2802757475

AF411471

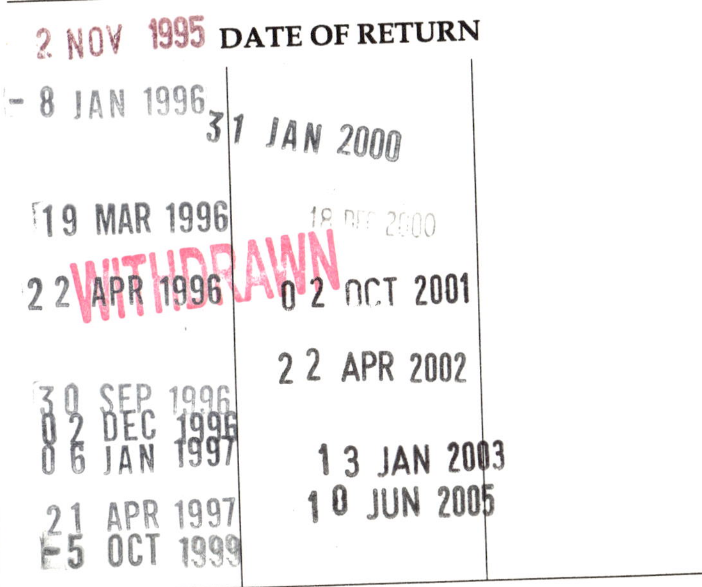

This item must be returned or renewed by the last date shown below.
The loan period may be shortened if it is reserved by another reader.
A fine will be due if it is not returned on time.

2 NOV 1995 **DATE OF RETURN**

- 8 JAN 1996   31 JAN 2000

19 MAR 1996   18 DEC 2000

2 2 APR 1996   WITHDRAWN   0 2 OCT 2001

2 2 APR 2002

30 SEP 1996
0 2 DEC 1996
0 6 JAN 1997   1 3 JAN 2003
10 JUN 2005
2 1 APR 1997
5 OCT 1999

UNIVERSITY COLLEGE LONDON
Gower Street London WC1E 6BT

LF4D   RLO168.22.8.95

ADVANCES IN BIOPROCESS ENGINEERING

ADVANCES IN BIOPROCESS ENGINEERING

Edited by

Enrique Galindo

and

Octavio T. Ramírez

Instituto de Biotecnología,
Universidad Nacional Autonoma de Mexico,
Cuernavaca, Morelos, Mexico

KLUWER ACADEMIC PUBLISHERS

DORDRECHT / BOSTON / LONDON

Library of Congress Cataloging-in-Publication Data

```
Advances in bioprocess engineering / edited by E. Galindo and O.T.
  Ramírez.
       p.  cm.
   Papers presented at the first International Symposium on
Bioprocess Engineering, held in Cuernavaca, Mexico.
   Includes bibliographical references.
   ISBN 0-7923-3072-2
   1. Biochemical engineering--Congresses.   I. Galindo, E. (Enrique)
II. Ramírez, O.T. (Octavio T.)  III. International Symposium on
Bioprocess Engineering (1st : 1994 : Cuernavaca, Mexico)
TP248.3.A3825  1994
660'.6--dc20                                              94-30302
```

ISBN 0-7923-3072-2

Published by Kluwer Academic Publishers,
P.O. Box 17, 3300 AA Dordrecht, The Netherlands.

Kluwer Academic Publishers incorporates
the publishing programmes of
D. Reidel, Martinus Nijhoff, Dr W. Junk and MTP Press.

Sold and distributed in the U.S.A. and Canada
by Kluwer Academic Publishers,
101 Philip Drive, Norwell, MA 02061, U.S.A.

In all other countries, sold and distributed
by Kluwer Academic Publishers Group,
P.O. Box 322, 3300 AH Dordrecht, The Netherlands.

UNIVERSITY
COLLEGE LONDON
LIBRARY

Printed on acid-free paper

All Rights Reserved
© 1994 Kluwer Academic Publishers
No part of the material protected by this copyright notice may be reproduced or
utilized in any form or by any means, electronic or mechanical,
including photocopying, recording or by any information storage and
retrieval system, without written permission from the copyright owner.

Printed in the Netherlands

This book consists of papers presented at the First International Symposium on Bioprocess Engineering organized by the Institute of Biotechnology of the National University of Mexico

Chairmen and General Coordinators

Enrique Galindo
Octavio T. Ramírez
Institute of Biotechnology, National University of Mexico (UNAM)

Scientific Committee

Eduardo Bárzana, Faculty of Chemistry, UNAM
Mayra de la Torre, National Polytechnic Institute
Enrique Galindo, Institute of Biotechnology, UNAM
Augustín López-Munguía, Institute of Biotechnology, UNAM
Adalberto Noyola, Institute of Engineering, UNAM
Rodolfo Quintero, Institute of Biotechnology, UNAM
Sergio Revah, Metropolitan University-Iztapalapa
Octavio T. Ramírez, Institute of Biotechnology, UNAM

Invited Authors

Arthur E. Humphrey, Pennsylvania State University, U.S.A.
Alvin W. Nienow, University of Birmingham, U.K.
Mattias Reuss, Universitat Stuttgart, Germany
Nicolaas W.F. Kossen, Gist-Brocades, The Netherlands
Murray Moo-Young, University of Waterloo, Canada
Andres Illanes, Catholic University of Valparaiso, Chile
Wei-Shou Hu, University of Minnesota, U.S.A.
Davis W. Hubbard, Michigan Technological University, U.S.A.
D. Grant Allen, University of Toronto, Canada
Irving J. Dunn, ETH-Zurich, Switzerland
Agustín López-Munguía, National University of Mexico, Mexico
Sergio Revah, Metropolitan University-Iztapalapa, Mexico
Mayra de la Torre, National Polytechnic Institute, Mexico

Sponsors:

Universidad Nacional Autonóma de México (UNAM)

Consejo Nacional de Ciencia y Tecnología (CONACyT)

- Institute of Biotechnology
- Coordination of Postgraduate Studies
- Coordination of Scientific Research

The financial support of the following entities is also acknowledged with thanks:

Biotechnology Regional Program for Latin America and the Caribbean,
PNUD/UNIDO/UNESCO

Centro de Investigación en Biotecnología, Universidad Autónoma del Estado de Morelos

The British Council

Saxa, S.A. de C.V.

Beckman Instruments de México, S.A.

Millipore S.A. de C.V.

Fisher Scientific Worldwide

Grupo Bacardí México, S.A.

Perkin Elmer, S.A./Analítica Representaciones S.A. de C.V.

Logistic Coordination

Leobardo Serrano
Juan L. García
Raunel Tinoco
Arturo Aguilar-Aguila
Lloyd Dingler

Fund Raising

Mayela Salinas
Elena Arriaga
Alfredo Martínez
Juan L. García

Administration

Ma. Elena Zamora

Secretary

Leticia Díaz

Assistants

Lorena Salazar
Roberto Cruz

Edition of Originals

Juan L. García
Laura I. Martínez
Celia Flores
Verónica Albiter

TABLE OF CONTENTS

x

Measurement, Control, and Automation of Bioprocesses

Metabolite Production, Physiology and Microbiology

Enzyme Engineering

Downstream Processing and Bioseparations

** denotes corresponding author.*

PREFACE

The exigent and rapidly growing markets of biotechnological products are demanding new and highly efficient bioprocesses. Molecular biology and immunology techniques have been crucial for achieving the important progress that the so called new biotechnology has made. However, improvements in actual process and industrialization of new ones depend more on how well a laboratory scheme is translated to production scale. In this activity, bioprocess engineering plays a key role.

Motivated by the belief that increased participation of engineering is needed in nowadays biotechnology, biochemical engineers at the Institute of Biotechnology of the National University of Mexico in Cuernavaca, Morelos, Mexico, organized the *First International Symposium on Bioprocess Engineering*. The Symposium aimed to be an International forum to present and discuss Bioprocess Engineering issues.

The Symposium consisted of invited and referred papers. Thirteen distinguished biochemical engineers were invited to give lectures at the Symposium. In addition, the response from the international community was very enthusiastic. About 120 abstracts were submitted for consideration. From the former, 80 were selected to submit full manuscripts, which then were sent to referees for reviewing. Finally, 55 papers were accepted for presentation in the Symposium and most of them are published in this book. The Symposium was truly international, 33 countries were represented from all over the world.

The papers address several topics of bioprocess engineering, namely: fundamentals of biochemical engineering, process development and optimization, control and automation of bioprocesses, design and scale up of bioreactors, enzyme engineering, waste treatment and downstream operations.

Basic bioengineering aspects such as transport mechanisms, models for mycelial growth, oxygen transport and structured models for bioreactors, rheology and mixing in fermentations and the utility of image analysis were addressed in some of the papers. Bioprocesses for the production of ethanol, gibberellic acid, lactic acid, proteases, β-galactosidase anakinra, streptokinase, penicillin-amidase, and lipases, as well as a bioleaching process, were described in various papers. Traditional and new processes (recombinant fermentations, plant and animal cell culture, and transgenic animals) are represented. Solid, liquid and gaseous substrates were considered in a variety of bioprocesses. The bioreactor modes studied included stirred tanks, air-lifts, percolation columns and packed and fluidized beds with immobilized cells. Interesting control problems are discussed for various model systems such as fermentors, enzymatic reactors and downstream operation systems. Authors from industry have also shared some of their experiences in process development and scale-up techniques. A good number of papers dealt with the treatment of solid, liquid or gaseous wastes using aerobic and anaerobic processes. Enzymatic engineering was well represented with examples of immobilized enzyme reactors, peptide synthesis and non-aqueous biocatalysis. Downstream operations papers dealt with the recovery of therapeutic bioproducts produced by recombinant microorganisms and transgenic animals.

The Symposium and the book would have been impossible to make without the participation of many people in various stages and aspects. We want to thank the invited speakers, the authors, the Scientific Committee and the referees for their enthusiastic response, their time and work in making this Symposium a valuable one. Our gratitude to our sponsors, who made possible the Symposium and the book. We thank the National University of

Mexico and particularly the Institute of Biotechnology for their support and encouragement in organizing this event. Many thanks to all the Organizing Committee for helping us in taking care of endless organization details and to our wives for their patience and support.

We hope this book can contribute to the development of Bioprocess Engineering, a key profession in turning biotechnology possibilities into biotechnology realities.

Enrique Galindo
Octavio R. Ramírez

Symposium Organizers and Editors
Institute of Biotechnology, National University of Mexico,
Cuernavaca, Mexico, 1994

Bioreactor Engineering

N.W.F. Kossen

Gist-brocades B.V., Research & Development, P.O. Box 1, 2600 MA Delft,
THE NETHERLANDS

This paper is not about formal design rules for bioreactors as such. There is plenty of literature available about this subject: Bailey and Ollis [1] ; Schügerl [2]; Van't Riet [3], Kossen [4]; as well as about its limitations: Zlokarnik [5]. The continuous thread through this paper is that the kind of product to be made in the bioreactor has a direct influence upon the time and the money that is available for the design of a new reactor, or the adaptation of an existing one. Time and money determine to a large extent the rules and methods that should be used. Designs with a built-in flexibility will appear to be essential. Finally the consequences for the education of those involved in bioprocess engineering (including bioreactor design) will be discussed.

THE DEVELOPMENT OF A BIOREACTOR AS A PART OF A PRODUCT/PROCESS DEVELOPMENT CHAIN (PDC)

The development of a bioreactor is an integrated part of a chain of events, the product/process development chain (PDC) **(fig.1)**. A number of remarks can be made regarding this PDC:

First of all the environment of the micro-organisms is different in every part of the chain. The greatest difference is between the environment during screening and production (any resemblance between these environments is a lucky coincidence). Due to the flexibility of biotechnology this problem can be overcome further down stream the PDC.

Second, the over all success of the development of a process in biotechnology depends on all the steps of this chain. Contrary to ordinary chains, however, weak links of this chain can be compensated by the contribution of other links.

If e.g. the productivity of a bioreactor is below standard this problem can be solved by a number of methods:

a. selection of a new strain
b. genetic improvement of the strain by either classical or molecular genetics
c. physiological improvements like changing the composition or the rate of addition of the substrate
d. improving the bioreactor as such
e. improving the yield of the down stream processing.

There is a definite difference between the methods mentioned in this list. The first three, a, b, and c, are "upstream" solutions (relative to the bioreactor). Furthermore they are "software" solutions (no investments in nuts and bolds are needed). Finally results in this area can often be obtained rapidly and with relatively few investments (2 or 3 man years of work and often much less). Solution d is a typical "hardware" solution, and the investments needed can be quite substantial.

Conclusions. The conclusions of this paragraph are that there is not just one solution to a problem. It even shows, "horribile dictu," that the solution can be achieved by disciplines different from your own, and that there is no "a priori"

*E. Galindo and O.T. Ramírez (eds.), Advances in Bioprocess Engineering. 1-11.
© 1994 Kluwer Academic Publishers. Printed in the Netherlands.*

argument why the contribution of one discipline is better than that of another.

THE RELATION BETWEEN THE DESIGN OF A BIOREACTOR AND THE PRODUCT.

There are a number of relations between the design of a bioreactor and the product:

a. Via the sequence: product ----> micro organism ----> reaction conditions (the "environment") ----> reactor design. This is the approach of the PDC and is mainly used for the design of a new process for a new product.
b. Via the influence of the process on the product:
   - yield
   - properties (composition, structure of enzymes etc.)
   This approach is often used when an existing process has to be optimized.
c. Commodities vs specialties:

commodities	specialties
-cost leader	-differentiation
-long life cycle (>10y)	-short life cycle (<5y)
-large market share (economy of scale)	-first to the market (economy of time)
-optimization	-"quickies"
-"hard/software" sol.	-"software" solutions
-patents an issue	-patents a big issue

The relations a and b are well known, and are well integrated in the teaching programs of the universities. This is not the case for relation c.
The focus will from now on be on the similarities and the differences between the development of bioreactors for commodities and specialties.

Commodities

The main point here is to realize a lower cost price than your competitors. This usually asks for economy of scale (for production and marketing). We will restrict our attention to the influence on the design of bioreactors.

A low cost price usually means (among others) a very sophisticated design of a reactor (low fixed and variable costs). This often results in a design where the process fits in the installation as a hand in a glove. If this fit is not proper, costly adaptations are necessary (as an example the adaptations in the ICI bioreactor for the production of "pruteen" to increase the mixing performance can be mentioned). Furthermore the collection of all the necessary data (especially those on kinetics) can be extremely elaborate. For the production of commodities in biotechnology this close fit is not necessarily the right philosophy as will be shown in the next paragraph.

Quite often the successes of the "software" solutions (improvements of strains or substrates) allow for the continuous increase of the productivity of bioreactors to such an extend that there is hardly any need for new bioreactors. The increase in productivity can often keep up with the increase in demand of the product. If the demand does not increase, the increased productivity often results in the availability of one or more of the existing reactors for new products. Often the main necessary changes in hardware are adaptations to the needs of the new strains (cooling and aeration capacity) to extend the "limits of growth".

When a company has decided to produce a new product in a new bioreactor the design of this reactor is often caught up by strain improvement: when the reactor is ready the strain is, as a result of ongoing improvements, already much more productive than the one used for the design calculations.

Also the quality of the substrate can vary enormously as a function of time (fluctuations due to changing seasons, geographical origin, or supplier).

The need for methods to improve the output of bioreactors increases when the product becomes a commodity and/or when it is at the end of its life cycle (cost cutting). The cost/benefit ratio for R&D for this purpose must be checked carefully and frequently (small margins).

<u>The development of bioreactor is a product/process development chain (PDC)</u>

Screening/selection

Strain improvement

- classical
- r-DNA

Laboratory fermentation

- physiology
- choice of substrate
- fermentation regime

pilot plant fermentation (scale up)

(reactor design)

down stream processing

formulation

Fig. 1: The product development chain

Flexible design/problem solving

A. General survey of methods to solve problems in   technology

	qualitative	quantitative
mechanistic	* ("mechanic")	("engineer") *
trial & error	* ("Kick & see")	("kick & measure") *

Figure 2.

The example mentioned above indicates that flexibility is an important property of a bioreactor, even for commodities. This need is strengthened by the tendency of product life cycles to decrease, also for commodities.

<u>Specialties</u>

The main point here is not "economy of scale" but "economy of time" (be on the market before your competitors are there). This means that there is usually no time for an optimized design of the reactor. Very frequently the process has to fit in an existing reactor, which is not likely to be a "hand-in-a-glove-fit". This development is driven by the ambition to realize a quick adaptation to customers demands. The result is an increased demand for bioreactors that are, above all, flexible (multi purpose). The solutions mainly have a "software" character.

<u>Conclusions.</u> The conclusion is that there are essential differences between the design of bioreactors for commodities vs specialties (optimization, time scale etc.). They have one thing in common however: the need for flexibility. A flexible design asks, mutatis mutandis, for a flexible and creative attitude of those involved in the development of processes in biotechnology, including the developers of bioreactors. In the next chapter we will see what the consequences are for the methodology of this development.

<u>FLEXIBLE DESIGN/PROBLEM SOLVING</u>

<u>General survey of methods to solve problems in technology.</u>

There are many methods to solve problems in technology. Two different distinctions are very essential however and can be used to illustrate in broad outline which choices can be made. These distinctions are between:

- trial & error versus mechanistic
- quantitative versus qualitative

A mechanistic method means that we know the essential mechanisms that contribute to the problem and that we use this knowledge to find a solution.

The "trial & error" method means that we do not know the mechanisms and that we try to solve the problem by what is at worst a series of random "experiments". It is also called the "hit & miss" method. "Kick & see" is probably more illustrative for this method. It is the behaviour most car drivers show when something is wrong with their vehicle (this comparison will be expanded later on).

Another method often mentioned is the "empirical" one. This is also a non mechanistic method, the approach is based on experience that has not (yet) been converted into mechanisms.

In other words, one knows by experience where to kick and how intense, but not why this works. The empirical method is therefore closely related to trial and error (an empirist is a trial & error man/woman with experience).

The distinction between qualitative and quantitative goes without further explanation.

The two distinctions just mentioned can be represented in a square **(fig.2)**. We shall use this square to illustrate four archetypes of methods and a number of very useful "in-betweens".

When confronted with a problem one can start in two different ways to solve it: with trial & error (including empirism) or mechanistically (or "know how" versus "know why"). A simple example: we all know how our car has to be managed in order to drive from the city of Mexico to Cuernavaca, but the vast majority of us does not know why the car responds as it does to our actions. This is not a problem until the car breaks down. Most of us then open the bonnet of the car, and after a puzzled look at the motor we start with the "kick & see" procedure or, when we met the problem before and know where to kick, with the empirical method. This is the lower left corner.

When this method does not work we ask for someone with "know why" in order to fix it. Usually this is a mechanic, with a good qualitative (but very effective) mechanistic feeling of what is going on. This is the upper left corner.

Mechanical engineers, specialized in the design of cars, are usually not the right people for this job. They are experienced in design, where sound quantitative mechanistic methods are a necessity (the upper right corner). They are not very good at solving down to earth problems with a qualitative mechanistic approach. It is usually also below their standing.

The trial & error quantitative approach ("kick & measure") is quite popular in technology, particular in the early stages of research when the object of research still has a black box character. One then measures in a quantitative way output vs input (frequency response methods and the like). It can be an effective method to unravel the mechanisms determining the behaviour of the object, but it can also be used as such. This is the lower right corner.

Let's have another look at the empirical or the trial & error method vs the mechanistic approach.

The empirical method is much older than the mechanistic one. In the case of empirism one has gained a collection of know how (not know why) that can be applied to a rather limited set of problems. One does not know why a particular proposal for a solution should work, but experience has shown that, for this particular set of problems it is a reasonable suggestion. However, there is no a priori reason to believe that it will work next time for a slightly different problem. Many problems in industry, are still solved that way, particularly in biotechnology. The advantage is that a problem can often be solved quickly (if it is in a particular set). The disadvantages are that problems outside this set are generally very difficult to solve, the knowledge consists of huge amounts of facts and it is difficult to transfer it to others. When this

is the case, one falls back to the pure trial & error method ("kick & see"). This is a straight nightmare for many industries (and a possible application area for expert systems). Finally, an overriding problem with any non-mechanistic model is that it cannot be used for extrapolation (scale-up!).

The second way, the mechanistic approach, tries to unravel the basic mechanisms that could explain the problem and that give a clue to its solution. Experience has shown that a rather limited number of mechanisms is sufficient to do the job (a kind of 80/20 rule: 80% of the problems can be solved with 20% of the known mechanisms). This makes this method very powerful. Its advantages are: applicable to many sets of problems, can be used for extrapolation (scale-up), limited amounts of facts, easy transfer to others. The disadvantages are; easy transfer to others ("other others") and the fact that many problems are still too complicated for this approach.

There is a general but very profound tendency, a guiding principle for all scientists (in industry or university, pure or applied): from empirical to mechanistic and from qualitative to quantitative.

However, empirism and other trial & error methods still are very valuable and should deserve attention during training. Furthermore there are extremely useful methods in between trial & error and mechanistic.

So far the archetypes. Let's now deal with "in-between" methods and stick to the knitting: bioreactor engineering in stead of motor cars.

Flexible design/problem solving in biotechnology

For a typical commodity like penicillin, the upper right approach (the "engineer") is very valuable. Intensive quantitative research after the metabolism, resulting in mathematical modelling of kinetics, and corresponding insight in the transport phenomena in the bioreactor, is necessary to stay

6

ahead of competitors. The success of this method can be quite impressive **(fig.3.)**. In the upper right corner you furthermore communicate mainly with people of the same, or closely related, disciplines.

However, when suddenly an unexpected drop in performance of the process in the bioreactor occurs after the introduction of a new substrate with a composition different from the previous one, an approach is necessary in between mechanistic and trial & error. A pure mechanistic approach is too time consuming for this problem. One therefore starts with a chemical analysis of the new substrate to find out which components present could be inhibiting, or which essential components are missing. This approach is semi-quantitative, and somewhere in between trial & error and mechanistic. Once the problem has been located there are many ways to solve it (add the missing component, eliminate the inhibiting one, change to the old substrate, use a mixture of substrates, change the micro-organism etc.).

A different problem is the rapid development of a new product, including the process for its production.

When this product is a new enzyme one has to go through the whole procedure of fig.1: screening & selection, strain improvement (including protein engineering), fermentation, down stream processing, formulation. Lots of questions have to be answered and decisions to be taken:
1. Has the enzyme the right properties (activity, stability, absence of toxic effects etc.). If not so, how can it be changed or can wrong properties (eg. insufficient stability) be compensated by finding a proper formulation.
2. Is the micro-organism acceptable as a production organism (do we have experience with it, is it GRAS, is it genetically stable, does it produce unwanted by-products, has the broth an acceptable viscosity, is the production level high enough etc.). If not so can we alter the

properties of this organism with classical or molecular genetics, or do we have to clone the gene coding for the enzyme in another micro-organism with well known and acceptable properties.
3. How pure should the product be and how do we have to separate, concentrate and purify the product.
4. What is the best formulation of the product.

The prevailing general questions are among others:
1. Does the product and its method of production meet the standards for health, safety and environment.
2. Are there problems to be expected due to legislation or the attitude of the public.
3. Do we have an acceptable patent position.
4. Do we have the necessary tools to make the product.
5. Do we know the market (potential clients and competitors)
6. Does the product meet the standards of potential clients
7. Can we be on the market in time.
8. Does the product result in an acceptable profit for the company.

Engineers are involved in this process right from the beginning, first in the role of consultant, later on as designers.

The formal upper right ("engineer") approach is out of the question for the majority of problems mentioned in this chapter (4.2) because it is far too time consuming and costly. Such an approach is also often not necessary. One needs experienced engineers with a flexible, holistic, multidisciplinary and creative attitude, that have the insight and the guts to use mechanistic rules of thumb to produce answers with an accuracy that is sufficient (and not more) and right in time for the problem at hand. In other words "good is good enough".

The approach is a typical in between one: somewhere between trial & error and mechanistic and between qualitative and quantitative. It asks for the ability to communicate with

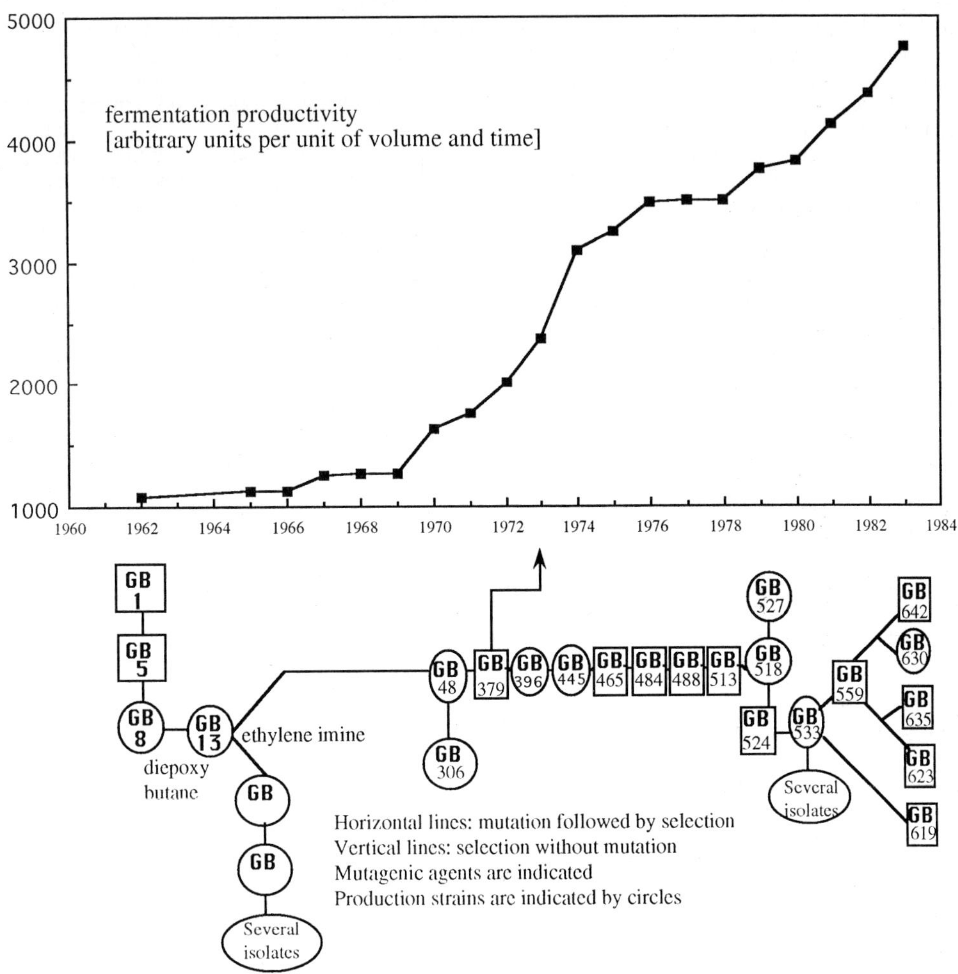

Figure 3. PHYLOGENY AND PRODUCTIVITY OF *PENICILLIUM CHRYSOGENUM* STRAINS AT GIST-BROCADES.

people from very different disciplines.

The essential point I want to make here is that such an approach asks much more from the skills of an engineer than the formal upper right approach. Everyone with the right classical engineering education can be successful in the upper right corner.
Only the very good ones, however, can be successful in the in-between approach. It asks for creativity and flexibility and an holistic approach and this is not taught at school.

<u>Conclusions</u>. Problems can be solved in a number of ways (trial & error vs mechanistic and quantitative vs qualitative). The choice of the method depends very much on the problem at hand.
The "upper right" approach has its own area of applicability. In biotechnology, however, "in between" methods are the rule rather than the exception. They ask for a very flexible and creative attitude. This attitude does not get the necessary attention at school.

<u>An example of the "in between" approach</u>

The aim of this example is to show how an "in between" approach has been used successfully for the design of a bioreactor for the aerobic treatment of waste water.

The example is part of the work of two chemical engineers, J.J. Heijnen and R. Weltevrede, at the time they worked for the R&D-organization of Gist-brocades. The subject is:

"Hydrodynamics of air-lift suspension reactors and scale-up/scale-down aspects".

The choice for the design was an air lift suspension reactor (see fig.4). Reasons are: no moving parts, good retention of the biomass on the suspended particles and the expertise we had with this type of retention.

Because the design had to serve too many aspects to be dealt with here we shall restrict ourselves to two of

them: the circulation velocity and the gas hold-up of the air/liquid/solid mixture.

<u>The circulation velocity</u>

The first approach is an airlift reactor without particles. The following variables have been considered:
1. diameter of the reactor
2. the gas distribution system
3. the distance between the central pipe (the riser) and the bottom
4. the distance of the central pipe to the surface of the liquid
5. the ratio of the diameters of the central pipe and the reactor
6. The coalescence properties of the liquid

The outcome of this consideration was that, within practical limits, these variables had little influence upon the circulation velocity within the reactor, provided the air is uniformly distributed in the riser.

A rather good description of the circulation velocity can be given by the following equation: Heijnen; van 't Riet [6]:

$$v_1 = 0.5 * (g * H * v_g)^{1/3}$$

v_1 = liquid velocity in the air lift (m/s)
g = acceleration due to gravity (m/s^2)
H = height of the reactor (m)
v_g = superficial velocity of the air (m/s)

Many other equations are available, but this one has a number of interesting properties: it has a mechanistic background, it has been checked at different scales and it has a simplicity that is almost elegant. Lots of these equations for the description of varying phenomena, relevant for the design of bioreactors, do exist in literature. A good survey is that of van 't Riet and Tramper [3].
Measurements at different scales supported the usefulness of this equation.

About the influence of the solids in the reactor very little literature

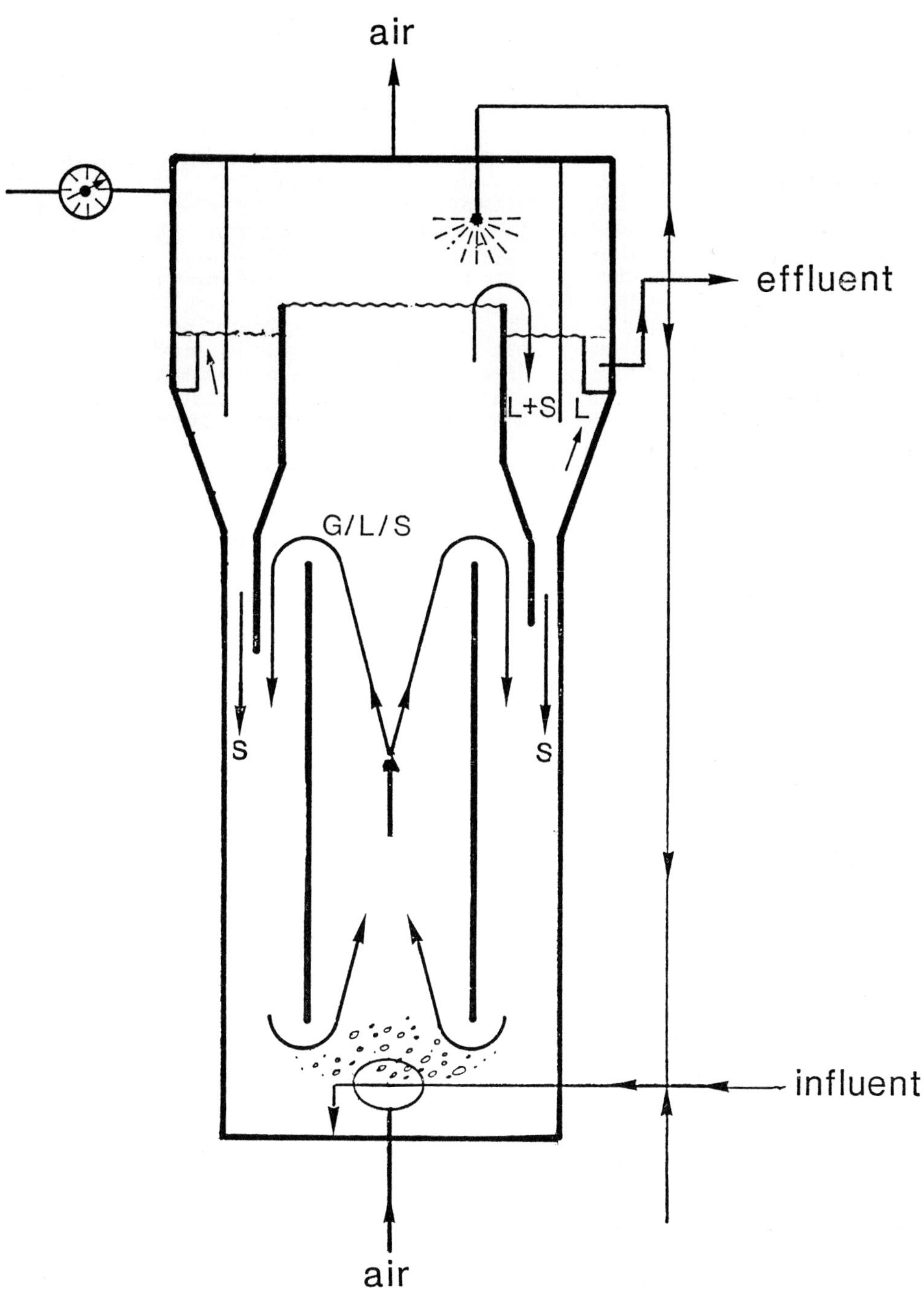

Figure 4   Air–lift suspension reactor

appeared to be available. The assumption was made that this influence was negligible for the particles we used (low concentrations and low falling-velocities).

<u>The gas hold-up</u>

The next point was the gas hold-up in the riser and the downcomer. This has been measured experimentally. The outcome is that the gas hold-up in the column as a whole is higher than predicted by the literature about bubble columns. Thus an estimate from the literature on bubble columns gives a conservative design.

With similar down to earth approaches the mass transfer and the conditions for suspension of the solids have been checked.

At the end the possible effects of scale-up or scale-down have been studied. When these effects were likely to occur, their magnitude has been estimated with equations of a simplicity similar to the one given above.

When the full-scale reactor was started up it worked as expected, right from the beginning, which is the best proof for the approach used for the design, a typical "in between" approach.

If we had decided to use the formal approach, setting up and solving the differential equations for the different phases and their interaction with each other, the design certainly would not have been better due to all the uncertainties involved (eg the influence of the particles upon the coalescence of the gas bubbles). It also would have caused unnecessary delay in the final design. Furthermore it would have consumed a lot of the capacity of the people involved, a capacity that could now be used to solve in a very creative way a number of details of the design.

<u>CONSEQUENCES FOR THE TEACHING OF (BIO)PROCESS ENGINEERING</u>

Bioreactor engineering is an essential part of biotechnology, and has its roots in chemical engineering. Chemical engineering is in itself already a multi disciplinary activity (chemistry and process technology).

The emphasis in the teaching of chemical engineering is on the development of mainly continuous processes. Furthermore hardly any attention is paid to the development of products, although there is an urgent need for chemical engineering input in product development.

In bioreactor engineering batch, fed batch and continuous processes are used, with a strong emphasis on the first two. The interest in product development is virtually absent although contributions from bioreactor engineering could be extremely useful.

The conclusion is that there is a partial similarity between chemical and bioreactor engineering regarding reactor design. There is a large similarity regarding the absence of interest in product development.

<u>CONCLUSIONS</u>

1. There is not just one solution to a problem in biotechnology. Often the solution can be achieved by disciplines different from your own. There is no a priori argument why the contribution of one discipline is better than that of another. This asks for a holistic approach of the problems at hand.

2. There are essential differences between the design of bioreactors for commodities vs specialties (optimization, investments, time scale, etc.). They also have one thing in common: the need for flexibility.

3. Problems can be solved in a number of ways (trial & error vs mechanistic and quantitative vs qualitative). The choice of the method depends very much on the problem at hand.
   The "upper right" approach has its own area of applicability. In

biotechnology, however, "in between" methods are the rule rather than the exception because of lack of time or of consistent data. Engineers should be better trained in using these "in between" methods in a correct way.

4. There is an urgent need for creativity, flexibility and the like of our engineers. This need is not fulfilled within the present educational system.

<u>REFERENCES</u>

1. Bailey, J.E. and D.F. Ollis, *Biochemical Engineering Fundamentals*, McGraw Hill, NY (1977).

2. Schügerl K., *Bioreaction Engineering* Vol.I, John Wiley & Sons N.Y. (1987).

3. van 't Riet K., *Basic Bioreactor Design*, M.Dekker Inc. NY (1990).

4. Kossen, N.W.F., *Bioreactors: Consolidation and Innovation*, Proceedings 3rd European Congr. Biotechnn, München 1984, 151, VCH Weinheim (1985).

5. Zlokarnik M., *Quo vadis Bioverfahrenstechnik?*, BTF-Biotech-Forum <u>5</u>, 3, pp160-165 (1988).

6. Heijnen J.J., K.van t' Riet, *Mass Transfer, Mixing and Heat Transfer Phenomena in Low Viscosity Bubble Column Reactors*, The Chemical Engineering Journal, <u>28</u>, B21-B42 (1984).

Design and Scale-up of External-Loop Airlift Bioreactor

Y. Kawase

Biochemical Engineering Research Center, Department of Applied Chemistry, Toyo University,
Kawagoe, Saitama 350, JAPAN

Hydrodynamics and oxygen transfer in external-loop airlift bioreactor were discussed. New models of gas hold-up and volumetric mass transfer coefficient were developed. They are based on the hydrodynamic model in which the similarity between the liquid circulation in airlift bioreactors and the natural convection is considered. The models include both cases of Newtonian and non-Newtonian fermentation media. The applicability of the proposed models was examined using the present experimental data obtained in two external-loop airlift bioreactors and the available results in the literature. Reasonable agreement suggests that the proposed models are very useful for the rational design and scale-up of external-loop airlift bioreactors.

Much attention has been given to airlift configurations among the bioreactor designs. The airlift bioreactor is classified into external-loop and internal-loop bioreactors according to the liquid recirculation type. There have been a number of experimental studies on internal-loop airlift bioreactors[1]. However, little generalized information is available on gas hold-up and mass transfer in external-loop airlift bioreactors. Although a relatively well-defined liquid circulation flow has been observed in external-loop airlift bioreactors, this is too complicated to be rigorously analyzed and there is little quantitative understanding of the liquid-phase mixing. Due to variations in the geometries of experimental bioreactors, the published results are sometimes inconsistent[1]. External-loop airlift bioreactors have a separate conduit for the downcomer. The injection of air into the bottom of the vertical tube (the riser) induces liquid circulation upwards and down the other vertical tube (the downcomer). External-loop bioreactor performance is significantly influenced by a variety of airlift configurations, particularly designs of connections between riser and downcomer. Clearly, additional experimental data for liquid-phase mixing and mass transfer in external-loop airlift bioreactors are necessary. In particular, more information is required on the influence of non-Newtonian flow behaviors on hydrodynamics in these bioreactors.

The aim of the present study is to examine gas hold-up and mass transfer in external-loop airlift bioreactors. Experimental data for gas hold-up and volumetric mass transfer coefficient are presented. Models for gas hold-up and volumetric mass transfer coefficient in external-loop airlift bioreactors are developed. They are based on hydrodynamics of liquid phase in external-

13

E. Galindo and O.T. Ramírez (eds.), Advances in Bioprocess Engineering. 13-19.
© *1994 Kluwer Academic Publishers. Printed in the Netherlands.*

loop airlift bioreactors. Their applicability is tested using present and available results.

Since the airlift bioreactor design includes applications for non-Newtonian broths, such as fermentation of filamentous fungi and production of xanthan gum, particular attention is paid to non-Newtonian flow behaviors of fermentation broths. The volumetric mass transfer coefficients significantly decrease with increasing viscous non-Newtonian flow behaviors. The non-Newtonian fluids used in this work were carboxymethylcellulose(CMC) and xanthan gum solutions.

EXPERIMENTAL

The schematic diagram of the model external-loop bioreactor used in this work is shown in Figure 1. Two different downcomers were used. The bioreactor was composed of a riser having an inside diameter of 0.155 m and a downcomer having an inside diameter of either 0.105 or 0.070 m. The risers were 1.37 m high. For reference, a bubble column bioreactor of inside diameter 0.155 m and height 1.17 m was also employed. The details of the bioreactors are given in Table 1.

Table 1 Airlift bioreactor dimensions

Reactor	D_r[m]	D_d[m]	A_d/A_r	H_d[m]	V_L[m^3]
Airlift 1	0.155	0.105	0.458	0.6	0.026
Airlift 2	0.155	0.070	0.204	0.8	0.023
Bubble C.	0.155	0	0	0	0.016

In order to avoid solids settling, the connection pipes between riser and downcomer were inclined. This connection pipe configuration has been used previously by other investigators [2,3].

Air was sparged into the bioreactor through a perforated plate with fifty seven 1.0-mm holes and the aeration rate was measured with a calibrated rotameter.

The physical properties of liquids used in this work are given in Table 2. The rheological properties of CMC and xanthan gum solutions were measured with a concentric viscometer and described by a power-law model(i.e. $\tau = K\dot{\gamma}^n$).

Table 2 Physical properties of solutions

Solution	ρ[kg/m^3]	n[--]	K[Pas^n]	σx10^3[Nm^{-1}]
(Newtonian fluid)				
water	999	1.0	0.00095	72.0
glycerin	1264	1.0	0.00147	60.9
Dextrose	1158	1.0	0.00380	61.2
(non-Newtonian fluid)				
CMC	998	0.870	0.041	69.9
Xanthan	1001	0.422	0.160	64.6

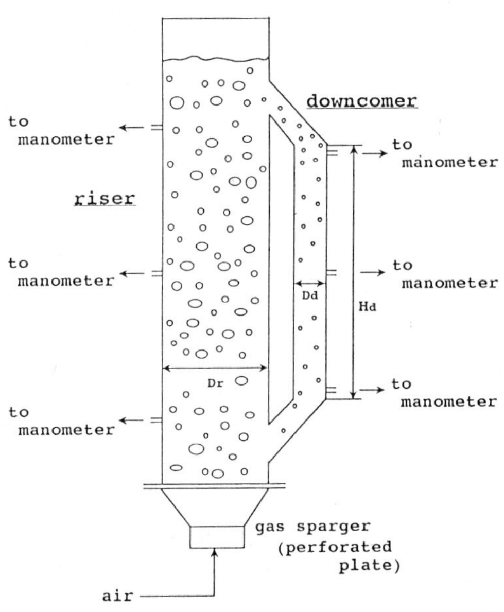

Figure 1. External-loop airlift bioreactor

RESULTS AND DISCUSSION

Gas hold-up

Recently, Kawase et al.[4] developed a model for gas hold-up in external-loop airlift bioreactors using Kolmogoroff's isotropic turbulence theory. The basic concept for the new model is the similarity between the liquid circulation due to the local variation of gas hold-up in airlift bioreactors and the natural convection due to temperature difference. The resulting correlation may be written as

$$\phi_{gr}/(1-\phi_{gr})=2^{-(3n+5)/(n+1)}n^{-(n+2)/2(n+1)}$$
$$(K/\rho_1)^{-1/2(n+1)}g^{-n/2(n+1)}(1+A_d/A_r)^{-3(n+2)/4(n+1)}$$
$$U_{sgr}^{(n+2)/2(n+1)} \qquad (1)$$

When $A_d/A_r=0$, Equation (1) reduces to that for a bubble column obtained by Kawase et al.[5]. For Newtonian fermentation media(n=1), Equation(1) becomes

$$\phi_{gr}/(1-\phi_{gr})=2^{-4}n^{-3/4}(K/\rho_1)^{-1/4}g^{1/4}(1+A_d/A_r)^{-9/8}$$
$$U_{sgr}^{3/4} \qquad (2)$$

The present model suggests the decrease in the riser gas hold-up with increasing downcomer-to-riser cross-sectional area ratio, A_d/A_r. In general, the gas hold-up in an airlift is less than the corresponding value in a bubble column due to the effect of higher liquid circulation velocity.

Table 3 shows the dependencies of ϕ_{gr} on $(1+A_d/A_r)$, U_{sgr} and η in the available correlations. The exponent of $(1+A_d/A_r)$ in the model is similar to that in the empirical correlation proposed by Popovic[6]. The exponential dependence of ϕ_{gr} on $(1+A_d/A_r)$ in the present model is rather stronger compared with that in the correlation of Chisti et al.[7] but it is weaker than that in the correlation of Bello et al.[8].

Table 3. Exponential dependency of ϕ_{gr} on $(1+A_d/A_r)$, U_{sgr} and η

	$(1+A_d/A_r)$	U_{sgr}	η
Bello et al.	$-5/3$	$2/3$	–
Chisti et al.	-0.258	0.603	–
Choi and Lee		0.504	
Popovic	-1.06	0.65	-0.103
Present work	$\dfrac{-3(n+2)}{4(n+1)}$	$\dfrac{(n+2)}{2(n+1)}$	$\dfrac{-1}{2(n+1)}$

Figure 2 compares the data for Newtonian media(water, glycerin aqueous solution, and dextrose aqueous solution) obtained in this work with the predictions of the present model. It is seen that Equation(2) agrees satisfactorily with the data. The average deviation is about 9.8%.

A comparison of the present model, Equation(1), with the data for non-Newtonian media(CMC and xanthan gum aqueous solutions) obtained in this work is shown in Figure 3. Satisfactory agreement between the predicted

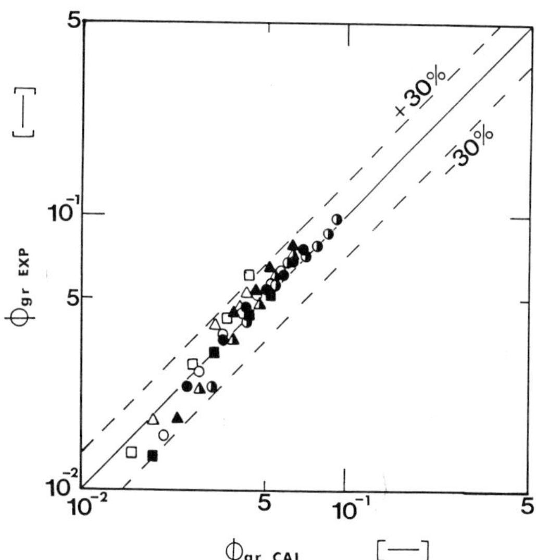

Figure 2. Comparison between Equation(2) and the present experimental data for Newtonian media(water, glycerin, Dextrose)

●■▲ Airlift 1 ○□△ Airlift 2
◐◭ Bubble column

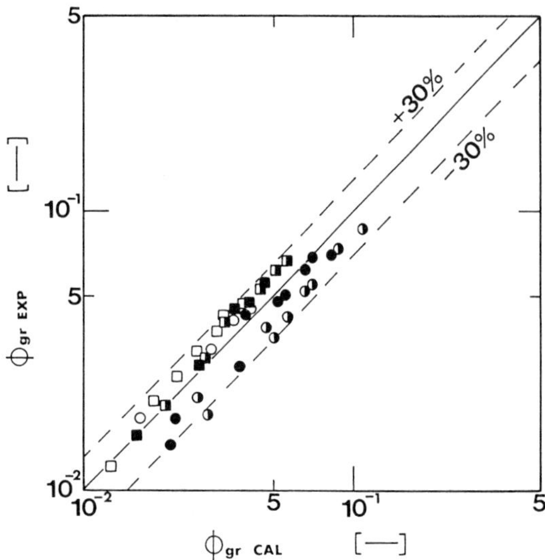

Figure 3. Comparison between Equation(1) and the present data for non-Newtonian media (Airlift 1 and 2, bubble column)

●○◑xanthan gum ■□◨CMC

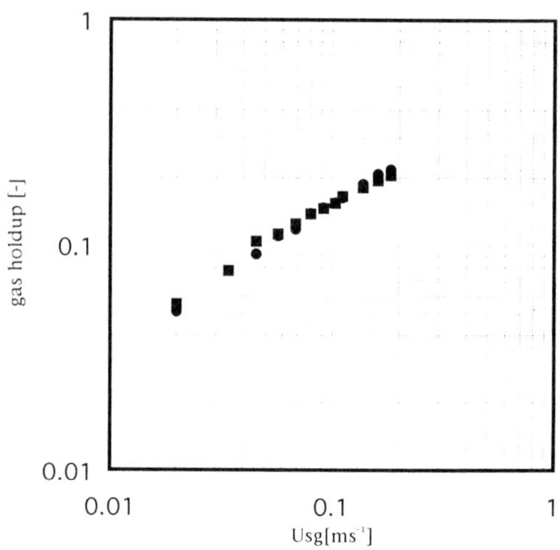

Figure 4. Gas hold-ups in small bubble column(D_r=0.065m)

■CMC ●water

and experimental results can be found. The average deviation is about 21%. Although it has been known that xanthan gum aqueous solutions have elasticity besides shear-thinning, the elasticity had only a little influence on the average riser gas hold-up.

We also measured gas hold-ups in a small bubble column(D_r=0.065m) with CMC aqueous solution. In large bubble columns, the viscous non-Newtonian flow behaviors reduce the gas hold-up. As shown in Figure 4, however, the gas hold-ups for CMC are very close to those in water. Similar results were reported by Vatai and Tekic[9]. These facts indicate that the results obtained in small bioreactors should be carefully used for scale-up of bioreactors with highly viscous non-Newtonian fermentation broths. In small bioreactors with viscous non-Newtonian media, gas hold-up is dependent of the reactor diameter.

Volumetric mass transfer coefficient

Kawase et al.[10] developed a theoretical correlation for volumetric mass transfer coefficients in bubble columns. It is based on Higbie's penetration theory and the Kolmogoroff theory of isotropic turbulence. Applying a similar approach in which Equation(1) is employed to estimation for gas hold-ups in external-loop airlift bioreactors, a correlation for k_La is derived. The resulting correlation may be written as

$$Sh = C\ Sc^{1/2}\ Bo^{3/5}\ Re*^{(2n+5)/4(n+1)}\ Fr^{(9n-6)/20(n+1)}$$
$$(1+A_d/A_r)^{-(31n+66)/40(n+1)} \qquad (3)$$

where

$$C = 0.142n^{1.63} \qquad Sh = k_LaD_r^2/D_L$$
$$Sc = (K/\rho_1)D_r^{(1-n)}/D_LU_{sgr}^{(1-n)}$$
$$Bo = gD_r^2\rho_1/\sigma \qquad Fr = U_{sgr}^2/D_rg$$
$$Re* = D_r^nU_{sgr}^{(2-n)}/(K/\rho_1)$$

In Figure 5, our data for water are compared with Equation(3). A satisfactory agreement was found.

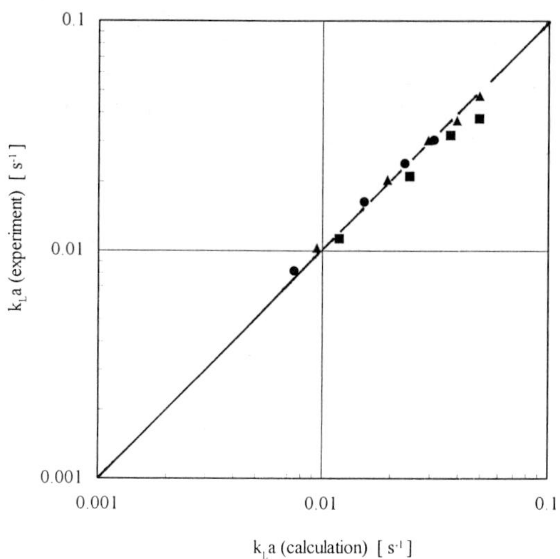

Figure 5 Comparison of k_La data with Equation(3) for water

■Bubble column ●Airlift 1 ▲Airlift 2

Figure 6 compares the predictions of the model with the data for 0.15wt%CMC solution. It can be seen that the correlation agrees reasonably with the data in CMC solution.

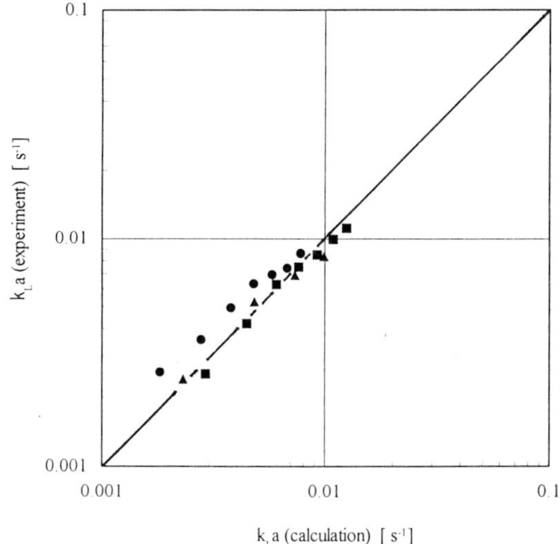

Figure 6 Comparison of k_La data with Equation(3) for CMC

A comparison of the model with the experimental data for xanthan gum is given in Figure 7.

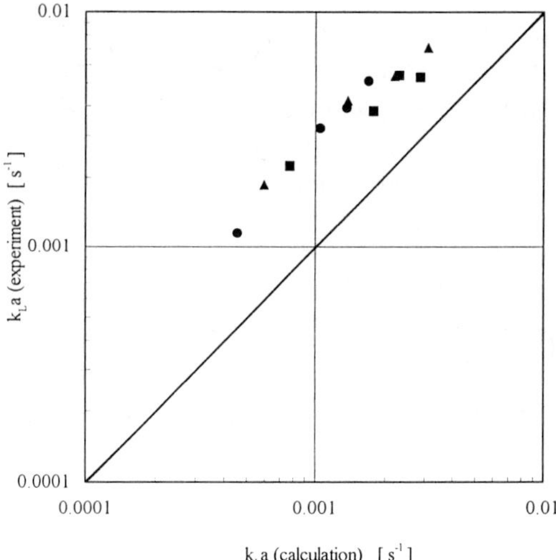

Figure 7 Comparison of k_La data with Equation(3) for xanthan gum

It is seen that the experimental data are somewhat larger than the theoretical predictions. In highly viscous non-Newtonian fluids, large bubbles formed by coalescence rose in the presence of a large number of tiny bubbles formed by break-up. This bubble size configuration led to a decrease in the specific surface area and as a result caused a significant decrease in the volumetric mass transfer coefficient. However, the k_La coefficients in xanthan gum solution, which presents viscoelasticity, are higher than the predictions of the model in which viscoelasticity is not taken account. The viscoelasticity may be responsible for the higher k_La data in xanthan gum compared with the theoretical predictions.

As described above, the oxygen mass transfer rate decreases with an increase in viscous non-Newtonian flow behaviors. For the improvement of the mass transfer rates in viscous non-Newtonian fermentation media, short draft tubes covered with perforated plates(Figure 8) were installed. As shown in Figure 9, an increase in the mass transfer rate was obtained in the whole range of gas flow rate. Bubble coalescence and break-up enhanced by the introduction of the perforated plates may be responsible for the increase in the k_La coefficients in viscous non-Newtonian fermentation media.

CONCLUSIONS

Reasonable agreement was obtained between the predicted and measured results for Newtonian and purely viscous non-Newtonian media except those for viscoelastic media, when correlations for gas hold-up and volumetric mass transfer coefficient which are based on hydrodynamics in external-loop bioreactors were compared with the experimental data. It may be concluded that the correlations for gas hold-up and volumetric mass transfer coefficient are useful for design and scale-up of external-loop airlift bioreactors.

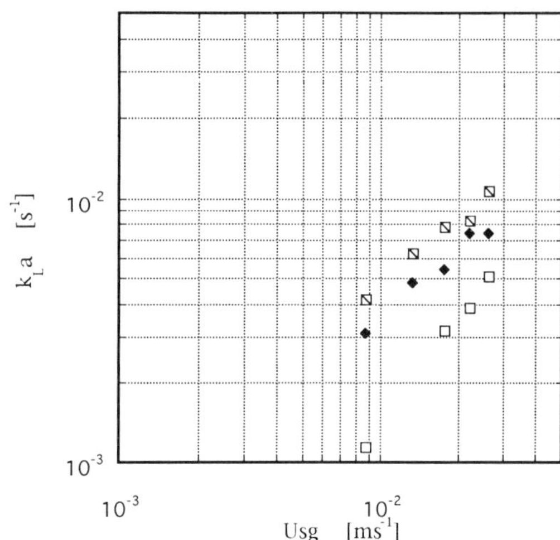

Figure 9. Effect of short draft tubes on k_La (xanthan gum)

□ no draft tube ◆ two draft tubes
◺ four draft tubes

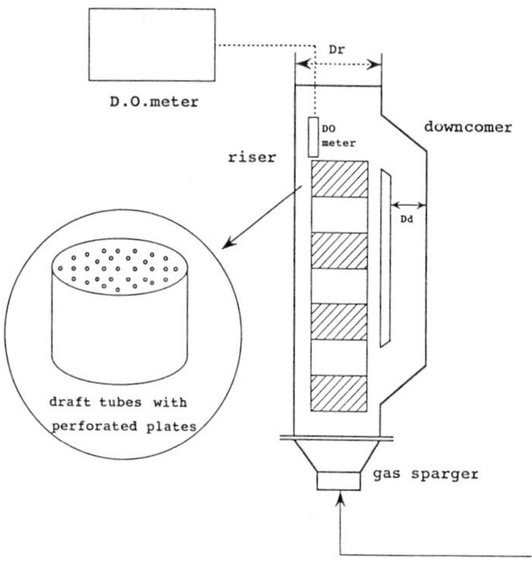

Figure 8. External-loop airlift bioreactor with short draft-tubes covered with perforated plates

NOMENCLATURE

A_d downcomer cross-sectional area
A_r riser cross-sectional area
a   specific surface area
Bo   Bond number
C   coefficient
D_L diffusivity
D_d downcomer diameter
D_r reactor or riser diameter
Fr   Froude Number
g   gravitational acceleration
K   consistency index
k_L liquid-phase mass transfer coefficient
n   flow index
Re* modified Reynolds number
Sc   Schmidt number
Sh   Sherwood number
U_{sgr} superficial gas velocity in the riser
V_L liquid volume

$\dot{\gamma}$ shear rate
η viscosity

ϕ_{gr} gas hold-up in the riser

ρ_l density

σ interfacial tension

τ shear stress

LITERATURE CITED

1. Chisti, M.Y. and Moo-Young, M., Chem. Eng. Commun., **60**, 195 (1987).

2. Posarac, D. and Petrovic, D., Chem. Eng. Sci., **43**, 1161 (1988).

3. Lindert, M., Kochbech, B., Pruss, J., Warnecke, H.-J. and Hempel, D.C., Chem. Eng. Sci., **47**, 2281 (1992).

4. Kawase, Y., Tsujimura, M. and Yamaguchi, T., Bioprocess Eng. in print.

5. Kawase, Y., Umeno, S. and Kumagai, T., Chem. Eng. J., **50**, 1 (1992).

6. Popovic, M. and Robinson, C.W., AIChE J., **35**, 393 (1989).

7. Chisti, M.Y., Fujimoto, K. and Moo-Young, M.,"Hydrodynamic and oxygen transfer studies in bubble columns and airlift bioreactors," presented at AIChE Annual Meeting, Miami (1966).

8. Bello, R.A., Robinson, C.W. and Moo-Young, M., Chem. Eng. Sci., **40**, 53 (1985).

9. Vatai, G. and Tekic, M.N., Chem. Eng. Sci., **44**, 2402 (1989).

10. Kawase, Y., Halard, B. and Moo-Young, M., Chem. Eng. Sci., **42**, 1609 (1987).

Engineering a Successful Scale-Up of a Process to Produce a rDNA Product

G.F. Slaff

Synergen, Inc., 1885 33rd Street, Boulder, Colorado 80301, USA

Synergen recently completed start-up of a new plant designed to produce 1500 kg per year of anakinra, a human therapeutic protein. Anakinra is a native human molecule that is used by the body to control the activity of interleukin-1. The product is produced in a recombinant E.coli fermentation and then purified using a combination of standard unit operations. This paper discusses the issues related to scaling up this process for use in the new plant.

Synergen recently completed start-up of a new plant designed to produce 1500 kilograms per year of Anakinra, a human therapeutic protein. The product, anakinra, also known as interleukin-1 receptor antagonist, is a protein which is ca. 18,000 MW. It is produced as a monomer with no intramolecular disulfides. Anakinra is a native human molecule that is used by the body to control the activity of interleukin-1. It is currently in clinical trials in a number of inflammatory diseases such as septic shock, rheumatoid arthritis, and asthma.

The product is produced in a recombinant E. coli fermentation and then purified using a combination of standard unit operations such as centrifugation, ultrafiltration and chromatography columns. The data used to design this plant was based on a process that was producing about 100 fold less product per batch that that anticipated for the new plant. This paper will discuss the issues related to scaling up this process for use in the new plant.

PLANT AND PROCESS DESIGN

The process used to produce anakinra is very straight forward (see Figure 1). The protein is produced by a recombinant strain of E. coli as a intracellular, soluble, active product. The cells are harvested using a disc stack centrifuge, resuspended in a breaking buffer and then lysed using a high pressure homogenizer. The resulting lysate is clarified using a disc stack centrifuge and the product recovered by using a ion exchange chromatography column. The resulting eluant is further purified to remove contaminants such as host cell proteins, DNA and endotoxins by use of a series of ultrafiltration and chromatography column operations. The final purified bulk material is then formulated and filtered into bulk storage containers which are then stored until shipment to an offsite fill and finish facility.

Early in the development of anakinra as a pharmaceutical product, Synergen made a business decision to take the steps necessary to ensure that the ability to produce this product in a licensable facility was not rate limiting. This meant that plant design and construction had to start early in the product development phase. Synergen had been producing clinical supplies of anakinra at a scale roughly 1/100 of that anticipated for the new plant. It was this small scale process data that was to be used as the basis for design of thos new plant.

The design and scale-up of each unit operation was approached in essentially the same way. An overall mass balance was written for the process as a whole by building it from

E. Galindo and O.T. Ramírez (eds.), Advances in Bioprocess Engineering. 21-23.
© *1994 Kluwer Academic Publishers. Printed in the Netherlands.*

the balances around each individual operation. Basic assumptions about the performance of each unit operation were made based on the small scale process data. Performance specifications formed the basis for the design and construction of the plant equipment. In the following section the approaches used to generate these specifications will be discussed for a few of the basic process operations.

Fermentation. Small scale data on growth rate, dissolved oxygen (DO) concentration and oxygen update rate (OUR) were used as a basis for the design of the large scale fermentor. Agitation power input, back pressure limits and aeration capacity were designed to ensure that the OUR value would be met at the same scale fermentors. Heat evolution rates were calculated and it was determined that additional heat transfer surface area in the form of internal coils was required to augment the jacket's cooling capacity in order to control the temperature adequately in the large scale fermentor.

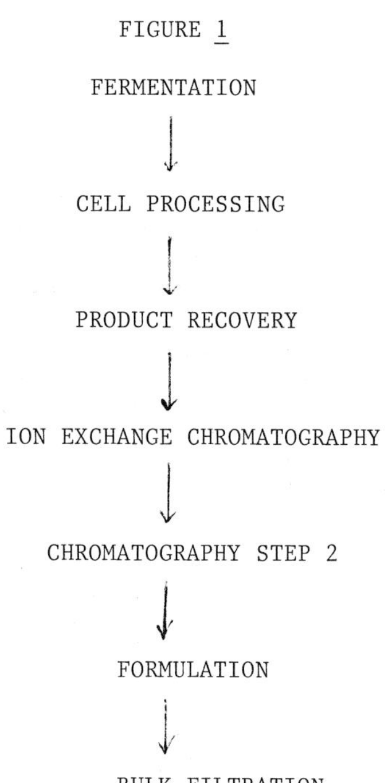

One problem encountered during the scale-up process was the need for additional antifoam. The use of additional antifoam however, initially effected product quality for reasons still not completely understood. Higher specific antifoam needs (antifoam amount per liter of fermentation broth) were necessitated by the increase in superficial air velocity as the fermentor scale was increased. A higher superficial air velocity for the large scale fermentor was required to maintain the same volumetric air velocity as for the small scale fermentor. Development work was required to identify a different antifoam which was more effective in breaking the foam and therefore was used in lower amounts, and did not cause the quality problem seen in the early scale-up runs.

Centrifugation. The design of the two centrifugation processes, cell harvest and debris removal, was based on data from a geometrically similar lab scale disc stack centrifuge which had a sigma factor ca. 200 fold smaller than that of the large scale machine. The initial large scale process flow rates for these two steps were determined by keeping the Q/sigma relationship constant. However, in order to obtain the same degree of cell recovery or clarification in the large scale system, it was necessary to use a flow rate equal to 85 to 90% of that predicted by the scale-up equation.

Ion Exchange Chromatography. The purification process contained a number of chromatography operations which were designed to reduce the level of host cell proteins, DNA, endotoxin and product variants to below specified levels. The design of the ion exchange chromatography columns to be used for this new plant was based on columns which were roughly 1/25 the volume of that to be used. Data from these small scale columns was used to determine the column dimensions as well as the critical operating parameters such as elution flow rate and gradient volume.

One basic assumption that was used to drive the design of these systems was that the load to each column would be divided into a specific number of sub-batches. Each sub-batch would be processed through the chromatography column using a complete cycle of equilibration, load, wash, elution and regeneration. The pools from each sub-batch would then be combined before processing on the next purification step.

The basic algorithm used to determine the column parameters is shown in Figure 2. Steps one through three are straight forward. Steps four, five and six require further explanation. Step four uses an equation based on the work by Yamamoto[1] to determine the elution flow rate and gradient volume required to achieve a resolution on the large scale column

FIGURE <u>2</u>

<u>SCALE-UP ALGORITHM</u>

1. DETERMINE TOTAL AMOUNT OF PACKING REQUIRED

   = <u>AMOUNT OF PROTEIN TO PROCESS PER BATCH</u>
       BINDING CAPACITY

2. CALCULATE THE MAXIMUM NUMBER OF CYCLES
   POSSIBLE

   = <u>MAXIMUM BATCH TIME</u>*
       HOURS PER CYCLE

3. COMPUTE THE REQUIRED BED HEIGHT

   = <u>TOTAL AMOUNT OF PACKING REQUIRED</u>
     CROSS SECTIONAL AREA X NUMBER OF CYCLES

4. CALCULATE OPERATING PARAMETERS (EG, FLOW
   RATES AND GRADIENT VOLUME) USING YAMAMOTO
   EQUATION

5. CONFIRM THERE ARE NOT PRESSURE DROP
   PROBLEMS USING VENDOR PRESSURE VERSUS
   FLOW DATA AND DARCY'S LAW

6. ITERATE AS NECESSARY

*Maximum Batch Time is the time available
to process the entire batch through the
chromatography step.

equivalent to that obtained with the small
scale column. Yamamoto's work is based on
equilibrium plate theory and relates column
resolution to easily measurable parameters.
Yamamoto showed the following:

$$R_s^2 \propto \frac{Z}{GH \times U \times D_p^2}$$

where:

   R_s = resolution

   Z = bed height
   GH = gradient slope x packing
        volume (excluding void space)
   U = linear elution flow rate
   D_p = packing particle diameter

If the resolution and packing particle diameter are held constant in the above equation,
a scaling equation relating the critical para-

meters of one size column (eg. column one) to
a column of another size (eg. column two) can
be developed as follows:

$$\frac{Z_1}{GH_1 \times U_1} = \frac{Z_2}{GH_2 \times U_2}$$

This scaling equation was used to determine
the elution flow rate and the gradient volume
required to give the same resolution in the
large scale column as that achieved in the
small scale column.

Once these parameters were determined it
was necessary to confirm that the pressure
drop across the column would not exceed the
design capacity of the packing or the column
itself. This was done by using Darcy's Law
to calculate the pressure drop across the
column for the computed bed height and flow
rate. This pressure drop was then compared
to the vendor supplied data for the packing
and the column.

Finally, steps two through five were repeated
until an optimal design was achieved.

CONCLUSION

The plant, and process, described in this
paper are currently in continuous operation.
The plant is operating successfully at greater than 90% of the design capacity and yield.
This sample demonstrates that a 100 fold
scale-up of a process to produce a biotach
product is possible as long as one uses
sound engineering fundamentals as a basis for
the design.

ACKNOWLEDGEMENTS

ABEC
Alfa-Laval
Biotage
Fluor Daniel
Pharmacia

Mr. Randy Hassler
Mr. John Hyde
Dr. Robert Kuhn
Mr. George Mackey
Mr. Phil Rayburg
Dr. Scott Rudge
Mr. Steve Schlager
Dr. Robert Seely
Ms. Kathleen Sniff
Mr. Mark Tomusiak
Dr. Mark Young

LITERATURE CITED

1. S. Yamamoto et al, "Ion-Exchange
   Chromatography of Proteins",
   Chromatographic Science Series,
   Volume 43, Marcel Dekker, Inc.,
   1988.

Invited paper

Slurry Bioreactor Design for Shear-Sensitive Mycoprotein Production

Y. Chisti and M. Moo-Young

Department of Chemical Engineering, University of Waterloo, Waterloo, Ontario,
N2L 3G1, CANADA

The effects of impeller associated agitation on fermentations of shear-sensitive mycelial microfungus Neurospora sitophila *are demonstrated. The impact of agitator type and the agitation rate on biomass protein production and cellulose utilization are shown in a relatively large 75 l fermenter. Results from cellulosic solid-substrate fermentations are presented. These fermentations tend to be unusual in that the broths contain solid-substrate particulates in addition to fungal mycelia and this leads to more complex rheological (and hence shear related) behaviour than in typical fungal cultures.*

All microorganisms, animal and plant cells are sensitive to various levels of hydrodynamic and mechanical forces. In fact, disruption of cells by fluid-shear, solid-shear or impact are established operations in bioprocessing [1]; however, in production of microbial or eucaryotic cells and their metabolites, exposure of the biocatalysts to high shear fields can be counterproductive [2,3]. Investigations of the safe limits of exposure of biocatalysts to fluid mechanical forces and design of bioreactors to not exceed the identified limits is of increasing importance especially because the newer genetically modified biocatalysts tend to be less robust than their wild counterparts [4,5]. Different morphological forms of the same organism may show different susceptibility to mechanical damage [6].

This paper demonstrates the effects of impeller associated agitation on slurry fermentations of the mycelial microfungus <u>Neurospora sitophila</u>. The effects of agitator type and the agitation rate are shown in a relatively large 75L fermenter. The food-grade fungus <u>N. sitophila</u> is of potential commercial significance in the context of a process developed for the conversion of lignocellulosic agricultural materials to protein-rich foods and feeds [2].

MATERIALS AND METHODS

Cultures and inocula

The microfungus <u>Neurospora sitophila</u> (ATCC 36935) was maintained at 4°C in submerged culture on glucose (10 kgm^{-3}) supplemented with yeast extract (Diffco) (2 kgm^{-3}) and the following nutrient salts (per litre): $(NH_4)_2SO_4$, 0.47 g; urea, 0.86 g; KH_2PO_4, 0.714 g; $MgSO_4 \cdot 7H_2O$, 0.2 g; $CaCl_2$, 0.2 g; $FeCl_3$, 3.2 mg; $ZnSO_4 \cdot 7H_2O$, 4.4 mg; H_3BO_3, 0.114 mg; $(NH_4)_6Mo_7O_{24} \cdot 4H_2O$, 0.48 mg; $CuSO_4 \cdot 5H_2O$, 0.78 mg; $MnCl_2 \cdot 4H_2O$, 0.144 mg. Inocula were grown at 26°C on the specified carbon source (5 kgm^{-3}) supplemented with 0.5 kgm^{-3} molasses (Hoffman Feeds Ltd, Heidelberg, Ontario) and the earlier specified salts.

The fermentation media contained a carbon source ("Solka Floc" wood cellulose or molasses). The Solka Floc cellulose (α-cellulose) was made from

25

E. Galindo and O.T. Ramírez (eds.), Advances in Bioprocess Engineering. 25-28.
© 1994 Kluwer Academic Publishers. Printed in the Netherlands.

wood pulp (James River Corporation, Berlin, New Hampshire). The KS1016 grade used in this work had average particle (fibre) length of 290 μm and a degree of crystallinity of 75-77% crystalline cellulose. Although the media were supplemented with the earlier specified nutrient salts, the entire complement of salts was not necessary for the naturally occurring substrates (e.g., corn stover); only ammonium sulfate and phosphates were essential.

Fermentation conditions

Fermentations were conducted either in shake flasks or in a 75 L (nominal) stirred tank fermenter (MBR Sulzer, Switzerland). The shake flask runs were performed in 250 mL flasks containing 100 mL medium including an specified carbon source and the nutrient salts. The flasks were sterilized at 121°C for 30 minutes, cooled to ambient, inoculated and held at the specified temperature on a gyratory shaker at 250 rpm. Unless otherwise indicated, the pH at inoculation was 6.0. At desired times, the flasks were rapidly cooled and stored at 4°C if necessary. The flasks were analyzed for total dry solids, crude protein and cellulose.

Crude protein and cellulose

For crude protein and cellulose determinations, the fermentation broth was filtered under suction through a 25 μm "Nitex" nylon cloth (Thomson Co., Scarbrough, Ontario), the filter cake was washed with several broth volumes of deionized water and dried overnight at 90°C. The dry biomass was ground to 1 mm particle size and a portion was analyzed for total nitrogen using a microKjeldahl technique [7]. The crude protein content of the biomass were calculated as 6.25 × total nitrogen, and percent (w/w) protein as gram protein per 100 g total dry solids. The cellulose content were determined by the spectrophotometric anthrone-sulfuric acid method [8]; percent cellulose was calculated on the same basis as crude protein.

Shear effects

The influence of shear on protein production was investigated in the 75 L fermenter (vessel diameter = 0.318 m) with a final working volume (after inoculation) of 50 L fermentation broth. The temperature and pH were controlled at 26°C and pH 6.0, respectively. The dissolved oxygen level was not allowed to drop below 20% of air saturation. Aeration rate varied (0.4 - 0.8 vvm) in response to the dissolved oxygen level. A 6-blade disc turbine was used for agitation (d_i/D = 0.57; C_i = 0.6·d_i; h_L/D = 1.9) at 250, 300 or 350 rpm corresponding respectively to tip speeds of 2.35, 2.82 and 3.29 ms^{-1}. N. sitophila was grown on KS1016 grade Solka Floc (5 kgm^{-3}) supplemented with molasses (0.5 kgm^{-3}), $(NH_4)_2SO_4$ (0.28 gL^{-1}); urea (0.52 gL^{-1}), KH_2PO_4 (1.0 gL^{-1}) and other, previously listed, nutrient salts at half the concentrations specified earlier.

RESULTS AND DISCUSSION

Agitation conditions

Mechanical agitation in stirred fermenters is known to damage mycelial biomass and affect the yield of the product [3,5,9]. Characterization of the influence of the impeller speed on N. sitophila protein production was required to identify the suitable operational conditions, any scale-up limitations and the sensitivity of this particular fermentation to impeller induced shear.

The protein production profiles at various agitation rates (tip speeds) are shown in Figure 1. For otherwise identical conditions, increasing tip speed of the Rushton disc turbine impeller lowered the rate of protein production (Figure 1), and the maximum protein yield. Thus, as shown in Table 1, the maximum specific protein production rate (μ) decreased from a high of 0.09 h^{-1} at 250 rpm to a low of 0.05 h^{-1} at 350 rpm. In relative terms, the protein production rate (μ_R) at the highest rpm was only 55% of that at the lowest agitation. Data on peak protein production and cellulose utilization (at 38 hours into the fermentation) are shown in Table 1 in absolute and relative terms. At the highest tip speed used (3.29 ms^{-1}) a distinct lag phase was noticed (Figure 1) in protein production compared to the results at lower agitation intensities. Clearly, the N. sitophila fermentations were quite sensitive to excessive agitation, and low agitation rates, consistent with adequate mixing and oxygen supply were indicated for the successful production process.

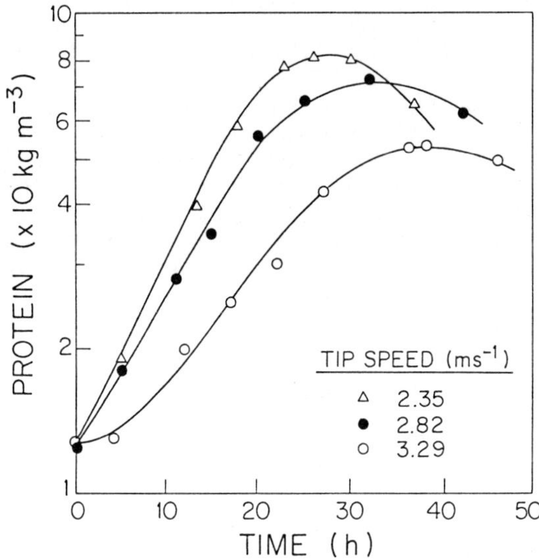

Figure 1. Effect of agitation on protein production. Impeller speed (rpm): (△) 250; (•) 300; and (O) 350. N. sitophila grown on KS1016 grade of Solka Floc wood cellulose (26°C, pH 6.0).

Table 1. Effect of agitation on protein production and cellulose utilization.

N rpm	μ h^{-1}	μ_R (-)	Crude Protein (%)	Cellulose Used (%)
250	0.09	21.0	31.1 (1)*	79.8 (1)*
300	0.07	0.78	27.7 (0.88)	69.0 (0.86)
350	0.05	0.55	21.2 (0.67)	55.6 (0.70)

* Values in parentheses are relative to the value at 250 rpm.

Identical fermentations conducted with the turbine replaced with an axial flow Prochem impeller (Figure 2) operated at 250 rpm gave marginally higher specific growth rate (μ = 0.11 h^{-1}) than was obtained with the Rushton turbines; however, the cellulose utilization was higher at ca. 86% with the Prochem impeller. Unlike the Rushton turbines, the axial flow impellers could be operated up to 400 rpm without any noticeable deleterious effects on the fermentations.

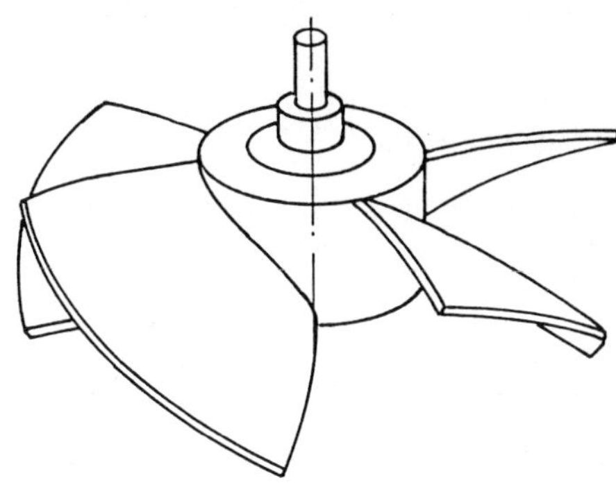

Figure 2. The Prochem Maxflo axial flow impeller.

CONCLUSIONS

Solid-substrate slurry fermentation systems for highly aerobic and shear-sensitive cultures of the mycelial fungus Neurospora sitophila present an unusually complex design problem. The fungal broths are highly viscous and non-newtonian because of the filamentous fungal solids and the cellulosic substrate fibres. Yet, the oxygen demand is high and the fungus is sensitive to agitation levels needed to provide sufficient oxygen transfer. Conventional, Rushton turbine agitated fermenters perform poorly for this system while simple airlifts have in the past been shown to be not particularly effective [10]. Redesign of stirred bioreactors with replacement of Rushton turbines with suitable axial flow devices improves bulk mixing in this rheologically complex fermentation. At the same time, the agitation-associated mechanical damage to mycelia is reduced. An alternative bioreactor configuration discussed elsewhere [10] and consisting of a concentric draft-tube airlift reactor with axial flow impellers located within the tube has also been shown to be effective for this fermentation at scales up to 1300L.

NOMENCLATURE

$\underline{C}_i$ Impeller clearance above bottom of tank (m)
$\underline{D}$ Diameter of fermenter (m)
$\underline{d}_i$ Diameter of impeller (m)
$\underline{h}_L$ Static liquid or slurry height (m)
$\underline{N}$ Impeller speed (rpm)
$\underline{\mu}$ Specific protein production rate (h^{-1})
$\underline{\mu}_R$ Relative protein production rate (-)

LITERATURE CITED

1.   Chisti, Y. and Moo-Young, M., <u>Enzyme Microb. Technol.</u>, **8**, 194-204 (1986). Disruption of microbial cells for intracellular products.

2.   Moo-Young, M., Chisti, Y. and Vlach, D., <u>Biotechnol. Lett.</u>, **14**, 863-868 (1992). Fermentative conversion of cellulosic substrates to microbial protein by <u>Neurospora sitophila</u>.

3.   Ujcova, E., Fencl, Z., Musilkova, M. and Seichert, L., <u>Biotechnol. Bioeng.</u>, **22**, 237-241 (1980). Dependence of release of nucleotides from fungi on fermenter turbine speed.

4.   Dunnill, P., <u>Chem. Eng. Res. Des.</u>, **65**, 211-217 (1987). Biochemical engineering and biotechnology.

5.   Moo-Young, M. and Chisti, Y., <u>Biotechnology</u>, **6**(<u>11</u>), 1291-1296 (1988). Considerations for designing bioreactors for shear-sensitive culture.

6.   Morimura, S., Kida, K. and Sonoda, Y., <u>J. Ferment. Bioeng.</u>, **74**, 129-131 (1992). The influence of shear stress and its reduction in the production of saccharifying enzyme from <u>shochu</u> distillery wastewater by <u>Aspergillus awamori</u> var. <u>kawachi</u>.

7.   Lang, C. A., <u>Anal. Chem.</u>, **30**, 1692-4 (1958). Simple microdetermination of Kjeldahl nitrogen in biological materials.

8.   Updegraff, D. M., <u>Anal. Chem.</u>, **32**, 420-4 (1969). Semimicro determination of cellulose in biological materials.

9.   Chisti, Y. and Moo-Young, M., <u>Biotechnol. Bioeng.</u>, **34**, 1391-1392 (1989). On the calculation of shear rate and apparent viscosity in airlift and bubble column bioreactors.

10. Moo-Young, M., Chisti, Y. and Vlach, D., <u>Biotechnol. Adv.</u>, **11**, 469-479 (1993). Fermentation of cellulosic materials to mycoprotein foods.

Invited paper

Strategies in the Design of a Penicillin Acylase Process

A. Gómez, M.Rodríguez, S. Ospina, R. Zamora, E. Merino, F. Bolívar,
O.T. Ramírez, R. Quintero, and A. López-Munguía

Instituto de Biotecnología, UNAM, Apdo. Postal. 510-3, Cuernavaca 62271, Mor., MEXICO

The development of an immobilized enzyme process for commercial application involves a series of steps and compromises, beginning with the selection of the enzyme and its source and ending with the downstream processing to obtain the product in a suitable form. Each step concerns one or various disciplines and is dependent on the other steps influencing the overall economics of the final process. In this paper such a situation is illustrated with the case of the 6-aminopenicillanic acid production process. Contributions concerning the production of the enzyme penicillin acylase are described. They involve molecular biology (the construction of a recombinant strain) and bioengineering (the overproduction of the enzyme in an exponentially fed-batch culture). The selection between an enzyme or a whole cell biocatalyst is discussed in terms of its technical and economical consequences. The kinetic behaviour, reactor type and operation strategy are also discussed, showing that specific activity of the biocatalyst and enzyme stability are the key properties affecting the performance of the process and therefore the overall economics. It is also shown that predictions made by modeling and simulation are important to define goals for genetic and protein engineering contributions to the process.

Penicillin acylase (PA) EC 3.5.1.11 hydrolyses penicillin G (PG) or penicillin V (PV) to yield 6-aminopenicillanicacid (APA) and phenyl acetic acid (PAA). The estimated worldwide production of 6-APA in 1992 was 7,500 tons [1] entirely used in the production of semisynthetic ß-lactam antibiotics such as ampicillin and amoxicillin. It is estimated that between 10 and 30 tons of immobilized penicillin amidase are now used globally per year, distributed in more than 10 commercial PA biocatalysts. It is one of the few immobilized enzymes in use at the industrial scale with annual sales only below those of glucose isomerase.

Although the enzyme is produced by a wide variety of microorganisms, the enzyme from Escherichia coli has been the most extensively studied and most industrial biocatalysts are prepared with this enzyme. In a review of 13 available industrial biocatalysts, only one was prepared with a PA from a different strain (Bacillus megaterium) and the rest with the purified enzyme from E. coli [2]. There are few reports concerning the production of biocatalysts using whole E. coli cells [3,4,5] and up to now there are no industrial applications, mainly due to the low specific activity of the wild strains. On the other hand, genetic engineering has allowed a 2-150 fold increase in PA production [6]. In a recent report, 100 U/g of cells of a recombinant E.coli were obtained in high density cultures [7] while Robas et al [8] reported 800 U/L with a strain obtained by transformation after chemical mutation, with a pBR322 derivative bearing the PA gene. Unfortunately the specific activity was not reported. However from their data it is infered that activities lower than 100 U/g of cells were obtained, with cell densities lower than 1 g/L. In spite of the advantage of a recombinant strain, its expression is often limited since translocation of the precusor through the cytoplasmic membrane and its processing in the periplasm are required to produce the active dimer, containing the α (24 kDa) and ß (62 kDa) subunits. Therefore, if a high productivity system is desired to increase the availability of the enzyme or to produce a high specific

E. Galindo and O.T. Ramírez (eds.), Advances in Bioprocess Engineering. 29-40.
© *1994 Kluwer Academic Publishers. Printed in the Netherlands.*

activity strain for whole cell PA biocatalyst, a combined strategy is required.

One of the main properties required for a successful biocatalyst is the stability. In the case of PA, the pH is the most important parameter affecting the half-life of the enzyme. Stable derivatives have been reported with half lifes between 1000 to 2000 hours. Stabilization procedures such as multipoint covalent attachment to agarose gels increase the stability 1400 fold compared to the wild strain [9]. The literature stability reports are usually obtained from storage experiments where the pH is constant and productivities are extrapolated from these data. However real productivities must include effects of macro and microenvironment pH shifts on the enzyme. Such shifts are caused by the PAA produced during the hydrolysis and the efficency of the pH regulation system, seldom considered in PA stability reports. In this particular case the pH-stability behavior is not only important in terms of the biocatalyst life but it is the base for the selection of the reactor configuration. Modeling is particularly useful for this purpose. It has been generally recognized that PA follows Michaelis-Menten kinetics, with inhibition by excess substrate, competitive inhibition by APA and non-competitive inhibition by PAA. The kinetic constants of the free enzyme, the immobilized enzyme and the enzyme in whole cells have been determined [2,3]. Models have also been proposed for the thermal and pH stability of the enzyme [2].

In the present paper an overview of different strategies followed in the development of a PA process is presented. All these activities are summarized in figure 1.

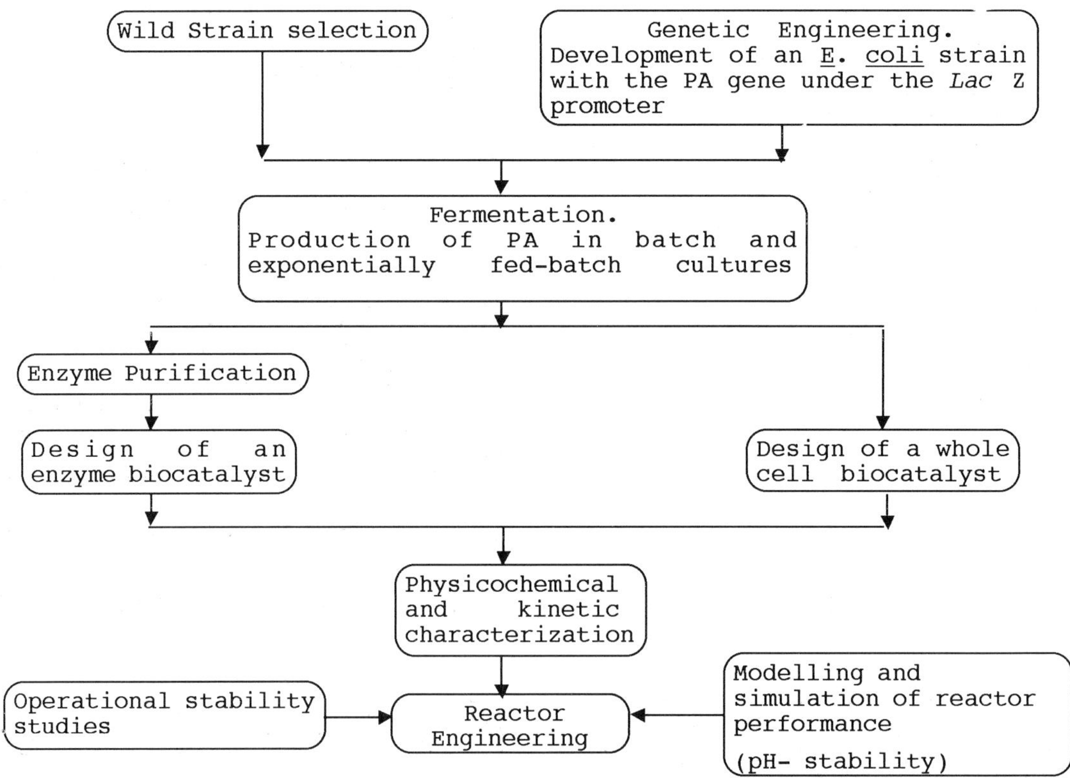

Figure 1. Strategies in the development of a penicillin acylase process.

ENZYME BIOCATALYST

Production and Stability

A PA process was developed by the company GENIN (Mexico) through a United Nations project using a mutant E. coli strain derived from ATCC 9637. This strain grows in a chemically defined medium in batch fermentations producing around 4 g of cells/L bearing 200 U/g. One activity unit is defined as the amount of enzyme producing 1 μmol/min of 6-APA at pH 7.8 and 37°C, using 2% (w/v) PGK. With these cells a biocatalyst was produced after successive extraction of PA in a Malton-Gaulin Homogenizer, ultrafiltration, salt precipitation, ion exchange chromatograhy and immobilization by covalent linkage to an epoxyacrylic resin. The biocatalyst obtained had an activity of 160-180 U/g catalyst (wet weight), particle size of 100-200 μm, density of 1.02 g/cm^3 and a porosity of 0.5. The purification process was optimized by scaling up an osmotic shock step instead of the extraction in the Malton Gaulin Homogenizer. This selective extraction resulted in a simple two step process with a 51.8 fold purification factor [10]. The enzyme displayed an optimum activity at 49°C, but due to the combined detrimental effect of temperature and pH on the enzyme stability it is often used at 37°C. At this lower temperature the activity is reduced in 40% but its half life is increased from 28 h to 2880 h at pH 7.5. The stability at 37°C as a function of pH has been described with an Arrhenius type model as follows [2]:

$$-dE/dt = k\,E \qquad (1)$$

with:
$$k = A\,\exp\,[B\,/\,pH] \qquad (2)$$
where:

E = enzyme activity (U/g catalyst)

$$A = 1.64 \times 10^{-6}\ h^{-1}$$
$$B = 37.1 \qquad \text{for pH} < 7.5$$

$$A = 3.27 \times 10^{6}\ h^{-1}$$
$$B = -173.98 \qquad \text{for pH} > 7.5$$

Equations 1 and 2 may be used simultaneously with kinetic equations to account for enzyme deactivation by pH in reactors. In this case the reaction rate is considered independent of pH as activity profile at 37 °C in the pH range 6.0 to 8.0 is constant [2].

Kinetic Properties

There is a general agreement that PA is inhibited by penicillin as well as by its two products. A mechanism that has been successfully used to describe the kinetic behavior of PA and the resulting kinetic model are presented in figure 2. The value of the kinetic constants obtained directly from initial rate experiments are given in table 1 [2,3].

Relationship between Penicillin Conversion and pH

If we consider the reaction catalyzed by penicillin acylase as:

$$[HP]_T \longrightarrow [HPAA]_T + [HAPA]_T \qquad (3)$$

where $[HP]_T$, $[HPAA]_T$ and $[HAPA]_T$ are the total concentration of penicillin, phenylacetic acid and 6-aminopenicillanic acid, respectively. Considering the dissociation equations of all the chemical species present in the reaction and their relation with penicillin conversion (defined as X= (So-S)/So), it is possible to write:

$$[HP]_T = [HP]+[P^-] = So(1-X) \qquad (4)$$

$$[HPAA]_T = [HPAA]+[PAA^-]= So\,X \qquad (5)$$

Table 1. Kinetic properties of penicillin acylase biocatalyst [2,3].

	T (°C)	pH	Km (mM)	K_{6-apa} (mM)	K_{paa} (mM)	Ks (mM)
Enzyme Biocatalyst	37	7.5	4.17	100.7	68.6	413
E. coli Cell	37	7.5	12	234.3	51.5	460

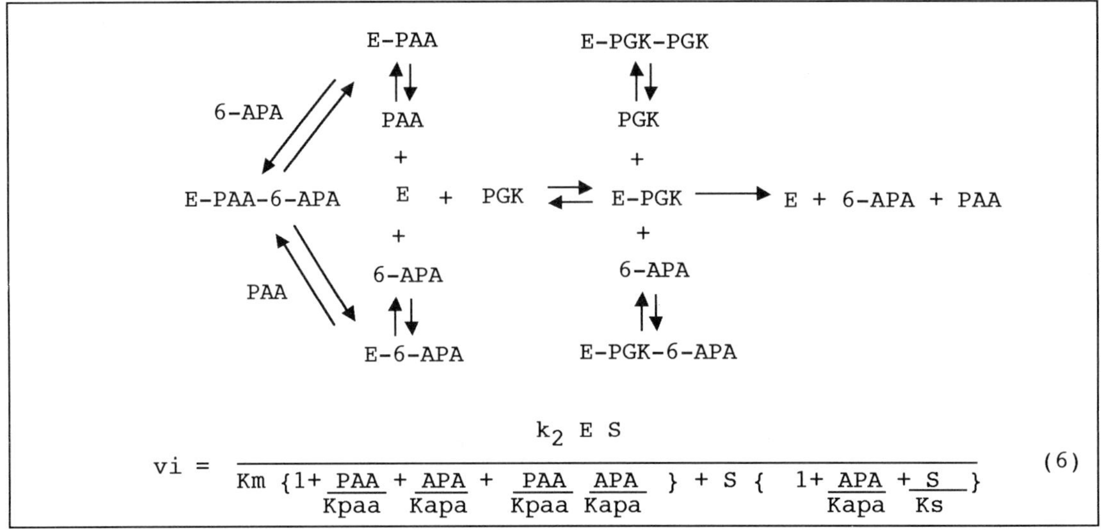

$$vi = \frac{k_2\ E\ S}{Km\ \{1+ \dfrac{PAA}{Kpaa} + \dfrac{APA}{Kapa} + \dfrac{PAA}{Kpaa}\dfrac{APA}{Kapa} \} + S\ \{\ 1+ \dfrac{APA}{Kapa} + \dfrac{S}{Ks} \}} \tag{6}$$

Figure 2. Reaction mechanism proposed for penicillin acylase and the corresponding kinetic model (adapted from [2]).

$$[HAPA]_T = [H_2APA^+]+[HAPA]+[APA^-]$$
$$= So\ X \tag{7}$$

$$[H_3PO_4]_T = [H_3PO_4] + [H_2PO_4^-]$$
$$+ [HPO_4^{2-}] + [PO_4^{3-}]$$
$$= C1 + C2 \tag{8}$$

$$CA = [NH_4^+] + [NH_3] \tag{9}$$

$$[H_2O]_T = [H_2O] + [OH^-] \tag{10}$$

where C1 and C2 are the molar concentrations of KH_2PO_4 and K_2HPO_4 respectively, and CA is the amount of ammonia required to maintain the pH at a given value. In order to relate the penicillin hydrolysis with the corresponding pH change and the ammonia required for its regulation, a model was constructed. The model is based on the electrochemical balance considering all dissociated and non dissociated forms of all the chemical species in the reaction. For each chemical species, the equilibrium constant defines the degree of dissociation. For instance, the fraction of dissociated penicillin from Equation (4) is given by:

$$\frac{[P^-]}{[HP]_T} = F_{10} = \frac{[P^-]}{[P^-] + [HP]} \tag{11}$$

$$F_{10} = \frac{[P^-]}{So\ (1-X)} \tag{12}$$

and

$$Ka1 = [H^+]\ [P^-]/[HP] \tag{13}$$

or

$$[HP] = [H^+][P^-]/Ka1 \tag{14}$$

substituting Equation (14) on equation (12) and introducing the definition of pH and pK, it is possible to obtain the fraction (in this case F_{10}) only in terms of pH and the pK of the chemical specie:

$$F_{10} = \frac{1}{1 + 10^{(pKa1-pH)}} \tag{15}$$

Similar expressions may be obtained for fractions F_{20}(PAA$^-$), F_{30}(APA$^-$), F_{32}(H$_2$APA$^+$), F_{40}(PO$_4$$^{3-}$), F_{41}(HPO$_4$$^{2-}$), F_{42}(H$_2PO_4$$^-$) and F_{50}(NH$_3$). Finally all the expressions are substituted in an electrochemical balance to give Equations (16) and (17), which relate conversion (X) with pH, and CA with X and pH respectively. These equations are presented in table 2, where K is the concentration of potassium ions (K = So + C1 + C2) and M is obtained from the water equilibrium constant:

$$M = 10^{(pH-14)} - 10^{-pH} \qquad (16)$$

A detailed description of the model has been published elsewhere[11].

The Batch Reactor

Due to the need for pH regulation during the reaction, industrial production of APA is carried out mainly in batch reactors. In order to model the process, we first validated the kinetic model in batch reactions, carried out at different substrate concentrations, considering constant temperature and pH. The integration of Equation (8) is compared with the experimental results (figure 3).

From data obtained after 150 batch reactions of two hours, the calculated half life of the enzyme was 1155 h. This is within the same order

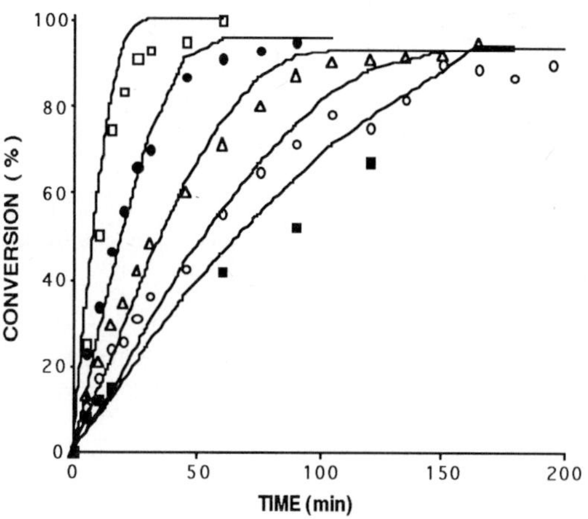

Figure 3. Hydrolysis in batch reactors with the PA biocatalyst at different substrate concentrations as described by the corrected triple inhibition model. Reaction conditions: 4.8 U/cm^3; 37 °C; pH regulation at 7.5 using 2M NH$_4$OH. PGK concentration (% w/v): - ☐- 2, - ●- 4; - △- 6; - ○ - 8; - ■ - 10. (adapted from [2]).

of magnitude reported for comercial biocatalysts, but it is half the time predicted from the storage stability data. To explain this behavior we carried out a second type of simulation based on the biocatalyst stability and considering the following sequence:

a) evaluation of one batch defining a time interval (Δt) and initial

Table 2. Equations that relate conversion with pH, and conversion with ammonia required to control the pH during penicillin hydrolysis [11].

$$X = \frac{F_{10}\,So + (C1 + C2)(3F_{40}+2F_{41}+F_{42}) - K + M + CA\,(F_{50} - 1)}{So(F_{10} + F_{32} - F_{20} - F_{30})} \qquad (17)$$

$$CA = \frac{F_{10}\,So + (C1 + C2)(3F_{40}+2F_{41}+F_{42}) - X\,So\,(F_{10} + F_{32} - F_{20} - F_{30}) - K + M}{(1 - F_{50})}$$

$$(18)$$

conditions (So, initial biocatalyst activity and pH= 7.5).

b) evaluation of reaction conversion by numerical solution of Equation (6) for each Δt.

c) evaluation of the pH change according to Equation (17).

d) when the pH reaches 7.4 or lower (lower regulation limit) the pH is reset to 7.6 (upper regulation limit), calculating the ammonia required using Equation (18).

e) evaluation of the residual activity according to Equations (1) and (2).

f) end the simulation when conversion reaches 95%.

The simulation from steps a) to e) may be repeated to evaluate the effect of the number of batch cycles on biocatalyst activity. Also, the values of pH in step d) may be set in a wider range to simulate the effect of defficient mixing and/or pH regulation systems in the stability of the biocatalyst. The results of this simulation are shown in figure 4. It may be observed that although the regulation system was set in ± 0.1, the actual stability results are equivalent to an oscilating pH value of ± 0.5. This may be due to inactivation by local acid or alkaline regions due to defficient mixing and regulation. It may be concluded that the operational stability is strongly influenced by the pH.

The Multicolum Recirculated Packed Bed Reactor (MRPBR)

Considering in one hand the absolute need for pH control during the reaction and in the other hand, the deactivating effect of the regulation system on the enzyme, a MRPBR system was proposed [12], (figure 5). It has the advantages of a pH regulation system which is not in contact with the enzyme, a short residence time (to avoid wide pH variations) and distribution of the biocatalyst in several columns (to

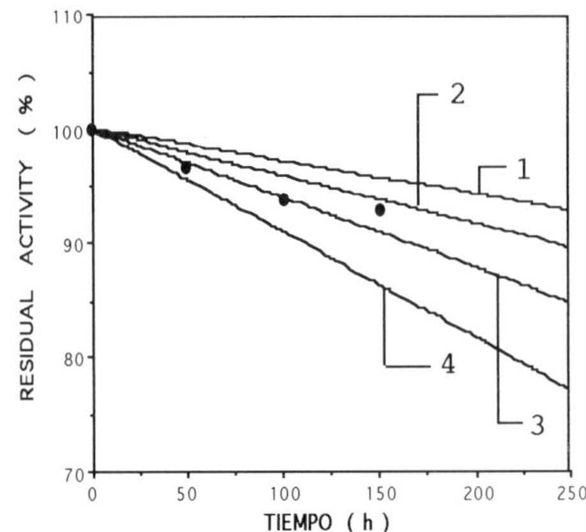

Figure 4. Residual activity of a penicillin acylase biocatalyst in a batch reactor. Continuous lines represent the simulation for the following conditions: 10% penicillin, 95% conversion, 120 units per gram of penicillin, 37 °C, in a 50 ml reactor with pH regulation using 2N NH_4OH at pH =7.5 ± Δ, with Δ =0.1 (1), Δ=0.3 (2), Δ=0.5 (3) and Δ=0.7 (4). Experimental results are also shown, for Δ = ± 0.1 (adapted from 11]).

avoid excessive pressure drop).

The design equation for a packed bed reactor is given by:

$$dX / dz = v \ \rho \ S \ \varepsilon /(Q \ So)$$

expressed in a non-dimensional reactor length:

$$dX/dz^* = v \ \rho \ S \ Z \ \varepsilon /(Q \ So) \qquad (19)$$
with $z^* = z/Z$

The substrate concentration in the neutralization tank shown in Fig. 5 is not constant as the inlet concentration (S_E) is a function of time. However it is assumed that both concentrations (S_T and S_E) are constant during the integration step (Δt). This allows the following mass balance in the neutralization tank [12]:

$$S_T(t+\Delta t)= [S_T(t) \ V_T + S_E \ Q \ \Delta t$$
$$- S_T(t) \ Q \ \Delta t]/V_T \qquad (20)$$

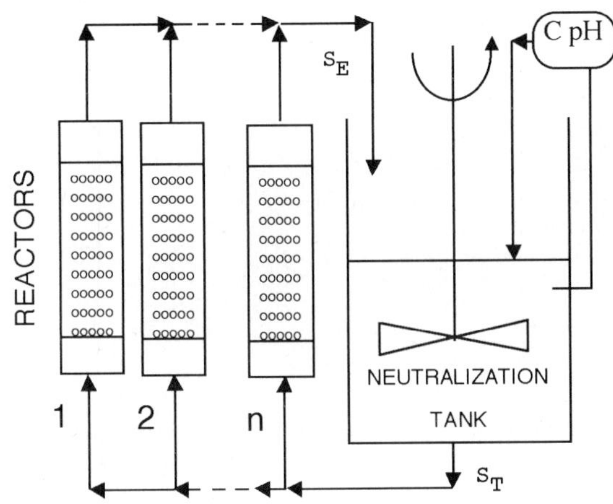

Figure 5. Multicolumn recirculated packed bed reactor (MRPBR) (adapted from [11]).

where:

$S_T(t)$ = Substrate concentration in the tank at time t.

$S_T(t+\Delta t)$ = Substrate concentration in the neutralization tank at time $t + \Delta t$ (Δt = time interval).

These equations, together with the pH deactivation model were used to simulate the experimental results obtained after 100 hours of use of two parallel reactors (figure 6). The extrapolation of these data resulted in a half life of 3430 h, three times the half life obtained in batch reactors. We may conclude that this configuration is particularly useful for the penicillin acylase system. Only one industrial process is actually operating with this type of reactors [13].

Whole Cell Biocatalyst with the Wild Strain.

Although it has been shown that enzyme biocatalysts are adequate from the technical point of view, their cost is high mainly due to purification and immobilization costs. In order to reduce these costs, the purification may be avoided and the immobilization procedure simplified using whole cells.

Among various alternatives of gel forming polymers, agar was selected due to its high stability, ease of handling and high entrapment capacity: up to 8.25 g of cells were immobilized per gram of agar. Gelatine was inadequate due to the proteolytic activity of the cells resulting in long term lysis of the biocatalyst.

The immobilization procedure has already been described [14]. In Figure 7, the effect of cell/agar loading ratio on the activity of the biocatalyst is shown for the wild strain. It is clear that the highest activity was obtained at the highest cell load, but there is a small effect of the particle size due to kinetic control.

As has already been mentionated, the low activity in the biocatalyst (a maximun of 11 U/g of biocatalyst) is due to the low activity of the E. coli cells (between 100 - 150 U/g of cells).

Due to its low density, there is a high load of biocatalyst required

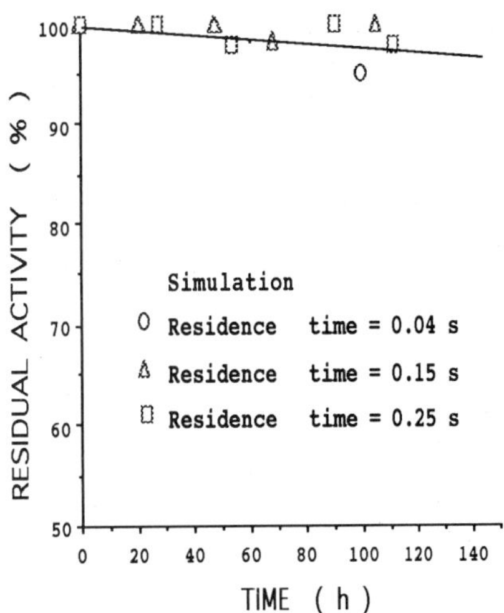

Figure 6. Residual activity of a penicillin acylase biocatalyst in two parallel packed bed reactors of 390 μl each, recirculating a total volume of 30 ml of 10% penicillin, with 44.5 units per gram of penicillin and operating at 37 °C. pH is regulated at 7.5 with 2N NH$_4$OH.

Recirculating flow rate in cm^3/min: -o - 273; -Δ - 85.5; -$\square$ - 47.2 (adapted from [12]).

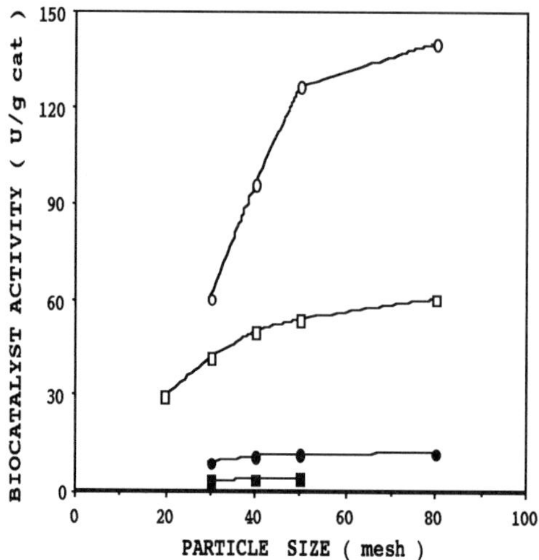

Figure 7. Effect of cell load of E. coli strains on a whole cell biocatalyst activity. Wild strain: – ■- 2 g cell/g agar, – ●- 8 g cell/g agar. JM102 strain: – □- 2 g cell/g agar, – ○- 8 g cell/g agar.

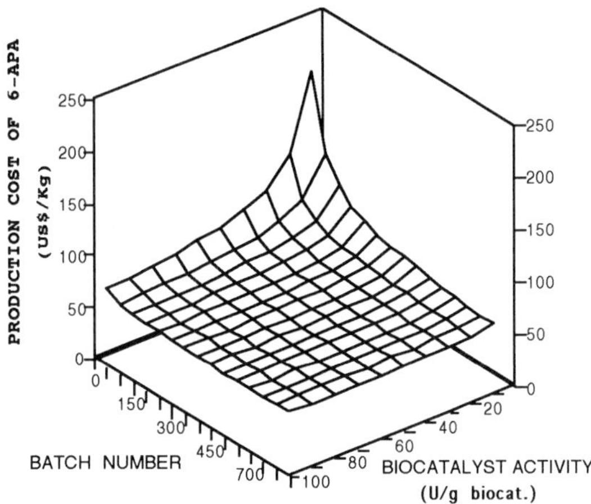

Figure 8. Effect of the number of runs of biocatalyst reuse and its specific activity on the 6-APA production cost (adapted from [3]).

for each reaction. It has also been shown that a minimum of batch cycles is required in order to minimize the influence of the biocatalyst cost in the 6-APA production costs. For a catalyst costing $2500 US/Kg, this minimum is estimated in 500 cycles, while for a whole cell catalyst with a calculated cost of $300/Kg, the number of batch cycles is a function of the specific activity [3]. In the later case, for a 10 U/g of biocatalyst, with only 50 batch cycles, the 6-APA production cost is higher than the actual selling price, while for a 100 U/g of biocatalyst this cost falls considerably. Details on the economical considerations are shown in figure 8 and reference [3]. It is therefore concluded that a whole cell biocatalyst requires a much higher specific activity strain.

CONSTRUCTION AND EXPRESSION OF A PA OVER PRODUCING STRAIN

Plasmid and Host

Plasmid pPA101 carries the E.coli *pac* gene whose transcription is under the control of the *lac* Z gene promoter and therefore can be induced with lactose as the sole carbon source or with the chemical inducer IPTG. In such a construction the amino terminal region of *lac* Z gene was positioned in the same translation phase of the *pac* gene. Thus, the *Hind* III restriction site between the two genes was digested, polymerized, and ligated to itself. Such modifications add four base pairs which avoid the formation of a *lac* Z-PA fusion product. The resulting plasmid, designated pPA102, was used to transform E. coli strain JM101 [15]. Figure 9 shows the construction and genetic maps of the molecular vehicles. The selection marker for the pPA102 plasmid was kanamycin. The strain JM102, shows a saturation type behaviour of PA production with respect to the inducer (isopropil- ß- thio-galactopyranoside, IPTG), with induction during inoculation resulting in the highest PA activity. The behaviour of PA specific production rate has been described as function of specific growth rate by simplification of an existing lumped mechanistic model [15]. Active PA results from processing of the polypeptide precursor. Acumulation of PA precursor protein suggests that postranslational processing and translocation through the cytoplasmic membrane may limit PA production. A typical batch culture of strain pPA102 results in 2 g/L of cells after 9 hrs, with 290 U/g of cells.

The use of exponentially fed-batch cultures (EFBC) was explored as an

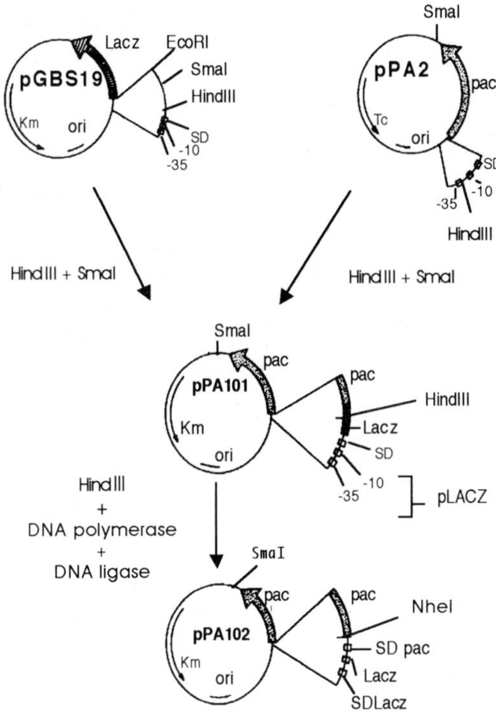

Figure 9. Construction of molecular vehicles for the over-expresion of *pac* gene. The dot-filled arrows represent the DNA fragments carrying the *pac* and the *lac*z genes, respectively (adapted from [15]).

alternative tool to chemostats for the study of the effect of growth rate on recombinant PA production. Its theoretical basis and the description of operation and control have been described elsewhere [16]. It is shown that after a transient period following initiation of the fed-batch phase, constant cell concentration and constant μ are reached. The same happens for glucose specific consumption rate and PA specific production rate. This situation is illustrated in Figure 10, for a $\mu = 0.01$ h^{-1}. The results suggest that the production of PA may be further increased if the fed-batch steady state is prolonged.

A summary of the fed-batch experiments is presented in figure 11. It is demostrated (11A) that at low dilution rates, the EFBC closely follows the behavior of a chemostat (μ=D). In figure 11B, the Luedeking-Piret kinetics describe the relation between the specific PA production rate and μ or D. Finally, figure 11C shows that the consequence of very low growth rates, is that PA accumulates at high concentrations, showing that the PA synthesis and/or processing rate is higher than the rate of cell division. This effect has also been shown for other recombinant proteins [17]. A maximum PA activity of 2000 U/g is obtained at μ of 0.01 h^{-1}. To our knowledege, such an activity is the highest reported in the literature, and is explained by the low growth rates reached with the EFBC system.

Finally, in figure 7, the activity of whole cell biocatalysts prepared with this biomass are reported. It may be observed that activities as high as 100 U/g of biocatalyst may be obtained using agar. For high cells loads, the biocatalyst is controlled by diffusion with effectiveness factors in the range of 0.6 to 0.9 depending on particle size. Nevertheless, the activity is high enough for an industrial reactor.

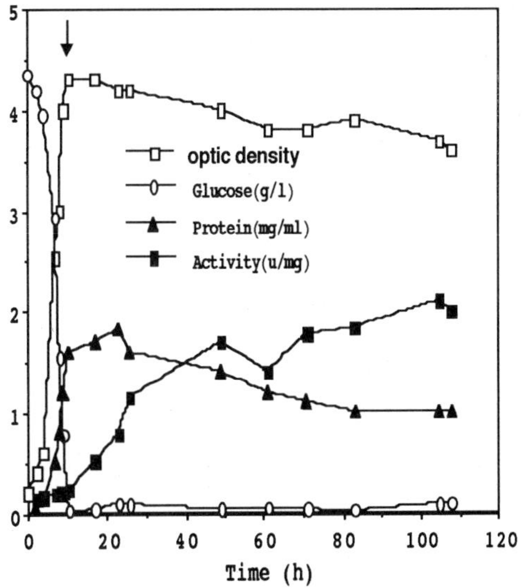

Figure 10. Penicillin acylase production in exponentially fed-batch cultures at a dilution rate of 0.01 h^{-1}. Medium addition starts when glucose concentration reaches zero (indicated by arrow). Initial and final culture volume were 4.3 L and 8 L, respectively.

38

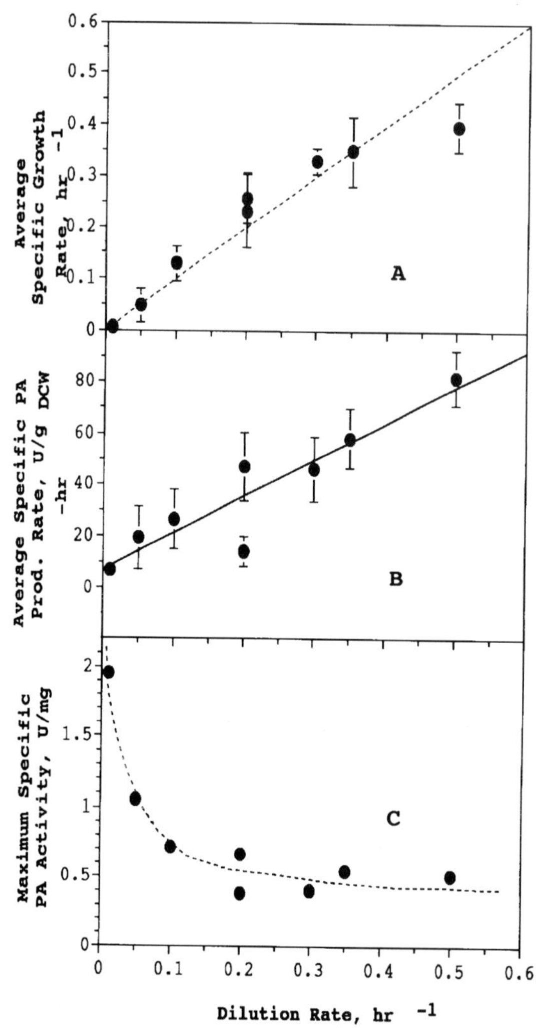

Figure 11. Summary of exponential fed-batch cultures. A, average specific growth rate. B, average specific PA production rate. C, maximum specific PA activity. The dotted-line in A correspond to the condition $\mu = D$, the continuous line in B is the linear fit, and the dotted-line in C is drawn to show the trend (adapted from [16]).

<u>CONCLUSIONS</u>

In this paper, various elements involved in the development of an enzymatic process were discussed. The interaction of molecular biology and bioingineering is illustrated, demonstrating the need for both disciplines in modern biotecnology. Specific requirements for enzymes or microorganisms such as stability and high productivity are defined in terms of process conditions. Such requirements may be the feed-back of information required for protein or genetic engineering programs. Also, the expression of new strains in fermentation process still require research in bioengineering. Combined strategies allowed the increse in specific activity of the wild <u>E</u>. <u>coli</u> strain 20-fold and the feasibility of a whole cell biocatalyst.

<u>ACKNOWLEDGEMENTS</u>

The authors wish to express their gratitude to Alfredo Martinez and Mario Caro for pilot plant support and to Fernando Gonzalez for technical assistence. The whole penicillin project was financed with a grant from DGAPA-UNAM.

<u>LIST OF SYMBOLS</u>

A	experimental constant for pH deactivation model (h^{-1})
APA or 6-APA	6 - aminopenicillanic acid
APA^-	dissociated form of phenylacetic acid
B	experimental constant for deactivation model
CA	ammonia concentration in the reaction mixture (M)
C1	concentration of KH_2PO_4 (M)
C2	concentration of K_2HPO_4 (M)
E	enzyme or biocatalyst activity (U/g)
ε	void volume in a packed bed
HP	non dissociated form of penicillin G

HAPA   non dissociated form of 6-aminopenicillanic acid

H_2APA^+ protonated form of the 6-aminopenicillanic acid

HPAA   non dissociated form of phenyl acetic acid

K   constant of deactivation model (h^{-1})

Ka1   equilibrium constant for dissociated and non dissociated penicillin forms.

Kapa   APA non-competitive inhibition constant (M)

Ks   excess substrate inhibition constant (M)

Km   Michaelis-Menten constant (M)

Kpaa   PAA competitive inhibition constant (M)

P^- dissociated form of penicillin G

PA   penicillin acylase

PAA   phenylxacetic acid

P   penicillin G

PGK   potassic salt of penicillin G

Q   recirculation flow rate (cm^3/min)

S   cross sectional area (cm^2)

S_E substrate concentration entering the neutralization tank (M)

So   initial substrate concentration (M)

S_T substrate concentration in neutralization tank (M)

t   time (h)

vi   initial reaction rate ($\mu mol/min/cm^3$)

V_T volume of neutralization tank (cm^3)

X   conversion

z   axial position in reactor (cm)

z*   non-dimensional axial position in reactor

ρ density of biocatalyst (g cat/cm^3)

LITERATURE CITED

1. Katchalski-Katzir, E., *Trends in Biotechnol.*, **11**, 471-478 (1993).

2. Ospina, S., López-Munguía, A., Gonzalez, R. L. and Quintero, R., *J. Chem. Tech. Biotechnol.*, **53** 205-214 (1992).

3. Rodríguez, M. E., Quintero, R. and López-Munguía, A., *Process Biochem.*, **29**, 213-218 (1994).

4. Klein, J. and Wagner, F., *Enzyme Eng.*, **5**, 335-345 (1984).

5. Sato, T., Tosa, T. and Chibata, I., *Eur. J. Appl. Microb.*, **2**, 161-168 (1976).

6. Shewale, J. G. and Sivaraman, H., *Process Biochem.*, August, 146-154 (1989).

7. Lee, Y. L. and Chang, H. N., *Biotechnol. Lett.*, **10**, 11, 787-792 (1988).

8. Robas, N., Zouheiry, H., Branlant, G. and Branlant, C., *Biotechnol. Bioeng.*, **41**, 14-24 (1993).

9. Alvaro, G., Fernandez, R., Blanco, R. M. and Guisán, J., *Appl. Biochem. Biotechnol.*, 181-195 (1990).

10. Rodríguez, M., Guereca, L., Valle, F., Quintero, R. and López-Munguía, A., *Process Biochem.*, **27**, 217-223 (1992).

11. Gómez Aguirre, A., Ospina, S., Queré, A., Quintero, R. and López-Munguía, A., "Modelling Simulation

of a pH-dependent Bioprocess: Enzymatic Conversion of Pen G to 6-APA", in *Process Computations in Biotechnology*, Ghose, T. K. (Ed.), Tata Mcgraw-Hill, New Delhi (in press).

12. Gómez Aguirre, A., Quintero, R. and López Munguía, A., *Bioprocess Eng.*, **9**, 147-154 (1993).

13. Poulsen, P. B., *Biotechnology and Genetic Engineering Reviews*, **1**, 121-140 (1984).

14. Castillo, E., Ramírez, D., Casas, L. and López-Munguía, A., *Appl. Biochem. and Biotechnol.*, **34** & 35, 447-486 (1992).

15. Ramírez, O. T., Zamora, R., Espinosa, G., Merino, E., Bolívar, F., and Quintero, R., *Process Biochem.*, **29**, 197-206 (1994).

16. Ramírez, O. T., Zamora R., Quintero, R. and López-Munguía, A., *Enzyme and Microbial Technology* (in press).

17. Bentley, W. E., Mirjalili, N., Andersen, D. C., Davis, R. H. and Kompala, D. S., *Biotechnol. Bioeng.*, **35**, 668-681 (1990).

Scaling-Up of a Lipase Fermentation Process: A Practical Approach

S.M.G. Geraats

Gist-brocades B.V., Bio Specialties Division, Research and Development, P.O. Box 1, 2600 MA Delft, THE NETHERLANDS

LIPOMAX™, a lipase for application in heavy duty detergents (HDD) and light duty detergents (LDD), obtained by a new Pseudomonas alcaligenes technology, will be introduced in 1995 by Gist Brocades BV. A newly developed fermentation process for the production of LIPOMAX™ was scaled up from 10 l to an existing 100 m³ fermenter. A loss of production during scaling up was observed. At 100 m³ scale, the maximum lipase concentration was reduced by 65% of the value at 10 l scale. The following most likely possible causes, with the highest priority for investigation, were put forward: a) gradients of soy-oil, pH and oxygen, b) raw materials, medium preparation, sterilization procedure and inoculation procedure, c) dissolved carbon dioxide concentration, d) shear and air-broth interfaces. To study the above items at lab and 100 l scale, practical downscaling experiments were designed. Medium and inoculum prepared at large scale was tested at lab scale. CO₂ was added to fermentations. Mixing effects were scaled down: pH, substrate and oxygen gradients were simulated at lab scale. Broth was subjected to high shear and large air-broth interfaces. Dissolved CO₂ was identified as the main cause for the scaling-up problem. An increased ventilation rate and a decreased head pressure combined with a lower pH, increased lipase production at 100 m³ up to values comparable to 10 l scale.

At Gist Brocades BV, the Bio Specialties Division has an important market position in enzymes for detergent applications. To obtain a broader product spectre and to strengthen the market position in detergent enzymes the lipase LIPOMAX™ will be introduced in 1995 for application in heavy duty detergents (HDD's) and light duty detergents (LDD's). An economic feasible process for LIPOMAX™ production is obtained by using new <u>Pseudomonas alcaligenes</u> technology.

A fermentation process for the production of LIPOMAX™ was scaled up from 10 l scale to an existing 100 m³ fermenter. A significant loss of lipase production during scaling-up was found. The downscaling experiments performed to investigate the possible causes for the discrepancy between small and large scale are described. A practical down to earth approach is chosen to get an indication of the sensitivity of the lipase production towards conditions representing large scale situations. A solution for the main cause is presented.

<u>DESCRIPTION OF THE LIPASE FERMENTATION PROTOCOL</u>

On 10 l scale, a fermentation process was developed for the production of lipase with <u>Pseudomonas alcaligenes</u>. <u>Ps. alcaligenes</u> strains obtained both by classical mutagenesis as well as with recombinant DNA technology were used for the production of lipase. The mode of fermentation was fed-batch. Temperature, pH and dissolved oxygen concentration were controlled. A mineral medium was used. Biomass and lipase production took place during a carbon limited feed phase on soy-oil.

<u>SCALING-UP TO 100 m³ PRODUCTION SCALE</u>

This newly developed fermentation process had to be scaled up to an existing production fermenter. For scaling-up to 100 m³, 100 l and 4 m³ fermenters were used for experiments. The maximum specific power input which can be achieved in the production fermenter was chosen at 100 l and 4 m³ scale. An airflow of 1 vvm was used at 10 l scale and 2 vvm at 100 l scale. At 4 m³ and 100 m³

41

E. Galindo and O.T. Ramírez (eds.), Advances in Bioprocess Engineering. 41-46.
© 1994 Kluwer Academic Publishers. Printed in the Netherlands.

scale 0.5 vvm was applied to preventexcessive foaming. Except at 10 l scale a head pressure of approximately 0.5 bar was applied. The composition and concentration of the medium was the same at all scales. The inoculum percentage as well as the temperature and pH did not change with scale. The specific soy-oil feed rate, expressed as kg soy-oil fed per m^3 broth volume per hour, was kept constant at all scales. The fermentation protocol, and the fermentation conditions as described above, at all scales was called the Standard Protocol (SP). The SP formed the basis for the lipase fermentation process development.

Production at all scales in the SP

10 l, 100 l, 4 m^3 and 100 m^3 scale fermentations were performed as described above. Figure 1 gives the relative lipase concentration as a function of time. It can be seen that the lipase concentration and therefore the production decreases with increasing scale. The production at 4 m^3 and 100 m^3 scale is much lower compared to 10 l and 100 l scale. At 4 m^3 scale the maximum lipase concentration was reduced by 50% of the value at 10 l scale and at 100 m^3 scale by 65%! It could be concluded that the lipase production loss was alarmingly high during scaling-up. Besides the lipase production, no major differences between the fermentations at the different scales were found, including the biomass development. The same scaling-up effect was found with a strain producing better at 10 l scale.

POSSIBLE CAUSES FOR THE LOSS OF PRODUCTION DURING SCALING-UP

From the characteristics of the lipase fermentation process, possible causes for the production loss can be deduced. The SP was used as the starting point. The possible causes are related to physical and physiological phenomena which can occur at larger scale (1, 2, 3, 4). The hypothesis is that if the changes in the immediate environment of the micro-organism do not influence the

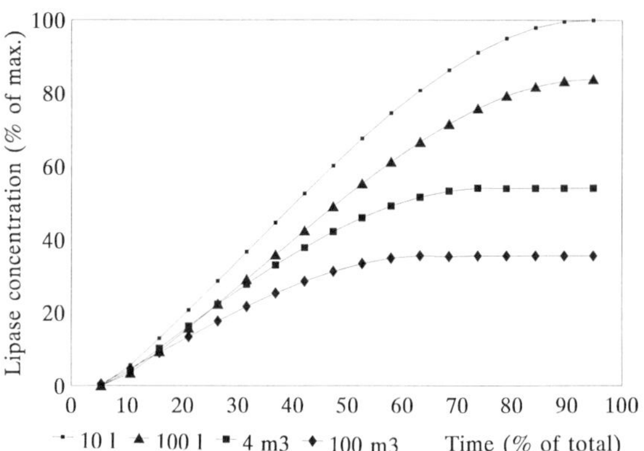

Figure 1. Relative lipase concentration (% of 10 l value) as a function of time at 10 l, 100 l, 4 m^3 and 100 m^3 scale.

behaviour of the micro-organism from lab scale towards large scale, scaling-up is successful. A classification was made in the type of effect and the priority for research. For the effects a distinction was made between the effects on the lipase-enzyme, on the micro-organism physically (vitality of the cells) and of the production of lipase by the micro-organism, (physiologically).
An extensive list of possible causes was set-up from which 5 main possible causes, with the highest priority for investigation, were put forward:
- Gradients of soy-oil, ammonia and pH caused by mixing effects
- Raw materials, medium preparation, sterilisation procedure and inoculation procedure
- Dissolved CO_2 concentration
- Shear and gas-liquid interfaces
- Local oxygen depletion and oxygen gradients caused by mixing effects

Except for shear and gas-liquid interfaces, the type of effect is believed to be physiologically. Shear and gas-liquid interfaces can affect the lipase enzyme and the vitality of the micro-organism which also can be expected from pH and ammonia gradients.
Lack of ideal mixing in larger scale fermentations causes gradients of components which are supplied to the fermenter like soy-oil, ammonia for pH control and oxygen. Fluctuation of the environment of the cells in a production fermenter may influence

the lipase production. The specific raw materials, medium preparation and sterilisation depend on the scale and the production location. At 10 l scale inoculation occurs with a shaking flask. At the other scales an aerated, stirred and pH controlled inoculation vessel is used. The ventilation rate at 4 m^3 and 100 m^3 scale was lower than at small scale. Both hydrostatic and head pressure increase with scale. The maximal dissolved carbon dioxide concentration and the maximum dissolved oxygen concentration are therefore scale dependent and do increase with larger scale. This can affect the lipase production (5). High velocity gradients which occur near the stirrer and splashing air-bubbles can have a negative effect by damaging the cells or the enzyme. At the gas-liquid interface the surface-active lipase enzyme is possibly inactivated. The gas is represented by the air bubbles, the liquid by the broth (air-broth interfaces).

DOWN SCALING EXPERIMENTS

To investigate possible causes for the production loss while scaling-up, down scaling experiments were performed. Down scaling experiments can give an indication of the sensitivity of the cells towards conditions representing large scale situations. The question is whether the cells (and the production) are hindered by the environmental conditions occurring at large scale (1, 2, 3, 4). The phenomena related to scaling-up, as described previously, were tested at 10 l and 100 l scale. In most experiments an attempt was made to change only one parameter at a time compared to SP conditions, which however can affect several fermentation conditions. In this way possible causes could be tested separately. Combinations and interactions of parameters, which are likely to occur at large scale, were not investigated.

Raw materials, medium preparation, sterilisation procedure

Medium prepared at 4 m^3 scale from industrial materials at industrial

conditions and inoculum prepared in an inoculation vessel were used both at 10 l and 4 m^3 scale. In this way a fermentation at 10 l and at 4 m^3 scale was performed at SP conditions with the scale as the only difference. The results were compared with the production in a 10 l fermentation completely prepared at the lab. No significant effect of the medium, the medium preparation at 4 m^3 scale and the inoculum was shown. Medium prepared at 100 m^3 scale also was used at 10 l scale. No significant effect on the production at 10 l scale could be seen.

Dissolved CO_2 concentration

At pH=7.0 CO_2 dissolves in the broth to fairly high concentrations mainly as HCO_3^- (pKa=6.3). The maximum dissolved CO_2 concentration at 4 m^3 and 100 m^3 was 3-5 times higher as at 100 l and 10 l scale. At 10 l scale CO_2 was added in the inflowing air, resulting in a CO_2 concentration comparable to large scale conditions. Figure 2 shows the relative lipase production as a function of time. The lipase concentration was reduced 35-60% compared to the SP. At 100 l scale an increase of the CO_2 concentration of a factor 3 can be obtained by increasing the head pressure and lowering the airflow. A 30% decrease of the lipase concentration was found.

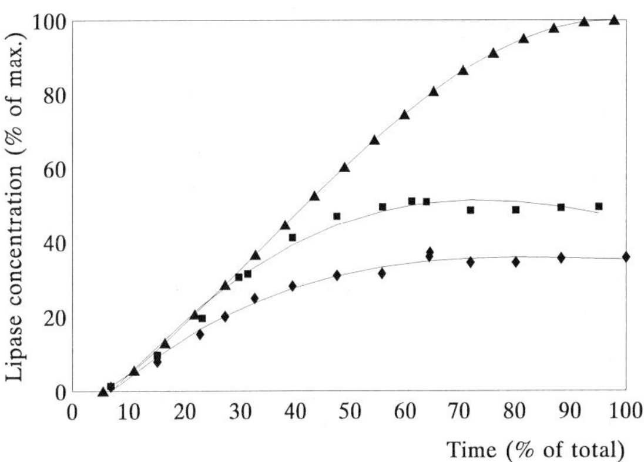

Figure 2. Relative lipase concentration (% of 10 l value) as a function of time at 10 l scale, addition of 5% and 10% CO_2 in the inflowing gas.

pH, soy-oil and oxygen gradients

At small scale gradients occurring at production scale can be simulated. As a model a volume element of broth which can be considered as completely mixed can be used. The volume element can be represented by a 10 l ideally mixed fermenter. The cells present in the fermenter (the volume element) experience fluctuating concentrations of soy-oil, ammonia, pH, oxygen, etc. as a function of location. The position of the element travelling through the production fermenter is replaced by the fermentation time in the 10 l down scaling experiment. Circulation times of 10-20 seconds are quite common on production scale. The circulation time represents the transport time of the volume of broth in the production fermenter. On basis of the fermenter geometry a well mixed compartment of 10 m^3 compared to 100 m^3 broth was assumed.

For down scaling of the soy-oil gradients an alternating soy-oil feed was used. The estimated circulation time gives the frequency of the alternating soy-oil feed. Two different soy-oil feed rates were tested. A circulation time of 20 seconds was simulated by a feed of 2 seconds followed by a pause of 18 seconds. To investigate an extreme condition, a total circulation time of 200 seconds was chosen with a feed of 10 seconds and no feed of 190 seconds. Surprisingly, no effect of the alternating soy-oil feeds was found.
To simulate pH gradients in a 10 l down scaling experiment a pH fluctuation of 6.5-7.5 and 6.7-7.3 was chosen. The pH was forced to the upper boundary by titration with ammonia. Metabolic activity caused the pH to decrease. No effect of the alternating pH was found.
The effect of local oxygen depletion and oxygen gradients was simulated at lab scale by not aerating a part of the broth. Due to poor mixing, stagnant zones may be present and high metabolic activity caused by a local high soy-oil concentration can give oxygen depletion. 10 m^3 out of 100 m^3 broth was assumed to be subjected to oxygen depletion. In a coupled two compartment system local oxygen depletion and oxygen gradients were simulated. One compartment was a completely controlled aerated and stirred fermenter, the second compartment was a non-aerated stirred and temperature controlled vessel. A residence time of 200 seconds in the non-aerated vessel was chosen. A significant decrease of the lipase production of 25% compared to the SP was found by subjecting a part of the broth to oxygen depletion.

It must be emphasized that the experimental set-up is an over-simplification of the production situation. The reality is far more complicated. Combinations of gradients and interactions are possible and also are likely to occur, also a circulation time distribution is more representative for large scale. However, with scale down an indication of the sensitivity of the production towards fluctuating pH, concentrations of soy-oil and dissolved oxygen can be obtained. It seems likely, that there are no major effects of mixing time concerning pH and soy-oil gradients which thus are not responsible for the discrepancy between lab and production scale.

Shear and air-broth interfaces

Description of shear is complicated. The effect of shear depends on the sensitivity of the micro-organism and the enzyme. The possible effect of shear and air-broth interfaces was investigated in 100 l experiments in which extreme physical conditions were created related to shear and air-broth interfaces. At production scale near the stirrers high shear rates and large gas-liquid interfaces are present. A four-fold increase of specific power input by increase of the stirring rate, an airflow of 3 vvm (giving a 50% increase of superficial gas velocity) and high oxygen concentration (by doubling the head pressure) were applied. A high oxygen concentration was applied to simulate the oxygen level near the stirrers. The broth volume was lowered to obtain an even more extreme situation. A significant lower lipase production of 20% was found. Similar experiments at lab scale also showed a 20% decrease of the lipase production. Although these extreme conditions do not occur constantly all the time at every location in a production fermenter, a negative effect on the production can be expected. The large observed gap between small and large scale can not be explained by shear and/or air-broth effects alone, a small contribution however can be expected. A possible effect of a high oxygen concentration and total pressure was tested separately

At 100 l scale the head pressure was increased to a level similar to the sum of the head pressure and the hydrostatic pressure at the bottom of the production fermenter at SP conditions. The dissolved oxygen concentration will increase, representing the higher maximum dissolved oxygen concentration at large scale. Due to the higher ventilation rate at 100 l scale, the increase of the CO_2 concentration can be neglected. A small negative effect of 10% on the lipase concentration was found.

Conclusions from the downscaling experiments

A major part (35-60%) of the production loss from 10 l towards production scale can be explained by the higher CO_2 concentration at larger scales. A smaller contribution (max. 25%) can be expected from local oxygen depletion and oxygen gradients, shear and interactions of air-broth interfaces. Interactions and combinations of fermentation parameters, which are likely to occur at large scale, were not tested.

No significant effect was found of raw materials, medium preparation (including the sterilisation procedure) and the inoculation procedure. Simulation of soy-oil gradients and also pH gradients showed no effect on the lipase production.

SOLUTION OF THE LOSS OF PRODUCTION DURING SCALING-UP

CO_2 has been shown to be the main cause for the production loss while scaling up. For the effect of shear and air-broth interfaces no simple solution is available. For aeration of a production fermenter a high local power input by the stirrers is inevitable. Also air-broth interfaces are not possible to avoid because of the need for aeration and the dispersion of air bubbles. To decrease local oxygen depletion as much as possible the oxygen transfer capacity must be as high as possible and the mixing time must be reduced. However, increasing the head pressure to improve the oxygen transfer causes an increase in the CO_2 concentration. Possibly the loss of production due to shear and air-broth interfaces can be compensated by an increased soy-oil feed.

If a negative effect of increase of the CO_2 concentration can be demonstrated, a positive effect at larger scales must be found by decreasing the CO_2 concentration.

To lower the CO_2 concentration at larger scales several methods are available:
- Decreasing the head pressure.
- Increasing the air flow.
- Lowering the broth volume will decrease the hydrostatic pressure and increase the ventilation rate, consequently the CO_2 concentration will be lowered.
- By decreasing the soy-oil feed rate, causing a lowering of the carbon dioxide production rate, the CO_2 concentration will be lowered.

These possibilities have been investigated in experiments at 4 m^3 and 100 m^3 scale. The lipase concentration could be doubled by decreasing the broth volume to 50 m^3. Decrease of the soy-oil feed rate by 50% at 4 m^3 scale gave a production comparable with the SP. At 4 m^3 scale an increased air flow and ventilation rate combined with a minimal head pressure resulted in a lipase concentration comparable to 10 l scale.

INCREASE OF PRODUCTION AT 100 m^3 SCALE

A reduction in the CO_2 concentration by almost a factor 3 at 100 m^3 scale can be obtained by lowering the head pressure and increasing the airflow. This is not enough to reach a CO_2 level comparable to 10 l scale. Decreasing the pH from 7.0 to 6.7 appears necessary. As the pKa of the CO_2 equilibrium is around 6.3, a small pH decrease has a large effect on the dissolved CO_2 concentration. At lab scale a low CO_2 level is present, lowering the pH slightly decreases the production at lab scale. By lowering the pH at larger scale the CO_2 concentration will be lowered, which has a large positive effect on the production. At pH=6.7 an optimum was found between the decrease of production by lowering the pH and the increase of production by lowering the dissolved CO_2 concentration.

An increased ventilation rate and decreased head pressure combined with decrease of the pH to 6.7 increased the lipase production at 100 m^3 considerably to values comparable to 10 l scale.

LITERATURE CITED

1. Oosterhuis, N.M.G., "Scale-up of bioreactors: a scale-down approach", PhD-thesis, Delft University of Technology (1984).

2. Sweere, A.P.J., "Response of bakers' yeast to transient environmental conditions", PhD-thesis, Delft University of Technology (1988).

3. Vardar, F. and Lilly, M.D., European. J. Appl. Microbiol. Biotechnol., 14, 203-211 (1982).

4. Feijen, J., Hofmeester, J.J.M., "Gradients in production scale bioreactors", presented at the 7th European Conference on Mixing, Bruges (1991).

5. Dixon, N.M., Douglas, B.K., Journal of applied Bacteriology, 67, 109-136 (1989).

Optimum Design of a Continuous Fermentation Unit of an Industrial Plant for Alcohol Production

S.R. Andrietta and F. Maugeri

Faculdade de Engenharia de Alimentos, UNICAMP Campinas, SP, CP 6121, CEP 13081/970, BRAZIL

The aim of this work was the optimum design of the fermentation unit of an industrial plant for alcohol production from sugar cane by Saccharomyces cerevisiae. *The kinetics was represented by a mathematical model having the growth rate as a function of the total reducing sugars, cell and alcohol concentrations and temperature. By means of computer simulations it was possible to obtain the best design for a fermentation plant under common Brazilian operational conditions. The result was a series of four CSTR's with medium feeding in the first reactor. The volume of each reactor was not necessarily the same. In fact, there was an optimal profile that gave the highest productivity. The optimum volume of each stage was obtained using the simulation results and the SIMPLEX method of optimization. The volume profile leading to the highest productivity was: 21% of total volume for the first reactor, 27% for the second reactor, 31% for the third one, and 21% for the last one. The equations used in the simulations were obtained from the mass and energy balances of the system and kinetic equations based on parameters taken from industrial data. The final result was implemented in an ethanol factory and the so far collected operational data showed good agreement with the theoretical results.*

The industrial production of ethanol started a long time ago, in the XIXth century. The large scale process has been much improved since then, and higher efficiencies have successively been achieved. The first breakthrough in ethanol production occurred in the 30's when the Melle-Boinot process was implemented. The main feature of this process was the cell recycle. The second breakthrough occurred in the 80's when the continuous fermentation with cell recycling was carried out in Brazilian factories.

According to Finguerut et al (1), the first continuous fermentation factories built in Brazil were very inefficient because they were rather adaptations from previous batch processes, leading to low-productivity units. More recently, new concepts for continuous alcohol fermentation plants have been developed, based on kinetics and process engineering. However, some of these failed mainly because they were developed under fermentation conditions other than the actual industrial and operational conditions. The understanding of process dynamics and the optimization studies are now much easier than some years ago with the today's faster computers, which simplified the task of simulating the fermentation processes. Therefore, computer simulation associated with a good knowledge of both industrial and environmental constraints and with the concepts of process engineering can be a very important tool for plant design.

In this work, the design of a continuous fermentation unit of an industrial alcohol fermentation plant was carried out, using mathematical models and computer simulations. This unit replaced a previous fed-batch unit in a Brazilian factory. Other industrial units, such as distilation and juice extration plants, remained unchanged.

KINETIC MODEL

The specific growth rate was considered to be affected by substrate, ethanol and cell concentrations, as stated by Lee et al. (2) in the model expressed by equation 1, and by the temperature (equations 2 and 3)

47

E. Galindo and O.T. Ramírez (eds.), Advances in Bioprocess Engineering. 47-52.
© 1994 Kluwer Academic Publishers. Printed in the Netherlands.

$$\mu = \mu_{max} \left(\frac{S}{S+K_s} \right) \left(1 - \frac{P}{P_{max}} \right)^n \left(1 - \frac{X}{X_{max}} \right)^m \quad (1)$$

$$\mu_{max} = A e^{\left(- \frac{E}{RT} \right)} \quad (2)$$

$$P_{max} = K_o e^{-aT} \quad (3)$$

According to van Uden (3), the ethanol tolerance by *S. cerevisiae* is constant between 12°C and 28°C and decreases at temperatures below 12°C and above 28°C. In a previous work, Andrietta (4) observed that the alcohol tolerance of the process microorganism decreases exponentially at temperatures above 32°C, as given by equation 2, where K_o and "a" are fitting parameters of the equation. According to Dale *et al.* (5), the effect of temperature on μ_{max} can be expressed by an Arrhenius-like equation as showed by equation 3.

KINETIC PARAMETERS

The kinetic parameters in Table 1 were adjusted on industrial scale plants by Andrietta and Stupiello (6) and (7). Parameters for equations 2 and 3 on Table 2 were determined from experimental results on a small scale.

PROCESS PARAMETERS

The process parameters considered in this work were the usual values used in Brazilian industrial plants, which are shown in Table 3.

REACTOR DESIGN

Figures 1 and 2 were constructed from the equations and tables above for a temperature fixed at 32°C. It can be seen in Figure 1 that a Plug Flow Reactor (PFR) is much more efficient(residence time is represented by the area below the curve) than a single Continuous Stirred Tank Reactor (CSTR), for which residence time is equal to the rectangular area, for alcohol production. However, a true PFR is

Table 1. Kinetic parameters determined in industrial scale

PARAMETER	VALUE
P_{max}	103 g/l
X_{max}	100 g/l
μ_{max}	0.41 g/l
n	3.0
m	0.9
K_s	1.6 g/l
$Y_{x/s}$	0.033
$Y_{p/s}$	0.445

Table 2. Kinetic parameters for equations 2 and 3.

PARAMETER	VALUE
K_o	895.6 g/l
a	−0.0676 °C^{-1}
A	4.50 x 10^{10}
E	1.54 x 10^4 cal/mol

practically impossible to obtain for large scale fermentation processes, so that a cascade of CSTR's could be a good solution since it has similar efficiency to a PFR and is much easier to obtain even in industrial scale. Figure 2 shows that the higher the number of CSTR's (up to eleven in the figure) connected in series, the lower the total volume, which means higher efficiency, for the same conversion of a PFR. However, eleven serial reactors in an industrial plant is not feasible, since operational and building costs are much higher than the gains represented by an improved productivity. Therefore, a compromise must be taken between the number of reactors and productivity gains.

The inclusion of one more serial reactor in a series can be considered economically feasible only if the total volume of the reactors decreases substantially. As a rule of thumb, 15% of the volume reduction is considered acceptable. In this present case, four reactors fulfilled the rule. The volume of each reactor was optimized using the SIMPLEX algorithm. The results are shown in Table 4, where the first column represents four different types of system.

Table 3.Process parameters normally used in Brazilian alcohol factories

PARAMETERS	VALUE
Substrate in the feed	180 g/l
Recycling rate	30%
Cell concentration in the cream	180 g/l
Cell concentration in the recycle	90 g/l
Cell concentration in the wine	3 g/l
Temperature of the cooling water	28°C

Types 1, 2 and 3 are existing industrial plants, already running in Brazil, and type 4 is the result of the optimization. Columns 2 to 5 represent the percentage of volume for each reactor related to the total volume, and column 6 the alcohol productivity calculated. It can be seen that the type 4 leads to the highest productivity. Although the difference between types 3 and 4 is less than 1%, it represents a gain of more than 700 m^3 of ethanol per year, for a similar plant.

The optimization was carried out with the following assumptions:

- according to the usual industrial practice in the alcohol factories, the susbstrate conversion in the first stage was taken as 55%, at least, in order to minimize

bacterial infection, and 99.5% at the fourth stage.
- all flow rates were considered constants, so that the residence time in each reactor was only

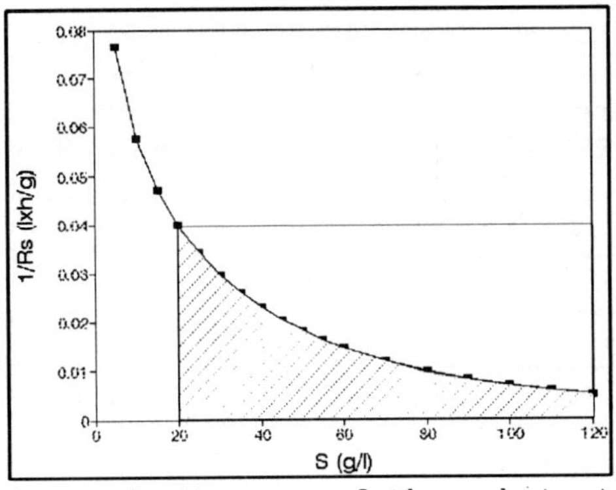

Figure 1. Inverse of the substrate uptake rate as function of the substrate concentration - Determination of residence time for PFR and CSTR.

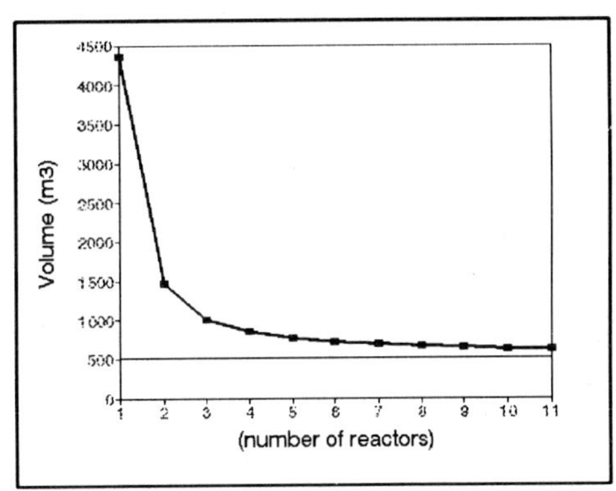

Figure 2. Total volume as a function of the number of CSTR's for a similar conversion of a PFR: —— PFR —█— CSTR

dependent on the reactor volume.
- the ethanol concentration in each reactor was only dependent on the substrate conversion, so that its concentration in the outlet of the forth stage was the same in all cases.
- all yield factors ($Y_{x/s}$, $Y_{p/s}$ and $Y_{x/p}$) were considered constants and equal in all reactors.
- the ethanol productivity, the parameter that was taken as the

objective function to be maximized, was only dependent on the total volume of reactors, since both the ethanol concentration in the outlet and the flow rate were constants.

Table 4. Comparation between four types of industrial plants.

TYPE	R1 %Vol	R2 %Vol	R3 %Vol	R4 %Vol	Φ
1	50	25	25	–	6.1
2	50	25	12.5	12.5	7.2
3	25	25	25	25	7.8
4	21	27	31	21	7.9

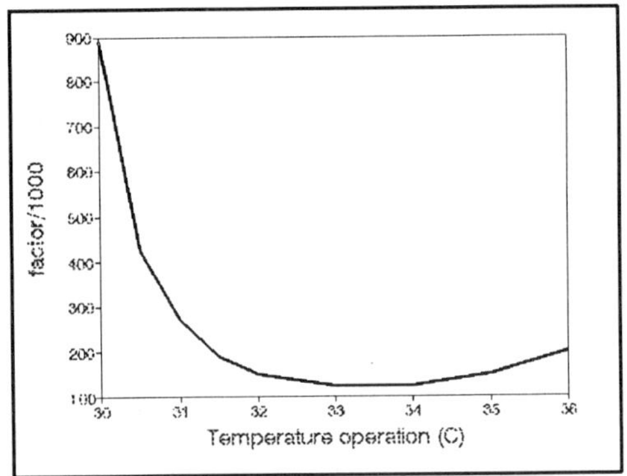

Figure 3. Factor "f" as a function of the temperature – Determination of the process temperature.

Other important parameters, most of them shown in Table 3 and used in this work, are traditionally used in the Brazilian factories and previously optimized by technical and economical analyses. For example, the cell recycling rate and cell concentration in the recycle cannot be much higher than the values considered here because otherwise they would increase the needs on centrifuges capacity. Centrifuges are very expensive and so are their maintenance costs.

THE PROCESS TEMPERATURE

The process temperature has a strong influence on reactor volumes and on the area of heat-transfer surface of the heat exchangers. Along with the centrifuges, the heat exchangers are the most expensive equipment in the plant, therefore the heat-transfer surface must be minimized according to the process temperature. From the energy and mass balance, it is possible to verify that the higher the medium temperature, the higher the reactor volume and the lower the heat-transfer surface. To obtain the processes temperature, a factor f that represents the product of the total volume of reactors times heat-transfer surface, was plotted against temperature in Figure 3. It can be seen that f decreases when the temperature increases, which means that there is a considerable decrease in the heat-transfer surfaces. The curve reaches the minimum at 33.5°C, considered the best value for the process temperature, since it represents a good compromise between heat-transfer surface and reactor volume.

THE INDUSTRIAL PLANT

Figure 4 shows the plant flowsheet. As state before, the plant has a series of four CSTR's (R_1 to R_4), with different volumes, according to the conversions values and to the production scale. There is a single feed (F_w) in the first reactor. The feed medium (F) is mixed on line with the yeast suspension stream (Recycling flow rate, F_r) coming from the acid treatment. The temperature is kept constant by means of four plate heat-exchangers (HE_1 to HE_4) whose colling water (F_j) comes directely from a nearby river. The fermentation broth flows (F_{ci}) to the heat exchangers are the same as the cooling water flows (F_{ji}). The fermentation broth coming out from the last stage goes to the centrifuges (C). The heavy phase (cream) or cell suspension flow (F_1) from centrifuges goes to the acid treatment tank (AT) and the light phase (wine) flow (F_{vt}) goes to the distillation columns. The ratio F_r/F_w is kept between 0.2 and 0.45, according to the production scale.

Overflow outlets for gases and foam

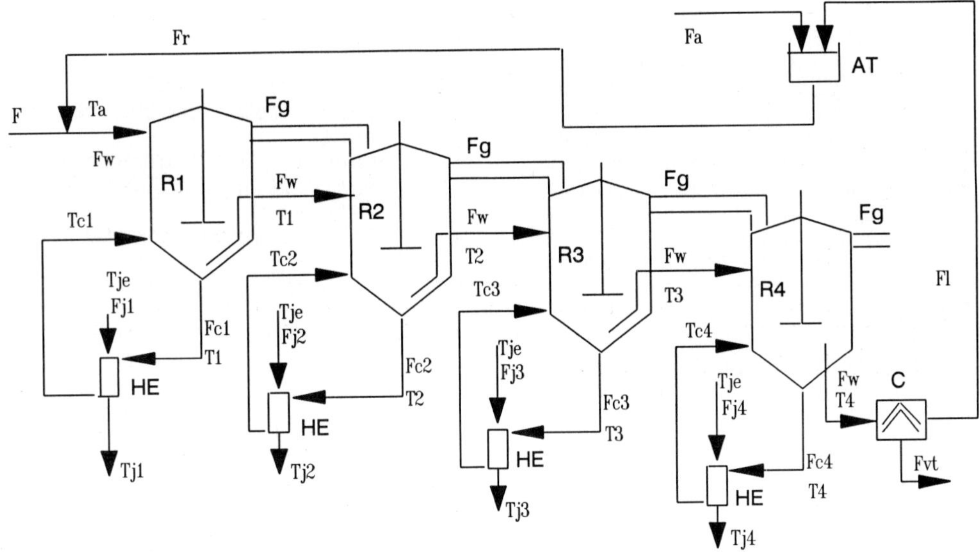

Figure 4. The plant flowsheet

are built on the top of reactors (F_G). These outlets turned out to be very efficient devices, because the foam passes throught the reactors, pushed by the gases and by the communicating vessels effect. Since the foam amount diminishes in the last stages, the anti-foam comsumption is minimized (25% of the traditional comsumption). The liquid outlets are in the bottom and the inlets on the top in order to enhance mixture and minimize hydraulic problems in the reactors. The agitation was promoted by liquid recycling at a very high flow with a tangential inflow.

The cooling system consists of an unit of plate exchanger for each reactor with flow rates for both cooling water and fermentation medium kept at similar values. The acid treatment unit consists of four serial tanks with aeration and pH control.

PRODUCTION DATA

This continuous fermentation plant was built to replace a previous fed-batch plant that consisted of 24 fermenters with 200 m^3 each, producing 400 m^3 of 96% ethanol per day. The continuous plant has a total volume of 2,500 m^3, producing about 440 m^3 of 96% ethanol per day. The productivity in the continuous plant is twice as high as the fed-batch plant. Also the theoretical prediction for the continuous plant production was 450 m^3 of 96% ethanol per day (a deviation of about 2%).

CONCLUSION

The process modelling approach with kinetics parameters adjusted from industrial data, a good knowledge of operational constraints, and computer simulations were efficient tools for the design of a large scale ethanol-producing plant. The inclusion of the temperature effect in the kinetic model allowed us to determine the area of heat-exchange surface for each reactor. Also, computer simulations associated with a method of optimization (SIMPLEX) led to the optimal design of the plant, considering Brazilian industrial conditions. The good agreement between theoretical and industrial data allows us to conclude that, for the alcoholic fermentation, the design of a large plant can be successfully achieved, skipping the traditional approach of Laboratory-Kilolab-PilotPlant demonstration previous to the Industrial Unit design.

The approach used in this work was very satisfactory and might be applied to fermentation process other than alcohol production, provided that a representative mathematical model is available and the adjustment of kinetic parameters can be accomplished from industrial data.

Currently, using the approach used here, an Expert System for the design of alcohol fermentation plants is under development.

NOMENCLATURE

a	Parameter in eq. (2)
A	Parameter in eq. (3)
E	Activation energy (cal/mol)
F	Inlet medium flow rate (l/h)
F_a	Dilution water flow rate (l/h)
F_{ci}	Reactant fluid flow rate in the ith heat exchanger (l/h)
F_G	Gas and foam exit
F_{ji}	Cooling water flow rate in the ith heat exchanger (l/h)
F_1	Yeast cream flow rate to acidic treatment (l/h)
F_r	Recycling flow rate (l/h)
F_{vt}	Wine flow rate to destilation (l/h)
F_w	Feedstream flow rate (l/h)
K_o	Parameter in eq. (2)
K_s	Substrate saturation constant (g/l)
m	Cell inhibition power
n	Product inhibition power
P	Product concentration (g/l)
P_{max}	Ethanol concentration limit (g/l)
R	Gas constant (cal K^{-1} mol^{-1})
S	Substrate concentration (g/l)
T_i	Temperature of the ith reactor (°C)
T_{ci}	Temperature of the reactant fluid in the ith heat exchanger (°C)
T_{ji}	Temperature of the cooling water in the ith heat exchanger (°C)
T_{je}	Temperature of the cooling water (°C)
X	Cell concentration (g/l)
X_{max}	yeast cell concentration limit (g/l)
μ	Specific cell growth rate (h^{-1})
$μ_{max}$	Maximum specific cell growth rate (h^{-1})
Φ	Ethanol productivity (G.l^{-1}.h^{-1})

<u>LITERATURE CITED</u>

1. FINGUERUT, J.; CÉSAR, A.R.P.; LEINER, K.H.; VAZ ROSSEL C.E. Fermentação contínua em múltiplos estágios. *Stab Açúcar, Álcool e Subprodutos,* **9**: 45-51, 1990b.

2. LEE, J.M.; POLLARD, J.F.; COULMAN, G.A. Ethanol fermentation with cell recycling: computer simulation. *Biotechnol. Bioeng.,* 25: 497-511, 1983.

3. VAN UDEN, N. Ethanol toxicity and ethanol tolerance in yeast. Use of extracelular acidification for rapid testing of ethanol tolerance. *Biotechnol. Bioeng.,* **28**: 1596-1598, 1985.

4. ANDRIETTA, S.R. Modelagem, Simulação e controle de fermentação alcoólica contínua em escala industrial. PhD Thesis, Unicamp/FEA - Brazil, 1994.

5. DALE, M.C.; CHEN, C.; OKOS, M.R. Cell growth and death rates as factors in the long-therm performance, modeling and design of immobilized cell reactor. *Biotechnol.Bioeng.,* **36**:983-92,1990.

6. ANDRIETTA, S.R. & STUPIELLO, J.P. Simulação e modelagem para processos de fermentação alcoólica (I) Batelada Alimentada. *Stab Açúcar, Álcool e Subprodutos,* **8**: 36-40, 1990a.

7. ANDRIETTA, S.R. & STUPIELLO, J.P. Simulação e modelagem para processos de fermentação alcoólica (II) Contínua. *Stab Açúcar, Álcool e Subprodutos,* **9**: 45-51, 1990b.

Scale-up

N.W.F. Kossen

Gist-brocades B.V., Research & Development, P.O. Box 1,
2600 MA Delft, THE NETHERLANDS

In scale-up as well as in carpentry, there is no single preferred tool to solve your problems. You need a whole tool box and you generally need more than one tool in order to do one particular scale-up job properly. After a general introduction about scale-up I will show you this tool box, a collection of the various methods that can be used to solve scale-up problems. Each of these methods has its advantages and disadvantages. They are arranged in a sequence going from very fundamental to trial and error. Apart from the tools you need a proper state of mind. My suggestion for the preferred state of mind is "scaling-down". This means that you work backwards, starting with your "minds eye" of the full scale. This is of great help to increase the efficiency of the scale-up job. I will show you a few examples that give some feeling for the fitness for use of the tools for different kinds of jobs. You will also find some introductory references to the literature.

The start-up of the new factory was not exactly what Tim had hoped for. In the beginning progression was not bad at all, but after three months this progression came to a stand still and lots of alterations had to be implemented. However, even now after six additional months of hard work and many nights without sleep, the production capacity was 20% below budget and the product quality was still below specs. The general manager and the sales people were furious at Tim.

Tim started as manager of the project two years ago. The research phase had just been finished. The results where so convincing that the general manager decided to invest the 50 million dollars needed for the further development of the product and for the new plant. The new product should give a badly needed boost to the companies' profits.

Tim was desperate.

SCALE-UP: THE CHALLENGE

The challenge as such

As you may have gathered from the title of this publication Tim's problem was a classical scale-up problem.

Very often scale-up is seen to be a problem but by the same token it can also be seen as a tremendous challenge and a source for new knowledge and insight.

It often implies working at the frontiers of your knowledge. It asks for creativity and imagination, for high professional skills, and for intellectual flexibility.

You need a number of professional skills like physical feeling, mechanistic insight (in genetics, kinetics of conversion, transport phenomena etc.) and you should master some mathematics.

These are necessary conditions, but they are not sufficient. For a succesful scale-up you also have to be aware of a number of important non-technical issues like:
1. A close collaboration between scientists, engineers, marketing and production.
2. The skills of personnel at different scales of operation and at parts of an organisation.

E. Galindo and O.T. Ramírez (eds.), Advances in Bioprocess Engineering. 53-65.
© 1994 Kluwer Academic Publishers. Printed in the Netherlands.

3. Changes in raw materials, instrumentation etc. during scale-up.

The emphasis in this publication is on the professional aspects (once more: they are only part of the game!). For the other aspects see Tong & Inloes [1] and Kossen [2, 3].

The scope of this publication

This publication is an introduction to scale-up for scientists and engineers in the (bio)process industry. The emphasis will therefore be on scale-up problems in the process industry and examples mainly refer to the production of (bio)chemicals.

After reading and digesting the material in this manual you should feel more comfortable when confronted with scaling-up because you understand the potential problems of scale-up and you are aware of the various methods to solve them.

It is assumed that you have some basic process engineering knowledge (not too much, don't worry).

Tim started to think what might have gone wrong and where and how similar problems could be avoided in the future.

SCALE-UP: WHAT IS IT AND WHAT KIND OF PROBLEMS CAN SHOW UP?

What is scale-up?

Is SU just an increase of size? Literally speaking the answer is "yes". Scale-up means designing and building a full scale airplane, oil tanker, refinery, a bridge, a fermentation plant etc. on the basis of calculations and small scale experiments.
Scale-up also means that present airplanes, oil tankers etc. are bigger than in the sixties.

At a higher level of abstraction, however, the answer is not so obvious. What we want to achieve with scale-up is generally an increase of

capacity of a device (eg a fermentor). An increase of the size of a fermentor is then one of at least three possibilities. Another possibility is to increase the capacity of an existing device (for an existing fermentor the capacity can often be increased considerably by strain-improvement). A third possibility is to use similar devices in parallel operation, like a second fermentor (parallel scaling-up).

The emphasis will be on the increase of size.

What kinds of problems can show up as a result of failed scale-up efforts?

The following inventory is not exhaustive but represents a number of often observed phenomena:
1. Delayed start-up (usually resulting in loss of market share).
2. Lower economic performance:
   a. lower yield on raw material
   b. lower capacity
   c. lower product quality
   d. frequent shut-down because of infections, fauling, corrosion etc.
3. Operational instability:
   a. mechanical instabilities like vibrations of stirrers
   b. loss of genetic capacity in a (continuous) fermentor
   c. unforeseen variations of flow, temperature and concentration

All of the above in short leads to loss of profit.

SCALE-UP PROBLEMS: THE CAUSES

Why do new wind generators and new airplanes sometimes crash, why do new bridges sometimes collapse, why do process plants quite often start up and become on spec much later than expected?
From a technical point of view there are a number of causes. The two most important ones will be given below:

A common cause

The common cause is a very fundamental one. In all sciences, also the natural sciences, we use models (verbal models, mathematical models etc.) and models can fail.

Why can models fail? What is a model?

"A model is a representation of certain aspects of a part of reality as we see it".

It is clear from this statement that a model is a strong reduction of the complex reality and it is definitely not reality itself. Therefore models can and sometimes do fail, also in scale-up. This is unpleasant but we have no alternative, we can not deal with reality as such. We can diminish the problem by improving the quality of our models but it will never disappear.

An example. Our models for microbial kinetics are always a tremendous reduction of the thousands of complex reactions within the micro-organism.

A trivial cause

There is another cause particularly for conventional ways of scale-up.
Every technical conversion process (eg fermentation) is based on a number of mechanisms (kinetics, mixing, mass transfer, heat transfer). Each of these mechanisms can be characterized with a time constant (t_c). This is the time that is characteristic for the speed of the mechanism. Usually t_c is equal to a capacity divided by a flow (see appendix 1 for a few examples) but also purely biological time constants exist.

If (the ratio's of) these time constants don't change during scale-up the system should behave in the same way at the different scales (see dimensional analysis in the section about the tools).
Unfortunately this is often impossible. For example: mixing times (t_{cm}) virtually always increase during scale-up in fermentors, while the kinetic time constant (t_{ck}) remains more or less constant.

By the way, this is a good reason to scale down (see the section about the tools): do your small scale experiments under the same mixing conditions as at the full scale.

Some other causes

There are a few other causes that frequently result in scale-up problems:
1. Surface phenomena:
   a. coalescence of bubbles and droplets $(k_1a$ etc)
   b. crystallization
   c. wetting, foaming, deposits etc.
   Surface phenomena can be a nuisance because even extremely small amounts of surface active materials can have a tremendous effect. These effects can show up if you use at production scale raw materials different from the ones at small scale.
2. Insufficient knowledge of the details of the kinetics of micro-organisms (eg of the formation of by- products).
3. Unpredictable fluid flow. In particular: two phase turbulent flow of non-Newtonion fluids (not uncommon in biotechnology!). In particular when the rheology is very dependent upon the morphology of the micro-organisms.
4. Other aspects of SU mentioned already like skills of personell, change of raw materials etc.

Tim had a number of discussions with the engineers in the development department about the rules used for the design of the apparatus. The majority of these rules had a proven track record in the company albeit under somewhat different conditions. Some of them were new, however, because the operations at hand were new. These new rules were based on models they had found in sound engineering literature. A number of the parameters needed in the design rules had been measured in R&D under conditions that where not clear to Tim but the chief engineer said: "The causes of your problems must be found elsewhere, we all did a proper job!" Tim was not reassured.

SCALE-UP PROBLEMS: AN INTRODUCTION TO THE TOOLS

The following sections deal with the tools of scale-up but mainly with the choice of the proper tools, the methodology. The tools as such will be used only as examples to illustrate the methodology. More about these tools can be found in the existing literature which is rather abundant. A number of references will be given. At the end you should have some feeling for the tools, their (combined) use and some pro's and con's.

Two important introductory remarks about scale-up:
1. **No scale-up surprises will show up if the micro-environment of the actors involved (molecules, micro-organisms, bubbles etc.) does not change during a change of scale (the actors are very short sighted).**
2. **Always, as soon as possible after the first laboratory experiments, start mentally with your image of the full scale (imagine what it should look like) and work backwards to the smaller scales**

These two remarks, when combined, give a clue to a very intersting method to solve scale-up problems: perform your small scale experiments in a way that is representative for the full scale (the representative small scale conditions: RSSC). This is scaling-down and will be delt with in the next chapter. Just two more remarks about scaling-down right now.

Sometimes you cannot ignore the full scale, even if you wanted to do so, because it is already there (if you have to fit a new process in an existing production unit, which is quite common in a.o. biotechnology and fine chemistry where frequently multi purpose plants are used).

Down-scaling can also be an excellent tool to solve problems in full scale production (to perform laboratory experiments under RSSC).

Tim tried to find out what kind of small scale experiments could be helpful to solve his present problems.

SCALE-UP PROBLEMS: THE TOOLS

The following list is a survey of the tools (methods) to solve scale-up problems. They will be delt with one after another.

A. From first principles: very fundamental.
B. "**A**ll **r**ubbish **t**o **o**ne **s**ide" ("ARTOS"): fundamental
C. Lumped models: semi-fundamental.
D. Dimensional-analysis (DA): semi fundamental.
E. Scale down: semi empirical.
F. Trial and error (T&E): empirical.
G. Mixed method (well choosen mixtures of A to F).

The amount of detailed physical and chemical knowledge, needed to solve the scale-up problems, decreases going from A to F.

Tim went to the R&D department and tried to find out what kind of scale-up methods had been used and the corresponding experimental conditions. He found out that they had used, as always, their preferred traditional method. The conditions were the normal ones they said.

From First Principles

This method is the most fundamental one. It consists of solving the (micro) balances for momentum, mass and heat, including conversion (kinetics). Essentially there are two methods: analytical and numerical.

A simple example of an analytical method is the calculation of the velocity distribution for the laminar flow of a fluid through a pipe. The solution is:

$$v_x = R^2/4\eta \, (\Delta P/\Delta x)(1 - r^2/R^2) \qquad (1)$$

This equation allows the calculation of the full scale velocity distribution straight away (without the need for small scale experiments). You should realise however, that the equation is only valid under a number of limiting conditions:
a. The flow is fully developed, one-phase and laminar.
b. The fluid is non-compressible and Newtonian.
c. The pipe is circular and horizontal.

Also for laminar flow of non-Newtonian fluids in a circular pipe analytical solutions exist although they are a little bit more complicated.

The calculation of laminar flow of a non-Newtonian fluid in a non circular pipe is a fairly complicated one. This scale-up problem can be solved with numerical analysis but also with method B ("all rubbish to one side").

For the calculation of turbulent flow into some detail numerical methods are the only tool.

Something to keep in mind. The calculation of the full scale (flow) phenomena from first principles will become more and more popular. The rapidly increasing calculation power of computers and the sophistication of our models ad considerably to this development.
This allows us to calculate how the full scale will behave. However, particularly the momentum balance (flow phenomena) is a difficult one and computational fluid dynamics (CFD) is at present not yet able to predict full scale flow under other than rather simple conditions. The combination with kinetics makes things even more complicated. However, this defenitely is an area where developments should be watched carefully in the future. See among others Bakker [4].

A critical attitude regarding computersimulations is very important. As mentioned before our models are always approximations of reality. Furthermore the computer is very non-critical. It accepts wrong models and wrong values of parameters without notice. It is always good to have a good overview of what you are doing. Therefore a warning:

"Never accept computer solutions as such. Always check the outcome by an independent method."

Such an independent method can be any of the other methods mentioned in this section. Also experience can be very valuable here (even when this is based on plain empirical results). Very important is also your "physical feeling".

A historical example. An "air-bridge", connecting two buildings, has been designed by computer calculations. The bridge is constructed in a workshop. The bridge is put in place with a crane and connected to the buildings while still hanging on the crane. When the crane slowly lets go its hold the connection between the bridge and the building slowly breaks down. The procedure is stopped and the bridge is put on the ground.
After careful analysis of what went wrong it appeared that during feeding the computer with parameters a decimal point had been put at the wrong place.
Obviously no one had noticed neither this nor the unusual constructional outcome. These things happen but can be avoided if you check your computer solutions.

"All Rubbish To On Side" (ARTOS).

This is a very useful and, contrary to the name, elegant technique. It starts with a fundamental (differential) equation (i.e. from first principles). ARTOS is very helpful when this equation cannot be solved in a straightforward manner.
The principle is as follows:

a. All the variables depending in some (unknown) way on one main variable (eg concentration or pressure or temperature) are brought together in an integral on one side of an equation.
b. If the value of the term on one side of the equation is constant, then the value of the term at the other side has to be constant as well.

A simple illustration. In biotechnology immobilized enzymes are often used in the form of spherical particles put together in an apparatus called a packed bed.
The flow of a liquid through a packed bed can be described with the so called equation of Kozeny-Carman. In a simplified form this equation is:

$$d\Delta P/dh = K\eta v_s \qquad (2)$$

v_s is the superficial velocity (i.e. the velocity of the liquid in a column without particles).

K contains properties of the particles (size and shape) and the bed (porosity).

This equation can only be solved in a straight forward way if K is a constant (i.e. constant particle size/shape and constant bed porosity). For rigid particles this is no problem.

However, in biotechnology sometimes particles are used that are compressible to some extend (enzymes immobilized in cross linked gelatine, alginate etc.). This results in particles with changing shape and a bed with a changing porosity, both as a function of the bed height (h).
The question now is: "how can we perform our small scale experiments under RSSC?"
The answer is: "bring all variables that depend in some way or another upon pressure (the "rubbish") to one side". The result is"

$$\int d\Delta P/K' = \int \eta v_s dh = \eta v_s H \qquad (3)$$

H is the bed height at zero pressure. The pressure dependance of h has been put in K'. A more thourough treatment of this problem is given by van Tilburg [5]). The right hand side of equation (3) now only contains quantities independent of pressure.

The RSSC is obtained by keeping $\eta v_s H$ at the same value as at the full scale. As a result $\int d\Delta P/K'$ will be the same too ad both scales. At a small scale the value of H is much lower. To keep $\eta v_s H$ at the full scale value either v_s or η (or both) can be increased.

Another powerful application of this method is the scale-down of the flow of fluids with unknown viscosity in pipes of non-circular cross section [6] but there are many more.

Lumped Models

Here so called lumped models are used, for example a number of mixers in series with exchange flow, as a model for stirred reactors with multiple impellors. The amount of lumping can be chosen at will and depends on the details of information wanted. A rather limited amount of

detail usually gives already enough information to use this method for scale-up. Examples are the work of Oosterhuis [7] and Baltes [8].

Dimensional Analysis (DA).

This can be a useful method and there are many succesful examples, particularly for flow phenomena. The method is based on the fact that all terms in a relationship that describes a physical/chemical process must have the same dimension. It is generally used for up-scaling but it is also useful for down-scaling. A good introduction is the book of Zlokarnik [9]. The method has a number of advantages (a.o. reduction of the number of independent quantities, relationships between quantities) but also a number of draw backs as mentioned among others by Dickey [10].

To use this method geometric similarity is needed. This means that the large size is a blown up version of the small one (or better: the other way around). In other words: all ratio's of the geometrical dimensions remain constant during scale-up. The physical basis of DA is described below.

Every physical/chemical process can be described with a (simplified) model. In process technology these models are usually presented as a balance equation, a balance based on the conservation of momentum, mass or energy. These equations have the following general form:

accumulation = Σ transport + Σ conversion

Often these balances are presented as a differential equation (DE). An example is the one dimensional mass balance for a chemical reactor:

$$\partial C/\partial t = - v\, \partial C/\partial x + ID\, \partial^2 C/\partial x^2 - r \qquad (4)$$
$$\underset{\text{accumulation}}{} \underset{\substack{\text{transport} \\ \text{by flow}}}{} \underset{\substack{\text{transport} \\ \text{by diffusion}}}{} \underset{\text{conversion}}{}$$

In appendix 1 it is shown how this equation can be transformed into a dimensionless form. The coefficients of this dimensionless DE (often called dimensionless numbers) can be interpreted as ratio's of time constants (as mentioned before), velocities or length's, or as a ratio between forces (only for the momentum balance).

If these coefficients and the boundary conditions are constant for the small scale and the production scale, and if these two scales are geometrically similar, the solution of the DE is the same for both scales. This means that the dimensionless concentration profiles are identical. The experiments at the small scale then result in concentration profiles that are geometrically similar to the full scale.

This possibility to obtain identical dimensionless concentration profiles at both scales suggests that the SU-problem has been solved. Unfortunately this is not always so because, as soon as a number of dimensionless coefficients are involved, it is often not possible to keep them all constant during scale-up under geometrical similar conditions (see the example below).

As has been touched upon in one of the preceding sections this problem is particularly manifest if chemical reactions are involved. At the small scale the mixing time is usually quite small. A a result the kinetics are often rate determining (micro kinetics). At the full scale, due to the much larger mixing times, mixing can be rate determining (macro kinetics). This phenomenon is called a change in regime (from a kinetic regime to a mixing regime). It can often, but not always, be solved by using different physical conditions (temperatures, viscosities etc.) at the two scales. Particularly mixed regimes (where eg kinetics and mixing rate are both determining) can cause problems during scale-up.

Two hints
1. You can easily check at laboratory scale if the decrease of the mixing performance at full scale could be important for the overall conversion by checking if a decrease of the stirrer speed has a negative influence on the conversion rate.

Many more similar experimental "tricks" can help you to elucidate potential scale-up problems.
2. Generally regimes can very effectively be traced by calculating the time constants of the mechanisms involved.

There are several ways to use DA. Most often used is the formal approach, see Zlokarnik [9], where the dimensionless numbers are obtained from a structured grouping of all the relevant variables and parameters. This asks for a lot of experience (the omission of one important variable can result in a very peculiar outcome).
Another way is to work with ratio's of time constants, forces, velocities etc. to be obtained from DE's or directly from the mechnisms involved. Because you then start with the mechanisms as such and not with a collection of variables and parameters this method is less prone to errors.

An example: the mixing time (t_{cm}) in a stirred tank. Problem statement: Find the mixing behaviour of an existing full scale tank from small scale experiments.

First find the relevant quantities that determine t_{cm}:
- stirrer speed N (s^{-1})
- a characteristic diameter D (m)
- viscosity η (kg m^{-1} s^{-1})
- density ρ (kg m^{-3})
- accelleration due to gravity g (ms^2)

After some fiddling around with these quantities (see Zlokarnik) the following relation is found:

$$N*t_{cm} = (\rho ND^2/\eta)^{\alpha} * (N^2D/g)^{\beta} = Re^{\alpha} * Fr^{\beta} \tag{5}$$

Re is the Reynolds number, an indication for the flow conditions and Fr is the Froude number, an indication for the relation between centrifugal forces and the forces due to gravity.
It is interesting to notice that it is not possible to keep Re and Fr constant if D decreases unless η is decreased as well (however, if D is decreased with a factor 9, η should decrease with a factor 27).

This is impossible if the liquid at the full scale is water because liquids with such a low viscosity don't exist.

There is one possible escape here: if the tanks are both fully baffled Fr is not important any more (neglegible rotation of the fluid) and Re = const is then the only criterium.

Above Re = 10^4 the value of $N*t_{cm}$ is more or less constant.

Equation $N*t_{cm}$ = constant describes the mixing process quite well, but only the so called "macro mixing" i.e. mixing at he scale of about 1 cm^3. Therefore there is a potential "fly in the ointment":

Macro mixing is relevant for slow conversion processes ($t_{cc} \approx$ 1 min). For fast conversions ($t_{cc} \approx$ 1 sec) micro mixing is important as well and therefore the size of the smallest eddy (where diffusion takes over from mixing due to flow). This size (λ) is given by:

$$\lambda = (\nu^3/\epsilon)^{.25} \tag{6a}$$

This results in another dimensionless number:

$$\lambda / (\nu^3/\epsilon)^{.25} \tag{6b}$$

As a consequence ϵ (the power consumption per unit mass) has to be constant (if ν does not change with scale).

If gas-liquid mass transfer is important as well P/V has to be constant during scale-up (i.e. N^3D^2 should be constant as well). Because the liquid density is always more or less constant this condition is similar with ϵ = constant. This results in an extra scale-up constraint.

In appendix 2 an overview is given of some similar problems.

Scale Down

Scale down is a method, but above all it is a state of mind. A necessary state of mind for successful scale-up.

As soon as a process looks "feasible" at the laboratory scale people start thinking what the full scale should look like. This is something all engineers (including process engineers) have in common. But somewhere down the development track often something goes wrong. People get stuck because they start scaling-up: enforcing small scale results and conditions upon the full scale installation.

Tim was surprised to find out that the crystallization process in the laboratory had been performed under rigorous stirring conditions, conditions that could never be met at the full scale. When he asked the technician why he had not deliberately tried to decrease the stirring rate the answer was: "but then my experiment could have failed". Tim tried to explain why this would have been the perfect experiment, without success. He felt like a missionary in a country where no one spoke his language.

Scaling down is a semi empirical method, heavily focussed upon the already mentioned representative small scale conditions (RSSC's). The method results in translatable empiricism (a constant micro- of macro environment for the molecule or microbe).

Sub-tools to obtain the RSSC's are:
* Time constants, forces, velocities etc. These quantities can be obtained from fundamental equations, rules of thumb, DA, "physical feeling", etc. In fact this is a kind of regime analysis (see above and appendix 1).
** Representative elements of the full scale scale equipment eg. one pipe from a multi tubular reactor (distribution problems can be very important then).

The time constants etc. can tell you what the regime will be at the full scale, whether or not you can expect a mixed regime or a change of regime, how you can eventually improve this situation and how you can simulate this improvement at the laboratory scale to find out what the effect is at the full scale.

Geometric similarity is not necessary. It is much more important to realize at the small scale the same time constants for the relevant processes as at the full scale. Just let your imagination go.

Scaling down is extremely useful in biotechnology. Kinetics in biotechnology can be very complicated. Small environmental changes can have a large effect upon yield, productivity, formation of by-products, morphology, foaming etc. If you simply scale-up here it is rather naive to think that these kinetic aspects don't change. RSSC through down-scaling can give you a good idea of what you might expect at full scale and how you could avoid this. For some literature in this area see Vardar [11], Oosterhuis [7] and Sweere [12].

<u>Pure Empiricism</u>

Pure empiricism is more or less equivalent with Trial and Error (T&E). Increase the size every time with about a factor ten and see what happens.
Many long existing processes have been developped by T&E (brewing, distilling, classical bakers' yeast production, etc.). If you have no other tools available it is the only method left.

An example: Based upon a number of laboratory experiments an idea is formed about the potential full scale fermentor. An existing fermentor is the preferred candidate (no investments needed).

Because mixing could be a problem dimensional analysis is used to set up a number of experiments in a small scale vessel that is more or less geometrically similar. Mixing is not very good indeed.

Now a down-scaled experiment is set up in the laboratory to simulate this mixing performance and its influence upon the macro-kinetics.

Meanwhile there are intensive discussions between engineers microbial physiologists (expert workshops). Empirical know-how from people with experience with mixing-sensitive fermentations is also used.

In the end the existing fermentor is modified to a limited extend with minor investments.

This method is very flexible and effective. Appendix 3 gives a picture of this process, just to get the taste of it.

Tim had a hard time to stop a number of people in the plant just trying to change operating conditions because "you never know whether it works or not". He introduced weekly meetings with all the foremen of the shifts and the technical staff to dicuss what could be done based on good engineering practice. The basis for this was a long term plan, discussed with his technical staff, people from engineering and sales and a few people from R&D. From that moment on the start-up made real progress.

<u>Mixed Method</u>

This method consists of a mixture of the methods mentioned in this chapter, choosen for the scale-up problem at hand. This is combined with regime analyses and a scale-down state of mind.

Now, two months later, Tim finally was convinced that he had the situation well in hand. Productivity was even better than planned and the product was on specs. He had made a report how the problems he had met could be prevented in the future. All people involved had contributed to this report (R&D, engineering, production, marketing and sales) and were convinced of the value of the proposed methodology. He got a compliment of the general manager.
It took some time to convince everyone in R&D that a "failed" experiment could be vital for a fast development but it worked: the next project started up within two months with only half the development costs compared with an estimate of what would have been the situation when the old procedures would have been used.

<u>NOTATION</u>

C	Concentration	(kg/m^3)
D	Diameter	(m)
ID	Diffusion coefficient	(m^2/s)
g	Acc. due to gravity	(m/s^2)
H	Height of a column	(m)
h	Height of a column	(m)
K	Penetration param.	(m^{-2})
N	Speed of rotation	(s^{-1})
P	Pressure	(N/m^2)
R	Radius (total)	(m)
r	Radius (partial)	(m)
t	Time	(s)
t_{cc}	Time count convers	(s)
t_{cm}	Time constant mixing	(s)
v_s	Superficial velocity	(m/s)
v_x	Velocity in x-direct.	(m/s)
x	x-coordinate	(m)
Δ	Difference	$(-)$
ϵ	Power per unit mass	(W/kg)
η	Dynamic viscosity	(Ns/m^2)
λ	Smallest eddy size	(m)
ν	Kinematic viscosity	(m^2/s)
ρ	Density	(kg/m^3)

<u>LITERATURE CITED</u>

1. Tong, G.E. and Inloes, D.S., *Chemtech*, 566, (Sept. 1990)

2. Kossen, N.W.F., "Scale-Up Strategy in Fermentation," in *Bioreactor Performance*, Mortensen, U. and Noorman, H.J. (Ed.), The Biotechnology Research Foundation, Lund (1993).

3. Kossen, N.W.F., "Some Remarks about Problem Solving in Biochemical Engineering", presented at the 3rd International Conference on Bioreactor and Bioprocess Fluid Dynamics, BHR, Cambridge (1993).

4. Bakker, A., "Hydrodynamics of Stirred Gas-Liquid Dispersions," PhD thesis, Delft University of Technology, Delft (1992).

5. Van Tilburg, R., "Engineering Aspects of Biocatalysts in Industrial Starch Conversion Technology," PhD thesis, Delft University of Technology, Delft (1983).

6. Reitsma, H., *De Ingenieur* (Dutch), **78**, W 247, (1966).

7. Oosterhuis, N.M.G., "Scale-Up of Bioreactors," PhD thesis, Delft University of Technology, Delft (1984).

8. Baltes, M., Winter, S., Jenne, N., and Reuss, M., "Optimal Experimental Design for Identification of Parameters in a Multiphase Compartment Model for Stirred Tank Reactors," presented at the 3rd International Conference on Bioreactor and Bioprocess Fluid Dynamics, BHR, Cambridge (1993).

9. Zlokarnik, M., *Dimensional Analysis and Scale-Up in Chemical Engineering*, Springer Verlag, Berlin (1991).

10. Dickey, D.S., "Dimensional Analysis, Similarity and Scale-Up," in *Process Mixing-Chemical and Biochemical Applications: Part II*, AIChE Symposium Series, **293**, 143, New York (1993).

11. Vardar, F., Lilli, M.D., *Eur. J. Appl. Microbiol. Biotechn.*, **14**, 203 (1982).

12. Sweere, A.P.J., Luyben, K.Ch.A.M. and Kossen, N.W.F., *Enzyme Microb. Technol.*, **9**, 386 (1987).

APPENDIX 1: TIME CONSTANTS

If you have 15 guilders (your financial capacity) and you spend 5 guilders a day (your financial flow) then your financial time constant is:

$$t_c = 15/5 = 3 \text{ days.}$$

Another example: the characteristic mixing time is t_{cm}. This is the ratio between the volume of the vessel (V = capacity) and the flow from the stirrer (Φ_s):

$$t_{cm} = V/\Phi_s$$

Because under geometric similar conditions Φ_s is proportional with the rotational speed of the stirrer (N, s^{-1}) and with the volume of the vessel the result is that $t_{cm} \times N$ = constant (provided the flow is turbulent).
For kinetics the time constant is concentration over conversion:
$$t_{ck} = C/r$$

For first order processes r = const $\times$ C and t_{ck} is a real constant then.

Time constants can be obtained either by simple mechanistic considerations as above or from the differential equation (DE) that describes the process concerned. This is demonstrated below with the help of the DE representing a mass balance:

$$\partial C/\partial t = - v\, \partial C/\partial x + ID\, \partial^2 C/\partial x^2 - r$$

accumulation   transport   transport   conversion
by flow   by diffusion

Time constants can be obtained from this equation by dividing all terms by C_0 and by making x dimensionless by dividing it by a caracteristic length (eg the main dimension of the apparatus). The result is:

$$\partial (C/C_0)/\partial t = -v/L\, \partial (C/C_0) / \partial (x/L) +$$

$$ID/L^2\, \partial^2 (C/C_0) / \partial (x/L)^2 - r/C_0$$

The terms indicated with an arrow each have the dimension of a reciprocal time. The relevant time constants (t_c) are then:
L/v t_c of the transport due to liquid flow
L^2/ID t_c of the transport due to diffusion
C_0/r t_c of the conversion process

APPENDIX 2: SOME (CONFLICTING) SCALE-UP RULES

	relative value at 10 m³ (small scale: 10 l)				
quantities ⟶	P	P/V	N	ND	R
SU criterion: const.value of ▼					
equal P/V	10^3	1	.22	2.15	21.5
equal N (or tcm⁻¹)	10^5	10^2	1	10	10^2
equal tip speed	10^2	.1	.1	1	10
equal Re	.1	10^{-4}	10^{-2}	.1	1

The tip speed (ND) is sometimes used as scale-up criterion for dispersion processes.

APPENDIX 3 : MIXED METHOD

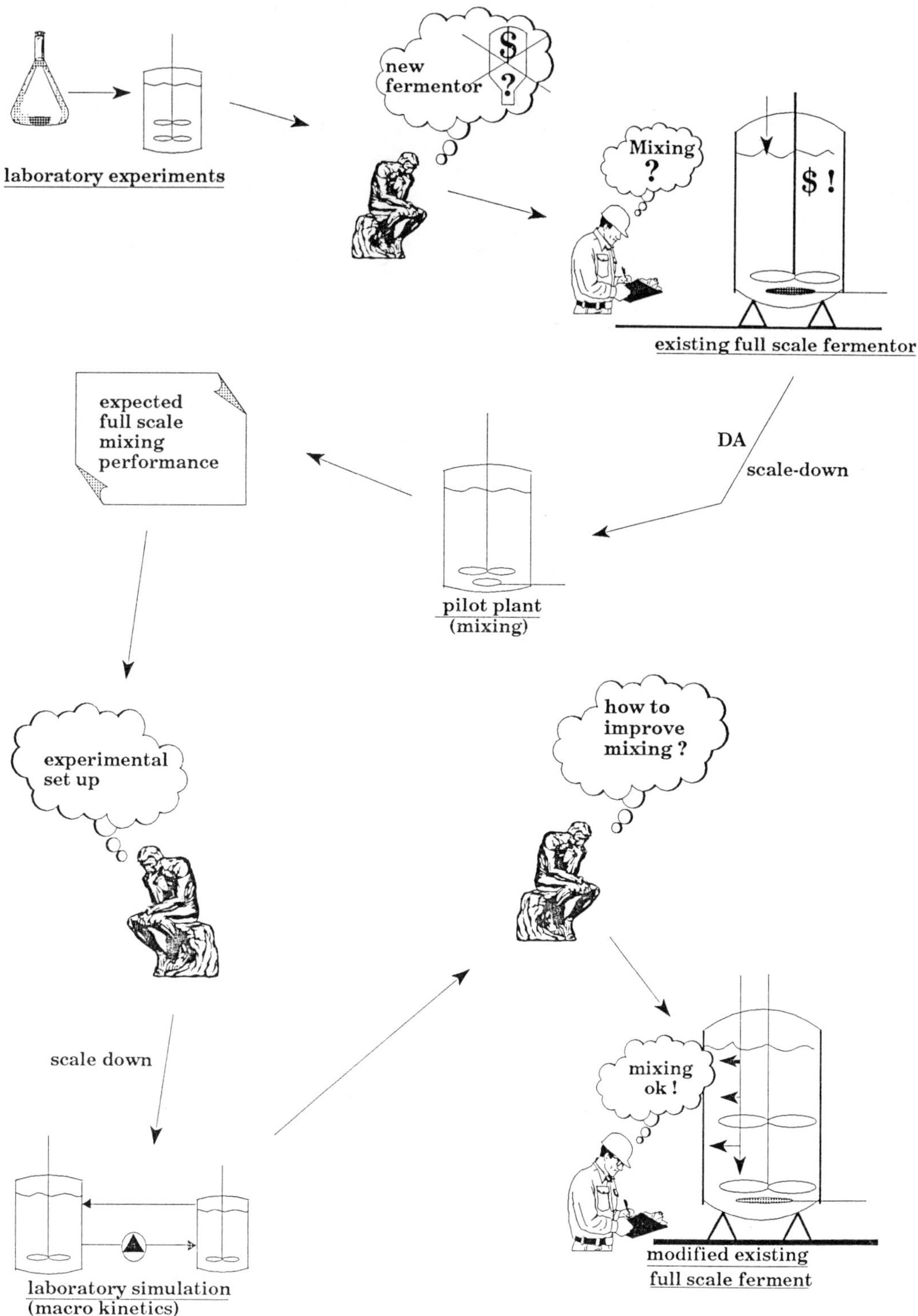

Invited paper

High Cell Density Yeast Production: Process Synthesis and Scale-Up

M. de la Torre, L. B. Flores, and E. Chong

Department of Biotechnology, CINVESTAV, P.O. Box 14-740, México 07000, D.F., MEXICO

A high cell density process for C. utilis *production on molasses was developed. A demonstration plant producing 70.8 kg h⁻¹ was operated in 1990 and 1991.* C. utilis *was continuously cultivated at cell densities of 80 to 110 kg m⁻³ in an especial 10.5 m³ jet loop fermentor having an oxygen transfer rate of 350 mol O₂ m⁻³ h⁻¹ at 1.1 VVM. The yeast was concentrated by a rotary filter yielding a 24% solids yeast cake, which is continuously thermolysed. The thermolysed product-which can be used as a natural flavor enhancer for food and fed - had a protein digestibility higher than 90%. The technology requires 20% less in capital investment and consumes only one third of the water involved in fermentation, if compared with conventional yeast production processes, being therefore a highly competitive alternative.*

A burst of activity appeared in the field of single cell protein (SCP) production in the 1950's and early 1960's. Much interest was aroused in the process developed by British Petroleum for production of yeast on hydrocarbons. By 1976 a full-scale 100,000 metric ton/year production plant had been built and was ready for operation. However, the oil crises and politics jeopardized the commercial viability of the project and subsequent events led to its end [1].

Later, ICI developed a process with methanol as an alternative raw material. A 50,000 metric ton/year plant was sanctioned in 1976 and started operating early in 1980. In this process both running cost and capital cost were reduced to a minimum in order to be competitive with protein supplements of vegetable origin, such as soya. The running cost was kept low by achieving a highly efficient conversion of methanol to protein, along with a minimum use of energy for fermentation and drying. The capital cost was kept down by operating a very large single-stream plant in continuous steady state [2]. No commercial plant currently uses this process.

Phillips Petroleum Co. chose a different approach.for SCP production. The products obtained are yeast-based natural flavor enhancers for food use and pet foods. The process is a continuous fermentation using high-cell-density technology and a special 25-m³ fermentor with an oxygen transfer rate of 800-1000 mol oxygen m⁻³ h⁻¹. In the process, liquid sucrose is added as the carbohydrate source and anhydrous ammonia as the nitrogen source. The broth contains 120 to 125 kg dry solids m⁻³. The plant is running since 1989 [3].

Since 1985, our research group has been developing a process for food-grade *Candida utilis* production from sugar cane molasses. A continuous 10.5-m³ jet loop fermentor was designed to take advantage of its intrinsic high oxygen transfer rate and energy efficiency [4]. This fermentor was provided with a computer system for online data acquisition and for controlling the molasses flow rate

67

E. Galindo and O.T. Ramírez (eds.), Advances in Bioprocess Engineering. 67-74.
© *1994 Kluwer Academic Publishers. Printed in the Netherlands.*

in response to inferred ethanol production rate [5]. The running cost was minimized because of the efficient conversion of molasses into biomass and of the low water consumption in the fermentation stage. The capital cost was kept down because of the high productivity achieved in the fermentation process. An overview of the most important aspects of the process developed is presented below.

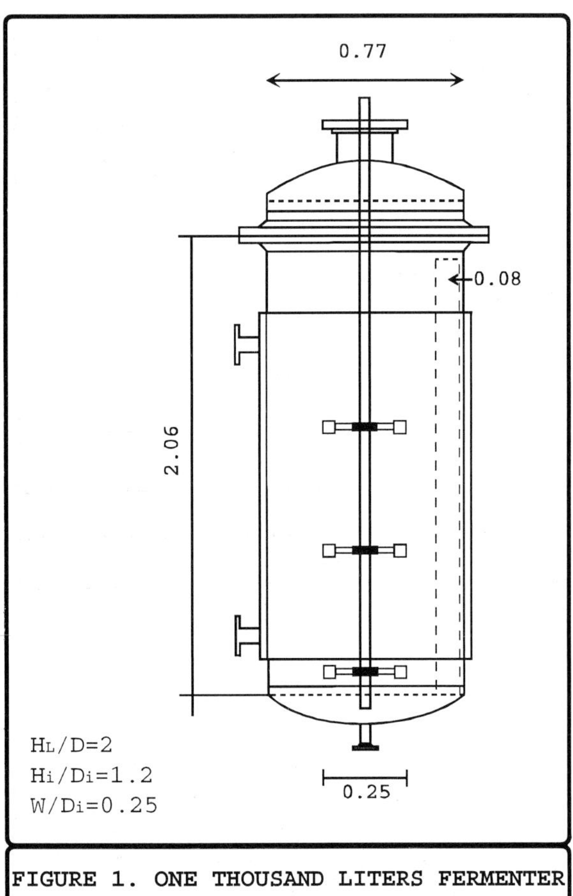

FIGURE 1. ONE THOUSAND LITERS FERMENTER (m)

EXPERIMENTAL.

Fermentors. Two stirred-tank fermentors (STR's) were used, a 0.030-m^3 fermentor (PEC-REACTOR, CHEMAP A.G., Switzerland) and a 1.0 m^3 made in México (Figure 1). The working volume of the STR's was 60% of the total volume.

In the 10.5-m^3 jet loop fermentor (Figure 2), the air was sparged at the base of the draft tube by three injectors having propulsion liquid nozzles of 0.008 m in diameter. The broth was pumped through the nozzles by a centrifugal pump (Labour A-30, U.S.A.).

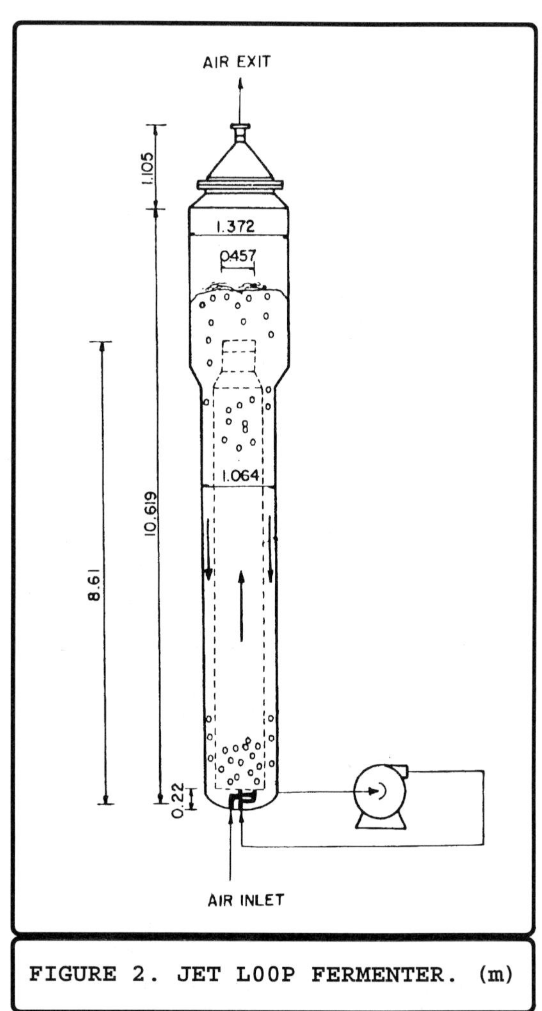

FIGURE 2. JET LOOP FERMENTER. (m)

Over time intervals of 45 seconds, average readings of 2000 signals from an oxygen analyzer (Servomex model 540A, U.K.) and a CO_2 infrared analyzer (Servomex, U.K.) were recorded by a computer (HP 1000 A600, U.S.A.) and used to calculate oxygen uptake rate (OUR), CO_2 evolution rate (CER) and respiratory coefficient (RQ). OUR and CER were used to calculate ethanol production rate (EPR) according with EPR= CER-C1*OUR [6].

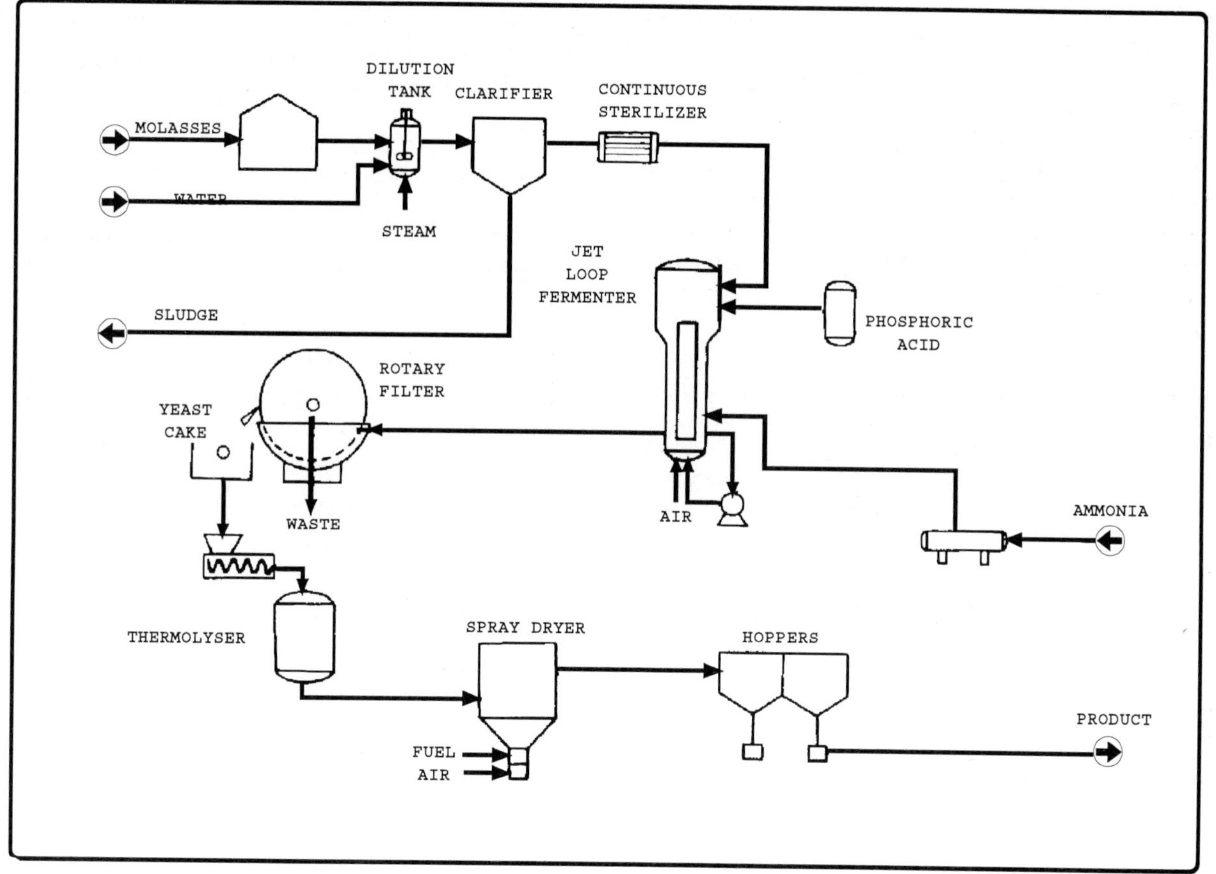

F I G U R E   3 .   S C H E M A T I C   F L O W S H E E T

To keep high conversion yields, the molasses flow was on-line adjusted by the computer in response to the EPR [5].

Microorganism and culture conditions.
Candida utilis NRRL-Y 900 was employed. The medium, procedures and operating conditions have been previously described by González et al. [7].

Analytical methods. Biomass was determined by dry weight, reducing sugars by Fehling's method, nitrogen by Kjelhdal's method, available lysine by the croceine orange method [8] and protein digestibility according to A.O.A.C. method 7-040/70.

PROCESS OUTLINE.

A flow sheet of the process developed is shown in Figure 3. Molasses are diluted with tap water 1:1 by volume and heated at 90 °C. The settled solids are separated from molasses. Clarified molasses are continuously fed to the fermentor through sterilizing equipment. Phosphoric acid is supplied by a metering pump while ammonia supply is continuously regulated by a pH controller. The fermentor operates under gas hold-up controlled conditions to optimize oxygen transfer [9].
The yeast is concentrated and washed using a rotary vacuum filter. The yeast cake obtained is fed to the thermolyser tanks

that the fraction of total reducing sugars utilized (FRSU) by the yeast varied from 85% to 94% and was dependent upon molasses composition. It was found too that, under the operating conditions, C/N ratios from 5 to 9.5 at dilution rates in the range of 0.14 h^{-1} to 0.25 h^{-1} did not affect the biomass yield based on sugars. These yield values showed a standard deviation of 0.05.

Under carbon substrate limiting conditions, the most efficient carbon substrate conversion to biomass is achieved [10]. Under these conditions biomass productivity is related to oxygen transfer rate (OTR), therefore, OTR was employed as criterion to scale up from bench to pilot plant (0.030-m^3 fermentor to 1.0-m^3 fermentor). Table 1 shows the average results of two sets of fermentation runs in STR's at steady state of at least 240 h each. The biomass volumetric productivity and the biomass yield were similar for both reactors, which indicates that OTR could be used as scale-up criterion.

Table 1. Biomass yield and volumetric productivity of *Candida utilis* for different fermentors.

FERMENTOR	C/N	D (h^{-1})	S_o (kgm^{-3})	FRSU $(\%)$	X (kgm^{-3})	Yx/s_c	Pv $(kgm^{-3}h^{-1})$
STR 0.030 m^3	5	0.25	62.6	90	24.1	0.43	6.0
STR 1.0 m^3	5.4	0.20	70.3	90	29.9	0.48	6.0
JET-LOOP 10.5 m^3	8.5 to 9.2	0.14	155.2	92	80	0.56	11.2

A jet-loop fermentor with increased OTR relative to STR's was designed to increase biomass productivity. The average results at steady state from four fermentation runs are shown also in Table 1. A significantly higher

to be liquified and increase its digestibility. Finally, the yeast is spray-dried.

RESULTS AND DISCUSSION.

Feed preparation. Molasses are the main contamination source, they usually contain wild yeasts that grow when diluted. Therefore, the molasses clarification must be a short residence time operation to avoid yeast growth and at the same time prevent browning reactions.

Continuous monoseptic yeast production requires that only sterile molasses are fed into the fermentor. The sterilizing equipment has to have temperature sensors at all important points to assure that the proper sterilization temperature is maintained, as well as bypasses for automatic recycling of non sterile molasses.

The sulfuric acid added to reduce the molasses Ca^{2+} content during the clarification step resulted in a severe fouling of the sterilizer. The encrustation was due to calcium sulfate precipitation under the sterilization conditions (130 °C to 135 °C). Fouling was minimized when sulfuric acid was omitted. No adverse effects in the fermentation were detected as a result of the remaining Ca^{2+} concentration.

Fermentation. Preliminary experiments were done at bench scale (0.030-m^3 STR) for strains screening and to define culture medium and operating conditions. Once the fermentation process conditions were stablished, work proceeded to pilot plant scale (1-m^3 STR) to scale up the process. Finally a demonstration plant with a 10.5-m^3 jet-loop fermentor was built and operated.

During four years of bench scale and pilot plant work, results of a total of 46 continuous runs in STR's indicate biomass yield was observed in the jet-loop reactor and yeast cells were larger and with less vacuoles than those cultivated in STR's. Reduced yields in STR's might be a result of ethanol production under oxygen limiting conditions. However, a physiological response of the cells continuously submitted to transient environmental conditions, such as recycling from a region of high turbulence to one of low turbulence in the jet-loop reactor, might be a contributing factor to enhance biomass yields. Katinger [11] reported, during continuous cultivation of *Candida tropicalis* on n-paraffin as substrate, that the biomass yields with respect to carbon and oxygen increased as a transient oxygen limitation was intensified in a tubular closed-loop fermentor. The oscillations in the dissolved oxygen tension due to the mixing characteristics of the closed-loop fermentor generated undamped short period oscillations in the respiratory

activity and ATP content in the *C. tropicalis*. This behavior could represent oscillations of allosteric feedback loops, which manifest themselves by some synchronizing action to the environmental transients in the fermentor.

In the 10.5-m^3 jet-loop fermentor, the average OTR was 350 mol O$_2$ m^{-3} h^{-1} at 1.1 VVM. However, we observed significant variations of OTR as a result of antifoam shots required for foam control. It is well known that bubble coalescence is naturally suppressed as surface-active materials released from microorganisms accumulate into the fermentation broth. Hence, fermentation broths may have a tendency to foam but also enhanced OTR relative to water. Excessive foaming causes a reduced liquid volume, foam overflow and wetted exit filters thus increasing contamination risk. During initial fermentation runs, foaming was controlled by sudden antifoam additions causing a sharp drop of OTR, severe oxygen limitation, and consequently reduced yields and productivities. Steady state conditions were lost because of the mentioned disturbances [5]. Afterwards, a novel gas holdup control scheme for gradual antifoam feeding was proposed in order to overcome the above mentioned drawbacks [9]. This control scheme was useful to produce a non foaming fermentation with high OTR while maintaining steady operating conditions.

Cell separation. Centrifugation tests were done in a disk-stack centrifuge (FESX 512S, Alfa-Laval). The yeast cream solids concentration after two centrifugation steps was 16% (dry weight). In addition, vacuum leaf filter tests were done and cotton fabric and Decalite 477 were selected as filter media and filter aid respectively. Trials with a rotary filter yielded a 24%-solids cake [12].

To get 24% final solids concentration, economic analysis showed that filtration alone is a better alternative than centrifugation followed by filtration at broth solids concentrations higher than 7% (Figure 4).

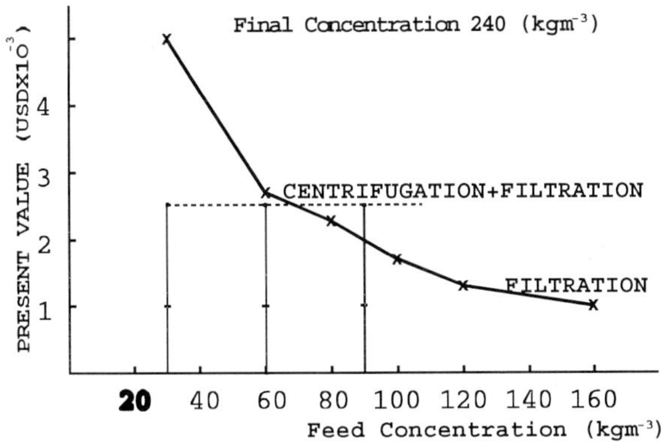

FIGURE 4. Comparison of different yeast- concentration processes for a 20 000 ton year plant

Thermolysis. In order to increase the availability of nutrients and to develop flavor notes, yeast used as food or feed is usually lysed. Three alternatives were investigated: autolysis, autolysis-thermolysis and thermolysis. Experiments were carried out to increase the protein digestibility while minimizing lysine and thiamine losses. The independent variables were temperature, pH and time, and the dependent ones were protein digestibility, thiamine and available lysine [13, 14, 15]. Results are shown in Table 2.

Table 2. Results of autolysis and thermolysis of *C. utilis*.

TREATMENT	AVAILABLE LYSINE (kg kg^{-1} of protein)	PROTEIN DIGESTIBILITY (%)
none	0.054	51.2
autolysis (24h at 50°C)	0.05	78.4
autolysis-thermolysis (24h at 50°C, 2h at 90°C)	0.046	90.2
thermolysis (1h at 70°C, 2h at 85°C)	0.053	95

The two-step thermolysis was the best alternative to increase protein digestibility with no adverse effects upon thiamine content and protein quality. The product has been used as flavor enhancer additive and to produce reaction flavors such as chicken notes.

ECONOMIC ANALYSIS

Table 3 shows an economical comparison among different yeast production processes. Speichim and Vogelbusch processes data were obtained from industrial facilities actually running in Cuba [16], while CINVESTAV process data were gathered from trials at the demonstration plant. Total investment and costs were estimated for new facilities in México (third quarter 1991).

Table 3. Comparative economic data for yeast production. Basis: 12 * 10^6 kg year^{-1}

PROCESS	TOTAL CAPITAL INVESTMENT (M USD)	*CAPITAL RELATED COST (USD kg^{-1})	RAW MATERIALS COST (USD kg^{-1})	UTILITIES COST (USD kg^{-1})	TOTAL PRODUCTION COST (USD kg^{-1})
VOGELBUSCH	22,500	0.412	0.439	0.202	1.053
SPEICHIM	21,150	0.388	0.440	0.183	1.011
CINVESTAV	18,000	0.330	0.358	0.215	0.903

* Annual capital related cost estimated as 0.22 of the total capital investment

The total capital investment for a new facility using the CINVESTAV process is about 80% to 85% the required for conventional processes. Raw materials cost is reduced nearly 20%, mainly as a result of a higher biomass yield on molasses and a minimal use of ammonia, which is utilized instead of ammonium sulfate. The specific energy consumption of the fermentors employed is 0.39, 0.55, and 0.53 kW-h kg-cell^{-1} for air-lift fermentor(Speichim), STR (Vogelbusch), and jet-loop fermentor (CINVESTAV) respectively, so energy consumption is very similar for the STR and jet-loop fermentor. On the other hand, as a result of culturing at high cell density, the water consumption is reduced by two thirds. However, utilities cost for the CINVESTAV process is slightly higher, because the Vogelbusch and Speichim processes employ combustion gases for yeast drying, while the CINVESTAV process uses indirect heating to improve product quality.

CONCLUSIONS

OTR is a suitable criterion to scale up yeast production processes employing either STR's or jet-loop fermentors. In addition, the jet-loop fermentor is suitable for culturing yeast at high cell densities. The transient environmental characteristics imposed on cells in this fermentor might have a positive effect on biomass yields. Filtration alone was a better separation alternative than centrifugation followed by filtration at broth solids concentrations higher than 7%. A thermolysis step (1 h at 70°C and 2 h at 85°C) increased the protein digestibility from 51% to 95% with no adverse effects on thiamine and product quality. The thermolysed yeast product has been used as flavor enhancer and protein supplement for food and feed as well as a raw material to produce reaction flavors.

Economic data showed that a computer controlled high cell density fermentation process carried out in a jet-loop fermentor can be a competitive alternative to conventional technologies.

Even when the fermentations in continuous culture with *C. utilis* are carried out at pH between 3.5 and 4.0, contamination with wild yeast is usual. Therefore, to achieve monoseptic fermentation in continuous mode at production scale, rigorous attention to minute engineering detail is essential. This means that sterile engineering technology is critical.

<u>NOMENCLATURE</u>

C/N		carbon:nitrogen ratio of fed broth
D	m	fermentor diameter
D_i	m	impeller diameter
FRSU	%	fraction of reducing sugars consumed
H_L	m	broth liquid height
OTR	mol m^{-3} h^{-1}	oxygen transfer rate
Pv	kg m^{-3} h^{-1}	biomass volumetric productivity
So	kg m^{-3}	total reducing sugar in fed broth
VVM	m^3 air m^{-3} broth min^{-1}	aeration rate
W	m	impeller blade width
X	kg m^{-3}	cell concentration
Yx/S_c	kg kg^{-1}	biomass yield based on sugar consumed

Acknowledgements:

We are very grateful to the staff of the Pilot Plant for their technical support, and to Dr.Eduardo Gutiérrez, M.Sc. Sergio García and M.Sc. Antonio Hernández for their collaboration during the realization of part of this work. This project was possible thanks to the financial support of Sindicato de Trabajadores de la Industria Azucarera y Similares de la República Mexicana, CONACYT and Secretaría de Educación Pública.

<u>LITERATURE CITED</u>

1. Guidoboni, G.E., The problems of large-scale yeast production, In: G.G. Stewart, I. Russell, R.D. Klein and R.R. Hiebsch (Eds.), *Biological Research on Industrial Yeast* vol. 1, p.47 CRC Press, Boca Raton, Florida (1987).

2. Westalake, R., *Chem. Ing. Tech.* (Germany) **58**, 934 (1986).

3. Staff article., *Food Technology* (USA) **43**, 7, 50 (1989).

4. Blenke, H., Biochemical loop reactors, In: H.J. Rehm and G. Reed (Eds.), *Biotechnology* vol. 2, p.465 VCH Verlagsgesellschaft, Weinheim (1985).

5. Flores, Z., L.B. Flores, S. García, J. Corona, O. Melchy and M. de la Torre, High cell density computer control of *Candida utilis* yeast production on molasses. In: N. Karim and G. Stephanopoulos (Eds.), *Modeling and Control of Biotechnical Processes and Computer Applications in Fermentation Technology*, p.227 Pergamon Press, New York (1992).

6. Wang, H.Y., C.L. Cooney and D.I.C. Wang, *Biotechnol. Bioeng.*, 21, 975 (1979).

7. González, A. and E. Chong, Producción de levadura en el reactor de 10 m^3 con control por computadora. Report No. PTE-054. Department of Biotechnology, CINVESTAV, Mexico (1990).

8. Hurrel, R.F., P. Lerman and K.J. Carpenter, *J. Food Science*, **44**, 1221 (1979).

74

9. Flores Cotera, L.B. and S. García Salas, U.S. patent appl. serial number 07/963,980 to CINVESTAV, (October, 1992).

10. Vasey, K.B. and K.A. Powell, *Biotechnology and Genetic Eng. Reviews*, 2, 285 (1984).

11. Katinger, W.D.H., *European J. Appl. Microbiol*, 3, 103 (1976).

12. Gutiérrez, E., Estudio técnico-económico sobre la separación de levadura forrajera por centrifugación, floculación, sedimentación y filtración. II Elección del arreglo óptimo. Report No. PTB-001. Department of Biotechnology, CINVESTAV, México.(1987).

13. Gutiérrez, E., Termólisis de la levadura *C. Utilis*. Report No. PTE-019. Department of Biotechnology, CINVESTAV, México.(1987).

14. De la Torre, M. and M. Luna, Termólisis de la levadura NRRL-Y 900. Report No. PTE-046. Department of Biotechnology, CINVESTAV, México.(1989).

15. De la Torre, M. and M. Luna, Evaluación de los procesos de autólisis y termólisis en cuanto al incremento en digestibilidad proteica y el efecto en el contenido de lisina de levadura tórula, tanto a nivel laboratorio como en planta piloto. Report No. PTE-058. Department of Biotechnology, CINVESTAV, México. (1991).

16. Estévez, R., Levadura forrajera a partir de las mieles finales de caña, In: ICIDCA (Ed.), *Los derivados de la caña de azúcar* p. 288 Editorial Científico-Técnica, La Habana, Cuba (1980).

Continuous Flow Cell-Recycle Fermentation of Biomass Hydrolysates

C.H. Choi[1] and A.P. Mathews[2]

[1]Department of Environmental Engineering, Kwandong University, Kangreung, KOREA;
[2]Department of Civil Engineering, Kansas State University, Manhattan, Kansas 66506, U.S.A.

Fermentation studies were conducted in batch and continuous flow cell-recycle reactors with glucose, xylose and biomass hydrolysates as substrates. Glucose fermentation at pH 6 indicates that Propionibacterium acidipropionici *rapidly consumes the substrate in an initial exponential growth stage producing large amounts of lactic acid. A slow growth stage follows, during which the lactic acid is further metabolized to produce acetic and propionic acids. The bacterial metabolism was altered when xylose was used as the substrate, and lactic acid was not produced during the fermentation. The product and cell yields obtained from the glucose fermentation were 0.76 g acid (g glucose)$^{-1}$ and 0.21 g dry cell (g glucose)$^{-1}$, respectively. Xylose fermentation gave about the same cell yield, but the total product yield was only 63% of that with glucose fermentation. Continuous fermentation with cell recycle using microfiltration resulted in a dramatic increase in cell concentration (X) and volumetric productivity (P). P ranged from 0.56 to 2.83 g dm^{-3}h^{-1} for dilution ratios (D) ranging from 0.036 to 0.23 h^{-1}. X and P at a D of 0.23 h^{-1} were 11 and 22 times the values for batch fermentation at pH 6. Cell-recycle fermentation with hydrolysates from bakery waste and wood chips gave P equivalent to 93% and 71%, respectively, compared to that with glucose.*

The recovery of valuable resources from low-grade biomass such as agricultural residues, and wastes from food processing operations and pulp production is an important aspect of solid waste and sludge management strategies. Many of these wastes contain starch and cellulosic materials that can be hydrolyzed to produce sugars. These sugars can be subsequently fermented to produce valuable feedstocks and useful end-products.

Wastes from agricultural and industrial operations involving biomass can be processed in several ways. In some cases, the structures of undegraded plant polymers can be modified to produce special types of resins, plasticizers and lubricants (1). Combustion and pyrolysis operations can be utilized for conversion of solid wastes to energy. However, a substantial amount of the energy available in the raw material will be lost or wasted. One of the most promising and challenging methods for energy conservation is the recovery of resources from biomass wastes and conversion to high value end-products by bacterial fermentation.

Anaerobic bacterial fermentations with high theoretical substrate-to-product conversion yields for production of organic acids are of great importance to the chemical industry. Proper microorganism and culture selection is essential due to the variety of metabolic pathways leading to the end product and the corresponding overall yield in fermentation (4). *P. acidipropionici* produces high concentrations of propionic and acetic acids using both glucose and xylose (5). Theoretically, two moles of propionic acid and one mole of acetic acid (2.47:1 mass ratio) would be produced from 1.5 moles glucose. Experimental studies with xylose gave similar results (6). However, wide variations in the ratios of organic acid products are observed in glucose fermentations. Propionic acid fermentations of glucose show propionic acid to acetic acid mass ratios of 1.5 to 1.8 and yields of propionic acid vary from 0.44 to 0.61 g propionic acid/g glucose (7).

Several systems utilizing dialysis, vacuum, flash and extractive fermentation have been proposed to increase volumetric productivity by selectively removing end-products that may be inhibitory. Continuous fermentation studies have also been conducted attempting to obtain high acid productivities by maintaining high cell concentrations in the reactor through crossflow filtration and cell recycle (9,10). Carrondo et al. (9) obtained a total acid productivity of 1.03 gdm^{-3}h^{-1} at 9.5 gdm^{-3} propionate and 1.2 gdm^{-3} acetate concentrations in an immobilized cell reactor with a starting sugar concentration of 25 g dm^{-3} at a residence time of 16.5 h. In a cell-recycle ultrafiltration system with pH at 6, they obtained volumetric productivities of 2.7 gdm^{-3}h^{-1} with propionate and acetate product concentrations of 18.0 gdm^{-3} and 4.0 gdm^{-3} respectively. This system was operated at a dilution rate of 0.12 h^{-1} (8.33 h residence time) and total sugar (glucose:xylose = 3:1) concentration of 50 gdm^{-3}.

75

E. Galindo and O.T. Ramírez (eds.), Advances in Bioprocess Engineering. 75-80.
© 1994 Kluwer Academic Publishers. Printed in the Netherlands.

The main objectives of this research were to stduy fermentation kinetics in the production of propionic and acetic acids from biomass hydrolysates and to examine methods to enhance acid productivities by continuous fermentation. This paper will describe the experimental studies and results from continuous fermentation runs with recycle of cells separated by membrane filtration.

MATERIALS AND METHODS

Microorganism and biomass samples. The microorganism employed in the study was *Propionibacterium acidipropionici*(ATCC 4875, American Type Culture Collection, Rockville, MD). Two types of low-grade biomass, bakery waste (BW) and wood chips (WC), were used as raw materials. Both samples were dried in the oven at 65°C for about one week. WC and BW were ground by using a roller mill and a hammer mill respectively, and separated using a Ro-Tap sieve shaker. The fraction that passed through a #20 sieve (0.85 mm) was collected and used in the experiments.

Acid hydrolysis. The biomass hydrolysis feedstocks were converted to fermentable sugars by acid hydrolysis using optimum conditions determined by Choi (11). BW and WC were dissolved in 2% H_2SO_4 solution to give 5% solid contents. The slurries were hydrolyzed at 132°C for 40 minutes. The hydrolysates were collected using a 1 μm Polypure capsule (Gelman Sciences Inc., Ann Arbor, MI) and neutralized with 5 N NaOH. Final solutions were diluted to 4% total sugar content.

Medium and inoculum preparation. Medium compositions used in this study contained glucose, xylose or mixtures, trypticase peptone (2.5%), yeast extract (0.7%), KH_2PO_4 and K_2HPO_4 (0.05% each), and Na_2CO_3 (0.4%). The sugar content was approximately 2% for all experiments. The medium from biomass feedstocks was prepared by mixing equal volumes of distilled water containing all nutrients but sugars and biomass hydrolysates with 4% sugar solution. The medium was adjusted to pH 6 with 5N HCl and autoclaved at 121°C for 17 min.

Fermentation. The batch fermentation system consisted of a Virtis Omni-Culture Bench-Top Fermenter (The Virtis Co., Inc., Gardiner, NY) and various control and monitor units. The impeller agitation speed was set at 175 rpm. The fermentation temperature was set at 30°C. A pH electrode (Ingold Electrodes Inc., Wilmington, MA) inserted into the medium was connected to a microprocessor controlled pH meter (Orion Research Inc., Boston, MA). Anaerobic grade nitrogen gas was supplied through a double-stage regulator. Dissolved oxygen (DO) in the medium was detected by a DO indicator module and kept at near zero for all experiments. An autoclavable galvanic type DO electrode connected to the monitor was inserted into the medium.

The continuous fermentation system consisted of several components to ensure continuous feed, product withdrawal and cell recycle. Two peristaltic pumps (Cole Parmer Instrument Co., Chicago, IL) were operated continuously at constant rate to pump the feed in and to remove the product out. A magnetic drive pump (Cole Parmer Co.) was connected to a filter capsule for continuous recycling of the medium. The flow rate of each line was metered using a Gilmont flow meter and manually adjusted when needed. An Acroflux crossflow filter capsule (Gelman Sciences, Ann Arbor, MI) was used in the cell recycling system. The normal recycle flow rate was 1.0 dm^3h^{-1}, and was increased to 1.5 dm^3h^{-1} when the membrane became plugged. Both 0.8 μm and 1.2 μm pore size filters were used in the study and they showed very similar performance. Membrane fouling was experienced in about two weeks of operation. When the on-line filter became plugged, the flow was diverted to a new filter and the fouled filter was removed and cleaned.

Analytical techniques. Cell mass was determined by monitoring absorbance with spectronic 710 spectrophotometer (The Bausch and Lomb, Rochester, NY). A 0.2 cm^3 sample was diluted in a corresponding amount of distilled water before measuring absorbance at 540 nm. Dilution factor was 50 for batch fermentation and increased up to 200 for cell recycling fermentation. Dry cell mass was determined by vacuum-filtering the samples through 0.45 μm nitrocellulose filters. The filters were then placed in an oven at 65°C for 72 h before weighing.

Substrate and product concentrations were assayed using High Performance Liquid Chromatography (HPLC, Varian Associates Inc., Palo Alto, CA). All separations were accomplished by an Aminex Ion-Exclusion Column (HPX-87H, 300 mm * 7.8 mm, Bio-Rad Lab, Richmond, CA) operated at 45°C. Components were eluted with 0.02 N aqueous sulfuric acid at a flow rate of 0.7 cm^3min^{-1}. Detection of sugars and organic acids was accomplished by UV absorption at 193 nm.

RESULTS AND DISCUSSION

Acid hydrolysis. BW and WC were hydrolyzed to maximize sugar conversion. BW was primarily derived from staled bread and contains about 66% starch and 15-20% carbohydrates other than starch (12). The chemical composition of WC depends on the source of wood. Gong et al.(13) reported that the quantities of cellulose and hemicellulose in soft wood chips are 45-50% and 25-35%, respectively. Hemicellulose is easily hydrolyzed and has very close thermal degradation points to starch. Kim and Lee (14) found that the yield of xylose exceeded 90% in about 50 minutes when hardwood hemicellulose was hydrolyzed at 130°C using 2.5% H_2SO_4.

The total fermentable sugars converted from acid hydrolysis is given in Table 1. BW has very high potential to be a candidate for feedstock in the production of chemicals or fuels, yielding almost 73 g glucose (100 g dry substrate)[-1]. With about 30% recovery of resources, hydrolysis of WC can also provide means for waste utilization while minimizing pollution.

Batch fermentation. Controlled batch fermentation studies were conducted at pH 6 using 19.1 g dm[-3] of glucose or xylose as the sugar source. For both cases, the production of acetic and propionic acids, and the production of cell mass during fermentation show similar trends as shown for xylose fermentation (Fig.1). Choi and Mathews(15) have reported that the optimum pH range for the genus *P. acidipropionici* is 5.5-6.5, and that the bacterial metabolism and fermentation pathway appear to be altered outside of this range. Glucose fermentation studies at pH 6 have shown that *P. acidipropionici* rapidly consumes all the glucose substrate in an initial exponential growth stage producing large amounts of lactic acid. A slow or non-exponential growth stage follows, and during this phase lactic acid is further metabolized to produce acetic and propionic acids. Lactic acid was not produced when xylose was used as the substrate, indicating that the bacterial metabolism is altered during xylose fermentation. The performance of the glucose and xylose fermentations are summarized in Table 2. The product and cell yields obtained from the glucose fermentation were 0.45 g propionic acid (g glucose)[-1], 0.30 g acetic acid (g glucose)[-1], and 0.21 g dry cell (g glucose)[-1], respectively. The total product yield was 0.76 g acid (g glucose)[-1]. In xylose fermentation, cell yield was about the same as glucose fermentation, but total product yield was only 63% of that obtained from glucose fermentation. The volumetric productivity based on the maximum productivity was 0.13 g acid dm[-3]h[-1] for glucose fermentation, and 0.15 g acid dm[-3]h[-1] for xylose fermentation. In the case of xylose fermentation, the maximum productivity was slightly higher than glucose fermentation because of the short fermentation time.

Continuous fermentation with cell recycle. Continuous fermentation runs were made in sequence with glucose, BW hydrolysate, and WC hydrloysate as the sugar source. The feed and permeate flow rates were adjusted to give a fixed dilution rate, D (h[-1]), calculated from D=(F/V), where, F (cm[3] h[-1]), is the volumetric flow rate, and V (cm[3]) is the total volume of media in the fermenter and the filter capsule.

The fermentation system was continuously run for about 35 days with 19.1 g dm[3] glucose as a sugar source. The dilution ratios under which the fermenter was operated ranged from 0.036 to 0.23 h[-1]. The fermentation run began with batch operation at pH 6, and cell recycle was initiated when the maximum cell concentration was reached in the fermenter. The pH was controlled between 5.5 and 6.3 during the continuous operation at 30°C. Cell mass increased continuously during cell-recycle fermentation until cell

Table 1. Conversion of sugars from low-grade biomass by acid hydrolysis[a] (g sugar (100 g sample)[-1]).

Biomass	Glucose	Xylose
Bakery Waste	72.80 ± 3.15	---
Wood Chips	2.59 ± 0.31	26.61 ± 0.95

[a]The values are based on 3 replications. Acid hydrolysis was conducted at 132°C for 40 minutes using 2% H_2SO_4.

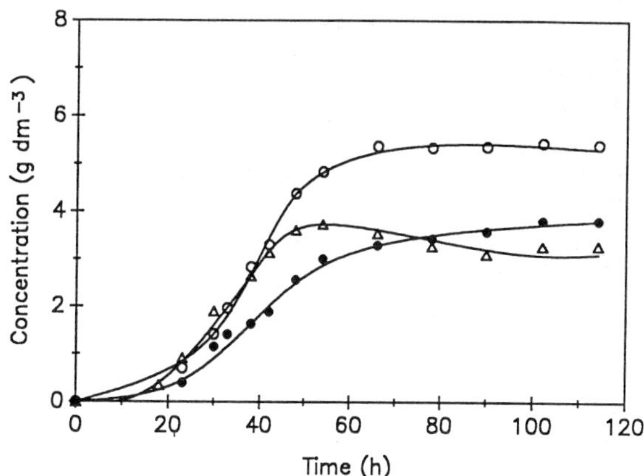

Fig. 1 Batch fermentation at 30°C and pH 6 using p. *acidipropionici* 19.1 g dm[-3] xylose as a sugar source. O-O, propionic acid; •-•, acetic acid; and ∆-∆ dry cell mass.

generation and lysis rates became balanced. To maintain high numbers of viable cells in the fermenter, 20-30% of the cells were bled and replaced with fresh medium through the sample port at the end of each cycle. The dilution rate was not changed until the cell mass and organic acids production reached a steady state. Fig.2. presents the final cell mass in the fermenter and total productivity of organic acids at each dilution ratio. Propionic, acetic and lactic acids are included in the total acid production. The cell mass and total acids productivity continuously increased as the dilution ratio increased, reaching 44.31 g dm[3] and 2.83 g dm[-3]h[-1] at 0.23 h[-1]. The application of cell recycle to fermentive organic acids production resulted in a dramatic increase in both cell mass production and volumetric productivity. In the ranges of operation (D = 0.036 h[-1] - 0.23 h[-1]), cell mass and total acids productivity increased 2.3 and 5.1 times, respectively. These results indicate that the performance of cell-recycle fermentation can be further improved by increasing the recycle ratio. Compared to batch fermentation at pH 6, the cell mass production increased by as much as 11 times and the volumetric productivity was increased by 22 times at the dilution rate 0.23 h⁻¹.

Table 2. Performance comparisons for controlled fermentation at pH 6 using glucose and xylose as sugar sources.

	Sugar Source	
	Glucose	Xylose
Total Sugar (g dm⁻³)	19.1	19.1
Maximum Cell Mass (g dm⁻³)	4.0	3.73
Maximum Cell Yield (g cell (g sugar)⁻¹)	0.21	0.20
Acid Concentrations (g dm⁻³)		
Propionic Acid	8.52	5.36
Acetic Acid	5.71	3.80
Lactic Acid	0.25	0.0
Total Acid Production (g dm⁻³)	14.48	9.16
Product Yield (g acid (g sugar)⁻¹)	0.76	0.48
Maximum Productivity (g prod. dm⁻³h⁻¹)	0.13	0.15

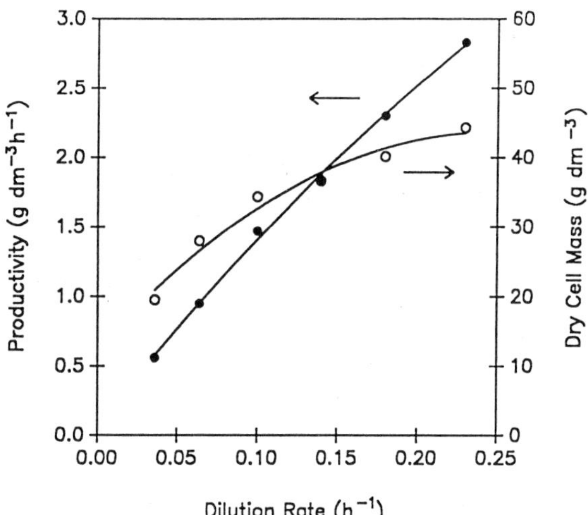

Fig. 2 Total acid productivity and cell mass production from cell recycle fermentation. O-O, dry cell mass; and •-•, total acid productivity (acetic, propionic, and lactic acids).

Another set of cell-recycle fermentation runs were made over a period of 20 days using glucose and sugars derived from BW and WC hydrolysis as substrates. The dilution rate was fixed at 0.1 h⁻¹. Data for the 15 days of fermentation are shown in Fig. 3, and performance comparisons are presented in Table 3. The cell mass in the fermenter

continuously increased reaching 34.31 g dm⁻³ at the end of glucose fermentation. This value is 8.6 times higher than the cell mass produced in batch fermentation at pH 6. Total acid yield was almost the same (0.77 g acid (g glucose)⁻¹) as compared to batch fermentation using glucose as a substrate. However, propionic and acetic acid concentration in the permeate decreased about 16-19% while lactic acid concentration considerably increased, indicating that the medium residence time (10 h) is not sufficient for the bacteria to convert all of the lactic acid to acetic and propionic acids. In the medium containing hydrolysates from BW, total sugar amount was 18.4 g dm⁻³ with 100% glucose. The medium from WC hydrolysates contained 19.6 g sugar dm⁻³ with 85.7% xylose and 14.3% glucose. Fermentation of BW hydrolysates produced excellent results comparable to that of glucose fermentation. In the case of WC hydrolysates, a lower organic acid production rate was expected due to its high xylose content. However, cell mass production was also significantly decreased, indicating the presence of potentially inhibitory compounds affecting cell metabolism. More acetic acid was produced than propionic acid in the fermentation of WC, suggesting that the inhibitory effect occurred in the late stages of the propionic acid pathway. The exact mechanisms of such effects need to be investigated further. The final cell mass production and product yields for BW fermentation were 30.84 g dm⁻³ and 0.75 g acid (g sugar)⁻¹ respectively, and 25.29 g dm⁻³ and 0.53 g acid (g sugar)⁻¹ respectively, for WC

Table 3. Results of continuous fermentation with cell recycle, using glucose and low-grade biomass as sugar sources[a].

	Glucose	Bakery Waste	Wood Chips
Total Sugar (g dm⁻³)	19.1	18.4	19.6
Glucose (g dm⁻³)	19.1	18.4	2.8
Xylose (g dm⁻³)	0.0	0.0	16.8
Final Cell Mass (g dm⁻³)	34.31	30.84	25.29
Final Acid Concentrations in Permeate (g dm⁻³)			
Propionic Acid	6.88	6.95	3.54
Acetic Acid	4.79	4.69	4.78
Lactic Acid	3.04	2.09	2.04
Total Acid Production (g dm⁻³)	14.71	13.73	10.36
Product Yield (g acid (g sugar)⁻¹)	0.77	0.75	0.53
Total Productivity (g acid dm⁻³ h⁻¹)	1.47	1.37	1.04

[a]Fermentations at 30°C and pH 5.5-6.3 with dilution ratio of 0.1 h⁻¹.

fermentation. The total productivity from BW (1.37 g acid dm^{-3} h^{-1}) was about 30% higher than that from WC (1.04 g acid dm^{-3} h^{-1}). The overall performance of biomass hydrolysates in cell recycle fermentation is, therefore, about 93% with BW and about 71% with WC, as compared to that of pure glucose.

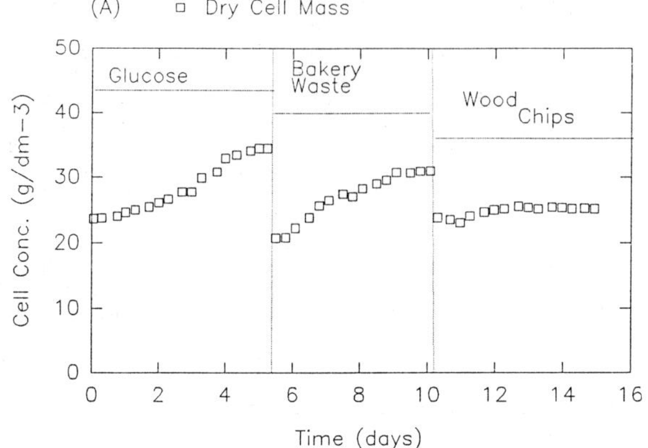

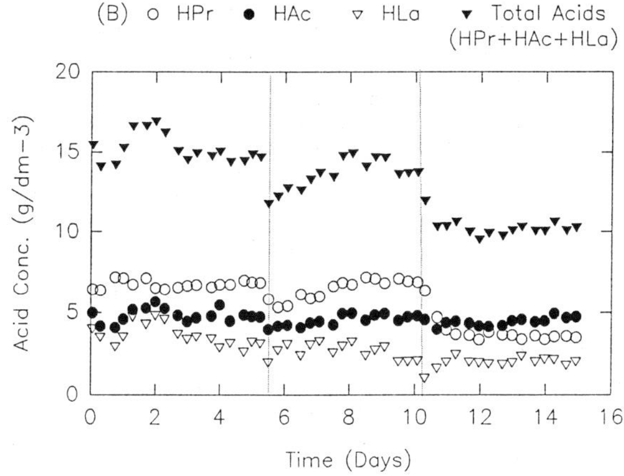

Fig. 3   Ccontinuous fermentation with cell recycling at dilution rate of 0.1 h^{-1} and pH 5.5-6.3 using glucose and sugars from biomass hydrolysates.

CONCLUSIONS

The genus *Propionibacterium acidipropionici* is an important microorganism in the production of organic acids, with an exceptionally high cell growth rate and over 75% substrate-to-product conversion yields. The separation and recycling of cells to the fermenter resulted in a dramatic increase in both cell concentration and volumetric productivity of acetic and propionic acids. The productivity of organic acids was evaluated from cell recycle fermentation with sugar sources from low-grade biomass such as bakery waste and wood chips. Excellent performance was achieved in the fermentation of bakery waste hydrolysates. In the case of fermentation with wood chip hydrolysates, organic acid production was lower due to the high xylose content in the medium. Fermentation with crossflow microfiltration of broth and cell recycle, is a promising alternative to petrochemical routes for the continuous production of organic acids. With a 0.71 to 0.93 product yield ratio with respect to pure glucose fermentation, cell recycle fermentation of biomass hydrolysates provides a valuable method to recover resources from waste materials and to reduce pollution load on the environment.

LITERATURE CITED

1. Pomeranz, Y., Symposium, in: *Cereals - A Renewable Resource. Theory and Practice*, Y. Pomeranz and L. Munck, Eds, The American Assoc. of Cereal Chemists, Inc., St. Paul, MN (1981).

2. Alter, H., *Environ. Conservation*, 4(1), 11 (1977).

3. Alter, H., *Science*, 189, 175 (1975).

4. Cheng, C.E., *"Development, modeling, and simulation of the anaerobic fermentation from glucose to acetic acid by Clostridium thermoaceticum"*, PhD Thesis, Rutgers The State Univ. of New Jersey, New Brunswick, 1987.

5. Buchanan, R.E. and N.E. Gibbons, *"Bergey's Manual of Determinative Bacteriology"*, 8th Ed., 633, The Williams & Wilkins Co., Baltimore,MD (1974).

6. Allen, S.H.G., R.W. Kellermeyer, R.L. Stjernholm, and H.G. Wood, *J. Bacteriol.*, 87(1), 171 (1964).

7. Wood, W.A., in: *The Bacteria*, Vol II, Gunsalus, C. and R.Y. Stanier, Eds., Academic Press, NY (1961).

8. Naser, N.F. and R. L. Fournier, *Biotechnol. Bioeng.*, 32, 628 (1988).

9. Carrondo, M.J.T., J.P.S.G. Crespo, and M.J. Moura, *Appl. Biochem. Biotechnol.* 17/18, 295 (1988).

10. Park, Y.S., H. Ohtake, K. Toda, M. Fukaya, H. Okumura, and Y. Kawamura, *Biotechnol. Bioeng.*, 33, 918 (1989).

11. Choi, C.H. *"Hydrolysis and Fermentation Kinetics in the Production of Road De-icing Salt from Low-Grade Biomass"*, PhD Thesis, Kansas State University, KS (1992).

12. Choi, C.H., *"Ethanol Production from Grain Dusts, Bread Waste, and Cake Waste with and without Brewers' Condensed Solubles (BCS)"*, MS Thesis, Kansas State University, KS (1986).

13. Gong, C.S., L.F. Chen, M.C. Flickinger, G.T. Tsao, *"Conversion of Hemicellulose Carbohydrates"*, Adv. Biochem.Eng., Vol. 20, Fiechter, A., Ed., Springer-Verlag, NY, 93 (1981).

14. Kim, S.B. and Y. Y. Lee, *Biotechnol. and Bioeng. Symp. Ser.*, Vol. 17, 71 (1986).

15. Choi, C.H. and A.P. Mathews, *Appl. Biochem. Biotechnol.* accepted for publication (1994).

Physiological and Technico-Engineering Aspects of Lignocellulose Solid-State Fermentation with Filamentous Fungi

U.E. Viesturs and M.P. Leite

Latvian State Institute of Wood Chemistry, 27 Dzerbenes iela, Riga LV-1006, LATVIA

The applicability of various solid-state fermentation (SSF) techniques, exemplified by the conversion of cellolignin- and starch- containing biomass is presented. Possible biopulping and protein biosynthesis in the SSF process with multizone air inlet, recirculation and conditioning has been shown. Some advances in process optimization based on the physiology of fungi have been reviewed. The established resistance of basidiomycetes to increased CO_2 concentration allowed to regulate biological heat formation by air recirculation and gaseous composition (O_2, CO_2) correction. The respiration intensity of cells and biological heat elimination depended upon the O_2 and CO_2 concentration in the aerating gas. The increased CO_2 concentration decreased the growth of bacterial pollution during SSF of basidiomycetes. The increased O_2 concentration promoted biological delignification. The technological parameters were improved in fed-batch culture. However, the feasibility of these processes on the industrial scale is rather debatable. SSF pro and contra have been discussed.

The analysis of the literature reveals rather a diverse treatment of the term "SFF" see: Pandey [1], Losane et al. [2], Losane et al. [3]. In our turn, we also suggest a possible definition of SOLID STATE FERMENTATION the microorganisms growth and/or product synthesis in a loose porous moist substrate from which water (liquid) is not separated any more under the effect of the gravitation force. A number of reports on the use and development of the laboratory type and pilot scale bioreactors are available: Pereira et al. [4], Schuchardt and Zadrazil [5], Villegas et al. [6], Zadrazil et al. [7], Zadrazil et al. [8].
It is known that at the first approximation SSF has several advantages:
- the possibility to realize the process at a relatively high substrate concentration;
- the absence of the free liquid phase, i.e. it is possible to work with natural polymers (starch, cellulose, etc) without the preliminary hydrolysis.
Hence, if the problem of conversion and degradation of the renewable products of photosynthesis are regarded from the global point of view (figure 1), it can be seen that the SSF process (or their analogues) occupy a definite important place. However, in the case of the controlled solid state process, the development engineer faces a series of problems.

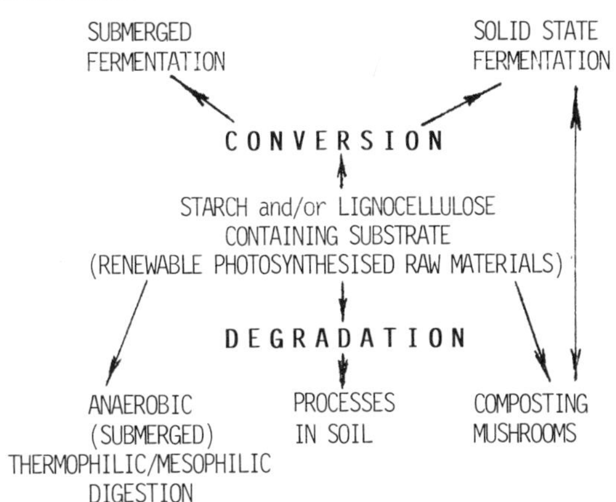

Figure 1. Conversion and degradation of the renewable products of photosynthesis.

The main problem in SSF is the contradiction: to ensure sufficient oxygen and heat removal, intensive mixing is required (both for SF and SSF), which, in its turn, results in lethal as well as partial (the state of

E. Galindo and O.T. Ramírez (eds.), Advances in Bioprocess Engineering. 81-86.
© 1994 Kluwer Academic Publishers. Printed in the Netherlands.

turbohypobiosis) cell damages.
We have chosen the term
"TURBOHYPOBIOSIS": Toma et al. [9],
Ruklisha et al [10], to describe a
state of the population in which the
physicomechanical interaction of the
cells and the external medium (shear
forces in turbulent liquids or solid
media), as well as the bioreactor units
causes a decrease in the growth
intensity, enzyme biosynthesis or other
functions of the population. In order
to grow intensively, the culture should
be mixed intensively. While growing
intensively, the culture releases a
considerable amount of heat (up to 3000
kcal from 1 kg assimilated substrate).
To ensure heat transfer, i.e.
homogeneity (t°, pH, etc.) in a
bioreactor, its content should be
stirred more intensively. However, it
results in a practical disintegration
of the culture and an actual
disturbance of the process.
The aims of the present work were:
- The development of the SSF technique;
- The development of the SSF process on
pretreated straw for protein production
(also the pretreatment technique
itself);
- The cultivation of wood materials for
the biopulping and biobleaching
purposes.
Different SSF and SF systems for
cultivation were developed and adapted:
1. The counterflow SF mixing system;
2. Rotary drums of different types;
3. The aerated column with liquid
circulation;
4. The aerated column with multilevel
air inlet and air conditioning (Figure
2).
The first three points were reported
earlier: Zeltina et al. [11], Zeltina
et al. [12], Viesturs et al. [13],
Viesturs et al [14]. In the present
study we report on our efforts to
improve the technological parameters
and biodelignification efficiency
during the _Coriolus versicolor_ 082 SSF
cultivation in an aerated column with
multilevel air inlet, aerating gas
recirculation and air conditioning
(CBMIC)

MATERIALS AND METHODS

Fungi. The white-rot species _Coriolus versicolor_ 082 and _Lentinus tigrinus_
from the culture collection of LS
Institute of Wood Chemistry were used
for this investigation.

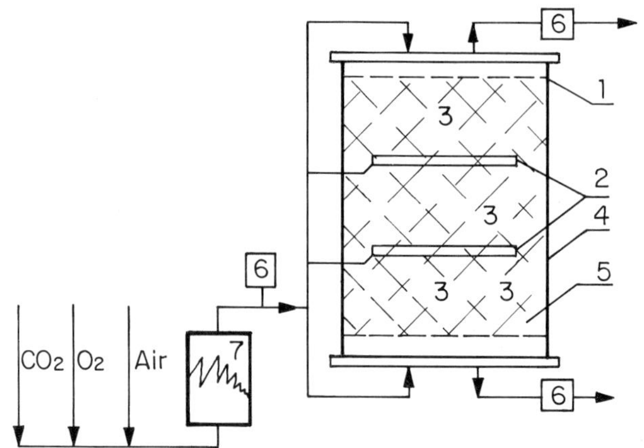

Figure 2. Aerated SSF column bioreactor
with multilevel air inlet and air
conditioning (CBMIC).
1. perforated plates, 2. air
distribution system, 3. temperature
control, 4. glass wall, 5. SS
substrate, 6. CO_2, O_2 control, 7. air
conditioning.

LC substrate. Untreated and alkali
pretreated winter wheat straw 1.5 to
2.0 cm fractions (1.5% NaOH, module
1:15, t 100°C, 20 min): Katke-vich et
al. [15], spruce chips (60 to 80 mesh)
and spruce thermo mechanical pulp (TMP)
were used as carbon sources.
Bioprocess. All the cultures were
maintained on 3% malt extract and 2%
agar slopes at 30°C. Non-shaken
cultures were grown in 750 ml flasks
containing 300 ml of a defined liquid
medium and 1% glucose as a carbon
source. The flasks were inoculated
with a single 10mm plug from 3-week-old
cultures grown on 3% malt extract and
2% agar plates. The culture medium was
harvested by filtration (Whatman No. 1
filter paper) and the biomass was
determined as the weight of the mycelia
dried at 70°C for 72 h. The mycelia
were fragmented in water for 30 sec
with a sterile Waring blender to
prepare mycelial suspensions. The LC
substrates were inoculated uniformly
with the mycelial suspension of each
fungus. Size of inoculum for all
treatment was 0.7% (DM) against dry the
LC substrate. Sterile water was added
to the pulp to give a final moisture
contents of 75% and 85% for straw and
spruce chips and spruce TMP,
respectively. In some experiments wood
substrates were moistened with a medium

containing mineral salts and macroelements necessary for the mycelial growth and enzyme synthesis: Viesturs et al. [13]. In protein synthesis experiments wheat bran was added (15% DM) as an easily digestible cosubstrate: Viesturs et al. [13]. Spruce TMP was also inoculated with a mycelial suspension containing small amounts of glucose (tentatively) according to the data Kirk et al. [16] reporting that Ph. chryzosporium and C. versicolor require glucose to metabolize ligning into CO_2 and/or to generate energy for the production of lignin degrading enzymes. Similarly, it has been established that the addition of energy rich substrates into the fed-batch culture of Trichoderma viride increases the rate of cellulose destruction and the energetic efficiency of the LC bioconversion into the feed protein: Viesturs et al. [14], and the ligning biodegradation rate into the fed-batch culture of C. versicolor and Cerrena maxima grown on wood substrates: Shvinka et al. [17]. The fermentor was equipped with a gas analysis system PGA (O_2, CO_2). The respiration intensity was measured by the gas balance method, using the online oxygen analyzer GTMK-16: Baburin et al. [18], Leite et al. [19].

Analytical aspects. The Klason ligning content from each treatment was measured by the method of Effland [20]; the protein content as described earlier: Leite et al. [19]. Cellulose was determined according to Updgraff [21].

Ligning and cellulose degradation rate was calculated as loss of dry weight of lignin or cellulose: g per gram biomass per day (g/gd) as described earlier: Shvinka et al. [17].

RESULTS AND DISCUSSION

The cultivation was performed in a column stationary layer fermentor with a working volume of 50 l. The height of the stationaty layer in the fermentor filled with the substrate was 800 mm. To provide the evenness of the culture growth in the whole volume of the substrate, a special aeration method - multiple zone aeration with aerating air recirculation and gas composition correction was applied. Special monitoring of the process was carried out to ensure the temperature of the fermented substrate within the physiological norm of C. versicolor and Lentinus tigrinus (30-70°C). It should be noted that in SSF one of the sprincipal problem is ensuring and omptimum temperature of growth during the logarithmic (intensive) growth phase, especially with such cosubstrates as wheat bran or glucose. The amount of the heat secreted during the fermentation process is stoichiometrically attached to the process of growth and enzyme production.

During the intensive growth phase, the amount of biological heat is directly proportional to the cell biomass. Due to the low thermal capacity of the gas phase as well as poor thermal conductivity of wood substrates and straw during the intensive growth phase of fungi, a remarkable gradient of temperature arises along the height of the mass in the direction of the air stream. Depending on the cosubstrates, the inoculum size etc., the temperature peaks appear exceeding the optimum 10-15°C in the upper part of the substrate within a short period of time (2-3 h). Upon 800 mm of the layer thickness it is the most difficult to monitor the midle part of the substrate. To overcome this constraint, the multiple zone areation was maked up (Figure 2). Since it is technologically impossible to ensure a sufficient homogeneity inside the reactor by stirring, the use of cold/hot water for temperatation is unacceptable, since it cause an unpermissible cooling/heating of the biomass close to the surfaces of heat-exchange. Due to the heat losses inside the fermentation system, only slow lowering of temperature takes place. More efficient mode for indirect temperature control was revealed based on the control of the respiration intensity of fungi by a purposeful alteration of the gas composition (O_2, CO_2) in the inlet air.

The heat evolution intensity depend on the respiration intensity, and was changed by correction of the gas composition in the inlet air. During the recirculation of the aerating gas, the O_2 concentration gradually went down, while the CO_2 concentration - went up. Accordingly, the respiration intensity as well as heat evolution decreased, and the temperature gradually leveled off. The established resistance of white-rot fungi C.

versicolor to the lowering oxygen: Thacker and Good [22], and raising partial pressures of CO_2, or O_2 as well: Sheffer and Livingston [23]. Reid and Seifert [24], allowed to carry out extensive manipulations with the gas composition during air recirculation or using different O_2/CO_2 mixtures. The lowest O_2 concentration in the aerating gas mixture used in our experiments with C. versicolor and Lentinus tigrinus was 4%, and the highest CO_2 concentration - 22%. The multiple-zone aeration and the correction of the inlet gas composition allowed us to succeed in ensuring comparatively good protein synthesis (biomass growth) (Table 1) as well as wood delignification (Figure 3).

Table 1

Protein biosynthesis during SSF cultivation of C. versicolor in a column bioreactor (CB) stanionary layer and CB with multizone air inlet and conditioning (CBMIC)

Substrate bioreactor	Time days	Prot. content %DM	Weight loss
Alkali treated wheat straw + wheat bran CBMIC	5	22.1	21.8
Untreated wheat straw + wheat bran	5	18.0	15.6
Alkali treated wheat straw CB, stationary layer	7	11.3	32.1

In an aerated column bioreactor with multizone air inlet and air conditioning, the lignin and cellulose biodegradation efficiency in spruce TMP by C. versicolor was 1.8 and 1.6 times higher, respectively, as compared to the control SSF in stationary layer of 800 mm, but 1.2 times lower as compared to submerged fermentation of C. versicolor on alkali treated straw, using a special stirring system and "C" substrate feeding. The lignin biodegradation rate in the experimental process reached 0.69 g/g d. against 0.38 g/g d. in the control process. The lignin biodegradation rate during C. versicolor submerged fermentation

with auxiliary energy sources (feeding with ethanol) reached 0.83 g/g d.: Shvinka and Leite [17].

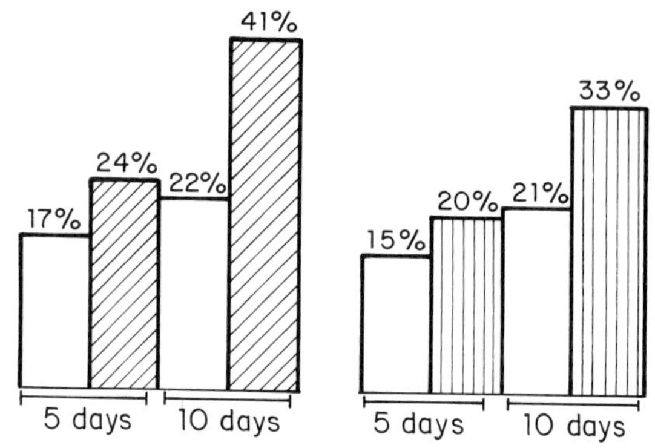

Figure 3. Comparison of lignin and cellulose weight losses during SSF of C. versicolor 082 on spruce TMP (glucose suplementation) in CBMIC and column bioreactor stationary layer (800 mm), monozone air inlet without air conditioning ( )
- Klason ligning loss;
- Cellulose loss.

In many treatments: Eriksson and Vallander [25], Jurasek and Paice [26], supplemental nutrients, such as glucose, were used to increase the fungal growth and lignin degradation. It should be pointed out that in the case of C. versicolor, the cellulose degradation rate was high - 0.6 g/g d. Improved technological parameters during the logarithmic growth phase (temperature, moisture, gas composition), and increased oxygen pressures during the secondary growth phase stimulated not only the lignin degradation but also the wood carbohydrates metabolism (Table 2).

Several wood substrates and several species of white-rot fungi vary substantially in their response to the lignin and wood carbohydrate degradation. Due to this specificity,

CBMIC is more or less suitable for the biopulping or biobleaching objectives. Among the fungi studied, Coriolus pubescens and Lentinus tigrinus should be pointed out as more selective biodelignificators. An additional advantage of CBMIC to be mentioned is that increased CO_2 concentrations in the inlet air at the initial growth phases prevent the growth of bacterial concomitants: Zadrazil and Peerally [27]. To sum up, the designed bioreactor and mode of aeration permit the flexible process monitoring and insuring the optimal conditions for the culture growth and LC substrate delignification. However, if compared with the fed-batch submerged fermentation, the lignin degradation rate is still lower, and the period of fermentation longer even in the best SSF process. The studies on the aerobic cultivation of filamentous fungi on wood substrates should be continued.

Table 2
Comparison of lignin and cellulose biodegradation during 10 days of C. versicolor cultivation in a column bioreactor (CB) (control) as well as a column bioreactor with multizone air inlet and conditioning (CBMIC)

Process	Spruce chips		Aspen chips		Wheat straw	
	Klason lignin loss,%	Cellulose loss,%	Klason lignin loss,%	Cellulose loss,%	Klason lignin loss,%	Cellulose loss,%
CBMIC	23.6	20.0	36.1	29.0	35.2	28.3
CB	20.0	16.4	30.8	22.0	19.5	17.6

CONCLUSIONS AND PROGNOSIS

The analysis of the literature and our own experiments demonstrates a technical possibility to realize the SSF or mixed SSF-SF process. However, the aerobic controlled SSF and SF processes of cellolignin raw materials, particularly prevailing for delignifying cultures, are energetically unfavourable ones for microorganisms. Therefore, to ensure an intensive growth, the addition of an energy rich substrates (ethanol, glucose, acetate, etc.) is necessary. This is evidently one of the reasons why a considerable delignification degree in SSF may be achieved only in long processes. Besides, the application of cellulose - or/and hemicellulose - assimilation cultures requires the complex treatment of raw materials (biomass delignification).

Hence, the direct bioconversion process (direct cultivation) appear, in general, to be rather complicated, and the end products rather expensive.

However, unfortunately, the acid hydrolysis of wood, agricultural waste, etc. is also currently rather expensive. The enzymatic hydrolysis is both expensive and hard to be realized on the industrial scale.
Several chemical processes are used, such as separation, gasification, liquefaction, etc. However, taking into account the fact that natural resources (oil, gas, etc.) are constantly decreasing, we should consider the photosynthesis in artificial plantation including and rational large-scale conversion of natural polymers (starch, hemicelluloses, cellulose, lignin and their composites) into monomers as the major way to obtain raw materials for the chemical, microbiological and other industries. In the future, the industrial-scale energy ethanol production under the enzymatic hydrolysis of special cultures grown in plantations, including fast-growing trees, is planned.
Our modest experience and the analysis of the world literature show that for such large capacity photosynthesized polymer conversion, SSF (as well as SF9 is doubtful to prove itself economical.
The major application of the SSF techniques may be in the output of comparatively expensive small amounts of microbial synthesis products for medicine, the food industry, agriculture, etc. For such purposes, the majority of the SSF-SF variants may be used for practical applications.
To summarize the afore mentioned, we consider mainly the following versions of the industrial aerobic SSF process:
1. Immobile or mixed like in the Koji proces thin layers for comparatively expensive products.
2. Mixed specially tailored bacterial cultures with a low shear sensitivity.
3. Bioscrubbers for gas decontamination.

LITERATURE CITED

1. Pandey, Ashok, Process Biochem., **27**, 2, 109 (1992).

2. Lonsane, B.K., Ghildyal, N.P., Murthy, V.S., "Solid State Fermentation and Their Challenges", presented at the Symp. of Assoc. of Microbiol.of India, Mysore (1982)

3. Lonsane, B.K., Saucedo-Castañeda, G., Raimbault, M., Roussos, S., Viniegra-González, G., Ghildyal, N.P., Ramakrishna, M., Krishniah, M.M., Process Biochem., **27,** 5, 259 (1992).

4. Pereira, S.S., Tores, E.F., González, V.G., Rojas, M.G., Appl. Microbiol. Biotechnol., **39,** 1, 36 (1993).

5. Schuchardt, F. and F. Zadrazil, Treatment of Lignocellullosics with White Rot Fungi, p. 77 Elsevier Appl. Sci., London, New York (1988).

6. Villegas, E., Aubague, S.S., Alcantara, L., Auria, R., Revah, S., Biotech. Adv., **11,** 3, 387 (1993).

7. Zadrazil, F., Diedrichs, M., Jansen, H. Schuchardt, F., Park, J.S. Advances in Biological Treatment of Lignocellulosics Materials, p. 43 Comiss. of the EC, Elsevier Applied Sciences (1990)

8. Zadrazil, F., Galetti, G.C., Picaglia, R., Chiavari, G., Francioso, O., J. of Anim. Feed Sc. and Technol., OECD Workshop (1990).

9. Toma, M.K., Ruklisha, M.P., Viesturs, U.E., Proceed, of Latvian AS, **8,** 1, 93 (1987)

10. Ruklisha, M.P., Vanags, J.J. Rikmanis, M.A., Toma, M.K., Viesturs, U.E., Acta Biotechnol., 9, 4, 347 (1989).

11. Zeltina, M.O., Leite, M.P., Vanags, J.J., Alpine, A.J., Viesturs, U.E., Acta Biotechnol., 7, 2, 157 (1987).

12. Zeltina, M.O., Leite, M.P., Treimanis, A.P., Viesturs, U.E., Egle V., Procceed. of the Latvian AS, 6 (539), 69 (1992).

13. Viesturs, U.E., Leite, M.P., Strikauska, S. V. "Bioconversion Alternatives and Technique for a Direct Cultivation of Microorganisms (on plant raw material)" presented at the VII Symp. 60, Espoo, Finland (1985).

14. Viesturs, U.E., Leite, M.P., Shvinka, J.E., Zeltina, M.O., Apine, A.J., "Energetic Efficiency of Complex Substrate Utilization by Morphological Forms of Fungi", presented at Soviet-Finnish Seminar on Bioconv., Riga, Latvia (1989).

15. Katkevich, J.J., Katkevich, R.H., Viesturs, U.E., Sakse, A.K., Leite, M.P., ActaBiotechnol., **8,** 6, 415 (1988).

16. Kirk, T.K., Schultz, Z.E., Connors, W.J., Lorenz, L.F., Zeikus, G.J., Arch. Microbiol. 117, 277 (1978).

17. Shvinka, J.E., Leite, M.P., Zeltina, M.O., Alpinee, A.J., Viesturs, U.E., "The Role of Auxiliarly Energy Source in Lignin Biodegradation by Lignolytic Fungi", presented at VII Symp. 122, Mustio, Finland (1990).

18. Baburin, L.A., Shvinka, J.E., Ruklisha, M.P., Viesturs, U.E., Acta Biotechnol., **6,** 2, 123 (1986)

19. Leite, M.P., Apine, A.J., Zeltina, M.O. Shvinka, J.E., Acta Biotechnol., 9, 6, 421 (1989)

20. Effland, M.J., Tappi, **60,** 1, 143, (1977).

21. Updegraff, D.M., Anal. Biochem., **132,** 1, 420 (1969)

22. Thacker, D.G. and H.M. Good, Can, J. Bot., **30,** 6, 475 (1952).

23. Sheffer, T.G. and B.E. Livingston, Am. J. Bot, **24,** 1, 109 (1937).

24. Reid, I.D. and K.A. Seifert, Can. J. Bot., **60,** 3, 252 (1982)

25. Eriksson, K.E. and L. Vallander, Svensk Paperstidn, **85,** 6, 33 (1982)

26. Jurasek, L. and M.G. Paice, "Direct Biological Bleaching of Pulps", presented at "Cellucon 90", Bratislava, Slovakia (1990).

27. Zadrazil, F. and A. Peerally, Biotechnol. Letters, **8,** 9, 663 (1986).

Growth of *Candida utilis* on Amberlite with Glucose and Ethanol as Sole Carbon Sources

P. Christen[1], R. Auria[1], R. Marcos[2], E. Villegas[2], and S. Revah[2]

[1]ORSTOM (Institut Français de Recherche Scientifique pour le Développement en Coopération), Cicerón 609, Col. Los Morales, 11530 México, D.F.

[2]UAM-Iztapalapa, Dpto. de Ingeniería de Procesos, Apdo. Postal 55-534, 09340 México, D.F., MEXICO

The results of Candida utilis *growth on an anionic resin (Amberlite) at high glucose concentration and using ethanol as the sole carbon sources are presented. The yeast consumed 240 mg glucose (g initial dry matter)$^{-1}$ (IDM) reaching a final population of 5.6x10^9 cells (g IDM)$^{-1}$ (initial inoculum size: 1x10^7 cells (g IDM)$^{-1}$. It was also shown that respirometry was a reliable on-line method for monitoring growth. The respiratory quotient (RQ) showed the changes in the metabolism of the yeast during glucose consumption, from a fermentative to an oxidative route. When* C. utilis *was grown on gaseous ethanol enriched air, a final population of 3.25x10^9 cells (g IDM)$^{-1}$ was attained. The importance of mineral salts concentration in the nutritive medium was clearly demonstrated. A two fold increase in the population was obtained when the mineral medium was not limiting. Small amounts of acetaldehyde and ethyl acetate were detected at the outlet of the reactor (1.88 μl l^{-1} and 0.87 μl l^{-1}, respectively). Ethanol accumulated in the reactor up to 120 mg (g IDM)$^{-1}$ (probably an inhibitory level for this yeast). RQ remained constant at around 0.6 during the fermentation.*

Solid state fermentation (SSF) is an old technique that has been recently revaluated and modernized to enhance protein content of agro-industrial wastes (1,2,3) or to produce enzymes (4), secondary metabolites (5), spores of fungi (6) or cheese or fruity flavors (7,8). Some reports deal more specifically with yeasts employed in SSF systems (9,10,11) and recently, the ability of a *Candida utilis* strain to grow on various substrate/supports was demonstrated (12).

Furthermore, yeasts from the genius *Candida* are known to be able to convert ethanol to ethyl acetate (13,14) or acetaldehyde (15). Both components present an economical interest in the food additives industry (16).

In this work, the effect of inoculum concentration and high glucose concentrations on growth of *C. utilis* on a synthetic support (Amberlite) under SSF conditions was evaluated in terms of kinetic and respirometric parameters. The ability of this yeast to grow and produce ethyl acetate and acetaldehyde when fed with air enriched with gaseous ethanol was also studied.

MATERIAL AND METHODS

Microorganism

The yeast *Candida utilis* ATCC 9950 (CDBB L245) was used in all experiments. It was periodically transfered on potato dextrose agar slants and stored at 4°C.

Culture media

The inoculum was grown on a liquid medium composed with glucose (20 g/l) and malt extract (20 g/l) in 150 ml Erlenmeyer flasks agitated at 200 rpm at 30°C for 20h. The yeast was then grown on Amberlite IRA-900 (Rohm & Haas), an anionic resin, prepared according to Auria *et al.*(17) and imbibed with the minimal salts medium described by Thomas and Dawson (18). This nutritive medium was used with different glucose concentrations from 40 to 240 mg/g Initial Dry Matter (IDM). For the case of cultures on ethanol, glucose (40 mg/g IDM) was used in the beginning of the fermentation and then gaseous ethanol was fed.

Solid State Fermentation procedure

Cultures were performed in the set up represented in Figure 1.

87

E. Galindo and O.T. Ramírez (eds.), Advances in Bioprocess Engineering. 87-93.

© 1994 Kluwer Academic Publishers. Printed in the Netherlands.

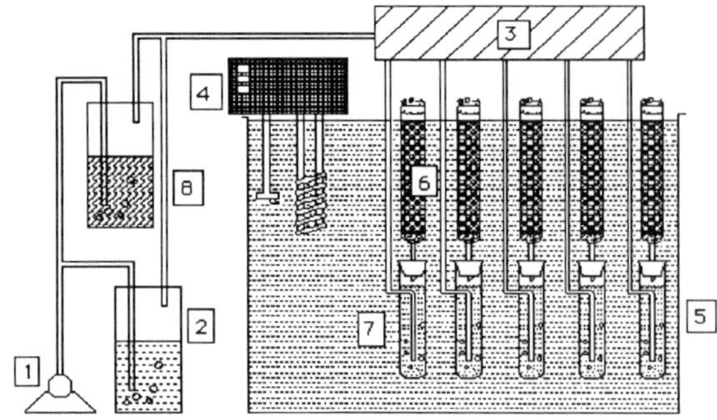

Figure 1. Experimental set up for SSF.

In the experiments run with ethanol, the alcohol was fed to the columns by bubbling air in pure solution as it was done for water in air.
The initial culture conditions were: pH, 6; Temperature, 30°C; Moisture content, 58% (w/w); yeast inoculum concentration, 1×10^7 cells/g IDM except when mentioned; aeration rate, 0.1 or 0.05 l/h.g IDM, ethanol rate feed, 5 µl/h.g IDM and packing density, 0.6 g/ml.

Analytical methods

Biomass was determined by direct cell count and viability with methylene blue coloration as described previously (12). In liquid culture, it was found that 10^7 cells correspond to 0.092mg (12). pH was measured with a Conductronic pH meter and Aw with an Aqualab CX-2 apparatus (Decagon, USA). With the same sample, glucose was determined by the dinitrosalicylic acid method (19) and residual ethanol was measured by gas chromatography. Headspace analysis of the exit air was achieved for acetaldehyde, ethanol and ethyl acetate determination. Gas chromatography analysis was made with a Hewlett-Packard chromatograph equipped with a flame ionization detector. Nitrogen was used as carrier gas at a 4 ml/mn rate. Split ratio was 1:50. Temperature were: injector and detector, 180°C; oven, 40°C. Separation was achieved with a Megabore HP-1 column (Length, 5m; Inner diameter, 0.53 mm). Concentrations were reported as µl liquid/l gas.
Respirometry (O2 and CO2 measurements) was realized with a Gow-Mac chromatograph equipped with a thermal conduc-

tivity detector and a concentric column CTR-1 (Alltech, USA). Helium was used as carrier gas (flow rate, 60 ml/mn). Carbon dioxide production rate (CDPR), oxygen uptake rate (OUR) and respiratory quotient (RQ) were calculated as follows :
CDPR = (%CO2 produced x F) / (100 x W)
OUR = (%O2 consumed x F) / (100 x W)
RQ = CDPR/OUR.

RESULTS AND DISCUSSION

Glucose as sole carbon source

.Influence of inoculum size
These experiments were run with an initial glucose concentration of 135 mg/g IDM and inoculum concentration of 1.4, 2.2 and 3.6×10^7 cells/ml respectively.

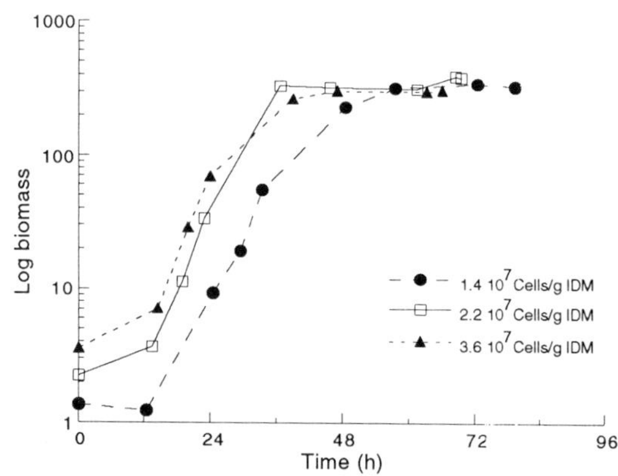

Figure 2. Logarithmic evolution of biomass vs. time. Influence of inoculum size.

In Figure 2, it can be seen that inoculum size did not have a strong influence on growth rate. There is not a significant influence on the final biomass (between 300 and 350×10^7 cells/g IDM) and all the glucose was exhausted after 50 hours. The same conclusions were obtained in submerged cultures.

.Influence of initial glucose concentration
Three different glucose concentrations were studied: 40, 135 and 240 mg/g IDM. It appeared that the phase lag and the fermentation time were longer, and the maximum biomass was reached later for higher glucose concentrations (Cf Figure 3). These maxima were related to the initial glucose concentration (more than 550×10^7 cells/g IDM

for 240 mg/g IDM).

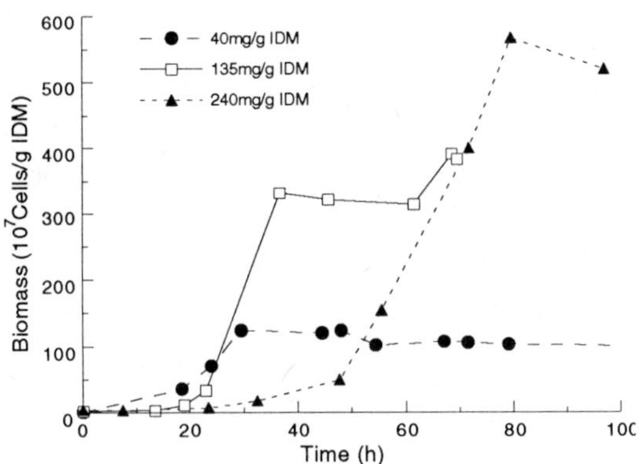

Figure 3. Evolution of biomass vs. time. Influence of initial glucose concentration.

Glucose was totally exhausted in all cases at different time courses (Figure 4). It can be seen that glucose concentration in SSF system has a strong influence on Aw (for the same initial moisture content). The relatively low value of Aw at the beginning of the fermentation for 240 mg/g IDM can explain the largest lag phase. When glucose was exhausted, Aw reached a value of 0.99 in both cases.

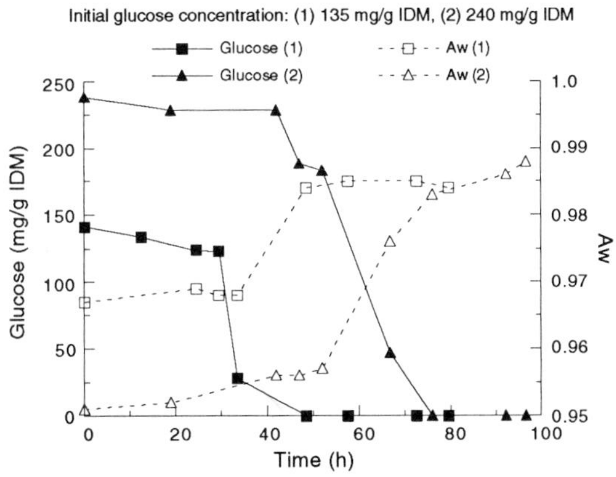

Figure 4. Evolution of glucose concentration and Aw vs. time. Influence of initial glucose concentration.

In Figure 5, it can be observed that despite the previous neutralization of the support (to pH 6), a dramatic drop in pH was observed (around 3.2 for low glucose concentration and 2.4 for high concentrations). This severe drop can

be explained by the excretion of organic acids (e.g. acetic) to the medium. These final values were reached faster for lower glucose concentrations which was related to growth and ammonium sulfate consumption and proton excretion. Mortality among the yeast cells was low except for the final time for the higher glucose concentration. This behavior was related to low pH (2.3) and lack of carbon sources.

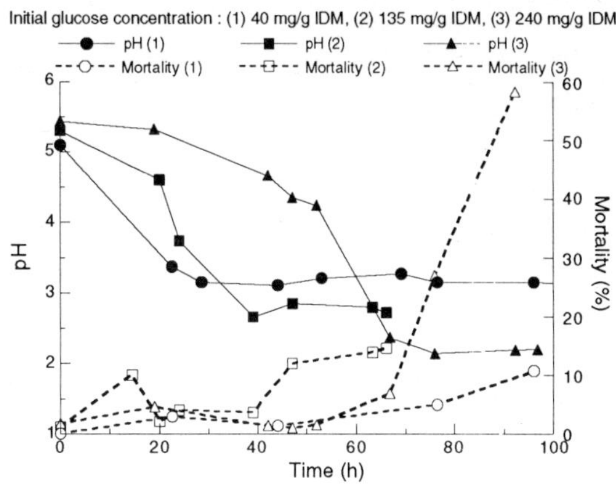

Figure 5. Evolution of pH and mortality vs. time. Influence of initial glucose concentration.

Figure 6 shows the CDPR evolution for different initial sugar concentrations. These curves present a maximum peak corresponding to the maximum volumetric growth rate. These maxima were reached in 28h, 34h and 55h respectively and are proportional to the initial glucose concentration.

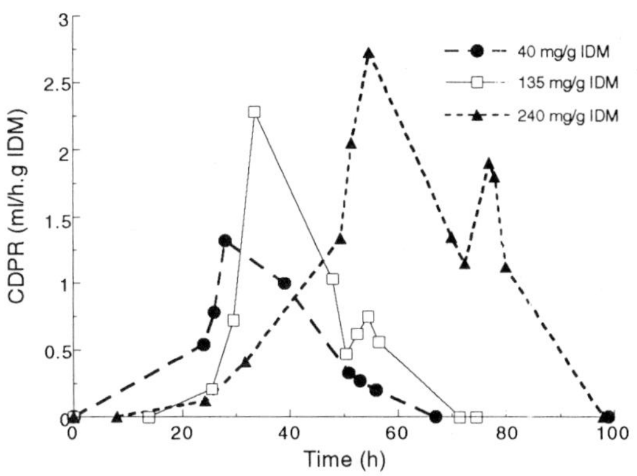

Figure 6. Evolution of CDPR vs. time. Influence of initial glucose concentration.

In the cases of 135 and 240 mg/g IDM, a second peak smaller and later was observed. It appeared when glucose was exhausted and was probably due to a diauxic effect from other substrates released previously in the medium (e.g. ethanol or organic acids). The maxima values of CDPR reached are comparable with those obtained by *Auria et al.* (17) for *Aspergillus niger* grown on the same support.

Table 1. Summing up of the results of the glucose fermentation by *C. utilis* grown on Amberlite. (* Cf Figure 10)

Initial glucose (mg/g)	40	135	240
X max	11.41	27.05	52.07
Fermentation time (h)	29.5	47	80.5
CDPR max	1.42	2.28	2.73
Δ R.Q.*	1.1/1.3	1.1/1.3	1.0/2.5
μ max	n.c.	0.55	0.58
Yx/s	0.283	0.201	0.225
Rx max	0.74	0.83	1.97

From table 1, it can be concluded that biomass production, in the range of the glucose concentrations studied, was proportional to the initial concentrations. The substrate conversion yields into biomass kept within a range of 0.2 to 0.28 with highest value for lowest concentration. There is no significant influence of glucose concentration on growth rate. On the contrary, the volumetric productivity (Rx) is greatly increased when substrate concentration is higher. There is also a clear influence of glucose level on the orientation of the metabolism. The respiratory quotient reached values superior to 2 (fermentative route) for 240 mg/g IDM, when these values kept near above 1 at lower concentrations (oxidative pathway). *C. utilis*, when grown on Amberlite, displayed a good tolerance toward high glucose levels, low pH value and seemed to be able to assimilate ethanol - or other metabolites released in the medium.

Ethanol as consecutive carbon source

.Influence of mineral medium concentration

Two experiments were made: one with mineral medium concentration corresponding to 40 mg/g IDM of glucose, the other one multiplied by a factor of 3.5 In both cases, the fermentation

was initiated with glucose (40 mg/g IDM) and ethanol vapor phase fed when glucose was totally exhausted (after 24 hours). Growth evolution is shown in Figure 7.

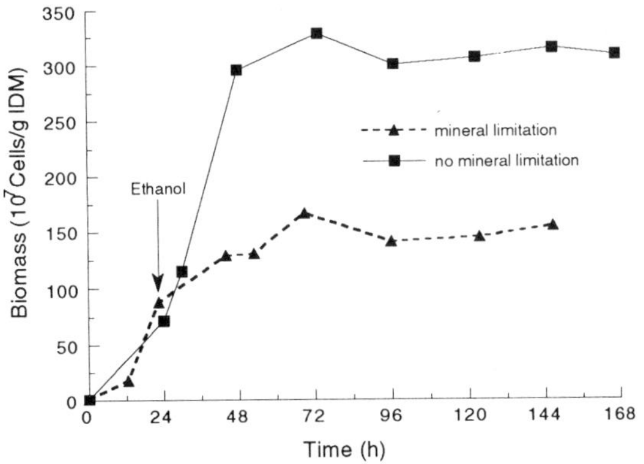

Figure 7. Evolution of biomass vs. time. Influence of mineral medium concentration. Growth on ethanol.

The interest of using the higher mineral salts concentration was clearly demonstrated since a final population of 330×10^7 cells/g IDM was reached against 150×10^7 cells/g IDM for the lower concentration. The maxima were reached in the same time (about 75h). The fact that the increase in biomass was not proportional to the mineral medium concentration implies that there is another limitation. Only the second experience (with the highest mineral salt concentration) is described hereafter.

.Growth and substrates consumption

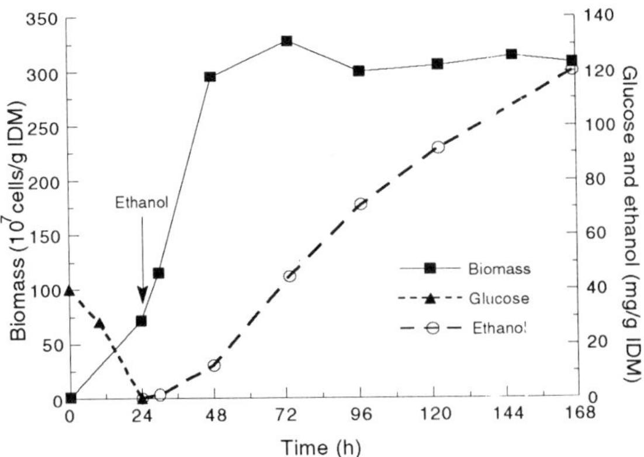

Figure 8. Evolution of biomass and substrates vs. time. Growth on ethanol.

From figure 8, it can be observed that the maximum population is reached after 75 hours when ethanol concentration in the medium was about 40 mg/g IDM and mortality yield was 16%. It can be seen that ethanol in the column increased constantly and with a higher velocity after 24 hours of culture on this substrate, which probably meant that the concentration reached was limiting but not lethal for the yeast (about 20 mg/g IDM). Meanwhile, the mortality yield increased up to 80% at the end of the fermentation. The yeast displayed a better tolerance toward ethanol than observed previously (13), which can be due to a protective effect of the support. The pH dropped to 2.5 in the first 30 hours and then kept constant around this value. Water activity was constant above 0.98 all along the experiment. It can be concluded from these data that SSF is an adequate system for the growth of *C. utilis* on gaseous substrate.

.<u>Ethanol, ethyl acetate and acetal-</u>
<u>dehyde evolution in the exit gas</u>
According to Armstrong *et al.* (13), ethyl acetate production from ethanol for *C. utilis* follows three steps:

1.Ethanol + O2 ----> Acetaldehyde
2.Acetaldehyde + O2 ----> Acetic acid
3.Ethanol+Acetic ac. --->Ethyl acetate + H20

These reactions are catalyzed by alcohol dehydrogenases (steps 1 and 2) and by an esterase (step 3).
The evolution of ethanol, acetaldehyde and ethyl acetate concentrations in the exit gas is plotted in figure 9.

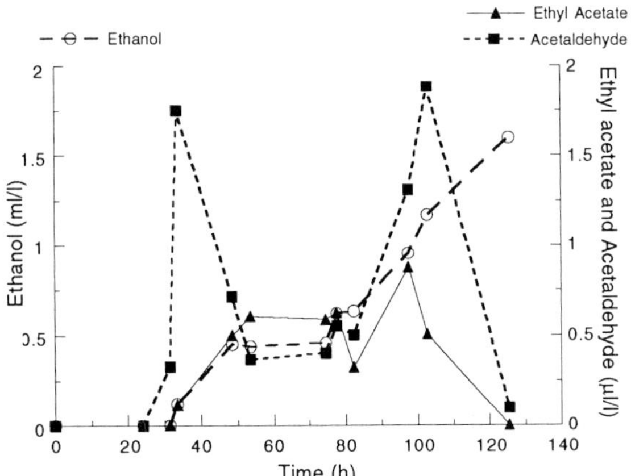

Figure 9. Headspace of solid state culture of *C. utilis* grown on ethanol.

The curves shown in figure 9 can be divided in three regions. From 24 to 55 hours, exit ethanol concentration increased as well as ethyl acetate and acetaldehyde, this one presented a maximum around 36 hours. From 55 to 75 hours, a steady state was set and the three concentrations remained constant. From 75 hours, when growth stopped, ethanol concentration increased again in relation to the concentration of the liquid phase, reflecting the fact that it was not consumed. No evidence was found that non-growing *C. utilis* cells produced ethyl acetate or acetaldehyde.

Most of the time, acetaldehyde levels were above those of ethyl acetate. Levels of both compounds were low probably because ethanol concentration was too high, pH low and hence inhibited redox enzyme system of the yeast. It should be observed that neither acetaldehyde nor ethyl acetate were detected in the solid state medium. By integration of the ethyl acetate and acetaldehyde curves, it was calculated that 23 mg and 38 mg were produced respectively.

Finally, a comparison of the respiratory coefficients in three cases is presented in figure 10.

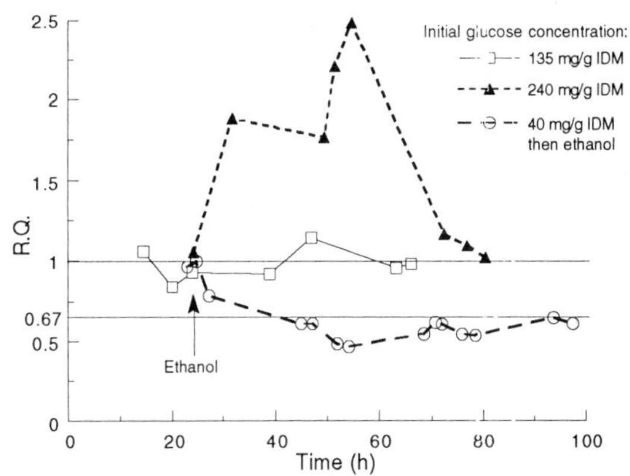

Figure 10. Evolution of the respiratory quotient vs. time on different glucose concentrations and on ethanol.

The evolution in each case is very significative:

. for 135 mg glucose/g IDM, RQ kept constant around a value of 1, characteristic of the oxidative metabolism of glucose, where :

Glucose + 6O2 ---> 6CO2 + 6H2O + **28 ATP**
     (theoretical RQ = 1)
. for 240 mg glucose/g IDM, RQ varied between 1 and 2.5. This is characteristic of a partial orientation of the metabolism toward a fermentative route *sensu stricto*, where :
Glucose ---> 2CO2 + 2 Ethanol + **2 ATP**
     (theoretical RQ ---> ∞)
. for growth on ethanol, RQ was constant around 0.6. This represents the oxidative metabolism of ethanol where:
Ethanol + 3O2 ---> 2CO2 + 3H2O + **11ATP**
     (theoretical RQ = 0.67)

Although *C. utilis* showed its capacity to grow on ethanol as sole carbon source, the overall results of the fermentation are lower than those obtained on glucose (μmax, Rx max) basically for the reason given above (cf Table 2).

Table 2. Kinetic parameters of the fermentation of ethanol by *C. utilis* on Amberlite.

X max	30.17
CDPR max	2.15
Δ R.Q.	0.5/0.72
μ max	0.13
Rx max	0.41

<u>CONCLUSION</u>

In this work, the influence of glucose concentration on growth was shown. It was demonstrated that total biomass, lag phase duration and yield of substrate conversion into biomass were dependent on initial glucose concentration as well as the type of metabolism (oxidative or oxidative/fermentative) as shown by the respiratory quotients study. Also, even at the higher glucose concentration, the relatively low water activity observed did not avoid the growth. The main limitation is finally due to the strong pH drop and the accumulation of ethanol in the water phase.
A SSF system was successfully designed to study the growth of the yeast with gaseous ethanol. It was shown that mineral salts concentration could be a limiting factor for growth and that *C. utilis* was able to use ethanol for its growth and to produce other metabolites. Moreover, the SSF system provided the advantage to increase the tolerance of the yeast toward ethanol. Although the conditions were not optimized, low excretion of ethyl acetate and acetaldehyde were obtained in the outlet gas phase.

R. Marcos was a post-graduate student from ENSBANA (Dijon-FRANCE). This work was achieved under research agreement between UAM (MEXICO) and ORSTOM (FRANCE).

<u>NOMENCLATURE</u>

Aw   Water activity (-)

CDPR   Carbon dioxide production rate (ml/h.g IDM)

F   Inlet aeration rate (ml/h)

OUR   Oxygen uptake rate (ml/h.g IDM)

R.Q.   Respiratory quotient (-)

Rx   Volumetric growth rate (mg biomass/g.h)

W   Initial dry matter weight (g)

X   Biomass (mg/g IDM)

Yx/s   Biomass yield on substrate (g biomass/g substrate)

μ   Specific growth rate (h^{-1})

<u>LITERATURE CITED</u>

1. Raimbault, M. and A. Alazard. *Eur. J. Appl. Microbiol. Biotechnol.,***9**, 199 (1980).

2. Baldensperger, J., J. Le Mer, L. Hannibal and P.J. Quinto. *Biotechnol. Lett.* **7**, 743 (1985).

3. Durand, A. and D. Chereau. *Biotechnol. Bioeng.*, **31**, 476 (1988).

4. Aidoo, K.E., R. Hendry and B.J.B. Wood. *Adv. Appl. Microbiol.*, **28**, 201 (1982).

5. Barrios-González, J., H. González and A. Mejía. *Biotechnol. Adv.*, **11** 539 (1993).

6. Desfarges, C., C. Larroche and J.B. Gros. *Biotechnol. Bioeng.*,**29**, 1050 (1987).

7. Revah, S. and J.M. Lebeault. *Lait*, **69**, 281 (1989).

8. Christen, P., E. Villegas and S.Revah. *Biotechnol. Lett.* Submitted (1994).

9. Grant, G.A., Y.W. Han and A.W. Andersson. *Appl. Env. Microbiol.*,**35**,

3, 549 (1978).

10. Rossi, J. and F. Clementi. *J. Food Technol.*, **20**, 319 (1985).

11. Saucedo-Castañeda, G., B.K. Lonsane, M.M. Krishnaiah, J.M. Navarro and M. Raimbault. *Process Biochem.*, **22**, 411 (1992).

12. Christen, P., R. Auria, C. Vega, E. Villegas and S. Revah. *Biotechnol. Adv.*, **11**, 549 (1993).

13. Armstrong, D.W., S.M. Martin and H. Yamazaki. *Biotechnol. Bioeng.*, **26**, 1038 (1984).

14. Corzo G., S. Revah and P. Christen. Effect of oxygen on the ethyl acetate production by *Candida utilis* in submerged cultures, In: G.Charalambous(Ed), *8th Int. Flavor Conf.* Cos, Greece. Submitted (1994).

15. Armstrong, D.W., S.M. Martin and H. Yamazaki. *Biotechnol. Lett.*, **6**,183 (1984).

16. Welsh, F. W., Murray, W.D. and Williams, R.E. *Crit. Rev. Biotechnol.*, **9**, 2, 105 (1989).

17. Auria, R., S.Hernandez, M.Raimbault and S. Revah. *Biotechnol. Tech.*, **4**, 391 (1990).

18. Thomas, K.C. and P.S.S. Dawson. *Can.J. Microbiol.*, **24**, 440 (1978).

19. Miller, G.L. *Anal. Chem.*, **31**, 426 (1959).

Invited paper

Scaling-up Aerobic Fermentation which Produce Non-Newtonian, Viscoelastic Broths

D.W. Hubbard, S.E. Ledger, and J.A. Hoffman

Department of Chemical Engineering, Michigan Technological University, Houghton,
Michigan 49931, U.S.A.

Gas liquid mass transport usually controls the rate of an aerobic fermentation. The effect of mass transport must be included when scaling-up this kind of fermentation. One method for scaling-up gas/liquid contacting systems is to keep the mass transfer coefficient constant. Both impeller speed, N, and gas flow rate, Q, must be determined for a successful scale-up. Equating mass transfer coefficients for similar systems of different volumes gives one relationship between unknown values of N and Q, but another relationship is needed so that both can be calculated. We measured mass transfer coefficients using a simulated fermentation broth. Carbon dioxide was reacted with sodium hydroxide in a dilute aqueous polysaccharide solution in a stirred tank. Mass transfer coefficients corrected for the effect of chemical reaction were calculated from measured reaction times. The work includes independent measurements of rheological and elastic properties and surface tension. Data from geometrically similar tanks of up to 2000 l were used to test the usefulness of mass transfer coefficient correlations for scale-up.

Scaling-up means reproducing laboratory results in an industrial-scale system. If a particular aerobic fermentation is accomplished successfullly at the laboratory scale, the values of the operating variables and the physical properties are known or can be measured. For equivalent organism growth on larger scales, the growth conditions observed in the laboratory unit must be maintained in the larger fermenter. The nutrient concentrations, temperature, and pH can all be controlled independently.

SCALE-UP STRATEGY

Oxygen uptake usually controls the rate of an aerobic fermentation in a stirred tank. The rate controlling step is often oxygen transport from the gas/liquid interface to the bulk of the liquid. The effect of mass transport must be included when scaling-up this kind of fermentation. This can be done by maintaining the oxygen uptake rate per unit volume, (OUR)/V, the same in each system.

$$(OUR)/V = k_L a (c^* - c_b) \qquad (1)$$

The bulk concentration of oxygen, c_b, is constant, because the organisms use oxygen rapidly, and the oxygen concentration at the gas-liquid interface, c^*, is determined by solubility. The scale-up strategy described is

$$(OUR/V)_1 = (OUR/V)_2$$

This leads to keeping the mass transfer coefficient the same for the laboratory-scale and industrial-scale systems.

$$(k_L a)_1 = (k_L a)_2 \qquad (2)$$

To implement this scale-up strategy, a correlation between the mass transfer coefficient and operating variables, fluid properties, and geometrical variables is needed. Selecting a particular $k_L a$ correlation means assuming that the correlation will apply to whatever process conditions are developed. Mass transport must be governed by the same mechanism at all scales.

Fermentation broths often contain macromolecules such as proteins and polysaccharides, so the mass transfer

95

E. Galindo and O.T. Ramírez (eds.), Advances in Bioprocess Engineering. 95-101.
© *1994 Kluwer Academic Publishers. Printed in the Netherlands.*

coefficient correlation selected as part of the scale-up procedure should apply to a liquid which has non-Newtonian and viscoelastic properties. Mixing in non-Newtonian liquids is often laminar, so a mass transfer coefficient correlation which applies for low viscosity liquids is not suitable. A dimensionless correlation can be devised using the dimensional analysis method described by Langhaar[1]. Perez and Sandall[2], Yagi and Yoshida[3], and Ranade and Ulbrecht[4] have presented such dimensionless correlations. These correlations were developed from measurements made in tanks having volumes of 2.7 to 21.2 liters. Yagi and Yoshida and Ranade and Ulbrecht include the effect of viscoelasticity in their correlations. The correlation by Yagi and Yoshida incorporating the effects of all the physical properties of the liquid has the following form.

$$N_{Sh}' = A(N_{Re})^{a1}(N_{Fr})^{a2}(N_{Sc})^{a3} \times$$

$$(\mu_e V_s/\sigma)^{a4}(ND_i/V_s)^{a5}[1+B(\lambda N)^b]^{a6} \quad (3)$$

The exponents, a1,...a6 and the constants, A and B, were evaluated from data and have values as follows.

$$
\begin{aligned}
A &= 0.06 \\
a1 &= 1.5 \\
a2 &= 0.19 \\
a3 &= 0.5 \\
a4 &= 0.6 \\
a5 &= 0.32 \\
a6 &= -0.67 \\
B &= 2 \\
b &= 0.5
\end{aligned}
$$

Effective viscosity, μ_e, is calculated according to the method of Metzner and Otto[5] as discussed by Calderbank and Moo-Young[6]. The power law model is used to describe the rheological behavior of the broth.

If all physical properties remain constant, equating the mass transfer coefficients for geometrically similar systems of two different volumes gives an equation relating impeller speed, N, and gas flow rate, Q, in the two systems. Both N and Q are unknown operating variables in the industrial-scale system, so another relationship between N and Q is needed. There are several possibilities for a second equation.

Constant gas flow number:

$$Q/ND_i^3 = constant$$

Constant superficial velocity:

$$4Q/\pi D_T^2 = constant$$

Constant volumetric gas injection ratio:

$$4Q/\pi D_T^2 H_L = constant$$

Constant impeller tip speed:

$$\pi ND_i = constant$$

Any one of these equations can be put into a form suitable for scale-up calculations. For example,

$$(Q/ND_i^3)_1 = (Q/ND_i^3)_2 \quad (4)$$

can be solved simultaneously with Equation (2)--the equation from the $k_L a$ correlation--to give values for N and Q to be used in the scaled-up system. To have a useful scale-up procedure, we need to determine which is the best of these equations to use.

SENSITIVITY OF THE $k_L a$ CORRELATION TO PHYSICAL PROPERTIES AND PROCESS VARIABLES

The dimensionless correlation given by Yagi and Yoshida--Equation (2)--was used to identify the factors which have the greatest effect on the value of the mass transfer coefficient predicted. The results of this sensitivity study appear in Table I. A base case was established for a 200-liter fermenter using the following values for the variables which appear in the $k_L a$ correlation equation.

$$
\begin{aligned}
V &= 2.0 \ m^3 \\
D_i &= 0.57 \ m \\
D_T &= 1.36 \ m \\
N &= 0.667 \ s^{-1} \\
Q &= 2.5 \times 10^{-4} \ m^3/s \\
n &= 0.32 \\
K &= 1.2 \ kg/m \ s^{n-2} \\
\rho &= 1000 \ kg/m^3 \\
\lambda &= 23 \ s \\
\sigma &= 0.047 \ N/m \\
D_{CO2} &= 2.0 \times 10^{-9} \ m^2/s
\end{aligned}
$$

Table I. Sensitivity of k_La to Variation in Physical Property Values and Operating Variables[*]

Variable	Common Range of Values	k_La Dependence (from the Y&Y correlation)	$k_La \times 10^5$ Range (1/s)	$|\Delta k_La|/k_La$
ρ	1000 (kg/m^3)	$\sim \rho^{1.0}$	---	---
D	1.9×10^{-9} to 2.5×10^{-9} (m^2/s)	$\sim D^{0.5}$	5.87 to 6.73	0.14
σ	0.04 to 0.07 (N/m)	$\sim \sigma^{-0.6}$	6.63 to 4.75	0.31
n	0.30 to 1.0	---	6.03 to 3.66	0.39
K	0.001 to 5.0 (0.5 to 5.0)[**] $(kg/m\ s^{n-2})$	$\sim K^{-0.4}$	102.8 to 3.41 (8.55 to 3.41)	16.4 (0.85)
λ	0 to 100 (s)	Figure 1	25.9 to 3.83	3.66
Q	10^{-5} to 0.04 (m^3/s)	$\sim Q^{0.3}$	2.45 to 25.0	3.74
N	0.01 to 5.0 (1/s)	---	0.0005 to 469	77.8

[*] Values calculated are based on the Yagi and Yoshida[3] (Y&Y) k_La correlation.

[**] Common values for non-Newtonian broths.

The mass transfer coefficient value calculated for this base case is $k_La = 6.03 \times 10^{-5}\ s^{-1}$.

The most probable range of variables for each physical property parameter was selected, and k_La was calculated for the extreme values of the range. The results are tabulated as values normalized by dividing by the base case k_La value. It is easy to determine the functional dependence of k_La on D, ρ, σ, K, and Q, so this dependence is indicated in the table.

Fermentation broths are mostly water, so the density is approximately the same as for water and is probably the same for all broths. As shown in the table. expected variations in diffusion coefficient, surface tension, and flow behavior index cause only a 14% to 40% effect on k_La over the range of most likely values. The value of the consistency index depends strongly on the concentration of dissoved macromolecular substances. The consistency index could vary by a factor of 5000 for water-like broths to broths containing substantial amounts of macromolecular solutes. For this situation, k_La could differ by a factor of 16. For a more restricted range of flow behavior index such as might be observed for non-Newtonian broths, our calculations show that a ten-fold variation in K causes a change of 85% in k_La.

The relaxation time could have approximately a 100-fold variation depending on the concentration of macromolecules in the broth. Our calculations show that this magnitude of variation in λ could cause k_La to change by a factor of 3.7. The effect of variations in λ depends on the impeller speed, the exponent of λN, the Deborah number, and, b, the coefficient of the Deborah number. This variation is shown for three cases in Figure 1. K_La decreases approximately as $\lambda^{0.3}$ if λ is greater than 1 second for N = 1.5/s or greater than 5 seconds for N = 0.667/s.

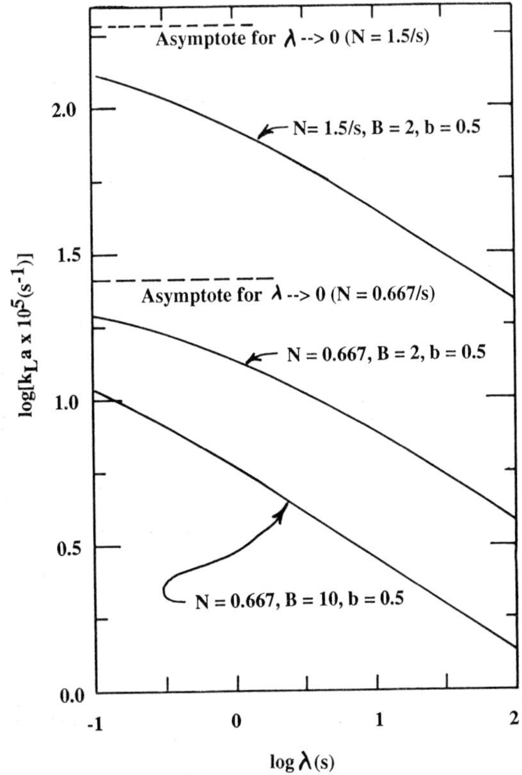

Figure 1. The Effect of Relaxation Time Variations on the Value of k_La Predicted by the Yagi and Yoshida Correlation

Different authors give different forms for the Deborah number term in the correlation. Ranade and Ulbrecht[4] measured k_La for solutions of polyacrylamide and carboxymethyl cellulose which had polymer concentrations of 100 to 1000 ppm. They used the Newtonian viscosity in their correlation which was as follows.

$$N_{Sh}' = A(D_T/D_i)^2 (N_{Re})^{a1} (\mu/\mu_w)^{a7} \times [1+B(\lambda N)^b]^{a6} \qquad (5)$$

where

$$
\begin{aligned}
A &= 2.5 \times 10^{-4} \\
a1 &= 1.8 \\
a7 &= 1.39 \\
a6 &= -0.67 \\
B &= 100 \\
b &= 1.0
\end{aligned}
$$

The most appropriate form for the Deborah number term is yet to be determined.

For predicting k_La, it is important to have the most accurate values possible for K and λ which are the physical property parameters which affect the value most strongly. The operating variables, Q and N, have the largest effects on the value of k_La predicted by the correlation, so these variables will be the best ones to adjust to accomplish an effective scale-up.

MASS TRANSFER COEFFICIENT MEASUREMENT

In our work, a fermentation broth was simulated using a dilute aqueous polysaccharide solution containing approximately 0.1 molar sodium hydroxide. Carbon dioxide was the gas transferred from the gas to the liquid, and carbon dioxide depletion in the liquid was caused by the reaction with sodium hydroxide.

Equipment

Pure carbon dioxide was sparged through an open pipe, a ring, or a tee sparger into a batch of sodium hydroxide solution where the reaction between sodium hydroxide and carbon dioxide took place. Geometrically-similar, cylindrical, flat-bottomed, stirred-tank reactors having batch volumes from 10.9 to 2000 liters were used for the experiments. Each tank had four baffles of width (0.1 D_T) spaced equally around the perimeter. Mixing was induced by a six-blade Rushton turbine for which $D_i/D_T = 0.4$.

Data Analysis

Danckwerts[7] discusses the kinetics of the reaction between carbon dioxide and sodium hydroxide and shows that the reaction is rapid. Mass transfer coefficients were calculated from the measured neutralization times and used to determine the exponents in equation (3). This correlation equation applies for mass transfer coefficients not affected by reactions, so the values of $k_L a$ calculated from the experimental data were corrected as suggested by Danckwerts[8]. A material balance for hydroxide ions in the reactor gives a first-order differential equation which is combined with the expression for mass transfer enhancement and solved to give an equation relating the measured reaction time to the mass transfer coefficient corrected for the effect of the chemical reaction. This equation is-

$$k_L a = \frac{\ln[1+(D_{OH}C_{OH}^{O}/2D_{CO2}C_{CO2}^{*})]}{t_R \; D_{OH}/D_{CO2}} \qquad (6)$$

Physical Properties

To apply the scale-up method described, we need to know the physical properties for the broth. The polysaccharide used in or work was Xanthan gum(Kelzan XC, Kelco Co., Houston), and the solutions contained 0.2 or 0.4 per cent by mass.

Rheological properties were measured using a Ferranti-Shirley cone-plate viscometer and a Bohlin VOR rheometer. The relaxation time was measured by a stopped rotation method. The surface tension was measured using a plate-withdrawal device (Tensimatic Model 03 161.02, LC Instrument Co.). The value of the diffusion coefficient for carbon dioxide used was that measured by Perez and Sandall[9]. The value of the diffusion coefficient for hydroxyl ions was taken from Newman[10]. We ignored the effect of other dissolved species on the diffusion coefficients of carbon dioxide and hydroxyl ions.

MASS TRANSFER CORRELATION RESULTS

Mass transfer coefficients measured at different gas flow rates for three types of sparger are shown in Figure 2. These results show that within the uncertainty of the measurements there is a negligible effect of sparger design on $k_L a$. The data plotted in Figure 2 can be correlated by

$$k_L a = 0.0017 \; Q^{0.266}$$

Table I shows that $k_L a$ should be proportional to $Q^{0.3}$, so the functional dependence on the volumetric gas flow rate expressed by the Yagi and Yoshida correlation is approximately correct.

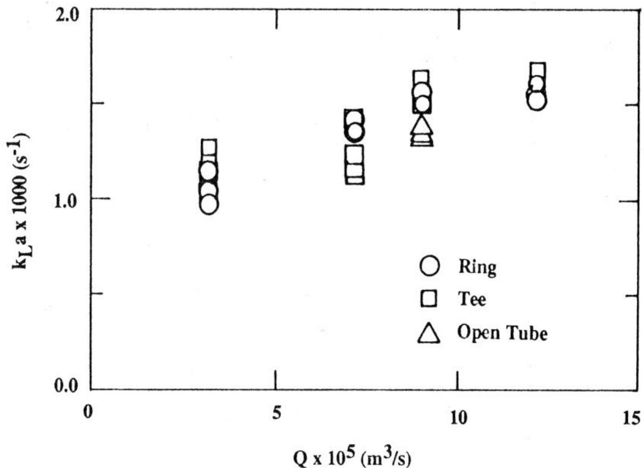

Figure 2. Effect of Sparger Design on $k_L a$ in a 10.9 liter tank with N = 2.9/s

Results for all three tanks-- 10.9, 102, and 2000 liter--are shown in Figure 3 where the ratio of the measured $k_L a$ to the value predicted from the Yagi and Yoshida correlation is plotted as a function of liquid volume. This figure shows that there are large effects of volume and polysaccharide concentration on this ratio. The Yagi and Yoshida mass transfer coefficient correlation does not appear to be a universal correlation, so it is not suitable for the scale-up strategy described earlier. We need to determine better exponents and the best form for the Deborah number term which represents the elasticity of the broth.

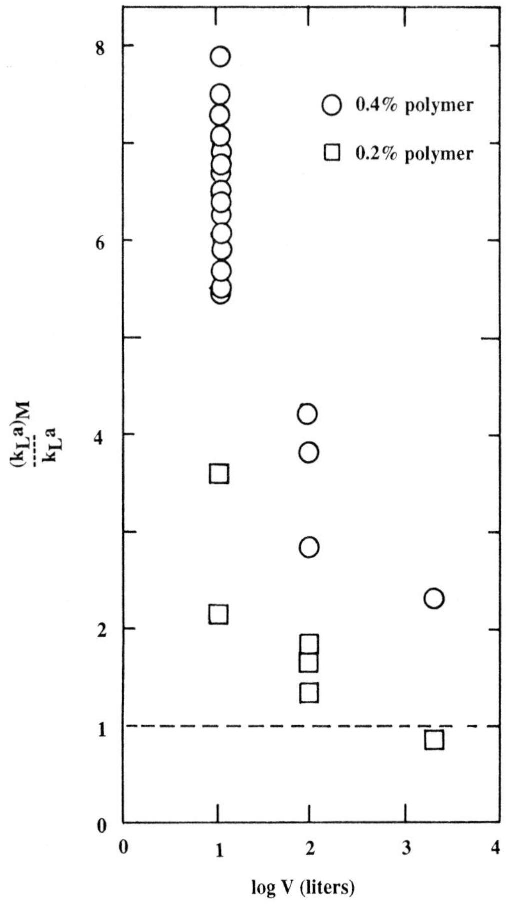

Figure 3. Effect of Liquid Volume and
Polymer Concentration on
$k_L a$

c_b dissolved gas concentration

c^* solubility of dissolving gas

c_{OH} concentration of OH ions

D_{CO2} diffusion coefficient for CO_2

D_{OH} diffusion coefficient for OH^-

D_i impeller diameter

D_T tank diameter

g gravitational acceleration

H_L liquid level in the tank

K consistency index

k_L mass transfer coefficient

$k_L a$ volumetric mass transfer coefficient

n flow behavior index

N impeller speed

$N_{Deb} = \lambda N$

$N_{Fr} = D_i N^2 / g$

$N_{Re} = D_i^2 N \rho / \mu_e$

$N_{Sc} = \mu_e / \rho D_{CO2}$

$N_{Sh}' = k_L a D_i^2 / D_{CO2}$

Q volumetric gas flow rate

t_R reaction time

V volume of fluid

$V_s = 4Q / \pi D_T^2$ superficial gas velocity

λ relaxation time for polymer in solution

μ Newtonian viscosity

μ_w viscosity of water

$\mu_e = K(11.5 N)^{n-1} (3/4 + 1/4n)^n$ effective viscosity

ρ density

σ surface tension

ACKNOWLEDGEMENT
This work was supported by a grant
from the EXXON Education Foundation.

We are grateful for the assistance of
Faith A. Morrison in guiding our
analysis of the relaxation time data.

NOTATION

a bubble area per unit volume

LITERATURE CITED

1. Langhaar, H. L., *Dimensional Analysis and the Theory of Models*, John Wiley and Sons, New York(1951) pp. 29-59

2. Perez, J. F. and O. C. Sandall, "Gas Absorption by Non-Newtonian Fluids in Agitated Vessels Liquids", *AIChE Journal,* 20, 770(1974)

3. Yagi, H. and F. Yoshida, "Gas Absorption in Newtonian and Non-Newtonian Fluids in Sparged Agitated Vessels", *Ind Eng Chem Process Des Dev,* 14, 488(1975)

4. Ranade, V. R. and J. J. Ulbrecht, "Influence of Polymer Additives on Gas-Liquid Mass Transfer in Stirred Tanks", *AIChE Journal,* 24, 796(1978)

5. Metzner, A. B. and R. E. Otto, "Agitation of Non-Newtonian Fluids", *AIChE Journal,* 3, 3(1957)

6. Calderbank, P. H. and M. B. Moo-Young, "The Prediction of Power Consumption in the Agitation of Non-Newtonian Fluids", *Trans Inst Chem Eng,* 37, 26(1959)

7. Danckwerts, P. V., *Gas-Liquid Reactions*, McGraw-Hill Book Co., New York(1970) pp. 239-251

8. Danckwerts, P. V., *Gas-Liquid Reactions*, McGraw-Hill Book Co., New York(1970) pp. 111-112

9. Perez, J. F. and O. C. Sandall, "Diffusivity Measurements for Gases in Power Law Non-Newtonian Liquids", *AIChE Journal,* 19, 1073(1973)

10. Newman, J. S. *Electrochemical Systems*, Prentice-Hall, Inc., Englewood Cliffs, N. J.(1973) p. 230

Invited paper

Plant Cells as Chemical Factories:
Control and Recovery of Valuable Products

A. E. Humphrey

Director, Biotechnology Institute, The Pennsylvania State University, University Park,
PA 16802, U.S.A.

Plant sources offer a great diversity of chemicals. More than 30,000 different chemical moieties are known to be produced by plants. The question is whether some or even any of these chemicals can be more economically produced by plant cell suspension culture than by traditional field grown plants. Suspension culture has a number of distinct advantages over field derived plant sources, including better control and year around availability. The concern, however, is expense. What are the major bottlenecks to economic plant cell suspension culture? Can they be significantly reduced?

Introduction

Plants are an important source of chemicals. Over 30,000 different chemical moieties are produced by plants. Because of rapid advances in plant culture techniques, there exists the potential to produce chemicals through plant cell suspension culture. This would be an alternative to extraction from whole plants. The question at the present time is whether it is economically feasible to produce these chemicals by plant cell culture. The answer is not a simple yes or no. It depends on the particular chemical, its value and market size. Fortunately, there are some rule-of-thumb relationships concerning chemical costs, market demand, and economical bioreactor productivities that can be used as guides in feasibility estimations. These will be discussed later.

A fair amount of literature exists on large scale production of chemicals by plant cell culture and the economics associated with such production systems (Fujita 1990, Lambie 1990, Drapeau et al. 1987, Goldstein 1987, Kargi and Rosenberg 1987, Sahai and Knuth 1985, Busche 1985, Curtin 1983, Shuler 1981, Tanaka 1981, Goldstein et al. 1980, Staba 1980, and many more). The first report of large scale (134 liters) production of plant cells was in *Science* in 1959 (Tulecke and Nickell 1959). In 1977, Noguchi et al., published a report on the production of tobacco cells on a 20,000 liter scale as a possible way of producing nicotine commercially. Since then, a number of chemicals have been produced using plant cell culture systems and their processes scaled up. This includes nicotine (Noguchi, et al. 1977), diosgenin (Drapeau, et al. 1986, Rokem and Goldberg 1985), rosmarinic acid

(Ulbrich, et al. 1985), ajmalicine (Drapeau, et al. 1987), shikonin and berberine (Fujita et al. 1986), and ginseng saponins (Ushiyama et al. 1986). In addition, taxol, vincristine and tropane alkaloids are under consideration. Pharmaceuticals such as codeine, morphine, scopolamine, atropine, L-dopa, hyoscyamine, quinine, and serpentine are possible candidates for production by plant cell culture (Curtain 1983). Fragrances such as jasmine, flavors such as vanilla, and sweeteners such as monellin are also candidates for production by plant cell culture.

What are the advantages and the disadvantages to plant cell culture in comparison to traditional agriculture for producing chemicals? The advantages are at least six in number. These include:

1. Can be carried out year long independent of environmental conditions such as climate and pests.
2. Isn't limited to production in a given local where political instabilities may exist.
3. Requires minimum use of land space.
4. Performed in a fully defined and contained system.
5. Limited to a single genetic line of cells.
6. Can achieve consistency in product yield and quality.

What are the disadvantages? These include:

1. Must compete with traditional agriculture having a well developed infrastructure.
2. May be difficult to achieve a high producing cell line without a lot of expenditure of research resources and time.

103

E. Galindo and O.T. Ramírez (eds.), Advances in Bioprocess Engineering. 103-107.
© 1994 Kluwer Academic Publishers. Printed in the Netherlands.

3. Slow growth leads to sterility problems and low productivities.

Economics

As a first estimate in considering the economic feasibility of various plant cell culture processes, there are some simple generalities that can be derived from existing cost/production data. Table I summarizes the cost of selected plant derived chemicals and drugs in comparison to their market size. Figure 1 is a logarithmic plot of these data along with those for 36 additional biochemicals and drugs. We see from this plot that there is a logarithmic relationship between product costs and market size.

Table I: Plant Derived Chemicals and Drugs: Selling Price and Market Size

Chemical/Drugs	$/Kg	Kg/yr	Market Value. $M
Ajamalacine	1,500	3,500	5.25
Codeine	650	75,000	50
Digitals	3,000	15,000	45
Jasmine	5,000	100	0.5
Pyrethrins	300	65,000	20
Quinine	100	500,000	50
Spearmint	30	3,000,000	90
Taxol	500,000 (est)	100 (est)	50
Vinblastine	5,000,000	4	20

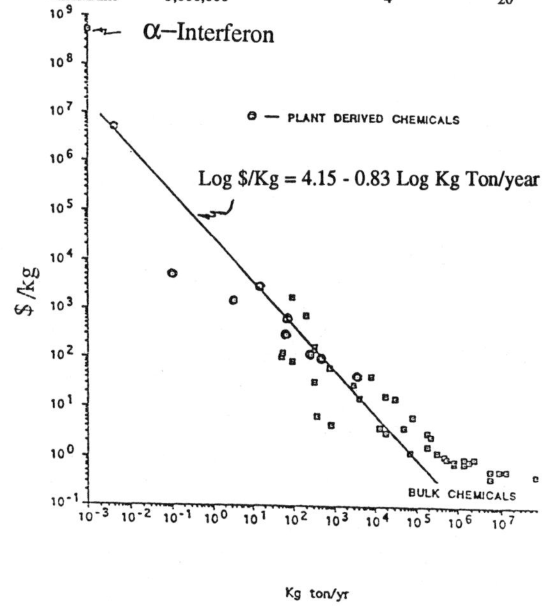

α–Interferon

O — PLANT DERIVED CHEMICALS

Log $/Kg = 4.15 - 0.83 Log Kg Ton/year

BULK CHEMICALS

$/Kg

Kg ton/yr

FIGURE 1: Relationship of Chemical Prices to Annual Production

Figure 1 suggests that a chemical with a 10,000 kg/yr market probably should sell for around $2,100/kg. Likewise, a product with a market of only 1000 kg/yr should sell for around $14,000/kg.

A similar generalization can be derived for the culture productivity in a bioreactor. In the fermentation field, there is a "rule-of-thumb" relationship that says the product value in the bioreactor when the fermentation is complete should be equivalent to 12 - 15¢/liter - day. (See Table II) These data assume typical recovery and purification costs, i.e., an extraction train followed by concentration and drying and only minimal validation and documentation steps necessary for product certification. Note that α-interferon, whose process requires considerable validation and documentation effort before product approval, falls well above the correlation line (Figure 1).

This relationship (12-15¢/liter-day value) has been used in setting yield goals for a new antibiotic fermentation. Supposing you are developing a new animal feed to be produced by fermentation. If it were projected to have a bulk sales value of $100/kg; then, within a 240 hour (10 day) fermentation period, the concentration that must be achieved for an economical fermentation would be 12g/liter. This is arrived at by the following calculation: ($100/kgx1kg/1000grx1 run/10 daysx$0.12/liter-day = 12g/liter).

How do potential plant products compare under these rule of thumb guidelines? Diosgenin with a 200kg-ton/yr market size selling at $674/kg is a chemical that could be considered for production by plant cell culture. Table III gives such a comparison for Diosgenin. As can be seen from the Table, present Diosgenin technology yields a bioreactor productivity value well below the 12¢/l - day goal for an economical process. It would appear that nearly a 6-fold combined improvement in concentration, yield, and productivity must be achieved on a large scale before diosgenin would be an attractive product for production by plant cell culture. Also, from the data in Figure 1, we see that for a 200kg - ton/yr market size, diosgenin should sell economically for around $162/kg in an established market. This means that an additional 4-fold improvement or a total 24-fold improvement in the existing plant cell culture technology for producing diosgenin might be necessary. This is not very attractive since it will could involve a high risk research undertaking.

Table II: Cash Flow Analysis of Fermentation Processes

Product	Concentration (g/liter)	Productivity (g/liter-day)	Price (¢/g)	Revenue (¢/liter-day)
Citric Acid	150	50	0.17	9
Ethanol	80	80	0.07	5
Glutamic Acid	100	33	0.40	13
Gluconic Acid	300	100	0.10	11
Lactic Acid	130	65	0.19	13
Lysine	73	33	0.50	12
Penicillin	30	4	4.0	15
Protease	20	7	1.0	7
Riboflavin	10	2	6.0	10
Streptomycin	25	3	1.0	15
Xanthan	25	3	1.0	15
Vitamin B_{12}	0.06	0.02	750	15

average: 12¢/liter-day)

Table III: Economic Potential for the Production
of Diosgenin by Plant Cell Culture

Selling Price	$674/kg (67¢/g)
Market Size	200,000 kg/yr.
Cycle Time	15 days
Cell Concentration	11.3 g/liter
Product Concentration in Cells	3.8% dry wt.
Product Concentration in Bioreactor	0.43 g/liter
Productivity	0.0287 g/liter -day
Productivity Value	1.9¢/liter -day
Selling Price from Figure 1 for 200,000 kg/yr Market Size	$162/kg

Let's examine another possible cell culture product, taxol, an exciting new anticancer agent. Hard data is difficult to come by because not all the potential use approvals are in. So, the taxol market size is only a guess at best. Also, virtually no productivity data has been published. Table IV gives some eduated guesses.

Based on a $500/g selling price, the estimated productivity value as seen in Table IV is considerably greater than 12¢/l-day. Thus, taxol at $500/g would appear to be an excellent candidate for production by plant cell culture. But, is $500/g a reasonable expectation for the selling price? Reference to Figure 1 suggests that a 100kg/yr market demand would only sustain a selling price of taxol greater than $141,000/kg. If this were to be the long-term selling price of taxol, then the achievable culture productivity value is only 240¢/l-day. Still, taxol is a very attractive candidate for production by cell culture.

Table IV: Economic Potential for the Production
of Taxol by Plant Cell Culture

Selling Price	$500,000/kg ($500/g)
Market Size	100 kg/yr.
Cycle Time	15 days
Cell Concentration	20 g/liter
Product Concentration in Cells	1.25% dry wt.
Product Concentration in Bioreactor	0.250 g/liter
Productivity	0.017 g/liter -day
Productivity Value	850¢/liter -day
Selling Price from Figure 1 for 100 kg/yr Market Size	$95,500/kg

Can one generalize from these "rule-of-thumb" relationships? Most plant cell culture systems can achieve secondary metabolite productivities fo 0.02 g/l-day. (Taxol = 0.017, Diosgenin = 0.0287 g/l-day) To achieve a cell culture productivity value of 12¢/l-day, any plant products to be economically produce by cell culture should have a selling price of 600¢/g or $6000/kg. Unfortunately, many of the plant chemicals and drugs sell in the range of $400 to 1000/kg. Obviously, an order of 6- to 15-fold improvement in productivity is necessary to achieve economic culture productivities.

What are the bottlenecks to achieving higher productiveness? They are mainly three:
1)   slow growth rates
2)   low cell concentrations in the biorecator
3)   low product yields per unit of cell mass

In recent years considerable progress has been made in achieving higher values for all three of factors.

Overcoming Low Growth Rates

Plant cells traditionally have long doubling times, particularly in comparison to bacteria such as E.coli. Plant cell doubling times are the order of 1-2 days. E. coli cell doubling times are the order of 15-20 minutes. Plant cells have nearly a 100-fold slower doubling time than bacteria. There are several reasons for this. Plant cells are several orders of volume larger than bacteria cells. They are usually grown on simple, chemically defined media, i.e. NH_4^+ and NO_3^- salts plus sugar.

Bacterial cells with fast doubling times are grown on complex media, i.e, a mixture of amino acids and/or peptides plus vitamins and sugars. Plants have to work longer and harder than bacteria to double. But there is some hope of generating plant cells with faster doubling times. Not all species of plants have large cells and corresponding large genomes. For example, Arabidopsis cells are only 20 microns in diameter. They have a relatively small genome - 100,000,000 base pairs. (Note: the genome for a tomato cell is 9x larger than Arabidopsis. Wheat is 40x larger). I suspect that as with bacteria, where E. coli became the preferred genetic vector for producing recombinant protein, so may Arabidopsis with its relatively small cell and genome size become the preferred vector for plant systems. Further, Arabidopsis is a common weed in the agricultural environment with no known toxic effects on humans and on other plants. So there is some hope of evolving more rapid growing plant cell systems. However, the smaller cells may have some loss of chemical function that exists generally in plants. I am convinced through genetic engineering a doubling in plant cells growth rate is achievable.

Achieving Higher Cell Concentrations

Plant cell concentrations of no more that 10 grams/liter are typcially obtained in bioreactors at present. However, this is by no means the limit. For example, Su (1991) and Asali (1992) have shown that by osmolarity control, concentrations of 50 grams dry weight/liter are achievable with Anchusa officinalis plant cells. The problem is that plant cells can swell 3 - 4 times their normal volume at low osmolarities. Such conditions, i.e. low osmolarities, typically occur near the end of a culture run when the sugar is used up. At that time the osmolarity may have decreased by at least an order of magnitude from the initial conditions. Continuous feeding to maintain a given sugar concentration or the use of a non-metabolizeable osmolyte can minimize the cell swelling problems. If one can routinely achieve 50g dry weight/liter then at least a 5-fold increase in density is possible in comparison with those conditions utilized for calculational purposes in Table III and IV.

Enhancement of Metabolite Recovery through Immobilzation

Encouraged by some early reports of enhancement of ajmalicine production by plant cells immobilized in calcium alginate beads, my group at Penn State (Ramakrishna, et al., 1993) investigated the enhanced production of solavetivone and lubimin by cells of <u>Hyoscyamus muticus</u> immobilized in calcium alginate beads. We were very encouraged to find that, compared with free cells, the immobilized cells, preinduced with a fungal elicitor, produced 53% more solavetivone than the freely suspended cells. We also found by controlling the initial viscosity of the culture medium nearly 2x higher yields were possible. Presumably, immobilization or higher viscosity provide shear protection.

<u>Enhancement of Metabolite Recovery through Metabolite Release</u>

About the same time of these observations, my group decided to investigate factors that enhanced the release of certain metabolites from cells. To study this, we selected transformed root cultures of <u>Nicotiana tabacum</u>. This was a nice model system, since tobacco hairy roots can be readily cultured in shake flasks. As such they form a kind of immobilized biomass that can be perfused under a variety of conditions. Also, tobacco roots produce nicotine intracellularly. Less than 1% of the total nicotine is excreted to the medium when the tobacco hairy roots are grown on B5 medium. The retention of nicotine is due to an equilibrium partiitioning process known as ion-trapping. Nicotine is an uncharged weakbase also present as its conjugated acid (pk=8.0) at physiological pHs. According to the ion-trapping theory, the uncharged weakbase freely diffuses across the plant membrane while the conjugate acid is held back. Our investigation (Larsen, 1993) showed that nicotine diffuses in and out of the <u>Nicotiana tabacum</u> roots following an equilibrium-partitoning. The equilibrium is related to the intracllular and extracelular pHs. Its release can be enhanced by almost a factor of 2x by changes in cation concentration and pH of the culture medium.

<u>Enhancement of Metabolite Recovery through Metabolite Entrapment</u>

The above observations suggested that enhanced nicotine production might be achieved not only through control of the culture medium composition but also by trapping out the acid form of the nicotine with an ion exhange material, thus driving the equilibrium towards excretion.

A two flask system was used for doing this in which the on-line removal of nicotine from batch culture was continuously accomplished with XAD-16, a hydrophobic polyamine exchange resin. In our experiments a 2.5-fold enhancement in metabolite recovery was achieved by continuous product removal (Gomez, 1993). It appears therefore, that some enhancement of metabolite release from plant cells can be achieved by control of culture conditions and by metabolite entrapment. Whether this, coupled with high cell densities and faster growing cells, will be sufficiently significant to make economical production of most plant derived chemicals feasible by cell culture remains to be seen.

<u>Feasible Goals for Productivity Improvement</u>

When one multiples all of these possible improvements, it is reasonable to expect that a 20-fold increase in productivity over existing technology is possible (See Table V). With a 20-fold improvement in productivity, then one is looking at a product productivity of 20x0.02 = 0.4 g/liter - day. This means to achieve a 12¢ liter - day productivity value in plant cell culture bioreactors, the economical plant product selling price could be as low as $300/kilogram. This represents the most optimisitc case for producing plant products by cell culture. However, it represents a risk or a need of a collectively 20-fold improvement in existing plant cell culture technology. So, one might conclude that some, but not all plant products are candidates for production by cell culture. There are opportunities for the economical production of chemicals by plant cell culture, but, improvement in cell density, growth rate and product recovery are necessary.

Table V: Possible Improvements to Plant Cell Culture Productivity

Item	Increase
Growth Rate	2X
Cell Concentration	5X
Product Yield	2X
Total Feasible Improvement	20X

<u>Conclusions</u>

From this analysis the following conclusions can be made:
1. Opportunities exist for the economical production of chemicals by plant cell culture.
2. The previous conclusion applies at the present time to chemicals having a selling price greater than $6000/kg plus a market demand greater than 4 kg ton/yr.
3. Since many plant derived chemicals sell in the range of $400 - $1000/kg, at least an 6- 15-fold improvement in existing productivity is necessary if they are to be economically produced by cell culture.
4. The bottlenecks to achieving higher productivity are
   a) low growth rates
   b) low cell concentrations
   c) low product yields and recovery
5. Improvements to overcome these bottlenecks are feasible. Collectively, as much as a 20-fold improvement appears to be an achievable goal.

<u>References</u>
1. Drapeau, D., H.W. Blanch, C.R. Wilke, "Economic Assessment of Plant Cell Culture for the Production of Ajmalicine, "Biotech. Bioeng. <u>30</u>, 946-953 (1987).
2. Fujita, Y., "The Production of Industrial Compounds" Chap. 11 in Plant Tissue Culture, ed. S.S. Bhojwani, Elsevier, New York (1990).

3. Sahai, O. and M. Knuth, "Commercializing Plant Tissue Culture Processes: Economic, Problems and Prospects," Biotech. Prog. $\underline{1}$, 1-9 (1985).

4. Curtin, M.E., "Harvesting Profitable Products from Plant Tissue Culture," Bio/Technology, $\underline{1}$, 649-657 (1983).

5. Kargi, F. and M.Z. Rosenberg, "Plant Cell Bioreactors: Present Status and Future Trends," Biotech. Prog. $\underline{3}$, 1-8 (1987).

6. Tanak, H., "Large-scale Cultivation of Plant Cells at High Density: A Review," Proc. Biochem., $\underline{22}$ 106-113 (1987).

7. Ulbrich, B., W. Wiesner and H. Arens, "Large-scale Production of Rosmarinic Acid from Plant Cell Cultures of Coleus blumei Benth," in Primary and Secondary Metabolism of Plant Cell Cultures, H. Neumann et al., editors, Springer-Verlag, Germany (1985).

8. Drapeau, D., H.W. Blanch, and C.R. Wilke, "Growth Kinetics of Dioscorea deltoidea and Catharanthus roseus in Batch Culture," Biotech. Bioeng. $\underline{28}$, 1555-1563 (1986).

9. Goldstein, W.E., L.L. Lasure, and M.B. Ingle, "Product Cost Analysis," Chapter 9 in Plant Tissue Culture as a Source of Biochemicals," E.J. Staba, editor, CRC Press, Boca Raton FL (1980).

10. Busche, R.M., "The Business of Biomass," Biotech. Prog. $\underline{1}$, 165-179 (1985).

11. Lambie, A.J., "Commercial Aspects of the Production of Secondary Compounds by Immobilized Plant Cells," Chapter 13 in Secondary Products from Plant Tissue Culture, B.V. Charluroodard M.S.C. Rhodes, editors, Proc. Phytochem. Soc. of Europe, Clarendon Press, Oxford (1990).

12. Rokem, J.S., and I. Goldberg, " Secondary Metabolites from Plant Cell Suspensions - Methods for Yield Improvement," Adv. Biotech. Proc., $\underline{4}$, 241-274 (1985).

13. Goldstein, W.E., "Large-scale Processing of Plant Cell Culture," Annals NY Acad. Sci. 295-408 (1987).

14. Shuler, M.L. "Production of Secondary Metabolites from Plant Tissue Culture - Problems and Prospects," Annals NY Acad. Sci., $\underline{369}$, 65-80 (1981).

15. Payne, G.F. and M.L. Shuler, "Selective Adsorption of Plant Products," Biotech. Bioengr. $\underline{31}$, 922-928 (1988).

16. Tulecke, W. and L.G. Nickell, "Production of Large Amounts of Plant Tissue by Submerged Culture," Science $\underline{130}$, 863-864 (1959).

17. Noguechi, M., et al., "Improvement of Growth Rates of Plant Cell Cultures," in Plant Tissue Culture and its Biotechnological Applications, W.Bars, E. Reinhad, and M.H. Zenk, eds., Springer-Verlag, Berlin, 85-94 (1977).

18. Su, W.W., "Production of Plant Production Secondary Metabolites from High Density Perfusion Cultures in a Membrane Aerated Bioreactor", Ph.D. Thesis, Lehigh University, May 1991.

19. Asali, E.C., "Characterization and Culturing of Plant Cell Supensions," Ph.D. Thesis, Lehigh University, December 1992.

20. Gomez, P.L., "Continuous Extractive Root Phytoconverstion: Monitoring Growth and Design of On-Line Product Removal Strategies for Transformed Root Culture," Ph.D. Dissertation, Lehigh University, June 1993.

21. Humphrey, A.E., "Bottlenecks to the Economic Production of Chemicals from Plant Cell Culture," Paper presented for Session 3, Frontiers in Bioprocessing, Boulder, CO, September 19-22, 1993.

22. Larsen, W.D., "The Characterization of Alkaloid Release from Nicatiana tabacum Hairy Roots," Ph.D. Thesis, Lehigh Univeristy, June 1993.

23. S.V. Ramakrishna, et al., "Production of Solavetivone by Immobilized Cells of Hyoscyamus muticus." Biotechnology Letters, 15:301-306 (1993).

Invited paper

Optimization of Fed-Batch Mammalian Cell Culture Processes

W. Zhou and W.-S. Hu

Department of Chemical Engineering and Materials Science, University of Minnesota,
Minneapolis, Minnesota 55455-0132, U.S.A.

Fed-batch cultures of hybridoma cells in a serum-free medium were carried out in a computer controlled bioreactor. Oxygen uptake rate (OUR) and base addition for controlling pH were employed to calculate key physiologically related variables. Using established stoichiometric relationships between measured variables and the physiological demand we were able to sustain optimal cell growth by feeding salt-free concentrated medium while also maintaining glucose at a prescribed range. A reduced lactic acid production and glucose consumption were observed. A high cell concentration with a very high viability was achieved through the prolongation of exponential growth stage. The results demonstrate the potential benefit of on-line instrumentation and control in mammalian cell culture.

Fed-batch operations and continuous cultures with cell retention are often used for cultivation of mammalian cells to increase cell concentration in the bioreactor. In comparison with perfusion cultures, fed-batch cultures allow both high cell concentration and product concentration to be achieved. However, these cultures are marked with a low viability as the metabolites accumulate in the culture. To achieve a high viability and cell concentration, one normally resorts to a perfusion culture with cell retention. The drawback is that the product concentration is also diluted.

It has been demonstrated that through controlled feeding of nutrient, the accumulation of metabolites at a high rate can be avoided[1,2]. With reduced metabolite accumulation, one can possibly achieve a higher cell concentration. However, in order to alter the cell metabolism to reduce the accumulation of metabolites the concentration range of nutrient (glucose) in a culture must be controlled at a rather low level. In a culture with a high cell concentration and thus a high nutrient consumption rate, the control of nutrient at a low level can be realized only by a continuous dynamic feeding. The adjustment of such dynamic feeding requires on-line determination of metabolic demands. Using on-line measurements of the nutrients concentration or metabolic parameters related to their consumption, the feeding of nutrients can be adjusted continuously to meet these demands.

We have established an instrumented bioreactor in which base addition and oxygen uptake rate (OUR) were monitored on-line[3]. Stoichiometric correlation between nutrient consumption and base addition for controlling pH, as well as between nutrient consumption and OUR were also established.

In this study nutrient feeding rates for fed-batch hybridoma cultures were estimated on-line via these stoichiometric correlation. Using either base addition or OUR as a metabolic indicator, we were able to manipulate the nutrient concentrations at a low level in fed-batch cultures for altering cell metabolism and reducing the specific

109

E. Galindo and O.T. Ramírez (eds.), Advances in Bioprocess Engineering. 109-113.
© 1994 Kluwer Academic Publishers. Printed in the Netherlands.

nutrient consumption rate and metabolite production rate. As a result of reduced metabolite accumulation and sufficient nutrient

supply, the exponential growth stage was successfully prolonged and a high viable cell concentration was achieved.

MATERIALS AND METHODS

Cells and culture conditions

The hybridoma cell line (MAK) used and its cultivation conditions have been described previously[3]. Experiments were carried out in a 750-ml reactor with a working volume of 500 ml. A serum-free DMEM/F12 medium with several additives was used. The initial glucose and glutamine concentrations were 0.5 g/l and 0.7 mM, respectively. The protein concentration in the medium amounted only 5 mg/l. The fed-batch was initiated about 25 hours after the glucose concentration decreased to 0.2 g/l. A 16 fold concentrated medium with a glucose and glutamine concentration of 47.4 g/l and 94.8 mM was used for the feeding. To avoid the increase of osmolarity, the fed medium contained no salts. During the cultivation, several culture parameters such as optical density, base addition for controlling pH and oxygen uptake rate were monitored on-line. Integrating OUR over time, cumulative oxygen consumption was calculated on-line and used to estimate glucose and other nutrients consumption.

Assays

Cell concentrations were estimated using a hemacytometer via trypan blue staining. Glucose and lactic acid concentrations were measured with an enzymatic analyzer (Yellow Springs Instruments, Ohio). Antibody concentration was measured by HPLC using a Protein G affinity chromatography column (Perseptive Biosystems, MA).

RESULTS

In a typical batch culture of MAK cells a viable concentration of 1.7×10^6 cells/ml with a high viability is achieved in approximately 50 hr. The exponential growth stage is about 40

hours. At the end of exponential growth phase about 17.0 mmol/l of glucose are consumed and 26.5 mmol/l of lactate are accumulated in the culture. The maximal antibody concentration amounted about 8 mg/l (Figure 1).

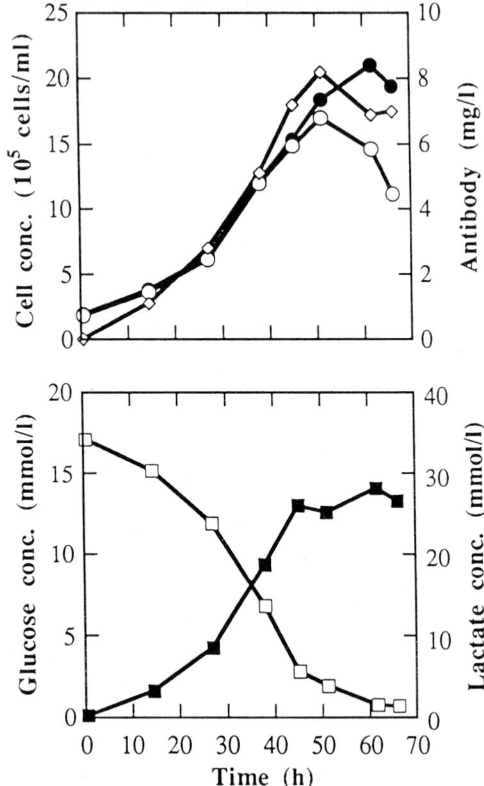

Figure 1. Batch kinetics of hybridoma MAK cells. (○), Viable cell concentration; (●),total cell concentration; (◇), antibody; (□), glucose concentration; (■), lactate concentration.

In the preliminary experiments the consumption of glucose, glutamine and other amino acids were measured. Some amino acids were consumed, while others produced in batch culture. In a fed-batch culture the accumulation of those amino acids produced by cells cannot be easily removed. However, those which are consumed can be replenished. A aim of our fed-batch culture is thus to prevent amino acid depletion. The specific consumption rate of glucose and those amino acids cannot be related by a single stoichiometric ratio. However, results accumulated from many experiments indicate that no amino acid depletion and no significant glutamine accumulation will occur if the ratios among the glucose, glutamine and other medium compounds in the feeding nutrient solution are adjusted as described in Materials and Methods. Two strategies of dynamic nutrient feeding were evaluated and are reported below as Case I and Case II.

Case I: Nutrient feeding coupled to base addition

Cells were inoculated at a viable concentration of 1.5×10^5 cells/ml from a 100 ml-spinner culture of late exponential growth stage. Feeding was initiated after glucose concentration decreased to the desired range. Since one mole of lactic acid produced requires one mole of NaOH for neutralization. The amount of base added is a good indication of the amount of lactic acid produced provided that the buffering capacity of the culture fluid remains relatively constant. We maintained CO_2 concentration in the gas stream at 5% during the cultivation to keep the buffering capacity constant. It was previously shown that the amount of glucose consumed and the amount of lactate produced is relatively constant during the exponential growth phase and the amount of base added can be used to estimated the amount of glucose consumed.

The stoichiometric ratio between base addition and glucose consumption was established from

previous experiments. During the cultivation the glucose concentration in the culture was measured off-line and the stoichiometric correlation between glucose consumption and base addition was confirmed and adjusted. During the fed-batch period, the nutrient feeding pump was coupled to the base addition pump. Whenever base pump was triggered for pH control, a stoichiometric amount of nutrient solution was also fed. Throughout the cultivation period glucose concentration was controlled below 0.5 g/l. Cells continued to grow exponentially to its maximal viable cell concentration of about 7.2×10 cells/ml and a total cell concentration 8.0×10^6 cells/ml at about 100 hours, while a maximal antibody concentration of 62 mg/l (6 times as high as that of

a batch culture) was obtained. Lactate reached about 70 mM at the end of cultivation (Figure 2). The maximal OUR

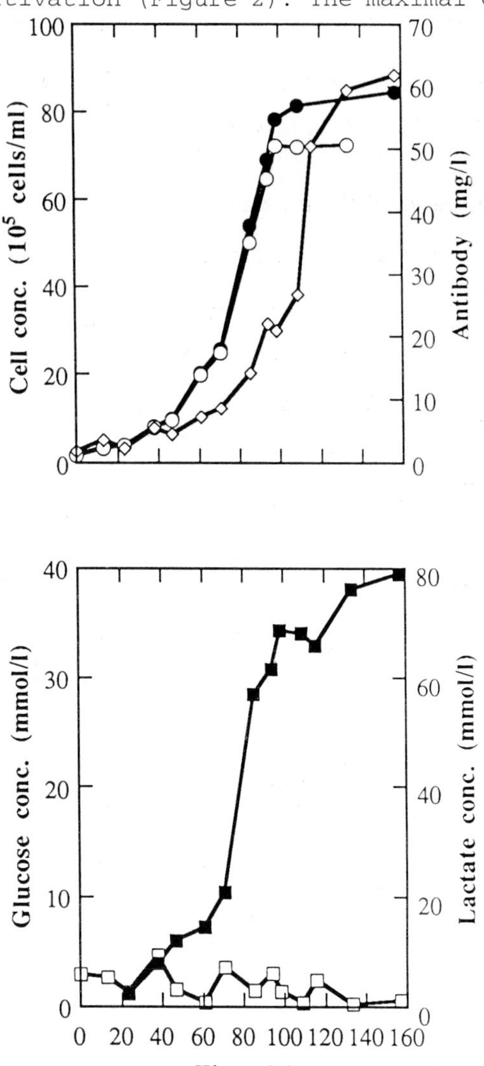

Figure 2. Dynamic feeding of nutrient in a fed-batch culture of hybridoma MAK cells. The feeding rate was coupled to base addition rate. Symbols are concentrations of : (○) viable cells, (●) total cells, (◇) antibody, (□) glucose, (■) lactate.

amounted about 2.5 mmol/l h. The conversion of glucose to lactate was 1.47 mmol lactate/mmol glucose in comparison with 1.65 mmol/mmol normally observed in the batch culture, while the ratio between oxygen and glucose consumption was increased from 1.06

mmol/mmol in the batch to 2.27 mmol/mmol (Figure 4).

<u>Case II: On-line adjustment of feeding rate by OUR</u>

Cells were inoculated at a viable concentration of 1.5×10^5 cells/ml from a 100 ml-spinner culture of late exponential growth stage. The conditions were the same as case I except that nutrient feeding was coupled to on-line OUR measurement. Integrating OUR over time, cumulative oxygen consumption was calculated on-line. This cumulative oxygen consumption was correlated to glucose and nutrient consumption. A stoichiometric ratio of 0.08 g glucose/mmol oxygen was used. During the cultivation the amount of glucose (and other nutrients) consumed was estimated from the cumulative oxygen consumption and replenished by feeding concentrated medium to maintain glucose concentration to be below 0.5 g/l.

After the initation of feeding cells continued to grow exponentially to a maximal viable cell concentration of about 5.6×10^6 cells/ml at about 96 hours, and thereafter. The total cell concentration reached 5.8×10^6 cells/ml. Subsequently oxygen limitation occurred and the experiement was terminated. The results show that glucose concentration was below 0.2 g/l . Lactate accumulated and reached only about 40 mM. The maximal OUR amounted about 2.0 mmol/l h (Figure 3). The conversion of glucose to lactate was 1.36 mmol lactate/mmol glucose, while the ratio between oxygen and glucose consumption was increased to 2.26 mmol/mmol (Figure 4) .

DISCUSSION

The above results show clearly, that both base addition and OUR can be correlated with nutrient consumption and applied to control the nutrient feeding rate. Through the dynamic adjustment of nutrient feeding rate of a salt-free concentrated medium, nutrient concentrations were controlled at low levels. Thus not only sufficient nutrients were supplied for cell growth and production formation, but also the nutrient consumption and metabolite

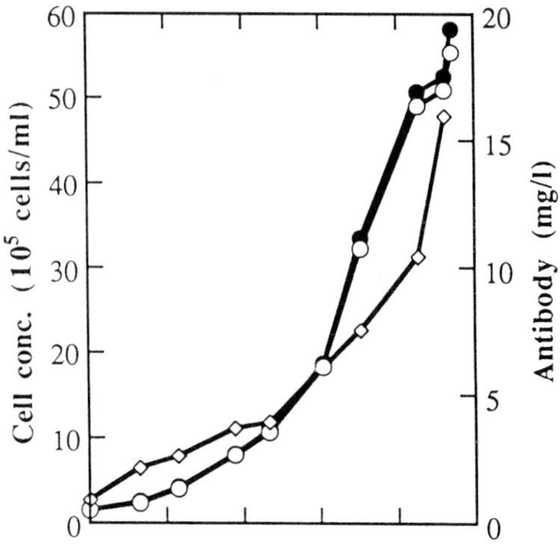

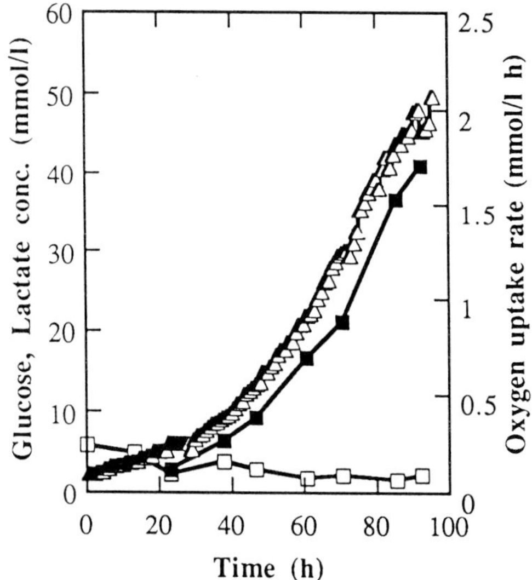

Figure 3: Fed-batch culture of hybridoma MAK cells with nutrient feeding coupled to OUR. Symbols are concentrations of (○): viable cells, (●): total cells, (□): glucose, (■): lactate, (◇):antibody, and (△): OUR.

production were reduced effectively. The conversion of glucose to lactate was decreased by keeping the glucose concentration at a low level, while the ratio between oxygen and glucose consumption was significantly increased in the fed batch cultures. A high cell concentration was obtained in both cases.

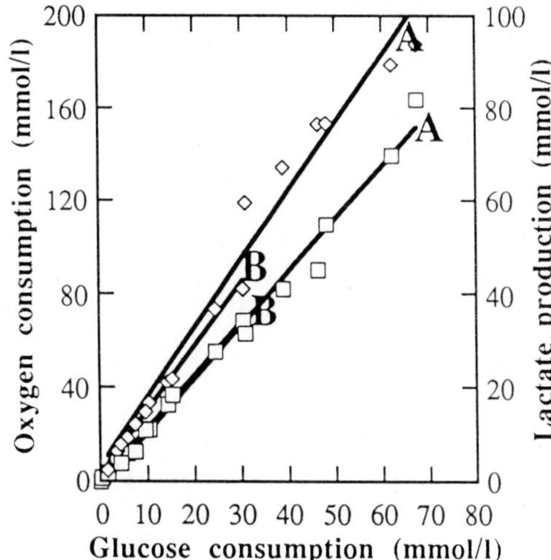

Figure 4. Stoichiometric correlation between lactate production and glucose consumption (◊) as well as between oxygen and glucose consumption (□). **A**: Case I; **B**: Case II.

Although the stoichiometric ratios were relatively constant in the cases presented, they are not necessarily constant when the growth stage changes. Furthermore measurement errors exist for both OUR and base addition causing uncertainty about the control actions and the fluctuation and deviation from the set point of nutrient concentration. Further improvement of the control strategy can be gained by combining base addition and OUR to assess the physiological state and demand of cells.

In comparison with the base addition, OUR is a better indicator of cell metabolic activities, especially when the rate of cell death is significant; the cellular materials released during cell lysis may cause an increase of the buffering capacity of the culture and change the realtion between the base addition and the lactate production.

REFERENCES

1. Hu, W-S., Dodge, T.C., Frame, K.K. and Himes, V.B. (1987) Effect of glucose and oxygen on the cultivation of mammalian cells. Dev. Biol. Standard. 66, 279-290.
2. Hu, W-S. and Oberg, M.G. (1990) Monitoring and control of animal cell bioreactors: Biochemical engineering considerations. In: <u>Large Scale Mammalian Cell Culture Technology</u>. Ed. A.S. Lubineicky, pp. 451-481, Marcel Dekker, Inc., New York, NY.
3. Zhou, W., W.-S. Hu. On-line characterization of a hybridoma cell culture process. Biotechnol. Bioeng., in press (1994)

A Fluidized Bed Reactor for the Cultivation of Animal Cells

J. Keller and I.J. Dunn

Biological Reaction Engineering Group, Chemical Engineering Department ETH, 8092 Zurich, SWITZERLAND

The development of a fluidized bed reactor for animal cell culture is described. Using both solid (Polysciences) and porous (Siran) glass beads, BHK cells have been grown to densities of $4x10^7$ cells ml^{-1}. Using external-loop oxygenation, the accurate measurements of online oxygen uptake rates were possible, which correlated well with the changing reactor conditions, as well as with the glucose and lactate uptake rates.

Fluidized bed reactors have been used in biotechnology for a number of applications related to biomass retention. These reactors include biofilm reactors for wastewater treatment, plant cell reactors which depend on cell aggregation tendencies and anchorage-dependent animal cell reactors using porous carriers. This work was started shortly before the development of a commercial fluidized bed reactor for animal cells. Although fluidized bed use for this application is logical, the importance of shear effects on animal cells would seem to rule out heavy, smooth-surfaced carriers. Initially, much effort was spent in finding a commercially-available carrier with the characteristics of high density and surface compatibility for the cell line used. It was known from work with biofilm systems that if the oxygen were supplied externally, the fluidized bed reactor would have the great advantage of oxygen uptake measurement by a dissolved oxygen difference measurement. This paper describes the development of the reactor and the experiments performed with two carriers, smooth and porous glass, that were finally selected.

MATERIALS AND METHODS

A fluidized bed reactor allows separation of the recycle loop into two compartments, one column section for cultivation on the suspended carrier and an aeration section. In this way the cells can be protected from the very high shear rates, which might be necessary for oxygen transfer. The carrier particles with adherent cells were kept in suspension in the reactor by an upwards stream of medium. Under these conditions each particle was suspended in a uniform shear field, which allows good diffusion rates and protects the cells from high shear forces. The medium from the reactor flowed through the oxygen measurement chamber and into the conditioning vessel (Fig. 1).

.The conditioning vessel was chosen to be a conventional 2 L fermenter (MBR BioReactor AG, Wetzikon, Switzerland) with instrumentation for the control of pH with carbon dioxide, and temperature. Temperature control was achieved by heating with warm water in a finger. At low cell densities, pH control was possible by adding $NaHCO_3$ to the medium and gassing with CO_2; at high cell

115

E. Galindo and O.T. Ramírez (eds.), Advances in Bioprocess Engineering. 115-121.
© *1994 Kluwer Academic Publishers. Printed in the Netherlands.*

densities the reactor was controlled with 0.15 M KOH. Oxygen transfer was achieved at low cell densities by surface aeration. Bubble-free and foam-free stirring up to 600 rpm was possible by replacing the upper turbine impeller with a propeller. Higher oxygen demands were handled by adding 2m of 3x4 mm diameter silicon tubing to the tank and controlling the pure oxygen supply with the PID dissolved oxygen controller. It was found necessary to improve the control by supplying a constant low nitrogen flow to remove the oxygen from the tubing when the DO had reached the set point. A condenser was added to retain the water removed by the surface aeration.

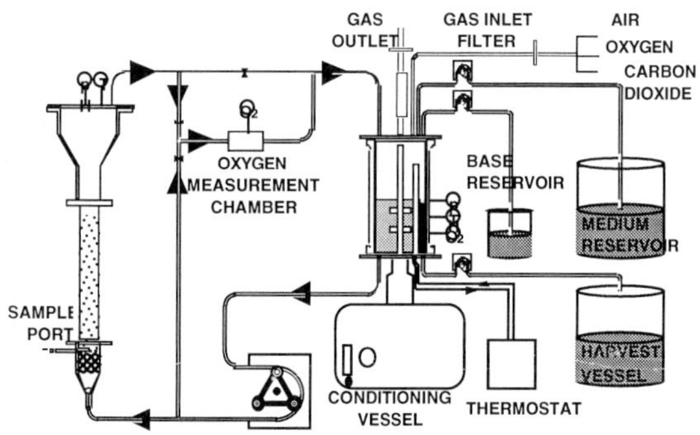

Figure 1. Fluidized bed reactor and equipment

The standard instrumentation of the fermenter allowed plotting the temperature, pH and DO in the tank and at the reactor outlet. Control was done with the built-in PI and PID controllers. All variables including rpm and the signal from an electronic feeding balance were collected and plotted using an Apple Macintosh II, with interface (MacADIOS II) and multiplexer (GW Instruments).

The system consists of the tank fermenter, and the reactor column connected together with tubing as a pumped recycle loop. An overflow kept a constant level in the column. For continuous operation, a feed pump and effluent pumps were used. Flow rate was determined by weighing the feed tank. The reactor column was built of glass with a stainless steel upper settling zone, which was not found necessary. The bottom of the column contained a static mixing element and above it a

polypropylene net to retain the carriers. Samples were taken at the end of a run from a port at the bottom of the column.

Since there were very few cells present in the conditioning vessel, the energy dissipation for aeration could be high, thus providing sufficient oxygen transfer necessary for high cell densities. The excellent mixing characteristics in this zone ensured also that the cells in the fluidized bed were supplied continuously with a well conditioned medium (pH, temperature, oxygen and other substrates), which was recycled back into the fluidized bed reactor by a peristaltic pump. The fluidized bed had a maximal expanded volume of 700 mL. The complete reactor system had a volume of 3.5 L.

The arrangement of the electrodes allowed an accurate difference measurement of the oxygen uptake rate, since the concentration was measured directly at the inlet and outlet of the reactor. A valve arrangement permitted their calibration against each other during fermentation.

Carriers

The choice of carrier materials for cell attachment was a critical point during reactor development. Beside the surface characteristics, also physical properties (size, density, porosity) are very important. Two different types of glass carriers were successfully used: nonporous glass beads with a diameter of 105-150 μm or 150 - 210 μm (Polysciences Inc. USA) and porous SIRAN® sintered glass carriers with a diameter of 0.4-1.0 mm and a pore size < 120 μm (Schott Glaswerke, Mainz, Germany).

Cell line and medium

The experiments were carried out with a BHK 21 (ATCC CCL 10) cell line using Glasgow Modification of MEM containing 10 % tryptose phosphate broth (Gibco-BRL, Basel, Switzerland) and in addition 1-10 % bovine serum (Gibco-BRL, Basel, Switzerland).

Characteristics of the reactor system

Table 1 summarizes some typical values used in the experiments and the achieved results:

Table 1. Reactor Conditions and Results

Volume of the fluidized bed	600 mL
Total volume	3.5 L
Medium throughput	up to 5 L/day (210 mL/h)
Glucose rate	max. 15 g/day (0.6 g/h)
Oxygen uptake rate	3 mmol/h
Max.cell density: Nonporous carriers Porous carriers	$2*10^7$cells/ml bed $4*10^7$cells/ml bed
Ratio of inoculum to cell number	approx. 5 % of final cell number

RESULTS AND DISCUSSION

For the inoculum, the BHK cells were grown for 2 days in a static culture on the carriers at 37 °C. After this period, the carriers with the cells were transferred into the fluidized bed reactor. For growth with the nonporous carriers, an initial expansion of 170 % and a superficial velocity of 3.4 mm/sec was used. During cultivation the expansion increased due to decreasing sedimentation velocity as the cells grew on the carriers (Fig. 2). The maximum cell density reached was $2*10^7$ cells/ml expanded bed volume.

For animal cell culture, the choice of carrier particle is a very critical aspect in the design of fluidized beds. The size and density of a carrier determines the superficial velocity of the medium which is used for optimal growth conditions. Good results were obtained with a nonporous glass carrier (105-150 μm Polysciences Inc. USA), whose sedimentation velocity was about 15 mm/sec. The oxygen uptake rate at high cell densities is the limiting factor for designing the bed height.

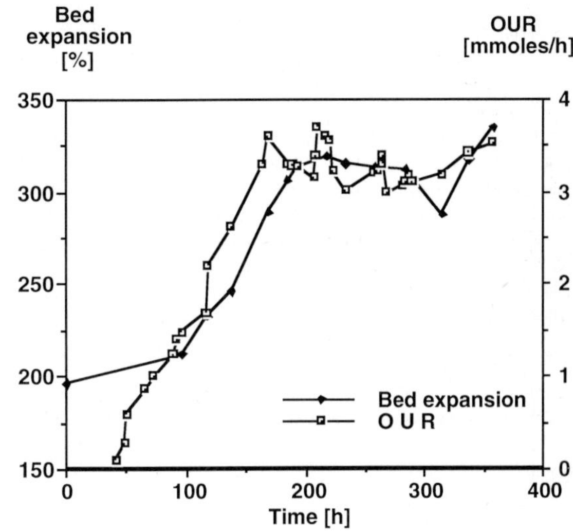

Figure 2. Bed expansion and oxygen uptake rate (OUR) during a 15-day cultivation

An interesting and perhaps useful characteristic of fluidized beds is that the less dense, overgrown carriers segregate at the top of the bed, permitting their observation and perhaps removal. Fig. 3 shows a typical cell distribution curve at the end of an experiment.

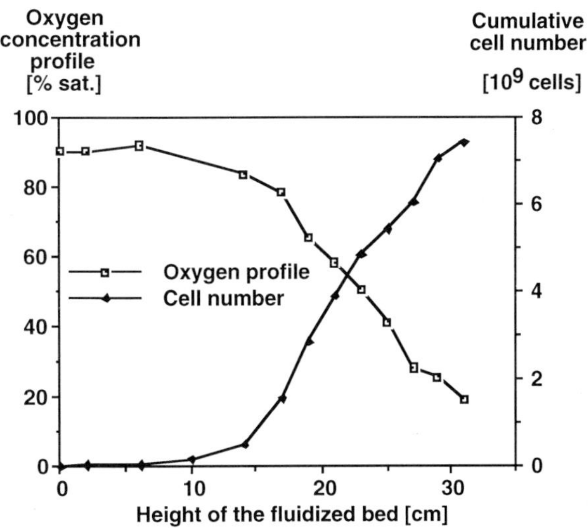

Figure 3. Oxygen concentration and cell distribution profile at the end of a cultivation period with the smooth glass carriers.

118

The oxygen uptake rate (OUR) could be easily determined using the special arrangement of the oxygen electrodes (Fig 4). The oxygen concentration difference between the inlet and outlet of the fluidized bed reactor permitted an instantaneous measurement of changes in cell metabolism. In Fig. 4 a fast response of the OUR during a pH change is shown.

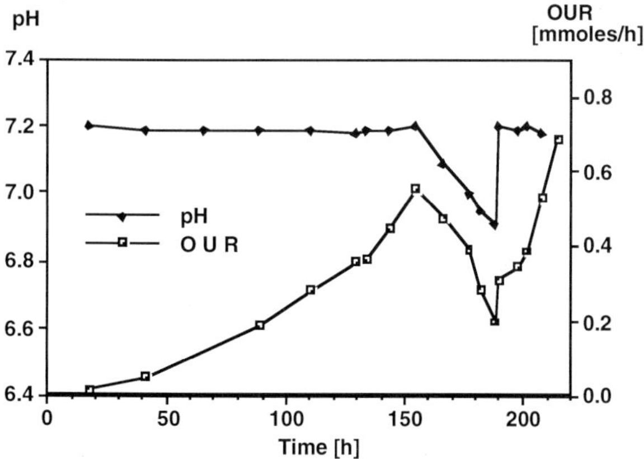

Figure 4. Response of the OUR caused by a pH-change.

Also the depletion of the medium could be rapidly observed by the change in the OUR (Fig. 5) This is useful for optimization of growth and culture conditions.

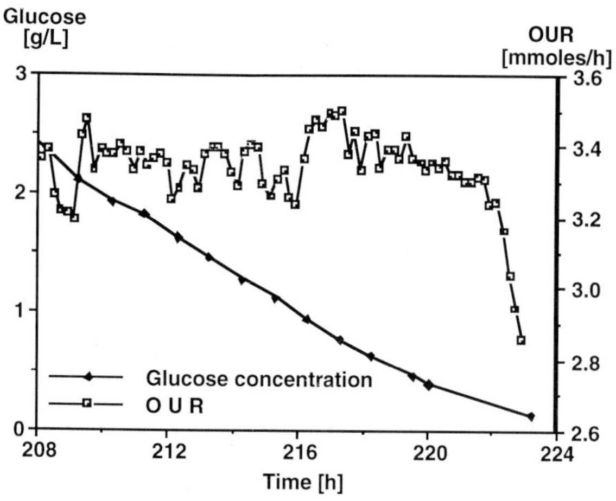

Figure 5. Rapid decrease of the OUR as a result of medium depletion during batch operation. Fluctuations in OUR are due to unsteady aeration conditions.

OUR is also important for process control and validation. It is shown in Fig. 6 that the OUR is correlated well with glucose consumption rate.

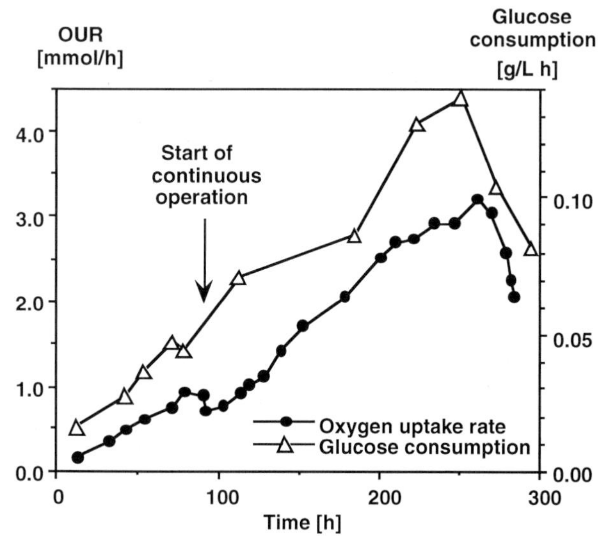

Figure 6. OUR correlates with glucose consumption rate during batch and continuous operation.

Figs. 7 and 8 show the glucose uptake rate in relation to the oxygen uptake rate.

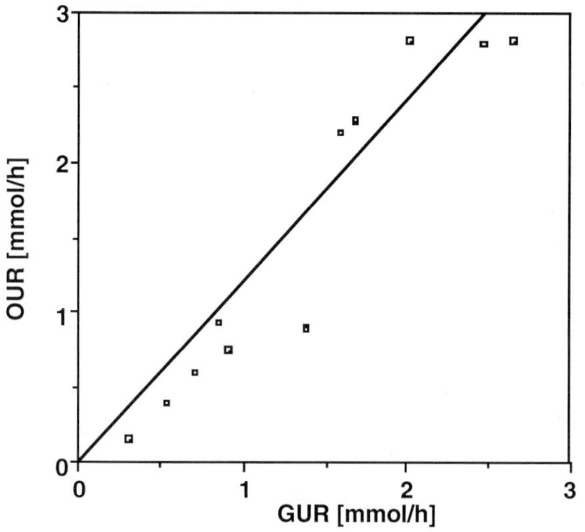

Figure 7. A linear relation between OUR and GUR

The dependency of OUR on oxygen as the limiting substrate is given in Fig. 8.

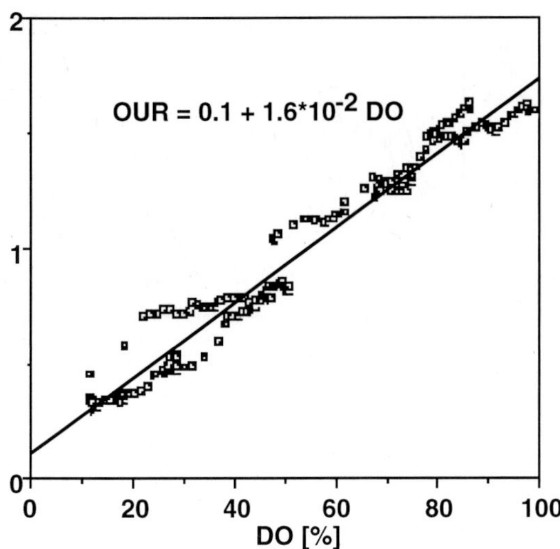

Figure 8. Dependency of OUR on the mean dissolved oxygen concentration (DO).

Fig. 9 demonstrates that the glucose uptake rate was also linearly related to the lactate production rate (LPR).

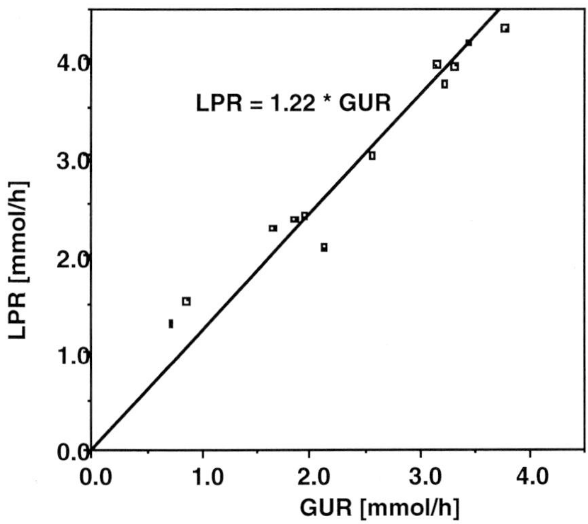

Figure 9. Linear relation between glucose uptake rate and lactate production rate (LPR)

A suspended cell inoculum was used for the porous glass carriers, which was recirculated through the bed. The rapid decrease in the free-cell concentration indicated rapid entrapment of the cells in the carrier pores (Fig. 11). This was a definite advantage over the smooth glass carriers, which had to be inoculated outside the reactor.

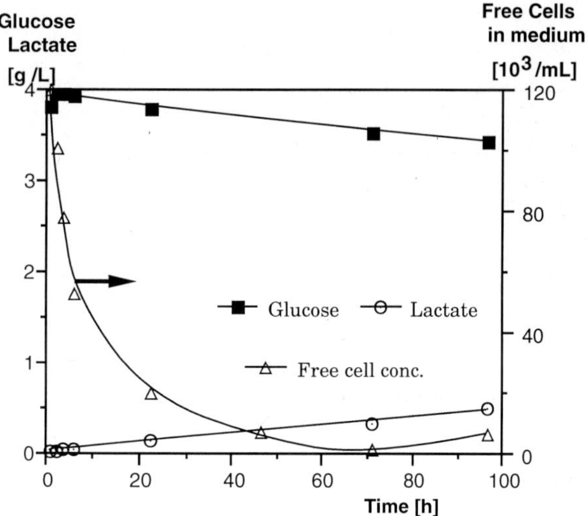

Figure 10. Free-cell concentration decreased during the inoculation period of the porous glass carrier

Medium optimization can be facilitated by OUR measurements, as indicated in Fig. 11. Noteworthy is the much higher OUR values during continuous culture and the fact that serum-free feeding did not influence the culture, after sufficient cell densities were reached. The fluctuations in OUR after 250 hours are due to other experiments.

The culture shown in Fig. 12 was run continuously after 150 h; the feeding was stopped during two period: for 17 hours and 72 hours. During the first feeding interruption the glucose concentration dropped to 0.25 g/L, which had a large effect on the rates. The longer feeding stop caused the rates to drop to zero and resulted in turbidity due to free cells and debris. In spite of this, the culture recovered rapidly after the feed was restarted.

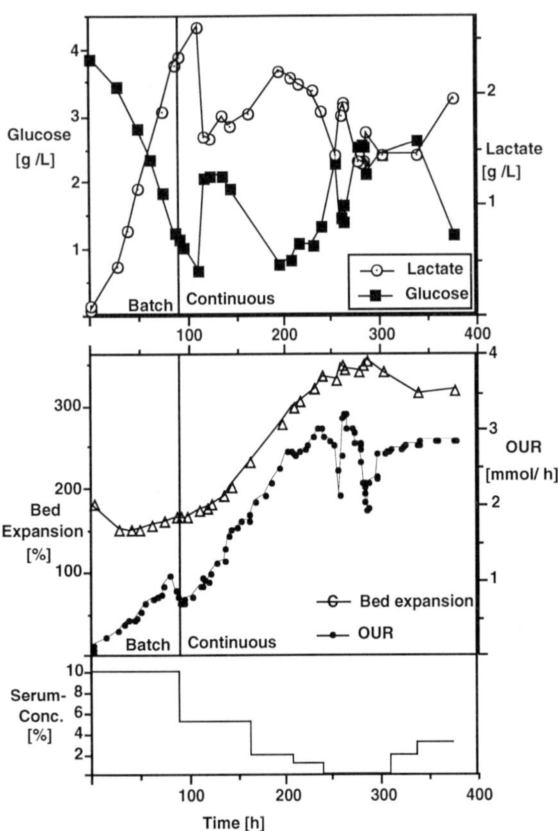

Figure 11. Switch from batch to continuous culture for the smooth glass carriers

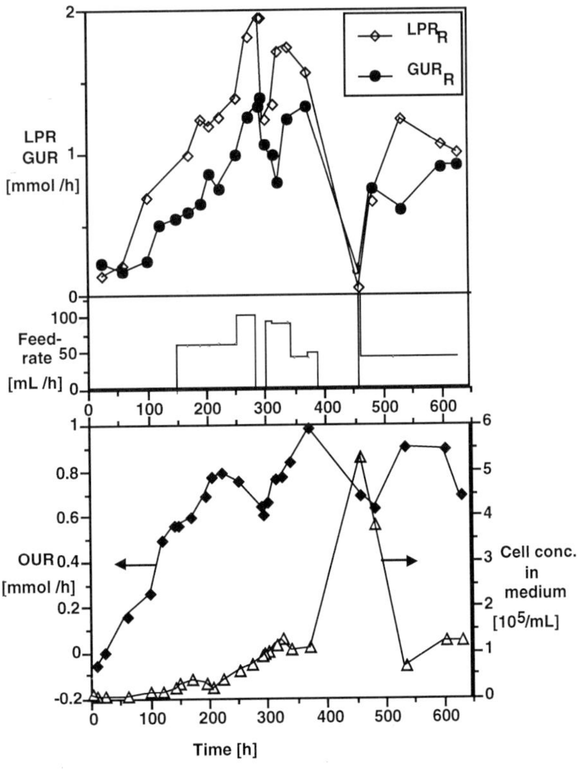

Figure 12. Response of the culture to two interruptions in the feeding.

Results from a culture using the porous glass carrier are shown in Fig. 13. Here it is seen that the development of the culture was considerably slower; this may be due to the inoculation method or to the glass surface properties. Due to insufficient pumping rate, these carriers were not completely fluidized. Unlike the smooth glass carriers, all of the carriers were populated. The final cell densities with the porous carrier were slightly higher than those with the smooth carrier.

CONCLUSIONS

Fluidized bed reactors are well suited for the cultivation of anchorage dependent cell lines. High density solid glass and porous glass carriers have successfully been used. Experiments with a BHK cell-line showed the advantages of this reactor system:

1. High cell concentrations (up to $4*10^7$ cells/mL).

2. High cell activity in a small reactor system.

3. On-line oxygen uptake measurements provide an excellent tool for cultivation control and optimization.

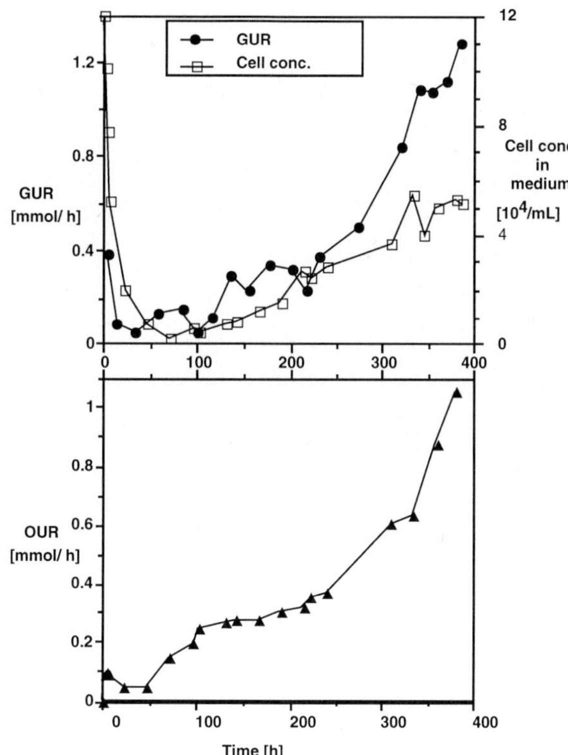

Figure 13. Continuous culture using the porous glass carrier.

4. Both porous and nonporous glass carriers were successfully used for cell retention. The porous glass had the advantage of easy inoculation, and the solid glass allowed cell inspection and harvest.

ACKNOWLEDGEMENT

Research grants from the Swiss Government (KWF) and the Sulzer Company, Winterthur, Switzerland as well as from the ETH-Zürich are gratefully acknowledged.

NOTATION

DO	Dissolved oxygen	% sat.
GUR	Glucose uptake rate	mmol/h
LPR	Lactate production rate	mmol/h
OUR	Oxygen uptake rate	mmol/h

LITERATURE CITED

1. Keller, J., M. Murkovic, I.J. Dunn, E.Heinzle,J.Prenosil, DECHEMA Biotechnology Conferences, **3**, 1989, 715

2. J. Keller, Dissertation, No. 9373, ETH-Zurich, 1991

Influence of Time and Multiplicity of Infection on the Batch Production of *Anticarsia gemmatalis* Nuclear Polyhedrosis Virus in Lepidopteran Insect Cell Cultures

G. Visnovsky and J. Claus

INTEBIO, Facultad de Bioquímica y Ciencias Biológicas, Universidad Nacional del Litoral, CC 530, (3000) Santa Fe, ARGENTINA

The influence of the time and the multiplicity of infection was investigated on the production of polyhedra of Anticarsia gemmatalis *nuclear polyhedrosis virus (AgNPV) in batch suspension cultures of IPLB-Sf-21 insect cells. Several cultures were infected at various multiplicities of infection, either in the early or in the late growth phase. While the dynamics of the infection process was clearly determined by the multiplicity of infection (affecting the cell growth, the viability evolution, the kinetics of virus synthesis and indirectly the virus yield), final virus yields were strongly influenced by the actual time of cellular infection. Early infected cells showed to be more permissive to replicate AgNPV than late infected cells. Such a higher permissivity of early infected cells determined the maximum polyhedra productivity obtained in early and synchronal infected cultures, despite the fact that late infected cultures showed to be more efficient than early infected ones to perform the final step in the process of polyhedra synthesis: the occlusion of enveloped viruses.*

The members of the **Baculoviridae** family are specific arthropod pathogenic viruses (<u>1</u>). Baculoviruses have a bi-phasic replication cycle, where two structurally distinct virus phenotypes, NOV (non-occluded virus) and polyhedron, are produced (<u>2</u>, <u>3</u>). NOVs are infective for susceptible cells, while polyhedra are the insecticide phenotype.

Anticarsia gemmatalis nuclear polyhedrosis virus (AgNPV) is a baculovirus widely used as insect control agent in soybean crops, both in Argentina and Brazil (<u>4</u>, <u>5</u>). AgNPV commercial production is currently performed by viral propagation in infected larvae. However, quality and safety considerations become highly desirable the development of an adequate technology to produce AgNPV in infected insect cell cultures. In addition, mass production in bioreactors is preferable than <u>in vivo</u> methods.

Batch multiplication of baculoviruses in infected insect cell cultures showed to be dependent on both cell culture factors and infection conditions. Oxygen level (<u>6</u>), nutrients and waste concentrations (<u>7</u>) and cell density (<u>8</u>, <u>9</u>) are critical cell culture factors to be considered. Both the multiplicity and the time of infection are parameters that are easily manipulated and that may be important in optimizing virus yields (<u>10</u>, <u>11</u>). The multiplicity of infection (MOI) is defined as the number of infectious units, measured as TCID50 (tissue culture infectious dose 50%), per cell that are added at the time of infection.

The aim of this work was to study the influence of both the time of infection and the MOI on the batch production of AgNPV in suspension cultures of infected IPLB-Sf-21 cells. In order to elucidate eventual relationships between both parameters, several combinations of them were assayed. Infected cultures were examined over time for cell density, cell viability, glucose concentration, titer of non-occluded virus and titer of polyhedra.

MATERIALS AND METHODS

<u>Cell cultures</u>. The IPLB-Sf-21 cell line (<u>12</u>) was obtained from Dr. V. Romanowsky, Facultad de Ciencias Exactas, Universidad Nacional de La Plata, R. Argentina. Cells were maintained as adherent cultures in 25 cm2 T-flasks. The incubation temperature was 28 C. TC-100 plus 10% fetal calf serum was used as culture medium. Subcultures were made every 4-5 days. Suspension cultures were done in 500 ml spinner flasks (Techne, U.K.). Culture volumes were 50 ml. The speed of the magnetic stirrer was adjusted to 60 rpm.

123

E. Galindo and O.T. Ramírez (eds.), Advances in Bioprocess Engineering. 123-128.
© *1994 Kluwer Academic Publishers. Printed in the Netherlands.*

<u>Virus.</u> The experiments were performed with a strain of AgNPV isolated from an infected larva of **Anticarsia gemmatalis** (<u>13</u>). Virus stock was prepared by amplification in suspension cultures of IPLB-Sf-21 cells, after two propagations of the original inoculum in cell cultures. NOVs particles from culture supernatants were titered by end-point dilution analysis (<u>13</u>). In brief, to determine the TCID50, cell suspensions (3 x 10^5 cells/ml) of IPLB-Sf-21 cells were seeded into 96-wells plates (50 ul per well) and then an equal volume of each viral supernatant dilution was added (each supernatant was processed in triplicate). After incubation at 27 C, plates were scored for infection and then the titer was calculated by the Reed and Muench procedure (<u>14</u>). For polyhedra quantification, aliquotes of whole infected cultures were harvested and centrifugated in a microfuge for 30 sec. The pellet was treated with 2% SDS during 1 hour at room temperature. The number of polyhedra was determined in both the supernatant and the treated pellet with a haemocytometer, counting three separate aliquotes and taking the mean.

<u>Analytical.</u> Cell counts were made with a Neubauer haemocytometer. Viability was assayed by 0.04% Trypan blue dye exclusion. Glucose concentration was measured in the culture supernatants with an enzymatic kit (Wiener Lab., R. Argentina).

RESULTS

Effects of the time and the multiplicity of infection on the cell growth and the cellular viability.

Several spinner flask suspension cultures of IPLB-Sf-21 cells were infected with AgNPV at various MOIs, either in the early or in the late growth phase. Figure 1 shows a typical growth curve of uninfected IPLB-Sf-21 cells and the growth curves of cultures infected at several sets of infection conditions. High MOI infection (MOI=5) provoked an immediate interruption of the cell growth. On the other hand, low MOI infection (MOI=0.05) only slightly interfered on the cell growth of early infected cultures, while no interference could be seen when cultures were infected in the late growth phase. Cultures infected at low MOI in the late growth phase reached maximum viable cell densities (2.9 x 10^6 viable cells/ml) at least as high as non-infected cultures. Later, viable cell density of the infected cultures decayed as a consequence of the increasing mortality.

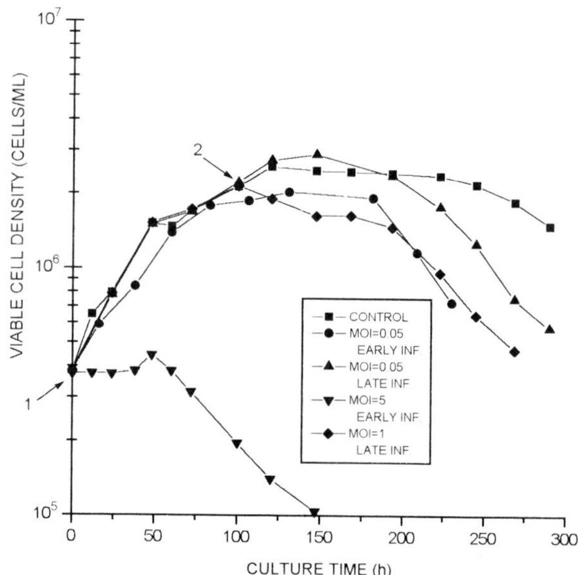

Figure 1. Evolution of the viable cell density in IPLB-Sf-21 cell suspension cultures infected with AgNPV at different conditions of infection. The arrows indicate the times of infection: 1- early infections; 2- late infections.

Viability percentages plotted against the time post-infection are shown in the Figure 2. Two well-defined phases could be seen: a constant viability phase (viability 90%) and a second phase with declining viability. The duration of the first phase appeared to be mainly determined by the MOI. While early infected cultures at high MOI (MOI=5) showed a short high viability phase (36 hours), this phase was significantly longer in cultures infected at low MOI (nearly 100 hours), and showed an intermediate duration in late infected cultures at MOI=1 (nearly 70 hours). The second phase was characterized by a first order death kinetics. However, while death rate remained unchanged in early infected cultures at high MOI, a final acceleration of the rate of cellular death was observed in the other cultures.

We have previously established that glucose is the first nutrient to be exhausted from high cell density IPLB-Sf-21 non-infected cultures in TC-100 medium (Beccaria and Claus, unpublished observation). Glucose exhaustion is followed by a rapid reduction of the cellular viability , that can be delayed by glucose addition. To inquire if the final acceleration of the death rate in

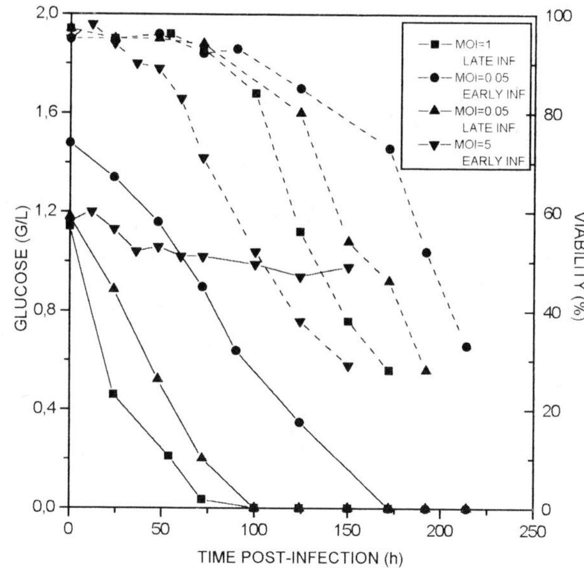

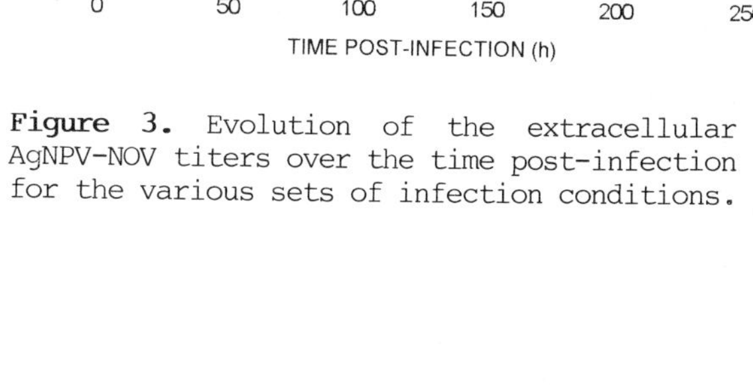

Figure 2. Viability and glucose concentration plotted against the time post-infection. Dashed lines show the viability evolution.

Figure 3. Evolution of the extracellular AgNPV-NOV titers over the time post-infection for the various sets of infection conditions.

high cell density infected cultures could be related to cultural factors, the evolution of the glucose concentrations in the infected culture media was followed. Figure 2 shows that infected cultures that reached high cell densities, where the final acceleration of the death rate was observed, exhausted glucose almost immediately before the beginning of the accelerated death phase. On the other hand, no glucose limitation could be observed in early infected cultures at MOI=5, where the maximum cell density was lower and the death rate remained unchanged.

Effects of the infection conditions on the AgNPV-NOV production.

Supernatant AgNPV-NOV titers plotted against the time post-infection are shown in Figure 3. First infectious progeny virus could be detected between 4 and 6 hours post-infection. Then, titers raised and the secreted NOVs accumulated in the culture supernatant. The kinetics of NOV synthesis appeared to be mainly influenced by the MOI. Cultures infected at high MOI showed a first period where NOVs were rapidly accumulated, followed by a period where NOV titer remained nearly constant. On the other hand, cultures infected at low MOI showed an initially slower but longer synthesis of NOVs.

On the other hand, the cellular efficiency for the NOV synthesis appeared to be mainly determined by the time of infection. Both the maximum titer and the specific yield of NOVs (table 1) were significantly higher in early infected cultures than in late infected ones, irrespective of the MOI. It must be noted that the specific yield of NOVs was specially high in early infected cultures at high MOI.

Effects of the infection parameters on the production of AgNPV polyhedra.

Figure 4 shows the evolution of the cell associated polyhedra titers plotted against the time post-infection. The first cell associated polyhedra were detected between 28 and 36 hours post-infection. The kinetics of polyhedra synthesis appears to be also influenced by the MOI. Cultures infected at high MOI showed a rapid accumulation of polyhedra, reaching the maximum titers between 100 to 120 hours post-infection. On the contrary, low multiplicity infected

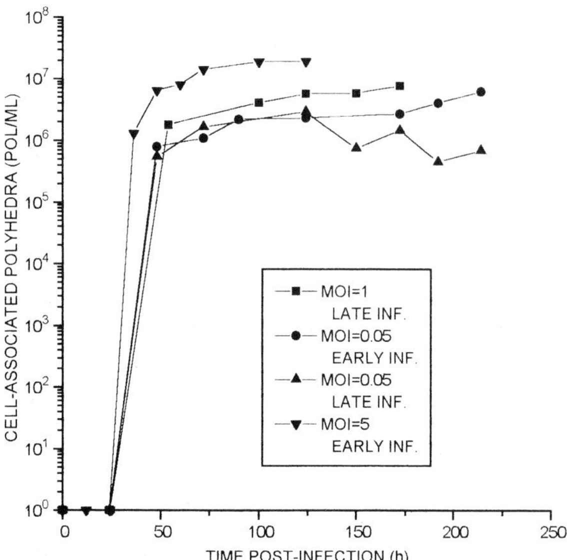

Figure 4. Evolution of the cell-associated polyhedra titers over the time post-infection for the various sets of infection conditions.

cultures synthesized polyhedra at an initially lower rate, but polyhedra synthesis was detected until the end of the cultures. The oscilating titer detected at the end of late infected cultures at low MOI was provoked by cellular lysis and polyhedra release. The released polyhedra accumulated in the culture medium (not shown).

With regard to the final polyhedra yields, early infected cultures at high MOI produced a significantly higher amount of occluded virus than cultures infected either at low multiplicity of infection or in the late growth phase (table 1). Such a difference is magnified when specific yields are considered due to the lower cell density reached in high multiplicity early infected cultures. While each high multiplicity early infected cell produced 52 polyhedra, no more than 3.5 polyhedra per cell were produced in the other cultures. Moreover, the time required to reach the maximum yield of polyhedra was substantially shorter in early infected cultures at high MOI (100-120 hours), than the time required in the other cultures (200-300 hours). Therefore, an early infection at

high multiplicity of infection appears to be essential to obtain a large productivity of AgNPV polyhedra in batch suspension cultures of IPLB-Sf-21 cells (table 1).

Table 1. Final results of batch suspension cultures of IPLB-Sf-21 cells infected with AgNPV at different infection conditions.

	MOI=5 early inf.	MOI=1 late inf.	MOI=0.05 early inf.	MOI=0.05 late inf.
Maximum viab. cell density (cell/ml) x 10^5	3.8	22.0	20.3	28.9
NOV max. titer (NOV/ml) x 10^6	11.2	1.2	6.0	1.2
NOV specific yield (NOV/cell)	29.5	0.6	3.0	0.4
Polyhedra max.titer (pol/ml) x 10^6	19.9	6.4	6.7	6.6
Polyhedra specific yield (pol/cell)	52.4	2.9	3.3	2.3
Productivity (pol/ml/h)	16.7	2.9	3.1	2.1
% Occlusion	64.0	84.2	52.8	84.6

DISCUSSION

Baculovirus replication is a complex process where two kinds of viral populations are produced. Early in the infection cycle the viral nucleocapsids are orientated to the synthesis of enveloped particles (NOVs) and then secreted to the extracellular environment, where they will be able to infect susceptible non-infected cells. Later, the nucleocapsids are enveloped in the nucleus of the infected cell and then occluded into the polyhedra. Therefore, polyhedra production in infected insect cell cultures can be considered as a two steps process where enveloped particles are first synthesized and then occluded.

The selection of the multiplicity of infection imposes to the process a characteristic dynamics. At high MOI (MOI=5), almost the whole cell population becomes infected at time zero (synchronal infection). Our experiments show that the culture growth is immediately arrested after infection at high multiplicity (Figure 1), indicating that cell division is incompatible with AgNPV infection. The same behaviour has been previously reported for other baculovirus-insect cell systems (6, 15). Licari and Bailey (10) reported that cell growth was not arrested until 48 hours post-infection in Sf-9 cell cultures infected at a theoretical MOI of 10 with wild-type or recombinant strains of **Autographa californica** nuclear polyhedrosis virus. However, their experiments were done infecting monolayer cultures and perhaps the actual MOI was substantially lower than the theoretical one. When cell cultures are infected at low MOI (MOI=0.05), only a small percentage of the cell population becomes infected at time zero. The growth of low multiplicity AgNPV infected cultures immediately after infection (Figure 1) is explained by the multiplication of non-infected cells. Late arrest of the cell growth would be provoked in part by secondary infection of the remaining non-infected cells. Such a dynamics of the infection process also explains the longer viability of cultures infected at low MOI, when compared with the rapid decrease of the viability in early infected cultures at high MOI (Figure 2). Nutrient limitation, particularly glucose, could be an additional factor to limit the viability of infected cultures where high cell densities are reached (Figure 2). It must be noted that the viability of late infected cultures showed a lesser sensitivity to the MOI changes, perhaps as a consequence of the reduced permissivity of older cells to the viral infection. Scott et al (6) have also suggested that an impairment of either the infectivity or the virus replication could be the reason for the relatively low cell death observed in late infected cultures of Sf-9 cells.

The dynamics of the infection process imposed by the MOI can also explain its influence on the kinetics of the virus production. The rapid accumulation of NOVs and polyhedra and the subsequent stability of the titers observed in high multiplicity infected cultures is a consequence of the synchronal infection of all the cells in the culture. In contrast, a slower and longer increase of the viral titers was observed in low multiplicity infected cultures, where several cell subpopulations become asynchronically infected by succesive generations of non-occluded viral progeny synthesized by precedingly infected cells.

The propagative nature of the viral infection assures that almost all the cells in a batch culture will become infected and productive along the time. If the cellular permissivity to the virus replication would be constant and independent of the physiological state of the cells at the time of infection, then cultures that finally reach high cell densities should yield larger amounts of virus than low cell density cultures. However, table 1 shows that virus yields were weakly dependent on the maximum cell density, but strongly dependent on the time of infection. Accordingly, the earlier the infection, the higher the virus yields. The requirement of an early infection to obtain high yields of either virus or recombinant products has been previously reported for numerous baculovirus-insect cell systems (6, 8, 9, 16, 17, 18, 19, 20, 21). Yield differences between early infected cultures at either low or high MOI can be explained as a consequence of the infection dynamics. In early infected cultures at high MOI most of the cells are effectively early infected, while in early infected cultures at low MOI, most of the cells are later infected, when they already are in a non-proliferating state. Therefore, cells infected during their active multiplication phase seem to be more permissive to virus replication than cells infected at the end of the growth phase. Such a reduced capability of late infected cells could be either an irreversible effect of the cellular ageing or a consequence of environmental changes. Baculovirus replication in late infected cultures showed to be restored by nutrients addition (9, 21) or medium replenishment (15). Thus, glucose exhaustion could be an important factor to limit AgNPV replication in IPLB-Sf-21 cells.

The comparison of the specific yields of both NOVs and polyhedra shows that the impairment of the virus replication in late infected cultures appeared to be preferentially exerted at the level of NOV synthesis, while polyhedra production was substantially less affected (table 1). A higher efficiency of later infected cells to occlude NOV particles could explain such a difference. Effectively, more than 80% of the non-occluded viral particles synthesized by late infected cells were derived to the production of polyhedra, while nearly 50-60% was the occlusion percentage in early infected cultures (table 1). Therefore, the higher efficiency of early infected cultures for the virus replication appeared to be partially compensated by the higher efficiency of late infected cultures for the occlusion process.

As a conclusion, early infection at high multiplicity of infection offers the highest productivity for the batch production of AgNPV polyhedra in insect cell cultures. Such a high productivity is mainly supported by the elevated level of virus replication in cells infected during their active multiplication phase. Viral replication significatively declines when cells are infected at the end of the growth phase, despite that late infected cells showed to be more efficient to perform the occlusion of the enveloped virions. The use of alternative culture strategies, such as fed-batch or perfusion, could aid to improve the polyhedra productivity of cultures infected either at low MOI or in the late growth phase.

NOMENCLATURE

MOI = multiplicity of infection.
NOV = non-occluded virus.
TCID50 = tissue culture infectious dose 50%.

LITERATURE CITED

1. Francki, R.I.B., C.M. Faquet, D.L. Knudson and F. Brown. Arch. Virol., 2 (Suppl.), 119 (1991).

2. Volkman, L.E., M.D. Summers and C.H. Hsieh. J. Virol., 19, 820 (1986).

3. Rohrmann, G.F. J. Gen. Virol., 73, 749 (1992).

4. Moscardi, Flavio and B.S. Correa Ferreira. "Biological Control of Soybean Caterpillars." in Proceedings of the World Soybean Research Conference III, Shiblis, R. (Ed.), Westview Press, London (1985).

5. Moscardi, Flavio, G.E. Allen and G.L. Greene. J. Econ. Entomol., 74, 480 (1981).

6. Scott, R.I., J.H. Blanchard and C.R.H. Ferguson. Enzyme Microb. Technol., 14, 798 (1992).

7. Wang, M.Y., V. Vakharia and W. Bentley. Biotechnol. Bioengn., 42, 240 (1993).

8. Lindsay, D. A. and M. J. Betenbaugh. Biotechnol. Bioengn., 39, 614 (1992).

9. Caron, A.W., J. Archambault and B. Massie. Biotechnol. Bioengn., 36, 1133 (1990).

10. Licari, P. and J.E. Bailey. Biotechnol. Bioengn., 37, 238 (1991).

11. Licari, P. and J.E. Bailey. Biotechnol. Bioengn., 39, 432 (1992).

12. Vaughn, J.L., R.H. Goodwin, G.J. Tompkins and P. McCawley. In Vitro, 13, 213 (1977).

13. Claus, J.D., G.E. Remondetto, S.A. Guerrero, A.M. Demonte, M. Murguia and A.J. Marcipar. J. Biotechnol., 31, 1 (1993).

14. Reed, L.J. and Muench, H. Am. J. Hyg., 27, 493 (1938).

15. Schlaeger, E.J., H. Loetscher and R.Gentz. Fermentation Scale Up: Production of Soluble Human TNF Receptors, In: J.M.Vlak, E.J. Schlaeger and A.R. Bernard (Eds.), Proceedings of the Baculovirus and Recombinant Protein Production Workshop, Interlaken, Switzerland, p. 201 (1992).

16. Maiorella, B., D. Inlow, A. Shauger and D. Harano. Bio/Technology, 6, 1406 (1988).

17. Hink, W.F., E.M. Strauss and W.A. Ramoska. J. Invertebr. Pathol., 30, 185 (1977).

18. Wood, H.A., L.B. Johnston and J.P. Burand. Virology, 119, 245 (1982).

19. McGlynn, E., J. Liebetanz, S. Reutener, B. Gay, M. Becker and N.B. Lydon. Large Scale Production of Protein Tyrosine Kinases in Insect Cells, In: J.M. Vlak, E.J.Schlaeger and A.R. Bernard (Eds.), Proceedings of the Baculovirus and Recombinant Protein Production Workshop, Interlaken, Switzerland, p. 209 (1992).

20. Shah, G., B. Crosset and R.S. Hale. Growth and Infection of Sf-9 Cells for Production of EGFR-Tk at 30 L Scale, In: J.M. Vlak, E.J. Schlaeger and A.R. Bernard (Eds.), Proceedings of the Baculovirus and Recombinant Protein Production Workshop, Interlaken, Switzerland, p. 216 (1992).

21. Reuveny, S., Y.J. Kim, C.W. Kemp and J. Shiloach. Biotechnol. Bioengn., 42, 235 (1993).

Invited paper

Design of Tubular Microporous Membrane Aerated Bioreactors for Plant Cell Suspension Culture

A.E. Humphrey

Director, Biotechnology Institute, The Pennsylvania State University, University Park, PA 16802, U.S.A.

Bubble-free aeration using immersed microporous membrane tubing holds great promise as an alternative to overcome the problems of culturing shear sensitive plant cells in a conventional air sparged bioreactor. In designing a membrane aerated system there are several key variables to be considered. These include the length, diameter and thickness of the membrane tubing, inlet gas composition and pressure, and the state of the growing culture, i.e., the dissolved oxygen and carbon dioxide concentrations in the culture medium. An analysis of the effect of these various factors on the oxygen transfer and the carbon dioxide removal is presented in this paper.

Introduction

Hydrodynamic shear damage of cells due to bubble breakup at the air-liquid surface in a sparged bioreactor is believed to be the major contributor to destruction of cells in destroying plant and mammalian bioreactor system (1). It can be avoided when a membrane aerated bioreactor is used for cell culture. Bubble-free aeration using immersed process tubular membranes in bioreactors has been proposed for more than a decade. Not only do membrane aeration systems avoid cell damage due to bubbles, but they sustain high gas transfer capabilities under high culture viscosity. This is due to the fact that tubular membrane systems have constant gas/liquid interfacial area. This is a very desirable feature for aerating commercial plant cell culture which with a cell density around 15-50/g dry wt./liter, can be very viscous. Also, membrane aeration systems eliminate foaming and cell floatation and, in turn, the undesirable wall growth and cell disruption.

Description of the System

The tubular membrane aeration system is illustrated in Fig. 1. Operating and design variables include the inlet pressure, P_e (which cannot exceed the bubble point pressure of the tubing in the liquid); the exit pressure, P_{ex}; the static pressure in the vessel, P; the tubing size or diameter, d_o; the tubing thickness, d_o-d_t; the length of submerge tubing, l, the dissolved concentration of the various gaseous components (c_i) the partial pressure of the various

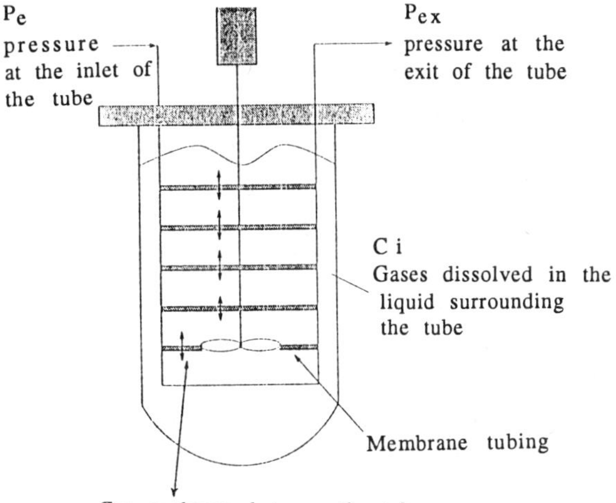

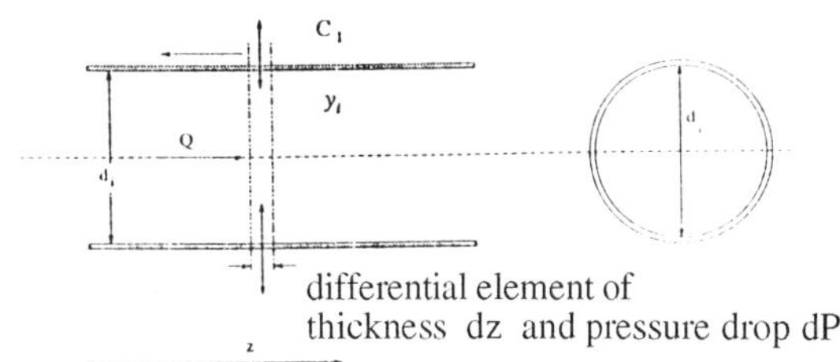

Figure 1: Membrane aerated bioreactor

E. Galindo and O.T. Ramírez (eds.), Advances in Bioprocess Engineering. 129-132.
© *1994 Kluwer Academic Publishers. Printed in the Netherlands.*

components in the gas, y_i; and the volume of the culture, V_r. Obviously, there are many control variables that can be manipulated such as the gas composition, the inlet and exit pressure, and the static (vessel) pressure to achieve varying rates of oxygen transfer and carbon dioxide removal. Further, as will be shown in the analysis, there exists a tubing length and/or size for any given set of operating conditions for optimizing the O_2 transfer to and CO_2 removal from the culture.

Thin walled silicone membrane tubing has been used to provide bubble-free oxygenation in small scale insect and animal cell cultures for a number of years [2,3]. Commercial bioreactors have been constructed using the silicon tubing to supply oxygen and to remove CO_2. Examples are: Invitron's maintenance reactor and B. Braun's Biostate MC fermenter. Silicone membranes have higher oxygen permeability than most dense polymeric membranes. However, the diffusion barrier associated with a silicone membrane is still quite significant. On the other hand, a low value of inherent mass transfer resistance (diffusion barrier) for the microporous polypropylene membrane has been demonstrated relative to the thin walled silicone tubine operating at similar conditions [4] . Moreover, the microporous polymer membrane shows better mechanical strength than the thin-walled silicone membrane. One restriction in using the microporous membrane, however, is its low bubble point (the operating gas pressure inside the tubular membrane, beyond which bubbles will form on the outer surface of the porous membrane). This limits the operating pressure inside the membrane tube. Also, for the microporous membrane the actual bubble point is much lower than the theoretical bubble point (< 50 mbar) mainly due to imperfect membrane wall wetting. This low bubble pressure limits the operating pressure inside the microporous tubular membrane and restricts the length of tubing one can use.

In this paper, a theoretical analysis of oxygen transport across the tubular microporous membrane will be described. This analysis provides some insight into the optimal design of the membrane aerator for plant cell culture where minimized shear is important.

<u>Problem Formulation</u>
In spite of a number of tubular membrane reactor systems described in the literature, none has dealt specifically with the optimal design of the system. The majority of studies simply assumed there was no detectable difference between the oxygen partial pressure at the gas inlet and outlet, i.e. a constant oxygen driving force was assumed along the entire length of the membrane tubing. For a small fermenter that requires only short length of membrane tubing, this may be a fair assumption.

Nevertheless, when scaling up the fermenter, much longer tubular membranes are required to meet the oxygen demand. Consequently, the constant oxygen driving force assumption is no longer valid.

For the tubular membrane aeration systems, the overall oxygen transfer rate can be obtained by integrating the oxygen molar flux, which is the product of the mass transfer coefficient and the oxygen transfer driving force, along the length of the membrane tubing, i.e.,

$$OTR_m = \left(\frac{\pi d_i k_{l,O_2}}{R_g T \cdot V_R}\right) \cdot \int_0^l \left(P \cdot \frac{n_{O_2}}{n_{N_2} + n_{O_2} + n_{CO_2}} - \frac{C_{l,O_2}}{H_{O_2}}\right) dz \tag{1}$$

where d_i is the inside diameter of the tubing, k_{l,O_2} is the oxygen mass transfer coefficient across the microporous membrane, R_g is the gas constant, T is the temperature, V_R is the culture volume, l is the length of the aeration tubing, P is the total pressure in the membrane gas phase, C_{l,O_2} is the dissolved oxygen concentration, H_{O_2} is the Henry's law constant for oxygen, and z is the axial coordinate, n_{O_2}, n_{CO_2} and n_{N_2} are the molar flow rates of oxygen, carbon dioxide and nitrogen, respectively. Similar transfer relationships can be derived for the CO_2 removal and N_2 transfer rates.

Upon examining eq. (1), it becomes apparent that the oxygen partial pressure in the membrane gas phase decreases along the length of the tubing, and the oxygen driving force may decline to an insignificant level at certain tubing length, beyond which there will be little oxygen transfer across the membrane tubing. In fact, as the pressure drops along the tubing, it is possible, for certain culture conditions, to get mass transfer reversals. As a consequence, there is an optimal length of tubular membrane for a given culture system, reactor size, and static pressure to maximize the volumetric oxygen transfer to the culture.

<u>Model Formulation and Solution</u>
The momentum balance for laminar flow inside the tubular membrane can be written as follows:

$$\frac{dP}{dz} = -\left(\frac{128 \mu}{\pi d_i^4}\right) \cdot Q \tag{2}$$

where μ is the gas viscosity, and Q is the gas volumetric flow rate inside the tubing.

By introducing the following dimensionless groups:

$$\chi = z/d_i$$

$$P_r = P/P_i$$

$$\Phi = \frac{n}{\dfrac{\pi d_i^3 P_i^2}{128 \mu R_g T}}$$

$$T_l = \frac{128 \mu k_l}{d_i P_i}$$

$$P_l = \frac{C_l}{H \cdot P_i}$$

where P_i is the gas pressure at the tubing inlet. Equation (2) can be rewritten in dimensionless form:

$$\frac{dP_r}{d\chi} = -\frac{1}{P_r}(\Phi_{N_2}+\Phi_{O_2}+\Phi_{CO_2}) \qquad (3)$$

$$\frac{d\Phi_{O_2}}{d\chi} = -T_{I,O_2}\left(P_r \cdot \frac{\Phi_{O_2}}{\Phi_{N_2}+\Phi_{O_2}+\Phi_{CO_2}} - P_{I,O_2}\right) \qquad (4)$$

$$\frac{d\Phi_{N_2}}{d\chi} = -T_{I,N_2}\left(P_r \cdot \frac{\Phi_{N_2}}{\Phi_{N_2}+\Phi_{O_2}+\Phi_{CO_2}} - P_{I,N_2}\right) \qquad (5)$$

$$\frac{d\Phi_{CO_2}}{d\chi} = -T_{I,CO_2}\left(P_r \cdot \frac{\Phi_{CO_2}}{\Phi_{N_2}+\Phi_{O_2}+\Phi_{CO_2}} - P_{I,CO_2}\right) \qquad (6)$$

The boundary conditions are

$$\chi=0, \quad \Phi_{O_2}=\Phi_{O_2,i}, \quad \Phi_{N_2}=\Phi_{N_2,i}, \quad \Phi_{CO_2}=0, \quad P_r=1$$

where $\phi_{O_2,i}$ and $\phi_{N_2,i}$ are the dimensionless inlet oxygen and nitrogen molar flow rate, respectively. When pure oxygen is fed continuously into the aeration tube, $\phi_{N_2,i}$ is equal to zero.

These coupled nonlinear ordinary differential equations can be integrated numerically using a fourth order Runge-Kuta method. Unfortunately, a discrete solution exists for each operating condition of entering gas pressure and composition, tubing size, and culture conditions. Dimensionless graphical solutions for these equations (3-6) have been reported by Bohtra (5).

<u>Results</u>

For a typical set of conditions, the transmembrane oxygen flux along the tubing results in an almost linear pressure drop along the tubing (Fig. 2). CO_2 partial pressure increases along the length of the tubing until it exceeds the level that is in equilibrium with the dissolved CO_2 in the culture (liquid phase). Beyond that point, CO_2 starts to diffuse back into the culture. Due to the low mass transfer driving force, the CO_2 molar flow rate is about two orders of magnitude lower than the O_2 molar flow rate. When aerating with air, for the molar flow rate of CO_2 into the tubing to approach that of O_2 out of the tubing, the dissolved CO_2 concentration must rise above 10% saturation of 1 atm CO_2 in the culture. The nitrogen transfer driving force increases along the length of the tube and then decreases. This is because in the front portion of the tubing, oxygen has a very high mass transfer driving force. The sharp decrease in oxygen mole fraction in the membrane gas phase results in an increase in the nitrogen partial pressure in the membrane gas phase. Since nitrogen is not utilized by the culture, there can be no net transfer of nitrogen.

For oxygen, the mass transfer driving force (and hence partial pressure), as well as the molar flow rate, decrease along the length of the tubing. For the tubing and conditions in Fig. 2, at a dimensionless length of 13377 or 24 meters for a 0.18 cm diameter tubing, the oxygen transfer driving force near the tubing outlet drops to only approximately 10% of that at the tubing inlet.

<u>Conclusions</u>

While the overall membrane oxygen transfer rate increases with increased tube length, the oxygen transfer rate per tube length always decreases with increased tube length. Because of this, when a large membrane surface area is required, the tubing should be divided into parallel segments in order to enhance the overall oxygen transfer rate. A manifold or a gas distributor can then be used to distribute gas into segments of tubing. Shorter tube segments give higher oxygen transfer rates per unit tube length.

This advantage has not been recognized in membrane aerated bioreactor design although it is the basis of the artificial lung (6). However, this advantage is counter balanced by the disadvantage that gas distribution into many parallel segments may not be uniform.

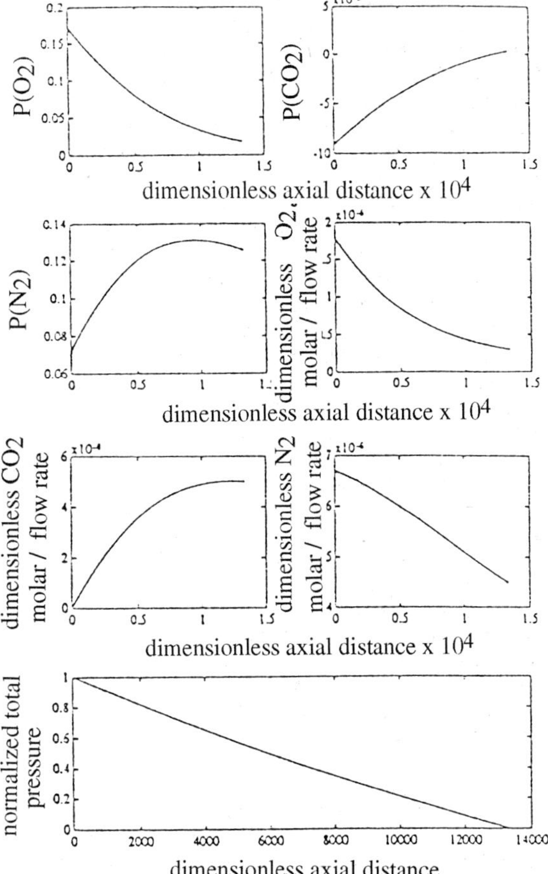

Operating conditions :
Air, P_i=1.1 atm, dimensionless tubing length=13377
Dissolved O_2 conc. : 20% saturation of 1 atm air
Dissolved CO_2 conc. : in equilibrium with 0.01 atm CO_2
Dissolved N_2 conc. : in equilibrium with 0.79 atm N_2

Figure 2: A typical set of simulation results for air being fed into the tubing inlet.

132

References

1. Yang, J.D., Wang, N.S., (1992) Cell inactivation in the presence of sparging and mechanical agitation, Biotech. and Bioeng., 40, 806-816.
2. Kilburn, D.G., Morley, M., (1972) A method for controlling dissolved oxygen tension and measuring respiration rate in suspension cultures of animal cells. Biotech. and Bioeng., 14, 499-504.
3. Eberhard, U., Schugerl, K., (1987) Investigations of reactors for insect cell culture. Developments in Biol. Stand. 66, 325-330.
4. Su, W.W., (1991) Production of plant secondary metabolites from high density perfusion cultures in a membrane aerated bioreactor. Ph.D. Thesis. Lehigh University.
5. Bohtra, D., (1993) Design of Bubble-free Membrane Aerated Bioreactors for Plant Cell Culture. Ph.D. Thesis, Lehigh University.
6. Ultman, J., (1993) Private communications, Penn State University.

Optimization and Scale-Up of High Density Cell Culture Bioreactors

S.S. Ozturk

Miles Biotechnology, 4th and Parker Streets, Berkeley, CA 94701, U.S.A.

High density cell culture systems offer more than a one-order of magnitude increase in reactor productivity. The design and scale-up of these systems are, however, more difficult. The challenges start with the mechanical design of the bioreactors; mixing and mass transfer are more critical in high density systems. In this paper, we analyze several methods of cell retention systems and discuss their operation. High density perfusion bioreactors can be operated in commercial scale production for more than 3 months at cell concentrations up to 10-40 million cells ml^{-1}. Optimization of these reactors involves altering feed rate and composition, reactor pH and DO. In some cases, local mass transfer gradients can also affect the cell physiology and increase cell specific productivity. This paper discusses our studies for industrial scale mammalian cell product development, optimization, and reactor scale-up.

Biomolecules of great investigative, clinical and commercial value have been produced by cultivating mammalian cells in bioreactors. Cellular macromolecules such as interferons, hormones, monoclonal antibodies, and physiological effectors already are produced as pharmaceuticals in large quantities. High density cultures are attractive for cost-effective production of these biologicals in large quantities mainly because of their higher volumetric productivities.

High density culture systems generally involve perfusion type operation with cell retaining or separating devices. Table 1 summarizes some of the important reactor types. In cell entrapment systems such as hollow fibers, flat sheet reactors, alginate beads, cells are retained as a separate phase with 100% retention. In cell separation systems cells are either attached to a solid support (porous microcarriers or other solid supports) or are separated by physical means such as filtration, centrifugation or sedimentation.

All these systems have several advantageous and drawbacks with respect to their usage in commercial production and laboratory research. Although they can reach very high cell densities (in the order of 100 million cells/ml total reactor volume) hollow fibers and flat sheet reactors are not practical for large scale production. Cell attachment systems and reactors with cell separation devices have been used successfully with cell densities greater than 10 million cells/ml (see Griffiths et al, 1992)

The choice of any high density system ultimately depends on the size of the production as well as operational and economical issues. Compared to simple batch cultures, these systems are more complex to design and operate. The need for cell retention and demand for higher oxygenation rates in these systems often require an external recycle loop. The media utilization efficiency of high perfusion systems needs to be improved. A fed-batch system for instance can reach higher product concentrations. However due to high cell densities obtained the reactor productivities are more than 10 fold higher making high density perfusion systems more economical in the long run (Griffiths, 1992).

In this paper, we discuss our experience in high density cell culture sys-

133

E. Galindo and O.T. Ramírez (eds.), Advances in Bioprocess Engineering. 133-139.
© *1994 Kluwer Academic Publishers. Printed in the Netherlands.*

134

Table 1. Examples of High Density
Perfusion Systems

**Cell Entrapment Systems
(100% Cell Retention)**

Immobilized Cell Systems
      Hollow-fiber, flat sheet
      Beads of alginate/Agarose

      Encapsulated Beads

**Cell Separation Systems
(Partial Cell Retention)**

Microporous Carriers
      Verax Collagen Microspheres
      Glass beads

Wall Attachment Systems
      Ceramic Cartridge
      Fiber-bed

Suspension Cultures with Cell
Separation Devices
      Membrane Filtration*
      Spin Filters
      Gravitational and Centrifugal
      Separation

---

* with cell purging

tems: (a) fluidized-bed reactor with
microporous microspheres, and (b) suspension culture with cell separation.
These systems are capable of producing
kilogram quantities of products from
mammalian cells. The performance and
optimization of these systems are illustrated for the production of monoclonal antibodies.

FLUIDIZED-BED REACTOR

Reactor system. Verax™ fluidized-bed
reactors were used for the cultivation
of human-mouse hybridoma cell line producing IgG1 class monoclonal antibody.
This reactor system developed about a
decade ago (Kalkare et al, 1985) has
been used successfully for many cell
lines (Runstadler et al, 1992). The
schematic description of fluidized-bed
system is shown in Figure 1A. This reactor system uses porous collagen microspheres for cell attachment or entrapment. Even for suspension cells
about 90% of the cells are confined to
the microspheres. The microspheres are

fluidized with recycle flow that eliminates external mass transfer resis-
tances (Figure 1A). The fermenter is
oxygenated by a gas-exchanger placed in
the recycle loop. Media can be added
and removed continually to the reactor
for perfusion.

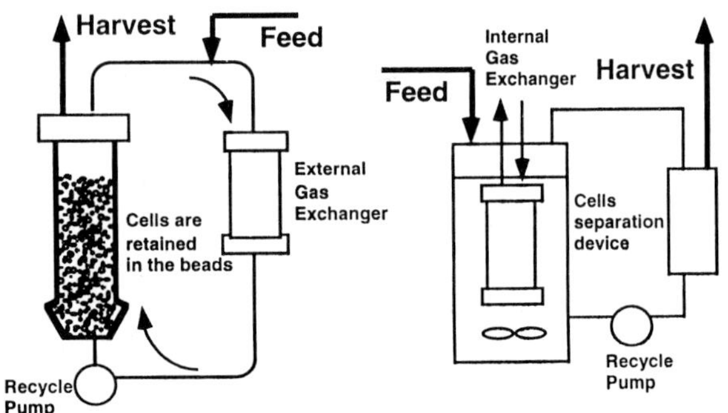

Figure 1. (A) Fluidized-bed and (B)
suspension culture with cell separation
system.

Reactor operation. Cells are cultured
in a proprietary protein-free medium.
The pH, temperature, and dissolved oxygen could be controlled at the specified set-points throughout the run. The
media feed rate (MFR) was constantly
adjusted to maintain a constant environment to the cells accessed by glucose levels in the reactor. Although
cell numbers in the fermenter could be
determined by microsphere sampling,
metabolic activities were used to estimate the cell concentrations continuously.

Figure 2 shows general trends in cell
densities. Metabolically estimated cell
densities are presented along with the
fraction of free cells. Cells were inoculated as a cell suspension and they
penetrated to the microspheres. Free
cell percentage dropped to 10% levels
after about 4-5 days. Cell number continued to increase and on day 27 a
steady-state was reached with 40 million cells per ml of fluidized bed. The
microspheres occupied about 25% of the
bed, so it was inferred that microspheres reached cell densities of 2.5
10^8 cells/ml. This value corresponds to
maximum occupancy of the beads and
aggress with real measurements obtained
by counting the cells after relapsing
the cells from the microspheres. On two
occasions microsphere were sampled and
as seen in Figure 2, the measurements

matched to estimated cell counts. The viability in these microspheres were greater than 70%. This particular fermenter run was terminated after 4.5 months of operation.

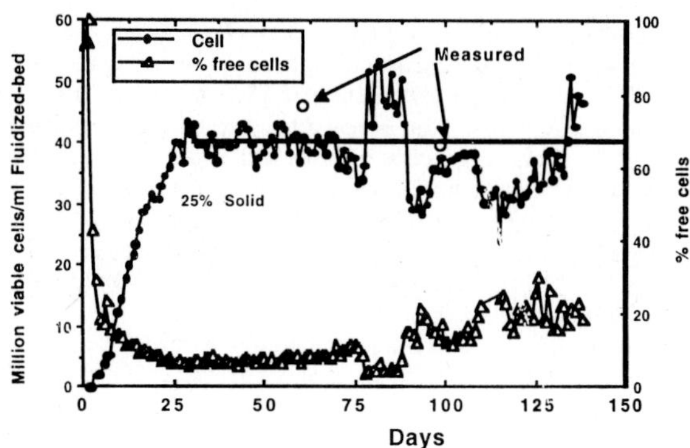

Figure 2. Characteristics of fluidized-bed reactor operation.

<u>Reactor optimization.</u> The fluidized-bed reactor maintained about 50 times more cells compared to batch culture (0.8 million cells/ml in the protein-free media). The increase in reactor productivity was even more; about 100-fold. Apparently cell specific productivity in this system was approximately 2 times higher compared to batch culture. The only concern about fluidized-bed reactor run was a 30% lower product concentration in the reactor. We then undertook a comprehensive optimization study to improve the product titer and further increase reactor productivity.

Figure 3 shows the results of these optimization studies. The fermenter was operated under conditions similar to the previous run. On day 134 an enriched medium was introduced to the reactor. The product titer was doubled as a result of these enrichments. Further improvements in the reactor performance were obtained by manipulating the reactor operating conditions and the use of secretion enhancers. Decreasing the medium feed rate, for instance, increased the product concentration without altering the reactor productivity. In fact the volumetric productivity continued to rise and approached levels 300-fold higher compared to batch cultures. On Day 182 the reactor conditions returned to the original conditions. The product titer returned to original values while reactor volumet-

ric productivity remained high. **The** reason for these higher productivities was mainly the gradual increase in the cell concentrations during this optimization period.

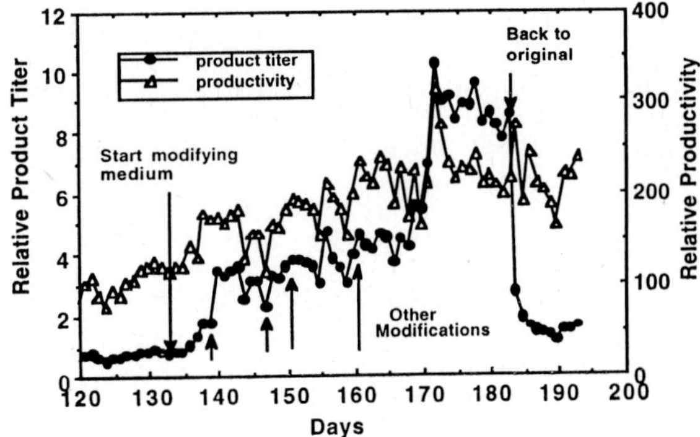

Figure 3. Optimization of reactor performance by manipulating medium and reactor operating conditions.

Cell specific productivities are presented in Figure 4. Before the optimization the specific productivity values were about 2 times higher in the fluidized-bed reactor compared to batch culture. During the modifications to the fermenter operating conditions they increased and about a 6-fold increase was obtained in cell specific productivities. The return to original conditions restored the specific productivities.

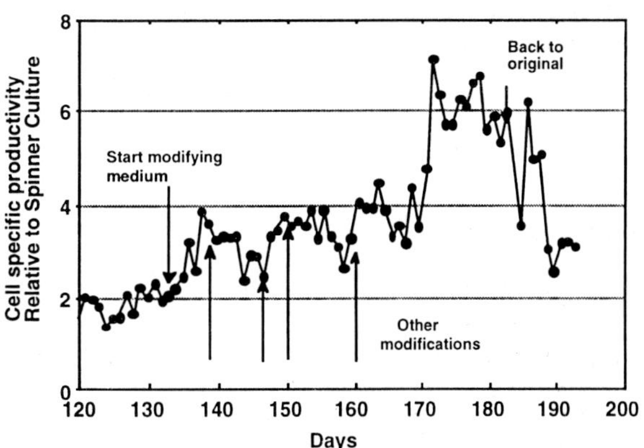

Figure 4. Cell specific productivities during the optimization of fluidized-bed reactor performance.

It is important to understand the physiological changes and the resulting increase in cell specific productivity observed during these optimization

136

studies. It is reported in the literature that reactor operating conditions,
such as osmolarity and pH, can alter
the secretion rates of monoclonal antibodies (Ozturk and Palsson, 1991a,b)
The cells used in this study responded
positively to low pH values and increasing osmolarities and about 2-3
fold increase in cell specific productivity could be obtained in suspension
cultures. The remaining 2-fold increase
should be related to the cells entrapment into the microspheres and its environment (Vournakis and Runstadler,
1989).

<u>Cell Physiology in Microspheres</u>. Better
cell performance in the microsphere
environment is not fully-understood.
Higher cell productivities were also
observed in other immobilized cell culture systems (Lee and Palsson, 1990).
The increased cell specific productivities could partially be attributed to
local concentration and pH gradients
(Ozturk et al, 1991). Mathematical models developed are used to simulate the
local gradients inside the microspheres. Figure 5 presents some of the
results obtained from the mathematical
analysis.

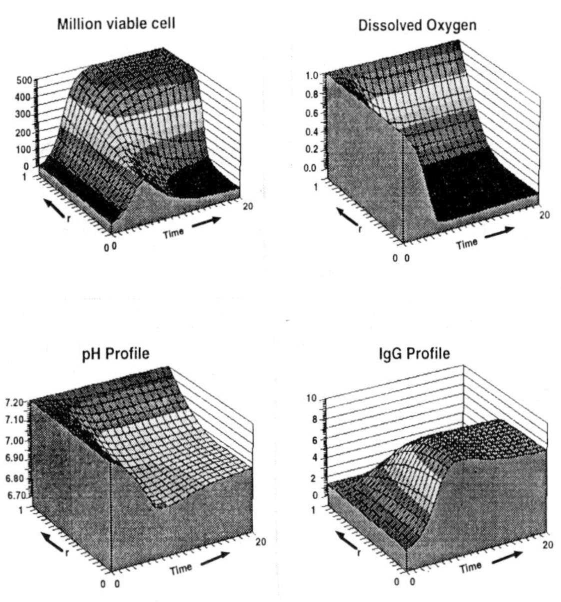

Figure 5. Mathematical analysis for
concentration and pH profiles in
immobilized beads.

Viable cell, oxygen, pH, and IgG profiles are simulated against time and
location in the bead. At the beginning
of the inoculation, cell density and
all concentrations in the bead are uniform in the radial direction. As the
time progresses cell numbers increase
and more nutrients are consumed. The
mathematical model considers the effects of medium composition and pH on
the growth, metabolic and production
rates. The growth rate is faster at the
surface of the bead (r=1) compared to
the center where nutrient levels are
lower. As the time progresses, oxygen
gets depleted at the center and cell
growth stops in the middle of the bead.
These cells eventually die and a
nacrotic core is established. This com-
prehensive model can also predict pH
variations in the bead. As seen in
Figure 5, the pH in the center of the
bead is about 0.3 units lower than the
bulk value. Also presented in Figure 5
is the profiles of IgG. The IgG concentration in the center of the bead is
about one-order-of-magnitude higher.
Due to very low diffusivities, similar
localization is expected in other protein concentrations. The apparent productivity of bead entrapped cells can
be about 2-fold higher when the secretion rate is pH dependent. All these
predictions are in accord with observations reported in the literature and
may well be the explanation for higher
productivities in the fluidized-bed reactors.

<u>Efficiency of media utilization</u>. One
disadvantage of perfusion cultures is
that the media utilization rates are
lower compared to batch or fed-batch
systems. In general, the product titers
are often lower for perfusion cultures.
In batch and fed-batch cultures the media is depleted completely and the
product continues to accumulate. The
reactor utilizes both growth and decline phases for production. Perfusion
cultures such as fluidized-bed reactor
cannot be used in the decline phase
and complete media utilization is not
allowed. Optimizing the reactor operating conditions can increase the productivity and improve the titers as presented above. Alternative to these
routes is the use of enriched media for
perfusion.

Traditionally cell culture media is
prepared for active growth of the cells

and can not support more than 1-4 10^6 cells/ml. The media can be enriched by maximizing the nutrient levels while minimizing waste by-product accumulation. Cell growth rates are normally lower in enriched media and their use in batch culture is limited to the stationary or decline phase. In perfusion reactor, however, low growth rates can be tolerated and medium enrichment can be used successfully as illustrated in Figure 3.

Development of enriched media requires a careful study of cell metabolic activities. A simpler solution is the use of concentrated medium. The concentrated media was prepared using NaCl/KCl-free powders. Osmolarites were then adjusted to the same levels in these concentrates.

Figure 6 presents the performance of a fluidized-bed reactor with concentrated media. In the first 10 days regular (1X) medium was used for perfusion. Medium feed rate was increased to maintain a constant glucose level in the reactor. Cell densities reached to the levels of 2 10^8 and media flow rate had to be increased to the levels of 50 volumes a day. Starting on Day 14, the medium was gradually switched to 2X and later to 3X concentrates. The feed rate was decreased in a stepwise fashion to 5 volumes per day by the end of the experiment.

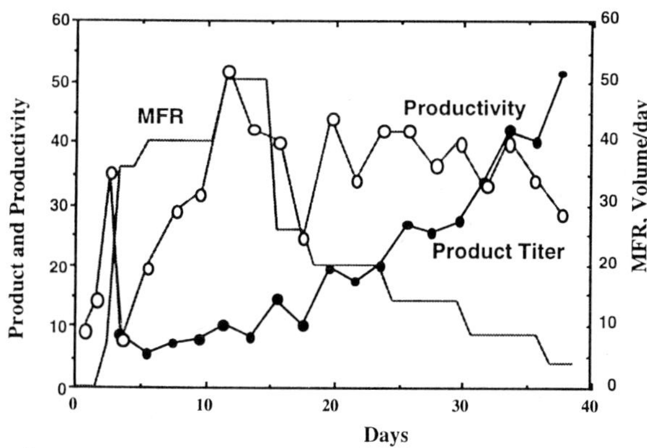

Figure 6. Use of concentrated medium to increase product titer and minimize medium usage. The product concentrations and productivities are scaled using the batch reactor values.

While the medium feed rate was decreased by a factor of 10, no significant change was observed in the reactor productivity. Apparently both cell number and cell specific productivity did not change during these alterations. As a result a 10 fold increase was observed in product concentration simply by increasing the nutrient value media and thus decreasing the feed rate. This was a significant gain for the process. The media utilization was increased without compromising reactor productivity.

<u>Scale-up of Fluidized-bed Reactors.</u> Cells were cultured in 3 different scale reactors. These reactors could be scaled up based on the number of microspheres, or volume in the fluidized-bed. System-one, the smallest size reactor, contains 25 ml of microspheres.

About 1000 times more microspheres are used in S-2000 reactor. The perfusion rate in these reactors can be up to 10-50 volumes per day and the S-2000 can produce kg quantities of product.

The reactor productivities are normalized using the number of microspheres and compared in Figure 7. It was observed that the reactors are scalable in size even with a scale-up ratio of 1000.

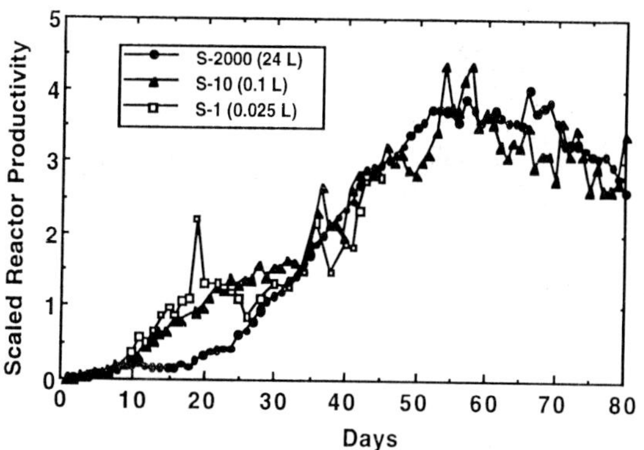

Figure 7. Scale-up performance of fluidized-bed reactor. The scaling is performed using the number of beads in different scales.

<u>SUSPENSION CULTURE WITH CELL SEPARATION</u>

<u>Reactor system</u>. Cell retention in conventional stirred tank reactors can be achieved by a variety of cell separating devices. These systems do not require a solid support for the cell separation and provide a homogeneous envi-

ronment for the cells (Hulscher et al., 1992).

Figure 1B depicts the reactor system used in this study. High density is achieved by separating the cells from the harvest stream in a external cell retention system and recycling them to the fermenter. Gas exchange was achieved by membrane aeration using silicone tubing. The reactor was used for the cultivation of a myeloma cell line producing monoclonal antibody. The cells are perfused by the exchange of medium between feed and harvest. The harvest flow rate is adjusted based on cell concentrations measured by direct sampling. A level controller controls the feed pump for constant volume in the reactor.

<u>Reactor operation</u>. Serum-free medium (DMEM:F12 based) was used for the fermentation. For most of the run temperature, pH, and DO were maintained at their set points of T=37°C, pH=7.3, and DO=50% air saturation, respectively.

Figure 8 presents the operation of cell separation system. The reactor was inoculated at a cell density of 10^5 cells/ml. Exponential growth was observed in the first 12 days and a cell density of 15 million cells/ml was obtained. This was a 160 fold increase or 7.2 doublings. Some mechanical problems caused a drop in cell density. The mixing in the reactor was not adequate; the base addition caused local precipitation or clumping of the cells. The base addition was stopped and the cells continued to grow. Oxygen supply reached maximum (pure oxygen in gas phase) on Day 20. After that the DO control was turned-off and the culture proceeded with oxygen limitations. Cell densities were maintained around 1.6-2 10^7 viable cells/ml. The viabilities remained around 90% .

<u>Oxygen consumption rate</u>. Oxygen mass transfer coefficients for the silicone tubing aeration were measured by a dynamic method as K_La=4.5 1/hr. Using the mass transfer coefficient and the cell numbers obtained, the cell specific oxygen consumption of these cells are determined to be 0.21-0.25 μmol/million cells/hr. These values agreed with the measurements using dynamic method during the growth phase at low cell densities.

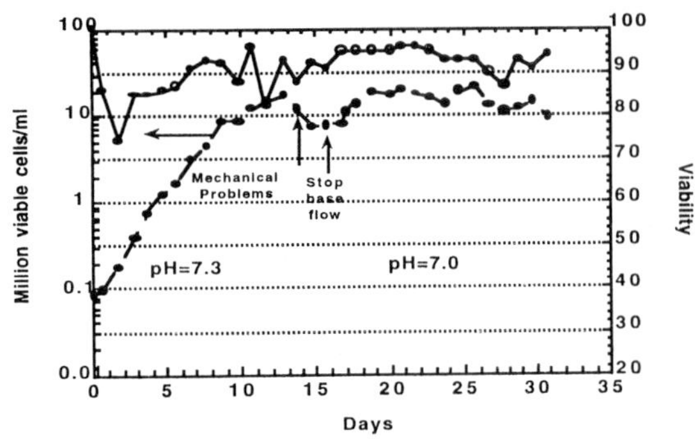

Figure 8. Performance of suspension reactor with cell separation.

<u>Reactor performance</u>. Volumetric and cell specific productivities are presented in Figure 9. These values are normalized using the values of a batch culture. About 20 fold improvement was observed for reactor productivity compared to batch culture. This was mainly because of higher cell densities. Unlike fluidized-bed reactors, high cell densities did not resulted in any increase in specific productivity.

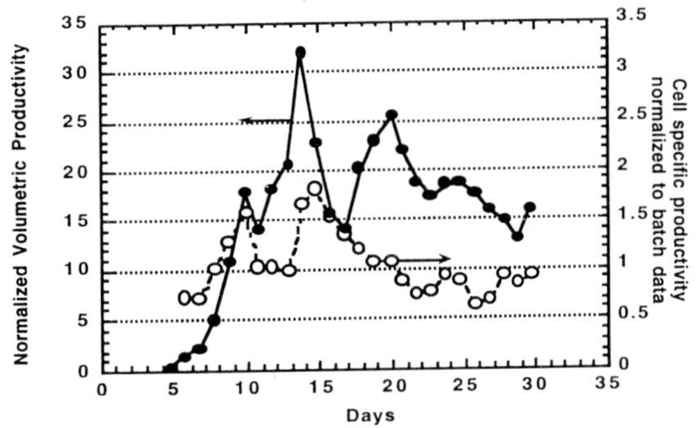

Figure 9. Volumetric and Cell specific productivities in suspension reactor with cell separation.

<u>CONCLUSIONS</u>

We have presented an overview of high density cell culture and shared our experience with two commercial reactor systems.

In the fluidized-bed reactor cell attachment capacity of the beads limited the cell density. Although local cell densities could reach more than 10^8 cells/ml, the effective cell density in the reactor was 4 10^7 cells/ml. Mathematical analysis predicted that local concentration and pH gradients could create a micro environment with enriched protein levels increasing the cell specific productivity. The volumetric productivities in the reactor was 200-300 higher because of higher cell concentration and higher cell specific productivity. Concentrated media was successfully used to increase product concentrations without sacrificing the reactor productivity.

Cell densities in a suspension reactor with cell separation were also high (2 10^7 cells/ml). Higher cell densities could be obtained by increasing the mass transfer rates, i.e., more tubing per volume, or sparging. About 20 fold increase in volumetric productivity was obtained compared to batch culture. Increasing cell concentration, from 10^6 (batch culture) to 2 10^7 cells/ml (perfusion), did not change the cell specific productivity.

ACKNOWLEDGMENT

The data on fluidized bed reactors were obtained when the author was affiliated with Verax Corporation. Experimental help by John Thrift and Jonathan Blackie on cell separation system is greatly appreciated.

LITERATURE CITED

1. Griffiths, J.B., Looby, D., and Racher, A.J., *Cytotechnology*, **9**, 3-9, 1992

2. Griffiths, J. B., *J. Biotechnology*, **22**, 21-30, 1992

3. Hulscher, M., Scheiber, U., and Onken, U., *Biotech. Bioeng.*, **39**, 442-446, 1992

4. Karkare, S.B., Phillips, P.G., Burke, D. H., Dean, R.C. and Runstadler, P.W.Jr., in: *Large-Scale Mammalian Cell Culture*, J. Feder and W.R. Tolbert, eds. Academic Press, New York., 127-149,1985.

5. Lee, G.M., and Palsson, B.O., **36**, , 1049-1055, 1990

6. Ozturk, S. S. and Palsson, B. O., Biotech. Progress, **7**, 481-494, 1991a.

7. Ozturk, S.S. and Palsson, B.O., Biotech. Bioeng., **37**, 989-993, 1991b.

8. Ozturk, S.S., Ray, N.G., and Runstadler, P.W.Jr., Symposium on Transport Processes in Bioreactors: Fundamentals and Applications, Annual Meeting of AIChE, Los Angeles, CA, 1991

9. Runstadler, P.W., Ozturk, S.S., and Ray, N.G., *Animal Cell Technology: Basic Applied Aspects,* Ed. H. Murakami, Kluwer, Nederlands, 1992

10. Vournakis, J.N., and Runstadler, *Bio/Technology*, **7**, 143-145, 1989

Recent Studies on Stirred Bioreactors at the SERC Centre for Biochemical Engineering at Birmingham

A.W. Nienow

SERC Centre for Biochemical Engineering, The University of Birmingham, Edgbaston, Birmingham B15 2TT, U.K.

The SERC Centre for Biochemical Engineering was established in 1987 by the Biotechnology Directorate of the Science and Engineering Research Council. A Rolling Grant of ~ £ 2.0 million was made for a Research Programme and another £1 million was given for fully refurbishing and extending the Biochemical Engineering Building to the highest modern standards. The building had been a state-of the-art development in 1956 and remains so today with the SERC Rolling Grant being confirmed every two years since 1987. The latest tranche of money awarded was £2.6 million for four years from October 1993. One of the main research topics of the Centre relates to stirred bioreactors and this paper considers recent developments. Much of the earlier work up to 1990 was covered in previous journal (1,2) and conference (3) reviews and, after a very brief introduction to set the scene, this article mainly presents work since then.

Until recently, most industrial-scale bioreactors were agitated by a set of Rushton turbines of about 1/3 of the fermenter diameter. However, fermenter performance is linked to agitation by interaction of the two following sets of parameters:

1) System-specific biological parameters (Table 1) which relate to the sensitivity of the bioreaction to factors such as high and low levels of dissolved oxygen and nutrients (especially in fed-batch operations), and pH. Mechanical stress may also cause damage from fluid dynamic or bubble burst effects.

2) Physical parameters independent of the biological system (Table 2). These parameters are not bioreaction specific as in Table 1 but are mostly specific to the scale of operation and the impeller/bioreactor geometry and are dependent on fluid dynamics and transport phenomena.

Without a good understanding of the parameters in Table 2, at different scales, with broths of different rheological properties, and with different agitators, there is little chance of linking mixing parameters to biological performance. This idea is the background philosophy to the work in the SERC Centre and both types of parameter are discussed here based on recent work in the Centre. Physical parameters are discussed first and, in addition to Rushton turbines, recent work on the Scaba 6SRGT, the Lightnin' A315, the Prochem Maxflo T and the Ekato Intermig agitators (Fig.1a to d respectively) are also briefly considered. All these agitators have lower power numbers than the Rushton turbine and therefore constant speed-power-torque retrofitting allows large D/T ratios to be used which makes homogenisation easier (1,2) (Table 3).

Table 1. Biological performance parameters

---

a)   *System specific parameters*

- Growth and product secretion
- Damage to microorganisms
- Morphology changes
- O_2 and CO_2 transfer rates
- Nutrient demands

b)   *Parameters which may be dependent on stirring conditions*

- Quality of homogenisation
  Oxygen toxicity
  Oxygen starvation
  CO_2 concentrations
  pH excursions
  Nutrient excursions

- 'Shear' damage stress from fluid dynamic and bubble burst phenomena

---

141

E. Galindo and O.T. Ramírez (eds.), Advances in Bioprocess Engineering. 141-147.
© 1994 Kluwer Academic Publishers. Printed in the Netherlands.

Table 2 Stirred bioreactor parameters that are not biological system-specific

- Bulk fluid mixing
- Air dispersion capability
- Heat transfer
- Power drawn
- Oxygen mass transfer coefficient
- Local fluid dynamics:
    - a) Close to the agitator
    - b) Energy Dissipation Rates
    - c) Bubble Burst

Table 3 Modern Agitators Studied

	Power No. $(Re>10^4)$	Fl	D/T
Scaba 6SRGT	~ 1.4	-	0.44
Prochem Maxflo T	~ 1.0	-	0.47
Lightnin A315	~ 0.85	~ 0.73	0.50
Ekato Intermig (a pair)	~ 0.60	~ 0.64	0.60

* Equal P-M-N retrofitting of a 1/3 T Rushton (Po~6)

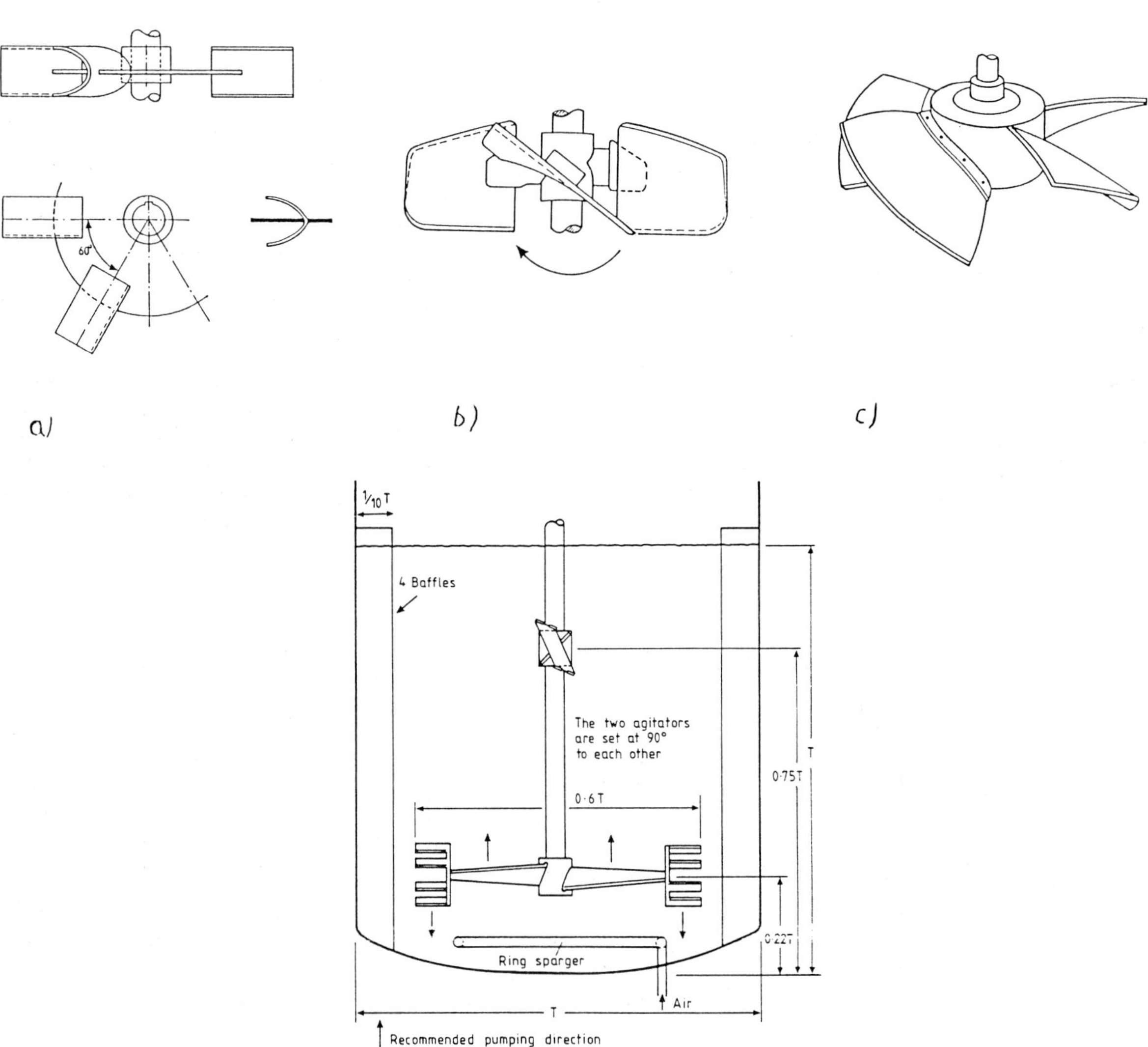

Fig.1   Recently-introduced low power number, air dispersing agitators.   a) Scaba 6SRGT   b) Lightnin' A315 c) Prochem Maxflo T d) Ekato Intermig (a pair)

Transport Phenomena

MIXING AND HOMOGENISATION

Theory ($\underline{1}$) suggests larger D/T ratios should give shorter mixing times in turbulent flow (Re > 10^4). Since that parameter is the one which changes most on scale-up, i.e., it increases significantly, and since it clearly has important consequences for biological performance, the large D/T retrofitting noted above should help reduce this problem. Thus, a large Prochem Maxflo T hydrofoil produced shorter mixing times than the standard radial flow Rushton turbine at the same energy dissipation rate ($\underline{4}$); and the radial flow Scaba was found to be superior to the Rushton, especially when aerated ($\underline{5}$). However, when multiple impellers are used, as are typically found in industrial fermenters, radial flow agitators produce very strong stratification or compartmentalisation ($\underline{6}$). This means mixing is very slow from the top to the bottom of a large fermenter. Though axial flow agitators are an improvement, they do not entirely eliminate this tendency so that some compartmentalisation was found even with A315 agitators ($\underline{7}$).

With a yield stress fluid, as found in high concentration polysaccharide and mycelial fermentation, agitators need to be placed closer together to ensure good motion of the fermentation broth. In such fluids, caverns form ($\underline{24}$), i.e., cylindrical regions with significant motion close to the agitator with stagnant or near-stagnant broth outside them. These caverns are roughly the same shape with an aspect ratio of ~ 0.5, whether radial flow agitators such as the Rushton turbine (24) and the 6SRGT ($\underline{8}$) or axial flow ones such as the A315 ($\underline{9}$) are used.

AERATED POWER DRAW AND AIR DISPERSION CAPACITY

Radial Flow Agitators. Rushton turbines lose power on aeration quite rapidly and in a complex way which is difficult to predict ($\underline{1}$) though typically $P_g/P{\sim}0.5$ on the large scale ($\underline{15}$). They also flood relatively easily. In fully turbulent conditions, this fall can be reduced in a variety of ways. The use of blades shaped like the Scaba ($\underline{5}$) (Fig.1) essentially eliminates the fall in power ($P_g/P{\sim}1$) and the use of Rushton's with more than 6 flat blades, e.g., 12 or 18 ($\underline{10}$) greatly reduces it ($P_g/P{\sim}0.75$). These modifications also allow more gas to be handled before flooding occurs ($\underline{5,10}$).

However, when the fluid becomes more viscous, gas filled cavities form on Scaba blades and P_g/P is about 0.8 ($\underline{5,8}$). With the 12 or 18 blade Rushton, they behave identically to the 6 blade one ($\underline{11}$) at Re $\sim$ 500 in that the cavity size becomes independent of aeration rate and P_g/P falls to about 0.4. With such high viscosity fluids, the use of the Stirring Intensity Measuring Device, developed in Latvia, has also shown, contrary to what has normally been believed to be the case, that more air can be handled before flooding occurs compared to the low viscosity case ($\underline{12}$).

Axial Flow Agitators and Intermigs. The axial hydrofoils when pumping downwards have a wide range of conditions where the fall in power is reduced ($\underline{13,14}$); and so too does the Intermig, though it becomes a radial flow impeller when aerated ($\underline{2}$). The axial hydrofoil also produce torque fluctuations ($\underline{13,14}$) as the flow pattern changes from axial to radial (Fig.2); and both they

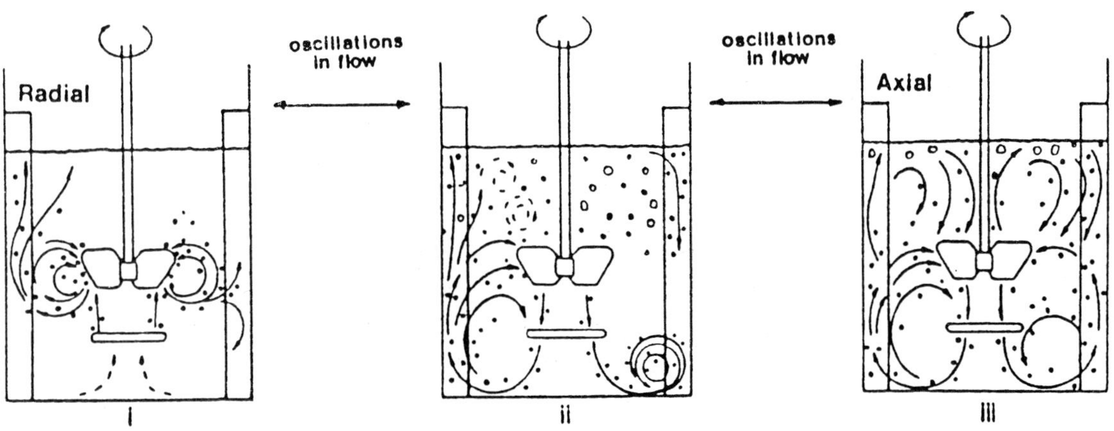

Fig.2   Typical flow pattern changes with axial flow hydrofoils (i to iii) with increasing N at constant Q_G or with decreasing Q_G at constant N (or vice-versa. iii to i)

(15) and Intermigs (16) may cause serious equipment vibrations. Changing to the upward pumping mode with pitched blade turbines (17), greatly reduces torque vibrations and the loss of power on aeration and enables more air to be handled before flooding occurs. A similar effect has been found with Maxflo T agitators (18). There appears to be substantial scope for upward pumping hydrofoils.

At high viscosity, the power draw, as with Rushton turbines, becomes independent of aeration rate for Intermigs (19) and A315's (9). Under these conditions, torque vibrations (9) are significantly enhanced with the A315 and vessel vibrations for both impellers (15,16). Torque vibrations are much less when the A315 is upward-pumping (9).

MASS TRANSFER AND HOLD-UP

Empirical correlations between mass transfer coefficients, $k_L a$, specific energy dissipation rate and air flow rate and between hold-up and these two parameters are often very similar. Thus, hold-up has been suggested as a good indicator of $k_L a$ potential (40). Our prior work (20) showed that Maxflo T impellers gave a much higher hold-up under conditions of heavily repressed coalescence than upward or downward pitched blade turbines or Rushton turbines at equivalent power and aeration conditions. However, since then it has been found that the high hold-up does not lead to an enhanced $k_L a$ (14). Using the steady state catalytic oxidation of hydrogen peroxide (14) as the measuring technique, the Rushton turbine and Maxflo T were equal in $k_L a$ performance. This result, i.e., enhanced hold-up without enhanced $k_L a$, obtained with polypropylene glycol (a typical antifoam in real fermentations), is of particular interest because high hold-up without enhanced $k_L a$ implies loss of productive capacity during a fermentation.

The precise relationship between $k_L a$ and aeration rate is still being discussed. A recent paper (21) has combined extensive data from the SERC Centre and the Fluid Mixing Processes Consortium of BHR Group. The data come from equipment up to 3m diameter and the paper compared correlations based on superficial gas velocity, $v_s (=4Q_G/\Pi T^2)$ with those based on vvm ($\alpha\ 4Q_G/\Pi T^3$). A very clear advantage for superficial gas velocity was found. This clarification is important because on scale-up at equal P/V and vvm, a correlation with vvm implies equal $k_L a$, whilst a correlation with v_s implies a significantly enhanced $k_L a$.

With shear-thinning fluids, the apparent viscosity is dependent on the apparent shear rate; which, in turn, is proportional to the agitator speed, N. Low power number agitators may be operated at higher speeds to give the same energy dissipation rate. Since $k_L a \ \dot\alpha\ \mu_a^{-1/2}$, it might be expected that the higher speed should give higher $k_L a$'s. When this concept was tested with Intermig agitators, the $k_L a$ values were indeed significantly higher (22).

HEAT TRANSFER

There is a paucity of heat transfer data obtained in real fermentation media. Heat flux probes placed on the inner surface of a jacketed 800 litre fermenter have been used to measure local inside film heat transfer coefficients (23). This technique is a very powerful one. For example, it has clearly demonstrated that in mycelial fermentations, the heat transfer coefficient is significantly greater than in a fluid of the same rheological properties without the mycelia.

Cavern sizes are significantly bigger with bigger D/T ratios at the same power input (24). Therefore larger impellers should produce a higher velocity at the wall than small ones. Thus, higher overall heat transfer coefficients have been measured in Xanthan broths during sterilisation and subsequent cooling with large D/T ratios compared to smaller ones (25).

SOLID-LIQUID SYSTEMS

Microcarrier are often used in animal cell culture. It has been reported that animal cells are easily damaged on solid supports (26) unless very low agitation intensity is used. The use of A310 and HE3 agitators, made by Lightnin and Chemineer respectively, together with cone-and-fillet contoured tank bottoms (27) leads to very low energy dissipation rates being required for microcarrier suspension (41), well below the level at which damage has been said to occur (26).

FLUID DYNAMIC STRESSES

Theoretical studies on the stresses associated with bursting bubbles (28) has shown that such stresses are much greater than those found due to turbulence at typical bioreactor agitation conditions (29). Measurement of the strength of animal cells by micromanipulation, strongly suggests that whilst the bubble bursting stresses would be expected to disrupt freely-suspended cells, those due to turbulence would not (29).

Biological Performance

XANTHAN GUM FERMENTATIONS (25)

Xanthan gum solutions and Xanthan gum broths both exhibit yield stresses at gum concentrations typical of those found at the end of a batch fermentation. As mentioned above, cavern theory (24) suggests that large D/T agitators give larger caverns at the same energy dissipation rate. Biologically, Xanthan gum fermentations should be able to give a higher yield of gum if they are operated as a two stage process. Thus the bulk of the glucose is added at the start (the tropo-phase) and, after it has fallen to a low value, the second shot of glucose should be added (the idiophase). In addition, the level of dO_2 should be maintained above a certain critical value.

The above concepts led to an operating strategy which involved progressively increasing agitator speed to keep dO_2 above 30%. This increase in speed was necessary to enhance k_La to meet the increasing oxygen demand due to the enhanced biomass level; and to overcome the fall in k_La due to the increasing viscosity as gum was produced. Power was measured throughout by strain gauge-telemetry and two D/T ratios were used, 0.42 and 0.63. 25 g/litre glucose were used in 2 ways; either all in at the beginning with the smaller D/T ratio or with 20 g/litre initially and the addition of a further 5 g/litre after the residual glucose had fallen to 5 g/litre with both ratios.

The improvement from the 0.43T with batch addition to the 0.63T, two-step addition was an increase in yield from 18 g/litre to 26 g/litre, a reduction in batch time from ~100 hrs to ~72 hrs and a reduction in total energy requirement of ~40%.

ANIMAL CELL CULTURE

By controlling dissolved oxygen by gas blending, it has been found that TBC/3 mouse hybridoma cells grow equally well with dO_2 from 5 to 100% with respect to air (30). Without sparging, agitation rates giving energy dissipation rates some two orders of magnitude greater than that required to satisfy oxygen demands did not damage cells (31). However, when bubbles were entrapped from the air interface (32) or introduced by sparging (30,31), cells died rapidly even at quite low aeration rates. This damage could be attributed to bursting bubbles leaving the bioreactor (32). The use of Pluronic as a protective agent essentially eliminated damage, mainly because they prevent cell attachment to the bubbles (33). Very similar results have been found with other cell lines.

With insect cells, the sensitivity to fluid dynamic and bubble burst stresses is very similar to hybridoma cells (34). On the other hand, the rate of insect cell infection is independent of the level of agitation which fits in with solid-liquid mass transfer concepts for dispersed particles of the size of such cells (35).

SEED PRIMING (42)

When vegetable or flower seeds are suspended in an osmotica, i.e., a fluid of controlled osmotic pressure such as a polyethylene glycol solution, then under the right conditions, they take up moisture sufficient to initiate the germination process but insufficient to enable germination to occur. Such a process is known as seed priming and initially, it was carried out on filter paper soaked with osmotica. Our work has shown that if the process is to be carried out in bulk, then it must be treated like other bioreactors, i.e.,

1. oxygen is required to meet the oxygen demand of the seeds and must be controlled
2. temperature control is important
3. the biomass is prone to damage

The process differs too, i.e.,

1. the biomass is large (the size of the seeds!)
2. the nutrients are in the seeds
3. the osmotica has a low O_2 solubility
4. the energy required to suspend the biomass is of the same order as that required for oxygenisation because a) the seeds have a very low oxygen demand (36) and b) a significant level of agitation is required to suspend the seeds

By using both air and enriched air and two, large pitched blade turbine agitators (D/T=0.5) which suspend seeds with a very low energy requirement at the low sparging rates needed to satisfy the oxygen demand (37), a wide range of vegetable seeds have been successfully primed. Such seeds, on sowing, germinate very much more rapidly and synchronously than untreated seeds and it is this behaviour which makes the treated seeds attractive to the seed trade and enables them to be sold at a premium price. Unfortunately, after about six months, the seedling that are produced are deformed, whether a laboratory or bulk technique is used. Therefore, a seasonal priming strategy should be adopted.

SMALL SCALE TESTS TO SIMULATE BIOLOGICAL PERFORMANCE ON THE LARGE SCALE

Mixing theory shows that the time required to achieve homogenisation (or, alternatively, the variation of concentration) is greater on the large scale than the small at realistic power inputs (i.e., broadly similar specific energy dissipation rates) (2). Model studies simulating this problem have been conducted (38) using two well-mixed fermenters side-by-side. Two types of inhomogeneity have been studied. Firstly, the dO_2 level has been held high in one and low in the other; and secondly, pH has been similarly controlled (39). The culture used, sensitive to both dO_2 and pH, was *Bacillus subtilis*. Though the technique has been used previously for dO_2 studies, pH has not been considered. In addition, high circulation rates between the two fermenters were used. As a result, the frequency of fluctuations more closely mimicked those found in real large scale stirred bioreactors than in previous work.

The technique is a powerful one for indicating the likelihood and form of changes in metabolism on scale-up. However, it is still unable to give much quantitative information on the effect of scale-up.

ACKNOWLEDGEMENTS

I would like to thank my many co-workers whose work, which is referenced, has enabled me to write this review; and especially Bob Badham and Hazel Jennings, senior technicians at Birmingham. Bob manages the "troops" who work on physical aspects and Hazel those concerned with biological ones. I would also like to thank the SERC for their support for the Centre for Biochemical Engineering at Birmingham and for other grants. Finally, I would like to thank the organisers for inviting me and for their generous support.

NOMENCLATURE

D	Impeller diameter (m)
Fl	Impeller flow number (-)
$k_L a$	Mass trnsfer coefficient (s^{-1})
N	Impeller speed (s^{-1})
P	Power draw (w)
Po	Power number (-)
Q_G	Air flow rate (m^{3}s^{-1})
T	Vessel diameter (m)
V	Volume of broth (m^3)
v_S	Superficial air velocity (m s^{-1})
vvm	(Volume of air/min) / Vol. broth (min^{-1})

Subscript

g	Under aerated conditions

LITERATURE CITED

1. Nienow, A.W., *Chem.Eng.Prog.*,**86**, 61 (1990).

2. Nienow, A.W., *Trends in Biotech.*, **8**, 224 (1990).

3. Nienow, A.W., In: C. Christiansen, L. Munck and J. Villadsen (Eds), *Proc. 5th Euro. Cong. on Biotechnology*, Vol.2, Munksgaard, Copenhagen, p 791 (1990).

4. Hass, V.C. and Nienow, A.W., *Chem.Ing. Technik*, **61**, 152 (1989).

5. Saito, F., Nienow, A.W., Chatwin, S. and Moore, I.P.T., *J.Chem.Eng. Japan*, **25**, 281 (1992).

6. Cronin, D.G., Nienow, A.W. Moody, G.W., *Trans.I.Chem.E.*, *Part C - Food and Bioproducts Processing*, accepted for publication (1994).

7. Manikowski, M., Bodemeier, S., Lübbert, A., Bujalski, W. and Nienow, A.W., *Can.J.Chem.Eng.*, accepted for publication (1994).

8. Galindo, E. and Nienow, A.W., *Chem.Eng.Technol.*, **16**, 102 (1993).

9. Galindo, E. and Nienow, A.W., *Biotech.Prog.*, **8**, 233 (1992).

10. Middleton, J.C., In: N. Harnby, M.F. Edwards and A.W. Nienow (Eds.), "Mixing in the Process Industries", Ch.15, Butterworths Heinemann, London, p 322 (1992).

11. Nienow, A.W., Kendall, A., Moore, I.P.T., Ozcan-Taskin, G. and Badham, R.S., *Chem.Eng.Sci.*, submitted (1994).

12. Vanags, J., Viesturs, U., Bujalski, W. and Nienow, A.W., CHISA 93, Prague, Czech Republic., Poster H3.111 (1993).

13. Martin, T., McFarlane, C.M. and Nienow, A.W., *The 1993 I.Chem.E. Research Event*, I.Chem.E., Rugby, U.K., p 648 (1993).

14. Martin, T., McFarlane, C.M. and Nienow, A.W., *Proc. 8th European Mixing Conf.*, accepted for publication (1994).

15. Nienow, A.W., Weetman, R.J., Hunt, G. and Buckland, B.C., In: A.W. Nienow, (Ed.), *Proc. 3rd Int.Conf. on Bioreactor and Bioprocess Fluid Dynamics*, BHR Group/MEP, London, p 505 (1993).

16. Kipke, K.D., *Proc. 4th European Mixing Conference*, BHRA, Cranfield, U.K., pp 355 (1982).

17. Bujalski, W., Nienow, A.W. and Liu Huoxing, *Chem.Eng.Sci.*, **45**, 415 (1990).

18. Hass, V.C., M.Sc. Thesis, University of Birmingham, U.K. (1988).

19. Ozcan, G.N. and Nienow, A.W., In: M. Bruxeelmane and G. Froment (Eds.), *Proc. 7th European Mixing Conf.*, (KVIV, Belgium, p 359 (1991).

20. Machon, V., McFarlane, C.M. and Nienow, A.W., In: R.King (Ed.), "Fluid Mechanics of Mixing" Kluwer, Dordrecht, Netherlands, p 91 (1992).

21. Whitton, M.J. and Nienow, A.W., In: A.W. Nienow, (Ed.), *Proc. 3rd Int.Conf. on Bioreactor and Bioprocess Fluid Dynamics*, Cambridge, BHR Group/MEP, London, p 135-145 (1993).

22. Dawson, M.K. Nienow, A.W. and Moody, G.W., *The 1993 I.Chem.E. Research Event*, I.Chem.E., Rugby, U.K. p 645 (1993).

23. Mohan, P., Emery, A.N., Bujalski, W. and Nienow, A.W., In: A.W. Nienow, (Ed.), *Proc. 3rd Int.Conf. on Bioreactor and Bioprocess Fluid Dynamics*, Cambridge, BHR Group/MEP, London, pp 33 (1993).

24. Nienow, A.W. and Elson, T.P., *Chem.Eng.Res.Des.* **66**, 5 (1988).

25. Zhao Zueming, Nienow, A.W., Kent, C.A., Chatwin, S. and Galindo, E., In: R. King, (Ed.), "Fluid Mechanics of Mixing" Kluwer, Dordrecht, Netherlands, p 99 (1992).

26. Croughan, M.S., Hamel, J.F. and Wang, D.I.C., *Biotechnol.Bioeng.*, **29**, 130 (1987).

27. Ibrahim, S., Nienow, A.W. and Chatwin, S., *The 1992 I.Chem.E. Res. Event*, I.Chem.E., Rugby, U.K., p 122 (1992).

28. Boulton-Stone, J.M. and Blake, J.R., *Proc. 3rd Int.Conf. on Bioreactor and Bioprocess Fluid Dynamics*, A.W. Nienow (Ed.), BHR Group/MEP, London, p 163 (1993).

29. Zhang, Z. and Thomas, C.R., *Proc. 3rd Int.Conf. on Bioreactor and Bioprocess Fluid Dynamics*, A.W. Nienow (Ed.), BHR Group/MEP, London, p 475 (1993).

30. Oh, S.K.W., Nienow, A.W., Emery, A.N. and Al-Rubeai, M., *J. Biotechnology*, **22**, 245 (1992).

31. Oh, S.K.W., Nienow, A.W., Al-Rubeai, M. and Emery, A.N., *J. Biotechnology*, **12**, 45 (1989).

32. Kioukia, N., Nienow, A.W., Emery, A.N. and Al-Rubeai, M., *Food Bioprod.Proc.*, (*Trans.I.Chem.E., Part C*), **70**, 143 (1992).

33. Bavarian, F., Fan, L.S. and Chalmers, J.J., *Biotechnol. Prog.*, **7**, 140 (1991).

34. Kioukia, N., Nienow, A.W., Emery, A.N. and Al-Rubeai, M., *The 1993 I.Chem.E. Research Event*, I.Chem.E., Rugby, U.K., p 126 (1993).

35. Kioukia, N., Ph.D. Thesis, The University of Birmingham, 1994.

36. Nienow, A.W., Kirk Othmer Encyclopedia of Chemical Technology, 4th Edition, Wiley, New York, Vol.1, p 645 (1991).

37. Chapman, C.M., Nienow, A.W., Cooke, M. and Middleton, J.C., *Chem.Eng.Res.Des.* (*Trans.I. Chem.E.*), **61**, 167 (1983).

38. Amanullah, A., Baba, A., McFarlane, C.M., Nienow, A.W. and Emery, A.N., In: A.W. Nienow, (Ed.), *Proc. 3rd Int.Conf. on Bioreactor and Bioprocess Fluid Dynamics*, Cambridge, BHR Group/MEP, London, p 381 (1993).

39. Amanullah, A., Ph.D. Thesis, The University of Birmingham, (1994).

40. Andrew, S.P.S., *Trans.I.Chem.E.*, **60**, 3 (1982).

41. Ibrahim, S., Ph.D. Thesis, The University of Birmingham, (1992).

42. Nienow, A.W., Bujalski, W., Maude, R.B. and Gray, D., *I.Chem.E. Symp. Ser.*, I.Chem.E., Rugby, U.K., accepted for publication, (1994).

Role of Turbulence in Fermentations

P.K. Namdev, E.H. Dunlop, K.Wenger, and P. Villenueve

Colorado Bioprocessing Center, Department of Chemical Engineering,
Colorado State University, Fort Collins, CO 80523-0002, U.S.A.

This paper raises the need for a systematic evaluation of non-linear interaction between various turbulent-driven physical processes in fermentors. Various turbulent/mixing studies published in the literature are formalized based on a multi-dimensional framework. This formalization leads us to the limitations of many existing approaches and scale-up rules to mixing problems. The results on the effects of micromixing and fluid shear on yeast cells will be presented as an example to show the possibility of a non-linear interaction. A micro-environmental approach is proposed whereby the effects of multiple turbulent-driven physical processes in the micro-environments of cells can be evaluated.

Fermentation systems are required to be scaled-up from laboratory-scale fermentors (few liters) to production-scale fermentors (5,000 L to 300,000 L) with reproducibility of the performance (1). To fulfill this requirement, the local conditions around each cell in the production-scale should be similar to that found optimum in the laboratory scale fermentor. These conditions are temperature, pH, dissolved oxygen, nutrient concentration, osmotic pressure and shearing forces. This requirement is rarely fulfilled and is the main cause of irreproducibility during the traditional step-wise scale-up. Other causes, which will not be discussed here, include strain degeneration during inoculum train, media degradation during sterilization, and constraints placed by the down-stream processing. The irreproducibility of local conditions on scale-up is due to the pronounced inhomogeneous distribution of mixing power or

turbulence in large fermentors and is termed here as "mixing effect/turbulence problem". The variation in local conditions on scale-up affects various types of fermentation processes to different degrees depending on the type of species, the medium composition and the *nature of turbulence*. The *biological effects* on cells typically measured are yield, productivity and lysis. Thus the simplest view of mixing problem can be represented by Figure 1. Most processes use a stirred tank bioreactor which is taken as the model system in this paper. The approach and results presented here, however, should be applicable to any other reactor format which employs turbulence-driven physical processes.

Traditional scale-up rules have been available for at least three decades which are based on maintaining a constant value of one of the following variables: power per unit volume, tip speed, mixing

149

E. Galindo and O.T. Ramírez (eds.), Advances in Bioprocess Engineering. 149-156.
© 1994 Kluwer Academic Publishers. Printed in the Netherlands.

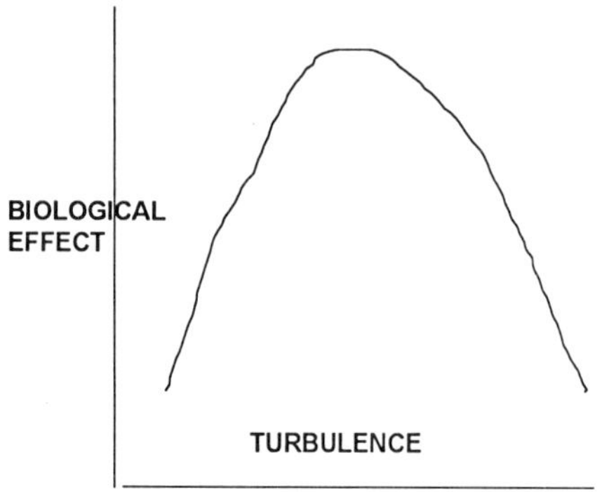

Figure 1. TURBULENCE IN FERMENTATION

time, and mass transfer rate(2).
Scale-up of fermentation, however,
still could be frequently
disappointing(3). For example,
scale-up based on a constant mass
transfer coefficient for vitamin B_{12}
fermentation shows an optimum value
of k_1a for operation. Results from
the scale-up of penicillin
fermentations based on constant
(P/V) suggest that product yields at
different scales will be
significantly different, both higher
and lower, depending on the value of
(P/V). The main cause of limitation
of these scale-up rules is that they
consider bulk average values for
physical processes and cellular
properties(4).

This limitation of the scale-up
rules has been the motivation for a
number of mixing/turbulence studies
published over the last two decades
to evaluate the effects of local
conditions on fermentations
(5,6,7,8,9,10,11). The premise for
these studies and traditional
scale-up rules has been that only
one of the local conditions is most
critical to one property of cells.
This approach, therefore, is
limited. The main reason is that the
physical processes driven by
turbulence can not be segmented into
"independent" physical processes
such as micro- and macro-mixing,
gas-liquid transfer, fluid forces,
and heat transfer. In addition, the
physiological effects on cells can
not be simply measured as an overall

yield. This overall effect could be
the cumulative effect of multiple
biological responses, each affected
by various physical processes to
different degrees. *Traditional
scale-up rules and mixing studies
fail to address such non-linear
interaction which may lead to
either amplification or cancellation
of effects by various physical
processes.*

A MULTIDIMENSIONAL PROBLEM

The turbulence-driven physical
processes are caused by five
intrinsic operating variables. These
are agitation (or power input),
aeration (or flow rate), scale of
equipment, internals and
cells/medium used. The physical
processes will map the operating
region based on their impact on the
physiology of cells. Figure 2 shows
how each of these individual
physical processes can restrict the
operating range on scale-up. For
example as the scale of operation
increases, the upper limits on the
agitation power and aeration will
decrease due to constraints placed
by the shear forces and the cost.
The lower limits on agitation and
aeration required for minimum

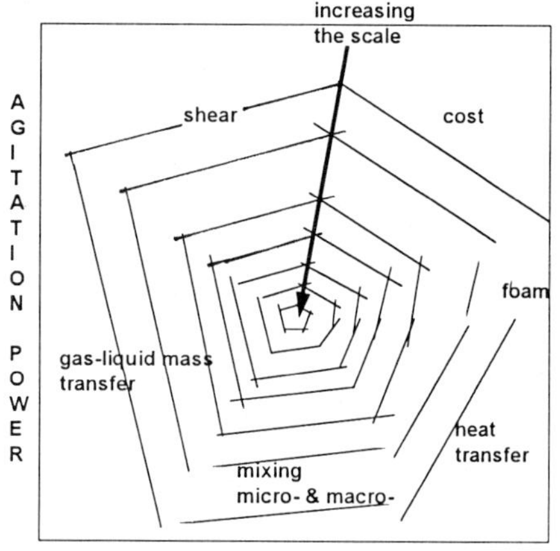

FIGURE 2. MULTI-DIMENSIONAL ROLE OF
TURBULENCE IN FERMENTATION

suspension and gas-liquid mass transfer will increase as the scale of operation increases.

The contours of the operating map in Figure 2 makes the turbulence problem to be a three dimensional with the scale of equipment as the third axis. In addition to the scale of operation, other intrinsic variables such as the internals and type of cells/medium used will make the turbulence problem to be *multi-dimensional*. For example, internals such as impeller, sparger and cooling coils would affect the physical processes. Also, if cells/medium show non-Newtonian rheology then depending on the local shear rate, the rheology of the suspension will vary. Such a multi-dimensional framework for the turbulence problem, of which Figure 2 is a simplified view will encompass all the mixing studies performed in the literature. These mixing studies can be broadly divided into three types (Figure 3):

(1) *Biological characterization of the fermentor*: In these studies, derived variables or biological effects of primary interest including growth, product yields and lysis are measured as a function of intrinsic variables. Such studies, where one or multiple intrinsic variables are varied simultaneously, are numerous in the literature. These studies could satisfy scale-up requirement for low volume fermentations but they would never elucidate the mechanisms at the levels of physical processes and its physiological consequences.

(2) *Physical Characterization of the fermentor*: In these studies, extrinsic variables quantifying the physical processes are measured as a function of intrinsic variables using cell suspension broth or simulated broth(12,13,14,15,16,17). Extensive studies, for example, have been performed to obtain correlations to predict k_la, and mixing times as a function of agitation, aeration, type of

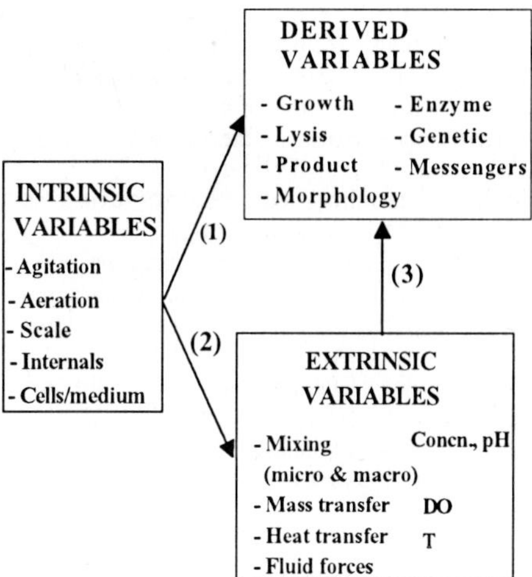

Figure 3. Relationships between various studies

impeller and nature of cells/medium. These studies represent a cross section of the multi-dimensional framework represented in Figure 2, for example k_la as a function of agitation and aeration. Very limited studies have been performed to determine distributions of circulation times, micro-mixing levels, k_la, fluid shear and heat transfer in production fermentors(12,18,19,20,21). These type #2 studies provide important data needed to perform mechanistic studies described below.

(3) *Mechanistic studies:* These studies, performed over the last three decades by many groups, determine the physiological effect(s) of a single extrinsic variable while others are maintained constant. For example, one of the earliest study attempted to evaluate the role of mixing time with respect to the feeding point, or micromixing, in the fermentor(22). The majority of these studies are based on a scale-down approach whereby small-scale strategies are designed to simulate a 'critical' local condition or extrinsic variable found in production fermentor. Examples include simulation of the spatial gradients in concentrations of the carbon

source(10,11), oxygen(10,11) and pH(1), micromixing at the feeding point(7) and fluid stresses(23). Many of these studies did not consider the data obtained from type #2 studies for physical characterization of fermentors to simulate realistic local conditions(8,9). Other major limitations of all these studies were that the possibility of the interaction between various extrinsic variables exist. We will illustrate this possibility by showing results from micromixing and shear studies using *Saccharomyces cerevisiae*.

INTERACTION OF EXTRINSIC VARIABLES

Saccharomyces cerevisiae, or baker's yeast, was chosen because of it is sensitive to glucose and dissolved oxygen levels. The biomass yield of yeast culture is reduced by more than half if the glucose level exceeds 0.05-0.2 g/L ("glucose effect") or the dissolved oxygen levels dips below 0.1 mg/L ("anaerobic")(24). To maintain sufficient mixing and mass transfer in production vessels, adequate agitation & aeration power (1-4 W/L)(1) is supplied without regards to the sensitivity of cells to fluid shear effects. Two sets of experiments were performed; one each to study the micromixing and shear effects with no previous indication of possible interaction. These studies were performed in a stirred and aerated bioreactor of volume 3 L operated in chemostat mode. Dissolved oxygen was maintained above 20 % in all experiments to avoid any interaction with mass transfer process.

(1) Micro-mixing effects- The goal of this experiment was to determine whether the micromixing of the incoming concentrated glucose (30 g/L) stream with the yeast cells will affect the biomass yield. Chemostat experiments were performed at dilution rates ranging from 0.16 h^{-1} to 0.32 h^{-1} at two agitation speeds of 800 rpm (1 W/L) and 1100 rpm (3 W/L) using one Rushton turbine. The feed was added at the injection point away from the impeller (IP#2).

(2) Shear effects- The goal of this experiment was to determine whether excess turbulent shear affected physiology of yeast cells. A chemostat was performed at a dilution rate of 0.27 h^{-1} and steady states were achieved at two agitation speeds, 650 rpm (3 W/L) and 1200 rpm (10.5 W/L) using two Rushton turbines on the shaft. The glucose concentration in the feed was 5 g/L and was added away from the impeller.

Micromixing effects

Figure 4 shows biomass yield ($Y_{x/s}$) as a function of dilution rate for two agitation speeds, 800 rpm and 1100 rpm. Data shows the onset of the *glucose effect* at a critical dilution rate and an *hysteresis effect*. Changing the agitation speeds showed these two metabolic effects to be sensitive to the mixing state. The onset of glucose effect occurs at a lower dilution rate, from 0.26 to 0.28 hr^{-1} for 800 rpm, compared to that at a dilution rate from 0.30 to 0.33 hr^{-1} for 1100 rpm during the forward direction of the experiment. Similar effects are observed in the reverse direction of the experiment. Consistently, the cytochrome aa_3 enzyme which is repressed by excess glucose showed higher activity at 1100 rpm compared to 800 rpm beyond the dilution rate of 0.24 hr^{-1} (data not shown).

Changing the agitation speed results in changes in both the macro- and micro- aspects of mixing, mass transfer and shear forces. The small volume of the fermentor and dissolved oxygen control ensures that any changes in the macromixing and mass transfer rates would be insignificant. The questions then remains is that whether the micromixing at the injection point is the only relevant

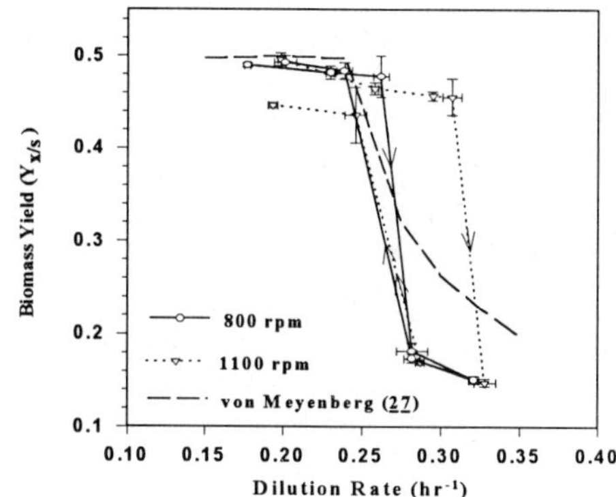

FIGURE 4. MICROMIXING EFFECTS
IN FERMENTATION

turbulence-driven physical process
or whether other physical processes
such as shear forces also play a
role. Measurements of micromixing
levels using a dye system(25) showed
that the Kolmogoroff's microscale at
the injection point can be changed
more effectively by changing its
location (from 15 μm near the
impeller IP#1 to 45 μm away from
the impeller IP#2 at 800 rpm) than
by changing the stirrer speed at the
injection point away from the
impeller (from 45 μm at 1100 rpm to
55 μm at 800 rpm). Additional
chemostat experiments at a constant
agitation speed of 800 rpm were
performed where the injection point
was changed from a location away
from the impeller (IP #2) to a
location in the impeller region (IP
#1) (data not shown). This shift in
the location of the injection point
produced effects on biomass yield,
cytochrome aa_3 enzyme activity and
hysteresis consistent with the
effects produced by changing
agitation speed. Thus additional
turbulence provided only at the feed
point (by better injection point in
the impeller region) or throughout
the reactor (by increasing the
stirrer speed) seems to have similar
effects on respiratory enzyme and
biomass yield. These results
indicate that micromixing at the
injection point may not be the only

turbulent-driven process controlling
the overall biological effect. Full
details of this experiment will be
published elsewhere.

<u>Shear effects</u>

Chemostat experiments were performed
at one dilution rate of 0.27 h^{-1} and
steady states were achieved at two
agitation speed, 650 rpm and 1200
rpm . Although the biomass yield
remained unchanged at these two
conditions, a number of
physiological responses representing
possible shear effects were
observed. A 20% increase in the
incorporation of glucan in the cell
wall was observed at higher
agitation speed. Correspondingly the
average cell size was increased and
surface area per unit volume was
reduced. Other biological assays
showed increased compression modulus
of walls, increased resistance to
enzymatic degradation of walls and
reduced elastic coefficients at the
higher agitation speed. The details
on method of assays and numerical
results are described by Villeneuve
(26). All these effects on the cell
wall indicate that yeast cells sense
and counter the increased exposure
to shear forces at 1200 rpm by
building mechanically stronger walls
and reducing effective surface area
per unit volume. To compensate for
the increased cell wall strength
related metabolic activity, other
metabolic activities are reduced at
higher agitation speed. These
reduced metabolic activities include
reduction in synthesis of proteins,
trehalose, glycogen and heat shock
proteins (HSP 90 & 26). These data
clearly demonstrates that yeast
cells, hitherto considered
insensitive to shear forces, are
responding to higher agitation
speed. This response, however, could
be result of either a complex chain
of events caused by
induction/repression of certain
genes or a simple case of selection
of shear resistant cell lines.

These multiple physiological
responses did not have measurable

154

impact on the biomass yield at the higher agitation speed of 1200 rpm (10.5 w/L). In contrast, the first set of micromixing experiments showed a higher biomass yield of 45% for 1100 rpm (3 W/L) compared to 15 % yield for 800 rpm (1 W/L). The potential causes for the differences in biological effects produced by shear- and micromixing- experiments are the total power input and inlet glucose concentration. The shear experiment had a very low inlet glucose level of 5 g/L compare to 30 g/L used in micromixing experiment. This could avoid the glucose effect and hence reduction in biomass yield as seen in the "shear effects" experiments. Whether increased agitation in micro-mixing experiments caused any physiological effects from shear forces, it is not known yet. Further experiments will be performed to confirm this interaction of micromixing and shear. Full details of this experiment will be published elsewhere.

MICRO-ENVIRONMENTAL APPROACH: NEW DIRECTIONS

The experimental data discussed above showed that an interaction between two or more turbulent-driven physical processes is probably non-linear. No systematic approach is yet available to impose defined and relevant values for more than one extrinsic variable simultaneously on cells. Here micro-environmental approach can be applied. The sizes of microscales (50- 300 μm) in a typical industrial fermentors are much larger than the size of microbial cells which range from 1 to 5 μm(7). As a microbial cell, embedded in its micro-environment, circulates in the fermentor, it would be exposed to a range of extrinsic variables in its microenvironment. The overall performance of the fermentor would depend on the response of individual cells to their respective microenvironment. The traditional rules of thumb for scale-up neglect

the interaction of the cells with their microenvironment by averaging the hydrodynamics through the use of the engineering variables. A rational strategy to aid scale-up studies, therefore, should be based on the evaluation of interactions of physiological consequences of various physical processes (extrinsic variables) present in the micro-environments of cells. Based on the list of extrinsic variables in Figure 2 and data from our laboratory we propose the following studies for interaction effects.

1. Micromixing and shear- A stirred bioreactor with a recycle loop will be employed where feeding point will be located in the loop region. Such a system is already being used in our labortory. The micromixing in the feeding region will be varied at defined levels by creating turbulence behind a wire mesh in the tube. The cell broth recirculates between a recycle loop and the stirred bioreactor. The turbulent shear can now be varied in the stirred tank by varying agitation and introducing additional turbines.

2. Micromixing and mass transfer- An annulus type tubular bioreactor will be employed with liquid flow inside and air/oxygen flow on the outside. The micromixing levels in the feeding region will be varied as described earlier. The mass transfer rates will be varied across the membrane by changing partial pressure of oxygen in the gas phase, membrane type and the liquid flow rate.

3. Micromixing and Macro-mixing- A stirred bioreactor will be employed with multiple injection point which will produce a range of micromixing levels. Macro-mixing will be imposed using a pulse feeding based on a Monte Carlo method to simulate the circulation time distribution(10). Similar strategies can be developed for other types of interaction, for example, mass transfer and shear for cell culture system, heat transfer and mass transfer for highly viscous

fermentation system, mass transfer and heat transfer for anaerobic fermentation. In addition to these experimental strategies, computer models are also being developed to evaluate the interaction, for example, interaction between mixing and mass transfer in fermentors(28).

CONCLUSIONS

A multi-dimensional framework was developed to comprehend the role of turbulence in fermentations. A review of published studies on this topic revealed that the non-linear interaction among various turbulence-driven physical processes has not been systematically evaluated. A comparison of yeast chemostat experiments showed a distinct possibility of interaction of micromixing with shear. Small scale experimental strategies based on a micro-environment approach were proposed to evaluate whether shear, mass transfer and bulk mixing would interact with micromixing in fermentation.

ACKNOWLEDGEMENTS

This study has been supported by funding from NSF grant #BCS-9022158, Anheuser Busch Co. and Colorado Bioprocessing Center for fellowships and experimental work.

LITERATURE CITED

1. Einsele, A., "Scaling Up of Bioreactors", *Process Biochem., July* **13**:13-14, (1978).

2. Oldshue, J.Y., "Fermentation mixing scale-up techniques", *Biotechnol. Bioengng.*, **8**, 3-24 (1966).

3. Bailey, J. and D. Ollis, *Biochemical Engineering Fundamentals*, McGraw-Hill, New York, 510-511, (1986).

4. Young, T.B., " Fermentation scale-up: Industrial experience with a total environmental approach." *Annal. New York Acad. Sci.* **326,** 165-180 (1979).

5. Amanullah, A., Baba, A., McFarlane, C.M., Emery, A.N., Nienow, A.W., "Biological models of mixing performance in bioreactors", *3rd Intnl. Conf. on Bioreactor and Bioprocess Fluid Dynamics*, Nienow, A.W. (Ed.), 381-400 (1993).

6. Bajpai, R.K. and M. Reuss, "Coupling of Mixing and Microbial Kinetics for Evaluating the Performance of Bioreactors", *Canadian J. Chem. Eng.*, **60:** 384-392 (1982).

7. Dunlop, E.H., Ye, S.J., "Micromixing in fermentors: Metabolic changes in Saccharomyces cerevisiae and their relationship to fluid turbulence", *Biotechnol Bioengng.*, **36**, 854-864 (1990).

8. Kristiansen, B., McNeil, B., "The design of a tubular loop for scale-up and scale down of fermentation processes", *Intnl. conf. on bioreactor-5 and biotransformations*, Moody, G.W. and Baker, P.B. (Ed.), Gleneagles, Scotland, UK, 321-334, (1987).

9. Larsson, G. and Enfors, S.-O, " Kinetics of microbial response to insufficient mixing in bioreactors",. *Second Eur. Congr. on Mixing*, Pavia, Italy. 443-450, (1988).

10. Namdev, P.K., Yegneswaran, P.K., Thompson, B.G., Gray, M.R., "Experimental simulation of large scale bioreactor environments using a Monte Carlo method", *The Can. Jnl. Chem. Engng.*, **69**, 513-519 (1991).

11. Sweere, A.P.J., Luyben, K.Ch.M., Kossen, N.W.F., " Regime analysis and scale down: tools to investigate the performance of bioreactors.", *Enzyme Microbiol Biotechnol.*, **9**, 386-398 (1987).

12. Bryant, J., "The Characterization of Mixing in Fermentors", *Advances in Biochemical Engineering*, A. Fiechter, ed., **5**: 101 (1977).

13. Calderbank, P.H., "Mass transfer in fermentation equipment", *Biochemical and Biological Engineering Science,* Blakebrough (ed.), **1**, (1967).

14. Mohan, P., Emery, A.N., Bujalski, W., Nienow, A., "Integrated heat transfer and mixing studies in a pilot-scale bioreactor", *3rd Intnl. Conf. on Bioreactor and Bioprocess Fluid Dynamics*, Nienow, A.W. (ed.), 35-45, (1993).

15. Nienow, A.W., Ulbretch, J.J., "Gas-liquid mixing and mass transfer in high viscosity liquids" in *Mixing of liquids by mechanical agitation*, Chapter 6, J.J. Ulbretch and G.K. Patterson (Eds.), Gordon and Breach, New York, 203-237 (1985).

16. Taguchi, H., "The nature of fermentation fluids", *Adv. Biochem. Engng.*, **1**, (1971).

17. Wang, D.I.C., Fewkes, R.C.J., "Effect of operating and geometric parameters on the behavior of Non-newtonian, mycelial, antibiotic fermentations", *Dev. Ind. Microbiol.*, Chapter 2, 39-56 (1977).

18. Buckland, B.C., Gbewonyo, K., Jain, D., Glazometzky, K., Hunt, G., Drew, S.W., *Proc. 2nd Intnl. Conf. on Bioreactor Fluid Dynamics*, BHRA/Elsevier, Cranefield, UK, 1-15 (1988).

19. Kossen, N.W.F., "Some remarks about problem solving in biochemical engineering", *3rd Intnl. Conf. on Bioreactor and Bioprocess Fluid Dynamics*, Nienow, A.W. (ed.), (1993).

20. Oosterhuis, "Scale-up of bioreactors, a scale-down approach", Ph.D. Thesis, Delft University of Technology, Delft, The Netherlands (1984).

21. Velasco, D., Martinez, A., Torres, L.G., Galindo, E., "Rheology and dual impeller of an industrial fermentation broth containing *Micromonospora purpurea*", *3rd Intnl. Conf. on Bioreactor and Bioprocess Fluid Dynamics*, Nienow, A.W. (ed.), 101-116 (1993).

22. Hansford, G.S., Humphrey, A.E., " The effect of equipment scale and degree of mixing on continuous fermentation yield at low dilution rates.", *Biotechnol. Bioengng,* **8**, 85-96, (1966).

23. Papoutsakis, E.T., *Trends in Biotechnol.*, **9**, 427-437 (1991).

24. Sonnleitner, B., and O. Kappeli, "Growth of Saccharomyces cerevisiae is Controlled by Its Limited Respiratory Capacity: Formulation and Verification of a Hypothesis", *Biotechnol. Bioengng.*, **28**: 927-937 (1986).

25. Wenger, K.S., Dunlop, E.H., MacGilp, I.D., "Investigation of the Chemistry of a Diazo Micromixing Test Reaction", *AIChE Jnl.*, **38**, 1105-1114, (1992).

26. Villeneuve, P.E., Ph.D. Thesis, Colorado State University, USA, 1994

27. von Meyenberg, K. "Catabolite Repression and the Germination Cycle of *Saccharomyces cerevisiae* (German)", Ph.D. Dissertation #4279, ETH, Zurich (1969).

28. Singh, V., Fuchs, R., Constantinides, A., "A new method for fermentor scale-up incorporating both mixing and mass transfer effects- I. Theoretical basis", in *Biotechnology Process: Mixing and Scale-up* , Ho, C.S., Oldshue, J.Y. (Eds.), 200-214, (1987).

Power Input and Oxygen Transfer in Fed-Batch Penicillin Production Process

A.C. Badino Jr., M. Barboza, and C.O. Hokka

Department of Chemical Engineering, Universidade Federal de São Carlos, Cx.P. 676, CEP 13565-905, São Carlos, SP, BRAZIL

Fed-batch experiments in a 20 l fermentor utilizing Penicillium chrysogenum *IFO 8644 were conducted using a medium containing sucrose and corn steep liquor as main components. The runs were interrupted at different stages of the process - 24, 48, 72, 96 and 120 h - and broths were transferred to an agitated vessel. Power consumption under gassed and ungassed broths and volumetric oxygen transfer coefficient, k_La, were measured for various agitation speeds and aeration rates. The rheological characteristics of each broth were also determined. Power number - modified Reynolds number plots were similar to those for other non -Newtonian fluids. Michel and Miller correlation was applied for predicting gassed power drawn for all broths. A general correlation for k_La, as a function of operating variables and apparent viscosity for broths of different cultivation ages was obtained.*

The penicillin production process is a classical example of a fermentation process producing highly viscous non-Newtonian fluids [1] and demanding large oxygen amounts. The penicillin broth rheological characteristics can be represented fairly well by the "power law" as a pseudoplastic fluid. Bongenaar et al. [2] observed that conventional rheometers are not appropriate to determine these characteristics and proposed the utilization of a turbine viscometer for such suspensions. Calderbank and Moo-Young [3] utilized the concept of average shear rate around an impeller in the laminar flow region and showed that it is proportional to the shaft agitation speed. Concerning power consumption in gassed systems, Taguchi and Miyamoto [5] working with pseudoplastic Endomyces sp. broths in a wide range of fermentor volumes verified the validity of the correlation proposed by Michel and Miller [6] for Newtonian fluids.

Ogut and Hatch [8] studied the effect of the rheological characteristics of various fluids on k_La. Based on Cooper's correlation. Their results indicated that, for Newtonian fluids the effect of P_g/V_L was significant, while for non-Newtonian fluids this term had only slight effect. On the other hand the superficial velocity, v_s, affected k_La of both types of fluids.

Zlokarnik [9] originally working with aqueous solutions, in cylindrical agitated vessels, proposed a correlation based on dimensionless groups, relating volumetric mass transfer coefficient with operation variables and the fluid kinematic viscosity as follows:

$$k_L a.V_L/Q = k_1.(P_g/(Q.\rho.(v.g)^{2/3}))^\beta.(N_{Sc})^\theta \tag{1}$$

Mention should be made here that this relationship is valid for the penicillin fermentation broth as it is a non-coalescent fluid.

Recently, studies on rheological characteristics of filamentous microrganisms broths were presented by Allen and Robinson [10] where rheological properties, measured by different methods, are related to cell mass concentration. These workers point out that, in spite of its importance, there are only few studies attempting to quantify the relationship between transport process and rheological properties of fermentation broths.

On the other hand, operating parameters, such as agitation speed, may also have a direct influence on penicillin productivity, and not just as a consequence of their interaction with rheology and transport properties. For instance, Smith and Lilly [11] examined the effect of agitation speed on penicillin productivity suggesting that there is a zone around the impeller in which mycelia may be damaged, impairing the production rate.

In this work the power requirements and the k_La dependence on operating variables and apparent viscosity were examined in broths of fed batch penicillin fermentation process during 120 hours.

157

E. Galindo and O.T. Ramírez (eds.), Advances in Bioprocess Engineering. 157-162.
© *1994 Kluwer Academic Publishers. Printed in the Netherlands.*

MATERIAL AND METHODS

Microorganism. Penicillium chrysogenum IFO 8644, kindly provided by I. C. Biotech., Osaka, Japan, and kept on glycerol-agar medium slants, was used throughout this work.

Seed and Main Media (g/ℓ): sucrose, 20.0; corn steep liquor, 30.0; KH_2PO_4, 7.0; $(NH_4)_2SO_4$, 5.0; $MgSO_4.7H_2O$, 4.0; $CaCO_3$, 5.0; soy-bean oil, 4.0; salt solution, 20.0; pH = 6.0.

Feed Medium (g/ℓ): sucrose, 96.0; corn steep liquor, 30.0; potassium phenylacetate (15% w/w), 12 mℓ; pH = 6.0.

Salt Solution (g/ℓ): $FeSO_4.7H_2O$, 3.0; $CuSO_4.5H_2O$, 2.5; $ZnSO_4.7H_2O$, 10.0.

Cultivation Procedure. Ten mℓ spore suspension from agar slants cultivated for eleven days were seeded into a one liter Erlenmeyer flask containing 150 mℓ seed medium. The flask was incubated, for germination in shaker, 250 rpm, 26 °C. After 48 hours, the contents of the flask were transferred to Bioflo C-32 fermentor (New Brunswick Scientific Co. Inc.) containing 1.35 ℓ of seed medium and the culture was incubated for 24 hours, 800 rpm, 26 °C, and 1 vvm aeration rate. Then 1.2 ℓ of this culture were transferred to a 20 ℓ fermentor (Equipamentos Cientificos Superohm, Piracicaba - São Paulo) containing 10.8 ℓ main medium and operated at 26 °C, 550 to 560 rpm and 0.8 to 1vvm. Figure 1 shows the geometrical dimensions of the vessel. Dissolved oxygen, monitored by means of an oxygen analyser with a galvanic electrode (B.E.Marubishi Co. Ltd., Tokyo), seldom tended to fall below 30% saturation. But, whenever it ocurred, agitation and aeration were slightly increased to keep it above 30%. For the fed-batch experiments, after 24 hours, culture feed medium was added by a peristaltic pump at a constant flow rate of 50 mℓ/h.

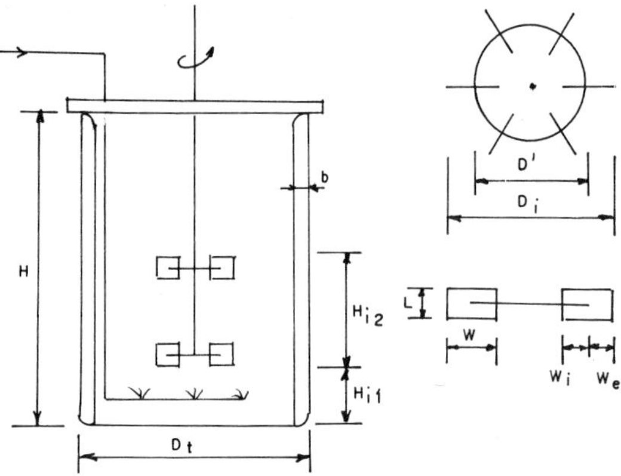

H=50.0 cm Hi$_1$=6.0 cm Hi$_2$=12.0 cm D$_t$=27.5 cm b=2.8 cm n$_b$=4 D$_i$=12.0 cm D'=7.4 cm L=2.3 cm W=2.9 cm W$_i$=0.6 cm W$_e$=2.3 cm

Figure 1. Geometrical dimensions of the vessel

Agitation and Aeration Experiments. Broths, from 24, 48, 72, 96 and 120 hours cultivation times were transferred to an agitated cylindrical vessel set, provided with two turbines and an air sparger, as shown in Figure 2. Their dimensions were exactly the same as the fermentor vessel. A 0.4 kW motor with Ringcone variable speed drive from Shimpo do Brasil Ltda, São Paulo-SP model RXMV-400, was placed over a vertical ball bearing allowing torque measurement with a digital dynamometer (KRATOS, São Paulo-SP). Aeration rate was adjusted through a flowmeter, and temperature was kept at 26 °C. Power input in aerated and non-aerated broths were determined for various agitation speeds and aeration rates. For each condition kLa was measured by the dynamic method as proposed by Taguchi and Humphrey [12] utilizing a fast response (first order response time, 3.6 s) galvanic electrode (B.E. Marubishi Co. Ltd., Tokyo). As the tank was open to the air, before each experiment the set was washed and rinsed with ethanol and boiled water.

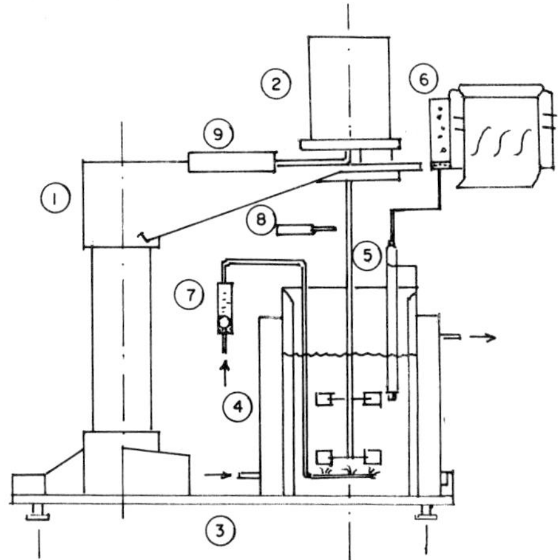

1.Metallic Structure 2.Motor 3.Agitated Tank 4.Water Jacket 5.Dissolved O_2 Electrode 6.Recorder 7.Flowmeter 8. Phototacometer 9.Dynamometer

Figure 2. Schematic view of the agitated tank set-up

Rheology. Paralelly, the broth rheological characteristics, based on the "power law", were determined in DVII series Brookfield viscometers (Brookfield Engineering Laboratories Inc., Stoughton, MA) adapted with a six blade Rushton's turbine (D$_i$ = 3.5 cm) according to the method developed by Bongenaar et al. [2] and previously calibrated with CMC solutions. The rheological characteristics of these solutions were determined with RVT DVII Brookfield viscometer with coaxial cylinders. The outer cylinder was the small sample chamber adapter (model 13R) and the inner cylinder was the spindle number 21.

Analysis. Cell concentration was evaluated as dry matter at 105 °C. The samples were centrifuged, washed twice with acetic acid (10 % v/v) to solve Calcium Carbonate, followed by washing with distilled water.

Sugar concentration was determined as reducing sugars, by the colorimetric Somogyi method after acid hydrolysis.

Penicillin was determined by bioassay according to Grove and Randall [13].

RESULTS AND DISCUSSION

Fermentation. A typical time course of the fermentation process is shown in Figure 3, corresponding to the run of 120 hour cultivation. The process behaves as a typical secondary metabolite fermentation, with higher production rates ocurring after the growth phase is completed. The other runs, interrupted at early times, showed very similar pattern, allowing to assume that the broth from each run can represent the broth of a whole process (120 hours) at that particular cultivation time.

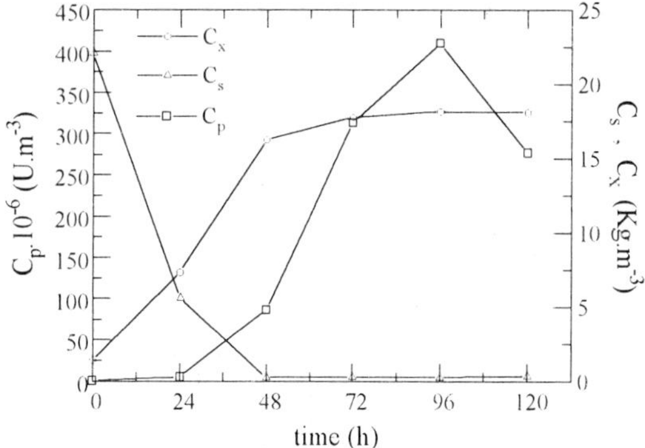

Figure 3. Time course of the 120 hours process

Power Requirements. After each cultivation, the rheological characteristics as well as the broth density were determined. The rheological equations obtained for each cultivation time are shown in Table 1. It can be observed that the broth behaved as a pseudoplastic fluid, showing different values of the constants, reflecting the differences in cell concentration and morphology. Both the consistency index, K, and flow behaviour index, n, could not be correlated with cell mass concentration as proposed by Allen and Robinson [10]. However, the K values calculated utilizing the equation proposed by the above authors lie in the same order of magnitude as those encountered in this work, though they have worked with a synthetic medium while here, a complex medium has been utilized. Besides, in the present work, rheological characteristics were evaluated using the whole broth at different cultivation times rather than diluted broth from batch experiments.

Aeration and agitation experiments were performed by transferring the whole broth to the agitated tank provided with two Rushton type turbine impellers as described in the previous section. For each cultivation time, power input in gassed and ungassed broth as well as k_La values were measured for various agitation speeds and air flow rates.

From the results obtained in ungassed conditions, Power Number, N_{Po}, and modified Reynolds Number, N_{Re} were calculated as:

$$N_{Po} = P_o/(N^3.D_i^5.\rho) \tag{2}$$

$$N_{Re} = D_i^2.N.\rho/\mu_{ap} \tag{3}$$

Table 1. Rheological equations and densities of the cultivated broths

Run	Rheological Equation	ρ (Kg.m^{-3})
24 h	$\tau = 0.54\,\gamma^{0.36}$	1,020
48 h	$\tau = 1.36\,\gamma^{0.12}$	1,000
72 h	$\tau = 0.92\,\gamma^{0.24}$	990
96 h	$\tau = 1.29\,\gamma^{0.03}$	990
120 h	$\tau = 1.43\,\gamma^{0.12}$	1,000

Apparent viscosity, μ_{ap}, was evaluated following the general procedure proposed by Metzner et al. [4], by the ratio between shear stress, τ, calculated from the equations shown in Table 1 and the average shear rate, γ_{av}, calculated as

$$\gamma_{av} = 11.5.N \tag{4}$$

The proportionality constant between average shear rate and agitation speed was evaluated as the mean value of 11.6 and 11.4, determined by Metzner et al. [4] for agitated vessels with one and two turbine type impellers respectivelly, since in our case working with an actual fed-batch process, the broth volume increases with the fermentation time. In Figure 4 the plot of N_{Po} and N_{Re} as proposed by Metzner et al. [4] for non-Newtonian fluid is shown. As it can be observed the data are fairly well fitted within the range of the relationship proposed by the above authors, though they have worked with homogeneous fluids rather than a suspension of filamentous organism.

The relationship between power requirements in gassed and ungassed systems was examined through the correlation proposed by Michel and Miller [6] as follows:

$$P_g = C.(P_o^2.N.D_i^3/Q^{0.56})^\alpha \tag{5}$$

The experimental results allowed the plot of Figure 5 and as it can be seen, though the volume variations, the correlation depicts closely the data obtained for different cultivation times. Furthermore, the value of $\alpha = 0.45$ was the same as calculated by the above authors.

Oxygen Transfer. The k_La values for each broth at various aeration rates and agitation speeds were measured by the dynamic method. Tables 2 to 6 shows the values of k_La calculated for each pair of N and Q in the different fermentation broths, as well as the power input and apparent viscosity.

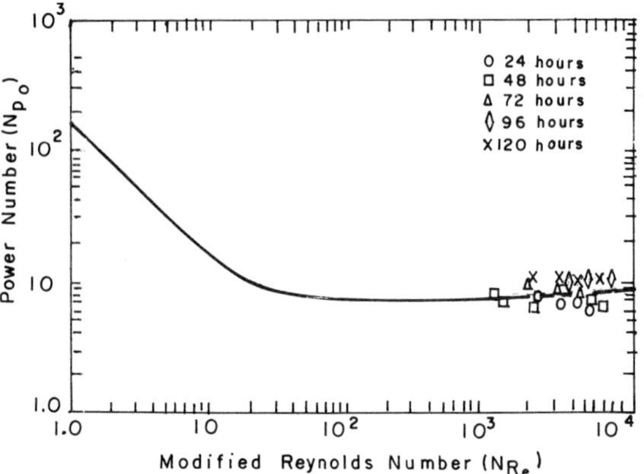

Figure 4. Experimental data adjusted to the diagram proposed by Metzner et al. [4]

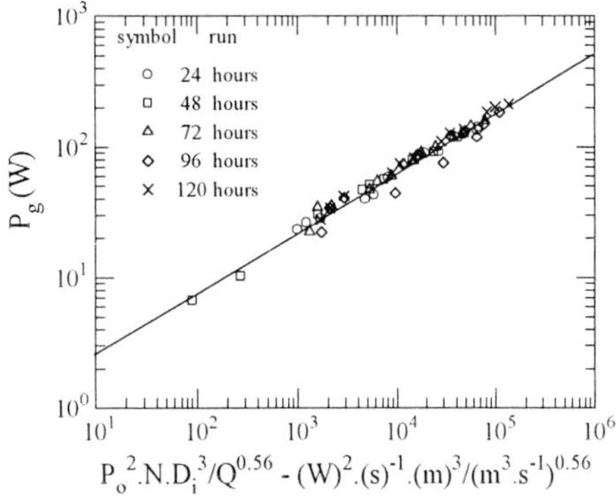

Figure 5. Correlation between power input in gassed system, P_g, and ungassed system, P_o

Table 2. Results from $k_L a$ measurements at different conditions for the 48 hour broth

N (s^{-1})	$v.10^5$ (m^2.s^{-1})	Q.10^4 (m^3.s^{-1})	P_g (W)	$k_L a.10^3$ (s^{-1})
5.50	3.72	0.94	28.5	81.0
		1.71	26.5	85.1
		2.43	23.5	80.6
7.00	3.19	0.94	50.0	95.0
		1.71	43.3	110.3
		2.43	40.0	97.7
8.33	2.85	0.94	93.0	134.0
		1.71	89.9	123.0
		2.43	84.2	123.0
9.83	2.57	0.94	136.2	162.7
		1.71	126.8	155.0
		2.43	120.8	154.0

Table 3. Results from $k_L a$ measurements at different conditions for the 48 hour broth

N (s^{-1})	$v.10^5$ (m^2.s^{-1})	Q.10^4 (m^3.s^{-1})	P_g (W)	$k_L a.10^3$ (s^{-1})
3.67	5.05	1.05	6.7	16.0
4.33	4.36	1.05	10.3	26.0
5.50	3.54	1.05	30.6	42.1
7.00	2.86	1.05	56.7	54.5
		1.94	51.9	62.2
		2.62	47.2	71.3
8.33	2.45	1.05	91.8	81.9
		1.94	86.9	69.8
		2.62	80.0	90.2
9.93	2.12	1.05	144.5	108.0
		1.94	130.4	102.0
		2.62	120.0	112.0

Table 4. Results from $k_L a$ measurements at different conditions for the 72 hour broth

N (s^{-1})	$v.10^5$ (m^2.s^{-1})	Q.10^4 (m^3.s^{-1})	P_g (W)	$k_L a.10^3$ (s^{-1})
5.50	3.97	1.11	35.0	35.1
		2.02	34.5	42.4
		2.80	22.4	53.9
7.00	3.31	1.11	60.1	51.2
		2.02	54.4	58.1
		2.80	47.1	56.6
8.33	2.90	1.11	99.9	64.5
		2.02	91.2	71.8
		2.80	78.3	70.2
9.83	2.56	1.11	158.0	82.2
		2.02	144.8	87.9
		2.80	126.5	86.8

Table 5. Results from $k_L a$ measurements at different conditions for the 96 hour broth

N (s^{-1})	$v.10^5$ (m^2.s^{-1})	Q.10^4 (m^3.s^{-1})	P_g (W)	$k_L a.10^3$ (s^{-1})
5.00	2.33	1.16	41.0	32.3
		2.15	34.4	39.8
		2.94	22.4	49.0
7.00	1.84	1.16	88.0	47.3
		2.15	75.2	54.1
		2.94	44.6	67.5
8.33	1.56	1.16	134.5	66.4
		2.15	122.8	69.9
		2.94	76.7	88.6
9.83	1.32	1.16	186.4	86.2
		2.15	150.6	100.0
		2.94	121.1	95.7

Table 6. Results from k_La measurements at different conditions for the 120 hour broth

N (s^{-1})	$v.10^5$ $(m^2.s^{-1})$	$Q.10^4$ $(m^3.s^{-1})$	P_g (W)	$k_La.10^3$ (s^{-1})
		1.32	42.7	23.2
5.50	3.72	2.35	34.0	24.6
		3.36	27.8	38.7
		1.32	83.3	29.6
7.00	3.01	2.35	74.6	33.0
		3.36	63.4	41.5
		1.32	138.0	43.2
8.33	2.58	2.35	129.5	50.0
		3.36	111.8	54.5
		1.32	214.6	58.3
9.83	2.23	2.35	203.7	66.7
		3.36	186.4	80.3

As it can be observed k_La is much higher at lower fermentation times, indicating the strong influence of viscosity, caused by cell mass, on this mass transfer coefficient. These data were related through Cooper et al. [7] relationship,

$$k_La = K_2.(P_g/V_L)^a.(v_s)^b \qquad (6)$$

For each broth, the values of K_2, a and b were estimated by the least squares nonlinear regression following Marquardt's procedure.

Table 7 presents the regression results and Figure 6 shows a typical plot of this relationship showing the data obtained for the 48 hours broth.

It can be observed that for the 96 and 120 hours broths, the influence of P_g/V_L and v_s is practically the same while for the 24, 48 and 72 hours broths, P_g/V_L affects more remarkably the mass transfer coefficient. We can presume that, as the morphology changes are less drastic from 96 to 120 hours fermentation time, the effect of morphology on the mass transfer rate is rather small as compared with that occurring at early fermentation times. These results are in disagreement with those found by Ogut and Hatch [8] who concluded that for pseudoplastic fluids k_La is weakly affected by P_g/V_L. However these authors, though working at similar aeration and agitation conditions, utilized different impeller type.

In order to incorporate the influence of apparent viscosity on k_La relationship, the dimensionless groups of Equation (1) were calculated through the data presented in Tables 2 to 6 for all fermentation broths. By nonlinear regression the values of K_1, β and θ were calculated as: $K_1=2.7 \ 10^{-7} \pm 7.8 \ 10^{-8}$, $\beta=0.68 \pm 0.08$, and $\theta=0.97 \pm 0.24$ with regression coeficient $r=0.75$. In Figure 7 the calculated line is plotted against the experimental values. It can be observed that this equation correlates, as a first approach , the operating variables with k_La, and probably the scatter of the data is due to morphology changes taking place during the fermentation which produces different broth characteristics. that can not be accounted for by the Zlokarnik's equation.

Table 7. Calculated values of the constants of the relationship proposed by Cooper et al. [7]

Run	K_2	a	b	r
24 h	0.07±0.05	0.41±0.08	0.03±0.12	0.996
48 h	0.08±0.08	0.57±0.13	0.18±0.16	0.993
72 h	0.12±0.09	0.43±0.10	0.23±0.14	0.996
96 h	0.58±0.84	0.51±0.18	0.53±0.27	0.987
120 h	0.23±0.26	0.53±0.14	0.46±0.21	0.991

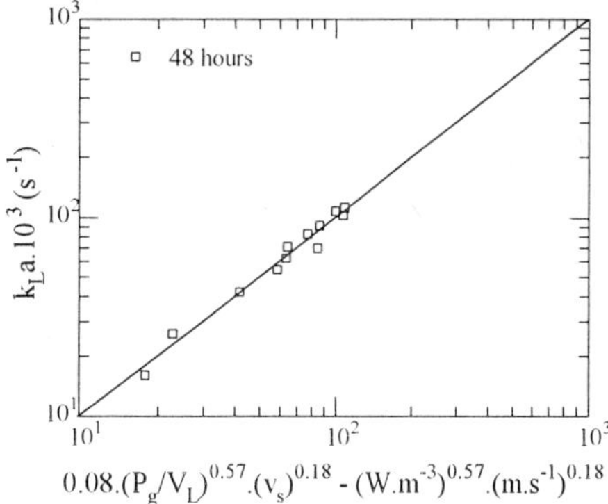

Figure 6. Experimental and calculated k_La values for the 48 h fermentation broth

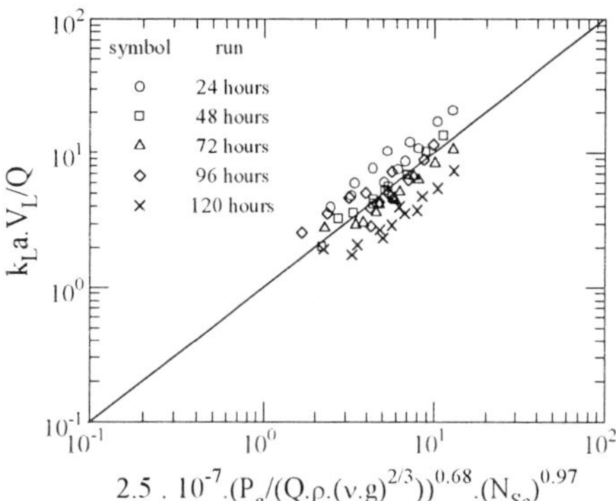

Figure 7. Experimental and calculated values of the dimensionless groups in Equation (1)

The results presented indicates that, though the difficulty in measuring certain variables, general correlations can be obtained relating operating variables with physical and transport properties even when dealing with complex non-Newtonian suspension of filamentous microorganism cultivated in variable volume fed-batch fermentors.

NOTATION

a, b	= constants, Equation (6)
C	= constant, Equation (5)
C_p	= penicillin concentration ($U.m^{-3}$)
C_s	= sugar concentration ($Kg.m^{-3}$)
C_x	= cell concentration ($Kg.m^{-3}$)
D_i	= impeller diameter (cm)
D	= tank diameter (cm)
g	= gravity aceleration constant ($m.s^{-2}$)
K	= consistency index ($Kg.m^{-1}.s^{n-2}$)
K_1	= constant, Equation (1)
K_2	= constant, Equation (6)
$k_L a$	= volumetric mass transfer coefficient (s^{-1})
N	= agitation speed (s^{-1})
n	= flow behaviour index (-)
N_{Po}	= Power Number (-)
N_{Re}	= modified Reynolds Number (-)
N_{Sc}	= Schmidt Number (-)
P	= power requirement (W)
P_o	= power requirement in ungassed systems (W)
P_g	= power requirement in gassed systems (W)
Q	= air flow rate ($m^3.s^{-1}$)
V_L	= broth fermentation volume (m^3)
v_s	= superficial air velocity ($m.s^{-1}$)
α	= constant, Equation (5)
β, θ	= constant, Equation (1)
γ	= shear rate (s^{-1})
γ_{av}	= average shear rate (s^{-1})
τ	= shear stress ($Kg.m^{-1}.s^{-2}$)
ρ	= broth density ($Kg.m^{-3}$)
μ_{ap}	= apparent dynamic viscosity ($Kg.m^{-1}.s^{-1}$)
ν	= apparent kinematic viscosity ($m^2.s^{-1}$)

ACKNOWLEDGEMENTS

The authors wish to thank Prof. T. Yoshida from I.C.Biotech., Japan, for his kind advise during his stay in S. Carlos, Brazil. Financial suport from PADCT/FINEP and CNPq-Brazil is also acknowledged.

LITERATURE CITED

1. Roels, J.A., Van Den Berg, J. and Voncken, R.M., *Biotechnol. Bioeng.*, **16**, 181 (1974).
2. Bongenaar, J.J.T.M., Kossen, N.W.F., Metz, B. and Meijboom, F.W., *Biotechnol. Bioeng.*, **15**, 201 (1973).
3. Calderbank, P.H. and Moo-Young, M.B., *Trans. Inst. Chem. Eng.*, **37**, 26 (1959).
4. Metzner, A.B., Feehs, R.H., Ramos H.L., Otto, R.E. and Tuthill, J.D., *AIChE J.*, **7**, 1, 3 (1961).
5. Taguchi, H. and Miyamoto, S., *Biotechnol. Bioeng.*, **13**, 43 (1966).
6. Michel, B.J. and Miller, S.A., *AIChE J.*, **8**, 2, 262 (1962).
7. Cooper, C.M., Fernstrom, G.A. and Miller, S.A., *Ind. Eng.Chem.*, **36**, 504 (1944).
8. Ogut, A. and Hatch, R.T.: *J.Chem.Eng.*, **66**, 79 (1988).
9. Zlokarnik, M., *Adv. Biochem. Eng.*, **8**, 133 (1978).
10. Allen, D.G. and Robinson, C.W., *Chem.Eng.Sci.*, **45**, 37 (1990).
11. Smith, J.J. and Lilly, M.D., *Biotechnol. Bioeng.*, **35**, 1011 (1990).
12. Taguchi, H. and Humphrey, A.E., *J.Ferment. Technol.*, **44**, 881 (1966).
13. Grove, D.C. and Randall, W.A., "Assay methods of antibiotics" chap. 2, Medical Encyclopedia Inc., New York (1955).

The Examination of Bioreactor Heterogeneity with Rheological Different Fermentation Broths

D.J. Pollard, A.P. Ison, P. Ayazi Shamlou, and M.D. Lilly

Advanced Centre for Biochemical Engineering, Department of Chemical and Biochemical Engineering, University College London, Torrington Place, London WC1E 7JE, U.K.

Bioreactor heterogeneity has been studied in a pilot scale airlift reactor ($0.3m^3$), which is multiconfigurable creating different degrees of heterogeneity. Hydrodynamic and oxygen transfer performance of two ring sparger configurations, annulus and draft tube, in combination with a marine propeller fitted at the base of the reactor, are compared using Newtonian baker's yeast and non-Newtonian Saccharopolyspora erythraea broths. In aerated only systems, cells experience a heterogeneous dissolved oxygen environment with both Newtonian and non-Newtonian fermentations. The use of the marine propeller when used in conjunction with the annulus sparger to draw liquid down the draft tube, was found to produce significant increases in the hydrodynamic and oxygen transfer performance of the vessel. This resulted in reduced dissolved oxygen heterogeneity with both fermentation broths, and increased erythromycin production from S. erythraea. The relationship between superficial gas velocity and gas holdup for different morphological S. erythraea fermentations were found to be significantly reduced when compared to Newtonian yeast broths, but similar to yeast when propeller and aeration operation was compared. Thus the effect of morphology, rheological properties and reactor design on overall reactor performance has been demonstrated.

It is well known that large industrial scale fermenters have circulation and mixing times ten times the values of lab scale vessels. At the scale of antibiotic production, a $100m^3$ vessel would have a mixing time in the region of 120–160 seconds. The extent of heterogeneity in a large scale vessel is known to effect growth and production yields, (Wang and Fewkes (1) and Vardar and Lilly (2)). However the way in which the heterogeneity is measured must also be considered. Oosterhuis and Kossen (3), showed that the dissolved oxygen tension varied along vertical axes in a $19m^3$ fermenter. Along one vertical axis they showed that a minimum value of 10% dissolved oxygen tension was measured. However at a second axis a small distance further away from the centre of the reactor the dissolved oxygen tension registered zero indicating pockets of oxygen depletion. Therefore in standard reactors the heterogeneity that the cells experience may be greater than that observed by the measurements.

MATERIALS AND METHODS

<u>Bioreactor configuration</u>.Experiments were performed in a multiconfigurable concentric airlift reactor, (Chemap AG,Volketswil, Switzerland), consisting of three stainless steel vessel sections clamped together to give a total vessel height of 4.11m. An internal draft tube of height 2.77m was used with a working volume of 250l, which gave an unaerated liquid height of 0.47m above the draft tube. The vessel diameter was 0.317m and the draft tube diameter was 0.211m. The height of the annulur gap between the draft tube and the base of the reactor was 0.06m. Aeration could be achieved by either a draft tube ring sparger,(0.16m diameter), or an annulus ring sparger, (0.16m diameter),and both spargers had 30 orifices of 0.002m diameter on the upper surface and four on the lower surface. In addition to the ring spargers a three bladed marine propeller,(0.16m diameter) was fitted at the base of the vessel and rotated in either direction inside the bottom of the draft tube, (fig.1). The propeller was operated to push liquid up the draft tube when used with the draft tube sparger and then rotated to draw liquid down the draft tube when operated with the annulus ring sparger.

<u>Hydrodynamic measurements</u>. Overall gas holdup was obtained from the volume expansion method involving measurements of the aerated and unaerated liquid heights. Gas holdup in the annulus of the vessel

163

E. Galindo and O.T. Ramírez (eds.), Advances in Bioprocess Engineering. 163-170.
© 1994 Kluwer Academic Publishers. Printed in the Netherlands.

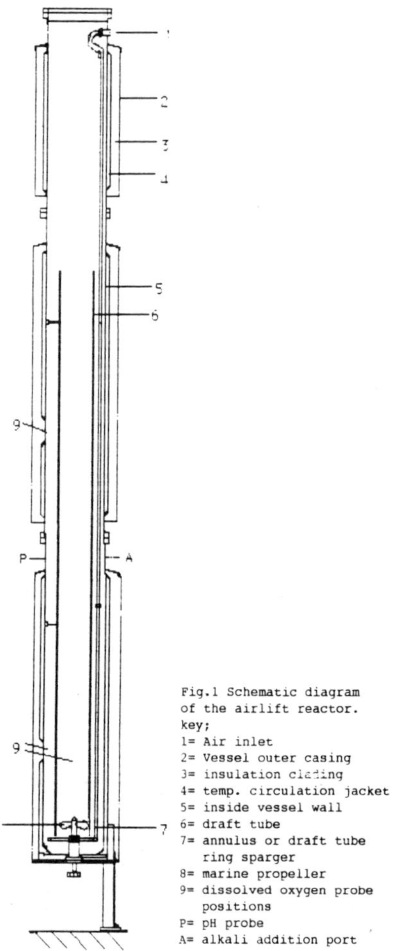

Fig.1 Schematic diagram
of the airlift reactor.
key;
1= Air inlet
2= Vessel outer casing
3= insulation cladding
4= temp. circulation jacket
5= inside vessel wall
6= draft tube
7= annulus or draft tube
   ring sparger
8= marine propeller
9= dissolved oxygen probe
   positions
P= pH probe
A= alkali addition port

was measured using the differential pressure measurement technique, by two Druck pressure transducers positioned at 0.53m and 2.11m heights. The operating conditions during the experiments were such that wall shear stresses had a negligible effect on holdup. Liquid circulation was measured by a pH pulse tracer technique. Addition of the alkali was by over pressure injection at a height of 1.48m into the annulus of the vessel and detection by a pH probe,(Ingold), at the same height in the annulus of the vessel.

Oxygen transfer measurements. The reactor was configured with three dissolved oxygen probes, placed at lower and upper annulus positions and the third penetrated into the lower draft tube. A fourth mobile dissolved oxygen probe was used to monitor dissolved oxygen tensions in the sections of the vessel during the non-sterile baker's yeast fermentations. The method of measurement for kla was that used by Sobotka *et al* (4), whereby the rate of oxygen transfer was measured by a gas balance method and was adapted for the airlift reactor by Russell(5). The inlet and exit gas compositions were monitored to obtain an estimate of the rate of oxygen transfer. The volumetric mass transfer coefficient could then be calculated with knowledge of the dissolved oxygen concentration at each probe position.

Saccharomyces cerevisiae fermentation. The airlift reactor was inoculated with baker's yeast *Saccharomyces cerevisiae*, (Distillers Company Ltd, Surrey, U.K.) at a concentrations of 10g dry cell weight per litre into a non-sterile medium containing (g/l),glucose,10;yeast extract, 10;ammonium sulphate,5; potassium dihydrogen orthophosphate, 2.5; polypropylene glycol, 0.25ml/l. The temperature was kept constant at 26^{0}C, and pH was automatically controlled at pH 7 by alkali addition.

Filamentous fermentation.
Saccharopolyspora erythraea, (NRRL B-2338) spores were harvested from a tap water agar medium after incubation at 29^{0}C for 11 days. 2ml of a spore solution, (10^6 spores/ml) was inoculated into a 2L baffled shake flask containing 500ml sterile medium. The medium contained, (g/l):bacteriological peptone, 4(Oxoid Ltd,Basingstoke);yeast extract, 6(Yeatex, Bovril Foods Ltd, Burton-on-Trent) ;glycine2; magnesium sulphate, 0.5;glucose, 10; potassium hydrogen orthophosphate, 1.36. The medium was adjusted to pH 7.0 and sterilised at 121^{0}C for 20 minutes. The glucose and potassium were sterilised separately and aseptically added following sterilisation. Flasks were incubated for 40h in order to provide an 8% (v/v) inoculum for the following seed stage. The fermenter (LH Fermentation Ltd,Reading) contained 25L of sterile medium and was agitated at 500 rpm by three Rushton turbine impellers. The airflow was set at 0.5vvm and the temperature controlled at 29^{0}C. Foam control was by addition of polypropylene glycol as required. The seed culture was grown for 19h and then transferred to the airlift reactor via a flexible tubing by overpressure in the seed fermenter. The airlift reactor was sterilised prior to inoculation at 121^{0}C for 45 minutes. The

medium was identical to that used in the seed fermenter except that the glucose concentration was higher, (30g/l). Carbon dioxide evolution rate, oxygen uptake rate, dissolved oxygen tension, pH and temperature were monitored throughout the fermentation. Samples were removed every four hours for viscosity, dry cell weight analysis and glucose analysis. The hydrodynamic measurements were performed when the biomass concentration reached 10g/l dry cell weight. Rheological measurements were performed at 26^{0}C using a Contraves rheomat 115 with a concentric cylinder viscometer.

<u>RESULTS AND DISCUSSION</u>

<u>Oxygen transfer characteristics of the aeration only operated bioreactor,(without propeller), with baker's yeast broth</u>

The mobile dissolved oxygen tension probe was used to measure the dissolved oxygen profile of the reactor in the direction of liquid flow with the baker's yeast broth. The dissolved oxygen profile for the vessel using the annulus ring sparger can be seen in figure 2, whereby air was sparged into the annulus of the vessel, which became the riser and the draft tube formed the downcomer. The dissolved oxygen tension in the riser can be seen to be low and independent of the airflow rate. In the top section where bubble disengagement from the vessel and entrainment into the draft tube occurs, only small rises in dissolved oxygen tension were observed at high airflow rates, (358L/min,0.136m/s). Dissolved oxygen was found to increase down the draft tube which was dependent on airflow rate. At all airflow rates the dissolved oxygen tensions rose down the draft tube with a dramatic reduction as the base of the vessel and subsequent flow direction change was reached. The position of maximum dissolved oxygen tension was found to move further down the draft tube with increasing airflow rate and hence a greater proportion of the downcomer had high dissolved oxygen tension values as the airflow rate was increased.
Even at the maximum airflow rate of 358L/min,(0.136m/s), where the mean oxygen uptake rate, (OUR), was 42mmol/L.h the dissolved oxygen tensions in the riser were still low which indicates that the oxygen demand, (OUR), was greater than the oxygen transfer rate to the liquid at this section of the vessel.

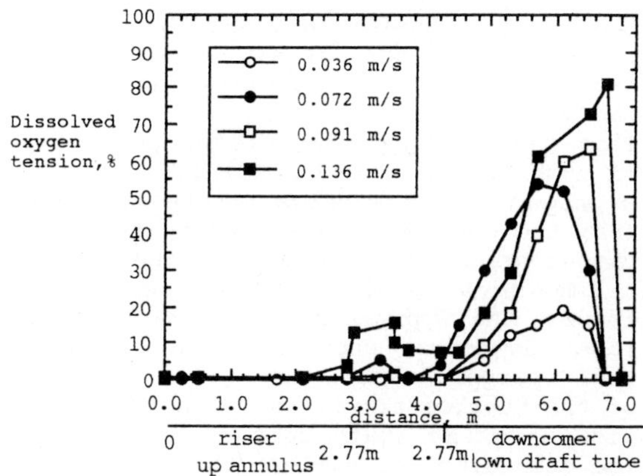

Figure 2. The dissolved oxygen tension profile of the reactor using the mobile dissolved oxygen probe with baker's yeast, annulus sparger, and four superficial gas velocities.

The high dissolved oxygen tension values in the downcomer indicate that the OUR was less than the oxygen transfer rate at all airflow rates but again, becomes equal to or greater than the oxygen transfer rate at the base of the reactor as the dissolved oxygen tensions reach low values. To discuss oxygen transfer around the vessel two assumptions must be made; there is uniform distribution of the broth and that the rate of oxygen uptake by the yeast can be regarded as being constant throughout the vessel. Abel *et al* (<u>6</u>), investigated the effect of periodical variation of dissolved oxygen concentration with *Saccharomyces cerevisiae*. They found that the cells metabolism was not affected when the cells experienced an oxygen limitation period of 10 s. However when the oxygen limited duration was 60 seconds the cells metabolism was affected. Sweere *et al* (<u>7</u>) also found that a cells metabolism was affected when the oxygen limited period was greater than 15 seconds. Therefore in the airlift reactor at low gas flow rates there may be an effect on metabolism by oxygen limitation as the circulation time was 34 s. Thus the cells spent at least half there time (17s) in an oxygen limited zone. However at all other airflow rates the duration of the oxygen limited zone that the cells experience, will be less than 15s and so the assumption of a constant oxygen uptake rate throughout the vessel is justified. As the rate of oxygen uptake can be regarded as being constant with position, the higher dissolved oxygen tensions in the downcomer and low values in the riser indicate that the oxygen transfer rate was the limiting parameter in the riser, with greater oxygen transfer to the liquid occurring in the downcomer.

This may indicate that of the population of bubbles introduced into the riser only a proportion of bubbles, those entrained into the downcomer, are more suitable for oxygen transfer. Bubbles in the region of 5-7mm have been observed descending through the sight glasses of the upper and lower downcomer sections with the draft tube sparger. Also, the actual velocity of the bubbles travelling with the liquid flow down the draft tube is a result of the counteraction against the bubble rise velocity, hence the difference between the gas and liquid phase velocities must be much greater in this section. This may provide a more conducive environment for oxygen transfer than in the riser. This shows the importance of liquid circulation time in relation to oxygen transfer, discussed with fig 5.

The values for the kla from the three fixed dissolved oxygen probe positions with the annulus sparger as a function of superficial gas velocity can be observed in figure 3. At low airflow rates the kla values were similar from the three probe positions, but as the airflow rate was increased the difference between lower downcomer and riser values becomes progressively larger, demonstrating the greater kla values obtained in the downcomer.

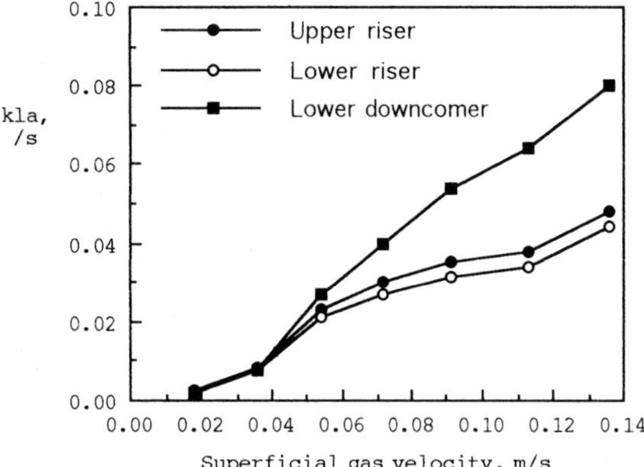

Figure 3. kla values measured at three dissolved oxygen probe positions with baker's yeast broth and the annulus ring sparger.

The circulation time is an important parameter when considering the position of greater oxygen transfer in the vessel. Figure 5 shows that at low gas velocities the circulation time was 30 seconds which

decreases with increasing gas velocity down to 11 seconds at the highest airflow rate, (358L/min,0.136m/s). To validate the liquid circulation results a typical probe response profile is shown in figure 4.

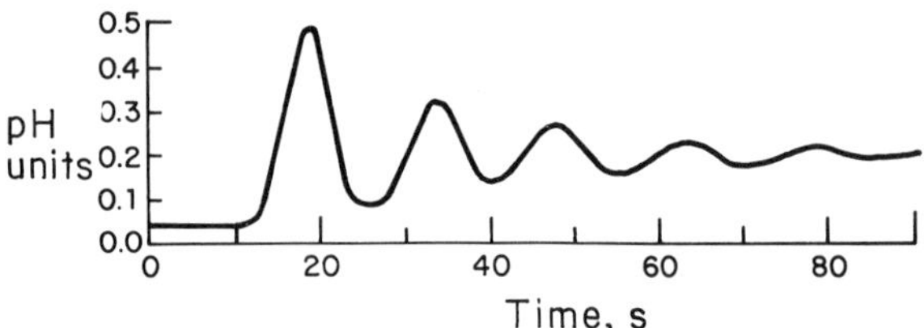

Figure 4. Probe response curve from the injection of an alkali pulse into the annulus of the vessel at a height of half the unaerated liquid height. Response measured with yeast broth at a superficial gas velocity of 0.072m/s .

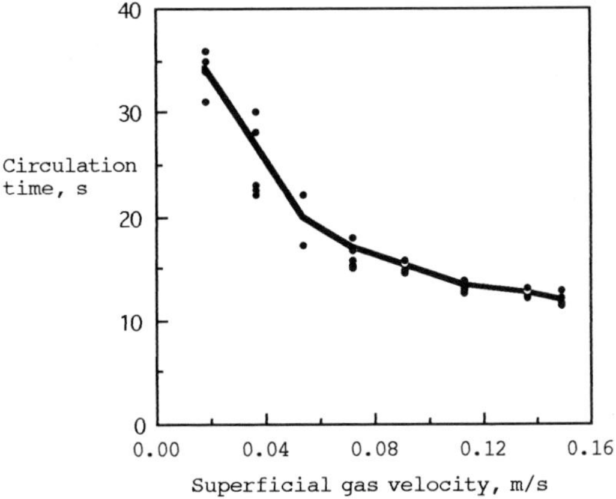

Figure 5. Circulation time with baker's yeast broth and the annulus ring sparger.

When the circulation times are considered with the profiles in fig.2, it can be observed that the cells experience extreme dissolved oxygen heterogeneity, in that the dissolved oxygen tension can change from 0 to 60% in 7 seconds at the highest airflow rate. Other researchers have observed dissolved oxygen heterogeneity in airlift reactors with biological fluids, (McNeil *et al*(8) and Lubbert *et al* (9). Lubbert *et al* (9) measured high dissolved oxygen tensions in the downcomer and low values in the riser, during yeast fermentations in a 78 L airlift reactor. Circulation times for the gas phase were found to be higher than the liquid phase, to the extent that bubbles moved slower in the downcomer, resulting in enhanced oxygen transfer in this region.

<u>Oxygen transfer and hydrodynamics of the aeration and propeller operated reactor with baker's yeast broth</u>

<u>Propeller operation with the annulus ring sparger</u> The reactor was operated with air sparged into the annulus,(riser), of the vessel and the propeller rotated to draw liquid down the draft tube,(downcomer). The effect of the propeller on aeration only operation can be seen in fig.6 using an high airflow of 358L/min. When the propeller was used in conjunction with the annulus sparger, the circulation time was found to decrease with increasing propeller speed and a more significant effect was observed at low airflow rates. Overall gas holdup was found to increase with similar increases at all airflow rates. Figure 6 also shows the kla at the lower riser position to have increased by three-fold with propeller operation which occurred at both high and low airflow rates. The point at 0rpm can be related to the individual points for the lower riser probe position on fig.3.

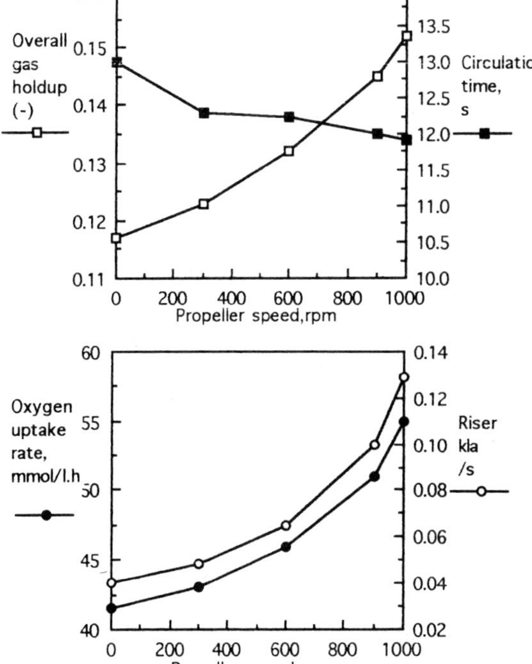

Figure 6. Effect of the propeller on the hydrodynamics and oxygen transfer performance of the annulus sparged reactor using an airflow 358L/min with yeast broth.

Thus the propeller improves the performance of the aeration only operation by increasing gas holdup, circulation time and kla which resulted in significant increases to the oxygen uptake rate with increasing propeller speed,(see fig.6). The effect on OUR was greater at low airflow rates. At the highest airflow rate, (358L/min), and propeller speed, (1000rpm), the propeller was found to raise the dissolved oxygen tensions from 0 in the riser to values in excess of 30%, see fig 7. However at all lower airflow rates the effect on dissolved oxygen tension was reduced and at low airflow rates the dissolved oxygen tensions were still zero but moderate increases in dissolved oxygen down the downcomer were observed.

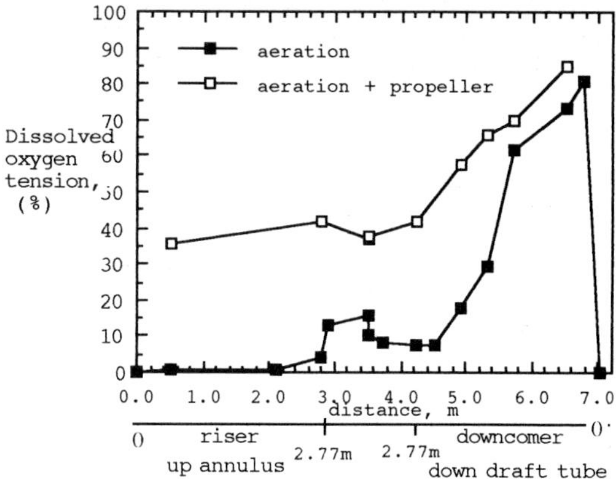

Figure 7. Effect of the propeller (1000rpm), on the dissolved oxygen tension around the vessel, airflow of 358 L/min with yeast broth and annulus sparger.

<u>Hydrodynamic and oxygen transfer characteristics of the aerated vessel with Saccharopolyspora erythraea fermentations</u>
The air sparged and propeller operated reactor has also been studied with the non-Newtonian S.erythraea, where vessel performance is complicated by rheology and morphology.

<u>Dissolved oxygen tension profile for the annulus sparged vessel with S.erythraea fermentation.</u> The dissolved oxygen profile for the two probe positions, lower riser and downcomer are shown in fig.8 for a fully dispersed mycelial fermentation. During the first nine hours the dissolved oxygen tensions fell rapidly with similar values around the vessel. At nine hours the airflow rate was increased to 358L/min and kept at this value for the remainder

of the fermentation. After nine hours the dissolved oxygen tension in the riser fell rapidly to zero and at the maximum OUR at 19 hours the dissolved oxygen at the lower downcomer position had increased, and remained high for remainder of the fermentation. Maximum biomass, (10g/l dry cell weight), was observed at 30 hours but the dissolved oxygen tensions remained low in the riser up to 49 hours whereby all the excess glucose in the media had been consumed. Maximum viscosity was observed at 49 hours. After 49 hours the dissolved oxygen tensions in the riser rose dramatically and reached 100% some 20 hours later.

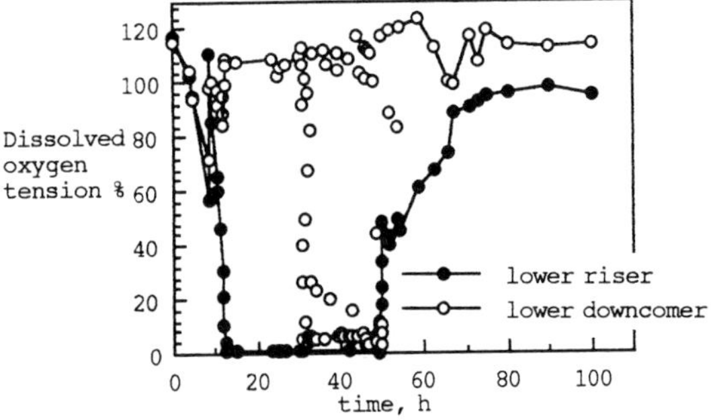

Figure 8. The dissolved oxygen tension profiles and operating conditions, for the *S.erythraea* mycelial fermentation in the annulus air sparged airlift reactor

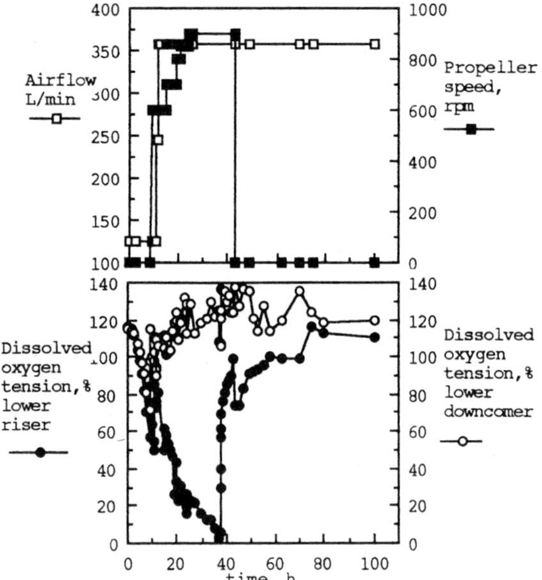

Figure 9. The dissolved oxygen tension profiles and operating conditions for the *S.erythraea* mycelial fermentation with the annulus air sparged and propeller operated reactor.

Dissolved oxygen profile for the annulus sparged and propeller operated reactor with *S.erythraea* fermentation

A fully dispersed mycelial fermentation was performed with the propeller operating as dissolved oxygen tension became limiting in the riser. The dissolved oxygen profile is shown in fig.9. During the first nine hours the dissolved oxygen tensions fell rapidly around the vessel to 45%. At this point a step wise increase of propeller speed and airflow rate was performed to prevent the dissolved oxygen tension in the riser from becoming limited. Maximum OUR occurred at 19 hours and biomass at 30 hours as with the aeration only fermentation. During this period the dissolved oxygen tension was maintained at 20% in the riser compared to zero in the aeration only operation.

After 30 hours the propeller speed was maintained at 900rpm and the dissolved oxygen tension in the riser fell gradually which coincided with an increase in viscosity. Maximum viscosity was observed at 37 hours where upon glucose consumption was complete resulting in the dissolved oxygen tension rising to 80% in the riser. The propeller operation was then removed and aeration operation only was maintained for the remainder of the fermentation.

Further differences between the aeration only and aeration, propeller operated mycelial fermentations were observed between the oxygen uptake rate,(OUR), profiles shown in fig.10. For the aerated only fermentation maximum OUR had been reached at 19 hours. The OUR then fell rapidly reaching a value of 11mmol/L.h which was maintained until 49 hours when the glucose consumption was complete. However operating the propeller with aeration allowed the OUR to be maintained at a maximum value of 23mmol/L.h until the glucose consumption was complete at 37 hours. Therefore operating the propeller during the growth phase and early stationary phase of a mycelial *S.erythraea* fermentation increased gas holdup,kla,and liquid circulation. This prevented low values of dissolved oxygen tensions in the riser for the majority of the fermentation. This resulted in the maintenance of a high OUR which concluded in the faster consumption of the excess glucose in the media, complete consumption occurring some 12 hours earlier than when using aeration only. These factors also contributed to increased production of the antibiotic erythromycin. The highest concentration at 50 hours with aeration and propeller operation was 108µg/ml compared to 75µg/ml with aeration only operation. Thus the dissolved oxygen heterogeneity observed in the aerated mycelial fermentation can be reduced by the utilisation of the propeller which enhances the hydrodynamics and oxygen transfer performance of the reactor.

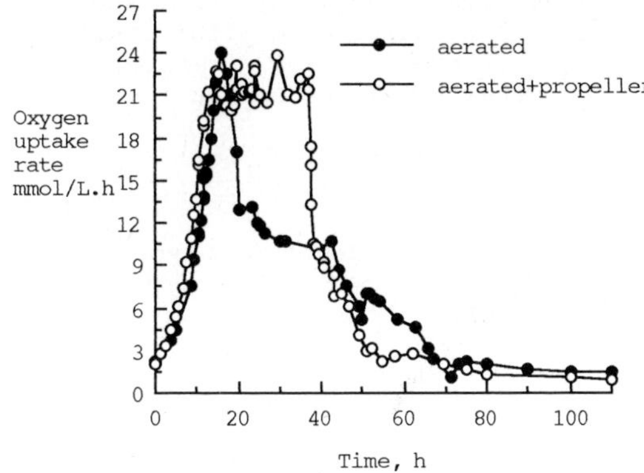

Figure 10. Differences in oxygen uptake rate from the *S.erythraea* mycelial fermentations using the aerated reactor and aerated + propeller operation.

Comparison of gas holdup between different rheological broths, morphologies, and reactor conditions

Figure 11 shows the overall and riser gas holdup profiles as a function of superficial gas velocity for baker's yeast, mycelial and pelleted *S.erythraea* aerated fermentations and mycelial fermentation operated with aeration and propeller. Gas holdup measurements for the *S.erythraea* fermentations were performed at 29 hours into the fermentation and the rheology of the broths can be seen from the table 1 to be considerably shear thinning compared to the Newtonian baker's yeast. Gas holdup varies significantly with superficial gas velocity for baker's yeast where a rapid rise in gas holdup was observed for gas velocities below 0.054m/s,(120L/min), followed by a reduced rate of increase at higher gas velocities. A similar situation was observed for the pelleted *S.erythraea* fermentation although the relationship was reduced and for a mycelial *S.erythraea* fermentation only small changes in gas holdup were observed for a large increase in gas velocity. The profile for the mycelial fermentation with aeration and propeller operation can be seen to be considerably above the mycelial aeration only profile, and only slightly lower than the Newtonian baker's

yeast profiles. Similar reductions in gas holdup with viscous non-Newtonian fermentation broths compared to Newtonian systems have been demonstrated in airlift reactors, (Moo-Young *et al* (10), Suh *et al* (11) and Kawase *et al* (12))

Reactor operation fermentation	n (−)	K (mPa.s^n)
Aeration only		
baker's yeast	1	30
S.erythraea mycelial	0.475	1200
S.erythraea pelleted	0.45	1195
Aeration +propeller		
S.erythraea mycelial	0.41	1240

Table 1. Rheology data for the Newtonian and non-Newtonian fermentations. Measurements for *S.erythraea* fermentations were made at 29 hours into the fermentation. The power law was used: flow behaviour,n, and consistency index, K.

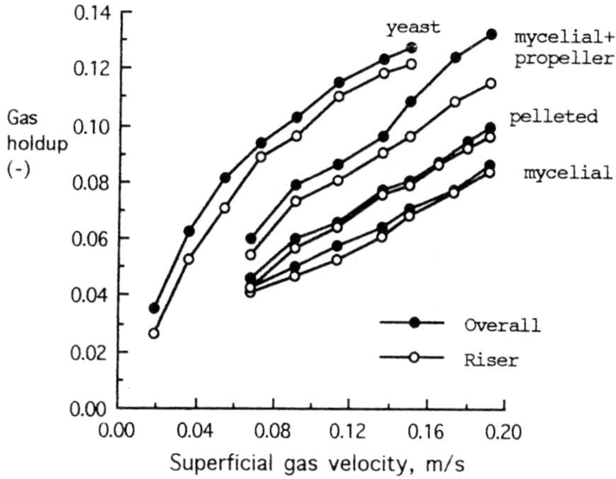

Figure 11 Comparison of the effect of gas velocity on gas holdup between yeast and *S.erythraea* mycelial and pelleted fermentations with aeration and aeration+ propeller operation.

CONCLUSION

It has been shown that extreme dissolved oxygen heterogeneity exists in the reactor with a simple rheological system of Newtonian baker's yeast and complex non-Newtonian mycelial fermentations. The hydrodynamic and oxygen transfer performance of the air sparged reactor can be improved when the annulus sparger is used in conjunction with the marine propeller. However only at the high airflow rates can the propeller be used to reduce dissolved oxygen heterogeneity. At lower airflow rates increasing the airflow is sufficient. These results demonstrate the effect of morphology and rheological properties of the fluids on the engineering environment, and the importance of reactor design on overall performance of the vessel.

ACKNOWLEDGEMENTS

University College London is the Biotechnology and Biological Sciences Research Council Interdisciplinary Research Centre for Biochemical Engineering and the Council's support to the participating University College London departments is gratefully acknowledged.

LITERATURE CITED

1. Wang, D.I.C., Fewkes, R.C.J. *Devel. Ind. Microbiol.* 18, 39-56 (1977).

2. Vardar, F., Lilly, M. D. *Eur. J. Microb.Biotechnol.*14,203-210 (1982).

3. Oosterhuis, N.M.G., Kossen, N.M.F. *Biotechnol. Bioeng.*26, 546-550 (1984).

4. Sobotka,M.,Prokop,A.,Dunn,I.J.,and Einsele,A.*Ann. Rep.Ferm.Proc.* 5, 127-210 (1982).

5. Russell,A.B.,"Hydrodynamics and oxygen transfer in a pilot scale airlift reactor,"PhD thesis, University College London (1989).

6. Abel,C.,Hubner,U.,Schugerl,K. *J.Biotechnology* 32,45-57 (1994)

7. Sweere,A.P.J.,Janse,L.,Luyben, K.Ch. A.M.,Kossen,N.W.F.*Biotechnol.Bioeng.* 31,579-586 (1988)

8. McNeil,B.,and Kristiansen,B. *Biotech Letts.* 12,1,39-44 (1990).

9. Lubbert,A., Frochlich,S., Larson,B., and Schugerl,K. *Proc.2nd.Int. Con. Bioreactor fluid dynamics.*p.379-393.

10. Moo-Young,M.,Halard,B.,Allen,G.D., Burrell,R.,and Kawase Y.*Biotechnol. Bioeng.* 30,746-753 (1987).

11. Suh,S.,Schumpe,A.,and Deckwer,W-D. *Biotechnol. Bioeng.*39,85-94 (1992).

12. Kawase,Y., and Moo-Young,M. *Can.J.Chem.Eng.* 70,2,48 (1992).

Invited paper

The Influence of Cell Concentration and Morphology on the Yielding Behavior of Filamentous Fermentation Broths

M. Mohseni[1], H. Kautola[2], and D.G. Allen[1]

[1]Department of Chemical Engineering and Applied Chemistry, University of Toronto, Toronto, CANADA,

[2] Seinajoki Institute of Technology, Seinajoki, FINLAND.

The influence of cell concentration and morphology on the yielding properties of filamentous broths of A. niger *and* S. levoris *was examined using the rotating vane technique and using dynamic measurements with a cone and plate viscometer. Cell concentrations ranging from 3 to 20 gdw l^{-1}, with corresponding volume fractions between 0.005 to 0.05, were generated via concentration / dilution of broths generated in stirred tanks and shake flasks. The yield stress values correlated with biomass concentration (X, gdw l^{-1}) using the equation $\tau_y = aX^b$ with "b" ranging between 2 and 3. Increasing particle length significantly increased the yield stress at a given biomass concentration and this morphology effect was well described using a morphology factor (range 0.016 - 0.11 Pa(gdw l^{-1})$^{-2.5}$), determined by setting the value of "b" to 2.5. The yielding strain was independent of cell concentration but increased with increasing cell size. The vane technique measures a combination of viscous and elastic properties and provides a realistic representation of the yield stress at zero shear rate that is significantly higher than the yield stress obtained by extrapolating shear stress-shear rate data.*

Fermentation broths of mycelial microorganisms are heterogeneous suspensions that are highly viscous, non-Newtonian, and possess a yield stress. This behaviour, that is due to the morphology of these organisms, leads to problems in mixing and mass and heat transfer. Several researchers have examined the measurement of broth rheology [e.g. 1,2,3] and its application to transport phenomena in bioreactors [e.g. 4,5,6,7]. However, few studies have attempted to relate broth characteristics such as organism morphology and concentration to rheological and transport properties. In particular, the role of broth characteristics on the yield stress has received little attention. Studies of fiber suspensions [8,9] suggest that the effect of particle morphology is most pronounced for low shear rheological properties (e.g. at or near the yield stress). Further study in this area is warranted since the yield stress of a broth influences mixing and mass transfer. Yield stress measurements may also provide a simple method for quantifying organism morphology.

The main objective of this study was to examine the influence of cell concentration (mass and volume fraction) and morphology on the yield stress of filamentous fermentation broths. What is unique about this work is that we directly measured the yield stress using a rotating vane technique that was recently developed in our lab [10]. Since the volume fraction of the particles is usually a key parameter in the study of the rheology of suspensions, the first step of the study involved developing a technique for volume fraction measurement in filamentous broths. Different cell morphologies were generated using two strains of filamentous organisms (*Aspergillus niger* and *Streptomyces levoris*) cultivated under different processing conditions. Cell concentration effects were examined by dilution of the broth.

Another aspect of the study involved the further examination of the vane measurement technique. In particular, we were interested in using viscoelastic measurements on broths to explain the significantly higher yield stress values obtained

171

E. Galindo and O.T. Ramírez (eds.), Advances in Bioprocess Engineering. 171-182.
© *1994 Kluwer Academic Publishers. Printed in the Netherlands.*

with the vane technique in comparison with those based on extrapolation of steady state shear data [10].

THEORETICAL BACKGROUND

The Vane Technique

The sensing element used in this technique consists of a vane with a number of blades (typically two to eight) arranged at equal angles around a cylindrical shaft (Figure 1). Although the technique was originally used for measuring the shear strength of soils, Nguyen and Boger [11,12] applied it to yield stress measurements of concentrated bauxite residue suspensions. Since that time it has been employed for other types of fluids including filamentous fermentation broths [10].

The constant rate vane method is based on monitoring the torque exerted on a slowly rotating vane of height, H_V, and diameter, D_V, that is immersed in the fluid. As the vane rotates, the torque increases until it reaches a maximum value (T_m), corresponding to the yielding point, and then decays to a steady state value. The method assumes that the vane plus the sample contained between the blades act as a solid cylinder and that the effects due to the upper free surface and the container boundaries are negligible. Thus, the torque (T) on the vane is related to the stress on the cylindrical wall (τ_w) and the two end surfaces of the vane by [10,11]:

$$T = \left(\frac{\pi}{2} D_V^2 H_V \right) \tau_w + 2 \left(2\pi \int_0^{D_V/2} \tau_e(r) r^2 dr \right) \quad (1)$$

where τ_e is assumed to be constant and equal to τ_w at the point of yielding. The maximum torque (T_m) is related to the yield stress τ_y by [10,11]:

$$T_m = \frac{\pi D_V^3}{2} \left(\frac{H_V}{D_V} + \frac{1}{3} \right) \tau_y \quad (2)$$

Demonstration of the validity of Equation (2) on filamentous broths and further details on the method are available elsewhere [10].

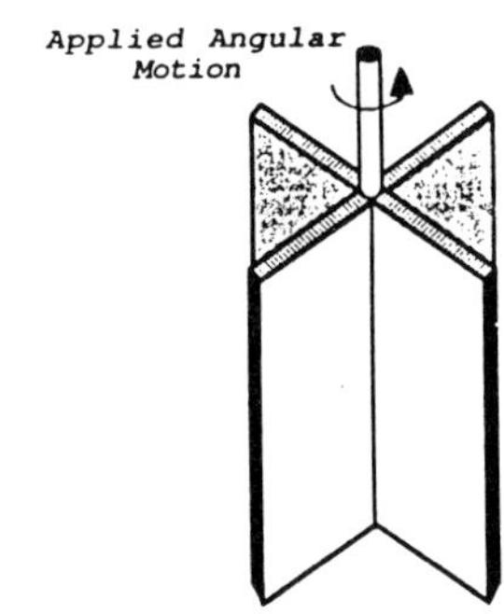

Figure 1. Schematic diagram of the vane.

EXPERIMENTAL

Organisms and Culture Conditions

Rheological data were obtained on broths of two different organisms: a mould, *Aspergillus niger* (ATCC 10864), and an actinomycete, *Streptomyces levoris* (ATCC 15876). The compositions of the media used are reported by Allen and Robinson [1].

To provide different cell morphologies, the broths were produced in stirred tank fermentors and shake flasks at a variety of processing conditions. Two stirred tank fermentors were used: (i) a 6-L stirred tank fermentor (4-L working volume; WHE Process Technology Group, Inc., Weston, ON, Canada); and (ii) a 5-L stirred tank fermentor (3-L working volume, Microferm fermentor, New Brunswick Scientific, NJ). Two liter Erlenmeyer flasks were used as shake flasks.

The fermentation of *A. niger* was carried out for about three days at an impeller speed of 400 to 500 rpm. *S. levoris* fermentations were carried out for about two days with an impeller speed of 500 rpm. A silicon antifoam (1520-US-Dow Corning Canada Inc., Mississauga, ON) was used for foam control, the temperature was maintained at 30°C, and the air flow rate was at 1 VVM (STP). The pH for *A. niger* was controlled at 5±0.1 with 1N NaOH solution. For the *S. levoris* the pH was not controlled and increased from 6.8 to 8 over the course of the fermentation. Controlled conditions for growth in the shake flask (shaker model G25-671542, New Brunswick Scientific, NJ) were at 30°C and 200 rpm.

In order to obtain various concentrations of the same broth (range 2 to 20 gdw/L), the broth was concentrated by either filtration or centrifugation and then diluted to the desired concentration with the supernatant.

Cell Concentration and Morphology

Biomass. The cell mass concentration was determined by filtering a known volume of the broths (about 20 ml) through a filter paper (1μm A/C glass fiber for _A. niger_ and 0.2μm Supor-200 Membrane filter for _S. levoris_; Gelman Science Inc.) and drying the filtered sample at 104°C for about 24 hours. The determination of biomass concentration of diluted broths was based on the dry weight of the concentrated sample and dilution by filtrate or centrifuged supernatant.

Volume fraction. Three different methods were applied for cell volume fraction measurements. The first two were based on the measurement of extracellular volume using a compound "excluded" from the cell membrane and the third method involved the measurement of the volume of a dewatered broth filter cake.

The first two methods were based on the techniques proposed by Reuss et al. [13], Arnold and Lacy [14], and Ho and Ju [15]. Naphthol Green-B (M_w=878.47, Fluka AG) and dextran (M_w=515000, Sigma B512) were the compounds tested in this study. A known amount of dye or dextran was added to a known volume of broth and centrifuged for about 15 min at 5000 rpm. Assuming that there was no adsorption of the compound on the biomass, the net volume of the cells, V_C, can be determined via mass balance on the dye or dextran [13,14,15]:

$$\phi = \frac{V_C}{V_t} = 1 - \frac{M}{C_s V_t} \qquad (3)$$

where ϕ is the cell volume fraction, V_t is the total volume of the broth, M is the amount of dye or dextran added to the broth, and C_s is the concentration of dye or dextran in the supernatant. The concentration of dye in supernatant was measured by spectrophotometry at 720 nm and a gravimetric method was employed for dextran.

The measurement of dewatered cake volume was based on a method reported by Oolman and Liu [16]. About 50 ml of broth was filtered with a constant pressure drop and the filter cake was rinsed with about 10 ml of 99% ethanol to displace interstitial water. The alcohol was evaporated for 1 hr at room temperature and the volume of dewatered cake was measured by displacement in a pycnometer (Kimble 1511524, Canlab Scientific Products). The cell volume fraction was determined by dividing the cake's volume by the total volume of the broth. Using the cell volume fraction and biomass dry weight, the hyphal density (gdw/ml) was also determined.

Morphology. In order to provide an approximate measure of cell morphology, several slides of each broth were examined under a microscope at 500X magnification. The length, diameter, and number of branches of 20 to 30 randomly selected free filaments were measured and averaged.

Rheological Measurements

Direct yield stress measurements were performed using a Haake Rotovisco RV12 viscometer (Haake Buchler Instruments, Saddle Brook, NJ) with an M150 measuring head. In order to cover a range of yield stresses, one of two 4-bladed vanes (D_V=5 cm, H_V=7.5 cm; or, D_V=4 cm H_V=8.8 cm) were attached to the measuring head. A 1-L glass beaker (H_C=15 cm; D_C=11 cm) or a 2-L glass beaker (H_C=17 cm; D_C=12.5 cm) was used to hold the broth sample. The vane was then immersed in the fluid, rotated at constant speed and the torque was recorded with time. The yield stress was calculated using the maximum torque value according to Equation (2). Details of the procedure are available elsewhere [10].

Dynamic tests were carried out on several *A. niger* broths using a Rheometrics RAA Analyzer (Rheometrics Inc., NJ). A cone and plate viscometer was used with a cone that was 50 mm in diameter at an angle of 0.04 radians. In order to find the linear viscoelastic region, all of the properties were measured as a function of strain amplitude at a fixed frequency. Frequency sweep tests were then performed within the linear strain values.

RESULTS AND DISCUSSION

Volume Fraction Techniques

The extracellular volume measurements based on dye and dextran exclusion methods were very erratic and sometimes produced negative volume fractions. The statistical analysis of the results obtained from replicate measurements on sub-samples from several broths showed that there was a significant difference between the sub-samples. Such a variation and the negative values may be explained based on the partial adsorption and/or partitioning of dye or dextran during the tests. A mycelial organism is made of both live and dead regions that vary in relative quantity from broth to broth and it is likely that the structure of the cell wall in the dead region is much more porous such that dextran or dye molecules are able to pass through it. The invalidity of the adsorption assumption. in this test and also the difference between the composition of the cell walls for molds and yeasts or bacteria can explain why other investigators [13,14] have been successful with this technique. Although Ho and Ju [15] reported volume fraction measurements on broths of the mould *Penicillium chrysogenum* using this technique, they also state that the data were somewhat scattered.

The results obtained on the volume fraction measurements using the filtration method for different concentrations of each broth are reported in Figure 2. The data points represent the average of two or three measurements at each biomass concentration. The hyphal density was calculated to range from 0.4 to 0.65 gdw/ml for both organisms. Statistical analysis of the data in Figure 2 indicated that the data for each broth was linearly correlated and these correlations were significantly different from each other and had an intercept which was not significantly different from zero. These two points (linearity and zero intercept) along with the good agreement of our hyphal density with those obtained by others [16,17] working on filamentous organisms, suggest that this method is suitable for cell volume fraction measurements.

The results also suggest that the hyphal densities for each broth were significantly different although variability in the dry weight measurement and differing broth dewatering/ethanol drying characteristics can also account for the measured differences. From a practical perspective, the volume fraction at a given biomass concentration is not radically different between the broths. Given the extra effort associated with the volume fraction measurement, dry weight fractions provide a more realistic measure of broth concentration for rheological studies. However, when comparing results from model slurries (e.g. pulp suspensions) and actual broths, a volume fraction approach would likely be required.

Characteristics of The Vane Method

A typical vane test curve for a filamentous fermentation broth is shown in Figure 3. As the vane starts rotating from rest, the region of suspension close to the edge of the vane blades deforms elastically, as is shown by the linear part of the torque-time response (region AB). Such linear behaviour is considered to result from the initial stretching and deformation of filaments and aggregates. As the deformation or strain increases, the elastic behaviour of the system becomes more complicated and the torque-time curve is non-linear (region BC). Finally, when all the network bonds are broken, additional deformation becomes irreversible, the material yields (point C), and the torque declines with time (region CD).

<u>Influence of vane speed.</u> The effect of vane rotational speed on the yield stress on two different *A. niger* broths is illustrated in Figure 4. The measured yield stress did not vary significantly over speeds ranging from 0.01 to 0.2 rpm but for some samples the yield stress declined significantly (but slightly) at higher rotational speeds (e.g. above 0.2 rpm). Since the vane method is based on slowly shearing to detect the yielding of the material, it can be concluded that the optimum operating range of vane rotational speeds is between 0.01 to 0.2 rpm.

For the *S. levoris* broths, at lower rotational speeds, the torque

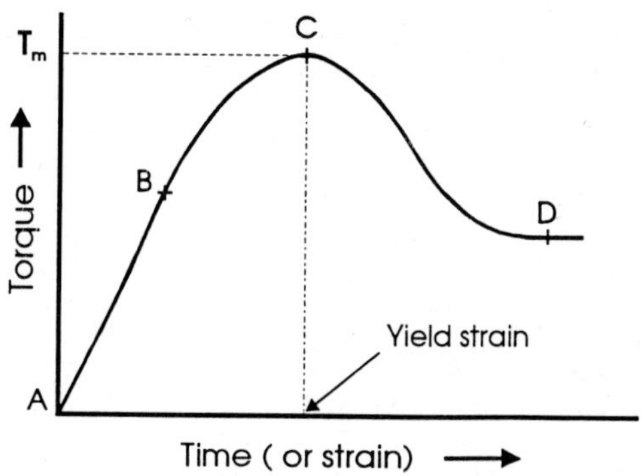

Figure 3. Typical torque-time response recorded during the tests on filamentous fermentation broths.

gradually approached a maximum, constant stress with time and did not show the characteristic yielding peak (i.e. point C on Figure 3). The same type of response was reported for concentrated Solka-Floc suspensions in carboxymethylcellulose solution by Leong-Poi and Allen [10]. Visual observation during these tests on *S. levoris* broths also showed that the whole system was rotating as a solid body and did not reach the yielding point. As the vane speed was raised to about 0.08 to 0.16 rpm, the momentum in the tangential direction increased, the network bonds interconnecting the structure elements were broken, and the material yielded coincident with a maximum in the torque time trace. These peak values were subsequently used to calculate the yield stress for these broths.

The contrast in the torque vs. time trace and the yielding behaviour with vane speed for *A. niger* on the one hand and *S. levoris* and CMC solutions on the other hand, is postulated to be a result of the different "particle" morphologies (Table 1) and the relative magnitude of the viscous vs. the elastic resistance of these materials. The larger, *A. niger*, particles provide for a more highly structured, solid-like matrix that requires a substantial stress to yield. While the yield stress is lower for the smaller, *S. levoris*, particles

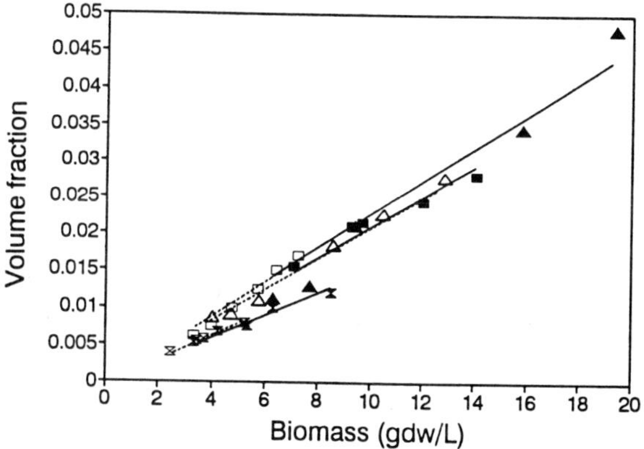

Figure 2. Cell volume fraction versus biomass concentration based on the filtration method; *A. niger* broths: (▲) stirred tank #1, (■) stirred tank #2, (X) shake flask. *S. levoris* broths: (△) stirred tank #1, (□) stirred tank #2, (Ⓩ) shake flask. Lines denote linear least square fits; (——) *A. niger*, (---) *S. levoris*.

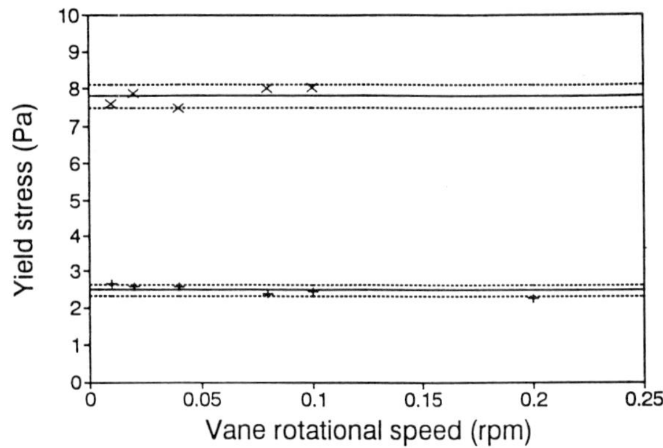

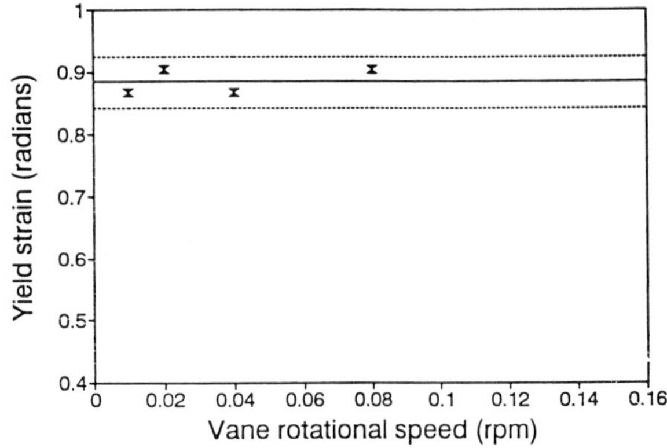

Figure 4. Variation of yield stress with vane rotational speed for *A. niger* fermentation broths; (×) 5.46 gdw/L shake flask, (+) 4.83 gdw/L stirred tank:
(——) mean value of yield stress.
(---) 95% confidence interval of mean τ_y.

Figure 5. Variation of yield strain with rotational speed for a 8.5 gdw/L *A. niger* shake flask fermentation broth (D_v=5 cm) ; (——) mean value of yield stress, (---) 95% confidence interval of mean yield strain.

(Figure 6) and the polymers in solution, the substantial relative viscous resistance of these materials leads to the movement of the entire fluid in the beaker prior to yielding and the absence of a characteristic peak. As the speed of the vane increases, the shear rate increases and the viscous resistance drops to the point where the *S. levoris* broths eventually yield at the vane edge and the characteristic peak is observed on the torque vs. time trace. Eventually at a higher rotational speed, a lower viscous resistance due to shear thinning of the broth together with inertial effects may introduce a small reduction to the maximum torque as was observed for the *A. niger* broths.

<u>Yield strain and rpm</u>. As illustrated in Figure 5, statistical analyses showed that the yield strain was not correlated with the vane rotational speed for the *A. niger* broths. This indicates that there is a specific breaking point for a broth and that the shearing surface is reasonably constant at all the speeds.

Unlike the *A. niger* suspensions, the yield strain of the *S. levoris* broths was influenced by rotational speed, with higher rotating speeds resulting in higher yield strains. This type of behaviour can be explained due to the movement of the broth at the sides of the vessel

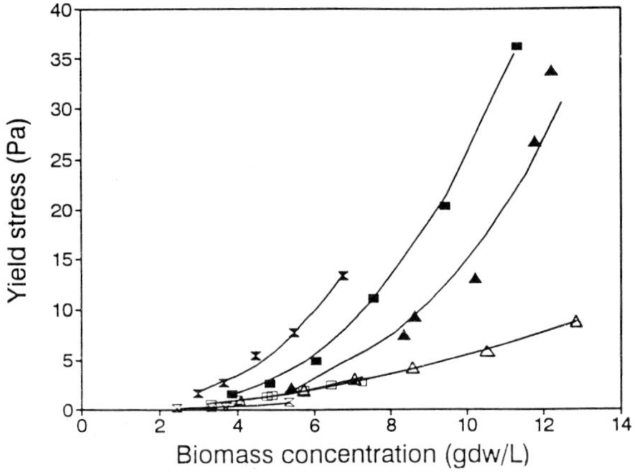

Figure 6. Yield stress as a function of biomass concentration with different organisms and processing conditions;
A. niger broths: (▲) stirred tank #1, (■) stirred tank #2, (✗) shake flask.
S. levoris broths: (△) stirred tank #1, (□) stirred tank #2, (✗) shake flask.
Lines denote least square fits of the data (Log-Log plot) to $\tau_y = aX^b$.

Table 1. Quantitative representations of the morphology of organisms used in rheological studies.

Sample	L_e*	D_e*	n*	Morphology Factor
A. niger stirred tank #1	420±120	5.5±3.5	2.9±2	0.055
A. niger stirred tank #2	540±150	5±1	2.7±2	0.08
A. niger shake flask	600±180	4±1.2	2±2	0.11
S. levoris shake flask	50±15	0.8±0.3	3.2±2	0.011
S. levoris stirred tank #1	45±14	0.8±0.3	4±2	0.016
S. levoris stirred tank #2	40±18	1.0±0.3	4.5±2.4	0.023

*L_e length of the main hypha (μm)
D_e diameter of the main hypha (μm)
n   number of branches from the main hypha
Error bars indicate ± 1 standard deviation.

during yielding; at higher speeds this would allow the vane to move farther relative to its initial starting point. In other words, the strain of the broth within the vane relative to the moving broth outside the vane may have stayed constant with speed.

Cell Concentration and Yield Properties

As expected, the vane measured yield stress for *A. niger* and *S. levoris* broths was highly dependent on biomass concentration (Figure 6). Plotting the data against either biomass concentration or volume fraction illustrated the same trends. The data in the Figure is well described by the correlation frequently used by others [1,2,3,18,19]:

$$\tau_y = a. X^b \qquad (4)$$

where *a* and *b* depend on the test procedure employed and other characteristics of the suspension. The value of *b* was in general agreement with those reported by others (Table 2) and was between 2.5 to 3.2 for *A. niger* broths, and between 2 to 2.5 for *S. levoris* broths.

Figure 6 also illustrates the influence of cell morphology on yield stress at a given cell concentration.

Table 1 provides the corresponding cell morphology measurements and values of the "morphology factor" as proposed by Roels et al.[3], which was determined by setting the exponent b in Equation (4) to 2.5. Although the morphological measurements are preliminary, a few general statements about the influence of morphology can be made.

Table 2. Comparison of the results for the correlation between biomass concentration and yield stress

Reference	Suspension	Yield Stress (Pa) [*]	b in Equation (4)
Roels et al. [3]	P. chrysogenum	τ_y^{C}	2.5
Metz et al. [2]	P. chrysogenum	τ_y^{C} τ_y^{B}	2.5-2.8 2.8-3.5
Allen and Robinson [1]	A. niger	τ_y^{B} τ_y^{C}	2.3 3.0
	P. chrysogenum	τ_y^{B} τ_y^{C}	2.1 2.5
Tucker and Thomas [19]	P. chrysogenum	τ_y^{B} τ_y^{C}	3.2±0.4 2.9±0.8
Bennington et al. [18]	Pulp fiber suspensions	τ_y	2.3-3.6
This study	A. niger	τ_y	2.5-3.2
	S. levoris	τ_y	2.0-2.5

[*] τ_y^{B} yield stress determined by extrapolation of steady state shear data using the Bingham model: $\tau - \tau_y^{B} = \eta_p \dot{\gamma}$

τ_y^{C} yield stress determined by extrapolation of steady state shear data using the Casson model: $\tau^{0.5} = (\tau_y^{C})^{0.5} + \eta^{C}(\dot{\gamma})^{0.5}$

τ_y yield stress measured using a direct method.

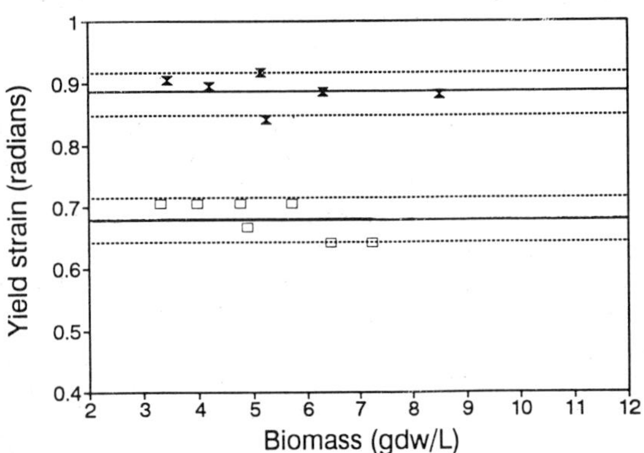

Figure 7. Variation of yield strain with biomass concentration for (X) A.niger shake flask, and (□) S. levoris stirred tank fermentation broths (D_v=4 and 5 cm); (——) mean value of yield strain, (---) 95% confidence interval of mean yield strain.

Larger particles such as those found in the A. niger broths tend to give a larger yield stress and therefore a higher "morphology factor". This effect is very pronounced when the two different microorganisms are compared to each other although it was also significant for the A. niger broths produced under different conditions. A more detailed morphological analysis coupled with image analysis is required before more quantitative conclusions can be made.

As shown in Figure 7, for all the broths tested (A. niger and S. levoris) there was no statistically significant correlation between the vane yield strain and biomass concentration at the 95% confidence level. The yield strain obtained for the S. levoris suspensions was also significantly lower than that of the A. niger broths and there was a small but significantly higher yield strain for the A. niger broths grown in shake flasks compared with those grown in stirred tanks. Similar to the yield stress, these differences were likely caused by cell morphology differences for the two strains of

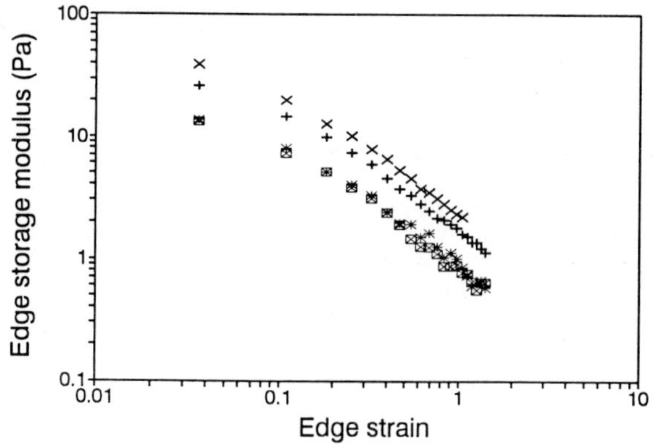

Figure 8. Storage modulus versus edge strain for *A. niger* stirred tank fermentation broths at a frequency of 10 rad/s; (×) 14.7 gdw/L, (+) 12.05 gdw/L, (∗) 9.7 gdw/L, (⊠) 7.11 gdw/L.

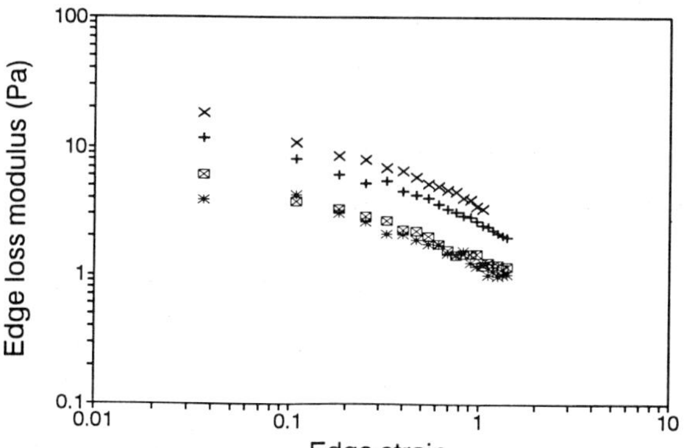

Figure 9. Loss modulus versus edge strain for *A. niger* stirred tank fermentation broths at a frequency of 10 rad/s; (×) 14.7 gdw/L, (+) 12.05 gdw/L, (∗) 9.7 gdw/L, (⊠) 7.11 gdw/L.

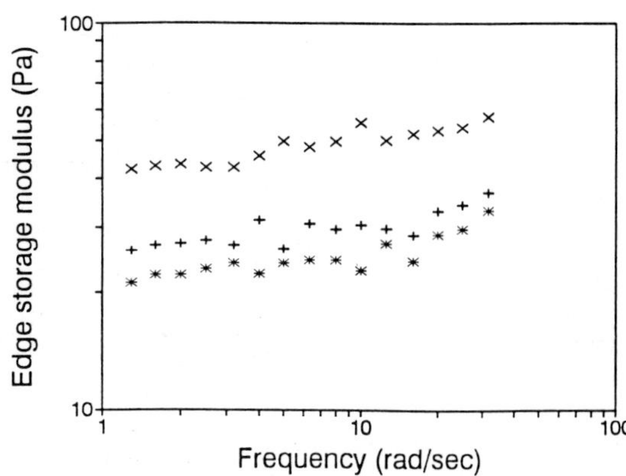

Figure 10. Storage modulus versus angular frequency for *A. niger* stirred tank fermentation broths at an edge strain of 0.04; (×) 14.7 gdw/L, (+) 12.05 gdw/L, (∗) 9.7 gdw/L.

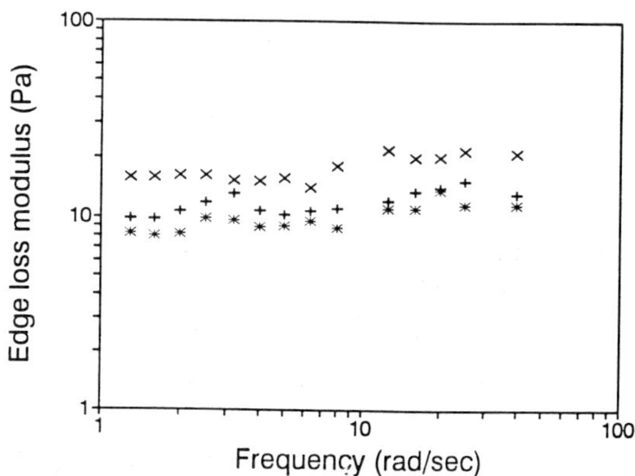

Figure 11. Loss modulus versus angular frequency for *A. niger* stirred tank fermentation broths at an edge strain of 0.04; (×) 14.7 gdw/L, (+) 12.05 gdw/L, (∗) 9.7 gdw/L.

organisms in that the longer the hyphae, the greater the deformation needed to stretch and break all the network bonds. This suggests that yield strain measurements may provide a rapid, indirect measure of cell morphology.

<u>Dynamic Properties and the Vane Measured Yield Stress</u>

Figure 8 shows the storage (elastic) modulus, G', as a function of edge strain for four different concentrations of *A. niger* fermentation broths. The data are presented at a constant nominal frequency of 10 s^{-1}. Despite the fact that the actual modulus values are likely higher due to wall slip on the smooth surfaces of the viscometer, they are qualitatively very similar to those found for pulp suspensions by Damani et al. [20] and are characteristic of a nonlinear viscoelastic material. G' decreased with increasing edge strain (i.e. strain at the edge of the cone), γ_e, and increased with increasing biomass concentration. Comparing the elastic modulus to the viscous component or loss modulus (G'') (Figure 9), the elasticity of the broths seems to be as, if not more, important as the viscous component at low strains but it declines much faster as the deformation increases.

The effect of frequency, ω, on dynamic properties was also investigated (Figures 10 and 11). This test was conducted at low edge strains (about 0.04) to ensure that we were as close as possible to the linear viscoelastic region. The storage modulus was nearly independent of frequency, with only a very slight increase that can be attributed to inertial effects.

Similar behaviour has been noted by Damani et al. [20] on pulp fiber suspensions. The loss modulus was also constant with frequency and was considerably lower than the storage modulus.

It is proposed that the quantity measured with the vane method is a relevant material property that includes both viscous and elastic properties. The non-linear portion of the torque vs. time plot (region BC in Figure 3) is consistent with the decreasing storage and loss moduli with strain (Figures 8 and 9) and is indicative of non-linear viscoelastic behaviour. The viscoelastic effect may be determined qualitatively from the ratio of the torque at the yielding point to that of the steady state situation (i.e. region D in Figure 3). This steady torque value is the result of viscous behaviour of the material, while, the vane yield stress is based on both viscous and elastic properties. The experimental results show that the ratio of the maximum torque to the steady torque value is a function of particle morphology and it rises rapidly as the mass concentration increases.

It may also be concluded that the significant viscoelastic nature of the broth is responsible for the greater yield stress values measured by the vane technique compared to those obtained from extrapolation of shear stress-shear rate data [10]. It would appear that the "true" flow curve for suspensions of broths like *A. niger* would show a high stress at zero shear rate (i.e. the true "static" yield stress) that rapidly declines with shear rate. Similar results have been reported in a review by Cheng [21] and by Sestak et al. [22] on bentonite suspensions in water. It is this "static" yield stress that is relevant when one is trying to initiate flow in a suspension (e.g. related to the formation of dead zones in bioreactor). Therefore, the vane measured yield stress is representative of the actual yield stress of the broth and is a potentially useful broth property for bioprocess engineering.

CONCLUSIONS AND RECOMMENDATIONS

1. Measurement of the volume of dewatered filter cake is preferable over dye and dextran exclusion techniques for measuring the cell volume fraction in filamentous suspensions. However, given the variability and complexity involved in making these measurements and the similar hyphal densities among broths, weight fractions are recommended for rheological studies except when making comparisons with other, non-biological, suspensions.

2. The vane yield stress is highly correlated with biomass concentration (X) and cell morphology. The yield stress increases with X^b with b between 2 and 3 and increases with cell length.

3. The vane yield strain is independent of biomass concentration but is affected by particle morphology in that it increases with increasing particle length.

4. It is recommended that further, detailed, morphological studies be carried out to quantitatively investigate the influence of morphology on the vane measured rheological properties. The potential use of the vane method to indirectly and rapidly quantify cell morphology should be explored.

5. The vane torque-time response is the result of the elastic properties of the broth. The vane measured yield stress is a good representation of the "true" yield stress at zero shear rate, this being considerably

higher than that obtained by extrapolation of steady state shear stress vs. shear rate data.

ACKNOWLEDGEMENTS

This work was supported from the Natural Sciences and Engineering Research Council of Canada. M. Mohseni also gratefully acknowledges the scholarship support from the Ministry of Culture and Higher Education in Iran. Dr. Kautola was supported with a fellowship from the Academy of Finland.

NOTATION

a	constant in Equation (4) $(Pa/(gdw/L)^b)$
b	constant in Equation (4) (dimensionless)
c_s	concentration of dye or dextran in supernatant (g/L)
D_e	diameter of the main hypha (μm)
D_v	vane diameter (m)
G'	storage (elastic) modulus (Pa)
G''	loss modulus (Pa)
H_v	vane height (m)
L_e	length of the main hypha (μm)
M	total amount of dye or dextran added to the broth (g)
n	number of branches of the main hypha
r	radial position along the vane (m)
T	torque (Nm)
T_m	maximum torque (Nm)
V_c	cell volume (L)
V_t	total volume of suspension (L)
$\dot{\gamma}$	shear rate (s^{-1})
γ_e	edge strain (dimensionless)
η_c	Casson viscosity (Pa.s)$^{0.5}$
η_p	plastic viscosity (Pa.s)
τ	shear stress (Pa)
τ_e	shear stress on the end surfaces of the vane (Pa)
τ_w	shear stress on the cylindrical wall outlined by the vane (Pa)
τ_y	yield stress (Pa)
ϕ	cell volume fraction (dimensionless)
ω	angular frequency (rad s^{-1})

LITERATURE CITED

1. Allen, D.G., and C.W. Robinson, *Chem. Eng. Sci.*, **46**, 37-48 (1990).

2. Metz, B., N.W.F. Kossen, and J.C. Van suijdam, *Adv. Biochem. Eng.*, **11**, 103-156 (1979).

3. Roels, J.A., J. Van der berg, and R.M. Voncken, *Biotechnol. Bioeng.*, **16**, 181-208 (1974).

4. Allen, D.G., and C.W. Robinson, *Can. J. Chem. Eng.*, **69**, 498-505 (1991).

5. Allen, D.G., and C.W. Robinson, *Biotechnol. Bioeng.*, **34**, 731-740 (1989).

6. Moo-Young, M., B. Halard, D.G. Allen, R. Burell, and Y. Kawase, *Biotechnol. Bioeng.*, **30**, 746-753 (1987).

7. Baker, M.R., A.N. Emery, and A.W. Nienow, "Mass Transfer and Power Characteristics of a Simulated Filamentous Fermentation Broth Exhibiting Biological Activity", in *Bioreactor Fluid Dynamics*, R. King (Ed.), Elsevier Applied Science Publishers, 79-93 (1988).

8. Goto, S., H. Nagazono, and H. Kato, *Rheol. Acta*, **25**, 119-129 (1986).

9. Goto, S., H. Nagazono, and H. Kato, *Rheol. Acta*, **25**, 246-256 (1986).

10. Leong-Poi, L., and D.G. Allen, *Biotechnol. Bioeng.*, **40**, 403-412 (1992).

11. Nguyen, Q.D., and D.V. Boger, *J. Rheol.*, **27**, 321-349 (1983).

12. Nguyen, Q.D., and D.V. Boger, *J. Rheol.*, **29**, 335-347 (1985).

13. Reuss, M., D. Josic, and M. Popovic, *Eur. J. Appl. Microbiol. Biotechnol.*, **8**, 167-175 (1979).

14. Arnold, W.N., and J.S. Lacy, *J. Bacteriol.*, **131**, 564-571 (1977).

182

15. Ho, C.S., and L. Ju, *Biotechnol. Bioeng.*, **32**, 95-99 (1988).

16. Oolman, T., and T. Liu, *Biotechnol. Prog.*, **7**,534-539 (1991).

17. Thomas, C.R., H.L. Packer, E. Keshavarz-poore, and M.D. Lilly, *Biotechnol. Bioeng.*, **39**, 384-391 (1992).

18. Bennington, C.P.J., R.J. Kerekes, and J.R. Grace, *Can. J. Chem. Eng.*, **68**, 748-757 (1990).

19. Tucker, K.G., and C.R. Thomas, Personal Communication (1993).

20. Damani, R., R.L. Powell, and N. Hagen, *Can. J. Chem. Eng.*, **71**, 676-683 (1993).

21. Cheng, D.C.-H., Rheol. Acta, **25**, 542-554 (1986).

22. Sestak, J., M. Houska, and R. Zitny, J. Rheol. **26**, 459-475 (1982).

A New Approach for Modelling the Kinetics of Mycelial Cultures

G. Viniegra-González, C.P. Larralde-Corona, and F. López-Isunza

Universidad Autónoma Metropolitana - Iztapalapa, Apartado Postal 55-535, C.P. 09340, México, D.F., MEXICO

A new approach for modelling the kinetics of mycelial cultures is presented. It involves the estimation of the microscopic branching frequency in terms of the rate of mycelial tip elongation and the critical length of mycelial leading segments. Branching frequency was then related to the macroscopic specific growth rate. Two models for estimating the braching frequency (zero and first order) as a function of segment length, were considered. Simultaneous predictions of the observed values of specific growth rates and critical branching lengths were made in order to compare those models. Also, a macroscopic kinetic expression for the evolution of a mycelial culture was developed, based in terms of a self inhibiting power function of biomass density. Such an expression was used to test the zero and first order models for predicting mycelial growth. Results indicated that the first order model was the more adequate one.

Mycelial cultures have become of great importance for industrial production of antibiotics, enzymes, organic acids and food protein. But unlike single cell cultures, their kinetics are much less understood because of the complex nature of the growth of hyphae having a set of rules for tip extension and branching that have been the subject of basic research for more than 25 years [1 to 6].

Until very recently, the approach often used to model the kinetics of mycelial cultures was based in the estimation of the specific growth rate, μ, a phenomenological coefficient with no specific reference to the fundamental mechanisms of mycelial growth. Due to the advent of computerized image analysis of mycelial cultures [7,8] there is the possibility of developing mechanistic models of mycelial growth which may correlate structural features such as hyphal lengths and diameters to the fundamental aspects of the branching mycelial mechanisms [9 to 13].

In this work, a new approach based on previous work [13, 15, 17] is presented for modelling the kinetics of mycelial growth in terms of the fundamental mechanisms already established for vegetative mycelia.

<u>THE SPECIFIC GROWTH RATE μ IS RELATED TO THE BRANCHING FREQUENCY ϕ.</u>

Mycelial tips are the only parts of the mycelia that grow [1,14]. Their number, N_t, increases in time with a binary branching frequency, ϕ

$$N_t = 2^{\phi t} \qquad (1)$$

Mycelial internodal segments can be taken as cylinders with diameter D_h [3,4], critical length, L_c, after which all branch [4,5,14], and average solid content, ρ, which is approximately constant along the hyphae [10]. Hence, in a population of N_o mycelia, the amount of dry biomass, X_c, for each distal segment of length, L_c, is given by

$$X_c = N_o(\tfrac{1}{4}\Pi D_h{}^2) L_c \, \rho \qquad (2)$$

183

E. Galindo and O.T. Ramírez (eds.), Advances in Bioprocess Engineering. 183-189.
© *1994 Kluwer Academic Publishers. Printed in the Netherlands.*

184

Assuming that the average number of tips of a single mycelium growing in a broth is N_t, with segment length, L_c, and, recalling that the number of segments N_s, according with [13] is:

$$N_s = 2N_t - 1 \qquad (3)$$

Then it follows that:

$$Ns = 2(2^{\phi t}) - 1$$

So

$$Ns = 2^{\phi t+1} - 1 \qquad (3a)$$

The total amount of mycelial biomass will be

$$X = N_o(\tfrac{1}{4}\Pi D_h^2)(2^{\phi t+1}-1)\, L_c\, \rho \qquad (4)$$

Considering that in a large population of mycelia, the average biomass increase is given by an increase in N_t, and using Equations 1 and 4, it is possible to calculate the biomass growth rate as a function of time if it is assumed that the overall average branching frequency ϕ is constant and also by noticing that

$$d(2^{\phi t+1} - 1)/dt = 2Ln2(\phi 2^{\phi t})$$

then

$$dX/dt = N_o(\tfrac{1}{4}\Pi D_h^2)\,2Ln2(\phi 2^{\phi t})L_c\, \rho \qquad (5)$$

Therefore, the specific growth rate of the biomass, μ_{obs}, is

$$\mu_{obs} \equiv (1/X)\,dX/dt$$

$$= 2Ln2(\phi 2^{\phi t})/(2^{\phi t+1} - 1) \qquad (6)$$

From equation 3a (or from 3) it is clear that, for mycelia having more than eight tips ($N_t > 8$), the ratio between the number of tips and segments (N_t/N_s) becomes close to $\tfrac{1}{2}$ as shown by

$$N_t/N_s = 2^{\phi t}/(2^{\phi t+1} - 1) \approx \tfrac{1}{2} \qquad (7)$$

Thus, using, Equations 6 and 7, the value of μ_{obs}, can be given approximately by

$$\mu_{obs} \approx \phi\, Ln2 \qquad (8)$$

This result is equivalent to say that the biomass is growing at a fixed exponential rate μ_{obs} between the range of time $t = 0$ ($X = X_o$) and $t = t$ ($X = X_t$)

$$X_t = X_o\, e^{(\mu_{obs}\, t)}$$

Then

$$X_t = X_o\, 2^{\phi t} \qquad (9)$$

which indicates that the kinetic constant μ_{obs} can be estimated if the average branching frequency ϕ of the mycelial population is known or *viceversa*.

<u>ESTIMATING THE BRANCHING FREQUENCY ϕ, IN TERMS OF TWO MICROSCOPIC MODELS OF MYCELIAL ELONGATION.</u>

Several workers [3, 4, 5, 10, 13, 14] have reported that, in a large population of mycelia, the parameter ϕ is proportional to the rate of tip elongation (u_r). Viniegra-González *et al* [13] tried to generalize this approach using a simple mechanism of symmetrically branching mycelia, by assuming that all tips elongate at exactly the same rate u_r until they reach a critical length, L_c, after which a new branch is formed. Hence, the estimated value for the branching frequency, ϕ, can be calculated as follows

$$\phi = u_r/L_c \qquad (10)$$

The value of u_r can be measured as the rate of radial extension of circular colonies of mycelia grown on solid media (K_r according to Trinci [3]) corresponding to the fastest rate of hyphal elongation and, L_c can be estimated by direct microscopic measurement of mycelia grown either in liquid suspensions [7], in microcultures [3,4] or by image analysis of the periphery of mycelial cultures grown on agar plates [3,4,15].

In order to justify Equation 10 at a mechanistic level, it is worth noting that Trinci [5] reported that, mycelia grown on cellophane extended over Petri dishes, had leading segments of length L, growing according to the following empirical law

$$dL/dt = u_{max} \, L/(K_L + L) \qquad (11)$$

where, u_{max}, is the maximum value of the hyphal rate of elongation and, K_L, is a saturation factor. That is, for $L \gg K_L$, $dL/dt \approx u_{max}$. Thus after integration of Equation 11, an expression is obtained (Equation 12) for the branching frequency taken as the inverse of the critical time, t_c ($\phi = t_c^{-1}$) necessary for reaching a length $L = L_c$

$$\phi = u_{max} \, /[K_L Ln(L_c/L_o)+(L_c-L_o)] \qquad (12)$$

where L_o is the initial branch length which has been found to be close to twice the hyphal diameter, D_h [16,17]

$$L_o \approx 2D_h \qquad (13)$$

Now, noting that $L_c \gg 2D_h$, Equation 10 is justified if $u_r \approx u_{max}$. For example, according to Trinci [5] *Geotrichum candidum* has L_c =330 μm, D_h =2.5 μm and K_L = 12 μm, therefore

$$\phi_1 = u_r/1.14L_c \approx u_r/L_c \qquad (14)$$

The model based on Equations 8 and 10, is called here the **zero order rate law of elongation**, or the **zero order model,**

For $L_c \ll K_L$ a **first order rate law of elongation** or **first order**

model is derived as follows, where α is a first order constant

$$dL/dt \approx \alpha L \qquad (15)$$

This in turn yields an exponential rate law for L [5,14]

$$L = L_o \, e^{\alpha t} \qquad (16)$$

For $L = L_c$, at time, t_c, the branching frequency, ϕ_2, will be given by

$$\phi_2 = \alpha/[Ln(L_c/L_o)] \qquad (17)$$

But at t_c, the value of dL/dt reaches its highest expected value which could be measured as the colony extension rate u_r and can be found from Equations 15 and 16 as follows

$$u_r = \alpha L_c \qquad (18)$$

Therefore, the branching frequency ϕ_2 can be estimated by

$$\phi_2 = u_r/[(L_c)Ln(L_c/L_o)] \qquad (19)$$

This later model (Equation 19) has been tested using experimental data from cultures (on agar plates) of *Gibberella fujikuroi* [17] and *Aspergillus nidulans* [4] as shown in Table 1. For example, in the case of *G. fujikuroi* it was observed that the critical length was in the range of 226μm<L_c<308μm, with initial length L_o = 6.2 μm and u_r≈100 μm/h, and the calculated values for the critical length, called, L_2, were found in the range of 310 μm< L_2 <420 μm (Table 1). For the zero order model the necessary values of L_c to fit the experimental values of the specific growth rates measured in a stirred fermenter (μ_{obs}) were $L_2 \approx$ 1,000 μm (results not shown in Table 1) and clearly above the experimental values

186

Table 1. Comparison of calculated (μ_{calc} and L_2) and measured (μ_{obs} and L_c) specific growth rates and apical lengths in cultures of *Gibberella fujikuroi*[1] and *Aspergillus nidulans*[2].

Organism and culture conditions.		u_r (solid) (μm/h)	$D_h{}^a$ (μm)	μ_{obs} (liquid) (1/h)	μ_{calc}# (1/h)	L_c (solid) (μm)	L_2# (μm)
G. fujikuroi[1]							
% of C from	25.0%	104	3.6	0.039	0.039	308	420
glucose in	37.5%	93	3.6	-----	0.048		
mixtures with	50.0%	103	3.6	0.058	0.072		
starch having	62.5%	102	3.6	0.079	0.074	240	310
total CH$_2$O=80g/L	75.0%	109	3.6	0.048	0.042		
	87.5%	102	3.6	0.046	0.045		
	100.0%	95	3.6	0.048	0.044	226	310
A. nidulans[2]							
Incubation	20 °C	86	2.5	0.090	0.089	---	184
temperatures	25 °C	146	2.5	0.148	0.148	299	188
(10 g/L glucose)	30 °C	215	2.5	0.215	0.215	300	190
	37 °C	297	2.5	0.360	0.359	327	164

1. Data from Larralde-Corona [17]; 2. Data from Trinci [4]; a = estimated average value; # = values calculated according to Equations 8 and 19.

Table 2. Macroscopic model for solid substrate fermentacion curve of *Aspergillus niger No.10* grown on cassava meal granules[a].

<u>Experimental data[a]</u>

u_r (μm/h)	276
N_o (spores x 10^9)	1.9
ρ (10^{-12}g/μm^3)	1.1
D_h (μm)	3.3

<u>Non-linear estimated parameters[b]</u> (calculated)

	<u>Logistic</u>	<u>Zero order</u>	<u>1st order</u>
L_c	850 ±23	2,308 ±159	274 ±2.9
X_m	2.65 ±0.12	2.99 ±0.22	2.69 ±0.038
p	1.00	25.3 ±138	3.17 ±0.49

<u>Statistical parameters</u>

SSR	0.4583	0.5947	0.0192

a) Experimental data from Larralde - Corona [16], Oriol [21] and Raimbault [22]. b) Parameters calculated from the models presented above, except for the logistic where, by definition X_c=0, p=1. Numerical calculations were done using the Marquardt technique with a fourth order Runge-Kutta method.

of L_c although the calculated values, μ_{calc} for the specific growth rates were very close to μ_{obs}

The results with *A. nidulans* [4] shown in Table 1 indicate that the calculated values of the critical length L_c, were smaller (164 μm $<L_2<$ 190 μm) than the experimental values (299 μm $<L_c<$ 327 μm) required by the first order model (Equations 8 and 12) in order to fit the experimental values of specific growth rates, measured in shake flasks (μ_{obs}). However the calculated values for L_2 using the zero order model, were around 700 μm (not shown in Table 1) in order to make Equations 8 and 10 to fit μ_{obs}.

The first order model did not fit exactly the experimental values of L_c which were taken as the apical cell length for *A. nidulans* [4], but, its predictions were good enough and can be considered as a better model for the estimation of μ_{obs} of mycelial cultures.

FITTING MACROSCOPIC FERMENTATION
CURVES OF MYCELIAL CULTURES.

A population of living organisms growing in a confined space until they reach a maximum level of biomass density X_m can be modelled by Equation 20, proposed by Bertalanffy [18] and Richards [19] for vegetal populations and lately, applied to microbial populations [13,20]

$$dX/dt = \mu_{max} [1-(X/X_m)^p]X \quad (20)$$

According to Viniegra-González *et al.* [13] the term $(X/X_m)^p$ can be considered as an empirical power law which measures the probability density function for organisms that stop growing at the fractional density level (X/X_m). If $0 < p \le 1$, it is assumed that such a probability increases very fast for relatively small biomass density levels, therefore, it corresponds to a process of early self inhibition. But for $p>1$, it is assumed that the

growing population is relatively resistant to self inhibition. In particular when p=1 this model becomes the so called **logistic equation**. Also, Viniegra-González *et al.* [13] showed that it was possible to generalize Equation 20 in order to agree with Equation 6 by using Equations 1, 3 and 4, yielding the following expression

$$dX/dt = \mu_{max}[\tfrac{1}{2}(X+X_c)][1-(X/X_m)^p] \quad (21)$$

Where, $\mu_{max} = \phi(2\,Ln2)$ and $X_c = N_o\,(\tfrac{1}{4}\Pi D_h{}^2)\,L_c\,\rho$ (see Equations 2 and 6) are parameters that can be estimated from morphometric analysis of mycelia as indicated above and X_m and p are phenomenological parameters to be estimated from this model. Again, as in the preceding section, two model choices, ϕ_1 (zero order) and ϕ_2 (first order), can be made to estimate the actual branching frequency ϕ (Equations 10 and 19). Equation 21 has the interesting property that two fundamental parameters ϕ and X_c share a common parameter, L_c, reducing the degrees of freedom of this model.

According to the data presented in Table 2, the model with the least SSR (sum of squared residuals) was the first order model (SSR = 0.0192) as compared to the logistic equation (SSR = 0.4583) and the worst case was the zero order model (SSR = 0.5947). The first order model had the smallest value for L_c = 274 ± 2.9 μm (1% error) as compared to a very large L_c = 2,308 ± 159 μm (7% error) for the zero order model and an intermediate L_c = 850 ± 23 (3% error) for the logistic model. Thus, the first order model was more consistent with available data (150 μm< L_c < 350 μm) measured by Larralde-Corona [16] and with a more accurate description of the fermentation curve.

DISCUSSION

The apparent good agreement between microscopic and macroscopic models of dense mycelial cultures seem to support the idea that the basic mechanisms for mycelial growth are very similar to those observed in the mycelial sparse cultures. That is, the dominant mechanism for mycelial branching seems to be related to a first order rate law in terms of the segment length increase.

The fact that the zero order model implied by Equation 10, does not seem to fit experimental data suggests that very few branches go to lengths well above the saturation constant K_L. Therefore, the average branching frequency is better represented by first order rate law than by zero order rate law.

Recent work [8] suggests that in a short while there will be fully automated image analyzers which will allow one to obtain quick estimates of average dimensions of mycelial structures. They will make possible the correlation of these measurements with others of mycelial biomass (gravimetric or photometric). Thus, the use of the models presented in this paper could help to analyze and control the evolution and metabolism of mycelial cultures. For example, it will be possible to predict the fermentation curves using early morphometric analysis of sparse mycelial cultures and this, in turn, will help to develop on-line and ahead of time control procedures for mycelial industrial fermentations.

ACKNOWLEDGEMENTS

This work was partially financed by the Universidad Autónoma Metropolitana and the Institut Français de Recherche Scientifique pour le Dévelopément en Cooperation (ORSTOM). Ms. C.P. Larralde-Corona was supported by a Graduate Fellowship from the Consejo Nacional de Ciencia y Tecnología (CONACYT, México). G. Viniegra-González had a Senior Fellowship (Level 3) as National Investigator.

NOTATION

α First order kinetic constant (h^{-1}).

ϕ Average branching frequency (h^{-1}).

ϕ_1 Average branching frequency of the zero order model (h^{-1}).

ϕ_2 Average branching frequency of the first order model (h^{-1}).

μ_{calc} Calculated specific growth rate (h^{-1}).

μ_{obs} Specific growth rate measured in shake flask or fermenter (h^{-1}).

μ_{max} Maximum specific growth rate (h^{-1}).

ρ Average biomass solid content $(g/\mu m^3)$.

D_h Hyphal diameter (μm).

K_L Saturation factor (μm).

L Length (μm).

L_o Initial branch length (μm).

L_c Critical length of hyphae (μm).

N_o Initial number of mycelia.

N_s Average number of segments in a mycelium.

N_t Average number of tips in a mycelium.

p Biomass self-inhibition exponent.

t_c Time to reach critical lenght (h).

u_{max} Maximum hyphal elongation rate $(\mu m/h)$.

u_r Colony extension rate $(\mu m/h)$.

X_o Dry biomass at time t_o (g/mL).

X_t Dry biomass at time t (g/mL).

X_m Maximum dry biomass (g/mL).

LITERATURE CITED

1. Bartnicki-García, S., F. Hegert and G. Gierz, *Protoplasma,* **153**, 46(1989).

2. Bartnicki-García, S. In "Tip growth in plant and fungal cells" (I.B. Heath editor). *Academic Press*, Cal. USA (1990).

3. Trinci, A.P.J., *Journal of General Microbiology* **57**, 11 (1969).

4. Trinci, A.P.J., *Journal of General Microbiology* **67**, 325 (1971).

5. Trinci, A.P.J., *Journal of General Microbiology* **81**, 225 (1974).

6. Prosser, J.I. and A.P.J. Trinci, *Journal of General Microbiology* **111**, 153 (1979).

7. Packer, J.I. and C.R. Thomas, *Biotechnol. Bioeng.* **35**, 870 (1990).

8. Thomas, C.R., *J. Chem. Tech* **56**(2), 204 (1993).

9. Aynsley, M., A.C. Ward and A.R. Wright, *Biotechnol. Bioeng.* **35**, 820 (1990).

10. Nielsen, J., *Adv. Biochem. Eng. & Biotech*, **46**, 187 (1992).

11. Yang, H., U. Reichl, R. King and E.D. Gilles, *Biotechnol. Bioeng.,* **39**, 44 (1992a).

12. Yang, H., R. King, U. Reichl, and E.D. Gilles, *Biotechnol. Bioeng.,* **39**, 49 (1992b).

13. Viniegra-González, G., G. Saucedo-Castañeda, F. López-Isunza and E. Favela-Torres, *Biotechnol. Bioeng*. **42**, 1 (1993).

14. Katz, D., D. Goldstein and R.F. Rosenberger, *J. Bacteriol*. **109**(3), 1097 (1972).

15. González-Blanco, P.C., C.P. Larralde-Corona and G. Viniegra-González, *Biotechnology Techniques* **7**(1), 57 (1993).

16. Larralde-Corona, C.P., "Acoplamiento energético en la germinación de *Aspergillus niger* CH4", M. Sc. Thesis. on Chem. Eng. Universidad Autónoma Metropolitana-Iztapalapa, D.F. México (1992)

17. Larralde-Corona, C.P., P.C. González-Blanco and G. Viniegra-González, *Biotechnology Techniques* **8**(4),261 (1994).

18. Bertalanffy, L. von, *Quart. Rev. Biol*. **32**, 217 (1957).

19. Richards, F.J., *J. Exp. Botany,* **10**(19): 290 (1959).

20. Mulchandani, A., J.H. Luong and A. Leduy, *Biotechnol. Bioeng*. **32**, 639 (1988).

21. Oriol, E., *These de Doctorat*. Univ. P. Sabatier, Toulouse, France (1988).

22. Raimbault, M., *These de Doctorat d´Etat*. Univ. P. Sabatier, Tolouse, France (1980).

On-Line Estimation of Yeast Growth Rate Using Morphological Data from Image Analysis

K. Zalewski, P. Götz, and R. Buchholz

Fachgebiet Bioverfahrenstechnik, Technische Universität Berlin, Ackerstr. 71-76,
13355 Berlin, GERMANY

The influence of process conditions on the microbiological growth was analysed by microscopical observation of Saccharomyces cerevisiae *cells, cultivated in a 2 l bioreactor. A time-dependent formation of characteristical cell-structures and the frequency of their distribution was found to be a criterion for the cell development. According to the cell cycle, different aggregates containing one, two, three or four cells are detectable. The peculiarity of the analysis is the four-cell-agglomerate, which appears only during the exponential growth phase with a distribution maximum at the end of the lag phase. As indicator for cell growth and vitality the morphological data can be used to model the growth rate.*

Yeasts are simple and robust eukaryotes without any demands on media. Typical substrates are cheap and mostly waste products of other industries as molasses, whey, alcanes and alcohols. Therefore the cultivation of yeasts is well known and because of their suitability for use under industrial environment the range of products is wide spread. Aside from the production of food and semi-luxury food in bakery and brewery, yeasts are interesting after modification as supplier of enzymes, proteins, flavours etc.(1).

For process control it is necessary to estimate the qualitative and quantitative development of biomass. Usually, biomass control is resticted to nephelometric measurements of the sample. This detection only determines indirect rather than direct properties of cells and the results are dependent on cell size and degree of clumping (2,3).

To obtain more detailed information the biosuspension has to be analysed on microscopic slides. Besides cell counting there is a possibility to give an estimate of cell vitality by the way cells appear in the microscopic field. The morphological data can be correlated to biomass and product yield (4).
In order to contribute to a better understanding of the physiology of microbial growth, there are different solutions to predict the course of the process. Some models use indirect measurements for on-line estimation of the biomass concentration e.g. as function of CO_2 generated during substrate metabolism (5) or the theoretical yield on substrate (6). For filamentous organisms

191

E. Galindo and O.T. Ramírez (eds.), Advances in Bioprocess Engineering. 191-195.
© 1994 *Kluwer Academic Publishers. Printed in the Netherlands.*

the active biomass and the growth rate can be determined by using morphological data (<u>7</u>).
The aim of this work was to describe the growth rate based on microscopically observed morphology of yeast cells. Using an image processing system the cells were counted and classified to four different structures.

MATERIALS AND METHODS

<u>Strain.</u> The yeast *Saccharomyces cerevisiae* Hansen, IFG SA-07180, strain 176/1 was used in all experiments.

<u>Analytical techniques.</u> Biomass was estimated as dry cell weight after separation by 12000 rpm, 12 min, twice with a centrifuge (Heraeus, Biofuge 13) as well as measuring the optical density at a wavelength of 560 nm in a spectrophotometer (Beckmann, DU-64). The blank was destilled water.

The microsopic observation was carried out with an inversmicroscope (Nikon, TMD) to record the distribution of the morphological structures (x100 objective lense) and to count the colony forming units, CFU, (x40 objective lense) by using a Neubauer counting chamber with a depth of 0,1 mm and image processing system. The main components for the digital image analysis are a transputer system with 2 IMNOS processores (T800, T222), camera interface (512 x 512/8 bit), black/white CCD camera (Pulnix, TM-6CN 2031) and flow-chamber integrated in the table of the microscope. Automatic sampling from the bioreactor consists of pumping broth into a cooled mixing chamber, diluting if neccessary and taking 20 samples from there for image analysis in the flow chamber. Total counts range from 30 CFU at the beginning of the

process to 250 CFU, the maximum value before dilution is activated. For determining the relation of dead to alive cells, methylene blue staining was used.

Glucose was detected in the cell-free sample with an enzyme testkit (Sigma Diagnostics, Tinder-Reagenz), the intensity of colour was measured at a wavelength of 505 nm. The variation of the ethanol concentration was also determined enzymatically.

<u>Fermentation broth.</u> The ingredients of the medium were 20 g/l glucose, 20 g/l peptone of caseine, 10 g/l yeast extract. The pH was adjusted to 4,8.
The composition of the media was equal for inocula and batch-cultivations.
The inoculum was grown in two steps on a horizontal shaker (Infors, TL-125) at a speed of 150 min^{-1}. At first the yeast was inoculated in a 100 ml shake flask containing 20 ml of medium and grown for 5 h at 30°C. Then, 2ml of the broth was transfered in a 1 l shake flask filled with 200 ml of medium and grown for 15 h at 26°C. The initial yeast concentration in the reactor was 0,5 g/l.

<u>Process conditions.</u> The yeast was cultivated in a 2 l bioreactor (Bioengineering, KLF 2000), equipped with two six-blade stirrers, PT-100, pH-electrode and oxygen probe, for 8-10h at an aeration rate of 2 vvm. The stirrer speed was 500 rpm. To avoid the formation of foam 0,2-0,3 ml PPG were added to the medium. The sampling volume amounted to 20 ml every hour.

RESULTS AND DISCUSSION

Batch-Process Features

To introduce the cell aggregates, Figure 1 presents yeast cells at the beginning of the exponential growth phase. The main structures are marked as tetrade, a cluster of four cells, and as bud cell, an association of a mother and one daughter cell.

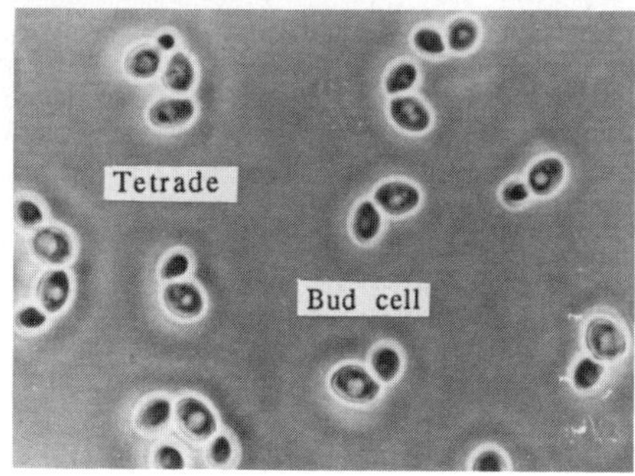

Figure 1. Microscopic picture of growing yeasts during batch process

Corresponding to the life cycle of *Saccharomyces cerevisiae*, a culture of growing yeasts shows a characteristical distribution of single cells and clusters of two, three or four cells. Depending on process time the compostion of aggregates and the frequency of the distribution alters. During the exponential phase, bud cells and four-cell-structures are dominant, whereas almost no single cells are detectable. On the other hand, at the stationary phase the major part of cells belongs to the group of single cells. To ensure reproducibility for the analysis, 5 fermentations were carried out under similar process conditions.

Process Model

In order to evaluate the results from the image processing system, first a process model is developed. Biomass growth, glucose consumption and ethanol formation are represented by a simple unstructured model using Monod-kinetics for describing the specific growth rate $\mu(S)$:

$$\frac{dX}{dt} = \mu(S) \cdot X \tag{1}$$

$$\frac{dS}{dt} = - \frac{1}{Y_{X/S}} \cdot \mu(S) \cdot X$$

$$- \frac{1}{Y_{P/S}} \cdot k_P \cdot X \tag{2}$$

$$\frac{dP}{dt} = k_P \cdot X \tag{3}$$

Model parameter identification from the 5 sets of data yields:
μ_{max}=0.41 1/h $\qquad$ K_S=0.013 g/l
$Y_{X/S}$=0.19 g/g $\qquad$ $Y_{P/S}$=0.678 g/g
k_P=0.5 g/(g·h)

A comparison of data from one batch run and the mathematical model is shown in Figure 2.

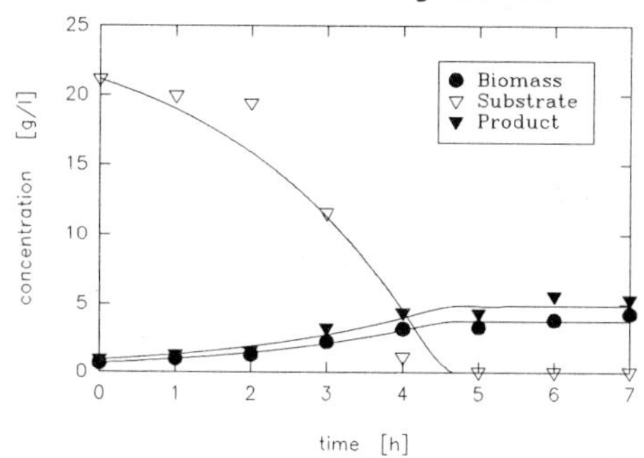

Figure 2. Batch process: Measurements and mathematical model

Evaluating the occurence of the different morphological states during the process gives a characteristic distribution

194

shown in Figure 3 (mean values from 5 fermentations).

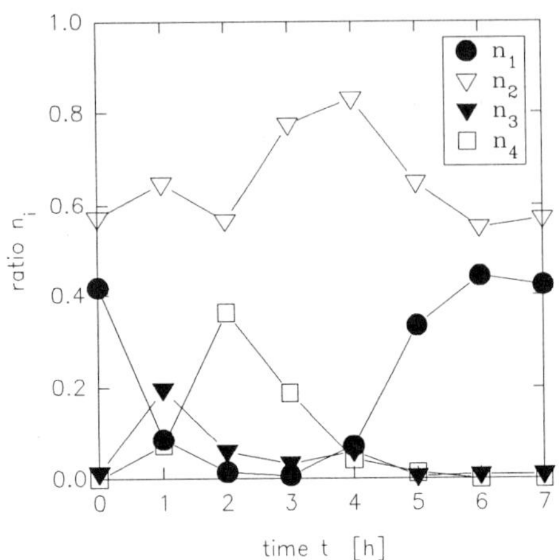

Figure 3. Ratio of different morphological structures during batch process

It can be seen, that a minimum of single cells coincides with a maximum of tetrades and after a time delay the bud cells having a maximum. The maximum value of the average deviation is 15 CFU. For process-relevant ratios n_i above 0.2 the relative average deviation is below 20 percent. Combining data from biomass growth and image analysis, a strong correlation between the tetrades and the observed specific growth rate is found (Figure 4).

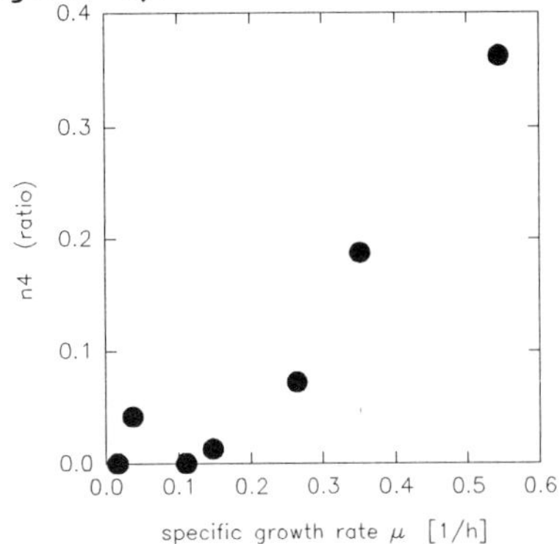

Figure 4. Correlation between specific growth rate and tetrades

This correlation can be used to give on-line information from image analysis about the growth conditions.

CONCLUSION

Image analysis is a powerful tool for delivering additional information about the growth state of a microbial population. Besides cell counting morphological data can be captured to estimate the physiological conditions of a yeast culture. The presented first results show the strong dependency of the growth rate from the distribution of tetrades as characteristical cell-aggregates indicating the phase of exponential growth during batch processes.

Further investigations are directed to the influence of different process parameters and media composition on morphology in order to get additional on-line information for process observation and control.

NOMENCLATURE

ATP	Adenosine triphosphate
CFU	Colony forming units
k_P	Product formation rate constant (1/h)
n_1	Ratio of single cells
n_2	Ratio of bud cells
n_3	Ratio of three-cell structures
n_4	Ratio of tetrades
P	Product concentration (g/l)
S	Substrate concentration (g/l)
t	Time (h)
X	Biomass concentration (g/l)
$Y_{X/S}$	Biomass yield coefficient (g/g)
$Y_{P/S}$	Product yield coefficient (g/g)
μ	specific growth rate (g/(g·h))

LITERATURE CITED

1. Sonnleitner, B.
Swiss Biotech 10, 7 (1992)

2. Siebert, K.J. and T.J. Wisk
ASBC Journal, 42, 71 (1984)

3. Monk, P.R. and P.J. Costello
J. Gen. Appl. Microbiol., 29,
467 (1983)

4. Cox, P.W. and C.R. Thomas
Biotechnol. Bioeng., 39, 945
(1991)

5. Whitaker, A.M. and S.R.
Elsden
ibid, 31, XXII (1963)

6. Abbott, B.J.
Process Biochemistry, 4, 13
(1973)

7. Reichl, U., R. King and E.D.
Gilles
Biotechnol. Bioeng., 39, 164
(1991)

Growth and Protein Formation of Recombinant *Aspergillus*: Utility of Morphological Characterization by Image Analysis

M. Carlsen, A. Spohr, R. Mørkeberg, J. Nielsen, and J. Villadsen

Center for Process Biotechnology, Department of Biotechnology, Technical University of Denmark, DK-2800 Lyngby, DENMARK

Three α–amilase producing strains of Aspergillus oryzae *used for recombinant protein production have been studied with respect to growth and protein production. The three strains were 1) a wild-type, 2) a transformant of the wild-type containing additional copies of the α–amilase gene and 3) a morphological mutant of the transformant. Carbon balances were set up for continuous cultures of the wild-type and the average recovery of carbon was 99 ± 2 % revealing that all major carbon sources were measured. Data from continuous and batch cultivations indicated that the α–amilase production was repressed at high glucose concentration. Image analysis was used to study the microscopic morphology of the fungus. Valuable information of the growth kinetics was obtained by comparing the three strains with respect to the tip extension rate and the branching frequency.*

Recombinant strains of *Aspergillus oryzae* have the ability to express high levels of heterologous proteins and this filamentous fungi is extensively used to produce industrial enzymes [1,2]. Industrial strains are screened to produce high yields of α-amylase [3], and recently, human proteins, e.g. human lactoferrin and human lysozyme, have also been produced in recombinant strains of *A. oryzae* [4,5]. Despite the extensive use of *A. oryzae* little is known about the mechanisms of growth and product formation, i.e. the basic cultivation physiology.

This paper presents data from a physiological and morphological study of three α-amylase producing strains of *A. oryzae*. The kinetics were determined by measurement of a number of culture variables, e.g. the exhaust gas composition and the concentration of medium components such as glucose and α-amylase. Steady-state data from continuous cultures of *A. oryzae* at different dilution rates are discussed and stoichiometric element balances are set up. Image analysis was used to study the microscopic morphology and to determine the average total hyphal length, the average number of tips and the average hyphal diameter in cultures of dispersed mycelia. By comparing the three strains with respect to the tip extension rate and the branching frequency valuable information about the growth kinetics was established. The synthesis and secretion of α-amylase was followed by measurement of the intracellular α-amylase level during cultivations.

<u>MATERIALS AND METHODS</u>

<u>Strains.</u> The three α-amylase producing strains of *A. oryzae* were donated by Novo Nordisk A/S. The wild-type is named A1560 and is derived from strain IFO 4177 obtained from the Institute for Fermentation, Osaka Japan. CF1.1 is a transformant of A1560 containing additional copies of the α-amylase gene. CF2.1 is a morphological mutant of CF1.1. The mutation was performed by adding nitrosoguanidin (NTG) to spores in the germination phase.

<u>Media.</u> All cultivations were carried out on a defined medium. The medium for the batch cultivations contained: 25.0 g/L glucose-monohydrate, 7.3 g/L $(NH_4)_2SO_4$, 1.5 g/L KH_2PO_4, 1.0 g/L $MgSO_4 \cdot 7H_2O$, 1.0 g/L NaCl, 0.1 g/L

197

E. Galindo and O.T. Ramírez (eds.), Advances in Bioprocess Engineering. 197-202.
© *1994 Kluwer Academic Publishers. Printed in the Netherlands.*

$CaCl_2 \cdot 2H_2O$, 0.5 mL/L pleuronic and 0.5 mL/L tracer metal solution. The tracer metal solution contained: 14.3 g/L $ZnSO_4 \cdot 7H_2O$, 2.5 g/L $CuSO_4 \cdot 5H_2O$, 0.5 g/L $NiCl_2 \cdot 6H_2O$ and 13.8 g/L $FeSO_4 \cdot 7H_2O$. The medium for continuous cultivations was identical with the batch medium except for the carbon and the nitrogen content (changed to 8.0 g/L glucose-monohydrate and 5.0 g/L $(NH_4)_2SO_4$).

<u>Cultivation conditions.</u> All cultivations were carried out in a 15 L MBR bioreactor. The temperature was 30°C and the pH was kept constant at 6.0 by adding 2M H_2SO_4 and 4M NaOH. The dissolved oxygen tension was kept above 60 % by controlling the agitation rate. The batch cultivations were inoculated with spores of approximately constant age to the concentrations listed in table 1.

Table 1. Inoculum of the three batch cultivations.

Cultivation identification	Strain	Inoculum (spores/m^3)
BAO16	A1560	$5.9 \cdot 10^{11}$
BA11	CF1.1	$4.3 \cdot 10^{11}$
BA21	C2.1	$1.0 \cdot 10^{11}$

<u>Analytical techniques.</u> The biomass concentration was measured by filtering the sample on a dry and preweighed filter followed by drying at 105 °C for 24 hours and measuring the weight gain. The exhaust gas was analyzed for oxygen and carbon dioxide by respectively paramagnetic (Magnos 6G) and infrared (Uras 3G) analysis (both from Hartmann & Braun, Germany). The glucose concentration was determined by a Flow Injection Analysis (FIA) using the glucose oxidase-luminol method as described by Benthin et al. [6]. A FIA method derived from Hansen [7] based on decolorization of the iodine-starch complex is used to measure the extracellular α-amylase activity. The α-amylase activity is given in FAU/mL (one FAU is the amount of α-amylase which at 37°C hydrolyses 5.26 g starch per hour). The morphology of *A. oryzae* was quantified using an image analysis system described in Nielsen and Krabben [8].

<u>Determination of intracellular α-amylase.</u> Filtered mycelium was washed with 0.9% NaCl, frozen in liquid nitrogen and stored at -80°C. 150 mg of frozen biomass was resuspended in 1.5 mL 0.1 M phosphate buffer, pH 5 and sonicated for 1 min with cooling (ice). The sample was centrifugated (10 min at 15,000 rpm) and the supernatant with the intracellular proteins was collected. The proteins were separated by SDS-polyacrylamide gel electrophoresis (SDS-PAGE) using 10% polyacrylamide gels according to the method described by Laemmli [9]. The separated proteins were transferred to nitrocellulose and incubated with rabbit antibody raised against the *A. oryzae* α-amylase. The immunoreactive proteins were visualized by treatment with peroxidase conjugated goat anti-rabbit antibodies. The detected proteins were quantified by a laser scanner (LKB UltroScan XL Densitometer, Sweden).

<u>RESULTS AND DISCUSSION</u>

<u>Continuous culture of *A. oryzae*</u>

A series of chemostat experiments with the wild-type A1560 was carried out at dilution rates varying from 0.025 h^{-1} to 0.167 h^{-1}. The average recovery of carbon in the experiments was 99 ± 2 % indicating that all the major carbon components were measured. The specific glucose uptake rate, the specific carbon dioxide formation rate and the specific oxygen consumption rate are all linear functions of the dilution rate. The stoichiometric coefficients and the maintenance coefficients are summarized in Table 2. About 95 % of the glucose fed to the bioreactor is converted to biomass and carbon dioxide and only a minor part is converted to α-amylase and metabolic products.

The residual glucose concentration was found to be 10 mg/L at a dilution rate of 0.167 h^{-1} decreasing to about 1 mg/L for the lowest dilution rates (see figure 1). The specific α-amylase production was almost constant 19 FAU/g DW/h for the dilution rate varying from 0.05 h^{-1} to 0.14 h^{-1}, dropping off outside this interval. The specific α-amylase production at high glucose concentration, obtained from several batch cultivations gives 3.0 FAU/g DW/h (data not shown). Since this is much lower than that obtained in the chemostat it strongly indicates that high glucose concentration repress the α-amylase production.

Table 2. Stoichiometric and maintenance coefficients estimated from continuous cultivations of *A.oryzae* (A1560). It is assumed that the biomass composition is given as $CH_{1.72}O_{0.55}N_{0.17}$ with an ash content of 7.5 % (w/w).

Stoichiometric coefficients
γ_{sx} = 0.64 C-mole biomass/C-mole glucose
γ_{sc} = 0.31 mole CO_2/C-mole glucose
γ_{so} = 0.23 mole O_2/C-mole glucose
m_s = 0.015 C-mole glucose/C-mole biomass/h
m_c = 0.017 CO_2 mole/C-mole biomass/h
m_o = 0.016 O_2 mole/C-mole biomass/h

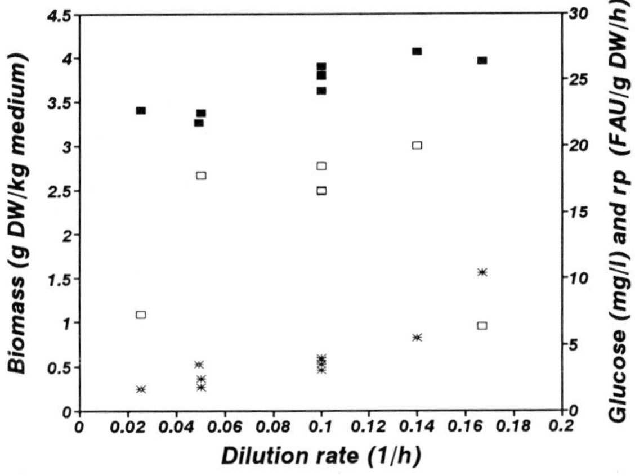

Figure 1 Steady-state values of the biomass concentration (■), the residual glucose concentration (*) and the specific α-amylase production (□) for *A. oryzae* (A1560).

Microscopic morphology

The microscopic morphology, i.e. the average number of tips (n_{av}), the total average hyphal length ($l_{t,av}$) and the average diameter (d_{av}) of the hyphal element was quantified during batch cultivations (inoculated with spores)

with each of the three strains. Assuming that the water content and the density in hyphal elements is constant, that the hyphal diameter is constant and that there is no fragmentation, one can estimate the specific growth rate (μ) as shown in [8]:

$$\frac{dl_{t,av}}{dt} = \mu l_{t,av} \qquad (1)$$

Nielsen and Krabben [8] found for batch cultivation with *P. chrysogenum* that the specific growth rate estimated from the average total hyphal length corresponds very well with the estimate from measurements of the biomass. Measurements of the average total hyphal length and the average number of tips during batch cultivation with the wild-type A1560 are shown in figure 2, and the specific growth rate is found to be 0.26 h^{-1} (the average hyphal diameter was constant 2.89 ± 0.10 μm).

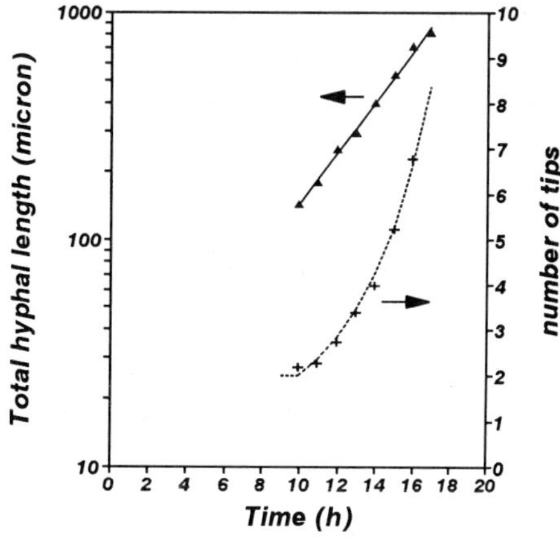

Figure 2 Measurements of the average total hyphal length (▲) and the average number of tips (+) during a batch cultivation with the wild-type strain A1560. The lines represent respectively exponential growth of the average hyphal length given by (1) (solid line) and the average number of tips given by (4) (dotted line).

Assuming that the spore germination and the hyphal fragmentation can be neglected the population balance equations for respectively the average hyphal length and average number of tips can be written as [8,10]:

$$\frac{dl_{t,av}}{dt} = n_{av} \cdot q_{tip,av} \qquad (2)$$

$$\frac{dn_{av}}{dt} = q_{bran,av} \qquad (3)$$

where $q_{tip,av}$ is the average tip extension rate (m/tip/h) and $q_{bran,av}$ is the average branching frequency (tip/h). From the measurements of the average number of tips it is observed that the rate of branching is low in the beginning of the cultivation. Significant branching starts at 10 hours where the hyphal elements have reached an average hyphal length of about 150 μm. Using the simple model proposed by Nielsen and Krabben [8] the rate of branching can be described as:

$$q_{bran,av} = \begin{cases} 0 & ; \ l_{t,av} < 150\mu m \\ k_{bran} \cdot l_{t,av} & ; \ l_{t,av} \geq 150\mu m \end{cases} \qquad (4)$$

The dotted line in figure 2 is the simulated average number of tips using the kinetics (4) with k_{bran} =0.0023 tip/μm/h together with equation (2) and (3).

By combining equation (1) and (2) the tip extension rate can calculated from the following equation:

$$q_{tip,av} = \mu \cdot \frac{l_{t,av}}{n_{av}} \qquad (5)$$

and in figure 3 the average tip extension rate is plotted as a function of the average total hyphal length. The data indicates that $q_{tip,av}$ can be correlated to the total hyphal length by the empirical saturation type kinetics:

$$q_{tip,av} = k_{tip} \cdot \frac{l_{t,av}}{K_t + l_{t,av}} \qquad (6)$$

where k_{tip} = 35 μm/tip/h and K_t = 148 μm. The expressions in (4) and (6) have no mechanistic basis [8], but the parameters, i.e. k_{bran}, k_{tip} and K_t are useful for comparison of the growth kinetics of different strains.

These kinetic parameters are very different for the three strains of *A. oryzae* (see Table 3). The specific growth rate for the transformant CF1.1 is about 15 % lower than the wild-type, whereas the mutated transformant has a 30 % lower specific growth rate than

the wild-type. The high value of k_{bran} for CF2.1 indicates that the mycelium for this strain is more branched than the wild-type A1560. In contrast to the dense mycelium for CF2.1 the mycelium of CF1.1 is more unbranched as represented by the low k_{bran} value.

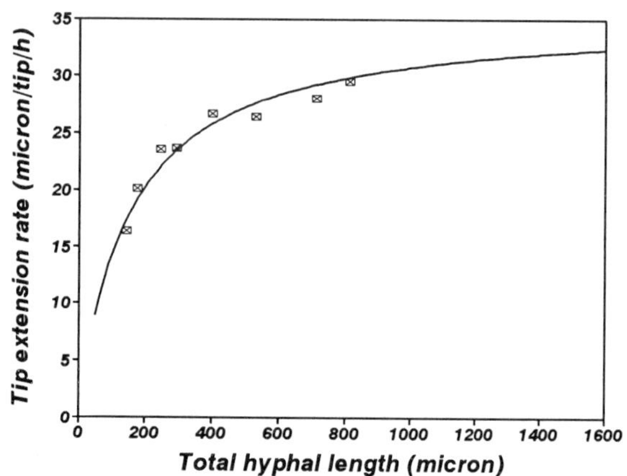

Figure 3 $q_{tip,av}$ as function of the average hyphal length for A1560. $q_{tip,av}$ is calculated from equation (5). The solid line is the model (6) with the parameters k_{tip} = 35 μm/tip/h and K_t = 148 μm.

The tip extension rate for CF2.1 is significantly lower than for the other strains. The maximum tip extension rate k_{tip} is higher for CF1.1 than for A1560 but at the same time the parameter K_t for CF1.1 is about twice the size of A1560. The transformant CF1.1 will therefore only have a higher tip extension rate than the wild-type for large hyphal elements, i.e. longer than 465 μm.

Table 3. Parameters estimated by equation (1), (4) and (6) for the three strains of *A. oryzae*.

Strain	k_{tip} (μm/tip/h)	K_t (μm)	k_{bran} (tip/μm/h)	μ (h^{-1})
A1560	35	148	0.0023	0.26
CF1.1	45	323	0.0010	0.22
CF2.1	28	304	0.0054	0.18

α-Amylase production

The α-amylase secretion was followed during batch cultivations with the three strains grown on glucose. For the transformant CF2.1 the extracellular α-amylase concentration increased exponentially with cultivation time (see figure 4), but at a lower rate than the biomass concentration. Thus the specific α-amylase secretion rate seems to decrease during the experiment. There is no obvious explanation for this observation, but a high production of α-amylase early in the cultivation, e.g. in the phase of germination would result in an initial high α-amylase concentration in the medium.

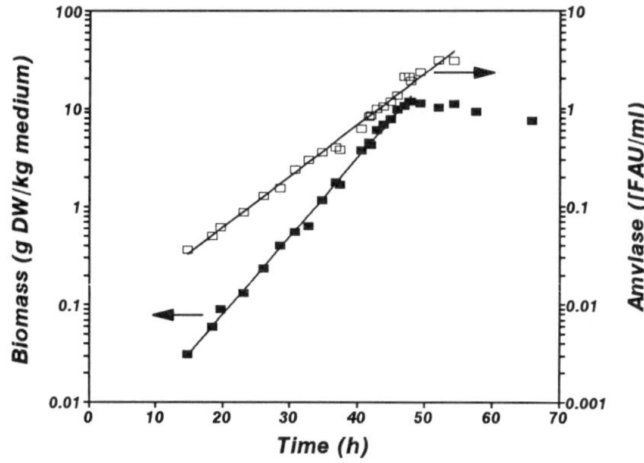

Figure 4 Biomass concentration (■) and the activity of extracellular α-amylase (□) for CF2.1 as functions of the cultivation time.

When the glucose is exhausted the biomass growth stops (see figures 4 and 5) whereas the secretion of α-amylase continues at almost the same rate. When the glucose concentration becomes very low the intracellular α-amylase level decreases rapidly (see figure 5). This indicates that synthesis of α-amylase is reduced due to depletion of energy sources. However, when the glucose is exhausted the cells may utilize other other energy sources, e.g. metabolic products or storage materials in the cells for α-amylase production, and since the chemostat experiments with the wild-type strongly indicates that the α-amylase production is repressed by glucose, there may be an enlarged production of α-amylase. This may explain the increase in the intracellular α-amylase concentration at 52 hours. Finally (after 55 hours) the intracellular α-amylase concentration decreases, probably due to cell lysis.

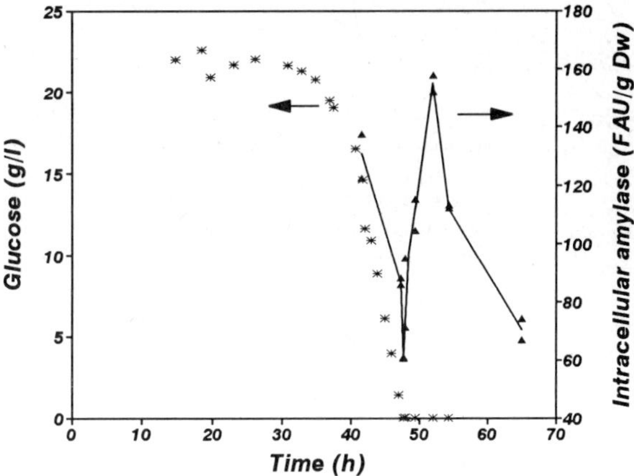

Figure 5 The Intracellular α-amylase concentration (▲) and the glucose concentration (*) for CF2.1 plotted as functions of the cultivation time.

CONCLUSION

Carbon balances were set up for continuous cultures of a wild-type strain of *A. oryzae* and the average recovery of carbon was 99 ± 2 % revealing that all major carbon sources were measured. For continuous cultivation of A1560 at glucose concentrations below 5 mg/L the specific α-amylase production is 19 FAU/g DW/h. Since the specific α-amylase production at high glucose concentration, i.e. during a batch cultivation was 3 FAU/g DW/h it is almost certain that the α-amylase production is repressed at high glucose concentration.

Wösten et al. [11] reports that protein secretion in *A. niger* occurs mainly at the tips. In high yielding protein strains the secretion procedure may be the rate limiting process. A dense mycelia with many tips may be able to secrete proteins at a higher rate than an unbranched mycelia. The growth kinetics can be evaluated from measurements of the microscopic morphology and when these variables are combined both measurement of biochemical variables such as substrate consumption and product formation new insight into the protein production is obtained.

NOMENCLATURE

d_{av}	Average hyphal diameter (m).
k_{tip}	Maximum tip extension rate (m/tip/h).
k_{bran}	Rate constant (tip/m/h).
K_t	Saturation constant (m).
$l_{t,av}$	Average total hyphal length (m).
n_{av}	Average value of the total number of tips.
$q_{tip,av}$	Tip extension rate (m/tip/h).
$q_{bran,av}$	Branching frequency (tip/h).
m_i	Maintenance coefficient (mole/mole/h).
γ_{ij}	Stoichiometric coefficient (mole/mole).
μ	Specific growth rate (h^{-1}).

Subscripts (i,j)

x	Biomass.
s	Substrate.
c	Carbon dioxide.
o	Oxygen.

LITERATURE CITED

1   Christensen, T., Woeldike, H., Boel, E., Mortensen, S.B., Hjortshoej, K., Thim, L. and Hansen, M.T.; Bio/Technology **6**, 1419 (1988).

2   Huge-Jensen, B., Andreasen, F., Christensen, T., Christensen, M., Thim, L. and Boel, E.; Lipids **24**, 781 (1989).

3   Erratt, J.A., Douglas, P.E., Moranelli, F. and Segligy, V.L.; Can. J. Biochem. Cell Biol. **62**, 678 (1984).

4   Ward, P.P., Lo, J.-Y., Duke, M., May, G.S., Headon, D.R. and Conneely, O.M.; Bio/Technology **10**, 784 (1992) .

5   Tscuchiya,K. Tada,S., Gomi, K., Kitamoto, K. Kumagai, C., Jigami, Y. and Tamura, G.; Appl. Microbiol. Biotechnol. **38**, 109 (1992).

6   Benthin,S. Nielsen, J. and Villadsen, J.; Anal. Chim. Acta **247**, 45 (1991).

7   Hansen, P.W.; Anal. Chim. Acta **158**, 375 (1984).

8   Nielsen, J. and Krabben, P.; Biotechnol. Bioeng. submitted (1994).

9   Laemmli, U.K.; Nature (London) **227**, 680 (1970).

10   Nielsen, J.; Biotechnol. Bioeng. **41**, 715 (1993).

11   Wösten, H.A.B., Moukha, S.M., Sietsma, J.H. and Wessels, J.G.H.; J. Gen. Microbiol. **137**, 2017 (1991).

Inducing Controlled Growth of *Penicillium chrysogenum* in Pelletized Form

S.M. Ratusznei and C.A.T. Suazo

Universidade Federal de Sao Carlos, Departamento de Engenharia Química, Rod. Washington Luiz, Km 235, Cx.P.676, CEP 13565-905, Sao Carlos, SP, BRAZIL

The aim of this work was to identify and quantify the relevant variables in inducing growth as pellets involved in the breeding of Penicillium chrysogenum *mold in complex medium. Experiments were carried out in an incubator shaker at $26^o C$ and pH=7.0 for 60 h, following by an experimental design that was defined by using response surface methodology. The independent variables tested were spore concentration in the inoculum, agitation speed and media composition. The dependent variables were bioparticle diameter and biomass. The results showed that by controlling two variables (spore concentration, $1x10^2$ - $2.0x10^3$, and agitation speed, 210 - 340 rpm) it was possible to grow the mold in pelletized form with predictable yield of biomass having a desired pellet diameter. A surface response represented by a third order model was found to be the best to correlate the studied variables.*

In the process of penicillin production when high biomass concentrations are reached, the culture broths are very viscous demanding agitation power as high as $5kw/m^3$, which may be considered very high [1,2]. Other strong stirring disadvantages are the heat generation, due to friction, and the destruction of the microorganism carrying out the biotransformation.

A critical factor in large scale bioreactors is the reduction of energy costs, which has been obtained by using nonmechanically stirred fermentors, such as the air-lift. This type of bioreactor presents the advantages of simple mechanical configuration, low operation cost and high efficiency in the transference of oxygen. Such advantages have turned them attractive to bioprocesses with low viscosity. In submerged cultures the morphology of molds is usually set between two extremes: filamentous mycelium with high viscosity characteristics, and spherical colonies or pellets. Although various commercial bioprocesses with filamentous mold have been implemented in new types of bioreactors, the oxygen transference is still inefficient, mainly due to the high viscosity of the culture broths. To eliminate these problems, the growth in pelletized form has been suggested as an interesting alternative by Metz and Kossen [3], since the most important advantage of a pellet suspension is a significant decrease in the broth viscosity; another advantage is the easy separation of the produced biomass.

The formation and growth mechanisms of the pellets are complex and not uniform, depending on the mold species,

inoculum size, growth medium, and physical conditions inside the bioreactor, among other factors.

Calam and Smith [4] studied some operational conditions on pelletization verifying the influence of the inoculum state in the penicillin productivity. Little information is found in the specialized literature about correlation of variables affecting mold pelletization. As a consequence, more information is extremely important to get mold growth selectively in pelletized form, either in laboratory or in industrial scale.

This study had the objective of identifying and quantifying the important variables in pellets formation aiming at the standardization of a methodology for their controlled production.

MATERIALS AND METHODS

Microorganism and Culture Media

Penicillium chrysogenum mold IFO-8644, was used. The strain was obtained from the Institute for Fermentation-Osaka, Japan.

The composition of the culture media used was as follow:

Sporulation - glycerol 7.5 g, glucose 7.5 g, corn steep liquor 2.5 g, peptone 5.0 g, $MgSO_4.7H_2O$ 0.05 g, KH_2PO_4 0.06 g, NaCl 4.0g, agar 20.0 g, distilled water ~1 ℓ, pH = 6.0.

203

E. Galindo and O.T. Ramírez (eds.), Advances in Bioprocess Engineering. 203-206.
© *1994 Kluwer Academic Publishers. Printed in the Netherlands.*

Germination - saccharose 30.0 g, corn steep liquor 20.0 g, $(NH_4)_2SO_4$ 2.0 g, $MgSO_4.7H_2O$ 0.25 g, $CaCO_3$ 5.0 g, soya oil 5.0 g, salt solution 20 ml, distilled water ~1 ℓ, pH = 6.0.

Salt solution - $FeSO_4.7H_2O$ 3.0 g, $CuSO_4.5H_2O$ 2.5 g, $ZnSO_4.7H_2O$ 10.0 g, distilled water ~1 ℓ, pH = 7.0.

Analytical Techniques

Cellular concentration: After centrifugation the cell mass was dried at $105^{o}C$ for 10 hours;

Spore counting: with Neubauer chamber on microscope;

Pellet size: the size of the pellets was obtained as the average diameter measurement of 50 particles. The measurement was achieved with a lens provided with a micrometric scale.

Experimental Procedure

A spore suspension was prepared after removing the spores from the surface of a slant of sporulation solid medium, followed by its suspension in a NaCl 0.9% w/w solution. Afterwards, 40 ml of germination medium was distributed in several 250 ml erlenmeyers flasks, and these were inoculated with 4 ml spore suspension. The flasks were agitated for 60 hours, in rotatory incubation chamber (shaker). The pellet size obtained with this time permitted an easier and more accurate measure of the dependent variables than in shorter times. The temperature was 26° C and the pH = 6.0. After that period of time the formed pellets were restrained in a 115 mesh (0.125 mm) screen, being washed with distilled water and treated with acetic acid, 5% v/v, to dissolve the residual $CaCO_3$ and determine the dried cell mass, afterwards, the size and biomass of the obtained pellets were determined.

Experimental design

On a preliminary stage the dependent variables studied by using a 2^3 factorial design were: spore concentration in the inoculum (x_1), agitation speed (x_2) and medium concentration (x_3). After a simplification, they were studied, by using a 2^2 factorial design, the variables spore concentration (x_1) and agitation speed (x_2). In both cases the independent variables were pellet size (y_1) and produced biomass (y_2). The (x_1) and (x_2) were varied in the ranges of $1.0 \cdot 10^2$ to $2.0 \cdot 10^3$ spores/ml and from 210 to 340 rpm, respectivelly. The empirical models were obtained by linear or nonlinear regression by using the least square method. To verify if they showed an adequate fitting to the experimental data it was used as parameters the regression correlation coefficient and the estimated value for the F test, according to the methodology proposed by Box, Hunter and Hunter [5].

RESULTS AND DISCUSSION

The regression analysis of the preliminary experiment showed that a first order model, for the response variable size of particle, was not statistically significant. The correlation coefficient found was 0.494. It was verified, in this case, that an increase in the inoculum concentration and an increase in the agitation speed helped the obtention of smaller bioparticles while the medium concentration had little importance.

In the case of the response variable produced biomass the first order model was statistically significative at a confidence level of 95%. The correlation coefficient found was 0.777. The three studied variables showed to be important in the obtention of higher values of produced biomass.

On a later stage a 2^2 factorial design was made, having as independent variables: spore concentration (x_1) and agitation speed (x_2) and as dependent variables: pellet size (y_1) and produced biomass (y_2). Even though the medium concentration showed an effect as significant as that on spore concentration and on agitation speed in produced biomass, it presented , by far, the lowest effect on pellet size. So, for the rest of the study the medium concentration was maintained constant at its original or undiluted value(see Culture media above).

Table 1 Values of the independent and dependent variables of the experimental design.

run	codified variables		response variables	
	x_1	x_2	y_1 (μm)	y_2 (g/ℓ)
1	-1	-1	1907	9.01
2	1	-1	1454	21.71
3	-1	1	1884	10.85
4	1	1	1128	22.26
5	0	0	1100	22.97
6	0	0	1073	22.29
7	0	0	1178	22.76
8	-1.414	0	2693	2.82
9	1.414	0	1129	23.40
10	0	-1.414	1689	16.17
11	0	1.414	1204	21.75
12	0	0	1116	21.60
13	0	0	1000	20.28

Initially, the results of the runs from 1 to 7 (see Table 1) were used to test first order models. Through the variance analysis it was observed that the linear models, for the response variables bioparticle size (y_1) and produced biomass (y_2) were not statistically significative. Additional experiments were carried out (runs from 8 to 13) to estimate the second and third order coefficients.

Through the variance analysis, for the response variable size of bioparticles, it was found a value of F(4,8) (at 4 (regression) and at 8 (deviation) degrees of freedom) equal to 27.96. In the table of values of F at significance level of 1% and associated with the same degrees of freedom, the value 7.01 was found. Therefore, the model describes the investigated region at a confidence level of 99%. The correlation coefficient obtained was 0.933. Figure 1 represents the second order model obtained.

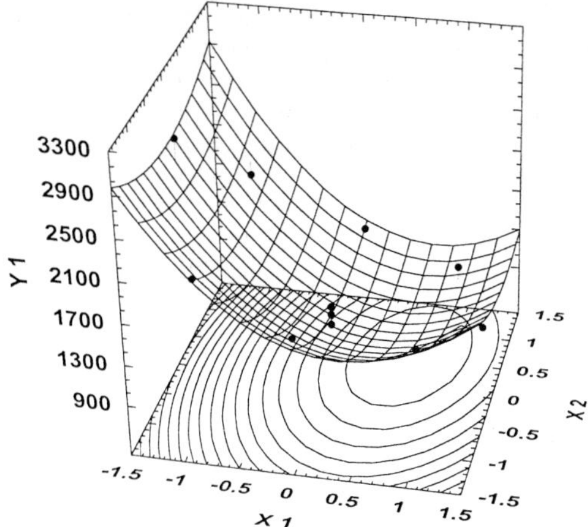

Figure 1 Surface referring to the second order model, y_1 = pellet size (μm) x_1 = codified spore concentration and x_2 = codified agitation speed, with disposition of experimental points and contour lines.

Analyzing Figure 1, it can be observed that for values of the codified variable x_1 over 0.5 there is an increase in the value of the pellet diameter, behavior not observed experimentally, since the increase of spore concentration should help the formation of the filamentous form or very reduced diameters. This increase in the diameter value may have been caused because the used model was not the most adequate to describe the data. So, a third order model was tested. The mathematical expression of the model, followed by the standard deviation of each parameter, is presented below. Only the important parameters in the model are presented, having the non significant parameters eliminated from it.

$$y_1 = 1093.4 - 75.8x_1x_2 + 387.4x_1{}^2 + 155.2x_2{}^2 - 281.7x_1{}^3$$

$$-86.0x_2{}^3 \qquad (1)$$

The variance analysis gave a value for F(5,7) = 119.9 while the value for F found in tables is 7.46. The value of the correlation coefficient was 0.988. The value of F and of the correlation coefficient obtained are higher than those for the second order model, indicating that the points are better represented by the third order model. Figure 2 shows the obtained model.

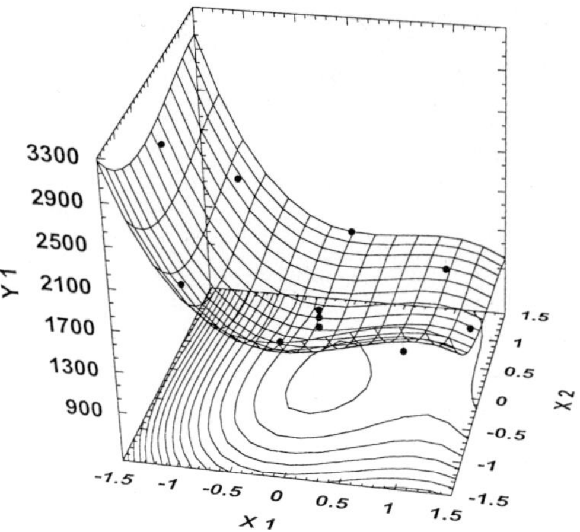

Figure 2 Surface referring to the third order model, y_1 = pellet size (μm), x_1 = codified spore concentration and x_2 = codified agitation speed, with disposition of experimental points and contour lines.

For the response variable produced biomass it was found that, through nonlinear regression, F(4,8) = 84.8, which is higher than the value 7.01 found in tables. The model describes the investigated region at a confidence level of 99%. The correlation coefficient was 0.977. Figure 3 shows the second order model.

It can be observed, on Figure 3, that for values of codified variable (x_1) over 0.5 there is a decrease in the value of produced biomass, tendency not observed experimentally, since the increase of spore concentration should produce an increase of the biomass. Such a decrease in the biomass value may have been caused, as in the case of pellet diameter, because the used model was not the most adequate to describe the data. Then, a third order model, whose variance analysis gave the value for F(5,7) = 128.8 and the correlation coefficient equal to 0.989, was tested. These values were higher than those obtained for the second order model, indicating that the points were better represented by the third order model. The mathematical expression for the third order model is shown below. Figure 4 shows the third order model.

$$y_2 = 21.98 + 4.77x_1 - 4.45x_1{}^2 - 1.53x_2{}^2 + 1.25x_1{}^3$$

$$+0.91x_2{}^3 \qquad (2)$$

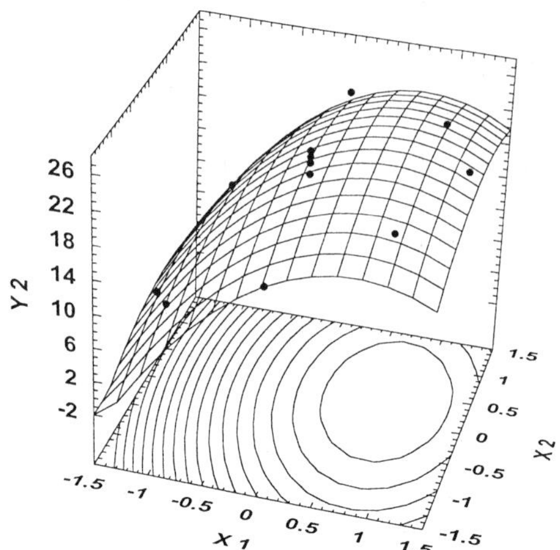

Figure 3 Surface referring to the second order model, y_2 = produced biomass (g/ℓ), x_1 = codified spore concentration and x_2 = codified agitation speed with disposition of experimental points and contour lines.

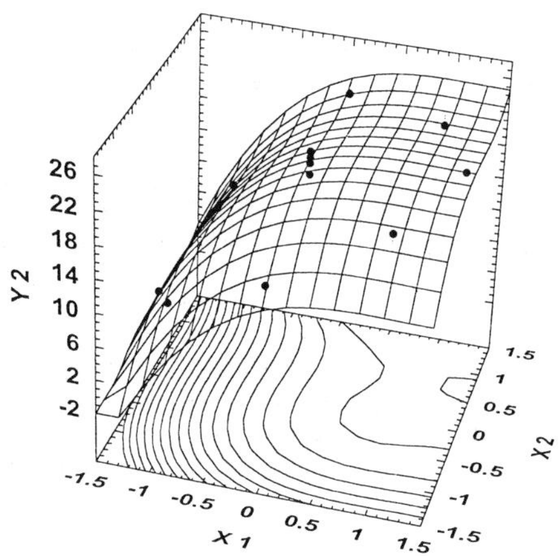

Figure 4   Surface referring to the third order model, y_2 = produced biomass (g/ℓ), x_1 = codified spore concentration and x_2 = codified agitation speed with disposition of experimental points and contour lines.

A desirable situation in industrial bioprocessing is the obtention of high quantities of biomass with very small particle diameter. By observing figures 3 and 4 it can be deduced that such conditions can be attained by using about $2 \cdot 10^3$ spores/ml in the inoculum combined with ~340 rpm agitation. It is important to notice that deductions based on these results in conditions different from the used in this work must be carefully drawn.

Even though optimal diameter was not obtained in the range of experimental conditions studied, in practice, it is possible to obtain it by controlling the fermentation time and agitation level. This strategy has been used in our laboratories for preparation of pelletizable inoculum with reliable results.

CONCLUSIONS

According to the results, we can conclude that:

- It is possible to grow the mold *Penicillium chrysogenum* in pellet form in rotatory incubator (shaker), by just controlling the variables: spore concentration in the inoculum and agitation speed.

- In the range of studied variables, an increase in the agitation speed and in the spore concentration favors the obtention of small bioparticles in increasing quantities.

- A third order response surface is the one which best correlates the studied variables. Dependent and independent variables are well correlated through empirical mathematical models given by Equations 1 and 2.

LITERATURE CITED

1. Aiba, S.; Humphrey, A. E.; Millis, N.F. *Biochemical Engineering*, University of Tokyo Press, 2.ed., 1973.

2. König, B.; Schügerl, K.; Seewald, C. *Biotechnology and Bioengineering*, 24, 259, 1982.

3. Metz, B.; & Kossen, N.W.F. *Biotechnology and Bioengineering* , 19, 781, 1977

4. Calam, C.T. & Smith, G.M. *Microbiology Letters*, 10, 231, 1980.

5. Box, G.E.P.; Hunter, W.G.; Hunter, J.S. *Statistics for Experimenters*, New York, John Wiley & Sons, 655p, 1978.

6. Schügerl, K.; Bayer, T.; Niehoff, J.; Möller, J.; Zhou,W. *Bioreactor Fluid Dynamics*, 229, 1988.

Structured Modelling of Bioreactors

M. Reuss, S. Schmalzriedt, and M. Jenne

Institut für Bioverfahrenstechnik, Universität Stuttgart, Allmandring 31
D-70569 Stuttgart, GERMANY

The aim of this contribution is to discuss some new aspects in the design of structured models for stirred tank bioreactors. The modelling strategy starts from a simple multiphase compartment model, which is based on an aggregation of well-mixed multiphase compartiments including mass transfer between the phases. In order to incorporate the available information of intrinsic fluid dynamics it is necessary to put finite volume elements in the place of these compartments. The fluxes from and into those elements can be estimated from the real fluid dynamics. Numerical solutions of the set of momentum equations for gas and liquid phase are shown along with application examples for a biological model (oxygen sensitive culture of Bacillus subtilis) and a physico-chemical model for pH-calculation.

During scale-up, an attempt is made to recreate a physiological and hydrodynamic environment in a large reactor as similar as possible to that established in bench scale and/or pilot plant vessels. Engineering solutions to this problem include maintaining geometric similarities, whenever possible, and also criteria such as constant power per unit volume, volumetric mass transfer coefficient, circulation time, shear rate, tip speed etc. The logic behind the different criteria is that preserving each of these singly represents maintaining the constancy of the corresponding characteristic of the extracellular environment at different scales. And, if this environmental property is the one that most critically influences the desired microbial productivity, a successful scale-up might result. Indeed, a number of microbial systems are in accord with these techniques and permit a production-scale operation reasonably in agreement with that established in the laboratory. However, it is easy to see that these criteria are mutually exclusive and, therefore, do not allow an exact replication of environmental similarity at any two different scales. Under such circumstances, the behavior of microor-

ganisms in different fermentors remains uncertain. As a matter of fact, the use of volumetric properties as scale-up criteria demands a guarantee of uniformity of properties throughout the system, something that may be impossible even in a small reactor.

However, it is well known that a uniform distribution of mass and energy becomes more and more difficult as the reactor volume increases. It seems, therefore, reasonable to asume that some non-uniformity exists at all scales of operations. If the reaction kinetics in question shows an interaction with the distribution, the different reactors would then have different performance.

The problem is particularly important for those processes in which nutrients are continously introduced into the broth. For specific nutrients such as oxygen and sometimes other nutrients such as carbon source, the time constant for their distribution (mixing-time) may be of the same magnitude as those of their consumption in any reasonable sized reactor beyond the bench-scale. If we accept that spatial variations exist we are faced with the problem that dyna-

207

E. Galindo and O.T. Ramírez (eds.), Advances in Bioprocess Engineering. 207-215.
© *1994 Kluwer Academic Publishers. Printed in the Netherlands.*

mically changing environmental conditions may result in drastic changes in metabolism and consequently the final outcome of the process. Although well known in classical fermentation processes, these problems may have even more serious consequences when dealing with recombinant microorganisms as, for example, during scale-up of high-density cultures. The long-term mathematical description of these phenomena requires flexible tools that can be easily adapted to different systems that integrate the process and the reactor. This, in turn, require the design of conceptual instruments for structuring both the abiotic and biotic phases of the system of interest.

As far as the abiotic (gas and liquid) phases of the bioreactor are concerned various tools are available to tackle problems of incomplete mixing and/or distribution of mass. The various models suggested for stirred tank reactors may be classified into two groups:

- reactor flow models

- turbulence models

The most important tools for modelling situations of incomplete mixing based on reactor flow models are:

- compartimental models
  (consisting of perfect mixing
   compartments and mixed
   models)

- recirculation models
  (considering various types of
   fluid circulation within
   the reactor)

The application of the various approaches to bioreactor modelling has been extensivley described by Reuss and Bajpai (1) and Reuss and Jenne (2). As a result of these critical reviews, several serious limitations of these modelling strategies have been outlined. One of the limitations is related to the omission of backmixing of the gas phase and therefore the disregard of the material balance of oxygen in the gas phase. Secondly, in many of the models the number of the compartments is connected to the intensity of mixing and, thus, the structures are virtual in space. With other words, the compart-

ments are without identity because of missing coordinates. It is easy to see that the two mentioned problems are closely related if attempts are made to couple mixing of gas and liquid phase.

MULTIPHASE-COMPARTMENT-MODEL

To overcome part of the aforementioned difficulties and incompabilities Ragot and Reuss (3,1) have proposed an alternative structure, the so called multiphase-compartment model. The approach was based upon an appropiate aggregation of well mixed multiphase-compartments including mass transfer between the phases (figure 1).

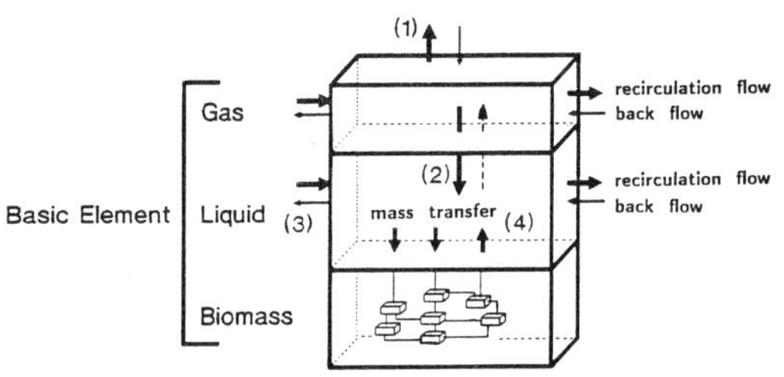

Figure 1: Single muliphase compartment

Aggregation is beeing performed by connecting the gas and liquid fractions of the compartments through circulation streams and back flow, thus trying to incorporate mixing of both, gas and liquid. Figure 2 illustrates a typical example of aggregation for a three impeller tank.

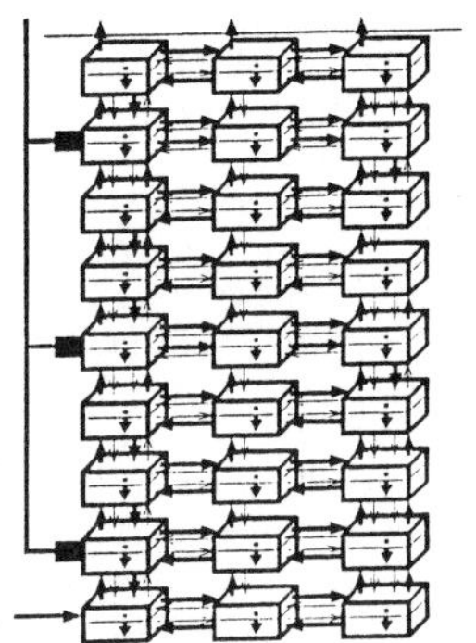

Figure 2: Aggregation of multiphase compartments for a three impeller tank

The structure was applied to simple Michaelis Menten type of oxygen consumption kinetics. The model simulation resulted in the distribution of oxygen concentrations, influenced by backmixing of gas and liquid as well as local hydrodynamic pressure.

In trying to understand the meaning of forward and backward flow from the physical point of view — this is essential for any further improvements of the model structure and meaningful parametrization — again we found some limitations which forced us to reconsider the concept. The problem is schematically illustrated in figure 3.

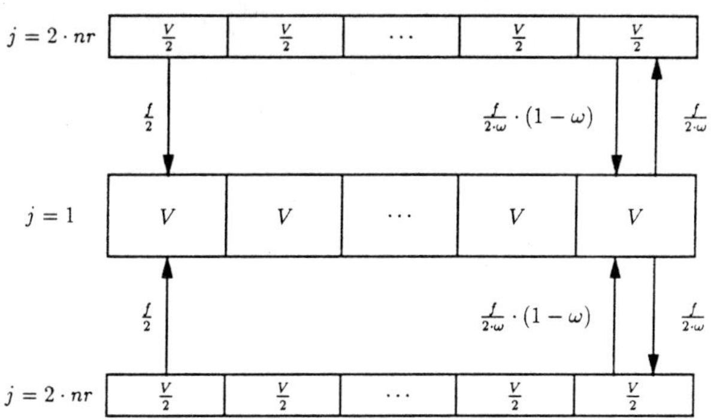

Figure 3: Limitations concerning the compartment model

Only in those cases in which the model structure consists of two symmetrical loops for a single impeller the parameter of the forward and backward flow can be interpreted as pumping capacity of the impeller and local dispersion coefficient as a measure for mixing in the loop. This represents a serious restriction as a result of which resolution in axial direction is beeing limited. The only way to overcome these difficulties is to put finite volume elements in the place of the compartments and, thus, following the concept of turbulence modelling.

<u>COMPARTIMENTATION WITH THE AID OF FINITE VOLUME ELEMENTS</u>

The two-dimensional material balance equations for a component characterized by its concentration c can be written as:

$$\frac{\partial c}{\partial t} + \frac{1}{r}\frac{\partial}{\partial r}(r\,u_r\,c) + \frac{\partial}{\partial z}(u_z c) = \frac{1}{r}\frac{\partial}{\partial r}\left(D_{eff}\,r\,\frac{\partial c}{\partial r}\right)$$
$$+ \frac{\partial}{\partial z}\left(D_{eff}\,\frac{\partial c}{\partial z}\right) + reaction \quad (1)$$

where:

r and z = radial and axial coordinates, respectively,

u_r and u_z = radial and axial components of the velocity vector

D_{eff} = eddy diffusivity or turbulent dispersion coefficient = ν_{eff} / Sc
with ν_{eff} = turbulent effective viscosity and Sc = Schmidt number

Next, a finite volume is introduced as shown by Patankar (<u>4</u>).

The finite volume element is located at point i,j in the coordinate system r,z. Integration of the material-balance equation between position i-1,j-1 and i,j leads to a set of 8 cell wall fluxes, which are schematically summarized in figure 4.

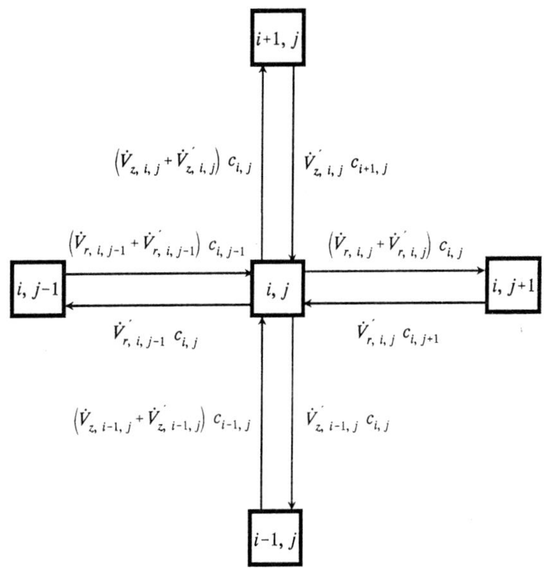

Figure 4: Cell wall fluxes

Thus, instead of two forward and two backward fluxes in the compartment model of figure 1, there are now 4 fluxes in each direction. The advantage, however, is the fact that these fluxes are related to intrinsic fluiddynamic parameters. As an example and with the assumption ϱ = const. the flow fluxes leaving the cell are beeing summarized in the following.

Convective fluxes:

$$\dot{V}_{r\,i,j} \;=\; 2\,\pi\,r_j\,(z_i - z_{i-1})\,u_{r\,i,j}$$

$$\dot{V}_{r\,i,j-1} \;=\; 2\,\pi\,r_{j-1}\,(z_i - z_{i-1})\,u_{r\,i,j-1}$$

$$\dot{V}_{z\,i,j} \;=\; \pi\,(r_j^2 - r_{j-1}^2)\,u_{z\,i,j}$$

$$\dot{V}_{z\,i-1,j} \;=\; \pi\,(r_j^2 - r_{j-1}^2)\,u_{z\,i-1,j}$$

Dispersive fluxes:

$$\dot{V}'_{r\,i,j} \;=\; 2\,\pi\,r_j\,(z_i - z_{i-1})\,\frac{D_{eff\,r,i,j}}{\delta r_j}$$

$$\dot{V}'_{r\,i,j-1} \;=\; 2\,\pi\,r_{j-1}\,(z_i - z_{i-1})\,\frac{D_{eff\,r,i,j-1}}{\delta r_{j-1}}$$

$$\dot{V}'_{z\,i,j} \;=\; \pi\,(r_j^2 - r_{j-1}^2)\,\frac{D_{eff\,z,i,j}}{\delta z_i}$$

$$\dot{V}'_{z\,i-1,j} \;=\; \pi\,(r_j^2 - r_{j-1}^2)\,\frac{D_{eff\,z,i-1,j}}{\delta z_{i-1}}$$

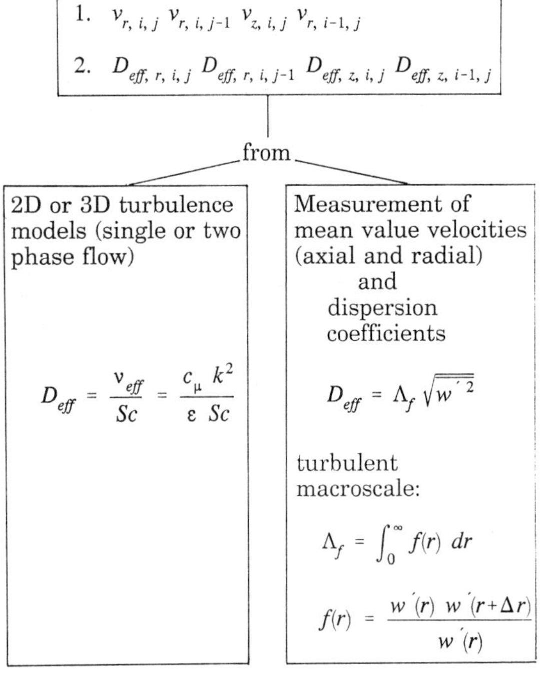

Table 1: Two routes to obtain fluiddynamical information

The values of the parameters (local values of u_r and u_z as well as local values of the turbulent dispersion coefficient D_{eff}) can be estimated from two different sources. Either simulation results from turbulence models are beeing used or the data are predicted from experimental observations of local values of mean axial and radial velocities and eddy diffusivities estimated from measured fluctuation velocities. The two routes are schematically summarized in Table 1.

Application of the first possibility – use of experimental data – has been shown in (2) for simulation of the dynamic change of the temperature field after pulsing a stirred tank reactor with hot water. In the following examples use of simulation results from the numerical solution of the momentum balance equations will be presented.

COMPUTATIONAL FLUID DYNAMICS SIMULATION

If we assume that the microbial reactions do not affect fluiddynamics – this assumption is valid for most applications – momentum and mass balance equations can be strictly seperated. This forms the basis for using numerical solutions of the momentum equations for parametrization of the fluxes in the finite volume elements for the mass balances (figure 4).

The governing equations based on the Eulerian concept and named the two-fluid method (Spalding (4)) are presented in the following for the general situation of two-phase flow (gas and liquid):

$$\nabla(\epsilon_L \underline{u_L}) \;=\; \nabla(\mu^L_{eff}\nabla\epsilon_L)$$

$$\nabla(\epsilon_G \underline{u_G}) \;=\; \nabla(\mu^G_{eff}\nabla\epsilon_G)$$

$$\nabla(\epsilon_L \rho_L \underline{u_L}\,\underline{u_L}) \;=\; \nabla(\mu^L_{eff}\nabla\underline{u_L}) - \epsilon_L\nabla p + \underline{F_f}$$

$$\nabla(\epsilon_G \rho_G \underline{u_G}\,\underline{u_G}) \;=\; \nabla(\mu^G_{eff}\nabla\underline{u_G}) - \epsilon_G\nabla p - \underline{F_f} - \epsilon_G\rho_L\underline{g}$$

with

$$\epsilon_L + \epsilon_G = 1$$

(ϵ_L, ϵ_G = liquid and gas hold up, respectively)

and

$$\mu_{eff} = \mu_{lam.} + \mu_t$$

($\mu_{lam.}$ = laminar viscosity, μ_t = eddy viscosity)

The pressure is considered to be common to the two phases, and both, gas and liquid are assumed to be incompressible.

The interphase friction term is calculated as:

$$\underline{F_f} = const. \, \epsilon_G \, \epsilon_L \, (\underline{u}_G - \underline{u}_L)$$

where const. = 5 x 10^4 kg m^{-3}s^{-1}.

This value corresponds to a slip velocity between the gas and liquid phases of 0.2 m/s. This is a good approximation to the experimentally determined terminal velocity in water of single air bubbles of diameter between 1 and 10 mm. The slip velocity is almost constant over this range of diameters because of increasing nonspericity. For this reason, equation (4) gives a much better representation of the friction over this range than does the formula for drag on a sphere.

In a first approximation of the complex turbulent flow we assume that the turbulent viscosity μ_t is constant throughout the entire reactor. Ongoing research is directed towards application of different turbulence models, for example the famous k - ϵ approach to two phase flow in stirred tank reactors (see for instance <u>5</u>, <u>6</u>, <u>7</u>) according to which the eddy viscosity μ_t is computed from:

$$\mu_t^i = c_\mu \rho_i \frac{k^2}{\epsilon}$$

where additional balance equations for the kinetic energy k and dissipation energy ϵ are required. Based on the aforementioned assumptions (μ_t = const.) the momentum balance equations have been numerically solved by applying the solution algorithm SIMPLER (Semi-Implicit Method for Pressure Linked Equations, Revised Version) suggested by Patankar and Spalding (<u>8</u>).

Figures 5, 6 and 7 summarize typical results for the velocity fields in gas and liquid phases as well as the local values of the gas hold up for a selected speed of agitation and aeration rate. The results indicate that even with the rather rough assumption μ_t = const.

reasonable results for the local fluiddynamic conditions can be estimated. These results will be used in the following for solving the mass balance equations for two selected examples.

<u>APPLICATION EXAMPLES</u>

<u>Bacillus subtilis culture for production of Acetoin/Butandiol</u>

Moes (<u>9</u>) and Moes et al. (<u>10</u>) have suggested a kinetic model to simulate the time course of glucose, biomass, acetoin, butanediol and dissolved oxygen in batch culture of *Bacillus subtilis*. In agreement with the experimental observations the model predicts a decrease of the ratio of acetoin to butandiol with decreasing oxygen concentrations in the region of microaerobic conditions. The system has been used several times as a model for an oxygen sensitive culture (<u>11</u>).

In addition to the balance equations for each reaction compound (see eqn. 1 and finite volume scheme figure 4) gas and liquid phases are to be coupled via oxygen transfer. This is schematically shown in figure 8.

For estimation of local volumetric mass transfer coefficient k_L has been estimated from the equation suggested by Kawase and Moo-Young (<u>11</u>)

$$k_l = 1.3 \left(\frac{\epsilon \, \nu_L}{\rho_L} \right)^{\frac{1}{4}} \left(\frac{\nu_L}{D_{O_2}} \right)^{-\frac{2}{3}}$$

and specific surface area from

$$a = \frac{6 \, \epsilon_G}{d_B}$$

with ϵ_G = f(r,z) (see for instance figure 7).

An example of the simulation is illustrated in figures 9a and 9b showing local oxygen concentration and local rate of butanediol in the tank.

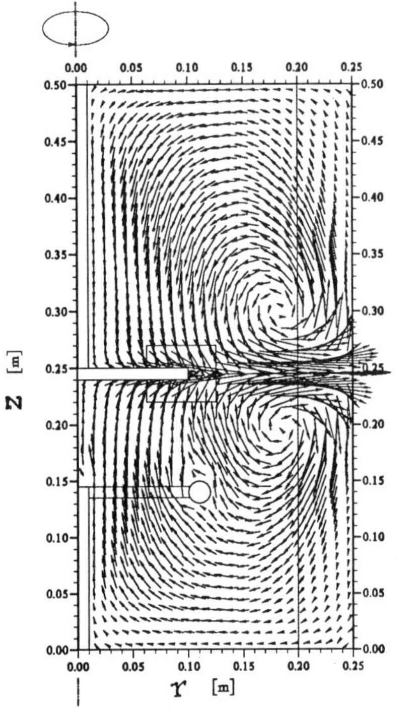

Figure 5: Velocity field in the liquid phase

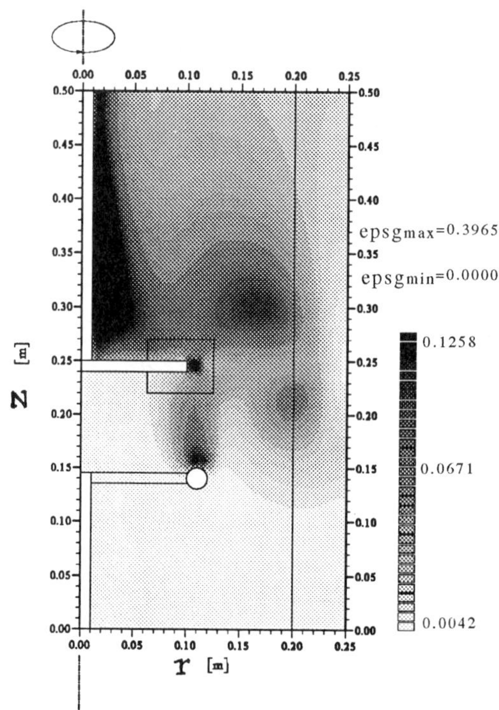

Figure 7: Local gas hold up

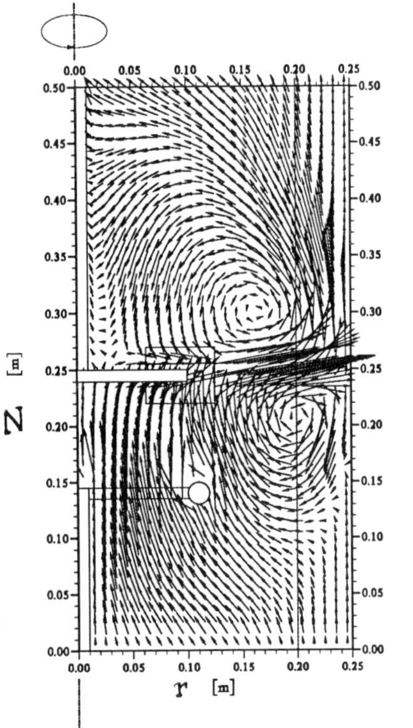

Figure 6: Velocity field in the gas phase

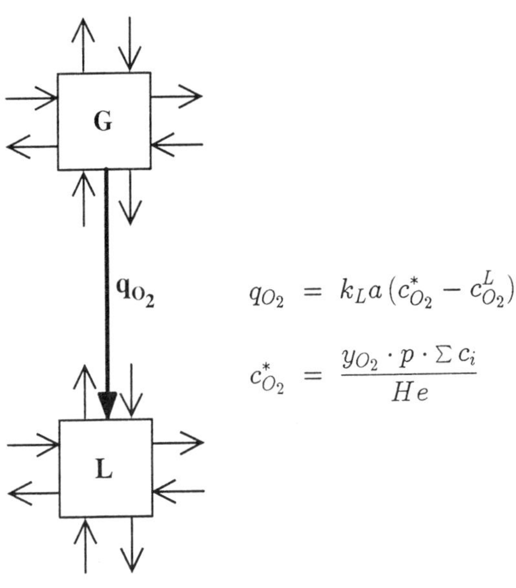

$$q_{O_2} = k_L a \left(c_{O_2}^* - c_{O_2}^L \right)$$

$$c_{O_2}^* = \frac{y_{O_2} \cdot p \cdot \Sigma c_i}{He}$$

Figure 8: Coupling of gas and liquid phases via oxygen transfer

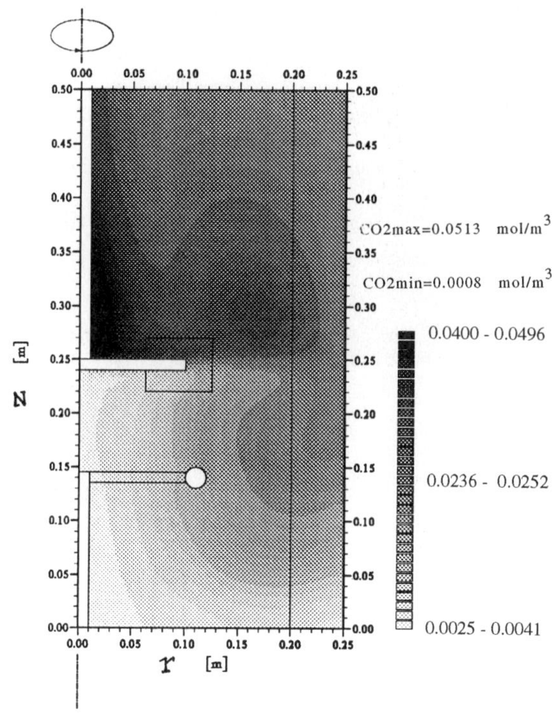

Figure 9a: Local oxygen concentration

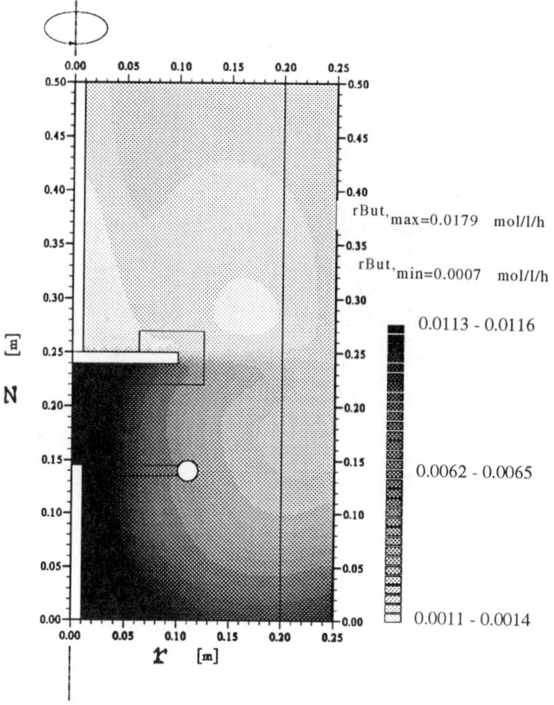

Figure 9b: Local rate of butanediol

Dynamics of pH-distribution

Knowledge of the dynamic response to pulses of ammonia is an important task to get a better inside into the scale-up of the dynamics of the control loop. The estimation of pH for mixtures of acids and bases requires the knowledge of the dissociation equilibra involved and the material and charge balances have to be calculated. The physico-chemical model applied in this paper has been originally suggested by Pons et al. (13, 14) and Belfares (15).

The model medium contains:

$(NH_4)_2SO_4$, KH_2PO_4, H_2CO_3, CH_3COOH, $MgSO_4$

The simulations were performed with the parametrization of the finite volume elements (figure 4) with the results from the fluiddynamic computations. The solution of this task requires the integration of the material balance equations of dissociated and non-dissociated ammonia (c_{NH3}+ c_{NH4+}) as well as the solution of the algebraic equation for estimation of c_{H+}. The numerical problem to be solved for a grid net of 25x50 finite volumes consists of 1250 coupled differential and an equal number of algebraic equations. Figures 10a, 10b and 10c illustrate the dynamics of the pH-distribution after a 1s-pulse of 0.125 mol ammonia into a 100 l stirred tank reactor. The initial pH value of 3.84 rises to a level of 4.1. After a time span of about 10 s the vessel is well mixed again.

CONCLUSIONS

Various recirculation and compartment models have been suggested in the past for simulating the effects of incomplete mixing in bioreactors. The reliability of these simulations was always confined due to the serious assumptions regarding the intrinsic fluid dynamics. Due to the advent of computational fluiddynamics it is now reasonable to envisage the use of these informations for application of modelling more complex biological reactions. The approach presented in this paper is based on a complete

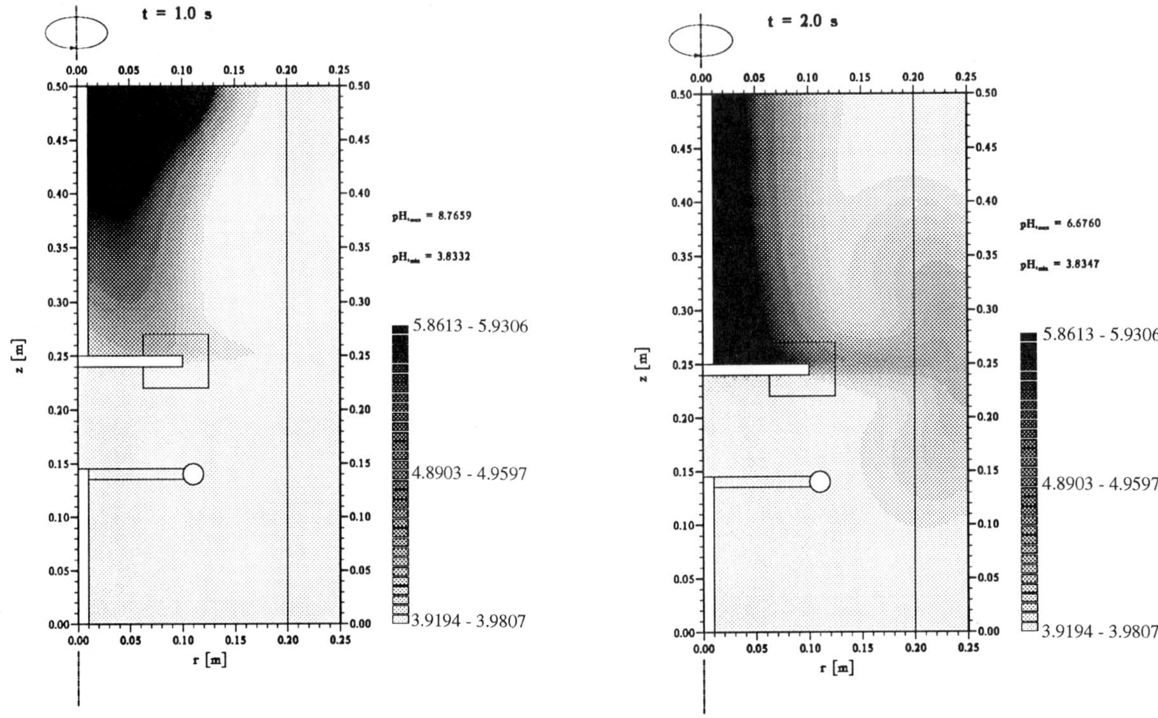

Figure 10a: pH-distribution in the stirred tank after 1s

Figure 10b: pH-distribution in the stirred tank after 2s

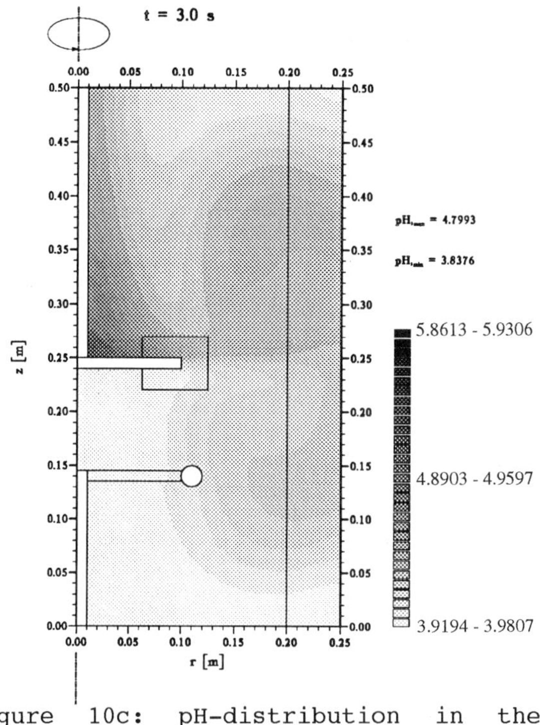

Figure 10c: pH-distribution in the stirred tank after 3s

separation of the momentum and mass balance equations. This assumption is essential to keep the effort for the numerical solution at a reasonable level for more complex reactions. Research in our group concentrates to the systematic development of more sophisticated turbulence models for the two-phase flow in stirred tank reactors. The results will be applied to various examples of structured models for microbial reactions. This should lead to further understanding of the interaction of transport and reaction for the scale up of bioreactors.

ACKNOWLEDGEMENT

Part of this work has been supported through a grant from the Deutsche Forschungsgemeinschaft (DFG). Thanks are due to Prof. Dr.-Ing. G. Eigenberger and Dr.-Ing. A. Sokolichin from the Institut für Chemische Verfahrenstechnik, Universität Stuttgart, for fruitful cooperation during software development for the Simpler-Algorithm.

LITERATURE CITED

1. Reuss, M. and R.K. Bajpai. Stirred tank models. In: H.-J. Rehm et al. (Eds.), Biotechnology, Vol. 4, VCH Verlagsgesellschaft, Weinheim-Deerfield Beach/Florida-Basel, p. 299 (1991).

2. Reuss, M. and M. Jenne. Compartment Models. In: U. Mortensen and H.J. Noorman (Eds.), Bioreactor Performance, Proc. The Biotechnology Research Foundation, IDEON, S-22370 Lund, Sweden, p. 63 (1993)

3. Ragot, F. and M. Reuss. A multi-phase compartment model for stirred bioreactors incorporating mass transfer and mixing. In: M. Reuss et al. (Eds.), Biochemical Engineering-Stuttgart, Gustav Fischer Verlag, Stuttgart, New York, p. 184 (1991)

4. Patankar, S.V. Numerical heat transfer and fluid flow. Mc Graw-Hill (1980)

5. Trägardh, C. A hydrodynamic model for the simulation of an aerated agitated fed-batch fermentation. In: Bioreactor fluid dynamics, Elsevier Applied Science Publ., 117 (1988)

6. Morud K. and B.H. Hjertager. Computational fluid dynamics simulation of bioreactors. In: Bioreactor Performance (Eds.: Mortensen, U., Noorman, H.). The Biotechnology Research Foundation, IDEON, S-22370 Lund, Sweden (1993)

7. Issa, R.I and A.D. Gosman. The computation of three-dimensional turbulent two-phase flows in mixer vessels. Numerical methods in laminar and turbulent flow, 829 (1981)

8. Patankar, S.V. and D.B. Spalding. Int. Journal of Heat and Mass Transfer, 15, 1787 (1972)

9. Moes, J. Diss. ETH Zürich, Nr. 7875 (1985)

10. Moes, J., M. Griot, J. Keller, E. Heinzle. I.J. Dunn and J.R. Bourne, Biotechnol. Bioeng., 27, 482 (1985)

11. Dunn, I.J. and E. Heinzle. Types of Understanding in Scaleing Down and Up, as Illustrated with an Oxygen-Sensitive Culture. In: U. Mortensen and H.J. Noorman (Eds.), Bioreactor Performance, Proc. The Biotechnology Research Foundation, IDEON, S-22370 Lund, Sweden, p. 189 (1993)

12. Kawase, Y. and M. Moo-Young. The Chem. Eng. J., 43, 19 (1990)

13. Pons, M.N., J.L. Greffe and J. Bordet. Talanta. 30, 205 (1983)

14. Pons, M.N., L. Garrido-Sanchez, P. Dantigny and J.M. Engasser. Bioproc. Eng. 5, 1 (1990)

15. Belfares, L. Diss. Inst. Nat. Polytechnique de Lorraine (1991)

Measurement and Modelling of Oxygen Transport into Biotechnical Immobilisates for Cell Entrapment

R. Wiesmann, P. Götz, and R. Buchholz

Fachgebiet Bioverfahrenstechnik, Institut für Biotechnologie, Technische Universität Berlin,
Ackerstr 71-76, D-13355 Berlin, GERMANY

The aerobic yeast Candida bombicola *was immobilized by entrapment of a cell suspension in a new, hollow spheres formig polymer system consisting of a cellulose derivate. As reference, this organism was entrapped in well known Ca-alginate beads. Using a microelectrode (based on Clark's principle) oxygen partial pressure was measured with step-sizes of 10μm inside cell-free and cell containing immobilisates. These pO_2-courses were simulated by a program calculating pO_2-values by numerical solution of Fick's second law. Oxygen transport properties of cell-free, non-stationary aerated polymer spheres could be characterized. A satisfying description of pO_2-curves inside cell-containing immobilisates was possible by using this program. Oxygen diffusion coefficients and O_2-penetration depth in both polymers were estimated. Further, it could be shown that the oxygen transport liquid/solid resistance in the liquid phase can not be neglected for high biomass concentrations inside the immobilisates.*

The immobilization of living cells offers a practicable possibility of cell retention in biotechnical processes. An effective cell retention is the essential condition for a higher productivity of continuous processes. Further a simplication of product recovery and stable operation in the case of high flow rates are the major advantages of immobilized cell systems. Any kind of cell entrapment however increases the mass transport resistance of substates into or of products from the immobilization matrix. Due to the low solubility of oxygen in aqueuos media, the oxygen supply of entrapped aerobic cells is the critical step in cultivating immobilized living cells. Some authors measured the oxygen partial pressure in immobili- sates (1). The oxygen transport in gel spheres was described by means of a model from other working groups (2-5).

MATERIALS AND METHODS

Strain and cutivation. *Candida bombicola*, ATCC 22214, was cultivated in 1000ml Erlenmeyerflasks containing 200 ml liquid medium according to Kosaric (6) at 30°C using a horizontal shaker (TL-125, Infors, Bottmingen, CH).

Polymers and immobilization.
Aqueous solutions of 2% Na-alginate and 2% $CaCl_2$ were both sterilized (30min, 121°C). A cell suspension is mixed with the Na-alginate solution and dropped in the $CaCl_2$-solution. Cell containing gel spheres were formed (7). These were washed twice, put into Erlenmeyer flask containing the described medium and cultivated for 5 days. The same procedure is used for aqueous solutions of 3% Na-Cellulose-sulfate (CS) (8) and 2,2% Polydiallyl-dimethylammoniumchloride (PDADMAC) (Fig. 1). Here the yeast cell

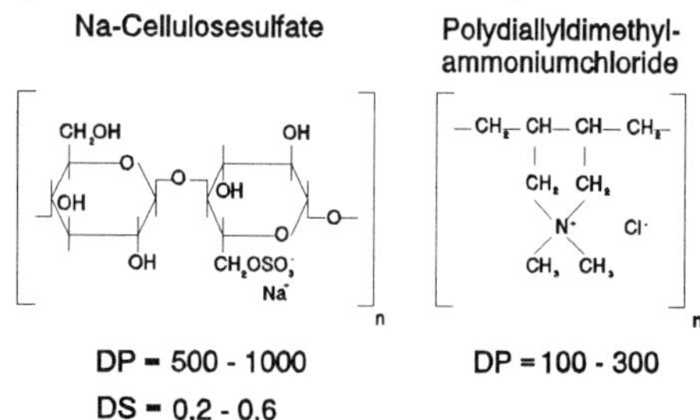

Fig. 1 Components for hollow sphere production (DP - degree of polymerisation, DS - degree of substitution)

suspension is added to the CS-solution and polymer hollow spheres with an

217

E. Galindo and O.T. Ramírez (eds.), Advances in Bioprocess Engineering. 217-220.
© *1994 Kluwer Academic Publishers. Printed in the Netherlands.*

asymmetric, microporous membrane (thickness $70\mu m$) (<u>9</u>) are formed.

<u>Oxygen detection and set-up.</u> The oxygen microelectrode (Fig.2) for the measurements is provided by the Max-

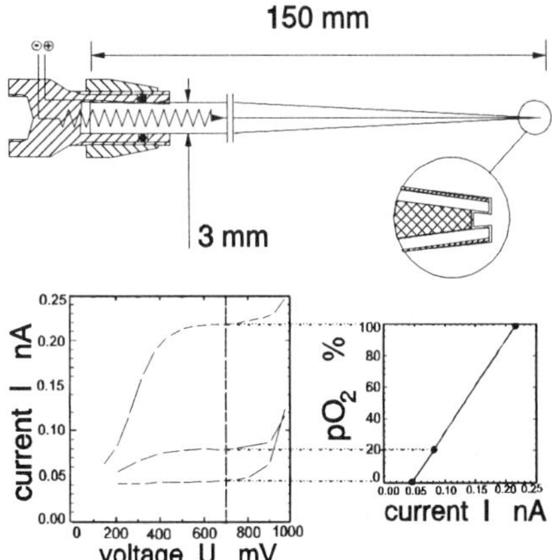

Fig. 2 Oxygen microelectrode

Planck-Institut für Molekulare Physiologie, Dortmund, Germany. The electrode is integrated in an airlift-type reactor (Fig.3). The insertion

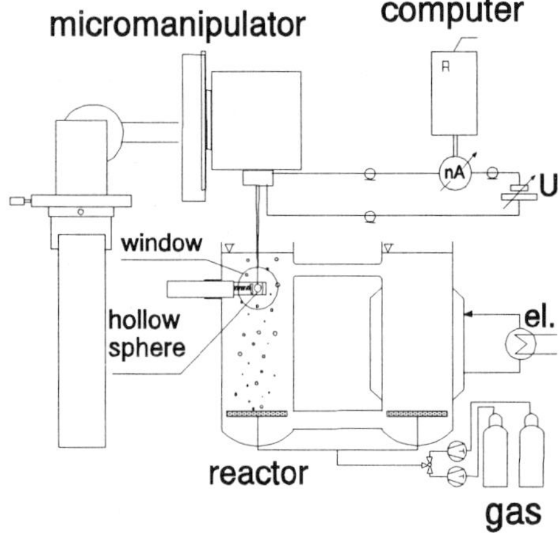

Fig. 3 Experimental set-up

into the fixed sphere is accomplished by a piezo micromanipulator. For an experimental run the system is saturated with nitrogen, aeration with air is started and the p_{O2} is recorded for different insertion depths, using a data aquisition unit and a personal computer.

<u>Mathematical model.</u> The O_2-transport in a sphere and the consuption inside the sphere can be balanced and presented in a dimensionless partial differential equation:

$$\frac{\partial c_{O_2}}{\partial t} = D_{O_2,H_2O}\left(\frac{\partial^2 c_{O_2}}{\partial r^2} + \frac{2}{r}\cdot\frac{\partial c_{O_2}}{\partial r}\right) - q_{O_2,max}\cdot X\cdot\frac{c_{O_2}}{c_{O_2}+k_0}$$

$$c_{O_2}(\ t=0,r\ ) = 0$$

$$c_{O_2}(\ t,r=R\ ) = 1$$

$$\left.\frac{dc_{O_2}}{dr}\right|_{r=0} = 0$$

This equation can be solved by means of an implicite difference method (Crank-Nicholson).

<u>RESULTS AND DISCUSSION</u>

<u>Cellfree immobilisates</u>

To characterize the employed polymer spheres p_{O2}-distribution in cellfree particles is measured by non-stationary aeration with alternating nitrogen and air. Fig. 4 shows the p_{O2}-profile in a Ca-alginate-bead.

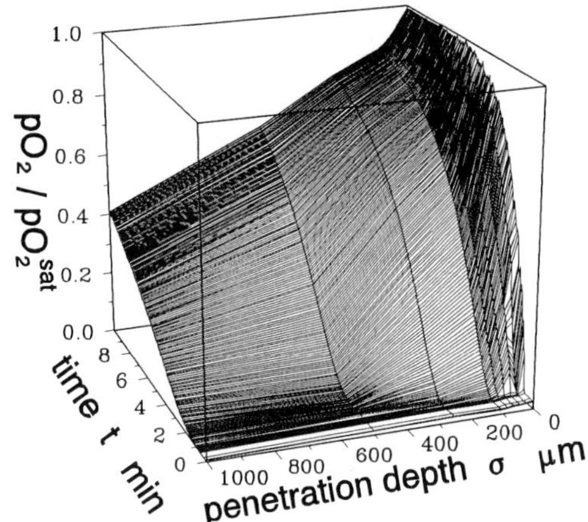

Fig. 4 Measured p_{O2}-profile in an non-stationary aerated Ca-alginate bead

Fitting the stationary model to this non-stationary measurement, the course of p_{O2} at the surface is calculated using eq. 5.

$$\left.p_{O_2}\right|_{r=R} = 1 - e^{(-t\cdot 0,89)}$$

The fit of the modelled profile (Fig. 5) is satisfying compared to the

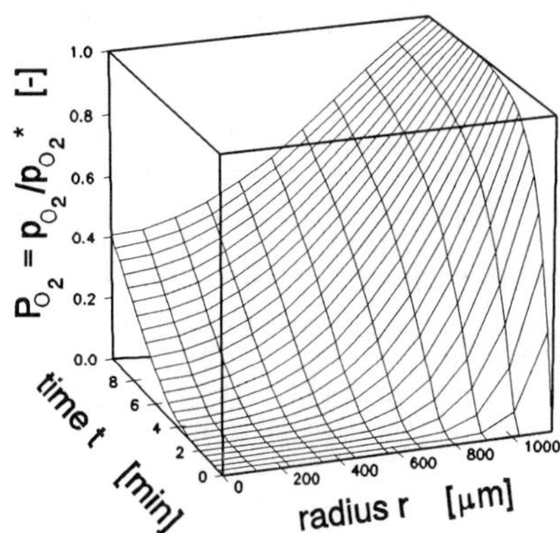

Fig. 5 Modelled p_{O2}-profile for a ca-alginate bead

measurements (Fig. 4). The D_{O2} in this simulation is 15% of the value estimated for water.

The same method of investigation is used for CS/PDADMAC-hollowspheres. In Fig. 6 the measured data are shown.

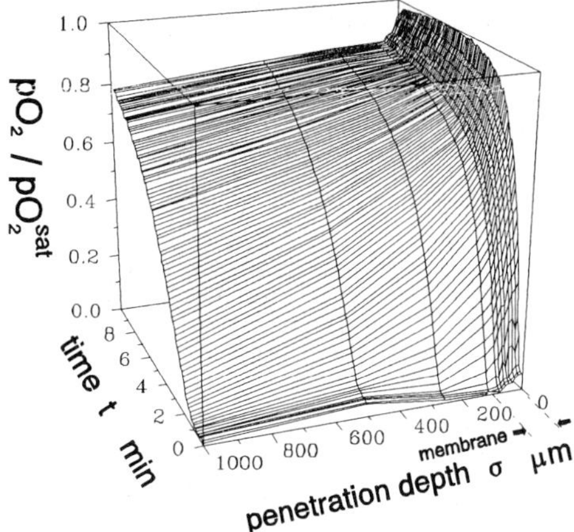

Fig. 6 Experimental data for p_{O2} in a CS/PDADMAC-hollowsphere

Hence the modelbased generating of p_{O2}-courses yields a profile being in good agreement to the measurement (Fig. 7). The D_{O2} for the 70μm membrane in this simulation is chosen to be 4.4% of the oxygen diffusion coefficient in water. Inside the liquid core, with a radius of 930μm, an D_{O2} of 90% of the water value has been assumed. The oxygen diffusion coefficient in the membrane is very small in comparison to the D_{O2} in Ca-alginate. It must be considered

however, that the D_{O2} of Ca-alginate is valid all over the radius of the homogenous sphere, while the low D_{O2} for CS/PDADMAC just belongs to the

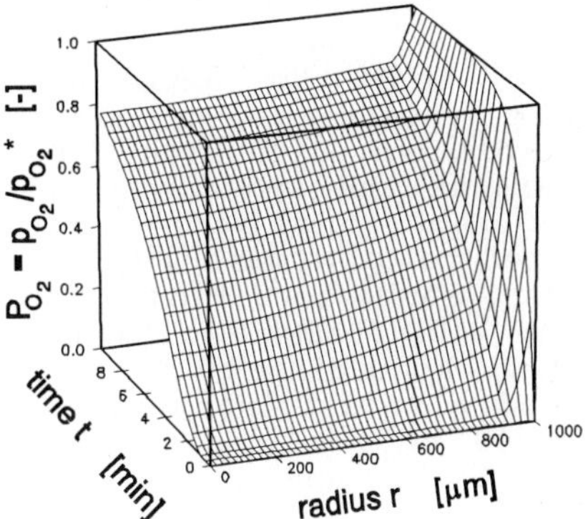

Fig. 7 Modelbased simulation of a p_{O2}-profile in a CS/PDADMAC-hollowsphere

70μm thin membrane. So the mean value for the D_{O2} of the hollowsphere is 84% of the oxygen diffusion coefficient in water.

Cellcontaining immobilisates

For these measurements immobilisates cultivated in batch process were taken from the Erlenmeyer flasks and fixed in the sphere holder inside a small airlift reactor. All conditions are equal to culture conditions, like temperature, medium, mixing. The electrode is moved fast in the center of the sphere and with constant

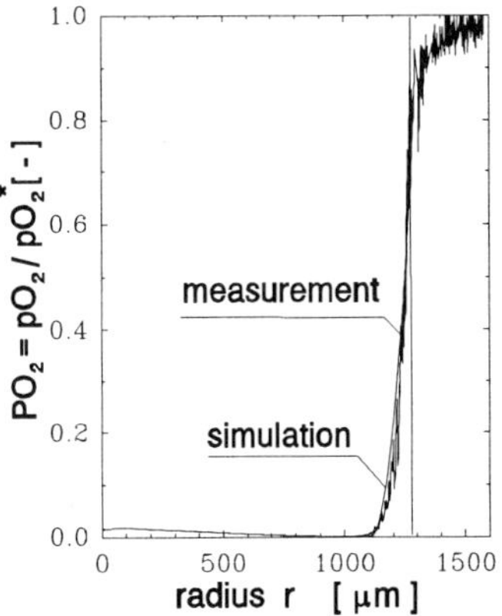

Fig. 8 Measurement an modelled data for p_{O2} inside a closely grown Ca-alginate bead.

velocity back to the surface. The measured data were recorded during the movement. It results in a course of p_{O_2} versus the radius. Fig. 8 shows this curve for a Ca-alginate bead measured at the end of the exponential growth phase. The D_{O_2} had to be enlarged to $0.3 \cdot D_{O_2,water}$ and the external mass transport resistance is not longer neglectable. A diffusion layer of $300 \mu m$ around the sphere is assumed. The oxygen consuption rate is determined to $Q_{O_2}=40mmol \cdot l^{-1} \cdot ^{-1}$.

Under same conditions the p_{O_2}-course of a CS/PDADMAC-hollowsphere is estimated. According to the Ca-alginate bead a satisfying curve fitting only could be realized by enlarging the D_{O_2} in the mem- brane. A oxygen diffusion coefficient of $D_{O_2,membrane}=0,15 \cdot D_{O_2,water}$ is determinated and a diffusion layer with a thickness of $300 \mu m$ is considered, according to the alginate bead. The oxygen consumption rate was $Q_{O_2}=40mmol \cdot l^{-1} \cdot h^{-1}$. In Fig. 9 the measured and modelled data in a densely grown CS/PDADMAC-hollowsphere are compared.

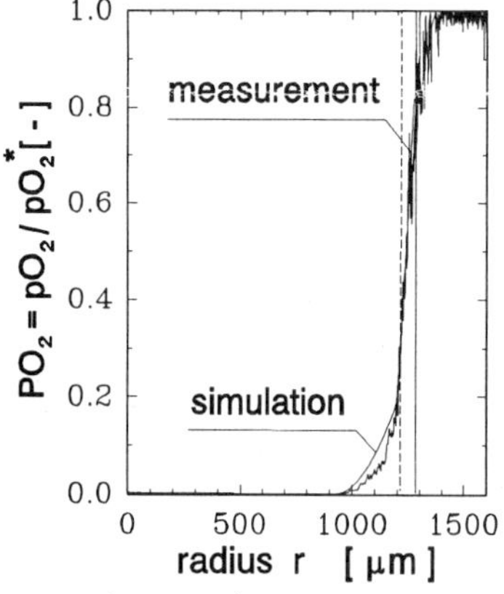

Fig. 9 Simulation and measured data for p_{O_2} in a cellcontaining CS/PDADMAC-hollowsphere

It is obvious, that sufficient oxygen supply exclusively in a thin layer at the surface of the sphere can be ensured. The core of the immobilisates is largely free of oxygen.

Conclusion

The employed oxygen measuring technique opens up the possibility of high performance p_{O_2}-detection inside biotechnical immobilisates. Oxygen

partial pressure profiles can be taken. From the modelbased simulation of these profiles oxygen diffusion coefficients can be calculated. Further the model allows a good description of p_{O_2}-courses in cell-containing immobilisates. With known hydrodynamic conditions and values of D_{O_2} and Q_{O_2} a prediction of oxygen penetration depth in the sphere is possible. For the investigated particles and organisms a critical sphere diameter of $520 \mu m$ for a Ca-alginate and $660 \mu m$ for a CS/PDADMAC-sphere could be estimated.

Nomenclature

A	surface of the sphere	m^2
c_{O_2}	oxygen concentration	$gO_2 \cdot g^{-1} \cdot h^{-1}$
CS	cellulosesulfate	
D_{O_2}	oxygen diffusion coefficient	$m^2 \cdot s^{-1}$
DP	degree of polymerisation	
DS	degree of substitution	
p_{O_2}	oxygen partial pressure	Pa
$p_{O_2}^*$	saturation oxygen partial pressure	Pa
P_{O_2}	normalized oxygen partial pressure	–
PDADMAC	polydiallyldimethylammoniumchloride	
q_{O_2}	specific oxygen consumption rate	$gO_2 \cdot g^{-1} \cdot h^{-1}$
Q_{O_2}	oxygen consumption rate	$gO_2 \cdot l^{-1} \cdot h^{-1}$
r	normalized radius	m
R	radius of the sphere	m
σ	insertion depth	μm
t	time	min

<u>LITERATURE CITED</u>

1. Beunink, J., H. Baumgärtl, W. Zimelka, H.-J. Rehm. *Experientia*, 45, 1041 (1989).

2. Hiemstra, H., L. Dykhuizen, W. Harder. *Eur. J. Appl. Mikrobiol. Biotechnol.* 18, 189, (1983).

3. Westrin, B.A., A. Axelson. *Biotechnol. Bioeng.*, 38, 439 (1991).

4. Toda, K., K. Sato. *J. Ferment. Technol.*, 63, 251, (1985).

5. Chang, H.N., M. Moo-Young. *Appl. Microbiol. Biotechnol.*, 29, 107, (1988).

6. Kosaric, N., G. Lu., "Removal of chlorinated aromatics in presence of sophorose biosurfactants." presented at the International Symposium Soil Decontamination using Biological Processes, Karlsruhe (1992).

7. Cho, M.-G., R. Buchholz. *Chem. Ing. Tech.*, 65, 1125, (1993).

8. Dautzenberg, H., F. Loth, K. Pommerening, K.-H. Linow, D. Bartsch, german patent DD-WP 160 393 (1980).

9. Wiesmann, R., W. Zimelka, H. Baumgärtl, P. Götz, R. Buchholz. *J. Biotechnol.*, in press, (1994).

Diffusion of Phosphate Ion in Xanthan Gum Solutions

G. Araiza, L.G. Torres[+], and E. Galindo

Departamento de Bioingeniería, Instituto de Biotecnología, Universidad Nacional Autónoma de México, Apartado Postal 510-3, Cuernavaca, 62271 Morelos, MEXICO

The diaphragm cell method was used in order to determine the diffusion coefficient of phosphate ion in xanthan gum solutions. The polymer concentration ranged from 0.125 to 2.5 kgm⁻³ and phosphate ion concentration from 12.5 to 50 mM. Potassium chloride solutions (0.01-5 M) were used in order to study the effect of ionic strength over the diffusion coefficients. All measurements were carried out at 29°C. An important decrease in the diffusional coefficients of phosphate was observed as the polymer concentration increased. Diffusion coefficients were influenced by the ion concentration. Phosphate diffusion coefficients in xanthan solution were a function of the phosphate concentration as well as of the ionic strength (modified by addition of KCl). The Ware model was employed successfully in order to describe the diffusion coefficients as a function of the polymer and salt concentrations.

Xanthan gum is a bacterial exopolysaccharide with a wide range of applications in the food, pharmaceutical, cosmetic and oil industries. It is produced by submerged culture of the bacteria *Xanthomonas campestris*. At the beginning of fermentation, the broth is a low viscosity Newtonian fluid; however, as fermentation evolves, the broth becomes highly viscous, showing strong non–Newtonian behaviour. At this stage, broth homogenization turns out to be very difficult and therefore, diffusional mass transfer process plays an important role.

Diffusion of sucrose (a culture's carbon source) and ammonium (a culture's nitrogen source) in xanthan solutions have been studied previously by our group (1,2). This work reports diffusionnal studies of a phosphorous source, the phosphate ion, in xanthan gum solutions.

Diffusion of small solutes in polymer solutions is a complex phenomenon. Few papers have been reported (3, 4, 5, 6, 7, 8) regarding diffusion of small molecules in high molecular weight-high viscosity solutions, like those formed by xanthan solutions. The study of these systems is particularly interesting besides the implications of substrate transport during xanthan fermentation.

The objective of this study was to determine the diffusion coefficient of phosphate in xanthan solutions. As the ionic strength plays an important role in the definition of the molecular conformation of polyelectrolites such as xanthan (9), diffusional studies were also conducted under different salt (KCl) concentrations in order to establish the effect of ionic strenght on phosphate diffusivity.

<u>MATERIALS AND METHODS</u>

Reagents. Potasium chloride and dibasic potasium phosphate were reagent grade obtained from J.T. Baker. Food grade xanthan gum ("Keltrol", Kelco, U.S.A.) was obtained from a local dealer. All other chemicals used in analytical determinations were reagent grade and obtained from J.T. Baker and Sigma. Deoinized water (milli-Q equipment, Millipore) was used in all experiments.

[+]Present address: Instituto Mexicano de Tecnología del Agua, Paseo Cuauhnáhuac 8532, Jiutepec 62550, Mor. México.

221

E. Galindo and O.T. Ramírez (eds.), Advances in Bioprocess Engineering. 221-225.
© 1994 Kluwer Academic Publishers. Printed in the Netherlands.

Diaphragm cell and diffussivity evaluations.
Diffusion coefficients were measured by the diaphragm cell technique as detailed before ([1]). Millipore HA filters, with pore diameters of 0.45 μm were employed as diaphragms. The cell was submerged in a water bath controlled at 29°C. To avoid boundary layer, two small magnets, one floating and one sinked, rotating at 60 rpm by using an external agitation system were used. The diaphragm cell constant, was evaluated by diffusing KCl. The diffusion coefficient of KCl was taken as (1.87 x 10^{-5} cm^2 s^{-1} at 25 $^{\circ}$C) as reported by Perkins and Geankoplis ([10]) and corrected to 29°C. Runs lasted between 2 to 35 h depending of the xanthan concentration used. The reported are the average of duplicates of experiments having a mass balance error lower than 6%. Experimental error among duplicates in D_{AP} determinations averaged 8.2%.

Analytical techniques. KCl concentration was determined following the Mohr's method as described by Brito ([11]). Phosphate ion concentration was measured using the phosphomolibdate method ([12]).

RESULTS AND DISCUSSION

Experimental diffusion results for phosphate in xanthan solutions as a function of xanthan concentration, ionic strenght and phosphate concentration are summarized in table 1. The average mass balance error and diffusion time are also quoted in table 1. Up to 35 h were required in order to diffuse at least 20% of the solute to the upper chamber. The mass balance errors of all experiments averaged 3.6%, which indicated accuracy in the experimental method.

Effect of xanthan concentration.

The effect of the polymer concentration over the diffusivity of phosphate relative to that in absence of xanthan (*i.e.* water or KCl solutions) (D_{AP}/D_{AW}) is shown in figure 1. The noticeable changes were found at the diluted regime. The D_{AP}/D_{AW} decreased as xanthan concentration increased. This drop was more pronounced below 0.5 kg m^{-3} of xanthan concentration. For higher xanthan concentrations, the D_{AP}/D_{AW} ratio showed very small differences.

As quoted in table 1, diffussion coefficients of phosphate in KCl solutions (containing no xanthan) were also affected by the KCl concentration. Clearly, diffusion coefficients of all electrolytes present in a solution are affected by the total ionic strength.

Table 1. Effect of xanthan and KCl concentration on the diffusion coefficient of phosphate ion.

Xanthan (kg m^{-3})	KCl (M)	$\bar{t}$ (h)	Average error in mass balance (%)	$D_{AP} \times 10^{-6}$ (cm^2 s^{-1})
0.000		2.27	5.1	11.47
0.250		4.56	5.7	2.35
0.500	0.01	5.00	3.3	1.58
1.250		20.50	1.2	1.06
2.500		35.50	3.2	1.03
0.000		2.54	1.9	10.01
0.125		3.03	0.1	8.70
0.250		4.94	6.1	3.81
0.500	0.05	6.09	5.5	2.01
1.250		20.27	3.0	1.09
2.500		20.11	4.0	1.02
0.000		3.00	4.8	9.02
0.125		5.81	4.5	3.48
0.250		7.50	3.9	2.65
0.500	0.10	7.83	4.9	1.91
1.250		20.00	3.1	0.93
2.500		25.25	1.4	0.12
0.000		2.77	2.3	7.15
0.125		5.17	2.3	6.34
0.250		5.03	3.9	3.76
0.500	0.50	7.85	2.2	1.39
1.250		20.10	5.9	0.89
2.500		21.73	5.5	0.92

Initial phosphate = 25 mM

Figure 2 shows a comparison of the results of this work with literature data ([1,2,7]). The D_{AP}/D_{AW} ratio for the phosphate ion showed the same trend if compared with those exhibited by other molecules: D_{AP}/D_{AW} decreases as the polymer concentration increases, and this behaviour is more pronounced for polymer concentrations below 1.25 kg m^{-3}.

Several authors (see a survery in reference [1]) have explained this behaviour by considering a blocking effect. The presence of a macromolecule in solution reduces the effective area for the diffusing solute, which must surround the macromolecule. An additional mechanism could be the binding of the small solute to the polymer. This phenomenon was not studied in this work but it has been reported as negligible for the system sucrose-xanthan (13).

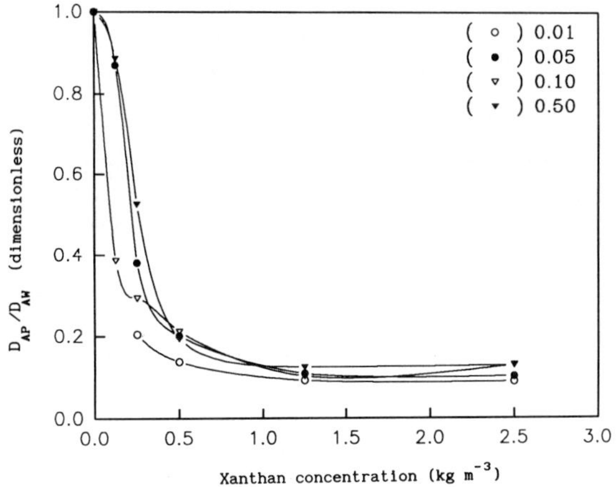

Figure 1. Normalized phosphate diffusion coefficients in xanthan solution as a function of xanthan and KCl concentrations.

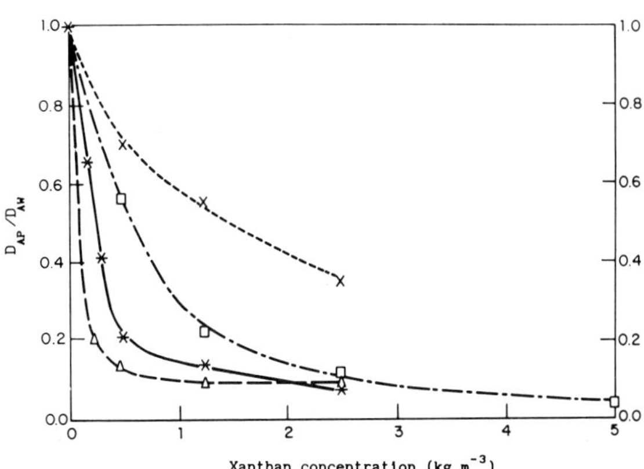

Figure 2. D_{AP}/D_{AW} as a function of xanthan concentration for oxygen (x) (7), sucrose (□) (1), ammonium (*) (2) and phosphate (Δ) (this work).

Effect of phosphate ion concentration.

The effect of phosphate concentration upon diffusivity in xanthan solutions was studied by varying the phosphate concentration and keeping constant the polymer concentration and ionic strenght. Only experiments with 0.01 M KCl and 0.5 kg m^{-3} of xanthan were carried out and the results are shown in figure 3.

As suggested by the data in figure 3, phosphate diffusivity is a function of phosphate concentration. This trend was the same observed by Torrestiana *et al* (1) when diffusing sucrose (5-15 kg m^{-3}) in 0.5 kg m^{-3} xanthan solutions. However, Torres (2) when diffusing ammonia 2.5-10 mM in 1.25 kg m^{-3} xanthan solutions, observed that D_{AP} practically did not vary with ammonium concentration. The fact that the experiments with sucrose were carried out in the diluted range, whereas the experiences with the ammonium ion were performed in the semi-diluted range, could explain the differences.

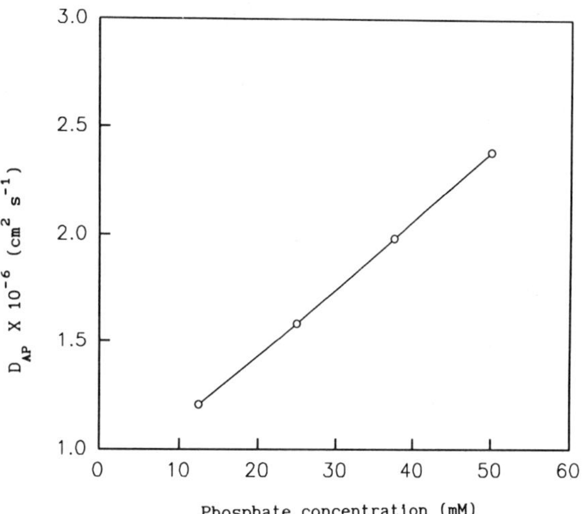

Figure 3. Phosphate diffussion coefficients as a function of the ion concentration. KCl = 0.01 M.

Effect of ionic strenght.

The diffusion coefficient of a small molecule in a polymeric solution, as a function of polymer concentration and ionic strenght, can be predicted using the expression of Gorti and Ware (14):

$$D_{AP}/D_{AW} = \exp(-a\,C^{\nu})$$

The experimental diffusivity data obtained in this work was used to determine the suitability of the Gorti and Ware expression in predicting diffusion coefficients of the phosphate ion, in terms of polymer and KCl concentrations.

The parameters a and ν of the model of Ware were calculated by fitting D_{AP}/D_{AW} data (table 1) to the former equation, including all xanthan concentrations. This was made for each KCl concentration. The calculated values of the Ware constants are quoted in table 2. The parameter a was a weak function of the ionic strenght, whereas the parameter ν increased as the ionic strenght was higher. However, no monotonic functions were evident nor the trends were similar to those reported by Gorti and Ware (14). Nevertheless, it should be pointed out that a and ν are empirical constants and the polymer studied by Gorti and Ware (polyestirene sulfate) is rather different to xanthan gum.

Table 2. Fitted values of a and ν of the Ware model as a function of KCl concentration.

KCl (M)	a (–)	ν (–)
0.01	2.15	0.18
0.05	1.8	0.37
0.10	1.83	0.25
0.50	1.63	0.47

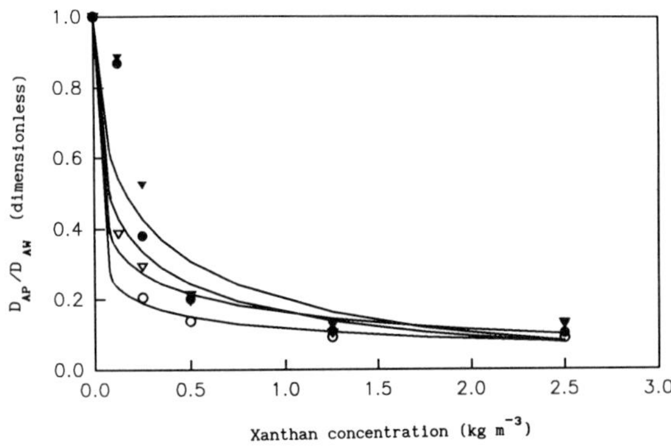

Figure 4 shows the experimental values of D_{AP}/D_{AW} for different xanthan and KCl concentrations as compared with the fitting of the Ware model. The Ware equation was able to describe well the trends in the experimental data, except for the lowest xanthan concentrations.

CONCLUSIONS

Diffusivity of phosphate ion in xanthan gum solutions was affected by the polymer concentration. The trend in the drop of D_{AP} was very similar to that reported in the literature for sucrose and ammonium.

When diffusing phosphate 12.5–50 mM in 0.5 kg m^{-3} xanthan solutions, the higher the ion concentration, the higher the diffusion coefficient.

The diffusion coefficients were affected by the total ionic strength (*i.e.* the KCl concentration). This phenomenon was more pronounced in diluted xanthan solutions and for KCl concentrations between 0.01–0.1 M.

For xanthan concentrations higher than 0.25 kg m^{-3}, the expression of Ware was able to describe well the dependance of the ratio D_{AP}/D_{AW} as a function of xanthan concentration and ionic strenght.

ACKNOWLEDGEMENTS

This work was supported in part by CONACyT (grant 1020-N9111). We thank E. Brito for helpful discussions.

NOMENCLATURE

a	Constant in the model of Ware (–)	
C	Polymer concentration (kg m^{-3})	
D_{AP}	Diffusivity of phosphate ion in xanthan gum solutions (cm^2 s^{-1})	
D_{AW}	Diffusivity of phosphate ion in water or KCl solutions (cm^2 s^{-1})	
$\bar{t}$	Average diffusion time (h)	
ν	Constant in the model of Ware (–)	

Figure 4. D_{AP}/D_{AW} values as a function of KCl and polymer concentrations. KCl (M): (o) 0.01, (●) 0.05, (▽) 0.10 and (▼) 0.50. Solid lines: Model of Ware.

LITERATURE CITED

1. Torrestiana, B., Galindo, E. and Brito, E. *Biopr. Eng.*, **4**, 265 (1989).

2. Torres, L.G. Difusión de amonio en soluciones diluídas de goma xantana. M.Phil. thesis, U.A.C.P.y P., CCH, Universidad Nacional Autónoma de México (1990).

3. Li, S. U, Gainer, J.L. *Ind. Eng. Chem. Fundam.*, **7**, 433 (1968).

4. Geankoplis, C.J, Okos, M.R., Grulke, E. A. *J. Chem. Eng. Data*, **23**, 40 (1978).

5. Nystrom, B., Roots, J. *Europ. Polym. J.* **16**, 201 (1980).

6. Oosting, E. M., Gray, J.I., Grulke, E. A. *AIChE J.*, **31**, 773 (1985).

7. Reuss, M., Debus, D. and Niebelschutz, H. Oxygen transfer in viscous fermentation broths. *Colloque Soc. Fr. Microbiol. Tolouse*, p. 46-70, (1980).

8. Navari, R.M., Gainer, J.L., Hall, K.R. *AIChE J.*, **17**, 1028 (1971).

9. Smidsrod, O. and Haug, A. *Biopolymers*, **10**, 1213 (1971).

10. Perkins L.R. and Geankoplis, C.J. *Chem. Eng. Sci.*, **24**, 1035 (1964).

11. Brito, E. Diffusion of sucrose and glucose in protein solutions with blockage and binding efects present. *M. Sci. thesis.* The Ohio State University, USA, (1982).

12. American Public Health Association, Methods for the Examination of Water and Wasterwater, p. 446-448, (1985).

13. Torrestiana, B., Galindo, E. and Brito, E. *Biotechnol. Prog.*, **11** (4), 14, (1988).

14. Gorti, S. and Ware, B.R. *J. Chem. Phys.*, **84** (12), 6449 (1985).

Modeling Protease Production by Immobilised *Serratia marcescens*

C. Quirós, L.A. García, and M. Díaz

Depto. de Ingeniería Química, Universidad de Oviedo, Julián Clavería s/n, 33071 Oviedo, SPAIN

Protease production and cell growth of Serratia marcescens, *when grown in free suspension in a synthetic medium, have been taken into account in order to model production when the cells were immobilised in a support such as calcium alginate beads. Microscopic analysis showed the breakage of the support and leakage of the microorganism into the medium. By substracting the effect of free cells from the overall production of protease in the system, the remaining production by the immobilised cells was successfully modelled by means of a diffusion-reaction equation, using kinetic parameters determined from free cells.*

There are several advantages in the use of immobilised microorganisms including increase in productivity due to higher cell density, relative ease of product separation, reutilization and regeneration of biocatalyst and reduced susceptibility to contamination[4]. The future application of immobilised microorganism techniques will depend on the development of systems which are technological applicable on an industrial scale. These techniques must permit high microbial concentrations and must allow mass transfer to take place with low diffusional limitations[11].

In this work the support chosen for cell immobilisation was one of the most widely employed, spherical calcium alginate gel beads[14]. Such beads are obtained from a natural polymer formed of D-mannuronic and l-glucuronic acid chains linked at the 1,4 position by glycoside bonds ionotropically gelled by calcium[6]. This particular matrix has the advantage of being compatible with the biomass and biochemically inert; however, it has some drawbacks such as low mechanical strength and poor chemical resistance to culture broths[7].

Proteases represent the world´s largest industrial enzyme market. Their most important application is as additives in detergents, although they are also employed in tannery, cheese making (rennin), bakery, brewery and food industry in general (enzyme modified soy protein, meat tenderisers)[1,15]. There are a great variety of proteases described in the literature and they can be classified into three groups depending on the characteristics of their particular active centre: Metallo, sulphydril or serine[16].

The protease production capacity of the enterobacteriaceae group has been little studied, because generally these microorganisms do not excrete their proteases but retain them inside the periplasmic space or they are firmly attached to the external membrane. *Serratia*

E. Galindo and O.T. Ramírez (eds.), Advances in Bioprocess Engineering. 227-232.
© *1994 Kluwer Academic Publishers. Printed in the Netherlands.*

marcescens is one of the few exceptions, having been identified as a producer of true exoenzymes[3]. Most of the strains seem to produce a mainly exometalloprotease and a little serinprotease[9].

In this work, we have studied protease production by *Serratia marcescens* immobilised in alginate beads in a batch reactor. The characteristics of the support were observed with time, including the leakage of cells and a diffusion model was tested using kinetic parameters from free cell experiments. This type of modelling is very common for some microorganisms but not for *Serratia*.

MATERIALS AND METHODS.

Microorganism and culture medium: *Serratia marcescens* (ATCC 25419) was the strain chosen because of its protease production ability[3]. This is an unusual microorganism when protease production is the process desired. Nevertheless, the strain employed has been shown to be a great producer compared with other more commonly used bacteria such as those of *Bacillus*, etc.

The strain was propagated and kept on nutritive agar. Afterwards, the microorganism was pre-incubated in 100ml of NBG medium, which was nutrient broth (Oxoid) plus 1% glucose, for 5 hours at 37°C and 250 rpm. Cells were harvested in 0.7% NaCl (0.4 g/ml final concentration) and transferred to 250 ml flasks containing 100 ml of TYE medium. TYE had the following composition: 5 g/l yeast extract, 10 g/l tryptone, 5 g/l NaCl.

Cell immobilisation. Cells were entrapped in calcium alginate under sterile conditions[14]. A volume of 5 ml cell suspension in 0.7% NaCl was mixed with 100 ml of a sodium alginate (Janssen) solution to give a 2% (w/v) final concentration. The mixture obtained was extruded dropwise through a 20-ml syringe (1.5 mm diameter) into a gently stirred 3% (w/v) $CaCl_2$ $2H_2O$ (Merck)

solution and hardened in this solution for 30 min. The calcium alginate beads containing cells were thoroughly washed with a 0.7% (w/v) NaCl solution and used as inoculum.

Fermentation conditions: Fermentations were carried out in 250 ml Erlenmeyer flasks containing 100 ml of the TYE culture medium in an orbital incubator Gallenkampf (N.B.S. Mod. G25) at 36°C and 250 rpm, under aerobic conditions, without pH to control. When cultures were of immobilised cells there were as many flasks as samples to be taken. Cultures were inoculated with 20 mg wet weight of cells in all cases. The growth of the microorganism was monitored by measuring the optical density of the culture at 587 nm in a spectrophotometer (Philips, Mod. PU8720).

Protease assay: Cell free supernatant fluids were obtained by centrifuging 1 ml culture samples at 19000 g for 10 minutes. A 120 μl portion of the sample was incubated at 37°C in a tube containing 480 μl of 2% (w/v) azocasein (Sigma) in 0.2 M Tris-glycin buffer pH 9.0 with 2 mM CaCl. After 60 minutes of incubation, the reaction was stopped by adding 600 μl of 10% trichloroacetic acid, and the mixture was centrifuged. The precipitate was removed and the supernatant added to a tube containing 200 μl of 1.8 M NaOH, and the absorbance at 420 nm was measured[8]. One unit of protease activity was defined as the amount of enzyme causing an increase in absorbance at 420 nm of 0.1 units in one hour.

In order to assay protease content inside the beads, these were dissolved into 2% Na_2HPO_4 (5.3 ml/g beads), centrifuged for 10 min at 19000 g, and the supernatant was tested. Dilution concentration and activity are considered to be proportional. Cell growth inside the alginate beads was determined by dissolving them in 2% Na_2HPO_4 and measuring afterwards the optical density of the resultant solution at 587 nm. A sample of alginate was used as blank. All determinations were performed in triplicate, and experiments at least in duplicate.

<u>Microscopic observance.</u>

Samples of immobilised cells in alginate beads were treated as follows for examination by scanning electron microscopy (S.E.M.). Firstly, samples were dehydrated with successively increasing concentrations, first of ethanol and then of acetone. The samples were dried in a critical point apparatus (Balzers Union) and subsequently were covered with gold in a metallizer (Balzers Union). The samples where then observed under an electron microscope JEOL JSM 6100 at 15 kv and photomicrographs were taken using AFGA 400ASA.

<u>RESULTS AND DISCUSSION.</u>

In first place the protease production was followed inside and outside the alginate beads, during some 24 hours. The results obtained are presented in Fig. 1.

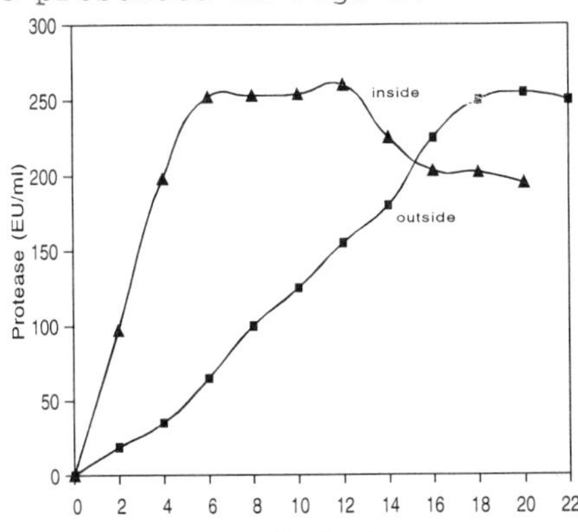

Fig.1. Protease production (EU/ml) by immobilised cells.

In this figure it can be observed that, when immobilised microorganisms are used for protease production, some of the enzyme produced appears outside the beads. The majority of the protease produced was initially achieved inside the beads, although after

some 24 hours of incubation the amount of protease outside and inside the beads were at the same level. This can be attributed to diffusional reasons[8,10] or the possibility of some degree of breakage of the particles resulting in leakage of cells into the medium.

In order to determine if the latter was the main reason for the extra-beads protease obtained, photomicrographs of the beads were taken throughout the incubation time. Visualisation of these (Fig.2) showed a certain level of breakage from first moment, which increased continuously as the fermentation proceeded (Fig.3). The degree of breakage was quantified and expressed as the number of holes (NH) achieved per bead surface unit. This level increased dramatically at longer times (11200 NH/mm^2 at total 48 h, 24 h after starting the second cycle of fermentation) than those used for protease production (120 NH/mm^2 at 24 h).

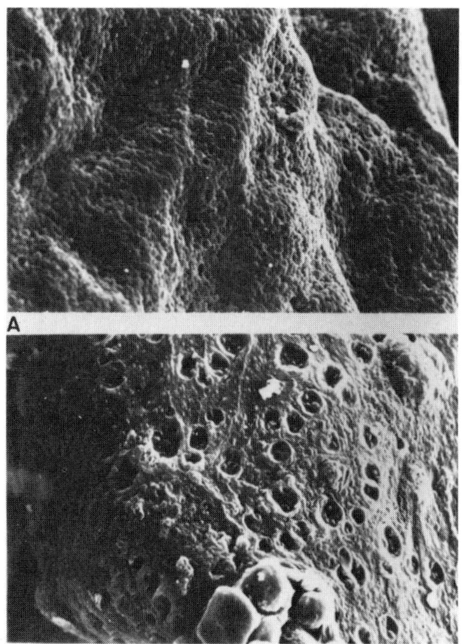

Fig. 2. View of calcium alginate particle surface by S.E.M. A: x190, 15 hours of fermentation. B:x1000 48 hours of fermentation (second cycle).

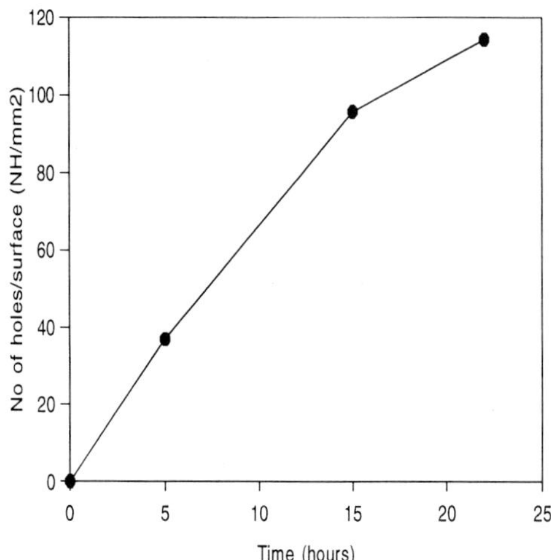

Fig. 3. Evolution of the number of breakages in the surface of the particles during fermentation.

The production of biomass was also monitored inside and outside the cells. As was mentioned before for protease production, some biomass was detected outside, Figure 4. In this case, the increasing levels of biomass were always kept at a low level compared with the biomass contents inside.

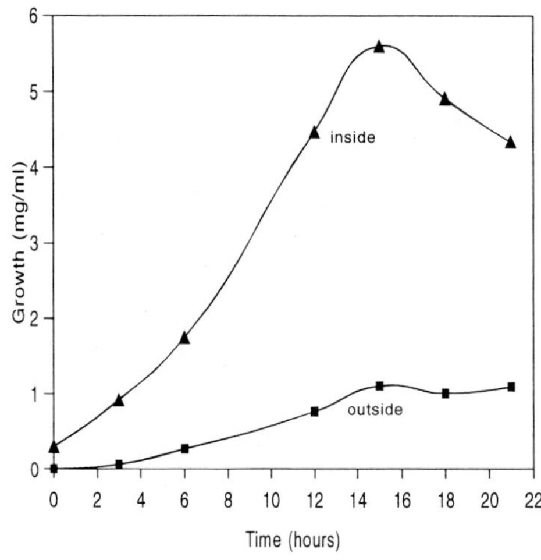

Fig. 4. Cell growth (mg/ml) in an immobilised system.

The protease production by the cells in the medium could be approximated, considering the production in previous experiments with free cells. Assuming that the production per cell unit was similar to those experiments, the contribution of the protease production of free cells in the medium in the immobilised cell experiments was less than 25% of the total during the first 15 hours. Therefore diffusion should be the reason for the presence of protease in the medium.

In order to model protease production during fermentation of immobilised *Serratia marcescens*, mass balances for the protease inside and outside of the particle were established. The mass balance inside the particle is:

$$\partial c_p / \partial t = r_p - \{\partial / r^2 \partial r [r^2 \varepsilon_i D_p (\partial c_p / \partial r)]\} \quad (1)$$

The same balance in solution gives:

$$\partial c / \partial t = (-3 D_p / r)[(1 - \varepsilon_L) / \varepsilon_L](\partial c_p / \partial r)|_{r=R} \quad (2)$$

The boundary conditions to consider are: $r = 0$: $\partial c_p / \partial r = 0$; $r = R$: $c_p = c$. The initial conditions are: $t = 0$: $c = c_p = 0$. To have a kinetic equation r_p to be introduced into the diffusional model (Eq. 1), experiments with free cells were done. In these experiments, the best fit for the protease production rate, was achieved using the following expression:

$$r_p = r_{ps} (1 - a_1 e^{-a_2 t}) \quad (3)$$

There are several expressions described in the literature which are quite similar to this one. For instance the empirical equation of Shu[13] was obtained working with free cells, on the basis of the assumption that r_p of individual cells is a genetically determined function of cell age t. This is an empirical approach and there is no clear biochemical reason that can be argued for explaining the meaning of the equation.

The parameters of the kinetic equation obtained from free cell experiments were also used to simulate the behaviour of immobilised cells. These were:

For $t<12h$: $a_1=1$; $a_2=0.39h^{-1}$.

For $12h<t<24h$: $r_{ps}=71.21$ EU/cm^3 h; $a_1=0.75$; $a_2=-0.0116$ h^{-1}.

For $t>24h$: $r_p=0$.

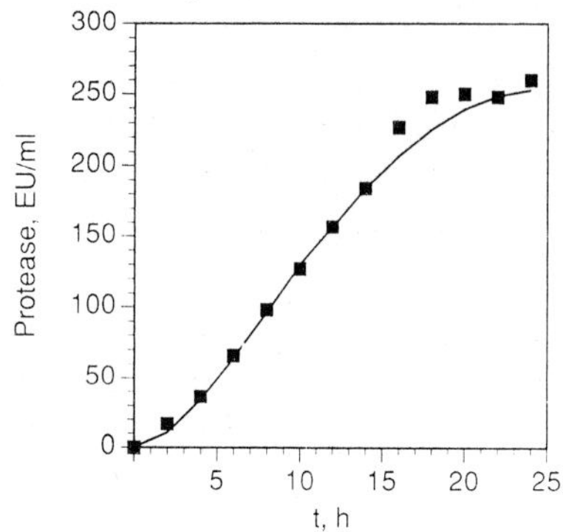

Fig.5. Fitting of the proposed model (straight line) to the experimental protease production data (dots).

In some experiments (Figure 5) the simulation for immobilised cells fits quite well with the experimental protease production after subtraction of the protease excreted by the free cells in the medium.

CONCLUSIONS.

Protease production by *Serratia marcescens* immobilised in calcium alginate beads, and in a batch reactor, can be considered as relatively high when compared with other microorganisms. Some of the protease detected is due to free cells leaking from the beads, which was increased during the experiments as determined by SEM.

Simulation of production in the beads excluding the free cells, has been made using a diffusion model and a kinetic equation from free cell experiments.

NOMENCLATURE.

c_p concentration of protease in the particles, EU/cm^3
c fluid phase concentration of protease, EU/cm^3

D_p effective diffusivity within the particles, cm^2/s
ε_i internal particle porosity
ε_L liquid void fraction in the system,
t,r time and spatial co-ordinates, s and cm
r_p rate of protease production, EU/cm^3 h^{-1}
r_{ps} rate of protease production during stationary phase, EU/cm^3 h^{-1}

<u>LITERATURE CITED</u>.

1. Aaslyng, D., Gormsen, E. and Malmos, H. "Mechanistic Studies of Proteases and Lipases for the Detergent Industry". J. Chem. Tech. Biotechnol. **50**, 321-330 (1991).

2. Ampon, K. "Distribution of an Enzyme in Porous Polymer Beads". J. Chem. Tech. Biotechnol. **55**, 185-190 (1992).

3. Braun, V. and Schmitz, G. "Excretion of Protease by *Serratia marcescens*". Arch. Microbiol. **124**, 55-61 (1980).

4. Büyükgüngör, H. "Stability of *Lactobacillus bulgaricus* Immobilized in K-carrageenan Gels". J. Chem. Tech. Biotchnol.. **53**, 173-175 (1992).

5. Galazzo, J.L. and Bailey J.E. "Fermentation pathway kinetics and metabolic flux control in suspended and immobilized *Saccharomyces cerevisiae*". Enzyme Microbiol. Technol. **12**, 162-172 (1990).

6. Guiseley, K. "Chemical and physical properties of algal polysaccharides used for cell immobilization". Enzyme Microb. Technol. **11**, 706-716 (1989).

7. Johansen, A. and Flink, J. "Influence of alginate properties and gel reinforcement on fermentation characteristics of immobilized yeast cells". Enzyme Microb. Technol. **8**, 737-748 (1986).

8. Longo, A., Novella, I., García, L.A. and Díaz, M. "Diffusion of proteases in calcium alginate

beads". Enzyme Microb. Technol. **14** 586-590 (1992).

9.   Miyazaki, H., Yanagida, N., Horinouchi, S. and Beppu, T. "Specific Excretion into the Medium of a Serine Protease from *Serratia marcescens*". Agric. Biol. Chem. **54** (10), 2763-2765 (1990).

10. Pu, H.T. and Yang, R.Y.K. "Diffusion of Sucrose and Yohimbine in Calcium Alginate Gel Beads with or without Entrapped Plant Cells". Biotechnol. Bioeng. **32**, 891-896 (1988).

11. Ruggeri, B., Gianetto, A., Sicardi, S. and Specchia, V. "Diffusion phenomena in the spherical matrices used for cell immobilization". The Chem. Eng. J. **46**, B21-B29 (1991).

12. Shreve, G. and Vogel T. "Comparison of Substrate Utilization and Growth Kinetics Between Immobilized and Suspended *Pseudomonas* Cells". Biotechnol. Bioeng. **41**, 370-379 (1993).

13. Shu, P. "Mathematical Models for the Product Accumulation in Microbiological Processes". J. Biochem. Microbiol. Technol. Eng. **3** (1), 95-109 (1961).

14. Vuillemard, J., Terré, S., Benoit, S. and Amiot, J. "Protease production by immobilized growing cells of *Serratia marcescens* and *Myxococcus xanthus* in calcium alginate gel beads". Appl. Microbiol. Biotechnol. **27**, 423-431 (1988).

15. Wiseman, A. "Designer Enzyme and Cell Applications in Industry and Enviromental Monitoring". J. Chem. Technol. Biotechnol. **56**, 3-13 (1993).

16. Zamost, B., Nielsen, H., and Starnes, R.L. "Thermostable Enzymes for Industrial Applications". J. Industrial Microbiol. **8**, 71-82 (1991).

A Mathematical Model for a Fermentation Process Carried Out in a Fluidized Bed Reactor

V. González-Alvarez[1], V. Alcaraz-González[1], and V. Zúñiga-Partida[2]

[1]Facultad de Ciencias Químicas, [2]Instituto de Madera, Celulosa y Papel "Karl Augustin Grellman", Universidad de Guadalajara, Blvd. Gral. Marcelino García Barragán 1451, Guadalajara, Jalisco 44430, MEXICO

A mathematical model describing the dynamics of a fermentation process carried out in a fluidized bed bioreactor with immobilised yeast is proposed. The model is a set of coupled partial differential equations, derived from the unsteady state mass balances of the species involved in the fermentation: substrate, biomass and product. The orthogonal collocation method is used to solve the resulting model. A series of experimental runs for fermentations using glucose was performed to validate the model. The model predicted satisfactorily the dynamic behavior of the fermentation species under various operating conditions.

Cell immobilization and fluidized bed reactors are often selected to increase fermentation yields. This selection is mainly due to the advantages that these two alternatives offer over traditional free cell systems and batch and continuous reactors. On one hand, cell immobilization leads to high reaction rates by virtue of high cell loadings in the support material, reduces contamination risks, keeps high cellular densities and dilution rates and eases product extraction (Mattiasson, [11]). Fluidized bed reactors, on the other hand, offer additional advantages: reduction of diffusional barriers, reuses of biocatalyst, reduction of product inhibition effects and that can be operated in continuous or semicontinuous manners. These advantages suggest that any fermentation carried out in a fluidized bed using immobilized cells will produce results which will be very difficult to improve with any other cell system and/or reactor configuration. One way to demonstrate this is by using a theoretical model of the system which will allow us to understand the behavior of the system and to identify the limiting steps and

hence focus on process improvement in the proper areas.

In this work, we propose a mathematical model for a glucose fermentation carried out in a fluidized bed bioreactor with immobilized yeast. The model is used to analyze the effect of process variables and reactor configuration on the yield. Reactor performance is evaluated on a basis of effluent ethanol concentration. The model is a set of coupled nonlinear partial differential equations which are solved by means of the orthogonal collocation method and an integration routine. A series of experimental runs for various operating conditions are also performed to validate the model.

MODEL DEVELOPMENT

Several considerations are taken into account to simplify the mathematical model and facilitate its numerical solution. The species considered in the model are; substrate, product and biomass. The model is based on the Levenspiel's dispersion model that includes convective and diffusive transport through a column. With these considerations, the unsteady state

233

E. Galindo and O.T. Ramírez (eds.), Advances in Bioprocess Engineering. 233-239.
© 1994 *Kluwer Academic Publishers. Printed in the Netherlands.*

mass balances of substrate and product are given by

$$\varepsilon \frac{\partial S_L}{\partial t} = D_{S_L} \frac{\partial^2 S_L}{\partial z^2} - u_o \frac{\partial S_L}{\partial z} - R_S$$

$$\varepsilon \frac{\partial P_L}{\partial t} = D_{P_L} \frac{\partial^2 P_L}{\partial z^2} - u_o \frac{\partial P_L}{\partial z} + R_P$$

$$(1)$$

where u_0 is the superficial velocity; ε is the bed porosity; z represents the axial direction , and t is time S_L, P_L are the substrate and product concentrations in the bulk liquid; and D_{S_L}, D_{P_L} are the effective diffusivities in the bulk liquid. R_S and R_P denote substrate and product rate expressions, respectively. For spherical particles, these expressions are given by:

$$R_S = \frac{3(1 - \varepsilon)}{R_o} D_{S_P} \left. \frac{\partial S_I}{\partial r}\right|_{r=R_o}$$

$$(2)$$

$$R_P = \frac{3(1 - \varepsilon)}{R_o} D_{P_P} \left. \frac{\partial P_I}{\partial r}\right|_{r=R_o}$$

where, S_I and P_I, are substrate and product concentrations inside the bead, D_{S_I}, D_{P_I}, are, respectively, the substrate and product diffusivities inside the bead, r represents the radial direction, and R_0 is the bead radius.

To evaluate R_S and R_P we assume: fermentation species in the bulk liquid and on the biocatalyst surface have the same concentration, homogeneous metabolite and substrate diffusion in the bead radial direction, uniform yeast distribution at the start of the fermentation, the amount of cells leaving the bead is negligible, and, constant concentration profile in the bead center.

With these considerations the transient behavior of substrate and product inside the beads is described by

$$\frac{\partial S_I}{\partial t} = D_{S_P} \left(\frac{\partial^2 S_I}{\partial r^2} + \frac{2}{r} \frac{\partial S_I}{\partial r} \right) - r_S$$

$$(3)$$

$$\frac{\partial P_I}{\partial t} = D_{P_P} \left(\frac{\partial^2 P_I}{\partial r^2} + \frac{2}{r} \frac{\partial P_I}{\partial r} \right) + r_P$$

Here, r_S and r_P denote substrate and product rates, respectively.

If the beads are very small, the quasi-steady state approximation can be applied to equations (3). This approximation allows us to incorporate R_S and R_P in (1) as an efficiency factor that can be expressed as

$$\eta_S = \frac{3 \, D_{S_P} \, dS_I/dr \big|_{r=R_o}}{R_o \, r_S (S_L)}$$

$$(4)$$

Thus, the equations (1), (2), and (4) are combined to give

$$\varepsilon \frac{\partial S_L}{\partial t} = D_{S_L} \frac{\partial^2 S_L}{\partial z^2} - u_0 \frac{\partial S_L}{\partial z}$$

$$- (1 - \varepsilon) \eta_S r_S (S_L)$$

$$(5)$$

$$\varepsilon \frac{\partial P_L}{\partial t} = D_{P_L} \frac{\partial^2 P_L}{\partial z^2} - u_0 \frac{\partial P_L}{\partial z}$$

$$+ (1 - \varepsilon) \eta_P r_P (P_L)$$

Here, r_S and r_P are, respectively, related to product formation and/or cell mass concentration (Wang, [21]). We use Monod kinetic expressions for r_S and r_P, since it is well known that fermentation with *Saccharomyces cerevisiae* follow that kind of kinetics. Furthermore, if we consider that the endogenous metabolism is negligible, the substrate concentration, product formation and cellular growth due to the fermentation can be described by

$$r_S = \frac{dS_L}{dt} = \frac{\mu_{max} S_L}{\left(S_L + K_S \right) Y_{XS}} X$$

$$(6)$$

$$r_P = \frac{dP_L}{dt} = \frac{v_{max} S_L}{\left(S_L + K_S \right)} X$$

$$\frac{\partial X}{\partial t} = \frac{\mu_{max} S_L}{\left(S_L + K_S\right)} X \qquad (7)$$

where X represents the cellular concentration per bead volume, μ_{max} is the maximum growth rate; K_S is the Monod constant and $v_{max} = \mu_{max} Y_{PX}$. Roels and Kossen [3] related Y_{XS} and Y_{PX} to stoichiometry. For ethanol production during anaerobic growth, they found that

$$Y_{PX} = \frac{1 - 1.22 Y_{XS}}{1.96 Y_{XS}} \qquad (8)$$

The boundary conditions are given by (Shügerl, [4])

$$S_L = S_0; \quad P_L = 0 \quad \text{at} \left(0, z\right)$$

$$\frac{\partial S_L}{\partial z} = 0; \quad \frac{\partial P_L}{\partial z} = 0 \quad \text{at} \left(t, L_e\right)$$

$$S_L - \frac{D_{S_L}}{u_0} \frac{dS_L}{dz} = S_L^i;$$

$$P_L - \frac{D_{P_L}}{u_0} \frac{dP_L}{dz} = P_L^i \quad \text{at} \left(t, 0\right)$$

$$(9)$$

Here L_e is the expanded bed length of the bioreactor and S_L^i and P_L^i are the input substrate and product concentrations, respectively. For cellular mass the initial condition is:

$$X = X_0 \quad \text{at} \quad t = 0 \qquad (10)$$

Our experimental set up also includes a recirculation stream and a tank for monitoring purposes (see **Figure 1**). The dynamic behavior for glucose and ethanol in the tank can be modeled as a unsteady state continuous stirred tank. Here we assume that no reaction is taking place in the tank. Hence, the species mass balances are:

$$V \frac{dS_L^i}{dt} = F\left(S_L^S - S_L^i\right)$$

$$(11)$$

$$V \frac{dP_L^i}{dt} = F\left(P_L^S - P_L^i\right)$$

where S_L^S and P_L^S are the substrate and product outlet concentrations for the bioreactor, while S_L^i and P_L^i are the substrate and product outlet concentrations for the recirculation tank. The initial conditions associated to equation (11) are:

$$S_L^i = S_0 \quad \text{at} \quad t = 0$$

$$P_L^i = 0 \quad \text{at} \quad t = 0 \qquad (12)$$

<u>EXPERIMENTAL SECTION</u>

The yeast used in our experiments was *Saccharomyces cerevisiae* ATCC-4126. The fermentation medium was a solution of glucose of 30 g/l and other nutrients. The detailed description of this medium is given in Alcaraz and Ortiz, [5]. The yeast was immobilized in calcium alginate. The yeast was mixed with a two mass percent sodium alginate solution, which was dropped into a two molar calcium chloride solution. The beads formed were cured in the solution for two hours. The fluidized bed reactor was a one meter glass tube with a 20 cm^2 cross sectional area. Ethanol samples were analyzed by gas chromatography.

Experiments were carried out for various flow rates (eigth to twenty ml/s) with a two liters of fermentation volume in the column. The recirculating tank had a volume of seven liters.

The modeling equations were solved by a numerical procedure that combined the spatial discretization by the orthogonal collocation method and a routine to integrate the resulting set of ordinary differential equations (ODE). The application of the orthogonal collocation method in many chemical engineering problems is very well documented in the literature (Villadsen and Michelsen, [6]; Finlayson, [7], [8]). In this work, we follow Finlayson's procedure to transform the modeling equations in a set of ODE's which are then solved by the MATLAB integration routine ode45 (Math Works, [9]). A detailed description of the numerical procedure is given in Alcaraz [10]. Chemical and physical properties used in the simulations were taken from other studies with the same type of immobilized yeast and they are reported in **Table 1**. The yield factor Y_{xs} was taken from the work of Aiba et al. [11], while Y_{ps} was calculated by stoichiometry (see Equation (8)).

<u>RESULTS</u>

The effect of the number of internal collocation points on the solution of the mathematical model is shown in **Figure 2**. No appreciable differences were found when two to six internal collocation points were used. No attempts were made to solve the modeling equations with more than 6 collocation points. **Figure 3** shows the simulation results of substrate and product concentrations for three collocation points and in the recirculation tank. We can see that the solutions at the collocation points generate practically the same curve. The maximum error in all cases was less than one percent. These results can be used to confirm that no reaction takes place in the tank.

We use **Figure 4** to confirm that the orthogonal collocation method did not exhibit instability problems during the solution procedure. This was probably due to the operating conditions selected in this work which allowed to avoid the effects of nonlinear kinetics and diffusional limitations.

In order to validate the model a series of experimental runs were carried out by varying the flow rate. **Figure 5** shows the comparison of both experimental and predicted ethanol concentrations as a function of time for several flow rates.

This comparison showed that the model yielded satisfactory results for the flowrates studied. We found a ± two percent error for the 20 ml/s flowrate which is acceptable within the experimental error (± five percent). However, the error was higher (± seven percent) for the eigth ml/s flowrate. This result may indicate that diffusional limitations have to be considered in our model for low flowrates (less than eigth ml/s).

<u>CONCLUSIONS</u>

A mathematical model was proposed to describe the dynamic behavior of a fermentation process carried out in a fluidized bed reactor with immobilized yeast. The model incorporated axial dispersion, convective transport, and Monod kinetics to predict the dynamic behavior of substrate, ethanol and biomass. The model was solved by means of a numerical procedure that combined the orthogonal collocation method and an integration routine. Experimental data and theoretical results were compared to validate the model and test the solution procedure. It was found that the model satisfactorily predicted the experimental data for a number of internal collocation points. The solution procedure was found to be fast and accurate and that can be extended to solve more complex cases (e.g. system with diffusional limitations, product and substrate inhibitions, etc.). The model can also provide a useful tool in deriving an optimal set of operating conditions after some key performance criteria have been decided upon.

<u>LITERATURE CITED</u>

1. Mattiasson B.; <u>Immobilized Cells and Organels,</u> CRC Press (1983).

2. Wang D. I. C.; <u>Fermentation and Enzime Technology,</u> Wiley, New York (1979).

3. Roels J. A.; y Kossen N. W. F.; "On The Modelling of Microbial Metabolism", in *Prog. Ind. Microbiol.*, **14,** 95 (1978)

4. Shügerl K.; <u>Bioreaction Engineering,</u> Wiley, N.Y. (1987)

5. Alcaraz G. V. and Ortiz F. J. R., "Diseño y construcción de un bioreactor de lecho fluidizado", B.S. Thesis in Chemical Engineering, University of Guadalajara; Jalisco, México (1991).

6. Villadsen J.; and Michelsen M. L., <u>Solution of Differential Equation Models by Polynomial Approximation.</u> Prentice Hall, New Jersey (1978).

7. Finlayson B. A.; <u>Nonlinear Analysis in Chemical Engineering</u>, McGraw-Hill, U.S.A. (1980).

8. Finlayson B. A.; "Orthogonal Colocation in Chemical Reaction Engineering", in *Cat. Rev.-Sci. Eng.*, **10(1)**, 69 (1974)

9. Math Works, Inc.; <u>Matlab for Macintosh Computers</u>, Math Works, Inc., Natick, MA (1985).

10. Alcaraz G. V.; "Estimación de estados de un proceso fermentativo realizado en un bio-reactor de lecho fluidizado", M.S. Thesis in Chemical Engineering, University of Guadalajara, Jalisco, México (1993).

11. Aiba S.; Shoda, M. and Nagatani, M.; *Bio. & Bio.* , **10**, 845 (1968).

12. Nakasaki K.; Murai, T.; and Akiyama, T.; *Bio. & Bio.*, **33**, 1317-1323 (1989).

13. Hannoun B. J. M. and Stephanopoulos, G; *Bio. & Bio.* , **28**, 829, (1986).

ACKNOWLEDGEMENTS

Financial support for this research provided by the University of Guadalajara is gratefully acknowledged. We also thank the National Council of Science and Technology (CONACyT) for supporting one of the authors (V.A.G.).

Table 1. Operating conditions and <u>typical parameters.</u>

PARAMETER	VALUE	REFERENCE
μ_{max}	0.113 h^{-1}	Nakasaki et al., <u>[12]</u>
K_s	4.21 Kg/m^3	"
Y_{xs}	0.121 Kg/Kg	Aiba et al., <u>[11]</u>
D_{S_L}	2.4×10^{-6} m^2/h	Hannoun et al., <u>[13]</u>
D_{P_L}	3.6×10^{-6} m^2/h	"
D_{S_I}	2.4×10^{-6} m^2/h	"
D_{P_I}	3.96×10^{-6} m^2/h	"
L_e	0.8 m	Alcaraz and Ortiz, <u>[5]</u>
R_0	1.5×10^{-3} m	"
ε	0.85	"
Residence time	5.45×10^{-2} h	"
S_0	30 Kg/m^3	"

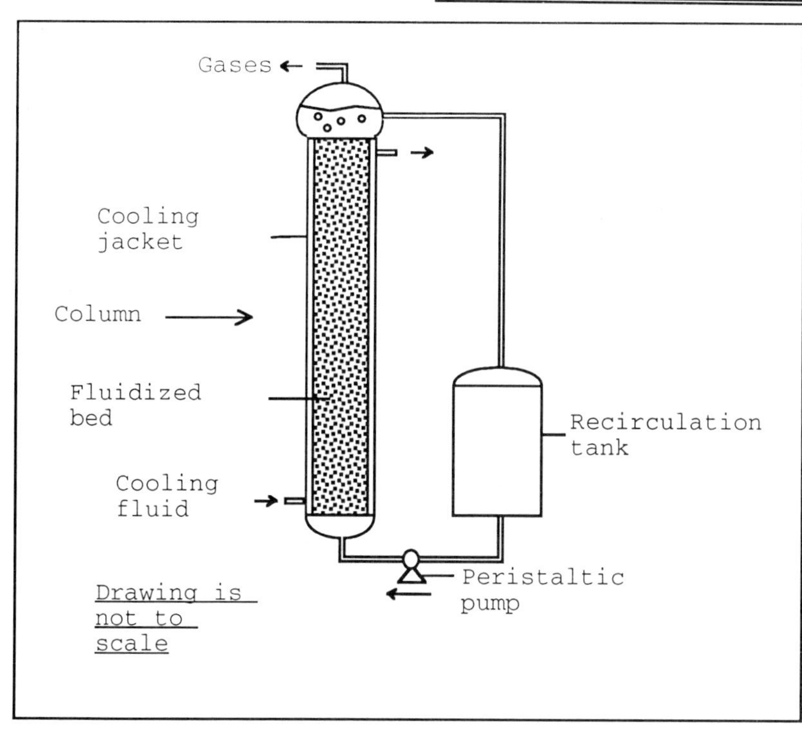

Figure 1. Fluidized bed bioreactor

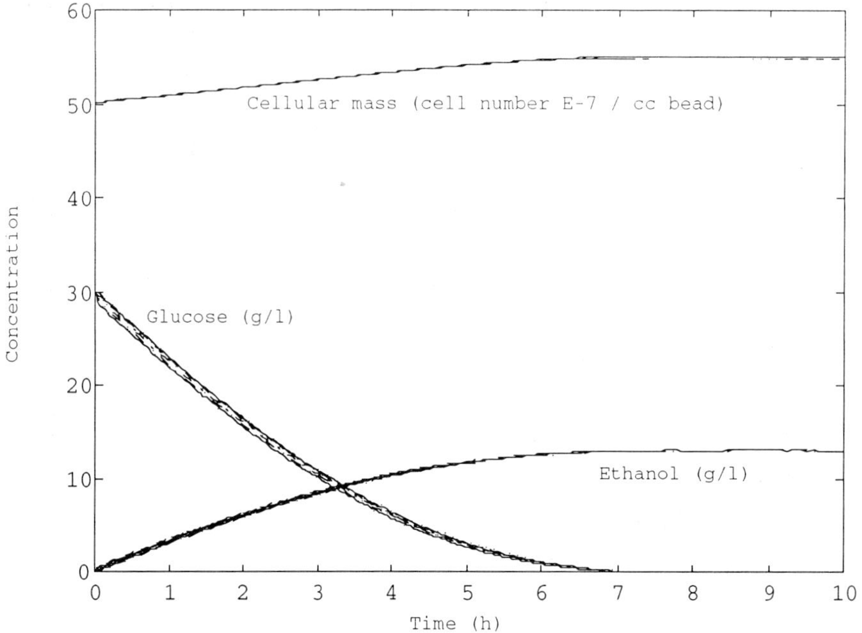

Figure 2. Simulation results using one, two, and three
internal collocation points

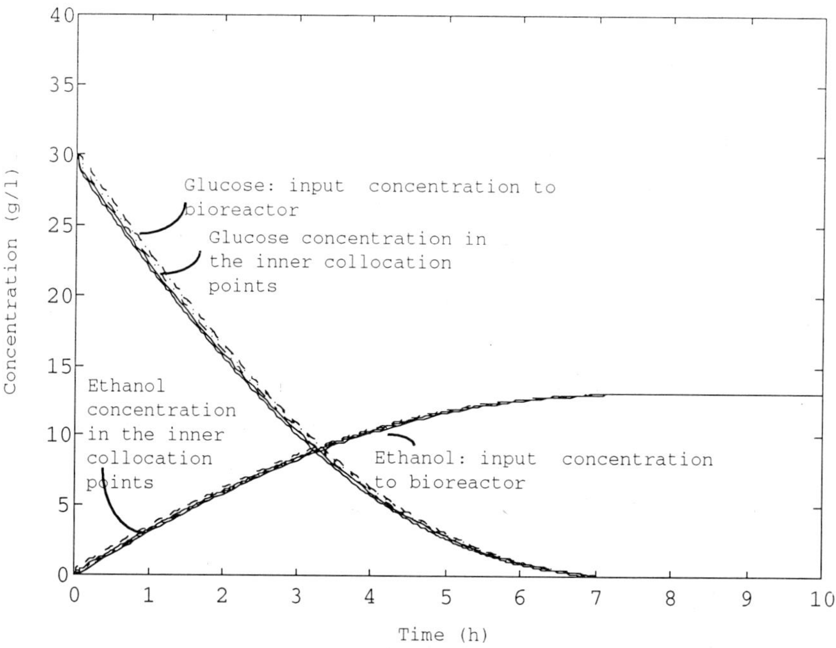

Figure 3. Simulation results·using one, two, and three
internal collocation points.   The concentration
differences in the bulk liquid in the axial direction is
negligible.

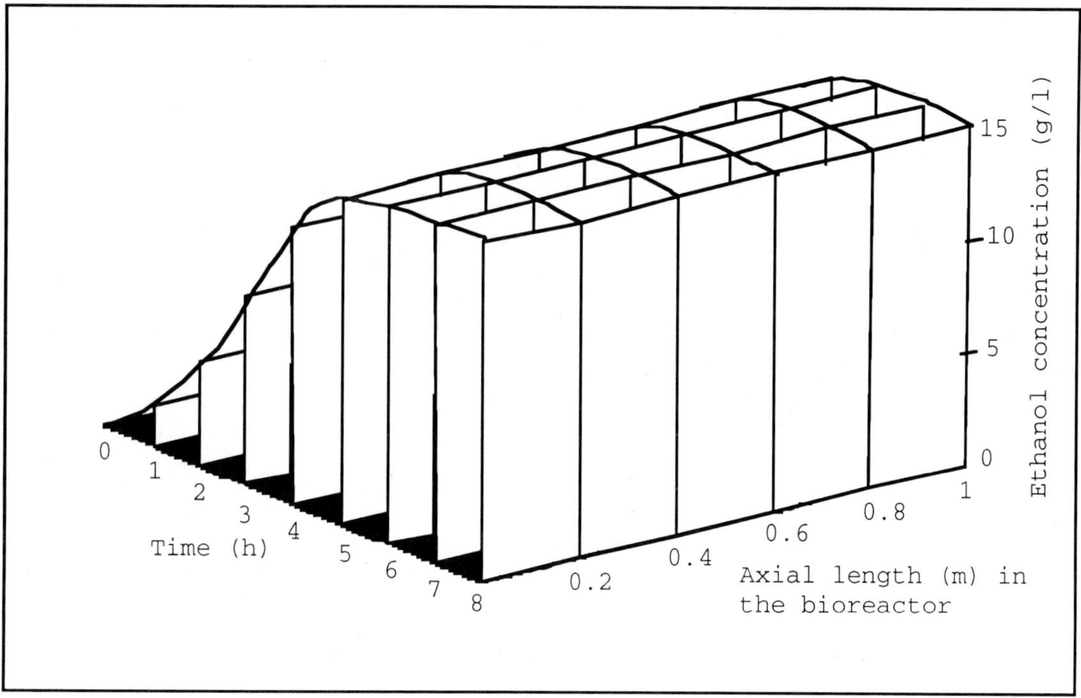

Figure 4. Simulation results for ethanol concentration using 5 internal collocation points: a tridimensional perspective.

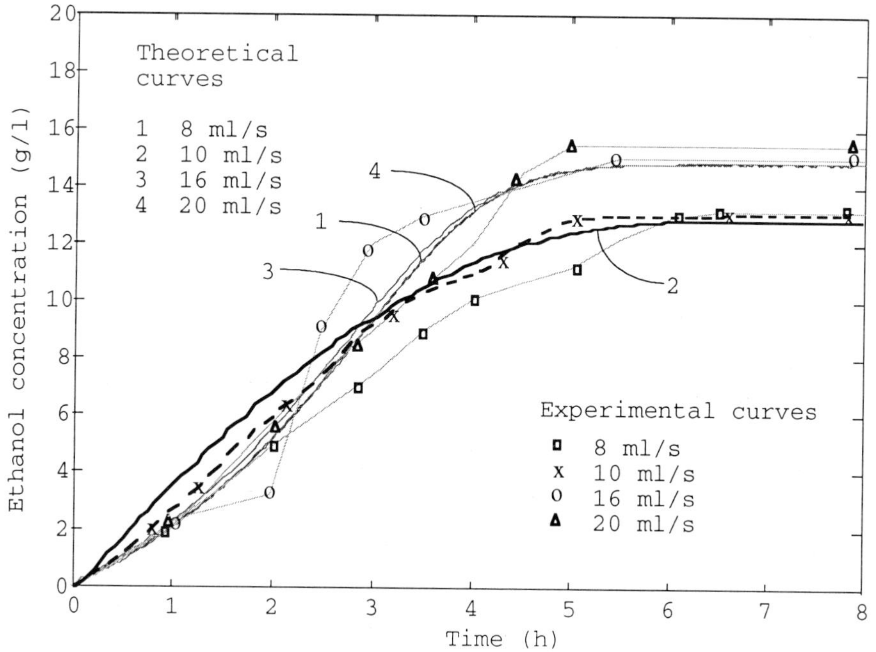

Figure 5. Comparison between simulation and experimental results for the ethanol fermentation from glucose using *Saccharomyces ceevisiae* ATCC-4126 immobilized in calcium alginate.

Invited paper

Modelling the Growth of *Acidothermus cellulolyticus*

D.W. Hubbard, T.B. Co, P.P.N. Murthy, and R. Mandalam

Department of Chemical Engineering and Department of Chemistry, Michigan Technological University, Houghton, Michigan, 49931, U.S.A.

Mathematical models of organism growth help in designing and scaling-up bioreactors and in predicting how microorganisms will respond to changes in growth conditions so that an appropriate process control system can be devised. Models also guide the analysis of experimental data and help in understanding biochemical growth mechanisms. Filamentous organisms which form pellets can be modelled using a cube root model or a modified logistic model. The cube root model allows detecting lag phases and stationary phases easily. The modified logistic model is a versatile model which can be used to model the entire growth cycle and which includes an adjustable inflection point for the rapid growth period. These two models were tested by studying the growth of Acidothermus cellulolyticus in glucose and cellulose media in batch culture. The pH was maintained at 5.3 and the temperature was controlled at 45, 55, or 65ºC. Cell biomass was determined by protein assay. The biomass productivity for the growth in a batch process shows that 55ºC is the optimum temperature for biomass productivity. The growth data were used to evaluate the model parameters for the two models.

Those who design bioreactors need mathematical models of the growth of the organism to be grown so they can predict the productivity of the process being designed. models of the growth are also useful in predicting how an organism will respond to changes in growth conditions so that an appropriate control system for the processs can be designed. Mathematical models can also guide the analysis of experimental data and help in developing an understanding of growth mechanisms.

MATHEMATICAL MODELS FOR ORGANISM GROWTH

For freely floating organisms growing in submerged culture, the Monod two-parameter model has often been adequate for describing cell growth. To describe the complete growth cycle for a microorganism including lag phases and the death phase, a more complicated model is needed. The logistic model described by Bailey and Ollis[1] has some of the features required, but modifications are needed if the death phase is to be included in the model. Metz and Kossen[2] describe a cube-root model which applies to the rapid growth phase of filamentous microorganisms which form pellets as they grow.

Generalized Logistic Model

Microbial growth data can be fitted to a generalized logistic model of the form

$$[1/(X-X_O)](dX/dt) = K_1[(1-K_2)X]^b \qquad (1)$$

When b equals one, the equation becomes the standard logistic equation. Introducing the parameter b allows for a better fit of growth data. An inflection point occurs between the initial concentration, X_O, and a stationary concentration, $X_{max}=1/K_2$. Using equation (1), the inflection point occurs when $X = X_i$, where

$$X_i = (1/K_2 + b X_O)/(1 + b) \qquad (2a)$$

or

$$X_i = (X_{max} + b X_O)/(1 + b) \qquad (2b)$$

The standard logistic model may be insufficient to model growth data

E. Galindo and O.T. Ramírez (eds.), Advances in Bioprocess Engineering. 241-245.
© 1994 *Kluwer Academic Publishers. Printed in the Netherlands.*

since it always fixes the inflection point exactly midway between X_O and X_{max}. This is the situation when $b=1$ in Equation (2b).

The parameters K_1, K_2 and b can be obtained by determining the stationary concentration, X_{max}, the infection point, X_i, and the slope at the inflection point, $dX/dt(X=X_i)$. In particular,

$$K_2 = 1/X_{max} \tag{3}$$

$$b = (X_{max}-X_i)/(X_i-X_O) \tag{4}$$
and
$$K_1 = dX/dt(X=X_i)(1-K_2 \; X_i)^{-b}/(X_i-X_O) \tag{5}$$

If the death phase is to be modelled, a second logistic equation is introduced as discussed by Hubbard and Co[3]. The two parts of the model are joined, and the six parameters are evaluated from experimental data.

Cube Root Model

Filamentous organisms may grow as spherical clumps or colonies of biomass. If the radius of a colony increases at a constant rate during the rapid growth phase, biomass is proportional to the cube of the elapsed time measured from the start of the rapid growth. This is described mathematically by the following equation.

$$X = [(4/3 \, \pi \rho)^{1/3} k_g t + M_0^{1/3}]^3 \tag{6}$$

To verify this model, one should plot growth data for the rapid growth period according to the following form.

$$X^{1/3} = K_{1/3} (t-t_{lag}) + X_0^{1/3} \tag{7}$$

Plotting the growth this way makes it easy to identify the true period of rapid growth and to distinguish this period from stationary or lag phases.

EXPERIMENTAL METHODS AND MATERIALS

Mohagheghi et al.[4] isolated and characterized a cellulolytic bacterium, *Acidothermus cellulolyticus*, from the hot springs of Yellowstone National Park. *A. cellulolyticus* are aerobic, acidophilic, thermophilic, and can break down microcrystalline cellulose completely. These characteristics make them attractive for converting lignocellulosic waste to useful products. We studied these bacteria to test the applicability of the modelling methods described above.

Acidothermus cellulolyticus was obtained from the American Type Culture Collection(ATCC 43068). The bacteria as received did not grow on cellulose, so they were grown in a glucose medium containing, in grams per liter, NH_4Cl: 1.0, KH_2PO_4: 1.0, $Na_2HPO_4 \cdot 7H_2O$: 0.1, $MgSO_4 \cdot 7H_2O$: 0.2, $CaCl_2 \cdot 2H_2O$: 0.02, yeast extract: 1.0, and glucose: 1.0.

To study growth of the bacteria on cellulose, the medium was prepared using D-cellobiose (Sigma Chemical Co.): 1.5, and Sigmacell 50(Sigma Chemical Co.): 5.0 instead of glucose. The pH was adjusted to 5.3, and 200ml of medium was transferred to 500 ml culture flasks where it was sterilized by placing the flasks in an autoclave at 121^OC for 20 minutes. Bacterial culture(20 ml) was added aseptically to the flasks, and the flasks were placed in an incubator-shaker at a constant temperature(45, 55, or 65^OC) operating at 200 RPM.

Growth was estimated by measuring the amount of protein in 10 ml samples of the culture broth using a modified Lowry's method as discussed by Markwell[5]. Because of insoluble cellulose, the growth in cellulose medium could only be measured by the protein assay method.

EXPERIMENTAL RESULTS AND MODEL VERIFICATION

The growth of *Acidothermus cellulolyticus* on cellulose medium at 55^OC is shown in Figure 1. This figure indicates the reproducibility of the results. Data from two different flasks are reported showing that flask-to-flask variation was small. A comparison of growth observed by Mohagheghi at al.[4] at 55^OC with our growth data is also shown in this figure.

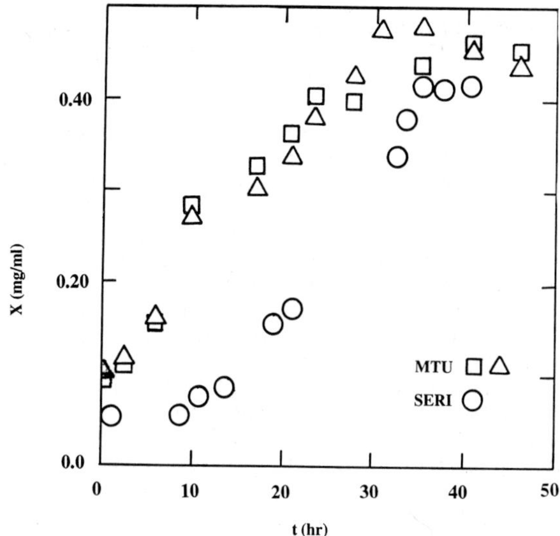

Figure 1. Flask-to-flask variation and comparison of growth curves for *A. cellulolyticus* growing on cellulose medium at 55°C and pH 5.3 for experiments at Michigan Technological University(MTU) and the Solar Energy Research Institute(SERI). SERI data are from Mohagheghi et al.[4]

Differences in the length of the lag phase could be due to differences in the metabolic state of the organisms in the innoculum. Our results were obtained using organisms which had been growing for at least five generations at the conditions at which the growth was measured. The magnitude of the maximum cell concentration and the slope of the growth curve during the rapid growth period are same within the experimental uncertainty.

Growth at three different temperatures is shown in Figure 2. These results are for actively growing organisms which had been growing at the designated temperature for at least five generations. Because of this method of innoculation, there is often only a short lag phase before the period of rapid growth begins. The maximum cell concentration occurs at different times depending on the temperature, and the magnitude of the maximum cell concentration also depends on temperature. When the maximum cell concentration is attained, a stationary phase begins.

The experiments were ended before the death phase began. The lines plotted in Figure 2 are lines which represent the generalized logistic model.

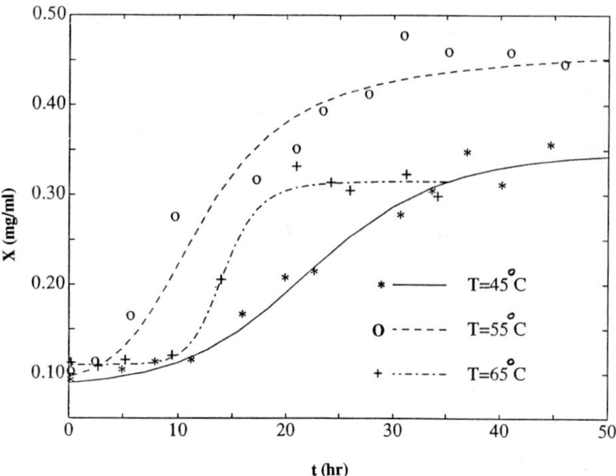

Figure 2. Comparison of the growth data with the modified logistic model

We determined the parameters from our growth data using the method of modulating functions described by Co and Ydstie[6]. These model parameters are summarized in Table I.

Table I. Parameters for generalized logistic equation

Temp(°C)	X_0	K_1	K_2	b
45	0.11	1.2	3.175	1.20
55	0.09	0.55	2.128	1.80
65	0.085	0.25	2.857	1.54

Certain observations can be made about Table I. First, the highest stationary-phase concentration occurs at 55°C--the temperature for which K_2 has the smallest value. Second, we note that K_1 decreases as temperature increases. This observation runs counter to what one might expect by considering Equation (5). Thus, the results show that the slope at the inflection point is not the dominant factor in Equation (5). Finally, the highest value of b occurs at 55°C. At this temperature, the inflection point occurs at the lowest value of cell mass.

The growth data plotted according to Equation (7) are shown in Figure 3.

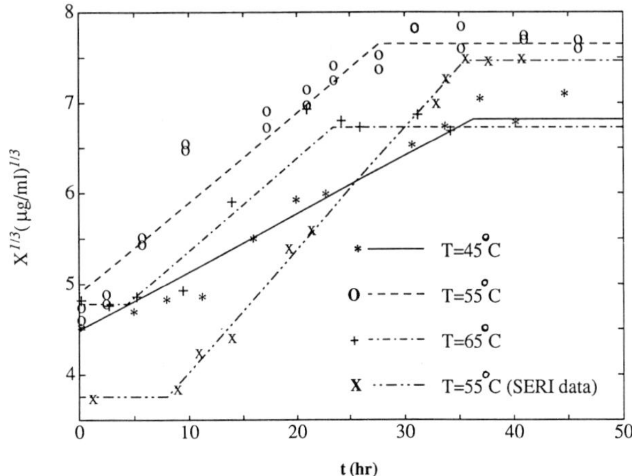

Figure 3. Growth data for *A. Cellulolyticus* represented by the cubic model

The model parameters and the times at the end of the lag and growth phases are listed in Table II for 45, 55, and 65°C. The model parameters are evaluated using only the data obtained during the rapid growth phase. Growth parameters calculated from the data measured by Mohagheghi et al.[4] are also presented in this table. These data show that if P is constant, the growth rate parameter increases with temperature over the range studied.

parameter. The equation

$$K_{1/3} = 550 \exp(-23,600/RT)$$

is obtained with a 0.98 correlation coefficient.

The growth data also show that there is no definite lag phase for growth at 45°C and 55°C. This is probably due to the way that the experiments were performed. In each case, the flasks in which the growth data were measured were inoculated with bacteria which were growing vigorously at the temperature at which growth measurements were being made. Since the organisms are fully adapted to the growth temperature, this may explain the absence of a discernible lag phase in these experiments. At 65°C, a short lag phase appears in spite of the fact that the innoculum contained vigorously growing organisms. Dilution or diffusion effects could explain the lag phase at the higher temperature. The data reported by Mohagheghi et al. show a definite lag phase. This is probably due to the organisms in the innocculum adjusting their metabolic state for new growth conditions.

Based on our growth results , the overall biomass productivity for a batch fermentation can be determined by the following relationship.

$$\text{Productivity} = (X_{max}-X_0)/(t_{max}+2) \quad (8)$$

Table II. Model Parameters for the Cubic Growth Model

T (°C)	X_{max} (g/ml)	$X_0^{1/3}$ (g/ml)	$K_{1/3}$	t_{max} (hr)	t_{lag} (hr)	r
45	317	4.49	0.075	36.3	0	0.992
55	448	4.90	0.101	27.7	0	0.966
65	305	4.78	0.102	23.4	4.3	0.961
55*	416	3.76	0.134	35.5	8.0	0.998

* Calculated from the results of Mohagheghi et al.[4]

Absolute reaction rate theory can be used to describe the temperature dependence of the growth rate

where the time for emptying, cleaning, sterilizing, and filling the fermenter is constant at two hours and X_0 is the initial concentration of biomass produced by the innoculation. The results for our data are shown in Table III. The biomass productivity has a maximum value at 55°C, so this temperature is the optimum for growth. Mohagheghi et al. noted that the maximum specific growth rate determined in chemostat experiments was also obtained at 55°C.

Table III. Biomass Productivity for the Growth of *A. Cellulolyticus*

T($^{\circ}$C)	Productivity(μg/ml hr)
45	6.2
55	11.1
65	7.7

In summary, growth of this filamentous bacteria, *Acidothermus cellulolyticus*, can be described by a model in which the cube root of biomass concentration is directly proportional to time. The cell growth can also be described by a modified logistic model containing an inhibition factor which depends on the biomass present. Both cell growth models apply for temperatures from 45°C, 55°C, and 65°C. The maximum biomass production in shake flask culture as determined by protein concentration occurs at 55°C.

ACKNOWLEDGEMENT

The authors gratefully acknowledge many helpful discussions with Dr. A. Mohagheghi of the Solar Energy Research Institute in Golden, Colorado.

NOTATION

b	Model parameter for the modified logistic model
K_1, K_2	Model parameters for the modified logistic model
k_g	Growth rate parameter
$K_{1/3}$	$[(4/3\pi\rho)^{1/3}k_g]$
M_0	Initial biomass
r	Correlation coefficient
T	Temperature
t	Time
t_{lag}	Time at the end of the lag phase
t_{max}	Time at which X_{max} is attained
X	Biomass concentration
X_0	Initial biomass concentration
X_{max}	Maximum biomass concentration
X_i	Biomass concentration at the inflection point in the growth function
ρ	Biomass density

LITERATURE CITED

1. Bailey, J. F. and D. F. Ollis, *Biochemical Engineering*, 2nd ed., pp.403-405 McGraw-Hill, New York(1986).

2. Metz, B. and N. W. F. Kossen, The Growth of Molds in the Form of Pellets--A Literature Review, *Biotechnology and Bioengineering*, **19**, 781(1977).

3. Hubbard, D. W. and T. B. Co, Modeling the Growth of Cellullase-Producing Organisms, in Hurst, C. J., ed., *Modelling the Metabolic and Physiologic Activities of Microorganisms*, pp. 89-113 John Wiley and Sons, New York(1992)

4. Mohagheghi, A., K. Grohmann, M. Himmel, L. Leighton, and D. M. Updegraff, Isolation and characterization of *Acidothermus cellulolyticus* gen. nov., sp. nov., a new genus of thermophilic, acidophilic, cellulolytic bacteria. *International Journal of Systematic Bacteriology*, **36**, 435(1986).

5. Markwell, M. A. R., S. Hass, N. Tolbert, and L. Bieber, Protein determination in membrane and lipoprotein samples: Manual and automated procedures, *Methods in Enzymology*, **72**, 296(1981).

6. Co, T. B. and B. E. Ydstie, System Identification Using Modulating Functions and Fast Fourier Transforms, *Computers in Chemical Engineering*, **14**, 1051(1990)

Bioreactor Coupled On-Line Measurements

M. Reuss, A. Riek, M. Schütz, and W. Mailinger

Institut für Bioverfahrenstechnik, Universität Stuttgart, Allmandring 31,
70569 Stuttgart, GERMANY

Although research on the development of on-line HPLC systems has been conducted for many years, there is still no satisfactory solution available in the area of fully automatic sampling and sample preparation during fermentation processes. Sample preparation, which is often complicated, is of great importance for a successful application. A particular problem, for example, is still the reliable and robust dilution of the sample. The proposed system fulfills the requirement of fully automatic sampling and sample preparation by use of the biofiltrator and an autosampler with micro-robotic-system. In addition to applications for on-line-HPLC, the same configuration can be applied to on-line FIA (Flow Injection Analysis). This double function leads to a new name: High-Pressure-Injection Analysis (HPIA). The selection of the individual components was largely based on convenient use and industrial suitability. Because of the modular structure of the system, it is possible to include other components at any time. Furthermore, it is even possible to use this application in other areas such as environmental analysis and process chromatography.

The last decade has seen the development of many approaches for improving the situation in the application of computer control of fermentation processes. As far as the observation part of control is concerned two seemingly distinct avenues have predominated these approaches: those based upon developments in the field of on-line measurements and those based upon state estimation algorithmus. Although model based measurements could achieve quite reasonable performances (1), the complexity of industrial fermentation processes often results in uncertainties of accurate model formulations and problems with the observability of the state vector from the available measurements variables. This is in particular true if we look beyond the horizon of simple processes for production of biomass. Stepping into the field of complex production processes we are often faced with the problem that important state variables are not directly observable from the simple on-line measurements most often available. It therefore seems that breakthroughs in the application of more sophisticated computer control of fermentation processes are unlikely without further increasing the amount of direct information from on-line measurements.

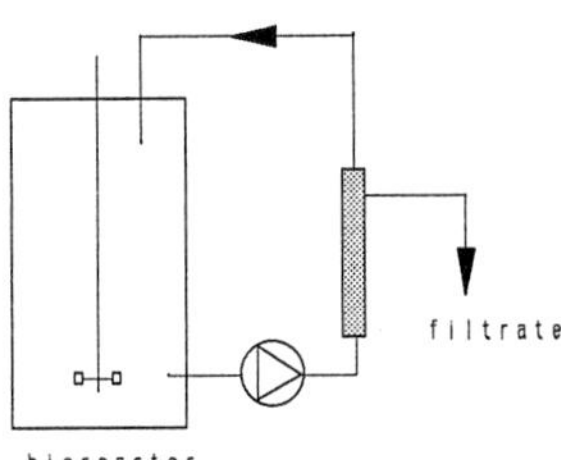

Figure 1 Coupling of automated analyzers via membrans separation in a bypass

Several developments started in the last twenty years which are all focussed on improving the unsatisfying situation in the on-line measurement area. There can be no doubt that developments of new on-line sensors for continuous measurements of various key variables are important contributions (2,3). In addition to the well-kown problems, however, regarding sterility, long-time stability, reliability, accuracy, and specifity of these sensors, it must be emphasized that these developments are rather time consuming, they require a lot of manpower, and, therefore, may result in very expensive solutions as far as the measurements of very specific compounds are concerned.

247

E. Galindo and O.T. Ramírez (eds.), Advances in Bioprocess Engineering. 247-253.
© 1994 Kluwer Academic Publishers. Printed in the Netherlands.

A more pragmatic, short-term solution for many of the problems is, therefore, still to envisage systems with greater flexibility, which can be applied to the measurement of more than a single compound. A powerful method is the coupling of various automatic analyzer systems to the bioreactor.

The major problem to be solved for coupling autoanalyzers and even more complex systems such as high performance liquid chromatography (HPLC) or flow injection analyses (FIA), is a reliable automatic sampling from the bioreactor and the on-line separation of the biomass and solid particles from the sample. In almost all of the solutions suggested so far, the separation has been performed with the aid of nonrenewable filtermedia (dialysis, ultrafiltration, crossflow filtration) installed in a by-pass of the bioreactor (figure 1) or directly introduced in the reactor with the aid of a membrane covered sensor. Due to the well-known problems of membrane fouling and pore blocking in biological systems theses separation techniques may create process-specific problems. These approaches also require some cautions in the environment of an industrial process because of the risk of mechanical damage of the membrane and a possible contamination of the fermentation.

This contribution describes an automatic sampling and filtration system which has been succesfully tested in pilot and large scale industrial plants. The concept of the system is schematically illustrated in figure 2.

The filter unit consists of a renewable filter media, positioned between media support and the filter body. The unit therefore operates in a cyclic batch mode. Due to the complete separation of the sterile region of the bioreactor and the cake filtration unit which are only connected through an automatically steam sterilizable valve. A very high degree of security against contamination of the fermentation can be guaranteed. In contrast to membrane separation techniques installed in a bypass of the bioreactor or directly introduced in a kind of in-line sensor where damage of the membrane will cause

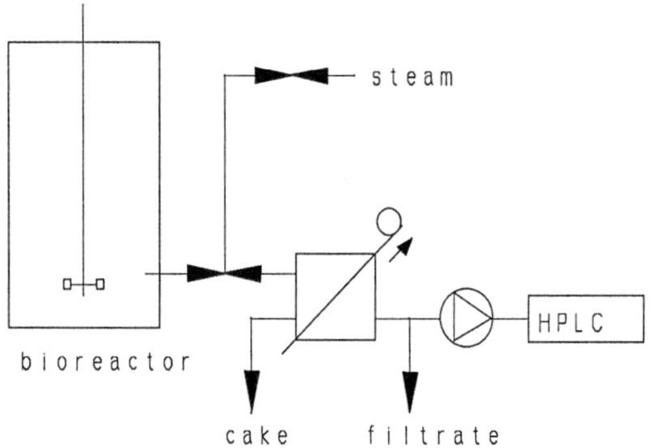

Figure 2 Coupling of automated analyzers via steam sterilizable valve and cyclic cake filtration (4,5,6)

the risk of process contamination, such malfunction in the filter unit will leave the fermentation unharmed. Also, the concept of renewable filter medium prevents any pore blocking and fouling up of the membrane. It therefore seems to be more plausible to envisage application for industrial fermentation processes, if industry is really willing to invest in better and more sophisticated instrumentation of their plants.

The system configuration as well as its computer controlled operation has been described in considerably more detail in earlier publications (4, 5, 6). In the same papers the concept of estimation of biomass concentrations from filtration characteristics of the biosuspension has been extensively discussed. The present contribution is mainly aimed at discussing new concepts for the solution of well-known problems in the automatic analysis of compounds in the filtrate. The configuration described in the following consists of a HPLC/FIA combination supported by a micro-robotic system which is under computer control.

<u>ON-LINE HPLC - THE "MINIMAL SYSTEM"</u>

The automatic sampling and filtration provide the first important step towards on-line analysis of the filtrate. In the most simple application injection into the HPLC then is performed with the aid of a pneumatic-electric Rheodyne-valve (figure 3). The complete set of control functions (start/stop-signals, signals for operation of the Rheodyne valve) is performed by the integrator via "time-events". The process computer is then only used for the purpose of data acquisition and analysis.

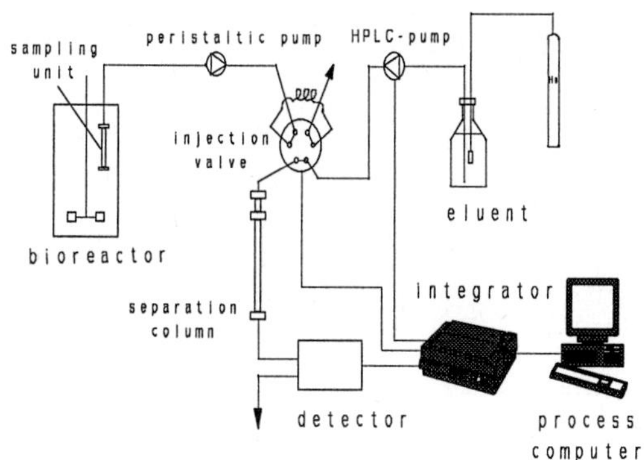

Figure 3 on-line HPLC "minimal system"

This "minimal system" has been applied to tasks of on-line measurements during various fermentations described in the above mentioned publications.

Disappointing experiences with this simple configuration can be summarized as follows:

(1) Blocking of columns due to macro-molecules

(2) Drifting retention times

(3) Overloading of columns at higher concentrations

(4) Limiting flexibility during various applications because of a lack of sample preparation.

It is to be supposed that most of the systems described in the literature show similar problems when applied at industrial conditions.

The major bottle neck of this con-figuration is the lack of an automatic dilution system in order to adapt the loading of the columns during a batch process. Simple dilution systems have been implemented using either variable speed peristaltic pumps or positive displacement-type syringe pumps (7). These systems show, however, low

flexibility and cannot be applied to more complex sample preparations like extraction, precolumn derivation, solid phase extraction and so on. A more

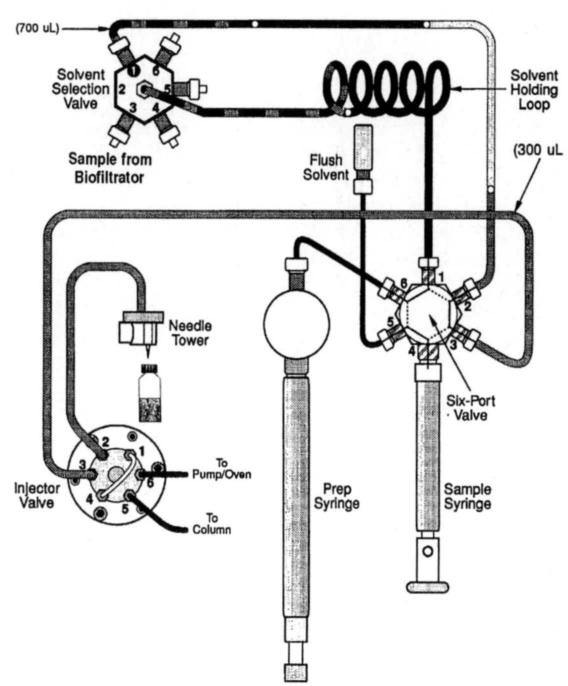

Figure 4 Major components of the autosampler AS 3000 (Thermo Separation Products, Darmstadt)

flexible solution to the problems of on-line analysis is the use of a robotic system which is described in the following.

<u>SAMPLING PREPARATION WITH THE AID OF A MICRO-ROBOTIC-SYSTEM</u>

A systematic analysis of the available systems (autosampler and various robots) leads to the selection of the Auto-sampler AS 3000 (Thermo Separation Products, Darmstadt) which consists of a three dimensional micro-robot. The system can be applied to automatic sample and standard preparations (including dilutions), mixing and extraction. The autosampler is equipped with a 6-port-solvent-selection valve which can be connected to 4 different solvents (figure 4). One solvent line is used as the transfer line to the small filtrate sampling vessel. The sample is then sucked in with the aid of the automatic syringe and injected through the septum of an empty sample vial. In the following, various analytical operations (dilution, adding of reagent solutions etc.) can be performed.

Particular attention has been paid to the development of a flexible user-friendly programming language. The result of this development is a new macro language (SYSMAK) which consists of simple commands for designing the neccessary sequential operations. Automatic control takes care of dilution, calibration and change of columns via switching if relative retention decreases below a predefined setpoint.

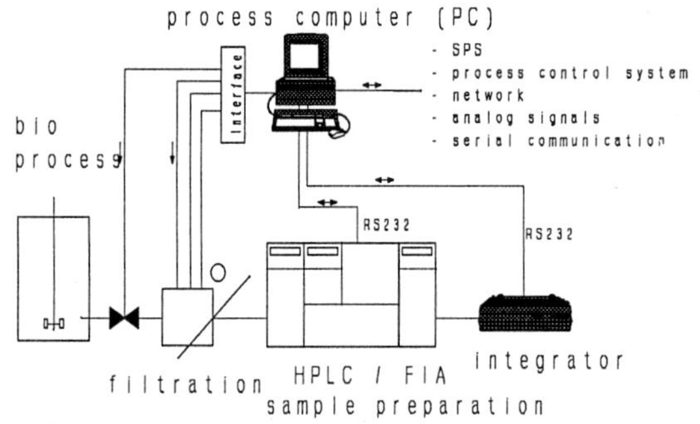

Figure 5 Complete configuration of computer controlled sampling device, sample preparation and HPLC/FIA

The complete configuration of the computer controlled sampling device, sample preparation and HPLC is illustrated in figure 5.

AUTOMATED SEQUENTIAL BATCH ANALYSIS WITH FIA

A particularly prominent feature of the automation concept is the fact that the same configuration used for tasks of on-line HPLC is applicable for enzymatic and wet chemical analysis of the sample. Automation for this kind of analysis is presently most often performed with the aid of flow injection analysis (FIA). In this method - the principle of which is schematically

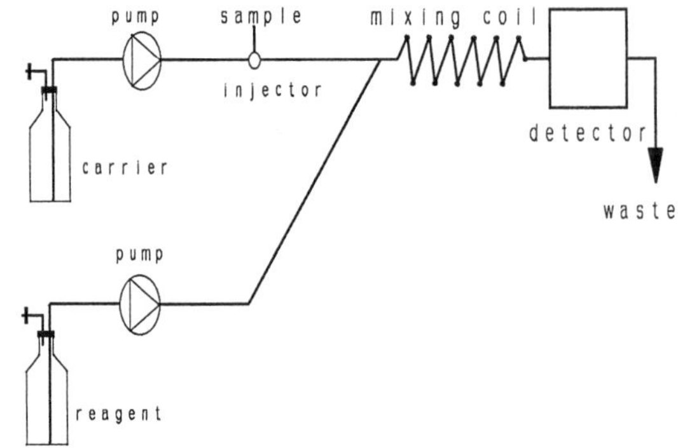

Figure 6 Schematic diagram of basic FIA system

illustrated in figure 6 - a sample zone is injected into a continuous stream, called the carrier. The latter may be either a non-reagent or a reagent fluid.

The automation concept suggested here is a mixture of a sequential batch analysis and the FIA concept (figure 7). The autosampler equipped with the micro-robot is able to replace part of the conventional FIA analysis. After automatic sucking of all the necessary reagents into the sample containing vial, the reaction mixture is being mixed under controlled temperature. After reaction has proceeded to the predefined extent, the mixture is automatically injected into the non-reagent fluid (carrier) which takes care of the transfer to the detector.

The advantage of this concept compared to conventional FIA analysis is two-fold: (1) The same equipment can be used for automatic wet chemical and chromatographical analysis and if required even in an alternating mode. This results in a remarkable reduction of costs for the instrumentation for on-line analysis. (2) Reduction of costs for reagents (particularly important for enzymatic analysis).

It should be also emphasized that the same programming tools (Macrolanguage SYSMAK) can be also used for the design of the templates for the different non-chromatographical methods.

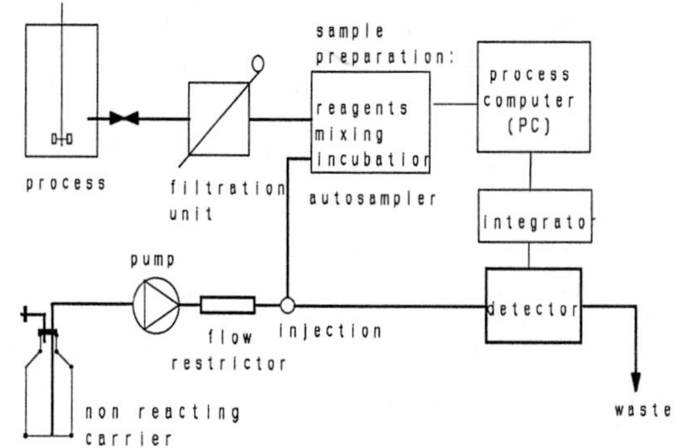

Figure 7 Micro-robotic supported autosampler for automated batch analysis and FIA

APPLICATION EXAMPLES

The system (BIOFILTRATOR for on-line sampling and filtration coupled to HPLC/FIA) has been succesfully applied to various tasks of on-line analysis during fermentation. The application examples include automatic HPLC as well as enzymatic and wet chemical analysis.

The first example is a HPLC-measurement of glucose during a batch fermentation of *Saccharomyces cerevisiae* (Figure 8).

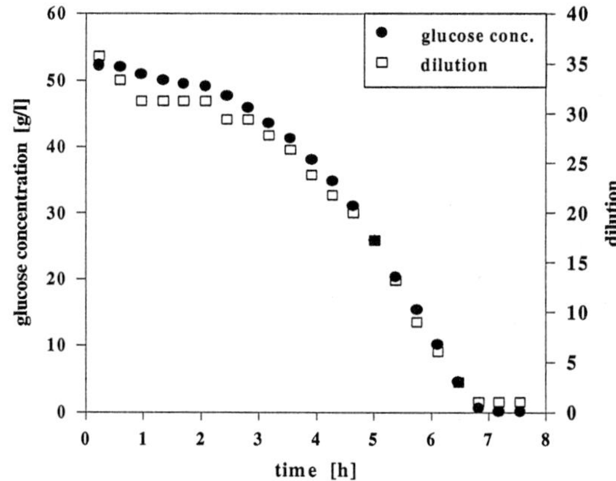

Figure 8 Glucose measurements with on-line HPLC during batch fermentation of *Saccharomyces cerevisiae*

This is a typical example illustrating the imperative necessity of automatic dilution of the sample which is indicated in the same figure by open symbols. The computer control of dilution guarantees the optimal operation of the HPLC-column and leads to very precise measurements at high as well as low glucose concentrations.

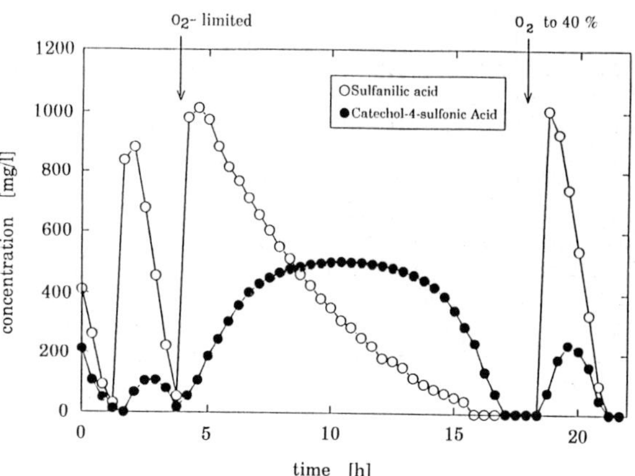

Figure 9 Sulfanilic acid and catechol 4-sulfonic acid measurements whith on-line HPLC during a biodegradation process with a mixed culture

The second example (figure 9) is another HPLC application performed for a biodegradation process. The results presented in this figure illustrate the measurements of sulfanilic acid (xeno-biotic substrate) and catechol-4 sul-fonic acid which is an intermediate excreted by strain 1 and consumed by strain 2 in a 2-species mixed culture community. Again, the rather large range of concentrations is easily handled with the aid of the automatic dilution system.

An example of application of automated enzymatic analysis making use of the same configuration is illustrated in figure 10. The figure shows the results of an automated calibration procedure for glucose measurements with the enzyme Glucosedehydrogenase. An interesting feature is the fact that even the preparation of the samples with the different concentrations is automatically performed via dilution in the autosampler. Thus, calibration can be performed in a fully automated mode.

A final example is demonstrated in figure 11. Shown therein are measure-ments of substrate phenol in a batch

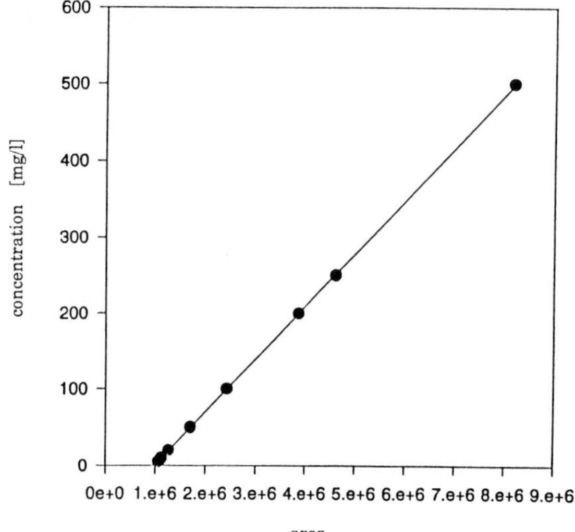

Figure 10 Automated calibration of glucose measurements with enzymatic analysis

fermentation of *Pseudomonas putida*. Analysis is performed according to the method suggested by Ochynski (8). The results presented serve as an example for application of automated wet chemical analysis with the systems configuration. The analysis consists of

a sequence of operations including a dilution step, addition of buffer, reagent A (4-aminoantripin), reagent B (ammoniapersulfate), mixing, 10 minutes of incubation and measurement of absorbance at 540 nm. The figure illustrates the reliability of the automated measurements during the batch fermentation of the bacteria.

<u>CONCLUSIONS</u>

We have shown a new concept for on-line analysis during fermentation. The system suggested in this contribution is able to fulfill simultaneously tasks of on-line chromatographic measurements as well as enzymatic and wet chemical analyis. The combination of on-line HPLC and FIA provides a powerful system for various tasks of on-line measurements during fermentation as well as routine analysis in the laboratory. The important features of the system are automatic dilution, calibration, change of columns and even possibility of automatic sample concentrating via solid phase extraction (not shown in this paper). Special attention has been paid to the development of a user-friendly flexible software (Macro language SYSMAK) which provides untrained users with all the necessary support for a quick transfer of manual lab analysis to a template for automated operation for on-line measurements.

<u>ACKNOWLEDGMENT</u>

We are grateful for the support of the company Thermo Separation Products (previous Spectra Physics), Darmstadt, Germany. Particularly we should like to acknowledge the cooperation with Reinold Scheffler.

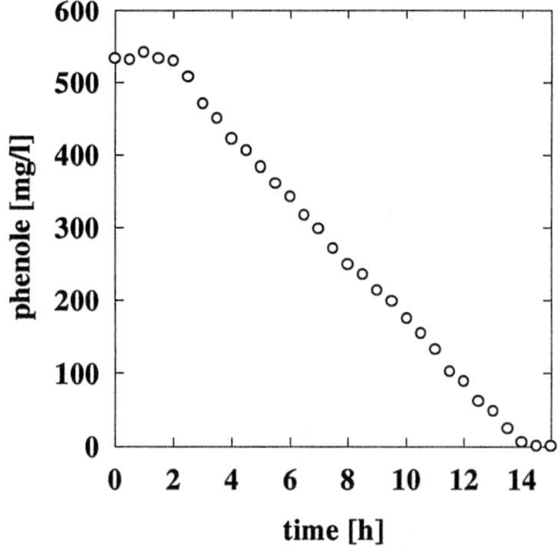

Figure 11 Wet chemical analysis of phenol during batch fermentation of *Pseudomonas putida*

<u>LITERATURE CITED</u>

1.   Stephanopoulos, G. and S. Park "Bioreactor state estimation" in Biotechnology, 2nd, completely revised Edition, Rehm et al. (Eds.), Vol. 4, VCH, Weinheim, New York, Basel, Cambridge (1991).

2.   Beardon, K.F. and T.H. Scheper "Determination of cell

concentration and characterisation of cells", ibid (1991).

3. Schuegerl, K., "On-line analysis of broth", ibid. (1991)

4. Lenz,R., C. Boelcke, U. Peckmann and M. Reuss, Proc. Modelling and Control of Biotechnological Processes, 1st IFAC Sym-posium, IFAC Publications, New York, 11 (1985)

5. Reuss, M.,C. Boelcke,R. Lenz and U. Peckmann, BFT-Forum, **4**, 2 (1987)

6. Reuss, M., "On-line fermentors sampling and HPLC analysis", in Computer Control of Fermentation Processes, Omstead (Ed.) CRC Press, Inc., Boca Raton, Florida (1990)

7. Reda, K.D. and D.R. Omstead "Automatic fermentor sampling and stream analysis", in Com-puter Control of Fermentation Processes,Omstead (Ed.), CRC Press, Inc., Boca Raton, Florida (1990).

8. Ochynski, F. W. Analyst, **85**, 278 (1960)

An Experimental Guide to the Relevant Aeration Rates in Microaerobic Bioprocesses

C.J. Franzén, G. Lidén, and C. Niklasson

Department of Chemical Reaction Engineering, Chalmers University of Technology,
S-412 96 Göteborg, SWEDEN

A new method for the study of microaerobic bioprocesses, called Oxygen Programmed Fermentation (OPF), was investigated theoretically and experimentally. Two major areas were identified, where the use of OPF can be advantageous. Firstly, it is possible to perform physiological studies under conditions that are not normally obtainable in chemostat cultures. This was illustrated by simulations of an OPF in a CSTR system, with a simple kinetic model of a purely fermentative yeast. Experimentally, by using OPF, it was shown that Candida utilis, *a yeast incapable of anaerobic fermentation and growth on xylose, actually produced ethanol from xylose during favorable microaerobic conditions. Secondly, it is also possible to use OPF in process optimization studies, in particular in the initial stages. Ethanol formation by the yeast* Saccharomyces cerevisiae *was studied by OPF, and it was shown that the optimal oxygen concentration in the inlet gas was below 1%. At this level, there was a minimum of byproducts formed. During the transitions between microaerobic and anaerobic conditions, fluorescence measurements indicated that the metabolism went through several phases. Based on these experiments, further optimization of ethanol production with* S. cerevisiae *should be made in the region below 1% oxygen in the inlet gas.*

Several bioprocesses require a microaerobic or semiaerobic environment for optimal performance. Examples are the fermentation of xylose by *e.g. Pichia stipitis* [1],the production of 2,3-butanediol by *Klebsiella oxytoca* [2], and ethanol production by *Saccharomyces cerevisiae* [3].

Traditionally, microaerobic processes have been optimized either by chemostat or batch (shake flask) experiments with aeration rates controlled at low and constant levels [1,3-6]. Xylose utilizing yeasts have very poor anaerobic growth capabilities [7-9]. This makes it difficult to perform chemostat experiments at very low oxygen transfer rates, since the biomass is washed out even at rather low dilution rates. Furthermore, chemostat experiments are very time consuming, since a steady state must be reached for every dilution rate and aeration rate tested.

Under anaerobic conditions, *Candida utilis* grows slowly on glucose and does not grow at all on xylose. However, it has been shown, that with glucose as carbon and energy source, both the growth and the fermentative capabili-ties of this yeast (and others) are greatly improved by oxygen limited conditions [10]. It is therefore inter-esting to investigate how this yeast reacts to microaerobic conditions also with xylose as the sole carbon and energy source.

Saccharomyces cerevisiae is able to grow, aerobically as well as anaerobi-cally, on glucose as sole carbon and energy source. However, for anaerobic growth, unsaturated fatty acids and ergosterol must be provided with the medium. The main product under anaero-bic conditions is ethanol. Glycerol, the main byproduct, is formed as a consequence of a net production of NADH from biomass formation [11]. Under aerobic conditions, a large part of the carbon is either respired or incorpo-rated into biomass. By adding an opti-mal amount of oxygen to the culture, it should be possible to decrease the amount of glycerol formed, without excessive biomass formation [3]. This is of potential interest for industrial ethanol production. However, it is very tedious and time-consuming to determine the optimal oxygen transfer rate in chemostat cultures. It would be very beneficial if such experiments could be

E. Galindo and O.T. Ramírez (eds.), *Advances in Bioprocess Engineering. 255-265.*
© 1994 *Kluwer Academic Publishers. Printed in the Netherlands.*

performed in a more efficient way.

Oxygen Programmed Fermentation (OPF) is a time-saving alternative and complement to previous methods of microaerobic experiments [12,13]. In an OPF, the oxygen transfer rate is slowly changed from one level to another. The change in oxygen transfer rate is brought about by changing the oxygen concentration in the inlet gas, while keeping the total flow rate constant. It is thus possible, using fast on-line measurements, to quickly "zoom in" on the aeration rates that are of real interest to the process.

In this work, experimental results from oxygen programmed fermentations with *Candida utilis* CBS 621, grown on a defined xylose medium, and with *Saccharomyces cerevisiae* CBS 8066, grown on a defined glucose medium, are presented. Simulations of the ramp response with a simple mathematical model of a purely respirative microorganism are also presented, to illustrate the important qualitative characteristics of the method.

THEORY

Oxygen programmed fermentation is an experimental method in which the oxygen transfer rate is changed in a predetermined way, after an initial steady state has been reached in a continuous culture. The change of oxygen transfer rate is preferably accomplished by controlling the inlet gas composition while keeping the total gas flow rate constant, since any change in gas holdup and stirring intensity will affect the response of optical probes. Here, we only consider linear changes (ramps) of the oxygen concentration in the inlet gas, although other types of changes, e.g. exponential, are possible.

Mathematical Model

The model is based on mass balances for oxygen, substrate, and biomass in a simplified CSTR system (Figure 1). It is possible to describe the whole reactor system with a system of dimensionless differential equations, if the dimensionless time is introduced according to

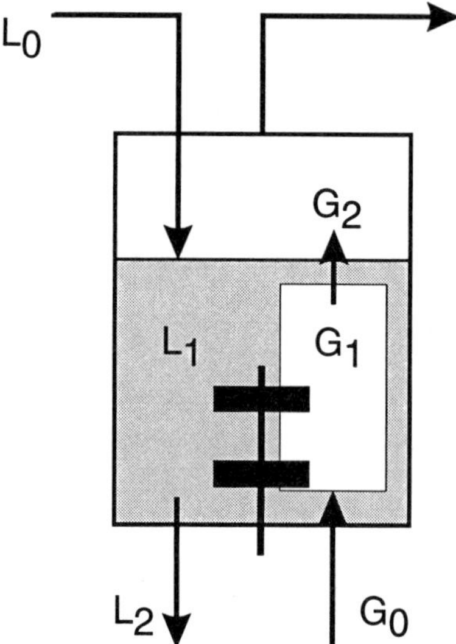

Figure 1. Gas and liquid phases in the bioreactor.

$$t^* = t / \tau_g \qquad (1)$$

where the residence time of gas in the headspace, τ_g, is given by

$$\tau_g = \frac{V_{G2} P}{F_0 R T} \qquad (2)$$

The dimensionless concentrations are consequently given by

$$c_0^* = c_0 \frac{He}{P} \; ; \qquad c_1^* = c_1 \frac{He}{P} \qquad (3)$$

$$S_1^* = \frac{S_1}{S_0} \; ; \qquad c_x^* = \frac{c_x}{S_0} \qquad (4)$$

To illustrate some important features of the OPF method, we will here concentrate on a simple, Monod type model for a purely respirative yeast. This case largely corresponds to the yeast *Candida utilis*, growing on xylose as the sole carbon and energy source. The macroscopic cell growth equation, for purely respirative growth, can be written

$$CH_2O + Y_{O/S}O_2 + Y_{N/S}NH_3 \rightarrow$$
$$Y_{X/S}CH_aO_bN_c + Y_{CO2/S}CO_2 + Y_{W/S}H_2O \qquad (5)$$

The kinetic model is based on the assumption that either the xylose uptake rate, or the oxygen uptake rate is rate limiting for the growth process. The rates of uptake of the nonlimiting compounds are determined by the stoichiometric coupling to the limiting uptake rate according to Equation 5 [14]. With the used parameters, the oxygen becomes rate limiting at dissolved oxygen concentrations below approximately 1 % of the air saturation value. No growth is possible under anaerobic conditions.

Steady state solutions were calculated for each oxygen mole fraction in the inlet gas, by using a nonlinear equation routine based on a modified Gauss-Newton method. The dynamic response of the system was simulated with a semi-implicit numerical method, since the differential equations were stiff. Further details of the modeling procedure are presented elsewhere [12].

MATERIALS AND METHODS

Strains and media. The yeast strains *Candida utilis* CBS 621, and *Saccharomyces cerevisiae* CBS 8066 (Centraalbureau vor Schimmelcultures, Delft, The Netherlands) were used throughout this study.

The defined xylose media used in the cultivation of *C. utilis* (Wickerham's medium) are described elsewhere [13]. For *S. cerevisiae*, the same medium was used, but with some modifications. The final medium for *S. cerevisiae* contained 30 g/l D-glucose and 0.1 g/l complex antifoam 289 (Sigma). The medium was supplemented with 10 mg/l ergosterol (USB, Cleveland, Ohio), 420 mg/l Tween 80 (MERCK-Schuchardt, Hohenbrunn, Germany), prepared according to Verduyn *et al* [11]. The medium was also supplemented with a filter sterilized vitamin solution, so that the final concentrations of all vitamins were twice that in ordinary yeast nitrogen base, YNB (Difco, Detroit, Michigan).

Bioreactor system. The bioreactor setup with surrounding equipment is schematically shown in Figure 2. The experiments were performed in a Chemap SG 3.5 bioreactor (Chemap AG, Switzerland). The liquid volume was 1.5 l in the *C. utilis* and 2.5 l in the *S. cerevisiae* experiments.

The change in the oxygen concentration in the inlet gas was accomplished by means of two mass flow controllers according to Figure 3.

Analytical methods. Determination of the major metabolites in the broth was made by an HPLC system, which was connected on-line. For *C. utilis*, xylose, xylitol and ethanol were separated on a Sugarpak column (Waters, Millipore Corp., Massachusetts). For *S. cerevisiae*, glucose, glycerol, ethanol, acetaldehyde and acetic acid were separated on a FAMpak column (Waters). Cell free samples for HPLC analysis were obtained on-line with a Biopem filtration unit (Braun, Germany). Biomass concentration was determined off-line gravimetrically.

Analyses of the inlet and exhaust gas streams were made with a QMS 420 capillary inlet quadrupole mass spectrometer (Balzers AG, Liechtenstein). Calibration was performed on-line at least every hour.
The carbon dioxide evolution rate (CER) and the oxygen transfer rate (OTR) were calculated by gas balancing [15].

On-line culture fluorescence measurements were made *in situ* with the FluoroMeasure System probe (BioChem Technology, Pennsylvania), tuned for NAD(P)H. On-line biomass estimation was made, also *in situ*, with a MEX-3 four-channel, alternating light beam absorbance probe, model OD 10/5 (BTG Källe Inventing AB, Sweden).

Cultivation procedures. During continuous cultivations, the temperature was maintained at 30°C, the pH was controlled at 4.5 by addition of NaOH, and the stirring rate was controlled at 400 rpm for *C. utilis*, and 500 rpm for *S. cerevisiae*. The bioreactor was continuously sparged with a mixture of air (100 Nml/min) and pure N_2. The total gas flow rate was 400 Nml/min for *C. utilis*, and 500 Nml/min for *S. cerevisiae*.

Fresh medium was continuously pumped into the bioreactor at a constant flow rate. The liquid volume in the bioreactor was kept constant by means of a weight controller, connected to the outlet pump. The resulting dilution rate was 0.043 h^{-1} for *C. utilis*,

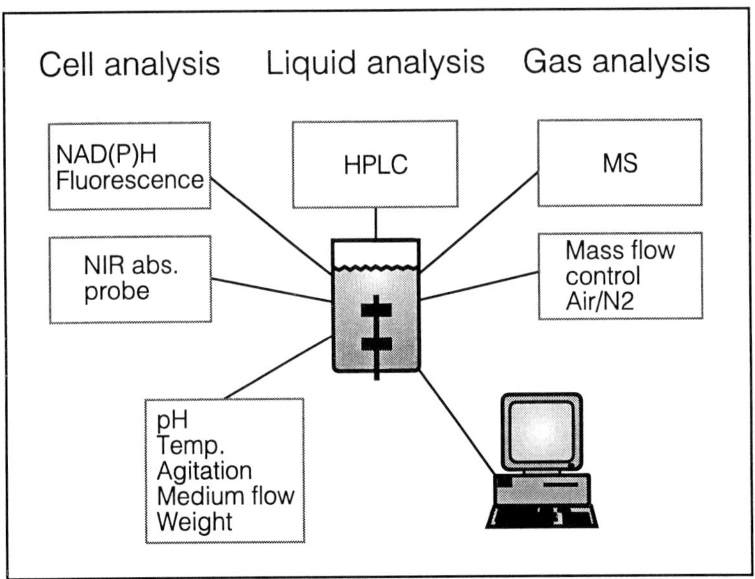

Figure 2. The experimental equipment.

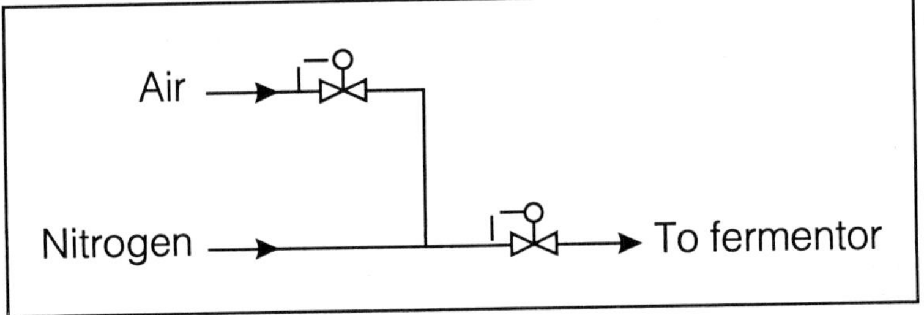

Figure 3. Mass flow control of the oxygen concentration in the inlet gas.

and 0.10 h^{-1} for *S. cerevisiae*.

The culture was allowed to reach steady state (verified by constant fluorescence and near IR signals, as well as by repeated HPLC analyses). The fraction of air in the inlet gas was then linearly decreased to zero. For *C. utilis*, the mole fraction of oxygen was reduced from 0.0525 to zero in 12 and three hours. Since *C. utilis* cannot grow anaerobically on xylose, the fraction of air was immediately increased with the same linear rate as before. *S. cerevisiae* readily grows under anaerobic conditions on the medium used here. Therefore, after reducing the inlet mole fraction of oxygen from 0.042 to zero in 12 h, the culture was allowed to reach steady state also under anaerobic conditions. The oxygen mole fraction was then linearly increased to 0.042 again, in 12 h.

RESULTS AND DISCUSSION

Simulations

The dynamic ramp response of the reactor system containing no cells, is shown in Figure 4. The three ramps consist of linear changes of y_0 from 0.21 to zero in the dimensionless times t^*. It can be seen, that as the ramp time is increased, the difference in oxygen concentration between the inlet and the outlet gas is decreased. At a ramp time of $t^* = 177.2$ (corresponding to 12 h with the used parameters), the difference between the dynamic and the steady state solutions is very small.

This suggests that with this or a lower ramp rate (larger t^*), analysis of the exhaust gas will not be complicated by any dynamic lag between the inlet and outlet gas. That is, the CER, the OTR, and the respiratory quotient (RQ) can be determined directly by exhaust gas analysis.

Figures 5 to 7 show simulated ramps from an initial y_0 of 0.0525 to zero, applied to a purely respirative organism (cf. Theory). In Figure 5, the effect of different ramp rates is shown. The cultures are substrate limited at the beginning of the ramps, but

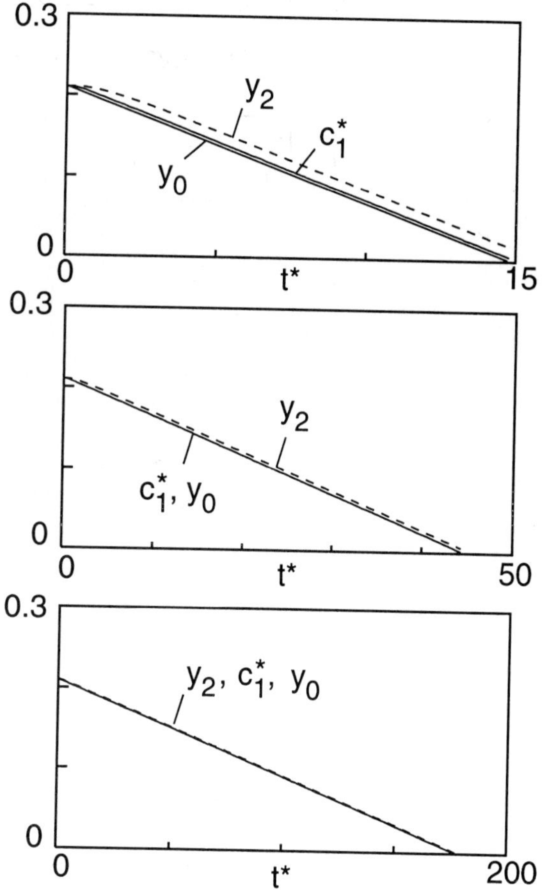

Figure 4. Simulations of three ramps in y_0, in a bioreactor system without oxygen consumption. Comparisons of solutions for three different ramp rates. Initial value of $y_0 = 0.21$, $k_La\tau_g = 12.18$, solid lines = c_1^*, dashed lines = y_2.

become oxygen limited when y_0 is reduced below approximately 0.03. This results in changes in the slopes of the steady state as well as the dynamic solutions.

In the oxygen limited region, the steady state solution for the biomass concentration decreases, and the substrate concentration increases, both linearly. The culture is washed out when y_0 equals zero. The steady state oxygen concentration, on the other hand, remains at a constant value until

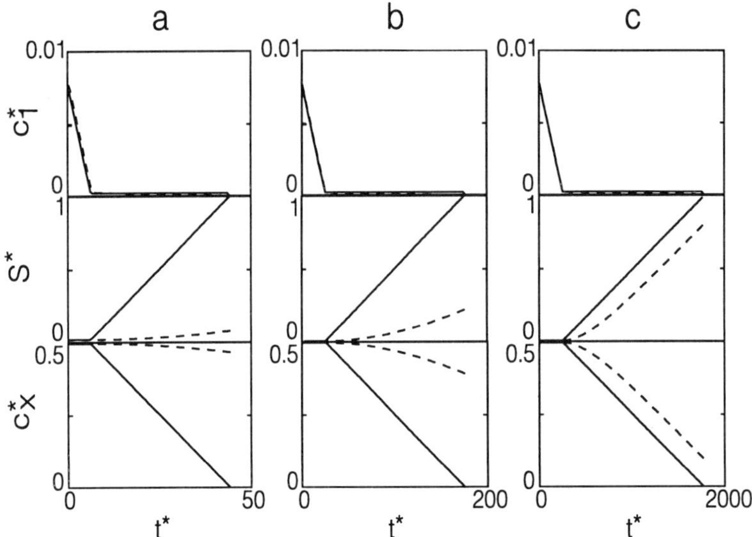

Figure 5. Simulations of three ramps in y_0, in a cultivation of a purely respiratory microorganism. Comparisons between steady-state chemostat solutions (solid lines) and dynamic solutions (dashed lines) for three different ramp rates. $D = 0.05$ h^{-1}, initial value of y_0 = 0.0525.

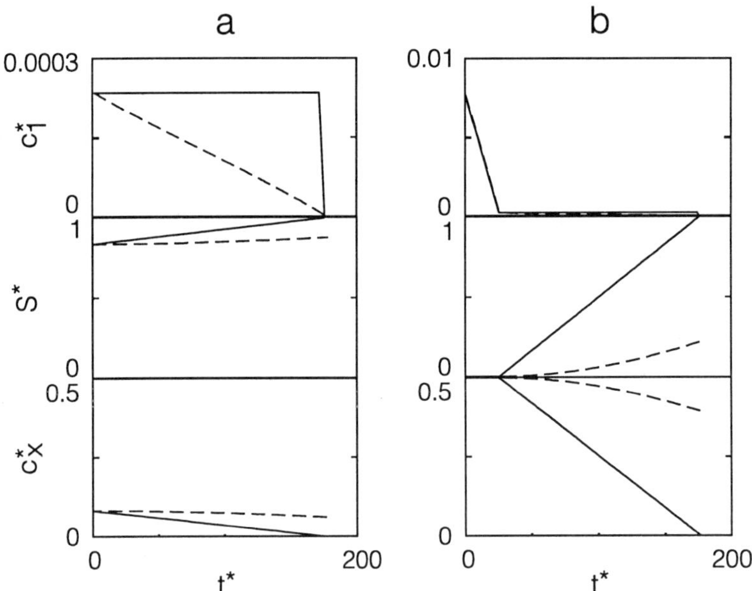

Figure 6. Simulations of two ramps in y_0, in a cultivation of a purely respiratory microorganism. Comparisons between steady-state chemostat solutions (solid lines) and dynamic solutions (dashed lines) for **a)** $k_L a = 0.005$ s^{-1}, and **b)** $k_L a = 0.05$ s^{-1}. $D = 0.05$ h^{-1}, initial value of y_0 = 0.0525.

the very last part of the ramp, where the inlet oxygen concentration gives an equilibrium concentration below this value.

The dynamic solution approaches the steady state solution at large ramp times, but at a ramp time of $t^* = 1772$ (120 h), there is still a substantial difference. At a ramp time of $t^* = 177.2$, the concentration of substrate and of biomass deviate relatively little from the initial steady state. However, the concentration of oxygen in the broth changes almost linearly, not only during the first substrate limited part but also during the oxygen limited part of the ramp.

Figure 6 shows how the solutions are affected by a change of the $k_L a$ value. For the lower $k_L a$ value the entire ramp is within the oxygen limited region, and the dynamic dissolved oxygen concentration decreases almost linearly also in this case.

Results from simulations at two different dilution rates are shown in Figure 7. At the higher dilution rate, the ramp is again entirely within the oxygen limited region. The dissolved oxygen concentration decreases linearly with y_0, whereas the biomass and substrate concentrations do not change very much from the initial steady state values.

These simulations show, that using oxygen programmed fermentation, changes in intracellular concentrations due to changing oxygen concentrations may be studied also in the oxygen limited region. This can be done by *e.g. in-situ* fluorescence measurements or rapid sampling and extraction of intracellular compounds. Head space gas analysis may furthermore be used to evaluate carbon dioxide evolution rates and oxygen uptake rates, provided that the ramp time is suitably chosen. It is important to remember that at the very

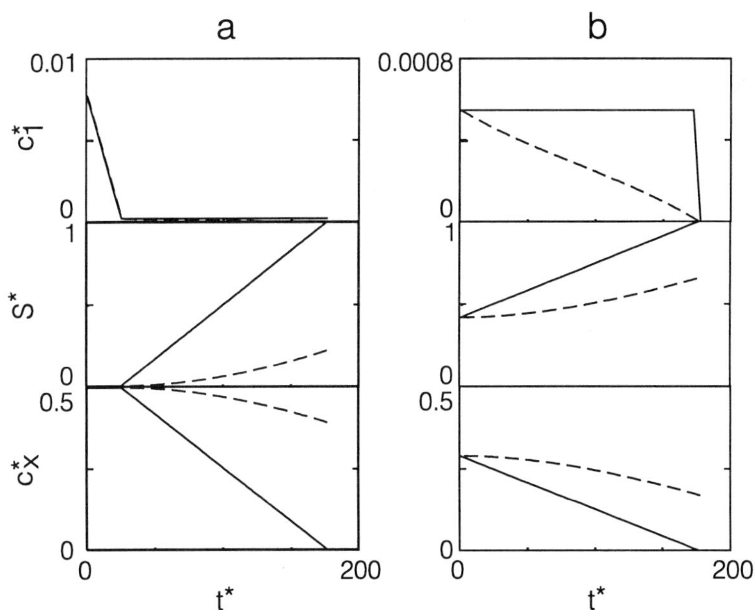

Figure 7. Simulations of two ramps in y_0, in a cultivation of a purely respiratory microorganism. Comparisons between steady-state chemostat solutions (solid lines) and dynamic solutions (dashed lines) for **a)** $D = 0.05$ h^{-1}, and **b)** $D = 0.10$ h^{-1}. $k_L a = 0.05$ s^{-1}, initial value of $y_0 = 0.0525$.

low oxygen concentrations towards the end of the ramp, a chemostat would be virtually empty of cells, making it very difficult to perform any kind of measurements. Additionally, in a chemostat, the only way of changing the steady state dissolved oxygen concentration for a purely respirative microorganism would be to change the dilution (growth) rate, which would change the whole basis of comparison.

Experimental Results

Candida utilis. Figures 8 and 9 show experimental results obtained with _C. utilis_, at two different ramp times. When grown in a chemostat with an inlet oxygen mole fraction of 0.0525, the culture was not oxygen limited during steady state conditions, and there were only minor amounts of xylose (< 0.1 g/l) and xylitol (< 0.2 g/l) in the broth. There was a dramatic increase in culture fluorescence when the DOT fell to zero, at both 12 h (Figure 8) and 3 h ramp time (Figure 9). After further reduction of the oxygen concentration, washout of cells occurred, and as a consequence, xylose and xylitol concentrations increased.

When the ramp time was shortened to only 3 h (Figure 9), the responses of the culture were similar, but not identical, to the results at a ramp time of

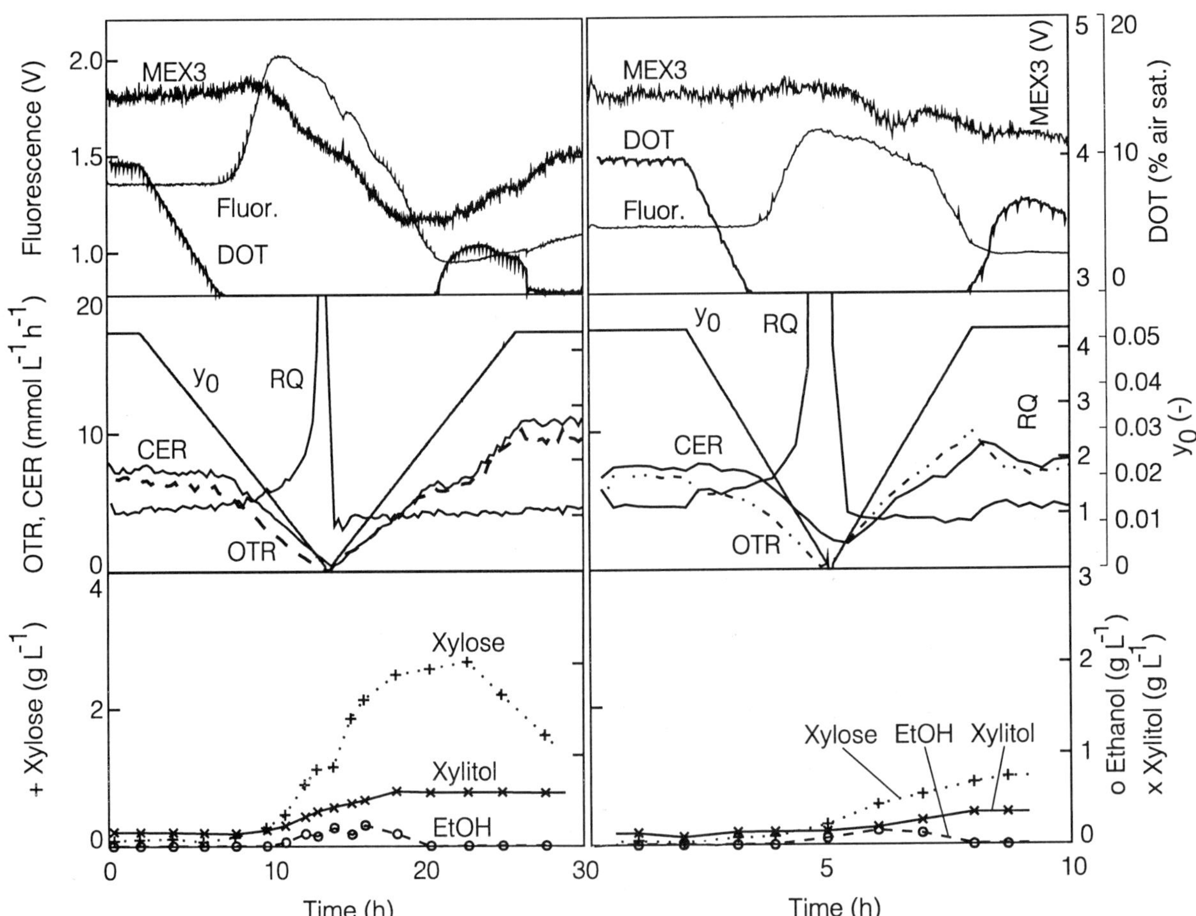

Figure 8. Results from an oxygen programmed fermentation with _Candida utilis_ CBS 621, with D = 0.043 h^{-1} and the ramp time = 12 h. y_0 is calculated from the air/N$_2$ ratio in the inlet gas.

Figure 9. Results from an oxygen programmed fermentation with _Candida utilis_ CBS 621, with D = 0.043 h^{-1} and the ramp time = 3 h. y_0 is calculated from the air/N$_2$ ratio in the inlet gas.

12 h. As predicted by the simulations, the biomass was to a larger degree retained in the bioreactor, and because of the very fast ramp, the calculated gas consumption and production rates as well as RQ suffered from dynamic lag problems.

Remarkably, ethanol was produced in small amounts also by *Candida utilis*, at very low inlet oxygen concentrations. This was seen at both 12 h and 3 h ramps, when the OTR was below 3 mmol L^{-1} h^{-1}. It has been suggested, that the incapability of *C. utilis* to produce ethanol under anaerobic conditions, is due to the inability of this yeast to maintain the intracellular $NADH/NAD^+$ ratio [7]. It is possible,

that the small amount of oxygen transferred to the culture in these experiments can, to some degree, sustain the fermentative pathway by the oxidation of NADH, while not being sufficient, however, for a complete oxidation of the substrate.

<u>Saccharomyces cerevisiae.</u> Results from two OPF experiments with *S. cerevisiae* are shown in Figures 10 and 11. At the first steady state (D = 0.12, h^{-1}, y_0 = 0.042, Figure 10), the culture was in a respiro-fermentative state. The RQ was approximately 5, and both acetate and acetaldehyde were produced. There was no residual glucose in the medium.

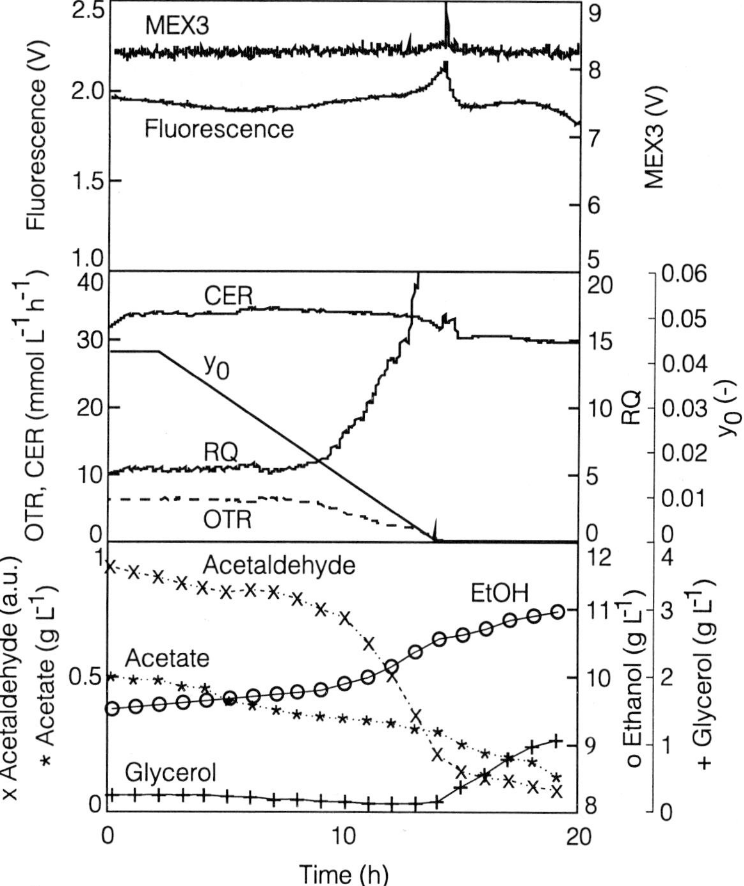

Figure 10. Results from an oxygen programmed fermentation with *Saccharomyces cerevisiae* CBS 8066, with an initial y_0 = 0.042, D = 0.12 h^{-1}, ramp time = 12 h. y_0 is calculated from the air/N_2 ratio in the inlet gas.

When the inlet oxygen concentration was reduced (Figure 10), the degree of fermentative metabolism increased. This became evident by the increasing RQ and ethanol concentration, and by the decreasing OTR and acetate and acetaldehyde concentrations. The biomass concentration changed very little. At the end of the ramp down, the fluorescence suddenly started to increase. This indicated, that at this point, the oxygen uptake rate was too low to maintain the intracellular NADH / NAD^+ ratio at a constant level. Just after the increase in fluorescence, the glycerol concentration started to increase, which had the immediate effect of decreasing the culture fluorescence to the same level as before, *i.e.*, alleviating the redox imbalance. It is interesting to note, that the glycerol concentration very close to the end of the ramp, actually reached a value which was smaller than the aerobic steady state value.

After the culture had reached steady state also under anaerobic conditions, y_0 was linearly increased again (Figure 11) to its original value. Both the ethanol and the glycerol concentrations eventually decreased, and acetate and acetaldehyde appeared in the medium. The fluorescence signal indicated that the cells could not maintain NADH / NAD^+ ratio properly during the whole transition from anaerobic to microaerobic conditions. On the contrary, the measurements indicated that the adaptation to microaerobic conditions runs through several phases.

From these experiments, it is clear that for maximizing the ethanol yield by minimizing the glycerol yield, the optimum oxygen concentration in the inlet gas should be very low. At y_0 greater than 0.01, the ethanol production rate and yield are reduced, and other byproducts apppear in the broth. Therefore, microaerobic chemostat studies should be performed with y_0 well below 0.01. This agrees well with the results of Grosz *et al* [3].

CONCLUSIONS

We have shown two major areas where oxygen programmed fermentation can be successfully applied. Firstly, it is possible to use OPF to study cell phys-

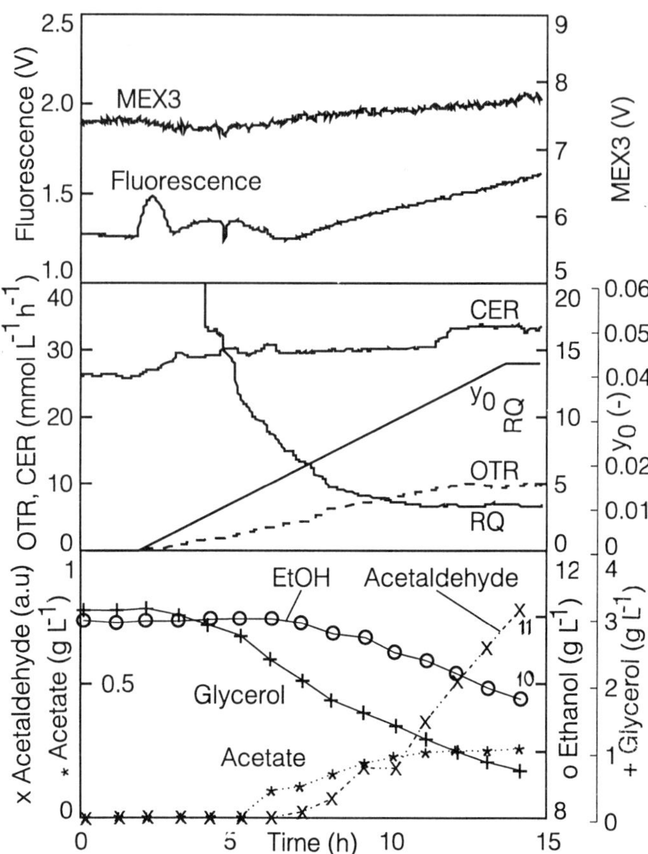

Figure 11. Results from an oxygen programmed fermentation with *Saccharomyces cerevisiae* CBS 8066, with an initial y_0 = 0, D = 0.12 h^{-1}, ramp time = 12 h. y_0 is calculated from the air/N_2 ratio in the inlet gas.

iology under experimental conditions not normally obtainable in chemostat cultures, *e.g.* at combinations of high dilution rates and very low dissolved oxygen concentrations. Secondly, it is possible to use OPF as an initial experiment, in which the regions of interesting oxygen uptake rates are identified. To gain a deeper understanding of the microbial physiology and kinetics, traditional chemostat cultures should then be performed, but these studies can focus on the oxygen transfer rates of real interest. Therefore, a significant reduction of the experimental effort should be possible. We believe, that the second area is where oxygen programmed fermentation will be of greatest value.

NOMENCLATURE

c_0	Concentration of oxygen in the inlet liquid flow L_0 (mol m^{-3})
c_1	Concentration of oxygen in the liquid phase L_1 (mol m^{-3})
c_x	Concentration of biomass in the liquid phase L_1 (C-mol m^{-3})
CER	CO_2 evolution rate (mmol L^{-1} h^{-1})
D	Dilution rate (h^{-1})
DOT	Dissolved oxygen tension (% air saturation)
F_0	Gas flow rate, inlet gas (mol s^{-1})
He	Henry's law coefficient (Pa m^3 mol^{-1})
k_La	Volumetric mass transfer coefficient (s^{-1})
OTR	Oxygen transfer rate (mmol L^{-1} h^{-1})
P	Pressure (Pa)
RQ	Respiratory quotient (-)
S_0	Concentration of carbon source in the inlet liquid L_0 (C-mol m^{-3})
S_1	Concentration of carbon source in the liquid phase L_1 (C-mol m^{-3})
T	Temperature (K)
t	Time (s)
t^*	Dimensionless time (t/τ_g)
V_{G2}	Volume of head space gas (m^3)
$Y_{CO2/S}$	Macroscopic yield of carbon dioxide (C-mol C-mol^{-1})
$Y_{N/S}$	Macroscopic ammonia yield (mol C-mol^{-1})
$Y_{O/S}$	Macroscopic oxygen yield (mol C-mol^{-1})
$Y_{X/O}$	Macroscopic biomass yield (C-mol mol^{-1})
y_0	Mole fraction of oxygen in gas phase G_0 (inlet)
y_1	Mole fraction of oxygen in gas phase G_1
y_2	Mole fraction of oxygen in gas phase G_2 (head space)

Greek letters

τ_g	Gas residence time in headspace (s)

Superscripts

*	Dimensionless variable

LITERATURE CITED

1. Skoog, K., Hahn-Hägerdal, B., *Appl. Environ. Microbiol.*, **56**, (11), 3389-3394 (1990).

2. Beronio, P.B., Tsao, G.T., *Biotechnol. Bioeng.*, **42**, 1270-1276 (1993).

3. Grosz, R., Stephanopoulos, G., *Biotechnol. Bioeng.*, **36**, 1006-1019 (1990).

4. Agrawal, P., Lim, H.C., *Adv. Biochem. Engng.*, **30**, 61-90 (1984).

5. du Preez, J.C., van Driessel, B., Prior, B.A., *Arch. Microbiol.*, **152**, 143-147 (1989).

6. Grootjen, D.R.J., van der Lans, R.G.J.M., Luyben, K.Ch.A.M., *Biotechnol. Tech.*, **4**, 149-154 (1990).

7. Bruinenberg, P.M., de Bot, P.H.M., van Dijken, J.P., Scheffers, W.A., *Appl. Microbiol. Biotechnol.*, **19:** 256-260 (1984)

8. Slininger, P.J., Bolen, P.L., Kurtzman, C.P., *Enzyme Microb. Technol.*, **9:** 5-15 (1987).

9. Ligthelm, M.E., Prior, B.A., du Preez, J.C., *Appl. Microbiol. Biotechnol.*, **28**, 63-68 (1988).

10. Visser, W., Scheffers, W.A., Batenburg-van der Vegte, W.H., van Dijken, J.P., *Appl. Environ. Microbiol.*, **56**, 3785-3792 (1990).

11. Verduyn, C., Postma, E., Scheffers, W.A., van Dijken, J.P., *J. Gen. Microbiol.*, **136**, 395-403 (1990).

12. Lidén, G., Franzén, C.J., Niklasson, C., *Biotechnol. Bioeng.*, Accepted (1994).

13. Franzén, C.J., Lidén, G., Niklasson, C., *Biotechnol. Bioeng.*, Accepted (1994).

14. von Stockar, U., Auberson, L.C.M., *J. Biotechnol.*, **22**, 69-88 (1992).

15. Buckland, B., *Bio/Technology*, **5**, 982-988 (1985).

Model Aided Multiple Correlation Analysis between Preculture and Main Fed Batch Culture

R. Guthke and W. Rausch

Hans-Knöll-Institute for Natural Product Research, D-07745 Jena,
Beutenbergstr. 11, GERMANY

For an industrial antibiotic process the course of physiological state of both, preculture and main fed batch culture, is characterized by the kinetics of pH and acid fed F_A to control pH, respectively. These kinetics could be described not by manageable causal dynamic models but by nonlinear functions of time t. A fast and global convergent algorithm was developed to identify the parameters of these nonlinear functions for a great number of fermentation runs (>30). The model parameters for the kinetics of pH(t) of the preculture as well as the carbon feed $F_C(t)$ and acid feed $F_A(t)$ of the main culture are correlated by multiple regression analysis. The result gives a data-based link between preculture and main culture. It provides a quite general approach for preculture dependent control of main fed batch culture.

Frequently the productivity of secondary product formation in the main culture depends on the preculture of fermentation. This fact can be explained by dynamical models, especially by models with bistable behaviour (1, 2). The simplest correlation between pre- and main culture can be formulated by the inoculum, i.e. the final biomass concentration of preculture and initial biomass concentration of main culture. It is more difficult to identify the physiological state of the inoculum, i.e. the state of induction of key enzymes of secondary product synthesis. This state is the result of course of partial limited preculture. The kinetics of pH or feeding rate of acid or alkali to control pH are very sensitive indicators for the course of carbon limited pre- or main culture and give rise for adaptation of carbon feed rate (3). Due to the interaction of serveral organic acids and puffers structured dynamic models are not applicable to formulate the sequence of limitation in complex media. Thus, for process identification a formal mathematical approach is necessary.

Artificial neural networks have been used to estimate the state variables in fermentation (4). In our contribution we does not apply neural network approach because we intend to exploit our knowledge about the qualitative shape of time courses of pH in the preculture and of acid feed rate in the main culture. The identified correlation will be used for the construction of preculture dependent control of carbon source dosage during the main fed batch culture using a pre-set acid dosage profile.

<u>MATERIAL AND METHODS</u>

<u>Experimental data</u>. For 34 runs of an industrial antibiotic fermentation three kinetic variables are taken into account:

- the uncontrolled pH
    of the batch preculture,
- the amount of acid feed F_A
    to control pH in the main
    culture,
- the amount of carbon feed F_c
    of the main culture.

267

E. Galindo and O.T. Ramírez (eds.), Advances in Bioprocess Engineering. 267-273.
© 1994 Kluwer Academic Publishers. Printed in the Netherlands.

These three measured variables as well as the time are presented in arbitrary units from zero to one. The experimental data are made equidistant with a fixed time interval Δt by averaging over the data measured in shorter intervals.

<u>Identification algorithm</u>. The function $f_H(t, \alpha_H, \beta_H)$ as well as the the time-integral over the funtions $f_A(t, \alpha_A, \beta_A)$ and $f_C(t, \alpha_C, \beta_C)$ with its vectors of nonlinear parameters α_H, α_A, and α_C as well as the vectors of linear parameters β_H, β_A, and β_C are fitted to the time series of experimental data of pH(t), $F_A(t)$ and $F_C(t)$ minimizing the sum of deviation squares (SDS) as follows: The components β_1, ..., β_L of the linear parameter vector β are estimated by HOUSEHOLDER's algorithm (<u>5</u>). This is done for each tested vector α of nonlinear parameters. The components α_1, ..., α_N of nonlinear parameters are estimated by four steps as follows. In the first step the parameter vector α is variated equidistributed within a preset region. In the second step the parameter vector α is variated normal distributed around the best result of the first step with a variance which decreases with the number of iteration linearly. The third step is an evolutive algorithm. The parameter α is variated as in step two but the variance will be increased by a certain factor after improvement of the objective functional SDS. The steps one to three are repeated three times. The 4th step is a systematical search within an N-dimensional interval around the best value of the third step. The interval will be enlarged if the best parameter α is on the border, else it will be reduced up to a minimum one.

RESULTS AND DISCUSSION

Models

<u>pH of the preculture</u>. It was found, that the pH profile of the batch preculture is characterized by two temporal minima and one maximum. A causal dynamic model to describe this kinetics pH(t) results in a set of differential equations for 3 organic acids and one base (<u>6</u>,<u>7</u>). This dynamic model includes 11 nonlinear parameters. To simplify the identification procedure instead of the dynamic model the following time-varying function with only 3 nonlinear and 6 linear parameters was fitted to the pH profile:

$$f_H(t, \alpha_H, \beta_H) = \beta_{H1} + \beta_{H2} \, t + \beta_{H3} \, t^2 +$$
$$+ \beta_{H4} \, g(t, \alpha_{H1}) + \beta_{H5} \, g(t, \alpha_{H2}) +$$
$$+ \beta_{H6} \, g(t, \alpha_{H3})$$

$$(1)$$

with

$$g(t, \alpha_{Hi}) = \begin{cases} \ln(\alpha_{Hi} - t) \\ \qquad \text{for } t \leq \alpha_{Hi} - \Delta t \\ \dfrac{(\alpha_{Hi} + \Delta t)^2}{t + 2\Delta t} - \alpha_{Hi} - \Delta t \\ \qquad \text{for } t \geq \alpha_{Hi} - \Delta t \end{cases}$$

The nonlinear function $g(t, \alpha_{Hi})$ decreases monotonously. The logarithmic part for $t \leq \alpha_{Hi} - \Delta t$ is used due to the definition of pH as the negative logarithm of concentration of H^+-ions. After the inflection point with $g(\alpha_{Hi} - \Delta t, \alpha_{Hi}) = 0$, i.e. for $t > \alpha_{Hi} - \Delta t$ the function $g(t, \alpha_{Hi})$ is continued smoothly by a hyperbolic term.

<u>Acid feed rate f_A</u>. The acid feed rate $f_A(t)$ is the time differential of the amount of fed acid $F_A(t)$. As process characteristics of the fed batch main culture it was found that the time course $f_A(t)$ has more than one maximum during the transient phase which is most sensitive for secondary product formation. Such maximum of acid feed indicates the instant of strong limitation, i.e. of minimal physiological availability of carbon source for the microbial metabolism. For modeling a piecemeal linear function $f_A(t)$ is used with 6 linear phases and 5 inflection points at α_{Ai} as shown in Figure 1, where the condition $0 = \alpha_{A0} \leq \alpha_{A1} < \alpha_{A2} < \alpha_{A3} < \alpha_{A4} < \alpha_{A5} < \alpha_{A6}$ is fulfilled:

$$f_A(t, \alpha_A, \beta_A) = \beta'_{Ak} + \beta''_{Ak} \, t \qquad (2)$$
$$\text{for } \alpha_{Ak-1} \leq t \leq \alpha_{Ak}; \; k = 1, \ldots, 6$$

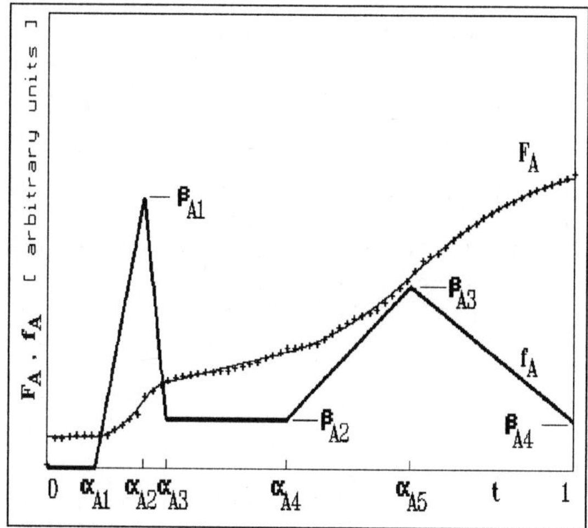

Fig. 1
Kinetics of acid Feed rate f_A and acid feed amount F_A: experimental data (+) and fitted function from equation (2)

To get a functional that is steady in time t only one of both, ß' or ß'' is a free parameter. Thus, instead of 12 parameters $ß_{Ak}'$ and $ß_{Ak}''$ as shown in equation (2) we have 6 free parameters, only. Once more, the number of free parameters is reduced to 4 parameters ($ß_{Ai}$, i=1,...,4) by two constrains: The function $f_A(t, \alpha_A, ß_A)$ is zero within the first interval $0 \leq t \leq \alpha_{A1}$ and it is constant with $f_A = ß'_{A4} = ß_{A2}$ within the 4th interval $\alpha_{A3} \leq t \leq \alpha_{A4}$ (i.e. $ß'_1 = ß''_1 = ß''_4 = 0$). It is linearly increasing or decreasing within the other periods (i.e. for k = 2, 3, 5 and 6). Here instead of the parameters $ß'_k$ and $ß''_k$ the extrem values are utilized: $f_A = ß_{A1}$ at $t = \alpha_{A2}$ and $f_A = ß_{A3}$ at $t = \alpha_{FA5}$. At the end of period considered for transient phase ($t = \alpha_{A6} = 1$), the acid feed rate is $f_A = ß_{A4}$.

Carbon feed rate f_c. The carbon feed rate $f_c(t)$ is the time differential of the amount of fed carbon source $F_c(t)$. To formulate this profile, which is one of the controlling inputs to the fermenter, a polynomial function is used:

$$f_c(t, \alpha_c, \beta_c) = \begin{cases} 0 & \text{for } t \leq \alpha_c \\ \\ ß_{C1}\,\tau + ß_{C2}\,\tau^2 + ß_{C3}\,\tau^3 & \\ \\ & \text{for } t \geq \alpha_c \end{cases} \tag{3}$$

with

$$\tau = t - \alpha_c$$

<u>Identification of parameters</u>

34 sets of 9 nonlinear parameters α_{H1}, ... α_{H3}, α_{A1}, ... α_{A5} and α_c as well as of 13 linear parameters $ß_{H1}$, ... $ß_{H6}$, $ß_{A1}$, ... $ß_{A4}$ and $ß_{C1}$, ..., $ß_{C3}$ are identified by fit of
- the function $f_H(t)$
    to the measured pH(t),
- the integral of $f_A(t)$

$$\int_0^t f_A(t')\,dt' \tag{4}$$

    to the measured $F_A(t)$,
- the integral of $f_c(t)$

$$\int_0^t f_c(t')\,dt' \tag{5}$$

    to the measured $F_c(t)$
of 34 fermentation runs. For 6 runs of them the result of this fit is shown in the Figures 2 to 7. The part "a" of these Figures shows the pH-kinetics of the preculture (measured values are indicated by small stars) and the fitted function $f_H(t)$. The part "b" of these Figures shows the kinetics of carbon feed. The measured values F_c and the integrated and fitted function $f_c(t)$ coincide very well for all runs as shown for the 6 selected runs. The integrated and fitted functions $f_A(t)$ for acid feed are not shown by Figures because the differences between the fitted kinetics and the predicted kinetics as shown in part "c" are small.

<u>Correlation analysis</u>

The course of the physiological states of the fed batch main culture -

- especially the kinetics of the acid feed F_A - depends on both,

(i) the profiles of the controlling inputs to the fermenter
- especially the carbon feed profile $F_C(t)$ - and

(ii) the history of the inoculum which is represented especially by the profile $pH(t)$ in the preculture.

Thus, the output parameter vector $Y=(\alpha_A, \beta_A)$ with its 9 components Y_i (= $\alpha_{A1}, \ldots, \alpha_{A5}, \beta_{A1}, \ldots, \beta_{A4}$; for i = 1, ..., 9) depends on the input parameter vector $X=(\alpha_C, \beta_C, \alpha_H, \beta_H)$ with its 13 components X_j (= $\alpha_C, \beta_{C1}, \ldots, \beta_{C3}, \alpha_{H1}, \ldots, \alpha_{H3}, \beta_{H1}, \ldots, \beta_{H6}$; for j = 1, ..., 13). The relation between the input vector and the output vector is nonlinear and can be approximated sufficiently by the following quadratic function:

$$Y_i = B_{i1} + \sum_{j=2}^{14} (B_{ij} X_j + B_{i\,j+13} X_j^2) \quad (6)$$

The 27 × 9 = 243 coefficients B_{ij} are estimated by multiple linear regression method from the 748 identified parameters α and β (34 parameter sets with 22 components). For this estimation 306 (= 34 × 9) equations of type (6) are used.

Prediction

The quadratic form of equation (6) with its 243 coefficients provides a functional for prediction of the course of acid feed. The parts "c" of the Figures 2 to 7 show the prediction of the amount of acid feed in comparison to the realized data F_A. For this prediction the parameters $\alpha_C, \beta_{C1}, \ldots, \beta_{C3}, \alpha_{H1}, \ldots, \alpha_{H3}, \beta_{H1}, \ldots, \beta_{H6}$ are put in the right side of equation (6). Before doing so these parameters have to be identified by model fit to the data F_C and pH of a certain fermentation run. From this equation (6) the output parameters $\alpha_{A1}, \ldots, \alpha_{A5}$ and $\beta_{A1}, \ldots, \beta_{A4}$ are calculated and the predicted acid feed rate $f_A(t)$ is estimated by equation (2). After integration according to equation (4) the result is comparable with the measured kine-

tics F_A as shown in Figures 2c to 7c. It is shown in these 6 Figures that the predicted acid feed is sufficiently correct for quite different profiles of $F_A(t)$.

CONCLUSION

The identification method by multiple regression analysis provides a sufficient approach to predict the physiological states of the fed batch main culture. Studies on model validation using new fermentation runs are under way. Analysing the coefficients in equation (6) the interesting fact was found, that acid demand during the main culture depends stronger on pH of preculture than on carbon feed rate of main culture.
On the bases of equation (6) the controller should be developed for carbon feed in dependence on the preculture course.

The results were unsufficient utilizing the parameters β'_k and β''_k of equation (2) instead of the transformed parameters $\beta_{A1}, \ldots, \beta_{A4}$, i.e. the extrem values of piecemeal linear function $f_A(t)$ are more appropriative than its slopes. This problem is caused by the fact that the identified slopes β''_{Ak} are disturbed more than the extrem values β_{Ai}.

The presented identification method is quite analog to the neural network approach. The nonlinear function approach, however, is more efficient (less time consuming) if a-priori knowledge is available about the qualitative shape of process kinetics. Thus, before using such identification method the qualitative kinetic analysis has to be done as demonstrated by cluster analysis previously ([8]). The most time consuming step of the approach presented is the identification of the vector α of nonlinear parameters. If this vector α can be choosen low-dimensional then the given method is very effective and should be applicable for real time control tasks. Whereas qualitative knowledge is applicable in expert systems and fuzzy control, the presented quantitative approach can be involved in conventional but nonlinear control circuits.

NOMENCLATURE

t   time
Δt   sampling time
pH   pH of the preculture (arbitrary
        units)
f_H funtion to be fitted to the
        kinetics of pH
F_A amount of acid fed to control pH
        in the main culture
f_A acid feed rate (after integration
        to be fitted to F_A)
F_C amount of carbon fed (arbitrary
        units) in the main culture
f_C carbon feed rate (after
        integration to be fitted to F_C)

LITERATURE CITED

1.   Guthke, R. and W.A. Knorre:
     Z.Allg.Mikrobiol. 20, 441 (1980).

2.   Hegewald, E., E. Wolleschensky,
     R. Guthke, M. Neubert and W.A.
     Knorre: Biotechnol.Bioeng. 23,
     1563 (1981).

3.   Pan, C.H., L. Hepler and R.P.
     Elander, Dev.Ind.Microbiol 13,
     103 (1972).

4.   Jalel, N.A., V. Hass, A.R.
     Mirzai, R. Boeker, A. Munack and
     J.R. Leigh. Modelling the cyathus
     striatus fermentation process:
     comparison of three methods. In:
     Proc. IFAC-Symp. Modeling and
     Control of Biotechnical
     Processes, Colorado, USA, p. 59
     (1992).

5.   Krausen, E. Numerische Mathematik
     mit Turbo-Pascal. Hüthig Verlag
     Heidelberg, 1989.

6.   San, K.Y. and G. Stephanopoulos.
     Biotechnol.Bioeng. 26, 1209
     (1984).

7.   Hatzinikolaou, D.G. and H.Y.Wang.
     Biotechnol.Bioeng. 37, 190
     (1991).

8.   Guthke, R. and R. Roßmann,
     Bioprocess Engineering, 6, 157
     (1991).

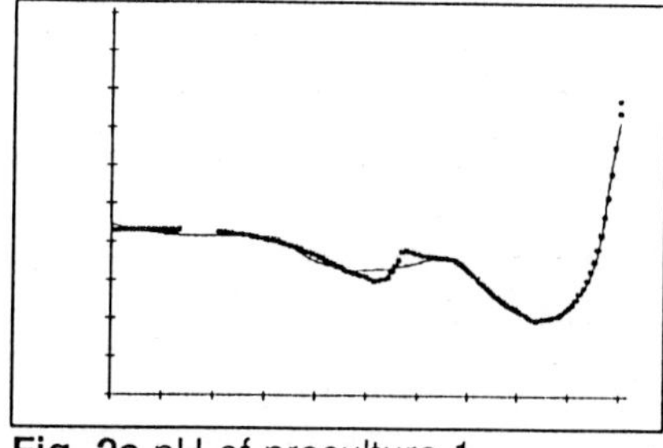

Fig. 2a pH of preculture 1

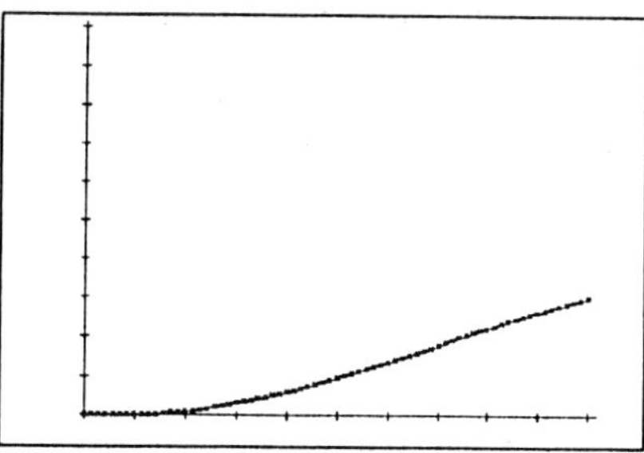

Fig. 2b Run 1: Carbon-Feed F_C

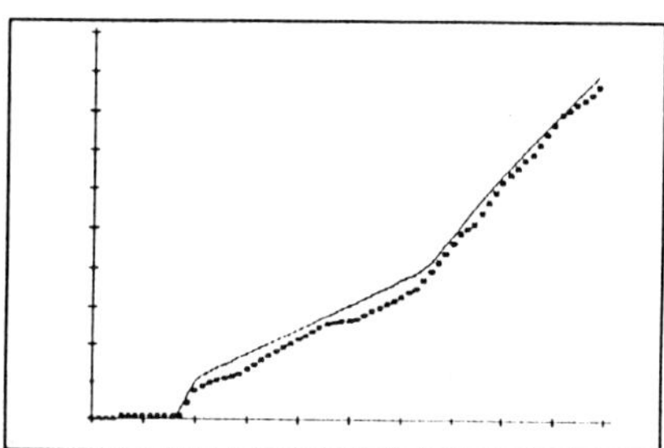

Fig. 2c Run 1: Acid Feed F_A
- measured and predicted

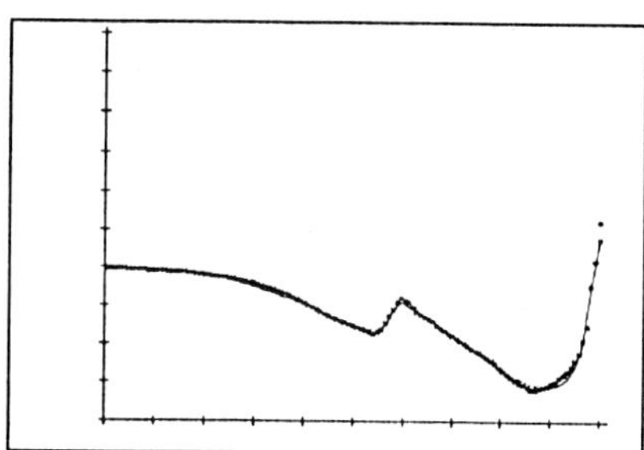

Fig. 3a pH of preculture 2

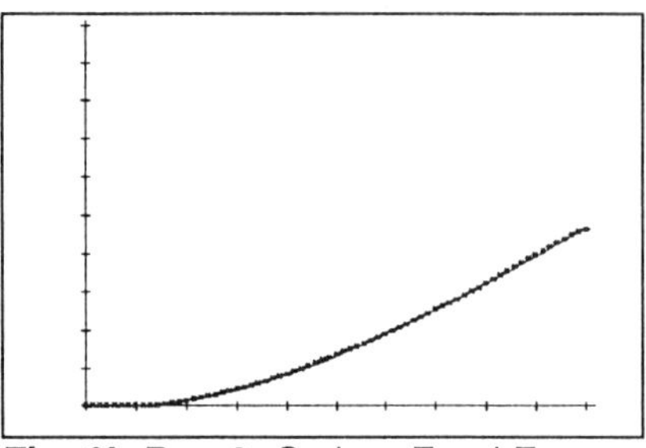

Fig. 3b Run 2: Carbon-Feed F_C

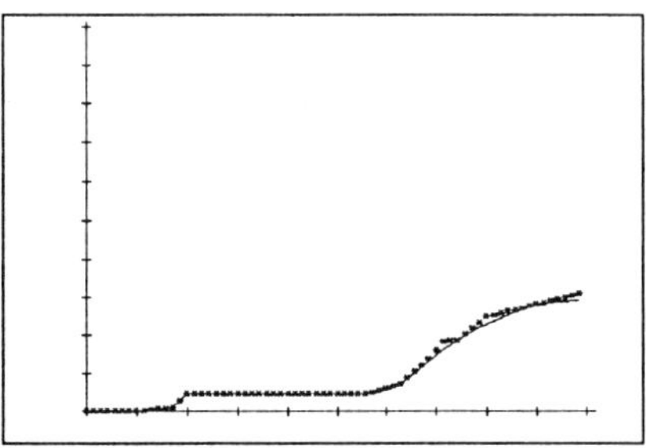

Fig. 3c Run 2: Acid Feed F_A
- measured and predicted

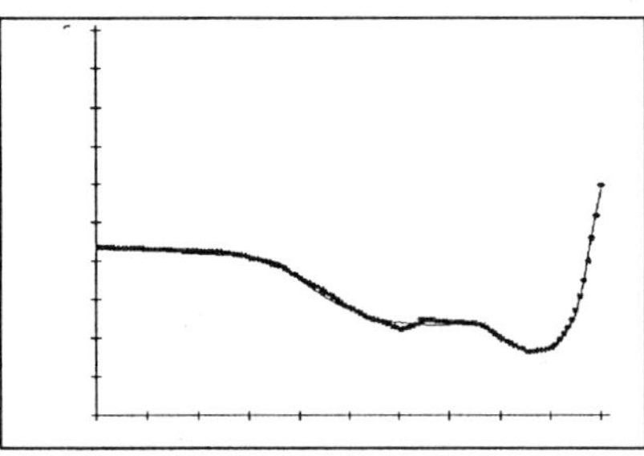

Fig. 4a pH of preculture 3

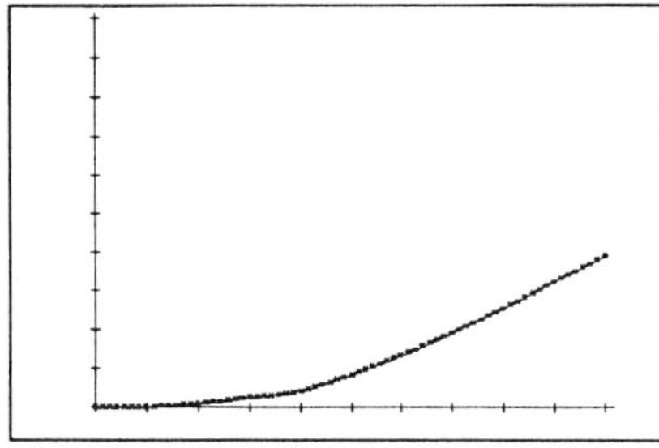

Fig. 4b Run 3: Carbon-Feed F_C

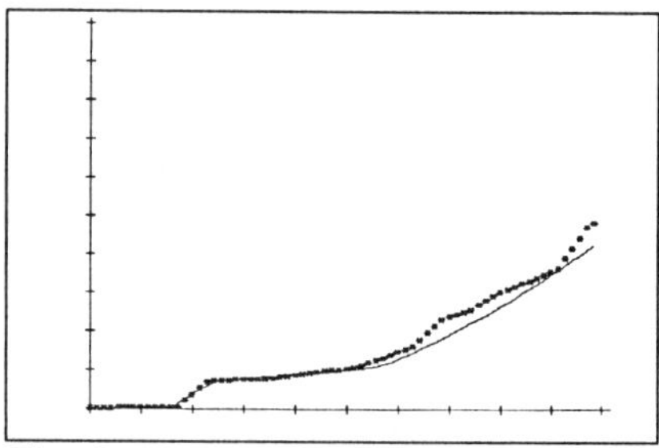

Fig. 4c Run 3: Acid Feed F_A
- measured and predicted

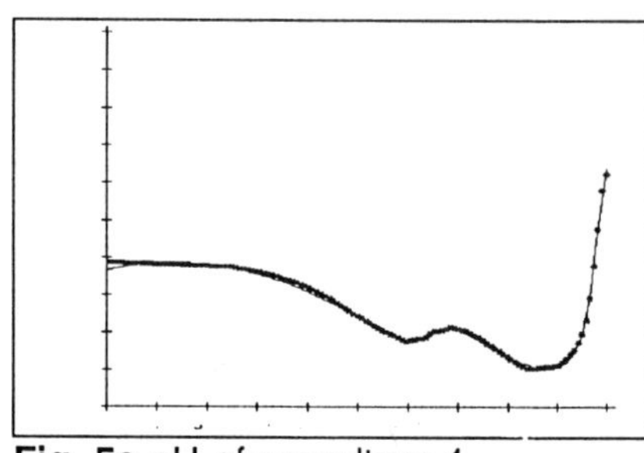

Fig. 5a pH of preculture 4

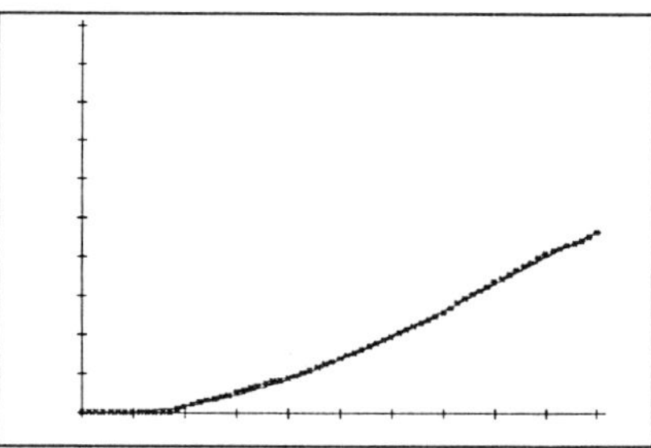

Fig. 5b Run 4: Carbon-Feed F_C

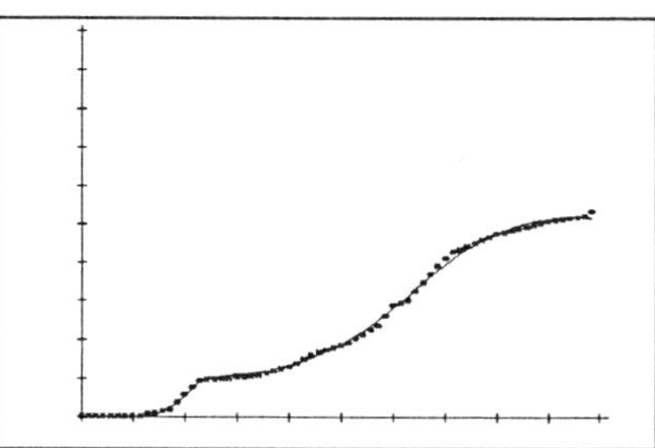

Fig. 5c Run 4: Acid Feed F_A
- measured and predicted

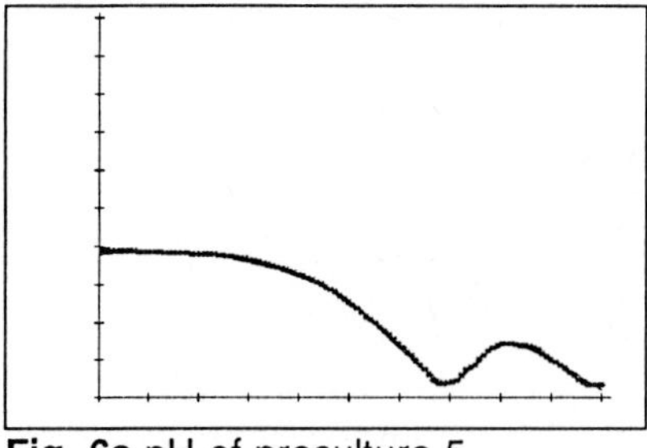

Fig. 6a pH of preculture 5

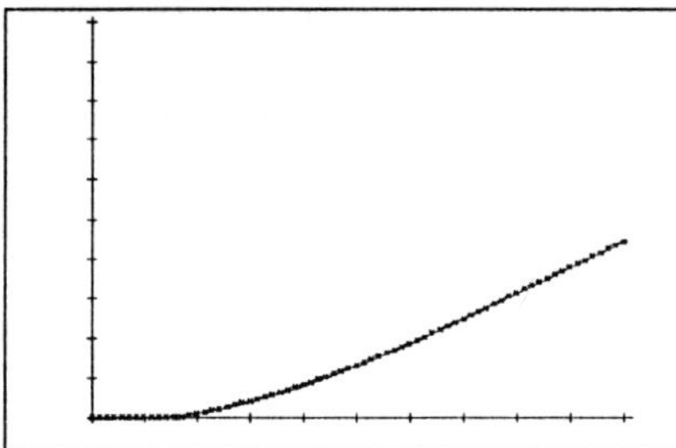

Fig. 6b Run 5: Carbon-Feed F_C

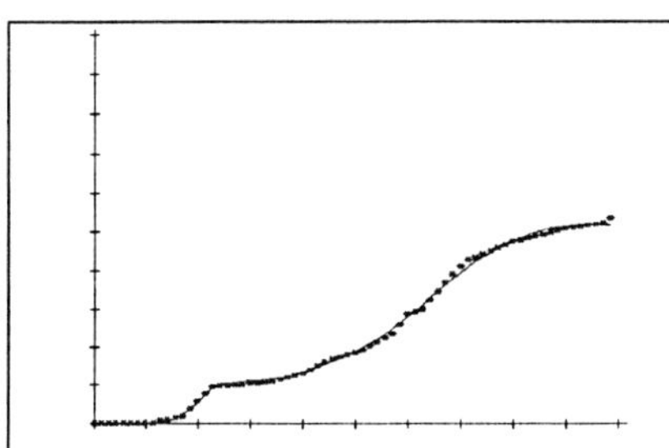

Fig. 6c Run 5: Acid Feed F_A
- measured and predicted

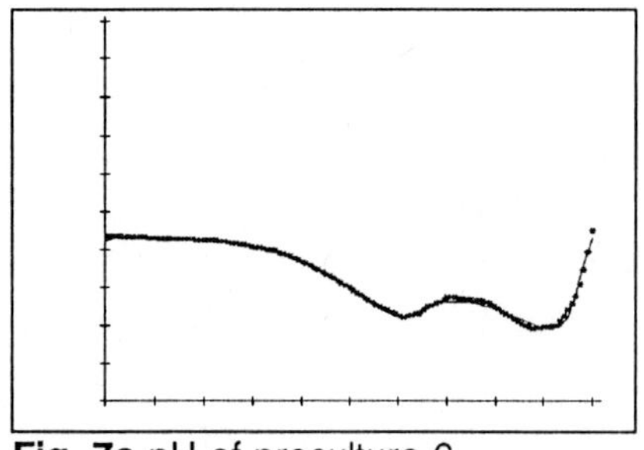

Fig. 7a pH of preculture 6

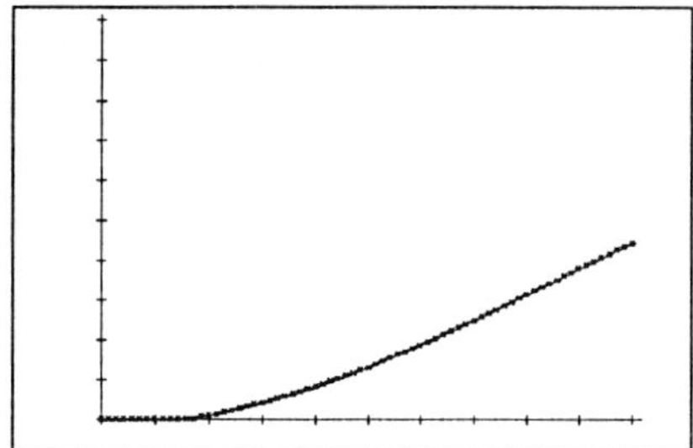

Fig. 7b Run 6: Carbon-Feed F_C

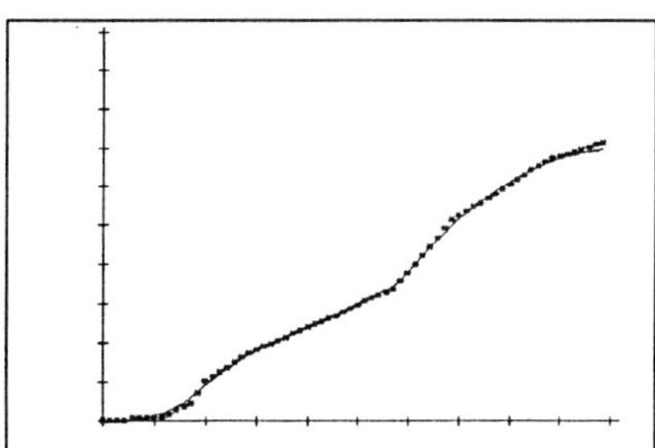

Fig. 7c Run 6: Acid Feed F_A
- measured and predicted

Identification Techniques for a Recombinant Fed-Batch Fermentation for Ethanol Production

V.M. Saucedo, B. Eikens, and M.N. Karim

Department of Agricultural and Chemical Engineering, Colorado State University, Fort Collins, Colorado 80523, U.S.A.

Optimization and control algorithms rely on mathematical models. Nonlinear models are necessary to describe biological system behavior. Two input/output modeling approaches are described for a fed-batch fermentation for ethanol production using recombinant Escherichia coli. *Volterra series shows a big improvement when an autoregressive term is added, leading to a more general nonlinear model such as NARX. Neural network models are also used for identification of the fed-batch fermentation experiments. In these experiments, neural networks outperform the estimation and prediction of Volterra series. Special attention is given to the design of experiments and modeling techniques.*

1. INTRODUCTION

System models are essential for optimization and control techniques. The accuracy, simplicity and reliability of the models used can determine the success of the operation. Many researchers disciplines are working towards the improvement of ethanol production. Ingram et. al. (1987) developed a strain of *Escherichia coli* containing plasmid from *Zymomonas mobilis* to produce ethanol from xylose, a five carbon sugar. More research needs to be done in order to make the industrial ethanol production more attractive. Classical first principle models which provide accurate predicting capabilities are sometimes difficult to obtain, and are not always conducive for implementation of advanced optimization and control strategies.

Volterra series is a tool for modeling nonlinear systems offering the advantage of estimating its parameters more accurately than classical kinetic models. NARX (Nonlinear Autoregressive with exogenous) model offers more possibilities for identification of

nonlinear systems (Chen, 1989). Statistical tools such as AIC (Akaike's information criterion) are used to select the best models, with respect to complexity and accuracy.

Neural networks is an area of Artificial Intelligence that has gained popularity for modeling and estimating nonlinear and time varying systems (Bhat and McAvoy, 1990; Karim and Rivera, 1992). In this research a multilayer perceptron (MLP) and a radial basis function network (RBFN) are applied to the identification of the fermentation process. Multilayer perceptron based on the back-propagation learning algorithm appears to be the traditional design methodology for practical neural network applications (Widrow and Lehr, 1990; Hunt *et al.* 1992). While the MLP belongs to the class of global mapping structures, radial basis function networks (RBFN) utilize local mappings. In this case, the input space is divided in localized receptive fields around center points (Poggio and Girosi, 1990). The architecture of a RBFN is similar to the feedforward neural network. These methodologies are studied and compared

275

E. Galindo and O.T. Ramírez (eds.), Advances in Bioprocess Engineering. 275-282.
© 1994 *Kluwer Academic Publishers. Printed in the Netherlands.*

for a fedbatch fermentation for ethanol production using recombinant *E. coli*.

An open loop experiment is used to generate the information necessary for model building. A series of random (white noise) manipulations yield the appropriate output. Comparative results for the modeling of the fermentation process and a discussion of the use of different identification techniques for applications in a fermentation process are given.

2. INPUT/OUTPUT MODELS

A dynamical system can be described externally by specifying how the system responds to different stimuli without specifying the internal properties. Kalman called this the external, input/output or empirical definition of a dynamical system. One basic tool employed in the external description of a non-linear system is the so-called Volterra series, which was developed early this century (Rugh, 1981).

A system described by a finite sum of homogeneous terms of the form

$$x(t) = \sum_{n=1}^{N} \int_{-\infty}^{+\infty} h_n(\sigma_1, \ldots \sigma_n) u(t - \sigma_1)$$
$$\cdots u(t - \sigma_n) d\sigma_1 \cdots d\sigma_n. \quad (1)$$

will be called a polynomial system of degree N, assuming $h_N(t_1, \ldots, t_N) \neq 0$. If a system is described by an infinite sum of homogeneous terms, then it will be called a Volterra system. The term $h_n(t_1, \ldots, t_n)$ is called the kernel associated with the system. The term homogeneous arises because the application of the input αn, where n, is a scalar, yields the output $\alpha^n x(t)$.

For nonstationary systems, the kernels are represented by $h(t, \sigma_1, \ldots, \sigma_n)$. From Schentzen theorem (Bendat, 1990), it is possible to represent a linear time varying system with a suitable transformation of a higher order system if and only if the time varying system is physically realizable.

The Volterra series representation is based on a very intuitive approach to nonlinear system description. It is natural to view the output $x(t)$ of a nonlinear system at a particular time t as depending (in a nonlinear way) on all values of the input at times prior to t. That is, $x(t)$ depends on all $u(t-\sigma)$ for all $\sigma \geq 0$.

For practical applications, the Volterra series in discrete form is as follows:

$$x(k) = \sum_{n=1}^{N} \sum_{i_1}^{\infty} \cdots \sum_{i_n}^{\infty} h(i_1, \ldots, i_n) u(k - i_1)$$
$$\cdots u(k - i_n) \quad k = 0, 1, 2 \ldots \quad (2)$$

The input signal $u(k)$ and output signal $x(k)$ are real sequences that are assumed to be zero for $k < 0$. The kernel $h(i_1, \ldots, i_n)$ is real and equal to zero if any argument is negative. The above system is stationary, causal and degree-n homogeneous. An autoregressive term is introduced to complement the output of the Volterra Series. This autoregressive term is the same input with one lag in the past, and is used as an input, i.e.

$$x(k) = a_1 x(k-1) + \text{Volterra series terms}$$

NARX models were initially introduced by Leontaritis and Billings (1985) and are defined as:

$$\mathbf{x}(t) = \mathbf{F}[\mathbf{x}(t-1), \ldots, \mathbf{x}(t-N_y),$$
$$\mathbf{u}(t-1), \ldots, \mathbf{u}(t-N_u)] + \varepsilon(t) \quad (4)$$

2.1. Kernel Identification

The determination of kernel values for an unknown system from a general input/output description is a linear problem, as long as only deterministic terms are considered (Ljung, 1987). This is more easily demonstrated for the case of a discrete time, polynomial system where for technical simplicity it is assumed that the system has a finite memory M described by the representation:

$$x(k) = h_0 + \sum_{i_1=0}^{M-1} h(i_1)u(k - i_1) +$$

$$\sum_{i_1=0}^{M-1}\sum_{i_2=0}^{M-1} h(i_1, i_2)u(k - i_1)u(k - i_2) + \cdots.(5)$$

In practice, the input-signal values $u(0),...,u(N)$ and the corresponding output signal values $x(0),...,x(N)$ are known. QR factorization methods are used for solving the LS problem, for determining the kernels of the Volterra series and the NARX models.

3. MODELING CRITERIA

For a certain value of N, there exists a model that fits a certain set of data. The "best" model represents reality as close as possible and is reproducible in similar situations. Two concepts that are helpful in selecting the best model for a particular case are the periodogram and the Akaike's Information Criterion or AIC.

The periodogram $I(\omega_j)$, a function of frequency ω_j, in an informal way, is defined for a set of n observations X_t in terms of the discrete Fourier transform by

$$I(\omega_j) = n^{-1}\left|\sum_{t=1}^{n} x_t e^{-itw_j}\right|^2 \qquad (6)$$

The resemblance of this definition with the spectral density function, suggests the importance of the periodogram as a means of estimating the spectral density function (Brockwell, 1992).

Ideally, a white noise signal, containing all the frequencies, guarantees that the system will be properly excited. In reality, few are white noise (purely random) processes, so the identification based on non-white noise as an input signal may not be accurate. One way of estimating how close a signal is to white noise is by calculating the periodogram defined above. The more the non-zero spectral values in the range of the frequency of interest, the more likely the signal will excite the system. For some systems, e.g. continuous

processes, it is possible to create a signal with proper non-zero values in frequency, but in some cases such as batch systems this may not be feasible as batch processes are basically "initial value" processes.

The problem of overfitting a model is addressed by the Akaike information criterion (AIC). This is a theoretical approach that can be applied to a variety of situations. The AIC is a function of a parameter ϕ which is the critical value of the chi-square distribution with one degree of freedom for a given significance level. Generally, for a given number of parameters q,

$$\text{AIC}(\phi) = -2log(\frac{\sigma_\varepsilon^2}{N}) + \phi q. \qquad (7)$$

A common value for ϕ is 4, according to the experience of the users (Chen and Billings, 1989). It penalizes models with many parameters. N is the number of effective observations and a σ^2 is the standard deviation of the error between the model and the real values (Priestley, 1981). If one plots AIC(ϕ) *versus* q the graph will, generally, show a minimum value, and the appropriate order of the model is determined by the value of q at which AIC(ϕ) attains its minimum value.

The AIC also tends to minimize the prediction error, i.e., it will try to minimize the error variance at $(t+h)$, $h = 1,2,...$ when the model is obtained using observations up to t.

4. NEURAL NETWORKS

As an alternative approach, neural networks are considered for modeling the fedbatch fermentation process. Two different neural network architectures have been used: the multilayer perceptron (MLP) and the radial basis function network (RBFN). The multilayer perceptron (MLP) consists of three layers of interconnected processing units. After summation of the weighted inputs, a nonlinear function, in this case the sigmoid function is used as the activation in each node. The structure of a three layer feedforward neural network consisting of three input and one output units is shown in Figure 1.

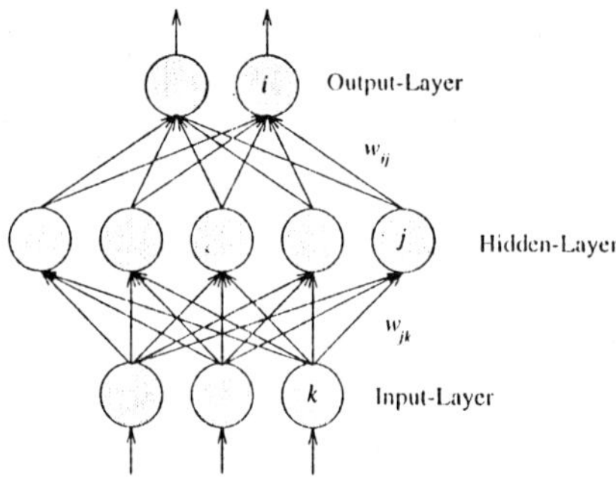

Figure 1.:   Neural network structure

In this research, the backpropagation algorithm including batching and step size optimization is utilized (Leonard and Kramer, 1990). The goal of the training procedure is to find proper weights of the network that minimizes the sum of the squared errors between the network prediction $\mathbf{y}' = [y_1', \ldots, y_N']^T$ and the desired output $\mathbf{y} = [y_1, \ldots, y_N]^T$. It has been shown that feedforward neural networks with one layer of hidden nodes and trained with backpropagation are universal function approximators (Hornik et al., 1990).

Radial basis functions (RBF) have been proposed as an alternative architecture for generating multivariate, nonlinear input-output mappings. They are traditional techniques for strict interpolation in multidimensional space. The generalized form of radial basis function is applicable to the network structure. The radial basis function network (RBFN) uses local mappings to construct the input-output space. It can be regarded as a feedforward neural network with three layers, namely an input, a hidden layer with the RBF nonlinearity and a linear output layer. With each node of the hidden layer a parameter vector, the center, is defined. Instead of evaluating just the weighted sum of the inputs, the Euclidian distance between the input pattern $\mathbf{x}$ and the center vector $\mathbf{c}$ is computed. The possible choices for input and activation functions for FNN and RBFN are shown in Table 1. The input of the hidden node is defined as q, $\Phi(q)$ denotes the nonlinearity associated with each hidden node and σ *is* the width or spread parameter for the Gaussian function.

Table 1: Comparison of a feedforward neural network and radial basis function network

Network	Node input	Activation
FNN (sigmoid)	$q = \mathbf{x}^T\mathbf{w}$	$\varphi(q) = (1 + e^{-q})^{-1}$
RBF (general)	$q = \|\mathbf{x} - \mathbf{c}\|$	$\varphi(q)$
RBF (Gaussian)	$q = \|\mathbf{x} - \mathbf{c}\|$	$\varphi(q) = e^{-q/\sigma}$
RBF (thinp.)	$q = \|\mathbf{x} - \mathbf{w}\|$	$\varphi(q) = q^2 \log q$

The centers are calculated using cluster analysis on the basis of Euclidian distance (Kaufman 1990). The initial centers $\mathbf{x}_c$ are selected based on the mean values of the input data. For each data-vector $\mathbf{x}_j$ of the input space, the dissimilarity d_{jc}, to each temporary center $\mathbf{x}_c$ is calculated. The vector $\mathbf{x}_j$, is assigned to the nearest cluster and the mean of that cluster is updated. The outline of this clustering algorithm is as follows:

1. Select a possible center candidate $\mathbf{x}_c$, from the input space.

2. Calculate the dissimilarity d_{jc} between a nonselected vector $\mathbf{x}_j$, of the input space and the center candidate $\mathbf{x}_c$, and the dissimilarity D_j, of $\mathbf{x}_j$, and the most similiar priviously selected center. As a measure of the dissimilarity, the Euclidian or Manhattan distance can be selected.

3. If the difference is positive, the input vector $\mathbf{x}_j$ will contribute to the decision to select $\mathbf{x}_c$ as a center.

4. Calculate the total gain $\Sigma_j C_{jc}$, by selecting $\mathbf{x}_c$ as a center for all vectors of the input space.

5. The center candidate which maximizes the objective function is chosen as the next center. Repeat this procedure until all k center vectors are found.

In the second step of the algorithm, the centers are adjusted to improve the clustering yielded by this set. Here the effect of changing one of the selected center vectors is calculated and a swap is carried out if the overall effect is positive. For the identification procedure in this paper, the thin-plate spline function has been chosen as the basis function.

The weights between the hidden and output layer are calculated using the gradient descent method.

5. EXPERIMENTAL SYSTEM

5.1. Materials and Methods

The strain *E. coli* ATCC 11303 containing plasmid PLOI297 was used in anaerobic conditions to convert xylose to ethanol. Tetracycline antibiotic is added to the medium to avoid instability of microorganism which do not carry the plasmid PLOI297 containing a genetic marker for tetracycline resistance.

Thus, segregation plasmid instability is avoided. The structural plasmid instability has been studied and found to be nonexistent in the strain. Cultures were grown in Luria Broth containing xylose (Hilaly, 1994; Ingram, 1988).

The fermentation system instrumentation was based on a 386 PC microcomputer equipped with an GPIB card and an HP3497A data acquisition and control system. Temperature, pH and CO_2 evolution were stored on-line. The CO_2 gas composition in the air outlet of the fermentor was measured with an infrared CO_2 gas analyzer. The amount of base added to keep the pH constant is recorded and digital signals sent to the data acquisition card. A small sample is injected on-line to a Waters HPLC unit with a R-400 Series Refractometer detector to measure the concentrations of xylose, ethanol and various acids. A Biorad HPX87H column running with a 0.01 N H_2SO_4 Solution as mobile phase was used. Optical densities were measured at 550 nm. Fermentations were carried out in a 7 liter Chemap fermenter with an initial volume of 2 liters.

5.2. Experimental Results

A random feed with 120 g/l xylose concentration was implemented during the first 50 manipulating intervals. The rest of the experiment was fed with a 200 g/l xylose concentration. The profile of feed implemented is shown in Fig. 2.

Approximately at 65 hrs, the xylose consumption rate is very high whereas the biomass and products do not follow the expected increase in concentration. The explanation is as follows. The xylose is used for cell and product formation. If we consider the volume changes shown in Figure 3, it can be explained that all components are diluted due to increasing feed rate. Hence, if the cell mass concentration shown in Figure 4 remains at a constant value, then it indeed represents an increase in biomass production. In the same manner the product concentration profiles can be explained.

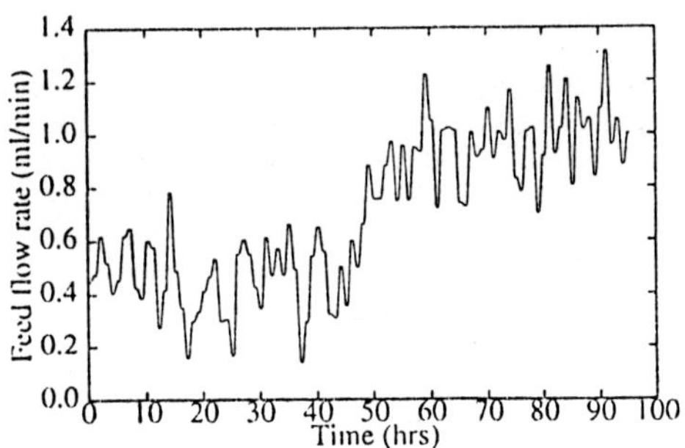

Figure 2.: Feed manipulation of the fed-batch fermentation

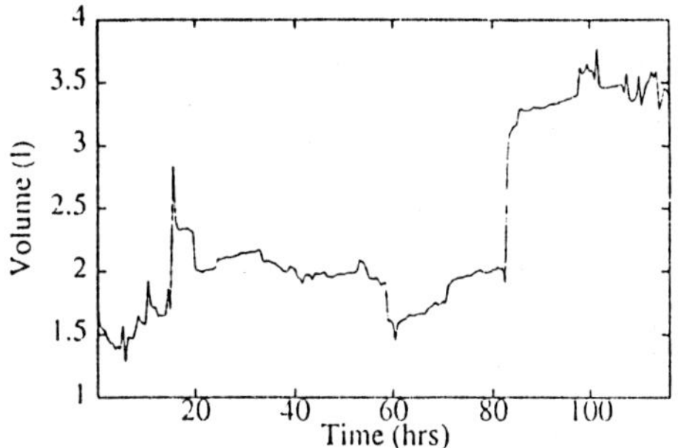

Figure 3.: Volume changes in fed-batch fermentation

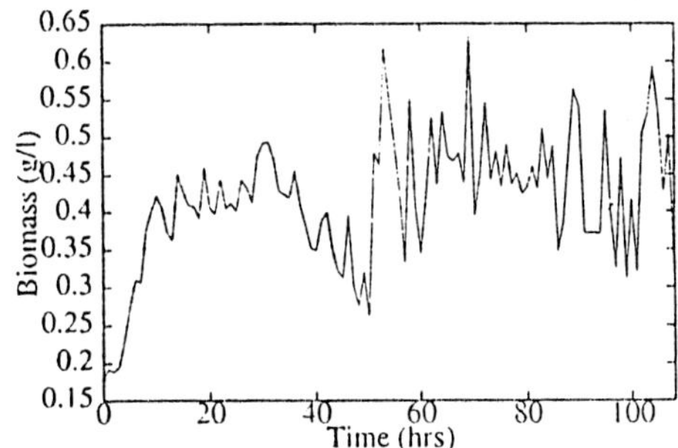

Figure 4.: Biomass changes in fed-batch fermentation

6. RESULTS AND DISCUSSION

The previously described identification techniques have been applied to the fedbatch fermentation process. The figures shown in this paper do not include the batch part of the experiment and present the obtained results for the fed-batch operation only. All models have been trained for $t< 70$ hrs. For 70 hrs $<t<$ 100 hrs, the obtained models are used as a predictor. All models implemented one-step ahead predictors. The NARX and the RBFN models show a better performance approximating the input/output training data. The sum of the squared errors is reduced significantly in these cases. Figures 5, 6, and 7 show the identification and prediction of the total acids. The radial basis function network shows a superior performance compared with the multilayer perceptron. This is not the case for the ethanol model as can be seen in Figures 8, 9, and 10. Again the NARX and RBFN based models are able to approximate the system during the training cycle accurately. The multilayer perceptron predicts the future outputs more closely. This is surprising since radial basis function networks are known to be a superior approximator. One reason for this behavior lies in the linear separability of the training and test sets. As shown in Figure 11 the centers of the RBFN are located in order to minimize the distance in each cluster of the training set. The test data are not represented by the training set and hence, the prediction is not very accurate. The multilayer perceptron is more robust with respect to linearly separable problems since it is based on a global training procedure. The situation is similar for the xylose model (Fig. 13 and 14).

Component	Algorithm	SSE
Acids	MLP	0.053333
	RBFN	0.058596
	NARX	0.052195
Ethanol	MLP	0.233381
	RBFN	1.223322
	NARX	0.248433
Xylose	MLP	0.005867
	RBFN	0.007448
	NARX	0.010127

Table 2: Comparison of the SSE for the identification procedures

The sum of squared errors (SSE) for these identification procedures are compared in table 2. It can be seen that the selected approaches give acceptable results. The overall performance in this application of the MLP is slightly superior.

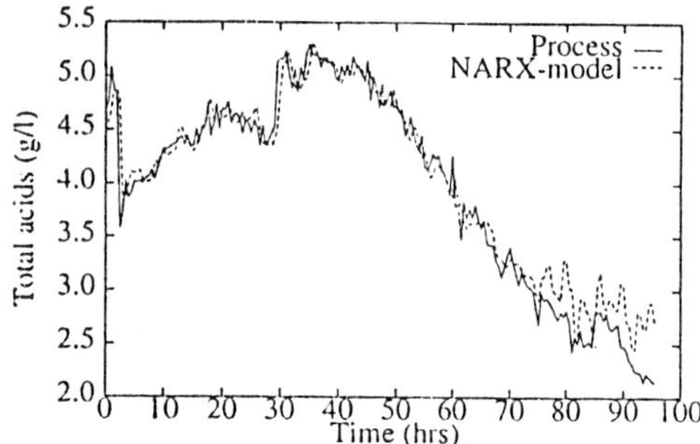

Figure 5.:   Total acids NARX model.

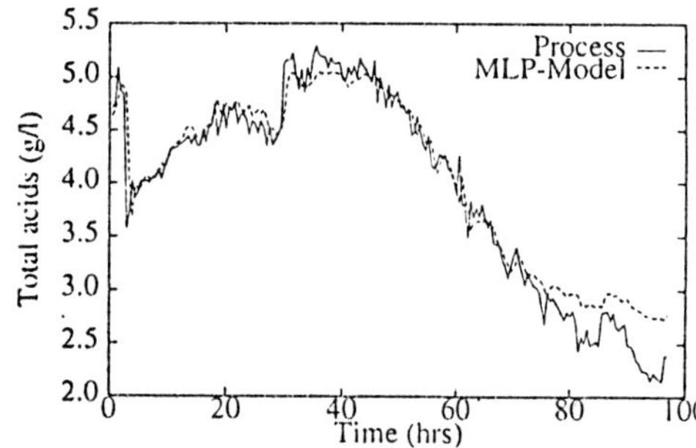

Figure 6.:   Total acids MLP model.

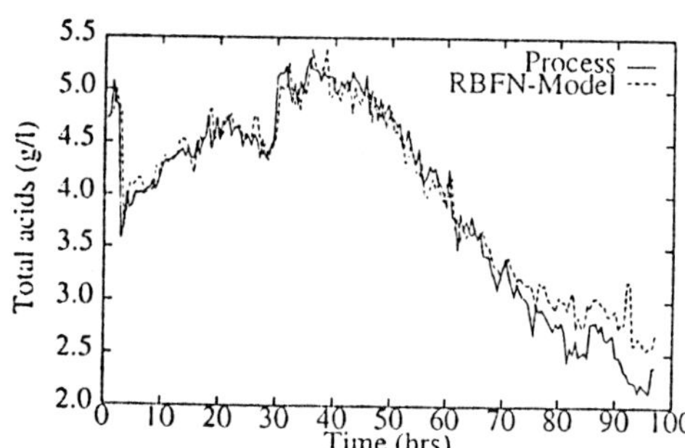

Figure 7.:   Total acids RBFN model.

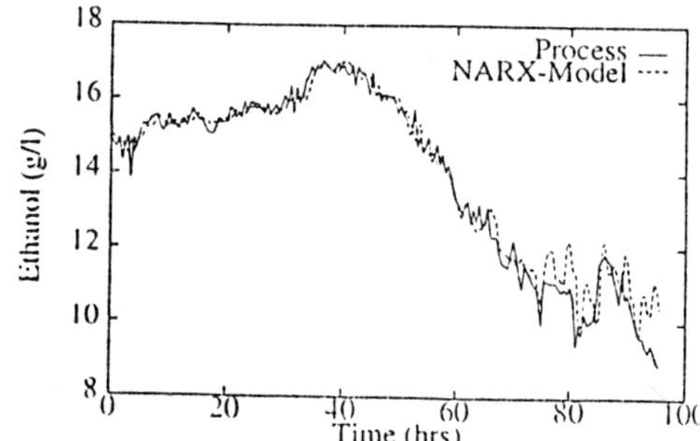

Figure 8.:   Ethanol NARX model.

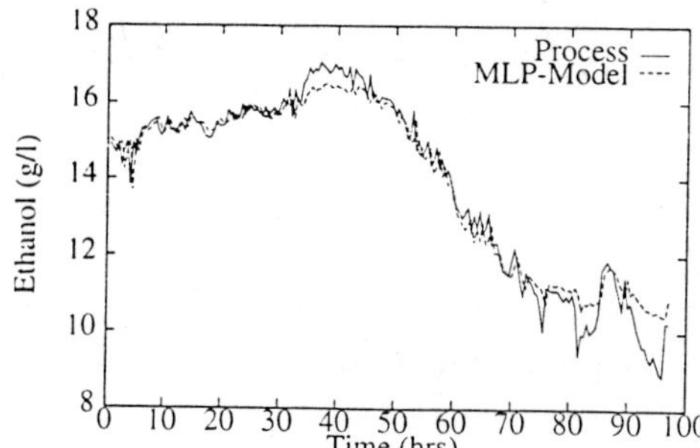

Figure 9.: Ethanol MLP model.

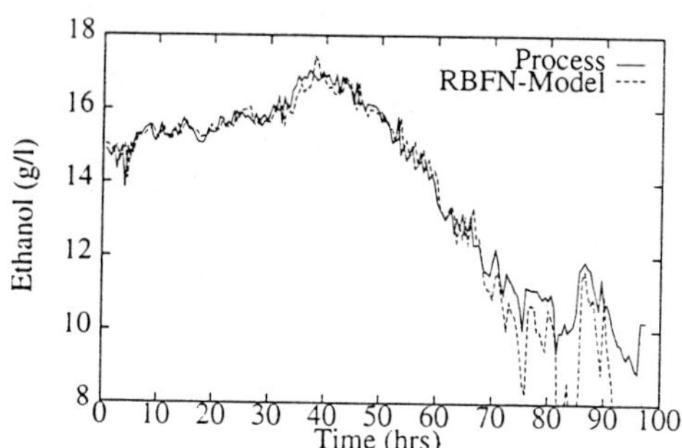

Figure 10.: Ethanol RBFN model.

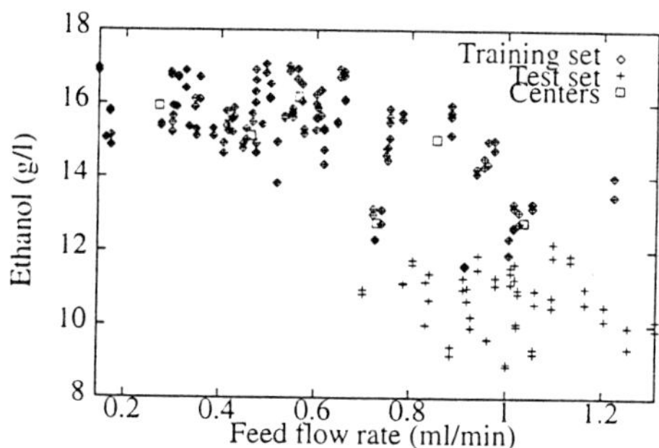

Figure 11.: Distribution of RBFN centers.

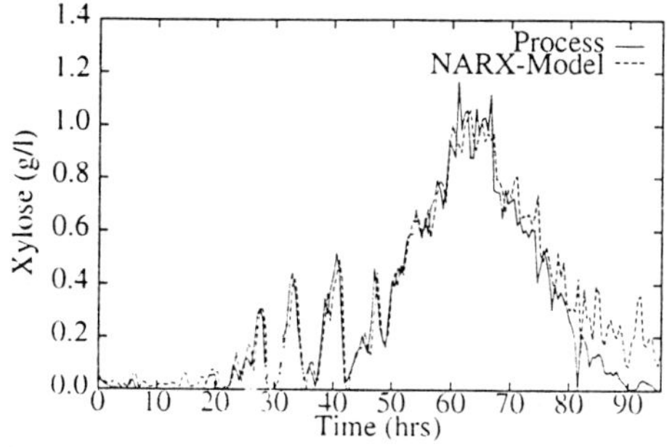

Figure 12.: Xylose NARX model.

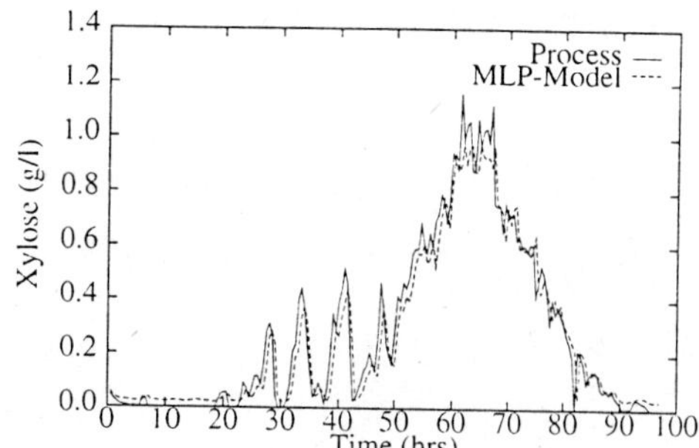

Figure 13.: Xylose MLP model.

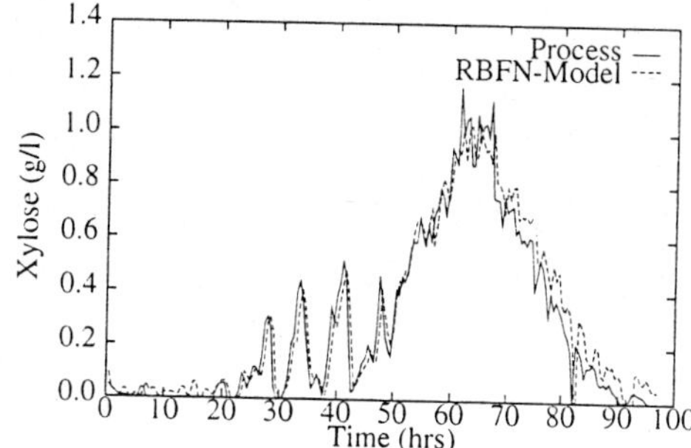

Figure 14.: Xylose RBFN model.

7. <u>CONCLUSIONS</u>

Neural networks and NARX models based on Volterra series are used in this study to effectively model a fed-batch fermentation process. The results of the identification procedures show a satisfactory performance of the neural network and the NARX approach. For this particular application the multilayer perceptron seems to be a better predictor. Nevertheless, the recombinant fed-batch fermentation is a challenging system for modeling and prediction studies. Further improvements of the prediction quality needs to be addressed in the future.

Acknowledgment: The authors wish to acknowledge partial financial support from the Colorado Institute for Research in Biotechnology (CIRB), the Colorado State Experiment Station, and the National Science Foundation (BCS - 9118955).

<u>REFERENCES</u>

Bendat, J.S. (1990). *Nonlinear Systems Analysis and Identification from*

Random Data. John Wiley & Sons, New York.

Bhat, N., and T.J. McAvoy (1990). Use of neural nets for dynamic modeling and control of chemical process systems. *Computers Chem. Engng*, **14**, 573--583.

Brockwell, P. and Davis, R.A. (1992). *Time Series: Theory and Methods*. Springer-Verlag, New York. Second edition.

Chen, S. and S.A. Billings (1989). Representation of nonlinear systems: the NARMAX model. *Int. J. of Control*, **51**, 1013--1032.

Chen, S., S.A. Billings and P.M. Grant (1992). Recursive hybrid algorithm for nonlinear system identification using radial basis function networks. *Int. J. of Control*, **55**, No. 5, 1051--1070.

Cooney, C.L., G.K. Raju and G. O'Connor (1991). Expert systems and neural nets for bioprocess operation. *International Symp. on Bioprocess Modelling and Control*, Newcastle, UK.

Hilaly, A.K., M.N. Karim and J.C. Linden (1994). Development of an automated system for state estimation of xylose fermentation by a recombinant *Escherichia coli*. *Journal of Ind. Microbiology*. (In Press).

Hunt, K.J., D. Sbarbaro, and R. Zbikowski (1992). Neural networks for control systems - a survey. *Automatica*, **28**, 6, 1083--1112.

Ingram, L.O., Conway, T., Clarck, D.P., Sewell, G.W. and Preston, J.F. (1987). Genetic Engineering of Ethanol Production in *Escherichia coli*. *Applied and Env. Microbiology*. **53**, No. 10, 2420--2425.

Ingram, L.O., Conway,T. (1988). Expression of different levels of ethanologenic enzymes from *Zymomonas mobilis* in recombinant strains of *Escherichia coli*. *Appl. Environ. Microbial*, **54**, No. 3, 397--403.

Hornik, K. (1991). Approximation capabilities of multilayer feedforward networks. *Neural Networks*, **4**, 2, 251-257.

Karim, M.N., and S. Rivera (1992). Comparison of feed-forward and recurrent neural networks for bioprocess state estimation. *Computers Chem. Engng*, **16** supp.

Kaufman, L., P.J. Rousseeuw (1990). *Finding Groups in Data - An Introduction to Cluster Analysis*. John Wiley & Sons, New York.

Leonard, J., and M.A. Kramer (1990). Improvement of the backpropagation algorithm. *Computers Chem. Engng*, **14**, 3, 337--341.

Leontaritis, I.J., and Billings, S.A.(1985). Input-output parameter models for nonlinear systems. *Int. Journal of Control*, **41**, 303--344.

Ljung, L. (1987). *System Identification: Theory for the user*. Prentice-Hall, New Jersey.

Narendra, K.S. (1992). Adaptive control of dynamical systems using neural networks. in *Handbook of Intelligent Control* (ID.A. White, D.A. Sofge, Ed), Van Nostrand Reinhold, New York.

Poggio, T., and F. Girosi (1990). Networks for approximation and learning. Proc. *of the IEEE*, **78**, 9, 1481-1497.

Priestley, M.B. (1981). *Spectral Analysis and Time Series*. Academic Press, London.

Rugh. W.L. (1981). *Nonlinear Systems Theory: the Volterra/Wiener Approach* Oxford University Press.

Widrow, B., and M.A. Lehr (1990). 30 years of adaptive neural networks: perceptron, medaline, and backpropagation. Proc. *of the IEEE*, **78**, 1415--1441.

Computer Controlled Enzymatic Reactor

L.A. Minim and R. Maciel Filho

Departamento de Procesos Químicos, Faculdade de Engenharia Química, C.P. 6066, CEP: 13081,
UNICAMP, Campinas, SP, BRAZIL

This work deals with the control and design of a medical reactor to be used in the extracorporeal treatment of leukemia in children. In this work, an adaptive controller based on state space approach is developed and tested for the control of a medical reactor, using L-asparaginase attached on the internal open nylon tube. As a medical reactor, the design must be simple and safe to be operated. It was shown that the control algorithm was adequate to maintain the system in a specified operational condition when the manipulated variable was the electric current, even when significant disturbances took place as well as set-point changes (servo control). This makes this system to be a promising method in the extracorporeal treatment of leukemia.

The enzyme L-asparaginase (L-asparagine amidohydrolase, EC 3.5.1.1) catalyses the hydrolysis of L-asparagine to L-aspartic acid and ammonia. Interest in this enzyme arose a few decades ago when it was discovered its value in the treatment of acute lymphoblastic leukemia in children (Crowter, [1]). Unlike normal cells, the malignant cells can only synthetize L-asparagine slowly and are dependent on an exogenous suply. The antineoplastic activity results from depletion of circulating polls of the aminoacid by the L-asparaginase (Gilman, [2]). The way of use has been through intramuscular applications of the free enzyme solution which must be pure. However, in many cases high doses are necessary for an effective treatment and as the enzyme is a foreign protein, troublesome side effects can arise such as nausea, vomiting and anaphylatic effects consequently limiting its use. This enzyme can be obtained from several microbial sources and recently was demonstrated the ability of *Erwinia aroideae* to produce the enzyme growing in supplemented cheese way (Minim, [3]).

Alternatively, the beneficial effects of the L-asparaginase can be used in this therapeutic without the side effects problems, if the extracorporeal treatment is used. It has been shown that the L-asparaginase covalently attached to nylon tubing are useful in clinical works. This tube method shows great promise for medical use for three main reasons: 1) it allows enzyme reactions to be carried out on a living system without the need to introduce into the systemic circulation (for example, blood could be perfused through an enzyme-activated nylon coil), 2) there is no loss of enzyme, so that a very limited quantity of enzyme can be used over extended periods of time, and 3) it may readily be incorporated into clinical systems, many of which already make use of tubes (Kobayashi and Laidler, [4]). Furthermore, we can design a system wich must be operationally simple and operated under conditions to provide high performance, stability and safe.

The ability to control enzymatic reactors at their desired point of operation accurately and automatically is of considerable interest in the design of medical reactors, since it can provide high safety performance. For the control system design of bioreactors, the following characteristics must be taken into account: 1) the lack of accurate mathematical models which describe the system, 2) the time-varying and nonlinear nature of the process and 3) the lack of reliable sensors which can detect the state variables such as the concentration of reactants and metabolites. With such a complex problem, simple control schemes without automatic adjustments in the parameters means that the efficiency of the process is low, so the use of more sophisticated adaptive control schemes

283

E. Galindo and O.T. Ramírez (eds.), Advances in Bioprocess Engineering. 283-287.
© 1994 Kluwer Academic Publishers. Printed in the Netherlands.

with a state and/or parameter estimator must be considered (Stephanopoulos and San, [5]).

Distributed parameter control problems can be treated through nonlinear stochastic techniques but the solution is very complex. On the other hand, it is possible to approximate the equations in such a way to enable linear techniques to be used, which can then use, for example, self-tuning regulators (STR). These form part of the general class of adaptive controllers dealing with uncertainties, nonlinearities and time varying process parameters (McGreavy and Maciel Filho, [6]; Maciel Filho and Minim, [7]).

In the present study we consider the problem associated with the maintenance of adequate operational conditions in an open tubular reactor with L-asparaginase attached in the internal surface, which can be used as a medical reactor in the leukaemia treatment. Using an appropriate mathematical model to represent the system, the process is simulated operating under control. A servo STR type, based in the state space approach is developed and is shown that this control algorithm is adequate to maintain the system in a specified operational condition, even when significant disturbances take place as well as a change in the set-point. Also, we have demonstrated the reconstruction of the states by the estimator, based on the Kalman filtering technique.

Process Model

The L-Asparaginase reactor consists of a nylon tube with the enzyme attached to the internal tube surface (Kobayashi and Laidler, [4]). The internal diameter is 1 mm and a variable eletric resistence is adjusted along the center of the reactor. An input eletric current is supplied at the terminals of the resistence to heat generation. The chemical reaction taking place in the system is the following:

$$L - Asparagine \overset{E}{\Longrightarrow} L - Aspartic\ Ac. + NH_3$$

where E is the L-Asparaginase enzyme. This reaction is weakly exothermic but very sensitive to temperature variation.

The mathematical equations which describe the dynamic behaviour of the system may be expressed as follow:

$$\frac{\partial C_i}{\partial t} =$$
$$D_i \frac{\partial^2 C_i}{\partial z^2} \quad v \frac{\partial C_i}{\partial z} \quad r_i \tag{1}$$

$$\frac{\partial T}{\partial t} =$$
$$\alpha_F \frac{\partial^2 T}{\partial z^2} - v \frac{\partial T}{\partial z} - \frac{4\alpha_F h_i}{k_F d_i}(T - T_T) + \frac{\alpha_F P_w}{k_F A_T dz} \tag{2}$$

with initial and boundary conditions:

$$t = 0, \ \forall\, z : C_i = 0 \qquad T = T0$$
$$z = 0, \ t > 0 : C_A = C0 \qquad C_B = 0 \qquad T = T0$$

The reaction rate is represented by the Michaellis-Menten model, which is valid for this system (Bunting and Laidler, [8]):

$$r_i = \frac{V_M(T, E)C_i}{K_M + C_i} \tag{3}$$

The maximal reaction rate (V_M) is a function of the enzyme concentration and the temperature (in the Arrhenius form). The enzyme is deactivated irreversible with the age time and temperature, expressed as the following manner:

$$E = E0 \exp\left(-K_d t\right) \tag{4}$$

$$K_d = K_{d0} \exp\left(-\frac{E_d}{RT}\right) \tag{5}$$

The system of differential equations is conveniently solved by a numerical algorithm based on a finite difference method. Details on the solution procedures are given elsewhere (Minim, [9]).

Adaptive Control Algorithm

A general strategy to design an adaptive control system is to estimate on line the model parameters and then adjust the controller parameters based on the model estimation (Seborg et al., 1986). This technique is often referred too as Self-Tuning Controller and is showed schematically in figure 1.

The control law in the STC algorithm and the adaptive estimation can be performed as follows: A

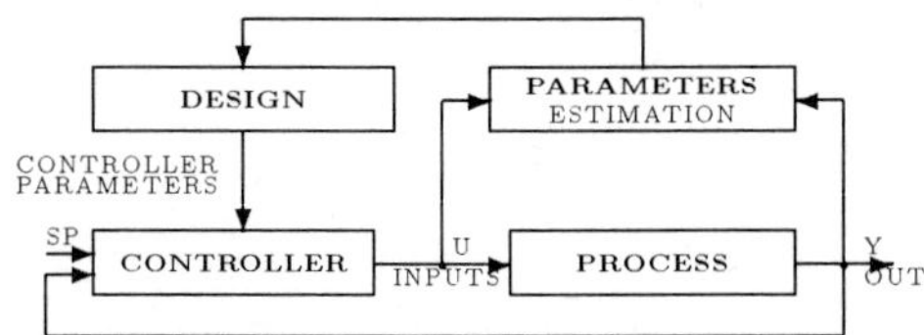

Figure 1: Self-Tuning control scheme

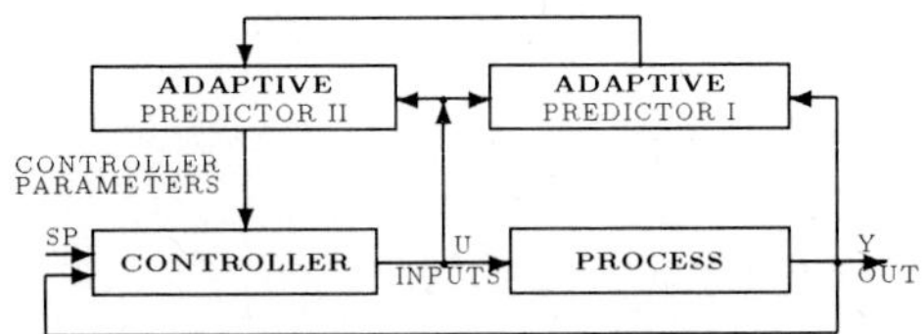

Figure 2: Improved self tuning controller

nonlinear system given by

$$x_{k+1} = f(x_k, \Phi_k, u_k) \tag{6}$$

can be modelled stochastically with varying parameters given by

$$\Phi_{k+1} = \Phi_k + w_k \tag{7}$$

so that the estimator stays permanently active if the process covariance is set to a positive number.

Linearisation of the system (eq. 6) leads to

$$x_{k+1} = \Phi_k x_k + \Gamma_k u_k + w_k \tag{8}$$

and the state measurements (y_k) are obtained through the equation

$$y_k = H x_k + \nu_k \tag{9}$$

where x are the states of the system and y are the observations.

The plant model (eq. 8), together with the state observations (eq. 9) can be used in a recursive Kalman Filter, and the following adaptive prediction is obtained:

$$\overline{x}_{k+1} = \overline{x}_k + G_{k+1}[y_{k+1} - H\overline{x}_k] \tag{10}$$

with $\overline{x}_k$ given by the linear model

$$\overline{x}_k = \Phi_k \overline{x}_{k-1} + \Gamma_k u_k \tag{11}$$

and with G_{k+1} being the Kalman Filter gain.

For the Kalman filter the calculation of G is done via the least squares method, minimizing an objetive function J which is a weighted sum of the variances of the estimation errors:

$$J = \sum_{j=1}^{n} a_{jj} P_{jj} \tag{12}$$

in which a_{jj} are weighting coefficients.

Thus, the equation for G can be written as

$$\begin{aligned} G_{k+1} = (\Phi P_k H^T \\ + \Psi N \Omega^T)(H P_k H^T + \Omega N \Omega^T + R)^{-1} \end{aligned} \tag{13}$$

with the actual value for the covariance P given by

$$\begin{aligned} P_{k+1} = (\Phi - G_{k+1}H) P_k \Phi^T \\ + (\Psi - G_{k+1}\Omega) N \Psi^T + \Gamma Q \Gamma^T \end{aligned} \tag{14}$$

The last equation indicates that the new variances are smaller than the prediction variance one step ahead.

In the same way it is possible to estimate the controller parameters by their inclusion in the form of an extended state:

$$X_k = \begin{bmatrix} x \\ x_{kc} \end{bmatrix}_k \tag{15}$$

where x_{kc} represents the elements of K_c (controller parameters).

An improved self tuning controller algorithm can be obtained if the error associated with the controlled variable and the desired value is considered as an input to the extended states. The controller gain K_c can be calculated by an embedded procedure within the Kalman filter which gives a minimum variance estimate (see Fig. 2).

For the implementation of the control strategy shown above to the system, it is necessary to find a simplified model of the form of eq. 8.

Empirical models are not suitable because they are not representative of all the phenomena taking place in the system, but a discretized linear approx-

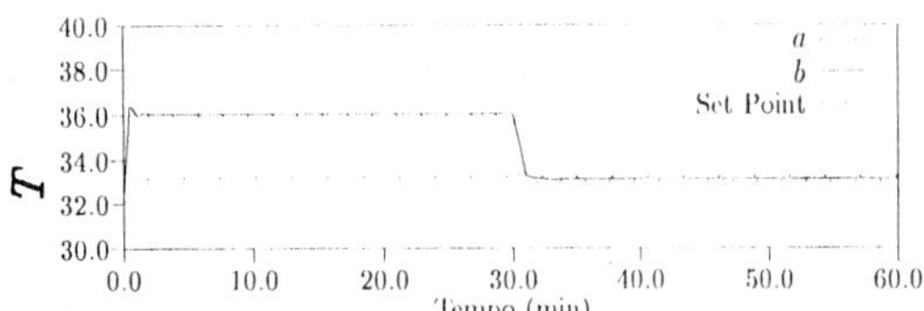

Figure 3: Fluid temperature at the reactor exit without control (a) and with control (b) using the eletric current as manipulated variable.

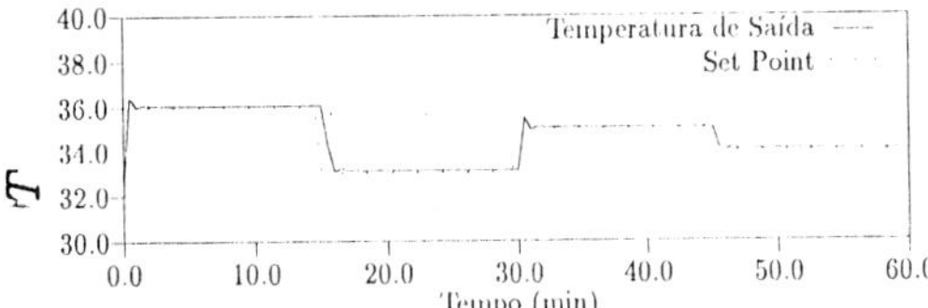

Figure 4: Fluid temperature at the reactor exit operated under control using a programmed set point change

imation can be developed which is satisfactory (McGreavy and Maciel Filho, 1989). Using a Taylor expansion series together with finite difference method makes it possible to generate a reduced model (resulting in a lumped parameter system), which is robust. Thus, the linear discrete time model then becomes

$$\frac{dx_i}{dt} = Ax_i + Bu + w \qquad (16)$$

where w is a white noise term, assumed to be a zero mean gaussian process which takes into account modelling errors, uncertainties in parameters and neglected nonlinearities.

Controller Performance

The control objective is to maintain constant the temperature at the reactor exit, despite disturbances in the input variables. The manipulated variable must be very well defined so that the system can be controllable. Only one manipulated variable was used and only the temperature was used as measured output variable so that a SISO controller was established.

The figure 3 shows the reactor exit temperature when the system is in open loop as well as when it is under control. The system is heated by an imposed

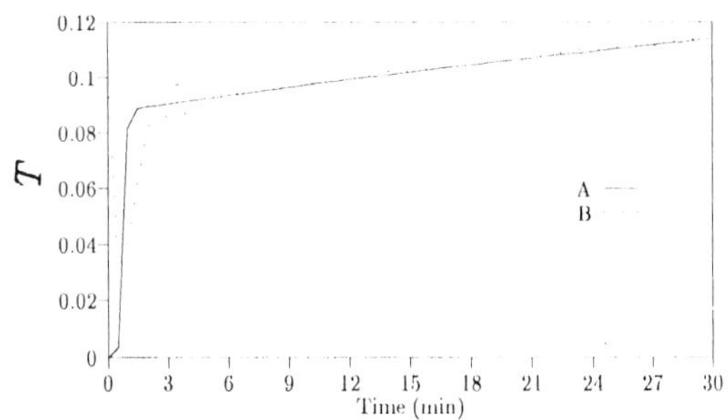

Figure 5: Reactant concentration of the process (**A**) and estimated (**B**) at the reactor exit.

initial eletric current, stabilising far from the desired value when it is without control. However, if the system is operated under control, using the eletric current as a manipulated variable, the desired condition is achieved fastly. Also, when the inlet fluid temperature is disturbed, the controller is able to drive the system to the desired conditions.

As a nonlinear and time varying system, its performance can be reduced during operation, mainly due to the enzyme deactivation. Thus, a new point of operation can be achieved and the controller must be able to track this new condition, acting as a servo controller. Figure 4 shows the performance of the controller acting as a servo controller, leading the system to the new condition fastly.

Figure 5 shows the process concentration of the reactant and the estimated by the filter. Initially, while the process outlet concentration was zero, the estimate was $0.1 moles/m^3$. After about 2 minutes the estimated concentration tracks the concentration at the process outlet. This is very important when the information of such state variables is desirable but the measurement device is time consuming or does not exists.

Conclusions

The problem of on line control of a medical reactor has been considered. The reactor is represented by a nonlinear model with time varying distributed parameters. It has been shown that the self tuning controller proposed in this work with the control strategy used here had a good performance, even under different operating conditions and using only one manipulated variable and only one measurement point (reactor exit temperature). The controller was also able to maintain the desired operational conditions when a programed set point was used. Also, good performance of the filter have been shown, making the system stable under control even under different operating conditions.

Nomenclature

C_i Reagent and product concentration $(moles/l)$;
D_A Reagent diffusional coefficient (m^2/min);
D_B Product diffusional coefficient (m^2/min);
v Reagent flow rate (m/min);
r_A Reagent reaction rate $(moles/lmin)$;
z Axial length (m);
t Time (min);
T Fluid temperature $(^\circ C)$;
α_f Thermal fluid diffusivity (m^2/min);
h_i Internal heat transfer coefficient $(cal/m^2 min^\circ C)$;
k_f Thermal fluid conductivity $(cal/mmin^\circ C)$;
d_i Internal tube diameter (m);
d_e External tube diameter (m);
P_w Electric power (cal/min);
C_0 Initial reagent concentration $(moles/l)$;
T_0 Initial temperature $(^\circ C)$;
E_0 Initial enzyme concentration$(UI/m^2$;
V_M Maximum reaction rate $(moles/lmin)$;
K_M Michaellis-Menten constant $(moles/l)$
A System coefficient matrix
B System input matrix
x States of the system
u Inputs of the system
y Output variable
Γ Input matrix
ν Measurement noise
Φ Transition matrix

Literature Cited

1. Crowter R.D. Nature,**229**, 168, (1971).

2. Gilman A.F., L.S. Goodman, T.W. Rall, and F. Murad, **As Bases Farmacológicas da terapêutica**, 7ª ed., p. 843, Editora Guanabara, (1987).

3. Minim, L.A. and R. Monte Alegre. Arq. Biol. Tecnol. (Brasil), **35**,, 2, 277, (1992).

4. Kobayashi, T. and K.J. Laidler. Biotechnology and Bioengineering, **16**, 99, (1974).

5. Stephanopoulos, G. and Ka-Yiu San. Biotechnology and Bioengineering, **26**, 1176, (1984).

6. McGreavy C., and R. Maciel Filho, "Dynamic Behaviour of Fixed Bed Catalytic Reactors", presented at the IFAC Symposium on Dynamics and Control of Chemical Reactors, Distillation Columns and Batch Processes, Maastricht, The Netherlands, (1989).

7. Maciel Filho, R., and L.A. Minim, "Performance of Control Algorithms in Industrial Chemical Reactors", presented at the IFAC Symposium on Dynamics and Control of Chemical Reactors, Distillation Columns and Batch Processes, Maryland, USA, (1992).

8. Minim, L.A., "Adaptive Control of High Performance Tubular Reactors: Applications to Enzymatic Reactors", DSc Thesis (in preparation). UNICAMP, Faculdade de Engenharia Química, Departamento de Processos Químicos, (1994).

9. Bunting P.S., and K.J. Laidler, Biotechnology and Bioengineering, **16**, 119, (1974).

10. Seborg, D.E., T.F. Edgar, S.L. Shah, AICHE J., **32:6**, 881, (1986).

Pressure Drop as a Method to Evaluate Mold Growth in Solid State Fermentors

R. Auria[1] and S. Revah[2]

[1]Departamento de Ingeniería de Procesos e Hidráulica, Universidad Autónoma Metropolitana-Iztapalapa, Apdo. Postal 55-534, 09340, México, D.F.,
[2]ORSTOM (Institut Français de Recherche Scientifique pour le Développement en Coopération), Ciceron 609, Col. Los Morales, 11530 México, D.F., MEXICO

The measurement of pressure drop (DP) across an aerated fermentation bed was used to follow Aspergillus niger *growth on Amberlite IRA-900, a synthetic resin, imbibed with a solution containing high concentrations of sucrose (Si=100, 200, 300 and 400 g l⁻¹). The DP allowed monitoring the germination, vegetative growth, limitation and sporulation for the four concentrations studied. A relation between the biomass and the relative intrinsic permeability established that the continuity of the gas phase was not broken and a regular mesh-like growth occurred. Relative intrinsic permeability as low as 0.0125 occurred at biomass concentrations of 103 mg dry biomass (g dry support)⁻¹ obtained with Si=400 g l⁻¹. Under these conditions the mold occupies 34% of the free space. The DP measurement was used to follow growth on cane bagasse and wheat bran, in both cases, the four growth phases (germination, vegetative, substrate limitation, and sporulation) were monitored.*

<u>**INTRODUCTION.**</u>

Solid State Fermentation (SSF) is an alternative cultivation method for microorganisms. Despite the increasing number of publications on applications of SSF there is little information about the engineering and technological aspects. Laukevics et al. (1) reviewed different fermentor configurations while Lonsane et al.(2) reviewed the equipment, the monitoring and control of the process, the automation and the modeling. In these papers are also included some recommendations as to the main engineering R&D areas needed for a more extensive use of SSF. Durand et al. (3) studied some of the constraints related to industrial applications. The most important limitations to SSF are process monitoring and control. Sargantanis et al.(4) studied some of the effects on the control of the SSF through the use of varying inlet air flow, humidity and temperatures.

The SSF process can be defined as a four phase system which consists of a gas, a solid support, a liquid containing the soluble substrates and a biotic phase. This multiphase heterogeneous system makes non destructive on-line monitoring more difficult than in liquid fermentation. Despite this some physical parameters can be measured directly on-line such as: temperature, gas composition, pH, and gas pressure drop. The biotic phase, either biomass or growth, has been estimated indirectly from physical measurements such as temperature, effluent gas composition (O_2 consumption and CO_2 production), composition changes as analyzed by infrared spectrophotometry and by variation of the dielectric properties and the variation in the pressure drop across the bed.

Temperature variations during the SSF are correlated to the metabolic activities of the mold and strongly determine the performance of the fermentation (4,5). Control of the temperature has been achieved by evaporative cooling (4, 6), by changing the inlet gas temperature, relative humidity (7) and by other methods (2).

Relevant information may be deduced from analyses of the gas phase. Water activity (Aw) of the medium is very important in SSF as it affects growth and germination rates for molds(5,8). It has been measured from the vapor pressure of the gas (9). Several methods have been proposed to control the water content in SSF

E. Galindo and O.T. Ramírez (eds.), Advances in Bioprocess Engineering. 289-294.

© *1994 Kluwer Academic Publishers. Printed in the Netherlands.*

($\underline{4}$,$\underline{10}$).

Oxygen uptake rate (OUR) and/or carbon dioxide production rate (CDPR) measurements ($\underline{11}$, $\underline{12}$, $\underline{13}$, $\underline{14}$) may be on- line and are directly linked to the microbial metabolism. They have been used extensively in SSF for growth estimation but to have a reliable data, yield coefficients should be evaluated for each microorganism, substrate and culture conditions ($\underline{12}$). Furthermore, these coefficients may change as a function of growth rate ($\underline{15}$).

Recently, it has been reported ($\underline{11}$) that fast biomass determination may be obtained by reflectance infrared spectrophotometry for the growth of *Beauveria bassiana* on clay granules. Another recently published technique shows a relation between growth and the change of the dielectric properties of the substrate in Tempeh fermentation ($\underline{16}$).

Changes in the pH result from the consumption of substrates and/or the production of primary or secondary metabolites and, thus, could be used as an indicator of metabolic activity. On-line measurements have been reported, ($\underline{7}$, $\underline{14}$, $\underline{16}$) but no relation to biomass or growth was established. Recently, it was shown ($\underline{14}$) that highly reproducible patterns in pH were obtained with an on- line measurement. Control has been achieved generally by initial medium formulation (increased buffer capacity, urea/ammonium sulfate relation, etc.) ($\underline{2}$, $\underline{15}$) or by addition of neutralizing agents during fermentation in agitated vessels ($\underline{7}$, $\underline{17}$).

The pressure drop (DP) defines the energy requirements for appropriate aeration. Durand and Chereau ($\underline{7}$) reported that the (DP) can be reduced through agitation. The relation between growth and pressure drop in static aerated fermentors was first reported by Auria et al. ($\underline{18}$) with a synthetic resin, (Amberlite IRA-900), used as a model support. Gumbira-Sa'id et al. ($\underline{19}$) followed the (DP) during the growth of *Rhizopus oligosporus* on sago beads. In a previous work, it was reported ($\underline{20}$) that DP across the fermentor could be used as an alternative on- line measurement for the qualitative, and under some circumstances, quantitative growth indicator. An increase in the DP was

correlated with the evolution of the different phases of *Aspergillus niger* growth: germination, vegetative growth, limitation and sporulation. With the above mentioned synthetic resin , the gas phase permeability of the bed was directly related to the biomass content up to 21.5 mg dry biomass/g dry support. Experiments with different initial sucrose solution concentrations showed that biomass could not be produced beyond this level.

In this article the DP measurement is used as an indicator of the growth of *A. niger* on SSF on a synthetic support under high cell densities and also on natural supports.

THEORETICAL BACKGROUND.

When a gas flows in a laminar regime through a porous bed, the pressure drop, (**DP**), is linked to the superficial fluid velocity (**U**) by the Darcy's equation ($\underline{20}$):

$$U = (k\ \rho_w\ g\ /\mu)\ \ (DP/L) \quad ..1$$

During microbial growth in SSF, the k varies due to the changes in the bed void fraction. The reduction of k can only be due to the increase in biomass which reduces the volume occupied by gas phase. In this case, the space utilization of fungal cells may be characterized by packing density Ψ, expressed the following way:

$$\Psi = V_x/V_p = (\rho_h\ (1-W)/\mathcal{E}_o\ )\ (X/\rho_s)....2$$

MATERIALS AND METHODS

Microorganism. *Aspergillus niger* No. 10, reported previously by Raimbault ($\underline{15}$) was used in this study. Conservation of the strain and spore production were reported ($\underline{18}$, $\underline{20}$). **Support.** Amberlite IRA- 900 (Rohm and Haas, Philadelphia, PA, USA) was used as a model support. The description and pre treatment of the support have been reported ($\underline{18}$). **Fermentation.** For all experiments the inoculum was 10^8 spores per gram dry support. The support was imbibed with the nutritive medium and the spore solution. The medium composition was: sucrose, 400g/L ; NH_4SO_4, 33.025 g/L; urea, 15.00 g/L; K_2HPO_4, 12.3

g/L; $MgSO_4$, 6.15 g/L; KCl, 6.15 g/L; $FeSO_4$, 0.123 g/L. For some experiments, the initial sucrose solution concentration (S_i) used were 100 g/l, 200 g/l and 300 g/l keeping the same $(NH_4)SO_4$/urea ratio and a **C/N** of 12. In addition, a solution of trace elements was used to provide the mineral nutrients with this composition: H_3BO_3 0.0199 g/L; Cu_2SO_4, 0.0199 g/L; KI, 0.0039 g/L; Fe_2Cl_3, 0.0799 g/L; $MnSO_4$, 0.0159 g/L; $ZnSO_4$, 0.0159 g/L; Na_2MoO_3, 0.0079 g/L which varies with sucrose concentration. The pH was adjusted to a 2.7 value. The initial water content and the temperature were respectively 58 % and 30 °C.

The experimental setup was presented earlier ([20]). One fermentor was provided with two side arms to measure the pressure drop through the bed with a simple **U** tube manometer that detects pressure variations of 0.5 mm H_2O. Pressure drop (**DP**) is expressed in mm H_2O height. Some fermentors, as those described by Raimbault ([15]) were used to evaluate biomass production. These fermentors had 2 cm I.D. and 8 cm bed height. Aeration was kept constant with a flow controller (SC440, Veriflo Corp. Richmond GA, USA). The aeration rates, used for the biomass and pressure drop measurements were respectively 1.3, 2.7 and 4.0 l air/L fermentor∗min.

Analysis.- Dry biomass was obtained from the protein measurement using a factor of 0.21 g protein/g dry biomass ([18]) measured by the Lowry method. For this study, an improvement of the technique for the biomass extraction by Auria et al ([20]) was made. One gram of humid resin was suspended in 7 ml distilled water. This mixture was agitated with an magnetic stirrer for five minutes. Twenty eight milliliters of glycerol was added to obtain a solution at 80 % (v/v) and the mixture was centrifuged at 4,500 rpm during 20 min. Under these conditions, a better separation between the biomass and the resin was obtained. For the determination of protein by the Lowry method, the liquid containing the biomass was recuperated. A NaCl solution at 1% was added to the mixture and centrifuged. After precipitation of the biomass it was wash three time with distilled water, filtered and dried. For the estimation of the dry biomass weight,

the liquid containing biomass was filtered and dried. Biomass, **X**, is reported as mg dry biomass/g dry support. Fermentations on cane bagasse and wheat ban were made as reported previously. ([14], [20])

RESULTS AND DISCUSSION

Figure 1 shows the evolution of the pressure drop (DP) and the dry biomass concentration (X) with time for the growth of *A. niger* N.10 on Amberlite IRA-900 with S_i= 200 g/L. A clear relation between the increase of these variables is observed. For the case of the DP the results shown correspond to two independent experiments made by duplicate. The DP behavior follows that found previously ([20]) for an S_i= 65 g/L: with an initial DP of 0.5 mm H_2O. The first 13 hours correspond to the germination phase and the DP is only due to the initial packing. A second phase, from 13 to 27 hours, corresponds to the vegetative growth where the free inter particular space is occupied by the mycelia, the DP increases as more biomass is produced. A third phase, from 27 to 31 hours corresponds to the limitation phase, the substrate is depleted and no growth is observed. In the fourth phase, beyond 31 hours, sporulation begins, the DP increases by the reduction of the free space by the formation of vesicle bearing conidia which can attain up to 100 μm. A fifth phase, which was not followed, occurs at longer times and signals the end of sporulation with a DP that stabilizes and then decreases.

Maximum values of X (55 mg d.b./ g d.s.) correspond to the DP (10 mm H_2O) of the limitation phase. This value of X was obtained from protein measurements and by direct dry weight. By comparison with previous results ([20]) with Si = 65 g/L the same trend is observed but higher final values of biomass were obtained (55 mg d.b./ g d.s. with S_i= 200 g/L *vs* . 21.5 mg d.b./ g d.s. with S_i= 65 g/L).

Similar experiments were made also with S_i= 100 g/L, 300 g/L and 400 g/L. The results of the evolution of the DP for these different initial Si is presented in **Figure 2**. The results are in DP units (mm H_2O) for the column but are shown in two scales to highlight the variations between the initial Si. The four curves show the

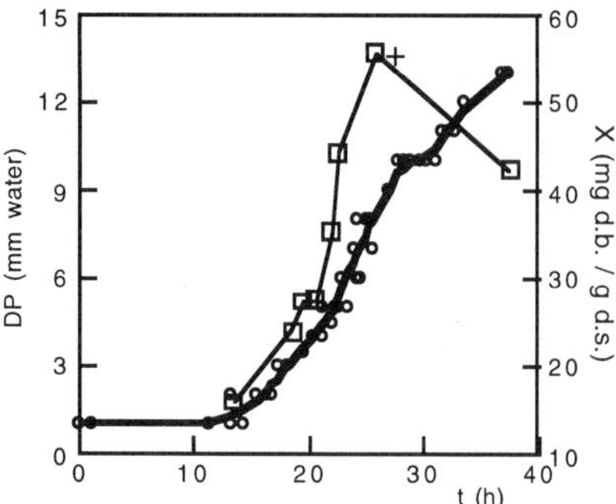

Figure 1.- Pressure drop (**DP**) and dry biomass concentration (**X**) variations with time for the growth of *A. niger* No. 10 on Amberlite IRA- 900 for initial sucrose solution concentration of 200 g/L.(°) Pressure drop; (+) Dry weight biomass; () Biomass concentration obtained from protein.

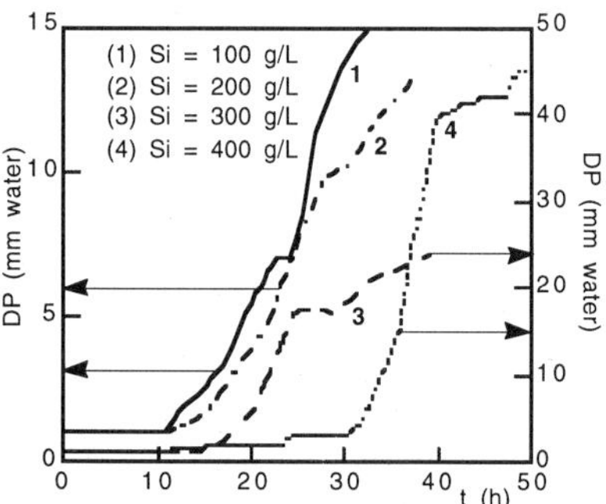

Figure 2.- Variation of pressure drop (**DP**) with fermentation time for the growth of *A. niger* No. 10 on Amberlite IRA- 900 for different initial sucrose solution concentration, S_i=100 g/L; 200 g/L; 300 g/L; 400 g/L.

same evolution corresponding to the phases described above. As it can be observed, the germination phase is retarded as the Si is increased. The DP at the onset of the limitation phase increases as Si is higher which corresponds to more biomass (17 mg db/gds. for Si= 100 g/L; 55 mg db/gds. for Si= 200 g/L; 74 mg db/gds. for Si= 300 g/L and 103 mg db/gds for 400 g/L).

Figure 3 shows the relation between the biomass and the relative permeability (k/kin). The (k/kin) was obtained from equation 1 as described previously (<u>20</u>) and using an initial DP of 0.5 mm H_2O. The data utilized for this graph is only from Si of 100 g/L, 200 g/L and 400 g/L. In the previous publication, (<u>20</u>) the maximum

d.s. and a straight line was used to establish the relation between (k/kin) and biomass. In **Figure 3** the experimental data beyond 21 mg d.b./ g d.s. showed an asymptotic trend. Even with a five- fold variation in the biomass, the continuity of the gas phase was not broken, this behavior correspond well to the mesh- like growth of the mycelia.

The relation between the dry biomass concentration and the biomass packing density is described by equation 2. The value of Ψ includes the effect of the density (ρ_X) and the biomass dry matter content (W_S). This two variables are very difficult to measure *in situ* , where they are relevant. In the literature the limits for these variables have been established. These limits are, for $0.105 < \rho_S < 0.405$ g/cm^3 with a mean of 0.24 g/cm^3. Under these circumstances the Ψ may have a very wide spread, for example for the maximum biomass achieved (103 mg d.b./ g d.s.) the mean value for Ψ is 34% but increases up to 78% for the high values and decreases to 20% for the low values. Although this spread is big, it is not probable that the mold changes its density that much along the growth phase. Despite this fact, the experiments proved to be highly reproducible. The value found (Ψ=34%) is similar to that reported by Laukevics et al.(17) as a maximum value before steric hindrances are found.

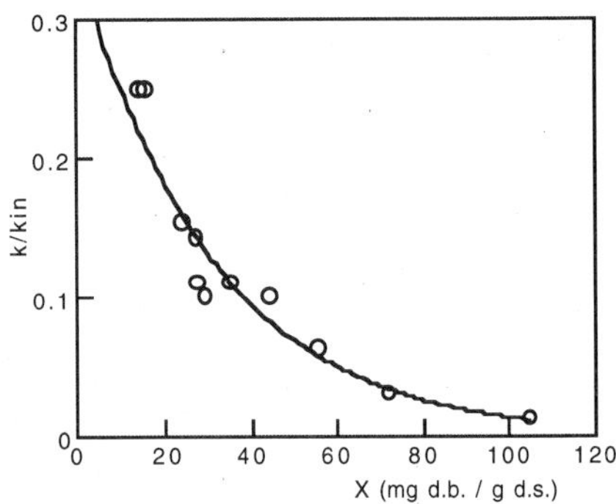

Figure 3.- Variation of the relative permeability, **k/k$_{in}$** (o) with dry biomass concentration (**X**) for different initial sucrose solution concentration, S_i=100 g/L; 200 g/L; 400 g/L.

From interpolation from **Figure 3** and using the DP values for the data of Si= 300 g/L, the values of biomass were calculated and are plotted in **Figure 4.** It can be seen that the DP can predict the biomass when used as an interpolating tool.

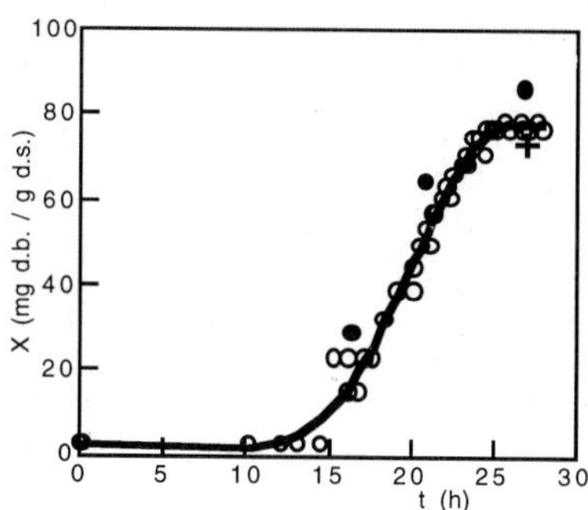

Figure 4.- Dry biomass concentration (**X**), variation with time for the growth of *A. niger* No. 10 on Amberlite IRA- 900 with initial sucrose solution concentration of 300 g/l. (o) Estimated values from pressure drop (**DP**); (+) Dry weight; (•) Dry weight from protein.

Figure 5 shows the evolution of DP for other supports. The behavior is similar for the three supports, the sporulation phase was shortened for the Amberlite while spore formation was still observed for the other two supports beyond 60 hours. The DP trend has been shown also for *A. niger* ([14]) under different gas environments. The three studied supports have an insoluble component that allows the porous structure to be maintained during growth. These components are the resin for the Amberlite and the cellulose-lignin complex for the wheat bran and the bagasse. When this structural component is missing, such as the work reported by Gumbira Sa'id et al. ([19]) who used sago, (a starch), two opposing actions occur: on the one hand, the mold occupies the free space by the hyphal growth and, on the other hand, the insoluble starchy substrate is consumed thus increasing the free space. Furthermore, the support looses structure and a collapse is observed. In their experiments Gumbira Sa'id et al. observed an initial increase in DP, up to 36 hours and a subsequent decrease while there was still growth. The authors explain this decrease by a reduction in the bed height and a contraction in the fermented biomass which favored channeling.

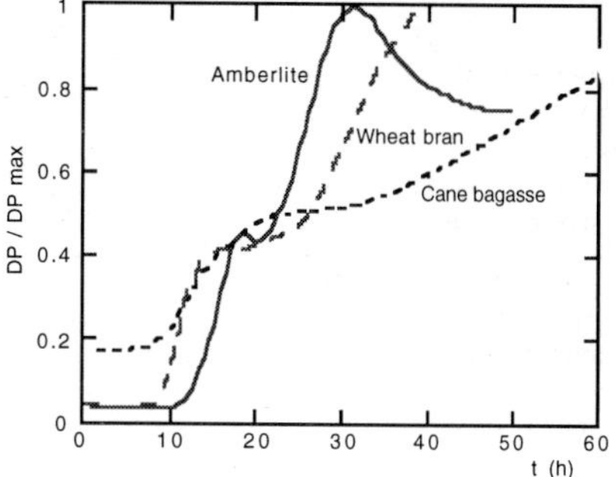

Figure 5.- Variation of relative pressure drop (DP/DP max) with time in the fermentation of *A. niger* No.10 on Amberlite and cane bagasse, and *A. niger* ANH-15 on wheat bran.

CONCLUSION

In this work the DP was used to monitor the germination, the vegetative growth, the substrate limitation and the onset of sporulation for *Aspergillus niger* under initial substrate solution concentrations ranging from 100 g/L to 400 g/L. The response of the DP followed the same behavior that was found earlier for lower substrate concentrations even with a five-fold increase in biomass. The relative intrinsic permeability decreased up to a value of 0.0125 as the biomass occupied 34% of the free inter particular space with a biomass concentration of 103 mg. d.b./ g.d.s. The technique has been also applied to other substrates that have a structural stability. This study shows that DP can be a valuable technique to study the growth of molds at high initial substrate concentrations and for a variety of substrates.

ACKNOWLEDGEMENTS

This work was performed under research agreements between the UAM (Mexico) and the ORSTOM (France). The authors acknowledge CONACYT for their contribution to the financial support of this study.

NOTATION

A: Bed cross sectional area (cm^2)
DP: **Pressure** drop between the inlet and the outlet of the porous bed (cm)
g: Acceleration due to the gravity (cm/s^2)
k: Intrinsic permeability(cm^2)
L: Bed length (cm)
Q: Volumetric flow rate (cm^3/s)
U= Q/A, superficial fluid velocity (cm/s)
V_p: Interparticular volume (cm^3)
V_x: Wet volume biomass (cm^3)
W: Support water content (g/g)
W_s: Biomass dry matter content (g/g)

Greek Symbols

ε_0: Initial interparticular porosity (cm^3/cm^3)

μ: Fluid dynamic viscosity (g/cm.s)

$\rho_s = \rho_x W_s$, apparent dry biomass density (g dry biomass/ cm^3 wet biomass)

ρ_h: Apparent wet support density (g/ cm^3)

ρ_x: Wet biomass density (g/ cm^3)

ρ_w: Water density (g/cm^3)

REFERENCES

1. Laukevics J.J., Apsite A., Viesturs, U. and Tengerdy R. (1984) Solid Substrate Fermentation of Wheat Straw to Fungal Protein.Biotechnol. Bioeng. **26;** 1465- 1474
2. Lonsane B.K., Saucedo-Castaneda G., Raimbault M., Roussos S., Viniegra-Gonzalez G., Ghildyal N.P., Ramakrishna M., Krishnaiah M.M. (1992) Scale-up strategies for solid state fermentation systems. Process Biochemistry. **27**: 259-273.
3. Durand A., De la Broise D., Blachère H. 1988. Laboratory scale bioreactor for solid state processes.J Biotech. **8**: 59-66.
4. Sargantanis J., Karim M., Murphy V., Ryoo D., Tengerdy (1993) Effect of Operating Conditions on Solid Substrate Fermentation. Biotechnol. Bioeng. **42;** 149- 158.
5. Narahara, H., Koyama, Y., Yoshida,T. Pichangkura, S., Ueda, R., Taguchi, H. (1982) Growth and enzyme production in solid-state culture of *Aspergillus oryzae*. J. Ferment. Technol. **60**: 311-319.
6. Barstow, L.M., Dale, B.E., Tengerdy, R.P. (1988) Evaporative temperature and moisture control in solid substrate fermentation. Biotechnol. Tech. **2**: 237-242.
7. Durand, A, Chereau, D. (1988) A new pilot reactor for solid state fermentation: Application to the protein enrichment of sugar beet pulp. Biotechnol. Bioeng. **31**: 476-486.
8. Oriol, E., Schettino, B., Viniegra-Gonzales, G., Raimbault, M. (1988) Solid-state culture of *Aspergillus niger* on support. J. Ferment. Technol. **66**: 57-62.
9. Gervais, P. (1989) New sensor allowing continuous water activity measurement of submerged or solid-substrate fermentations. Biotechnol. Bioeng. **33**, 266-271.
10. Sato K., Nagatani, M., Sato, S. (1982) A method of supplying moisture to the medium in a Solid-state culture with forced aeration. J. Ferment. Technol. **60**:607-610.
11. Desgranges, C., Georges, M., Vergoignan, C., Durand, A. (1991) Biomass estimation in solid state fermentation. Appl. Microbiol. Biotechnol. **35**: 206-209.
12. Sato, K., Nagatani, M. , Nakamura, K.I., Sato S. (1983) Growth estimation of *Candida lipolytica* from oxygen uptake in a solid state culture with forced aeration, J. Ferment. Technol. **61**: 623-629.
13. Weng, Y., Hotchkiss, J.H. (1991) Headspace gas composition and chitin content as measures of *Rhizopus stolonifer* growth. J. Food Sci. **56**: 274-285.
14. Villegas E., Aubague S., Alcantara L. Auria R. y Revah S. (1993) Solid State Fermentation: Acid Protease Production in Controlled CO_2 and O_2 Environments. Biotech. Adv., **11**, 387- 397.
15. Raimbault, M. (1980) Fermentation en milieu solide: Croissance de champignons filamenteux sur substrat amylacé, Thèse de Doctorat, U.P.S. Toulouse, France.
16. Davey, C.L., Peñaloza, W., Kell, D.B., Hedger, J.N. (1991) Real-time monitoring of the accretion of *Rhizopus oligosporus* biomass during the solid-subtrate tempeh fermentation. World Journal of Microbiology. and Biotechnology. **7**: 248-259.
17. Laukevics, J., Apsite, A.F., Viesturs, U.S., Tengerdy, R.P. (1985) Steric hindrance of growth of filamentous fungi in solid substrate fermentation of wheat straw. Biotechnol. Bioeng. **27**: 1687-1691.
18. Auria, R., Hernandez, S., Raimbault, M., Revah, S. (1990) Ion exchange resin: A model support for solid state growth fermentation of *Aspergillus niger*. Biotechnol. Techn. **4**: 391-396.
19. Auria R., Morales M., Villegas E. and Revah S.(1993) Influence of Mold Growth on the Pressure Drop in Aerated Solid State Fermentors. Biotechnol. Bioeng. **41**, 1007-1013
20. Gumbira Sa'id E., Greenfield P. F., Mitchell D. A and Doelle H. W. (1993) Operational parameters for packed beds in solid- state cultivation. Biotech. Adv. **11**, 599- 610

Computer-Aided Design of Integrated Biochemical Process

D.P. Petrides[1] and J. Calandranis[2]

[1]Dept. of Chemical Engineering, New Jersey Institute of Technology, Newark, NJ 07102;
[2]Intelligen, Inc. 2326 Morse Ave., Scotch Plains, NJ 07076, U.S.A.

As a result of the advances in molecular biology and genetic engineering, the scientific community has come to realize the great potential for developing new products and systems through novel use of microorganisms and enzymes. The challenge for the biochemical industry is now to scale-up and commercialize those products. This is a difficult task, especially for small corporations, because it is a complex problem that requires coordination of a large number of activities across many disciplines. Computer-aided process design tools have been successfully used in the chemical process industries for over three decades to scale-up and optimize integrated processes for the production of petrochemicals and other products. Similar benefits can be expected from the use of such tools in the biochemical industries. This paper describes the architecture and key components of computer-aided bioprocess design tools with particular emphasis on BioPro Designer. A case study on β–galactosidase production illustrates how such tools can be used in an industrial environment to analyze and optimize integral biochemical processes.

WHAT IS COMPUTER-AIDED PROCESS DESIGN?

Design is the creative process that leads from the identification of a need to an end product, system, or process that satisfies that need. Process design consists of two main subactivities, *process synthesis* and *process analysis*.

Process synthesis deals with the selection and assembly of a set of process steps capable of producing the desired product and satisfying a number of product quality, economic and environmental/safety constraints. The result of process synthesis is a number of feasible flowsheets.

Process analysis deals with the analysis and evaluation of individual or integrated processes. The results of process analysis for a flowsheet commonly include the calculation of material and energy balances, estimation of size and cost of equipment, economic evaluation, process scheduling and environmental and product/process safety evaluation.

The two subactivities of process design are not independent. Process design is an evolutionary activity that requires a chain of synthetic and analytic tasks. Based on the results of process analysis, for instance, one may want to modify either the structure or the operating conditions of a flowsheet aiming at a better solution. This flowsheet modification is a synthetic activity.

The automation of process synthesis and process analysis using computers is the focus of computer-aided process design. Automated design tools have found limited applications in the bioprocess industry till now because: 1) the biotech industry is relatively new and primarily dominated by scientists that are not process oriented; 2) most biological products, especially the high value ones, are complex and labile materials (e.g., proteins and polysaccharides), poorly

295

E. Galindo and O.T. Ramírez (eds.), Advances in Bioprocess Engineering. 295-303.
© *1994 Kluwer Academic Publishers. Printed in the Netherlands.*

characterized with respect to physical properties (e.g., diffusivity, solubility, viscosity, etc.) and these values are difficult to predict from thermodynamic and micro transport models; 3) several important bioprocess unit operations are poorly understood and predictive models which could be used for design do not exist; 4) most bioprocesses operate in batch or semicontinuous mode introducing time as an independent variable.

Benefits of computer-aided bioprocess design?

Computer-aided process design tools can play a key role at various stages of product and process development. The benefits from the use of such tools vary depending on the type of product, the size of the investment, the location of the facility, and other parameters.

Product and project selection. During project selection, they can be used to rapidly evaluate a large number of project ideas and discard the non-promising ones. This can drastically reduce engineering time because experience has shown that less than 5% of new project ideas are ever commercialized (1).

R&D planning. In research and development they can be used to plan and prioritize pilot plant work. More specifically, before serious process development is initiated, the engineer develops on the computer a conceptual model of the entire process that requires scale-up. Through the initialization steps, the user identifies all the missing information (e.g., partition coefficients for extractors, rejection coefficients for membrane filters, etc.) that is needed to develop an accurate model of the entire process. Then, based on the missing information, the engineer designs the appropriate experiments to gather that data and complete the modeling work. Having a good model of the entire process, the engineer starts asking "what-if" questions

and carrying out sensitivity analysis and optimization with respect to key design variables. If simulation yields any unexpected results, experiments are planned to test them. This systematic procedure of designing and running pilot plant experiments saves development time and yields more efficient and robust processes.

Environmental and safety issues. Environmental and process safety problems are best solved when such issues are considered during the early stages of process development and built into the initial process design. Computer-aided process design tools, when equipped with the appropriate features, enable engineers to consider the environmental impact of new products and processes at the early stages of process development when it is still easy to make process modifications (2). This is very important especially for therapeutic biological products because after a new product is approved as safe and effective by the Food and Drug Administration (FDA) it is very costly to make process modifications. Using such tools, for instance, the engineer can systematically select extraction solvents or chaotropic agents for solubilization of inclusion bodies that meet all process requirements and at the same time are environmentally benign. The engineer also can readily evaluate process modifications (i.e., addition of recycle streams, replacement of chemicals, etc.) that minimize waste generation.

Improved communication. When such tools are used by all groups involved in process development and design, they introduce a common language of communication that drastically reduces duplicate work. In a sense, their use constitutes a first step towards concurrent engineering and integrated product development with reduced cost.

<u>Retrofit of existing facilities.</u>
Computer-aided process design tools are not only useful for designing new processes but also for retrofitting existing ones. More specifically, they can be used to evaluate process modifications of existing manufacturing facilities in order to increase capacity, reduce production cost, reduce waste generation (and consequently cost of waste treatment and disposal), improve process safety, etc.

Historical review

Work in computer-aided process design began some 40 years ago. In the 1960's and most of the 1970's, research focused on the analytic aspects of design resulting in the development of powerful steady-state process simulators, such as FLOWTRAN developed at Monsanto (<u>3</u>) and ASPEN developed at MIT (<u>4</u>). In the 1980's, there also was development of dynamic process simulators, such as SPEED-UP at Imperial College in England (<u>5</u>).

Process simulators are mainly used to analyze and evaluate complex processing systems, carry out sensitivity analysis, and optimize given flowsheets. They have been successful in minimizing production cost of existing plants and improving the design of new plants. These tools, however, lack synthetic capabilities.

Work on automation of synthesis of entire flowsheets began in the late 1960's with Prof. Rudd and his students at the University of Winsconsin (<u>6</u>). They automated a methodology for the synthesis of initial flowsheets, based on heuristics and the *means-ends* analysis technique. Their program did not find industrial applications.

Recent approaches make use of Knowledge Based Expert Systems (KBES). KBES capture experiential design knowledge that is primarily used in process synthesis to constrain the solution space. However, KBES do not offer the best environment to handle algorithmic and "number crunching" tasks. Conventional programming languages are more appropriate for those tasks. To combine the best of the two approaches, hybrid systems are often developed. In such systems, a KBES that handles synthetic tasks is interfaced to a number crunching program that handles the analytic tasks of design.

Substantial research and development efforts in computer-aided design of integrated biochemical processes began in the mid 1980's following the development efforts and commercialization of new high value bioproducts.

<u>BioProcess Simulator (BPS)</u>. BPS is an extension of the Aspen Plus process simulator (<u>7</u>). It uses the infrastructure and the facilities provided by Aspen Plus. BPS, for a given flowsheet, carries out material and energy balances, estimates the size and cost of equipment, and carries out an economic evaluation.

<u>BioSep Designer</u> (<u>8</u>,<u>9</u>). BioSep Designer is a Knowledge Based system that carries out synthesis of protein separation systems. It is written in Common Lisp and runs on the Symbolics computer. Based on a given set of input data, BioSep Designer generates possible flowsheets and employs an evaluation technique to find the best one. Its synthetic knowledge base contains information mainly focused on the production of proteins by *E.coli*. It has facilities to estimate a number of protein properties from their amino acid sequence. It provides a database to store protein properties. BioSep Designer also includes some analytic and economic evaluation capabilities.

<u>BioDesigner</u> is a design tool that combines synthetic with analytic capabilities and the emphasis is on

interactive analysis (<u>10</u>,<u>11</u>,<u>12</u>,<u>13</u>,<u>14</u>). It carries out synthesis using a KBES based on the Nexpert *Object* (from Neuron Data, Inc., Menlo Park, California) expert system shell. The knowledge utilized for process synthesis is based on the properties of the producing microorganism, product, contaminants, etc. BioDesigner runs on personal computers and features an interactive and user friendly interface. The analytic component of BioDesigner was further developed and made commercially available (under the name of BioPro Designer) by Intelligen, Inc. (Scotch Plains, NJ).

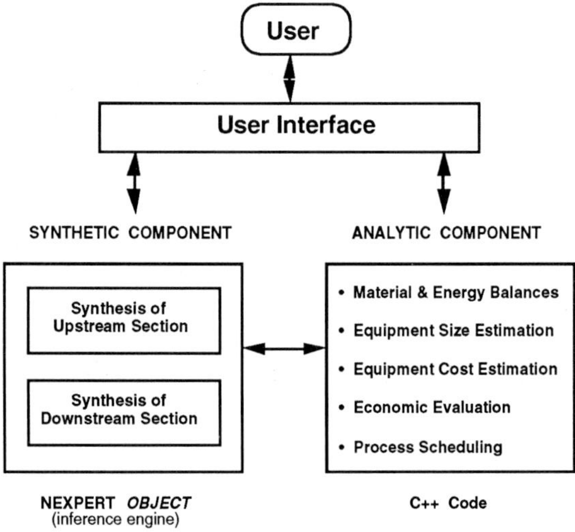

Figure 1. BioDesigner's system architecture.

COMPONENTS OF A BIOPROCESS DESIGN TOOL

Figure 1 shows the typical architecture of a computer-aided bioprocess design tool that combines synthetic with analytic capabilities. Such tools usually consist of three components, the user interface, the synthetic component and the analytic component.

<u>User interface</u>. Contemporary computer-aided process design tools use graphical user interfaces that facilitate the communication of the user with the computer. Graphical interfaces eliminate the need for development of FORTRAN-like input files and shorten the learning period. Figure 2, for instance, shows the main window of the analytic component of BioPro Designer. The user initializes a flowsheet interactively through appropriate dialog windows. Most contemporary computer-aided process design tools also feature on-line "Help" facilities.

<u>Synthetic component</u>. Process synthesis deals with the selection and assembly of integrated processes that are capable of producing, recovering, and purifying the desired product(s) while meeting financial and environmental constraints. Synthesis of entire flowsheets is a considerably more difficult problem compared to process analysis. This explains why the development of process synthesis tools is still a challenging research problem and why no synthetic tools (for entire flowsheets) are currently commercially available. As mentioned previously, one of the most successful techniques for process synthesis makes use of Knowledge Based Expert Systems (KBES). KBES offer a natural way to capture and utilize experiential design knowledge that is primarily used in process synthesis to reduce the solution space. Knowledge in the form of heuristics is stored in KBES mainly in the form of production rules. An example of the typical syntax of a rule is "*IF the type of microorganism is an animal cell, THEN use membrane filtration for media sterilization*". The synthetic component of BioDesigner (Figure 1) consists of two such knowledge bases. The first deals with synthesis of upstream sections (media preparation and bioreaction) and contains forty rules. The second knowledge base deals with synthesis of downstream sections (product

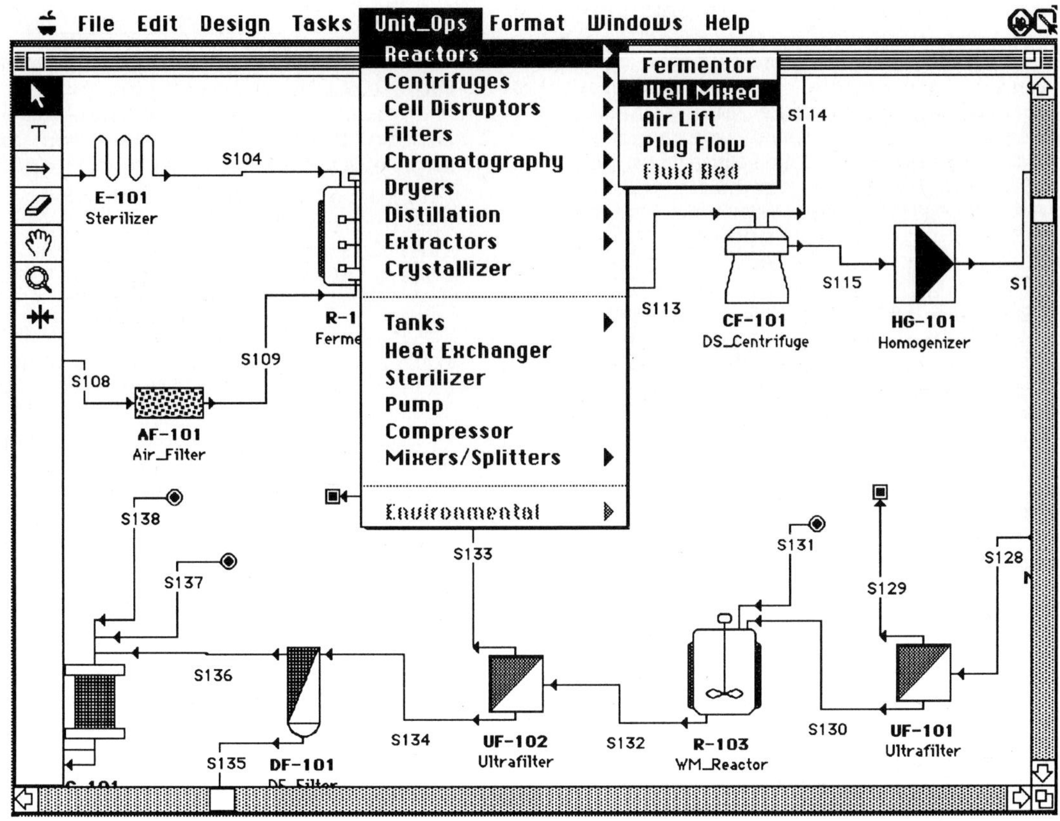

Figure 2. Main dialog window of the analytic component of BioPro Designer.

recovery and purification) and contains ninety rules. Industrial practice, common sense, and rules reported in the literature (15,16) constitute the main sources of synthetic knowledge in BioDesigner.

Analytic component. The analytic (simulation) component of a computer-aided process design tool calculates the material and energy balances of an entire flowsheet, estimates the size and cost of equipment and does the economic evaluation. The heart of any analytic component is the unit operation models. Models for traditional as well as for unit operations specific to biochemical processes should be included in

bioprocess design tools. The analytic component also generates a number of reports, such as the stream report, the economic evaluation report(s), the input/output data report, etc. The analytic component of the IBM version of BioPro Designer is written in Visual C++ (from Microsoft, Inc., Seattle, Washington). The material and energy balances around unit operations are calculated in sequential-modulal approach (17).

Process Scheduling. A number of unit operations in the biochemical industries operate in batch or semicontinuous mode. Also, many biochemical plants are multiproduct facilities that utilize the same equipment-line for the production of multiple products. To accommodate this time-dependency of equipment activity, a bioprocess design tool must be equipped with process scheduling capabilities and dynamic models for several unit operations. Process scheduling information is also needed for the accurate sizing of utilities and the estimation of volatile organic compound (VOC) emissions from process equipment.

An important issue in the development of a process simulator is validation of the unit operation models. In the case of BioPro Designer, this was done in collaboration with industrial partners over a period of six months before the release of the first version. A similar procedure is followed whenever new modules are introduced or the current ones are upgraded.

AN EXAMPLE - SIMULATION AND EVALUATION OF β-GALACTOSIDASE PRODUCTION

BioPro Designer was used to analyze and evaluate the production of an intracellular enzyme, β-galactosidase, by *E. coli*. The analysis of the production of any other intracellular protein that does not form inclusion bodies would be very similar. β-galactosidase has found limited industrial applications till now.It is mainly used in the utilization of cheese whey to convert lactose into glucose and galactose. An annual productionof about 5 tons of pure enzyme is considered as the design basis in this case study. The production rate has been selected to

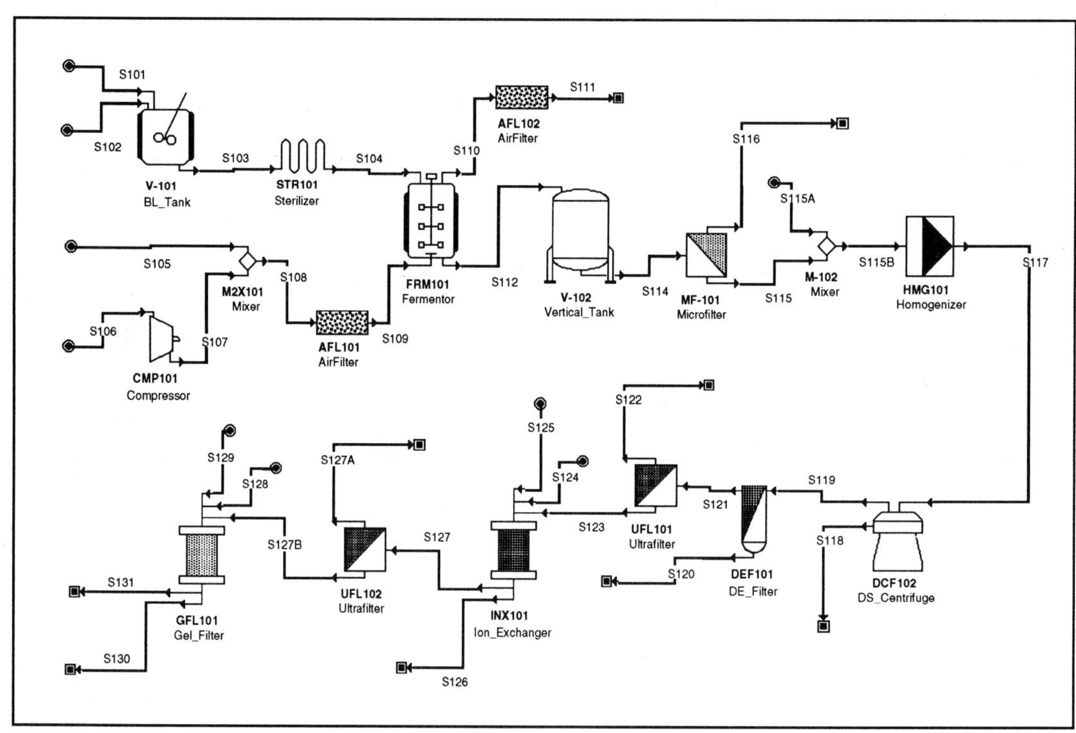

Figure 3. Production of b-galactosidase - Process flowsheet.

allow the size of the plant and the corresponding cost values to be high enough and representative of the values of a relatively large plant that produces an intracellular recombinant protein. It is assumed that the plant batch time is 72 hours and that 96 batches are processed per year.

Process Description (Figure 3)

Fermentation media are prepared in the stainless steel tank (V-101) and are continuously sterilized (STR101). The filling of the production fermentor with sterilized media takes 4 hours. The axial compressor CMP101 and the absolute filter AFL101 provide sterile air to the fermentors at an average rate of 0.4 VVM. The fermentation time in the production fermentor is about 18 hours, while the turnaround time is 12 hours. The final concentration of *E. coli* in the production fermentor is about 35 g/liter (dcw). The first step of the downstream section is cell harvesting and is carried out by a hollow-fiber membrane microfilter (MF-101) of 0.45 μm pore diameter. A high pressure homogenizer (HMG101) is used to break the cells and release the intracellular product. The broth undergoes two passes under a pressure drop of 800 bars. The removal of cell debris particles, created by homogenization, is accomplished by a disk-stack centrifuge (DCF102) in a period of 8 hours. The dead-end polishing filter (DEF101) removes any remaining particulate material. The dilute protein solution is concentrated by a hollow-fiber membrane ultrafilter (UFL101) of 100,000 molecular weight cut-off to a total protein concentration of 5-6% w/w. The INX101 ion exchanger is used for high resolution purification. Anion exchange is used because b-galactosidase is negatively charged in the range of pH that is stable. The type of resin is Fast Flow Mono Q (from Pharmacia). Four cycles are used each of 6 hours. The UFL102 ultrafilter is used to further concentrate (after its dilution by the ion exchanger) the protein solution to a final concentration of total protein of approximately 5% w/w. Finally, purification is completed by the GFL101 gel filter unit. The type of resin is Sephadex G25 or Sepharose CL (from Pharmacia). The final product purity is approximately 99.5%.

Economic Evaluation

For the plant of the design basis (5 tons of β-galactosidase per year) the equipment cost is $12.3 million. Fermentors and compressors are the most expensive pieces of equipment in the upstream section, while centrifuges and chromatography units dominate the downstream section. The total fixed capital investment is about $79.5 million. The total annual operating cost, including depreciation, is about $27.2 million. The production cost of β-galactosidase is about $5,500/kg.

Sensitivity Analysis

After a model for the entire process is developed on the computer, tools like BioPro Designer can be used to ask and readily answer "what if" questions and carry out sensitivity analysis with respect to key design variables. In this case study, the effect of two parameters on the production cost was examined.

Effect of dilution after cell harvesting on process economics.
Dilution after cell harvesting (stream S115A) and prior to cell disruption is done to mix the concentrated broth with a buffer solution that will ensure the stability of the product after its release by cell disruption. Dilution plays a negative role on process economics because the higher the dilution, the higher the equipment size and cost further downstream and consequently the higher the production cost. However, there is also a positive effect associated with dilution. More specifically, in

the removal of cell debris particles by the DCF102 centrifuge, the particles in the heavy phase of the centrifuge comprise only about 20 - 30% of total volume. The rest of the volume is liquid, with product protein dissolved in it. Consequently, the higher the extent of dilution, the lower the product concentration in stream S117 (feed to centrifuge) and as a result the lower the product loss in the heavy phase of the centrifuge. The results of the effect of dilution on process economics are shown in Figure 4a. As can be seen, the combined effect of dilution initially is positive but after a certain level (20,000 liters/batch) becomes negative.

<u>Effect of protein solution concentration on process economics.</u> After cell debris removal, the dilute protein solution is concentrated by ultrafiltration (UFL101 and UFL102) to reduce the volume of material that needs to be processed by the expensive

the concentration factor. The y-axis on the right shows the total protein solution concentration (as % w/w) in the feed stream to gel filtration. Chromatography manufacturers recommend that the total protein concentration in the feed stream to a chromatography unit should not exceed 5% w/w. This information can be used to set an upper bound to the extent of concentration.

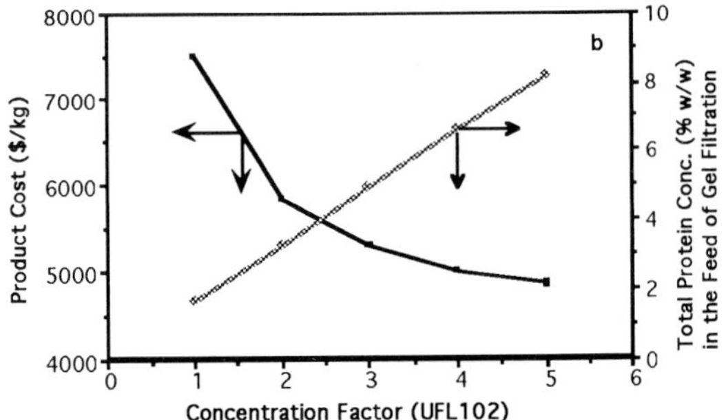

Figure 4b. Effect of protein solution concentration by the UF102 ultrafilter on process economics.

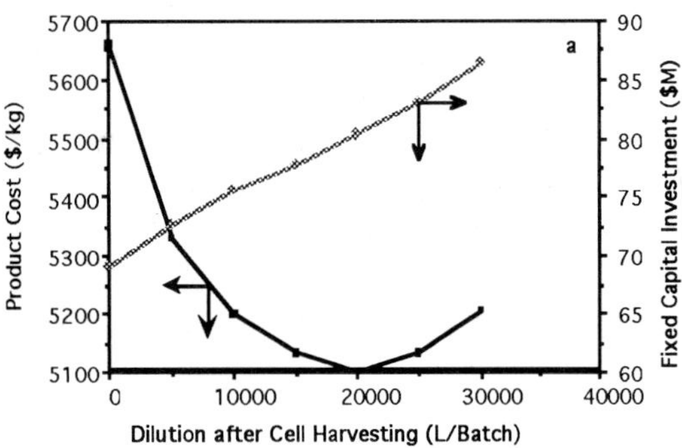

Figure 4a. Effect of dilution after cell harvesting on process economics.

chromatography units. In this case study, the effect of protein solution concentration by the UFL102 unit on process economics was analyzed. The higher the concentration factor, the lower the capital and operating cost of the expensive gel filtration unit (GFL101). The results of the analysis are shown in Figure 4b. The y-axis on the left shows the product cost that is reduced drastically by increasing

CONCLUSIONS

Computer-aided bioprocess design tools can play an important role in the development of biochemical processes and commercialization of biological products. At the early stages of project selection, such tools can be used to screen the large number of projects ideas from a profitability point of view and help focus development efforts on the most promising projects. During process development, such tools can be used to analyze and evaluate alternative processing schemes, reduce the impact of the whole process on the environment, interpret experimental results, and help design experimental protocols. During final design and plant construction, such tools can be used to optimize the entire process from a total-systems point of view. All these benefits substantially reduce cost and development time associated with a biological product.

LITERATURE CITED

1. Douglas, J.M., *Conceptual Design of Chemical Processes*, McGraw-Hill, New York (1988).

2. D. Petrides, K.G. Abeliotis and S.K. Mallick, "EnviroCAD: A Design Tool for Efficient Synthesis and Evaluation of Integrated Waste Recovery, Treatment and Disposal Processes," in *Comp. & Chem. Eng.*, **18** Suppl., S603 (1994).

3. Rosen, E.M. and A.C. Pauls, *Comp. & Chem. Engr.*, **1**, 11 (1977).

4. Evans, L.B., J.F. Boston, H.I. Britt, P.W. Gallier, P.K.Gupta, B. Joseph, V. Mahalec, E. Ng, W.D. Seider and H. Yagi, *Comp. & Chem. Engr.*, **3**, 319 (1979).

5. Perkins, J.D. and R.W.H. Sargent, "SPEEDUP: A Computer Program for Steady State and Dynamic Simulation of Chemical Processes," in *Selected Topics on Computer-Aided Process Design and Analysis*, Mah, R.S.H. and G.V. Reklaitis (Ed.), AIChE Symp. Ser., **78**, 1, 214 (1982).

6. Siirola, J.J. and Rudd, D.F., *I&EC Fundamentals*, **10**,353 (1971).

7. Evans, L.B. and R.P.Field, *Bio/Technology*, **6**, 200 (1988).

8. Siletti,C.A. "Computer-Aided Design of Protein Recovery Processes," Ph.D. thesis, Mass. Inst. of Technol., Cambridge (1988).

9. Siletti,C.A., "Design of Protein Purification Processes by Heuristic Search," in *Artificial Intelligence in Process Engineering*, Mavrovouniotis, M.L. (Ed.), Academic Press, New York (1990).

10. Petrides, D.P., "Computer-Aided Design of Integrated Biochemical Processes; Development of BioDesigner," Ph.D. thesis, Mass. Inst. of Technol., Cambridge (1990).

11. Petrides, D.P., "BioPro Designer: An Advanced Computing Environment for Modeling and Design of Integrated Biochemical Processes," in *Comp. & Chem. Eng.*, **18** Suppl., S621 (1994).

12. Petrides, D., C.L. Cooney, L.B. Evans, R.P. Field, and M.Snoswell, *Comp. & Chem. Eng.*, **13**, 553 (1989).

13. Cooney, C.L., J.E. Strong, Petrides, D. and L. Evans,"Computer-Aided Design of Biochemical Recovery Processes,"in *Biochemical Engineering - A Challenge for Interdisciplinary Cooperation*, Chmiel, H., W.P. Hammes, and J.E. Bailey (Ed.), Gustav Fischer Verlag, Stuttgard - New York. (1987).

14. Cooney, C.L., D. Petrides, M. Barrera, L. Evans,"Computer-Aided Design of a Biochemical Process," in *The Impact of Chemistry on Biotechnology*, Phillips, M., S.P.Shoemaker, R.D.Middlekauff, and R.M.Ottenbrite (Ed.), ACS Symposium Series 362, Washington (1988).

15. Wheelwright, S., *Bio/Technology*, **5**, 789 (1986).

16. Marston, F., *Biochemical Journal*, **240**, 1 (1986).

17. Westerberg, A.W., H.P. Hutchinson, R.L. Motard, and P.Winter, *Process Engineering*, Cambridge University Press, Cambridge, 1979.

A Strategy for the pH Control in Acidic Wastewaters

O. Galán-Domingo, J. Alvarez-Ramírez, and J. Alvarez-Calderón

Depto. de Ingeniería de Procesos e Hidráulica, Universidad Autónoma Metropolitana-
Iztapalapa, Apdo. Postal 55-534, 09000 México, D.F., MEXICO

Most wastewaters contain different acidic species, such as sulfuric and hydrochloric acids. Poorly known composition and concentration are central features of wastewaters. In this work, the pH control (in most cases, neutralization) of acidic wastewaters was addressed. In order to implement a feedback control, the process was modeled via the dynamics of fictitious polyprotic acidic and buffering species whose parameters can be estimated on-line. The robustness of the resulting control was tested with numerical simulations.

One critical problem in pollution control is the treatment of wastewaters derived from petrochemical processes (Bush, 1976). The mean feature of these waters is their high concentration of acidic and basic chemical species, which troubles their treatment by means of conventional biological methodologies (both aerobic and anaerobic sludges) (Davis and Cornwell, 1991). In fact, most wastewater treatment processes work efficiently at conditions with pH close to neutrality. For waters with a high load of acidic and/or basic chemical species, it is necessary to carry out a preliminary chemical treatment to regulate the pH in values ranging from 6.75 to 7.25 (Mavinic and Anderson, 1984; Ripley and Converse, 1986).

The aim of this work is to propose a strategy to control the pH in acidic wastewaters. It is assumed that the composition and concentration of the different species are unknown.

The control strategy is implemented by means of on-line measurements of (input and output) flowrates and pH. The central idea is to fit these measurements to aprocess modeled via the dynamics of fictitious polyprotic acidic and buffering species (specifically, by fitting the so called tritiation curve). Finally, departing of this model, a nonlinear feedback control is derived, and its performance is evaluated by means of numerical simulations.

Statement of the problem of pH control.

The main problem in pH control of wastewaters is to design a control law, which accounts for persistent changes in composition and concentration of flows. On the other hand, the control law has to be able to handle the high sensitivity (Okey, et. al., 1978) of

305

E. Galindo and O.T. Ramírez (eds.), Advances in Bioprocess Engineering. 305-312.
© 1994 *Kluwer Academic Publishers. Printed in the Netherlands.*

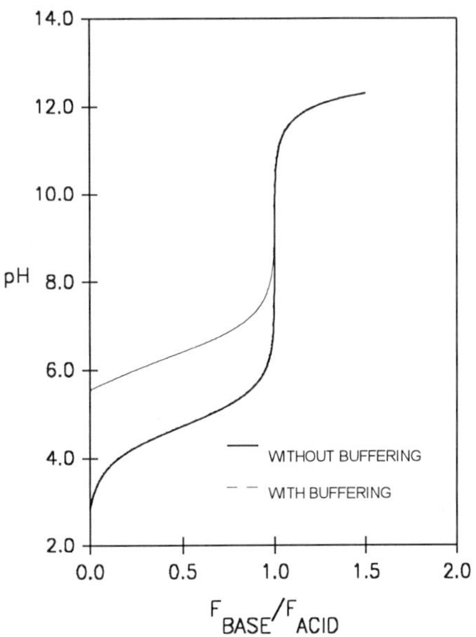

Figure 1. Tritation curves with and without buffering action.

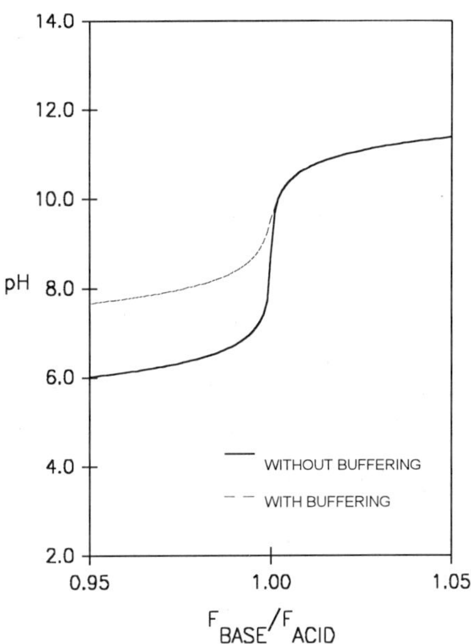

Figure 2. Enlargement view of the neutralization region in figure 1.

the process to small changes in the neutralizing agent in operation condition close to the neutralization point (pH = 7.0).

In the figure 1, one can observe that the tritiation curve possesses a large value of the slope (the inverse buffering factor) for values in a neighborhood of the neutralization point. This implies that small change in the flow of neutralizing agent (a basic chemical specie) induces large changes in the pH value.

A solution to this sensitivity problem is to reduce the value of the slope by using buffering species (in most cases, carbonate and bicarbonate salts) (see the figure 2).

The mathematical model of acid/base reactions and neutralization reactors.

The modeling of acid/base reactions taking place in a continuous stirred tank departs from mass balances and electroneutrality and chemical equilibrium expression. It is assumed that the temperature is constant and that all the chemical species are in the range of complete solubility in water (there is not problem with solubility equilibriums) (Stumm and Morgan, 1970; Lowenthal and Marais, 1978). The key for the modeling is the introduction of certain linear combinations of ionic species that are invariant under the presence of any set of ionic reactions. These combinations are related to the existence of irreducible ions. In what follows, such ion combinations will be referred to as chemical invariants. The final model is a set of first order differential equations that govern the dynamics of the irreducible ion, and an algebraic nonlinear equation that relates the irreducible ions concentrations to the instantaneous pH of the process. (Waller and

Makila, 1981).
The invariants are chosen as mass balances of irreducible ions associated to the bases (irreducible cations) and acids (irreducible anions). In this way linear combinations of concentrations lead to a mixing dynamic among ionic species, which are independent of the chemical reactions taking place into the stirred tank (Gustafsson y Waller, 1983; Gustafsson, 1985).
We illustrate the modeling of a strong monoprotic acid being neutralized with a strong base (McAvoy et. al. , 1972), to show the derivation of the final model. The following steps apply for more general systems.

Acid/Base Reactions.

$$HA \xleftrightarrow{K_a} H^+ + A^-$$

$$BOH \xleftrightarrow{K_b} B^+ + OH^-$$

$$H_2O \xleftrightarrow{K_w} H^+ + OH^-$$

Chemical Equilibrium.

$$K_a = \frac{[H^+][A^-]}{[HA]}$$

$$K_b = \frac{[B^+][OH^-]}{[BOH]}$$

$$K_w = [H^+][OH^-]$$

Chemical Invariants

$$U_A = [HA] + [A^-]$$

$$U_B = [BOH] + [B^+]$$

Observe that U_A is the total concentration of the chemical specie A^-, which is distributed between the cation A^- and the nondissociated acid HA. For example, if the strong acid to be neutralized is the clorhidric acid then

$$U_A = [HCl] + [Cl^-].$$

Electroneutrality.

$$[H^+] + [B^+] - [OH^-] - [A^-] = 0$$

The combination of the above equations leads to the following expression (the so called pH equation):

$$[H^+] + \frac{K_b[H^+]U_B}{K_w + K_b[H^+]} - \frac{K_a U_A}{[H^+] + K_a} - \frac{K_w}{[H^+]} = 0$$

The dynamics of the invariants.

The dynamics of the neutralization are governed by the mass balances of the chemical invariants. The following equations describe the mixing process of the such chemical *species*:

$$\frac{dU_A}{dt} = \frac{1}{V}[fU_{Af} - (f+m)U_A]$$

$$\frac{dU_B}{dt} = \frac{1}{V}[mU_{Bf} - (f+m)U_B]$$

Parameter estimation of a flow with unknown composition and concentration.

The neutralization of an acidic wastewater flow requires the knowledge of its concentration and composition. Due to the fluctuating nature of a wastewater flow, it is practically impossible to have on-line access to these operation variables. Furthermore, wastewater flows contain a great variety of corrosive chemical substances and suspended solids that trouble to carry out analytical procedures (with exception of pH and flowrate measurements).

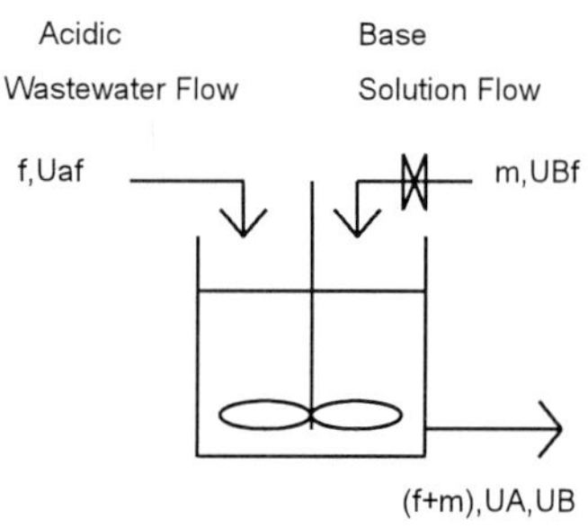

To overcome this problem, if on-line measurements of pH and flowrates are available, the dynamics of the process are fitted to a model with fictitious polyprotic acid and buffering species. In other words, all the acids present in the wastewater flow are lumped into an equivalent (maybe, polyprotonic) acid.

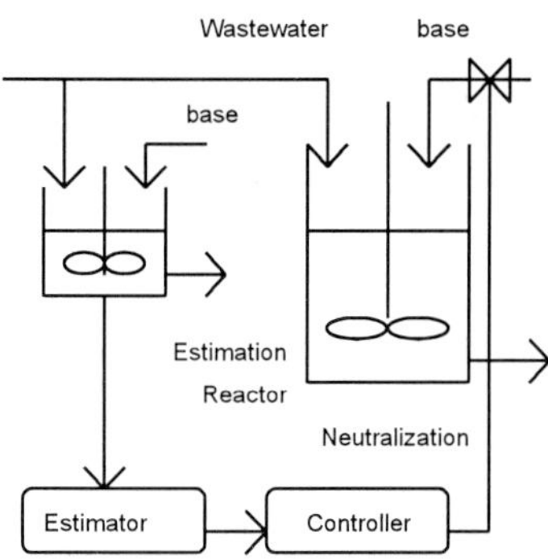

This is a common practice in modeling chemical processes where a complex network of reactions is taking place (for instance, in fuel cracking catalytic process).

Since the resulting equivalent model is nonlinear (the pH equation in the above section), in this work we have used a Marquart-Leverberg

algorithm to fit the on-line measurements to the equivalent model.

The model for representing the composition and concentration of the wastewater flows is:

Equivalent fictitious monoprotic acid,

$$[H^+]+U_B - \frac{K_{a,est}U_A}{[H^+]+K_{a,est}} - \frac{K_w}{[H^+]} = 0$$

Equivalent fictitious di-protic acid,

$$[H^+]+U_B - [HA^-](1+2\frac{K_{a2,est}}{[H^+]}) - \frac{K_w}{[H^+]} = 0$$

where,

$$[HA^-] = \frac{K_{a1,est}U_A}{K_{a1,est}(1+\frac{K_{a2,est}}{[H^+]})+[H^+]}$$

with

$$U_A = (\frac{1}{1+r})U_{Af,est}; U_B = (\frac{r}{1+r})U_{Bf}; r = \frac{m}{f}$$

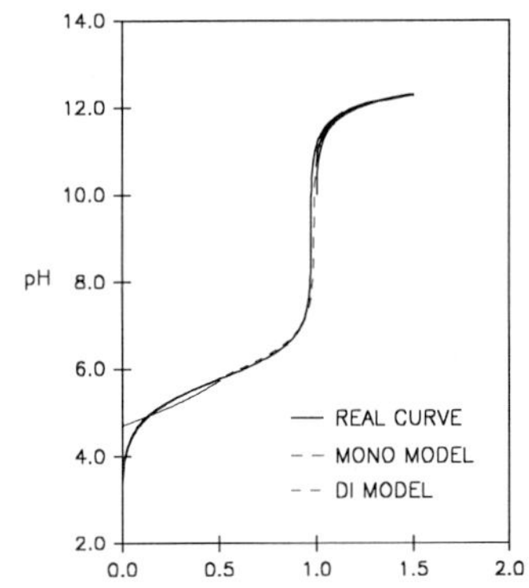

Figure 3. Approximation of a typical tritation curve (Okey *et. al.* , 1978) via ficticious monoprotonic and diprotonic acid models. The true flow is $U_{A,f}$=0.1 mol/l solution of acetic acid (CH_3COOH, K_a=1.85x10^{-5}), and the neutralizing agent is a $U_{B,f}$=0.1 mol/l solution of sodium hidroxide. The estimated parameters for the monoprotic acid model are $U_{Af,est}$=0.973x10^{-1}, $K_{a,est}$=1.81x10^{-5}, and and for the diprotonic acid model are $U_{Af,est}$=0.493x10^{-1}, $K_{a1,est}$=0.408x10^{-5}, $K_{a2,est}$=0.676x10^{-6}.

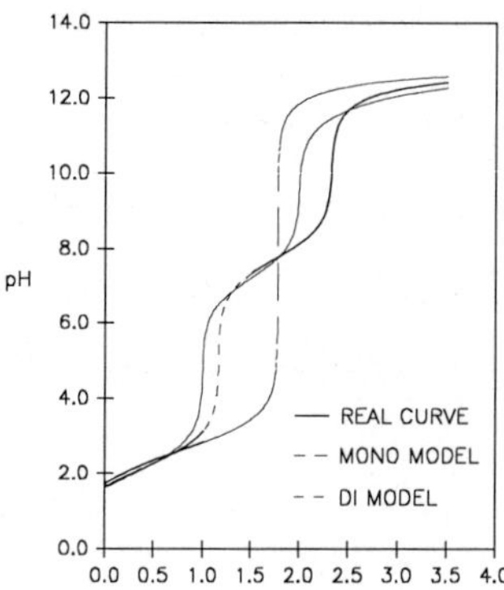

Figure 4. Approximation of a multiple inflection points tritiation curve. This case corresponds to the neutralization of phosphoric acid (H_3PO_4), [U_{Af}=0.1 mol/l and K_{a1}=7.5x10^{-3}, K_{a2}=6.2x10^{-8}, K_{a3}=1.0x10^{-12}] with sodium hidroxide U_{Bf}=0.1 mol/l. The estimated parameters for a monoprotic acid model are $U_{Af,est}$=0.177 mol/l, $K_{a,est}$=0.209x10^{-2} and for a diprotonic acid model are $U_{Af,est}$=0.116 mol/l, $K_{a1,est}$=0.539x10^{-2}, $K_{a2,est}$=0.190x10^{-7}.

Dynamics of the neutralization process.

Consider the neutralization of a strong monoprotic acid with a strong base, as described in above sections. For given inlet conditions for the chemical invariants, the mixing equations for the invariants U_A and U_B are integrated, resulting the following two expressions:

$$U_A(t) = [U_{A0} - \gamma_1 U_{Af}]\exp(-\theta t) + \gamma_1 U_{Af}$$

$$U_B(t) = [U_{B0} - \gamma_2 U_{Bf}]\exp(-\theta t) + \gamma_2 U_{Bf}$$

which , together with the pH equation:

$$\psi(U_A, U_B, h) = h^3 + h^2(K_a + U_B)$$
$$+ h(K_a[U_B - U_A])$$
$$- K_w] - K_a K_w = 0$$

where,

$$\gamma_1 = \frac{f}{f+m}; \gamma_2 = \frac{m}{f+m}; \theta = \frac{f+m}{V}$$

define the dynamics of neutralization. In the stationary case $(t \to \infty)$, and taking as parameter the base flowrate, the above equations define the *tritation curve*. The pH equation defines implicitly the pH dynamics. We will need the pH dynamics (dpH/dt) explicitly to design a control law. We will show that this procedure is always possible.

Let us apply the Chain Rule of derivation to the pH equation:

$$\frac{d\psi}{dt} = (\frac{\partial \psi}{\partial h})\frac{dh}{dt} + (\frac{\partial \psi}{\partial U_A})\frac{dU_A}{dt}$$
$$+ (\frac{\partial \psi}{\partial U_B})\frac{dU_B}{dt}$$

by solving for (dh/dt):

$$\frac{dh}{dt} = \frac{-(\frac{\partial \psi}{\partial U_A})\frac{dU_A}{dt} - (\frac{\partial \psi}{\partial U_B})\frac{dU_B}{dt}}{(\frac{\partial \psi}{\partial h})}$$

where the surface

$$\Gamma(U_A, U_B, h) = (\frac{\partial \psi}{\partial h})$$

satisfy the condition:

$$\Gamma(U_A, U_B, h) \neq 0$$
$$\forall U_A, U_B, h \in \mathfrak{R}_+$$

Therefore, the dynamics of the ion hydrogen concentration are governed by:

$$\frac{dh}{dt} = (hK_a(fU_{Af} - f_sU_A) - (mU_{Bf}$$
$$-f_sU_B)(h^2 + hK_a))/(3h^2 + 2h(K_a$$
$$+U_B) + K_a(U_B - U_A) - K_W)/V$$

with $f_s = f + m$. Obviously $dh/dt = -\ln(10)hdpH/dt$. Thus, the pH dynamics satisfy an algebraic equation:

$$\Phi(h, dh/dt, U_A, U_B, m) = 0$$

Derivation of the control law

To obtain a control feedback to regulate the pH (or equivalently, the ion concentration h) it is assumed that the manipulated variable is the base flowrate m(t). Solving for m(t) in the last equation, we obtain:

$$m(t) = \Theta(h, dh/dt, U_A, U_B)$$

where

$$\Theta = (dh/dtV(3h^2 + 2h(K_a + U_B)$$
$$-(K_a(U_A - U_B) + K_w))$$
$$-fh(hU_B + K_a(U_{Bf} + U_B - U_A)))$$
$$/\Sigma(U_A, U_B, h)$$

and

$$\Sigma(U_A, U_B, h) = h(h(U_B - U_{Bf})$$
$$-K_a(U_A + U_{Bf} - U_B)) \neq 0$$

Let pH_{ref} be a desired pH set-point. To have closed-loop (the system plus the control m(t)) stability is sufficient that the following equality be satisfied:

$$\frac{d(pH - pH_{ref})}{dt} = g_1(pH_{ref} - pH)$$
$$+g_2\int_0^t(pH_{ref} - pH(s))ds$$

where $g_1, g_2 > 0$ are known as the gains of the controller. So therefore,

$$m(t) = (\ln(10)(dpH/dt)V(K_{a,est}(U_A$$
$$-U_B) + K_w - 3h^2 - 2h(K_{a,est} + U_B))$$
$$-f(hU_B + K_{a,est}(U_{Af,est} + U_B - U_A)))$$
$$/\Sigma(U_A, U_B, h)$$

This is a nonlinear feedback control with proportional and integral actions. Observe that all the parameters of the controller are known or estimated on-line (via the equivalent system model and on-line measurements).

A numerical example.

To test the performance of the above controller, we are going to consider the neutralization of a strong monoprotic acid (of unknown concentration (U_{Af}) and equilibrium constant (K_a)) with a strong base, into a stirred tank . It is assumed that the input acid concentration follows a periodic variation around a nominal value ($U_{Af,av}$). The set of parameters is the following:

$$K_a \to \infty; K_w = 10^{-14}$$
$$U_{A0} = 0.1mol/l; U_{B0} = 0.0mol/l; pH_0 = 1.0$$
$$U_{Af,av} = 0.5mol/l; U_{Bf} = 0.4mol/l$$
$$U_{Af} = U_{Af,av} + 0.3\sin(10t)$$
$$V = 2000l; pH_{ref} = 7.0$$
$$g_1 = 10.0; g_2 = 20.0$$

The controller gains g_1 and g_2 were chosen in such a way that the closed-loop system be stable althoug, without a specific tuning criteria. The figure 5 shows that the pH is stabilized close the reference value $pH_{ref}= 7$, despite the sustained oscillation in the input flow concentration and parametric errors.

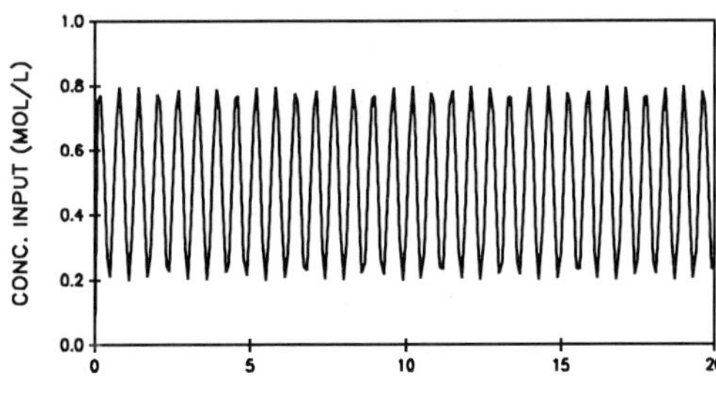

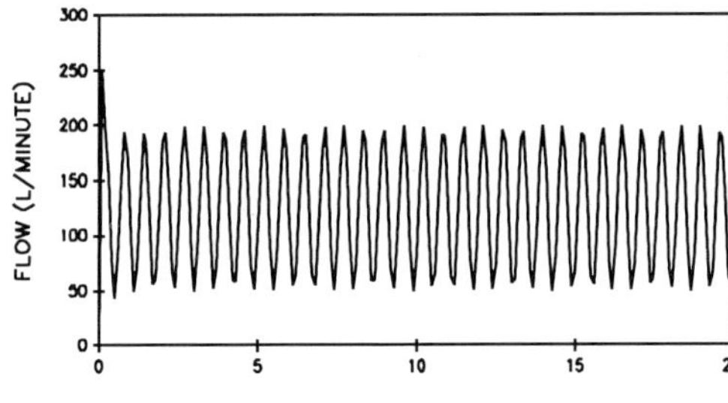

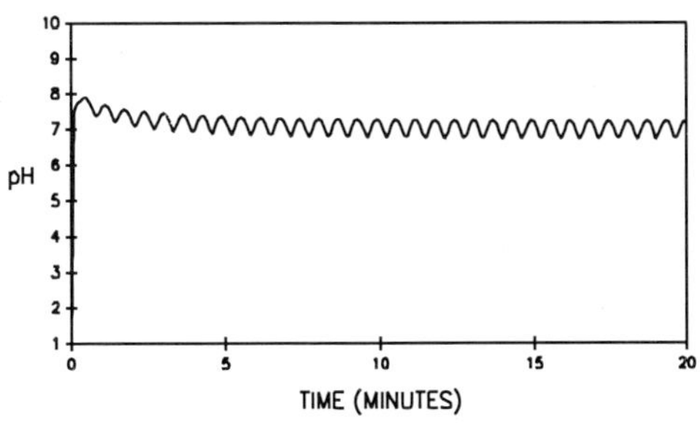

Figure 5. Control actions (base flow) for the case where the concentration of the input flow follows a periodic variation.

Conclusions.

In this work we have approached systematically the pH control for acidic wastewaters. Taking in account that the concentration and composition of pollutants in residual waters are highly fluctuant, we have modeled the dynamics of process as the dynamics of and simplified equivalent system. This approach allows us to derive a control law that is robust against perturbations in the composition , concentration and flowrates.

Nomenclature

A^- anion associated to the acid

B^+ cation associated to the base

H^+ hydrogen ion, mol/l

K_a dissociation acid constant mol/l

K_w dissociation water constant, mol/l

$pH \quad =-\log_{10}(H^+)$

t time min.

U_A invariant associated to the acid, mol/l

U_B invariant associated to the base, mol/l

V reactor volume, l

f acidic wastewater flowrate, l/min

m base solution flowrate, l/min

Subcripts

av average

A refers to acidic species

B refers to base neutralizer

f designates feed conditions

est refers to estimated values

Literature cited.

Bush, K.E., 1976, "Refinery Wastewater Treatment and Reuse", Chem. Engng., **12**, 113-118.

Davis, M.L. and Cornwell, D.A., 1991, "Introduction to Environmental Engineering", McGraw-Hill International Editions, U.S.A.

Gustafsson, T.K. and Waller, K.V., 1983, "Dynamic Modeling and Reaction Invariant Control of pH", Chem. Engng. Sci., **38**, 3, 389-398.

Gustafsson, T.K., 1985, "An Experimental Study of a Class of Algorithms for Adaptive pH Control", Chem. Engng. Sci., **40**, 5, 827-837.

Lowenthal, R.E. and Marais, G.v. R., 1978, "Carbonate Chemistry of Aquatic Systems: Theory & Application", Ann-Arbor-Science, Michigan, U.S.A.

Mavinic, D.S. and Anderson, B.C., 1984, "Aerobic Sludge Digestion with pH Control-Preliminary Investigation", J. Water Pollut. Control Fed., **56**, 7, 889-897.

McAvoy, T.J., Hsu, E. and Lowenthal, S., 1972, "Dynamics of pH in Tank Reactor", Ind. Eng. Chem. Process Des. Develop., **11**, 1, 68-70.

Okey, R.W., Chen, K.Y. and Sycip, A.Z., 1978, "Neutralization of Acid Wastes by Enhanced Buffer", J. Water Pollut. Control Fed., **50**, 1841-1851.

Ripley, L.E., Boyle, W.C. and Converse, J.C., 1986, "Improved Alkalimetric Monitoring for Anaerobic Digestion of High-Strength Wastes", J. Water Pollut. Control Fed., **58**, 5, 406-411.

Stumm, W. and Morgan, J.J., 1970, "Aquatic Chemistry", Wiley-Inter-Science, New York, U.S.A.

Waller, K.V. and Makila, P.M., 1981, "Chemical Reaction Invariants and Variants and Their Use in Reactor Modeling, Simulation and Control", Ind. Eng. Chem. Des. Dev., **20**, 1, 1-11.

Modeling and Control Strategies for the Transformation of D-Sorbitol to L-Sorbose on a Laboratory Bioreactor

A.T. Fleury[1], A. Bonomi[2], E.F.P. Augusto[2], L.H.C. Quiroz[2], M.F. Barral[2], and P.S. Pereiralima[1]

[1]Agrupamento de Sistemas de Controle-Dme, [2]Agrupamento de Biotecnologia-DQ, IPT, Instituto de Pesquisas Tecnológicas do Estado de S. Paulo S.A., Caixa Postal 7141, CEP 01064-970, São Paulo, SP, BRAZIL

This work shows the main aspects relative to modeling and control design of the microbial oxidation of D-sorbitol to L-sorbose. Based on experimental investigations in laboratory bioreactors an unstructured model is proposed. This is a five state variables model, with 13 adjusted parameters and it is linearized around an optimized operating point, taking into account productivity, yield, product and substrates concentrations in the effluent as well as the oxygen transfer requirements. Design of a control strategy for such a system is complicated by the facts that only few variables can be measured on-line and even so, some of them with large delays. The Singular Perturbation Method (SPM) is adapted to generate the regulator and observer parts of a Linear Quadratic (LQ) control system. Results obtained through digital simulation of the controlled system show that the described approach is a feasible solution to the control problem. Current efforts are focused on implementing the controller on an automated laboratory bioreactor.

In order to study the formulation and adjustment of a mathematical model and the design of a control strategy for a fermentation process, the bioxidation of D-sorbitol to L-sorbose, key step in the Reichstein chemical process to produce vitamin C, was chosen. The Reichstein procedure to convert glucose into L-ascorbic acid is a combination of several stages of chemical reactions, physico-chemical operations and a single biochemical step – the microbial oxidation of D-sorbitol to L-sorbose – performed by a bacterial strain of *Gluconobacter oxydans* (Boudrant [1]).

This paper proposes an unstructured mathematical model which includes the effects of the major biochemical phenomena on the process kinetics. A global non linear analysis was performed, in order to fit the model to the experimental data for a good estimation of the model parameters.

The resulting model was used for the selection of an optimal steady-state point, through the optimization of an objective function that takes into account important process parameters like productivity, yield, oxygen supply and product and substrates concentrations. The non-linear model is then linearized around this point in order to allow the design of a control strategy that keeps the bioreactor operation near the selected fermentation conditions.Although based on a linear model, design of a control strategy is not an easy task because of two factors: first, the presence of slow and fast variables in the model destroys system observability; second, this observability is also affected by the available instrumentation that only provides measurements with considerable delays (Fleury et al. [2]). To circumvent these problems, some of the topics developed inside the Singular Perturbation Theory (Kokotovic et al. [3]) are used in the achievement of a LQ Regulator combined with a Luenberger Identity Observer. This strategy was computer simulated with very promising results. It is being implemented on one of IPT's laboratory bioreactor.

MATERIALS AND METHODS

Microorganism. All the experiments were performed utilizing a strain of the bacteria *Gluconobacter oxydans*, ATCC-621, maintained lyophilized in the IPT's Culture Collection.

313

E. Galindo and O.T. Ramírez (eds.), *Advances in Bioprocess Engineering. 313-320.*
© 1994 *Kluwer Academic Publishers. Printed in the Netherlands.*

Culture Media. The culture media used in the laboratory experiments was prepared with sorbitol, yeast extract and salts (Mori et al. [4]).

Analytical Methods. Cell mass concentration was measured by the dry-weight method. The sorbitol and sorbose concentrations were measured with an HPLC (Waters-600E) with a RI detector (Waters-410), using a Shodex SC1011 Column (72°C) and with water as eluent (0.65 ml min⁻¹). The yeast extract was quantified only during the medium preparation. The dissolved oxygen was measured continuously with a galvanic probe mounted in the reactor.

Experimental Conditions. All the experimental runs were batch operations performed in a 7 l bioreactor (New Brunswick Microferm MF-104) with a 3 l working volume. The system was operated at constant temperature (30°C) and with the pH controlled (4.5). Agitation speed and aeration flow rate were variables depending on the experimental run conditions. The control strategy will be implemented in a 14l bioreactor operated continuously (Braun Biostat ED).

EXPERIMENTAL RESULTS AND MATHEMATICAL MODEL FORMULATION

Table 1 presents the nominal conditions of the experimental runs considered for the mathematical model formulation and the parameter estimation of the bioxidation of D-sorbitol to L-sorbose.

Figure 1 presents the experimental results obtained in one of the runs (number 16).

The following set of equations represents the proposed unstructured model for the bioconversion of sorbitol to sorbose:

Mass Balance Equations:

In the proposed model:

x biomass concentration (g/l)
p sorbose concentration (g/l)
s_1 sorbitol concentration (g/l)
s_2 yeast extract concentration (g/l)
o dissolved oxygen concentration (mg/l)

$$\dot{x} = r_x \tag{1}$$

$$\dot{p} = r_p \tag{2}$$

$$\dot{s_1} = -r_{s_1 x} - r_{s_1 p} \tag{3}$$

$$\dot{s_2} = -r_{s_2} x \tag{4}$$

$$\dot{o} = -r_{ox} - r_{op} + Na \tag{5}$$

Table 1 — Nominal Conditions of the Performed Experiments

run	Concentrations				$K_\ell a$
	s_1 (a)	s_2 (a)	o (b)	salts (c)	
1	50.0	1.25	3.75	4x	–
2	50.0	5.00	3.75	4x	–
3	50.0	5.00	3.75	4x	–
4	50.0	7.50	3.75	4x	–
5	50.0	5.00	3.75	4x	–
6	100.0	5.00	3.75	4x	–
7	200.0	5.00	3.75	4x	–
8	50.0	5.00	0.375	4x	–
9	50.0	5.00	1.50	4x	–
10	50.0	5.00	6.00	4x	–
11	50.0	5.00	6.00	4x	–
12	50.0	5.00	–	4x	60.0 (d)
13	50.0	5.00	–	4x	100.0 (d)
14	50.0	5.00	–	4x	150.0 (d)
15	50.0	5.00	–	4x	300.0 (d)
16 (e)	50.0	5.00	3.75	4x	–
17	100.0	10.00	3.75	8x	–
18	150.0	10.00	3.75	12x	–
19	175.0	10.00	3.75	14x	–
20	200.0	20.00	3.75	16x	–
21	50.0	5.00	3.75	4x	–
22	50.0	5.00	3.75	4x	–

(a) Initial concentrations in the medium.
(b) Dissolved oxygen concentration kept constant (PID control upon agitation/aeration) throughout the experimental run.
(c) Salts composition of the medium compared to the basic medium.
(d) $K_\ell a$ value kept constant throughout the experimental run.
(e) The inoculum for this experimental run was prepared in a bioreactor in order to compare the adapting conditions.

Kinetic Equations:

$$r_x = \mu_x x \tag{6}$$

where

$$\mu_x = \frac{\hat{\mu}s_1}{K_{s_1}+s_1+{s_1^2}/{W}}\frac{s_2}{K_{s_2}+s_2}\frac{o}{K_o+o}e^{-K_p p} \qquad (7)$$

$$r_p = \mu_p x \qquad (8)$$

where

$$\mu_p = \alpha\mu_x+\beta \qquad (9)$$

$$\beta = \frac{\beta_m s_1}{K_{\beta s_1}+s_1}\frac{o}{K_{\beta o}+o} \qquad (10)$$

Yield Equations

$$r_{s_1 x} = \frac{1}{Y_{xs_1}}r_x \qquad (11)$$

$$r_{s_1 p} = \frac{1}{Y_{ps_1}}r_p \qquad (12)$$

$$r_{s_2 x} = \frac{1}{Y_{xs_2}}r_x \qquad (13)$$

$$r_{ox} = \frac{1}{Y_{xo}}r_x \qquad (14)$$

$$r_{op} = \frac{1}{Y_{po}}r_p \qquad (15)$$

Transport Equation.

$$Na = K_\ell a\left(c_g - o\right) \qquad (16)$$

The impossibility of representing the adaptation period (lag phase), observed in the majority of the performed experimental runs, is the main limitation of the proposed unstructured model, with 5 state variables and 13 adjusted parameters (Bonomi et al. [5]).

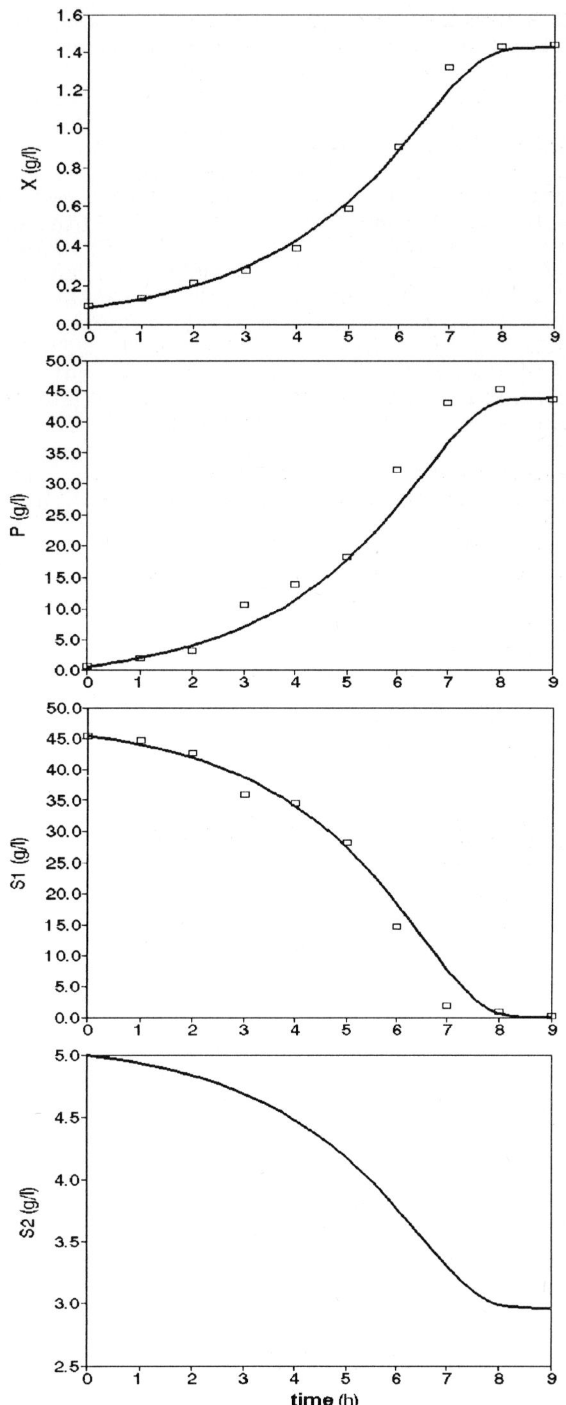

Figure 1: 16th run simulated with parameters from table 2.

The global estimation of the model parameters was performed applying the flexible geometric simplex method proposed by Nelder & Mead and known as the flexible polyhedron search (Himmelblau [6]). This technique, a derivative-free method of estimation, is especially effective in comparison to derivative methods when the number of parameters to be estimated becomes large (Himmelblau [7]). Different versions of the software "OTIMPAR", developed in our laboratory to adjust the model parameters using the flexible polyhedron search and the subroutine DGEAR from the IMSL MATH/PC - LIBRARY (IMSL [8]) to integrate the set of ordinary differential equations, had to be tested due to the use of different formulas to calculate the residue (based on the difference between the estimated and the experimental values) (Bonomi et al. [9]).

Table 2 presents the obtained set of model parameters. The curves in Figure 1 represent the results obtained simulating the experimental run 16, using the set of parameters from Table 2.

Table 2 — Final Estimation of the Model Parameters Set

parameter	value	parameter	value
$\hat{\mu}$	0.664	β_m	6.81
K_{s_1}	3.93	$K_{\beta s_1}$	4.91
W	204.	$K_{\beta o}$	0.544
K_{s_2}	0.994	Y_{xs_1}	0.795
K_o	0.158	Y_{xs_2}	0.656
K_p	0.00554	Y_{xo}	0.000664
α	18.6		

THE CONTROL MODEL AND THE BIOREACTOR SYSTEM

For control purposes of the continuous fermentation process, the 16 equations given in item 2 can be synthesized into 5 non linear differential equations with parameters given in Table 2.

$$\dot{x} = D(-x) + \mu x \tag{17}$$

$$\dot{p} = D(-p) + [\alpha\mu + \beta]x \tag{18}$$

$$\dot{s}_1 = D(s_{1e} - s_1) + \left[\left(-1/Y_{xs_1} - \alpha/Y_{ps_1}\right)\mu + \left(-1/Y_{ps_1}\right)\beta\right]x \tag{19}$$

$$\dot{s}_2 = D(s_{2e} - s_2) + \left[\left(-1/Y_{xs_2}\right)\mu\right]x \tag{20}$$

$$\dot{o} = D(c_g - o) + \left[\left(-1/Y_{xo} - \alpha/Y_{po}\right)\mu + \left(-1/Y_{po}\right)\beta\right]x + K_\ell a(c_g - o) \tag{21}$$

In consequence, there are five state variables and four variables are chosen to compose the control vector: D, specific flow rate (h^{-1}); s_{1e}, feed sorbitol concentration (gl^{-1}); s_{2e}, feed yeast concentration (gl^{-1}) and $K_\ell a$, oxygen transfer rate (h^{-1}).

The automation and control scheme for implementation of this process is shown in Figure 2.

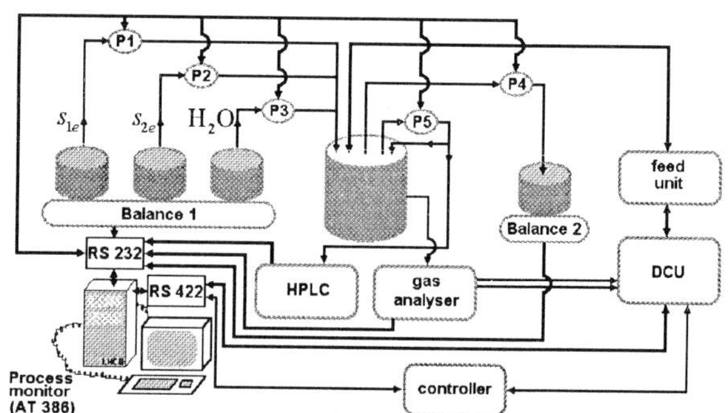

Figure 2: Control scheme implemented in the laboratory.

The bioreactor is the heart of the system. It incorporates a Digital Control Unit (DCU) and the feed unit. The agitation/aeration system, related to $K_\ell a$, is not shown in the figure. The other control variables are given by the pumps P1, P2 and P3 and are assured by the balances. The measurement system is composed by a gas analyzer and a High Performance Liquid Chromatography (HPLC). This measurement scheme imposes severe restrictions to the design of an adequate control strategy.

OPTIMAL OPERATING POINT

The achievement of an optimal steady-state point corresponds to the static optimization of the fermentation process described by the model above. The objective function to be minimized incorporates, through economically defined weights w_i, i=1,..,6, productivity, yield, final product concentration, residual substrates concentrations and oxygen supply. The first three parameters are to be maximized whereas the last three must be minimized. The resulting objective function is then given by (IPT [10]):

$$FO = -w_1 Dp - w_2 p / (s_{1e} - s_1) - w_3 p + w_4 s_1 + w_5 s_2 + w_6 K_\ell a \tag{22}$$

Minimization of the function FO was obtained through the use of the package OPT3 (Gabriele et al. [11]) which employs the Generalized Reduced Gradient Method constrained to the algebraic restraints given by Equations (17) to (21) with null dynamics. The optimal point and corresponding weights are:

Table 3 — Optimal Operating Point

state variables	values	control variables	values	weights	values
x	3.238	D	0.040	w_1	30
p	392.9	s_{1e}	422.1	w_2	10
s_1	20.79	s_{2e}	7.975	w_3	1.2
s_2	3.038	$K_\ell a$	269.6	w_4	1.0
o	1.597			w_5	30
				w_6	0.5

DESIGN OF A CONTROL STRATEGY

Once the optimal steady-state point was obtained, the non linear model (17) to (21) is linearized around this point in order to allow the design of a control strategy The resulting linear system is sumarized as:

$$\delta\dot{x} = A\delta x + B\delta u \tag{23}$$

$$y = C\delta x \tag{24}$$

where $\delta x \equiv x - x_e$, $\delta u \equiv u - u_e$, $x = [x \, p \, s_1 \, s_2 \, o]$ and $u = [D \, s_{1a} \, s_{2a} \, K_\ell a]$ are state and control deviations relative to the operating point.

In Equation (24) y is a 3 dimensional measurement vector, representing product p, substrate s_1, both given by the HPLC, and the dissolved oxygen concentration o measured with the probe. The other state variables, x and s_2, are not accessible (the response of the gas analyzer will be utilized to estimate x, in the future). Another difficulty imposed by the measurement system is that some of the observed variables are only available with large delays: 20 min for product, p, and substrate s_1. Even if these measures were available in real time, the linearized model (23)-(24) results controllable but not observable. The observability matrix is rank 3 for the chosen conditions. This fact is not linked to the measurement system but to the presence of slow and fast variables in the model (Fleury et al. [2]). The eigenvalue associated to oxygen concentration o is 4 orders of magnitude greater than the eigenvalues associated to the other variables (10^{+2} to 10^{-2}).

In order to preserve the aim of a control strategy based on the tested model of item 2, techniques of the Singular Perturbation Method (SPM) (Kokotovic et al. [3]) were adopted. SPM proposes the partition of the original system into slow and fast subsystems and the design of state observers and/or optimal controllers for each subsystem. The resulting global controller is suboptimal, with a deviation of order ε^2 relative to the optimal, where ε is a small parameter used for system partition (Kokotovic and Haddad [12]). For the case of D-Sorbitol to L-Sorbose transformation, ε was chosen as the ratio between the lowest and highest eigenvalues of the linearized model (Equations (23) and (24)), $\varepsilon=8,314E-5$. Taking this into account, the system is partitioned in two state vectors, $x_1 = [\delta x \, \delta p \, \delta s_1 \, \delta s_2]^T$ for the slow variables and $x_2 = [\delta o]$ for the fast one. The new equations are given by:

$$\dot{x}_1 = A_{11} \cdot x_1 + A_{12} \cdot x_2 + B_1 \cdot u \tag{25}$$

$$\varepsilon \cdot \dot{x}_2 = A_{21} \cdot x_1 + A_{22} \cdot x_2 + B_2 \cdot u \tag{26}$$

$$y = C_1 \cdot x_1 + C_2 \cdot x_2 \tag{27}$$

An identity observer associated to the system (25) to (27) can be obtained from the work of Porter ([13]):

$$\hat{\mathbf{x}}_1 = \mathbf{A}_{11}\hat{\mathbf{x}}_1 + \mathbf{A}_{12}\hat{\mathbf{x}}_2 + \mathbf{B}_1\mathbf{u} + \mathbf{P}_1\{\mathbf{y} - \mathbf{C}_1\hat{\mathbf{x}}_1 - \mathbf{C}_2\hat{\mathbf{x}}_2\} \qquad (28)$$

$$\varepsilon\hat{\mathbf{x}}_2 = \mathbf{A}_{21}\hat{\mathbf{x}}_1 + \mathbf{A}_{22}\hat{\mathbf{x}}_2 + \mathbf{B}_2\mathbf{u} + \mathbf{P}_2\{\mathbf{y} - \mathbf{C}_1\hat{\mathbf{x}}_1 - \mathbf{C}_2\hat{\mathbf{x}}_2\} \qquad (29)$$

where $\mathbf{x}_1$ and $\mathbf{x}_2$ are state vectors $\mathbf{x}_1$ and $\mathbf{x}_2$ estimates. Observer gain matrices P_1 and P_2 were determined in order to satisfy all observability, detectability and controllability conditions required by the method (Porter [13]). Poles were alocated according to the order of magnitude of the fastest pole, giving the following values for P_1 and P_2:

$$P_1 = \begin{bmatrix} 8.53E+0 & -3.74E+3 & -9.25E+2 \\ 2.65E+3 & -1.09E+2 & -7.75E+1 \\ -2.98E-1 & 2.40E+2 & 7.94E+1 \\ 1.62E+1 & 5.54E+4 & -9.88E+2 \end{bmatrix} \qquad (30)$$

$$P_2 = \begin{bmatrix} 0.00E+0 & 0.00E+0 & 6.39E-1 \end{bmatrix} \qquad (31)$$

The design of LQ (linear-quadratic) controllers for slow and fast modes follows the steps established by Chow and Kokotovic ([14]). The objective is the minimization of:

$$J = \int_0^\infty \{\mathbf{x}^t \cdot \mathbf{Q} \cdot \mathbf{x} + \mathbf{u}_c^t \cdot \mathbf{R} \cdot \mathbf{u}_c\}dt \qquad (32)$$

where $\mathbf{Q} = \mathbf{C}^T\mathbf{C}$ and $\mathbf{R}$ are state and control variables weighting factors. The composed control vector $\mathbf{u}_c$ contains the contributions of both individual controllers for the slow and fast subsystems, that is:

$$\mathbf{u}_c = k_0 \cdot \mathbf{u}_s + k_f \cdot \mathbf{u}_f \qquad (33)$$

In Equation (33) κ_0, κ_f are constant gain matrices obtained through the solutions of the Riccati equations associated to the two LQ problems and are given by:

$$k_0 = \begin{bmatrix} 5.25E-3 & 6.47E-4 & -1.28E-2 & -1.06E-5 \\ 1.65E+1 & -6.43E+0 & -1.48E+2 & 1.95E+0 \\ -2.96E+0 & 1.97E-2 & 1.39E-3 & -4.94E-1 \\ 1.09E+1 & -1.77E-1 & 2.63E+0 & 9.24E-1 \end{bmatrix} \qquad (34)$$

$$k_f = \begin{bmatrix} -3.61E-6 \\ 0.00E+0 \\ 0.00E+0 \\ -1.09E+1 \end{bmatrix} \qquad (35)$$

SIMULATION AND RESULTS

It must be clear that the LQ controller requires on-line, real time measurements and instantaneous control actions. As this is not the case, two different restraints were included to obtain a more realistic simulation:

1) measures of p and s_1 are available with 20 min delays, whereas measures of o have a 15s delay. These measures are included into the simulation after an actualization cycle through a discretized linear model. They are considered perfect measurements (no noise contamination), although a Kalman Filter is under development for future inclusion on the simulation scheme.

2) Control actions involving feed pumps for s_{1e}, s_{2e} and D never occur in time intervals below 1 min. They use mean values for these periods. Dead bands are included in order to fit pumps minimal technical specifications.

The simulation scheme results as indicated in Figure 3.

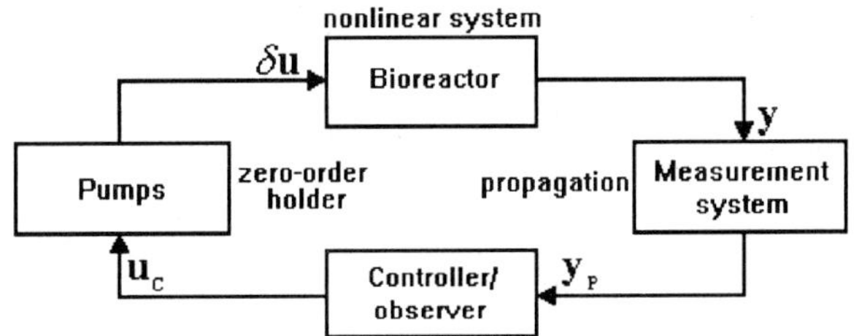

Figure 3 - Simulation scheme for the controlled bioreactor.

A program was developed at IPT to simulate the controlled bioreactor. Some other topics were also included, as a delay before starting control actions in order to allow a minimal convergence of the state observer to values near the "real" state vector.

Results achieved through computer simulation are shown in Figures 4 and

5.

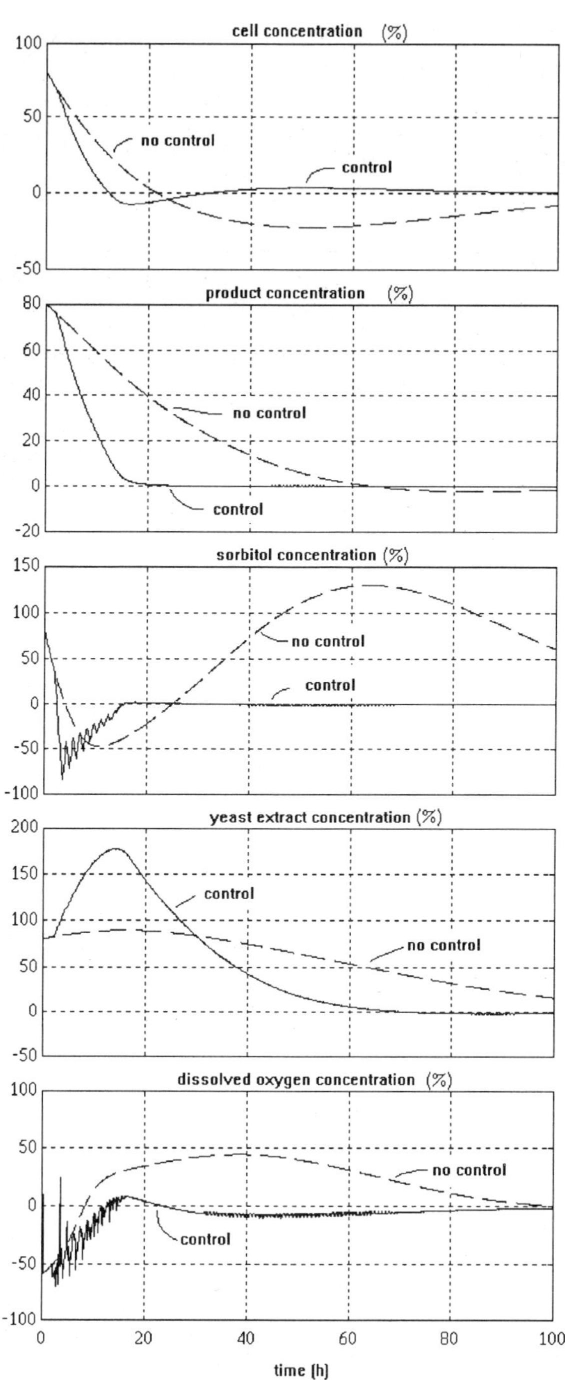

Fig. 4 - Non linear model simulation: controlled and non controlled systems.

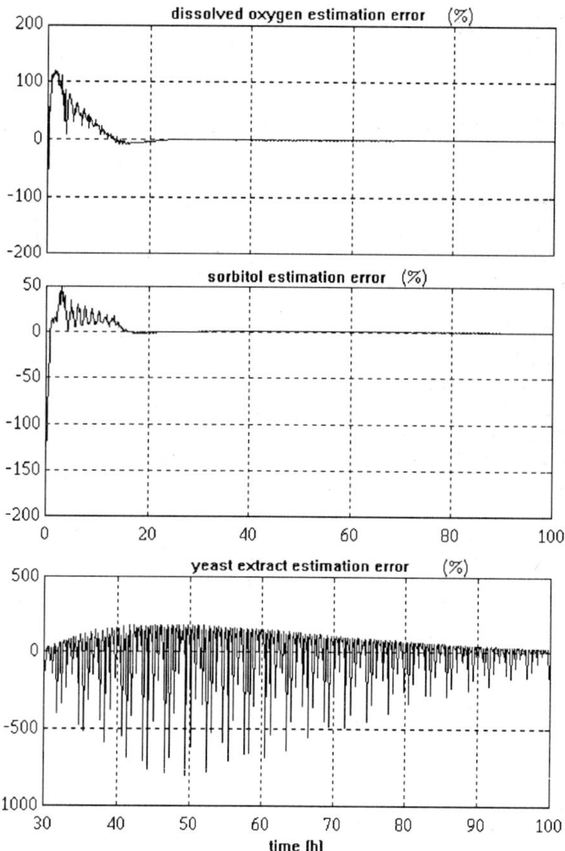

Fig. 5 - Estimation errors for o (fast), s_1 (slow) and s_2 (not available measure).

It can be seen from Figure 4 that the designed controller, although based on linear models, substantially improves bioreactor performance (non linear model) even for initial errors of 80% relative to the steady-state point for each of the state variables and even taking into account all the measurement delays. Curves for p and s_1 are particularly agreeable and very similar to the curves obtained in the ideal case of no measurement delays. This case was simulated but not included in the paper. The state observer also gives good results as can be seen from Figure 5. It is important to point out that the observer was simulated with initial errors of 160% relative to the real values. Large errors were used in the simulation bearing in mind that parameter variations and noise measurements are not yet included in the model.

CONCLUSION

The proposed unstructured mathematical model (Equations (1) to (6)), as well as the estimated parameters (Table 2), represented satisfactorily the microbial conversion of sorbitol to sorbose in a batch system in the range of the tested operating variables.

The Singular Perturbation Method seems to be an adequate tool for bioreactor controller design. Even for severe conditions in the simulation loop, where initial errors of 80% for each state variable, of 80%, in the other direction, for the state observer and all measurement delays, the designed controller was able to bring the system to the steady-state optimal point in time intervals compatible with the process dynamics.

This controller is being implemented on IPT's laboratory bioreactor. Next steps include the achievement of a new set of model parameters for continuous fermentation (batch process values were used throughout the paper), a deeper analysis on the weighting values of the matrices Q and R used for LQ controller design and the inclusion of probabilistic estimatiors (Kalman Filters).

LITERATURE CITED

1. BOUDRANT, J. (1990) "Microbial processes for ascorbic acid biosynthesis: a review" Enzyme Microb. Technol. 12, p. 322-329.

2. FLEURY, A. T.; QUIROZ, L. H. C.; LIMA, P. S. P. (1994) "Uma Aplicação do Método das Perturbações Singulares Ao Controle de Processos Fermentativos" Paper submited to the 10th Brazilian Congress Of Automatic Control, Rio de Janeiro, R. J., Brazil

3. KOKOTOVIC, P. V.; KHALIL, H. K.; O'REILLY, J. (1986) Singular Perturbation Methods in Control: Analysis and Design Academic Press, London.

4. MORI, H., KOBAYASHI, T. and SHIMIZU, S. (1981) "High density production of sorbose from sorbitol by fed-batch culture with DO-stat" J. Chem. Eng. Japan 14, p. 65-70.

5. BONOMI, A., AUGUSTO, E.F.P., BARBOSA, N.S., MATTOS, M.N., MAGOSSI, L.R. and SANTOS, A.L. (1993) "Unstructured model proposal for the microbial oxidation of D-sorbitol to L-sorbose" Journal of Biotechnology, 31, p. 39-59.

6. HIMMELBLAU, D.M. (1972) Applied Nonlinear Programming, McGraw-Hill, Inc., New York.

7. HIMMELBLAU, D.M. (1970) Processes Analysis by Statistical Methods, John Wiley & Sons, Inc., New York.

8. IMSL (1985) MATH/PC - LIBRARY. "Fortran subroutines for mathematical applications on a personal computer - User's Manual"

9. BONOMI, A., BARBOSA, N.S., AUGUSTO, E.F.P., BARRAL, M.F. (1993b) "Modelagem matemática de processos fermentativos - diferentes estratégias para a estimativa de parâmetros" In: Annals 14th Congr. Ibero Latino-Americano sobre Métodos Computacionais para Engenharia, 2, p. 1006-1015.

10. IPT (1992) "Desenvolvimento de Tecnologias para Automação e Controle de Processos Fermentativos" IPT Report 30.764/92, São Paulo, Brazil

11. GABRIELE, G. A.; RAGSDELL, K. M. (1989) "OPT 3.2: A Non Linear Programing Code in FORTRAN Implementing the Generalized Reduced Gradient Method - User's Manual" University of Missouri - Rolla-Rolla.

12. KOKOTOVIC, P. V.; HADDAD, A. H. (1975) "Controlability and Time-Optimal Control of Systems with Slow and Fast Modes" IEEE Trans Aut Control, V. AC-20, p. 111-113

13. PORTER, B. (1977) "Singular Perturbation Methods in the Design of Full-Order Observers for Multivariable Linear Systems" Int. J. Control, 26, p.589-594.

14. CHOW, J. H.; KOKOTOVIC, P. J. (1976) "A Decomposition of Near Optimum Regulators for Systems with Slow and Fast Modes" IEEE Trans Aut Control, V. AC-21, 5, p. 701

The Control Problem for Protein Separation by the Continuous Affinity Recycling Extraction Process

M.I. Rodrígues[1], R. Maciel Filho[2], and F. Maugeri[1]

[1]Facultadade de Engenharia de Alimentos, UNICAMP; [2]Faculdade de Engenharia Química, UNICAMP; Campinas, SP, CP 6121, CEP 13081/970, BRAZIL

Process modelling and simulation are important tools for better knowledge of systems, narrowing the ranges of experiments and leading to lower costs of process implementation. Both recovery yield and productivity in biochemical processes have been improved by genetic manipulation of microorganisms and by new techniques leading to better process control. It is important for the process economics to maximize yields, while maintaining high recovery efficiency and productivity. In this work, the dynamic modelling of the Continuous Affinity Recycling Extraction (CARE) process was carried out and its performance was studied against different disturbances. The parametric analysis and parameters sensibility studies were performed considering the steady state behavior. The appropriate manipulated variable was determined from the dynamic responses of the system, so that the control structure could be defined. The classical PI and PID feed back controls strategies were studied and both of them showed good performance when a time-delay of 8 minutes in the on-line analysis of the control variable was considered. The non-conventional feedback-feedforward control strategy was also studied and the results showed that the performance was good with time delays up to 24 minutes. In particular, this work showed that the CARE process under control can be a very attractive way to achieve continuous protein separation in industrial environments.

Generally, the cost of enzyme purification process is very high, especially due to the required level of separation efficiency and normally low productivity associated with the conventional processes. Many of the industrial enzyme separation processes operate in batch and this means that output conditions change with time.

On the other hand, continuous processes have the advantage of being able to operate at stable conditions as long as an adequate control strategy is employed. In fact, it is still possible to have optimized yield and productivity which makes some continuous separations process very attractive. This is the case of the continuous affinity recycling extraction (CARE) process considered in this work. It is based on the principle of cromatography by affinity and was originally proposed by Pungor *et al.* (1987).

The process, shown in Figure 1, consists of a continuous two stage reactor with recycle of adsorbent beads for the protein purification.

The adsorbing stage takes place in the adsorber (the first reactor) where the liquid feed from a fermentor (F_1), containing enzyme and contaminants, is contacted whith the adsorbent, so that the enzyme is linked to the suport. Desorption of the protein to be purified is obtained in the second reactor (desorber) by maintaining appropriate conditions, and particularly, by the action of a suitable eluting solvent. The adsorbent beads are recycled to the first reactor while the poor product is continuously removed (C_1). The second reactor is fed with a flow rate (F_2) containing an adequate solvent and leads to the desired output of a concentrated product (C_2).

The success of a such protein separation tecnique depends strongly on a good control of the exit stream (C_2) in spite of changes in feed conditions. In this work the performance of conventional control schemes with PI and PID control laws as well as feedback-feedforward algorithm are considered.

321

E. Galindo and O.T. Ramírez (eds.), Advances in Bioprocess Engineering. 321-328.
© 1994 Kluwer Academic Publishers. Printed in the Netherlands.

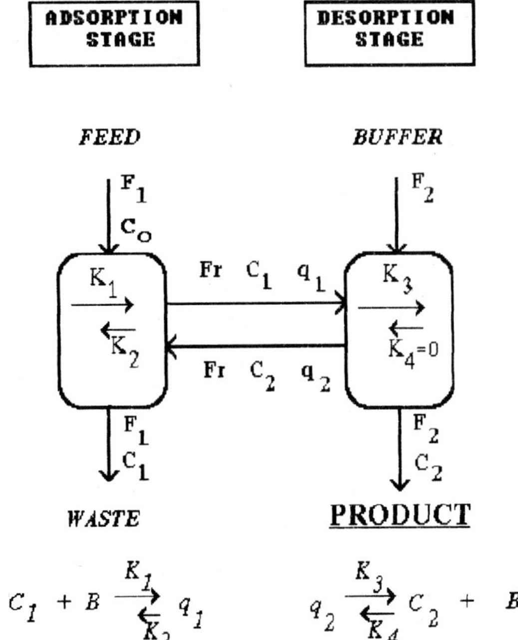

Figure 1 - The continuous adsorption recycle extraction system

Mathematical Modelling

A mathematical model of the continuous adsorption recycle extraction described before was developed by Rodrigues *et al.* (1992). It is based on the mass balance and kinetic equations describing the adsorption and desorption phenomena. In the model the two reactors are assumed to be perfectly mixed with the adsorption process being regarded as reversible second-order reaction whereas the desorption stage is considered to be a first-order irreversible reaction. For the protein concentration in liquid phase (C) and in gel phase (q) in the adsorber and desorber, respectively, the four governing equations are:

$$\frac{dC_1}{dt} = \frac{C_0 - C_1}{\tau_1} + \frac{\psi.\varepsilon\ (C_2 - C_1)}{\tau_1} + (k_2 q_1 - k_1 C_1 (q_m - q_1))\frac{1-\varepsilon}{\varepsilon} \quad (1)$$

$$\frac{dq_1}{dt} = \frac{\psi.\varepsilon.(q_2 - q_1)}{\tau_1} + (k_1.C_1.(q_m - q_1) - k_2.q_1) \quad (2)$$

$$\frac{dC_2}{dt} = \frac{\psi.\varepsilon.(C_1 - C_2)}{\tau_1} - \frac{C_2}{\gamma.\tau_1} + k_3.q_2.\frac{1-\varepsilon}{\varepsilon} \quad (3)$$

$$\frac{dq_2}{dt} = \frac{\psi.\varepsilon.(q_1 - q_2)}{\tau_1} - k_3.q_2 \quad (4)$$

The system of ordinary differential equations describing the dynamic behaviour was solved numerically using an algorithm based on the method of RungeKutta 4th order. The steady state solution is obtained by setting the accumulation terms (i. e. time derivatives) equal to zero, giving rise to a set of non-linear algebraic equations which can be solved analytically.

The model based on equations 1 to 4 is general enough to describe a number of CARE bioseparation process, since the kinetic parameters of the model are depend upon the pair adsorbent/enzyme.

Conventional Feedback Control Structure

Preliminary Considerations

Conventional feedback control is largely used in chemical and biotechnological industries. Indeed, it is cheaper and simpler to implement when compared with advanced control strategies, so that they have to be used when conventional controllers do not result in acceptable system performance.

In order to design an adequate control structure it is necessary to understand the system behaviour. To do that an extensive steady state simulation was carried out. The parametric analysis was based on the effect that changes in feed conditions as flow rates (F_1, F_2 and F_r), initial enzyme concentration (C_o) and system porosity (ε) caused on the major system variables, namely, free enzyme concentration at adsorption (C_1) and desorption (C_2) stages as well as the concentration of adsorbed enzyme at adsorption (q_1) and desorption (q_2) units, productivity, yield and concentraction factor (C_2/C_o).

The parameters used in the simulation come from Cowan *et al.* (1986), Chase (1984) and Pungor *et al.* (1987) and are shown in Table 1. The kinetics data to the lysozyme adsorption in sepharose gel are those obtained by Chase (1984).

Table 1 - Baseline Values

C_o	7.10×10^{-6}	[mol/l]
C_1	8.4×10^{-7}	[mol/l]
C_{2SP}	5.0×10^{-5}	[mol/l]
F_1	400	[ml/h]
F_2	50	[ml/h]
F_{2SS}	50	[ml/h]
F_r	15	[ml/h]
k_1	1.0×10^6	[ml.mmol^{-1}.h^{-1}]
k_2	1.8	[h^{-1}]
k_3	1.0×10^3	[h^{-1}]
q_m	1.0×10^{-3}	[mol/l]
q_1	3.1×10^{-4}	[mol/l]
q_2	1.9×10^{-8}	[mol/l]
T	0.20	[h]
V_1	100	[ml]
V_2	100	[ml]
ϵ	0,4	

Rodrigues (1993) found that the feed flow rate at the first (F_1) and second (F_2) stage in addition to the recycle flow rate (F_r) were indicated to be manipulated variables in order to control the process. Through the analysis of the dynamic behaviour of the system based on positive and negative 20% step changes around nominal values presented in Table 1, the time constant (τ_p) were determined and are in Table 2.

Table 2 - Characteristic time constants (h)

Input \ Output	c_1	c_2	q_1	q_1
F_1	15.0	18.1	15.8	15.8
F_2	17.0	**2.2**	17.6	17.8
F_r	16.9	____	16,9	____

These data lead to conclude that the new steady state as far as the purified enzyme (C_2), the controlled variable, is concerned is reached more rapidly for perturbation in (F_2) with τ_p being equal to 2.2 h. Therefore this variable (F_2) was choosen to be the manipulated variable.

Simulation Results

A primary control objective is to maintain the purified enzyme concentration (C_2) at desired value (set-point) at the exit of the 2nd reactor in such way that the enzyme concentration at the exit stream of the adsorber (C_1) is minimized, despite variations in the feed conditions.

For the process considered in this work, the product concentration is strongly affected by all inputs, so that appropriate monitoring and control schemes need to be designed. The implemented algorithm is a single-input single-output (SISO) feedback-control scheme based on the on-line measurement of protein concentration in the product stream, using the eluent flowrate as the manipulated variable. Two control laws, PI and PID, will be employed. The following variable classification was found to be adequate:

— input variables: F_1, F_2 and C_o

— output variables: C_1 and C_2

— disturbed variables: F_1, F_r and C_o

— controlled variable: C_2

— manipulated variable: F_2

Design parameters (ϵ, V_1, V_2) and kinetic data (k_1, k_2, k_3 and q_m) are essentially constants. For the feedback controller with a PID law, the eluent flowrate, in discrete velocity form, can be written as:

$$F_{2_i} = F_{2_{i-1}} + \left[K_c \left(1 + \frac{T}{\tau_1} + \frac{\tau D}{T} \right) E_i \right] - \left[K_c \left(1 + \frac{2\tau D}{T} \right) E_{i-1} \right] + \left[K_c \frac{\tau D}{T} E_{i-2} \right] \quad (5)$$

Obviouslly, the PI controller is easily obtained by setting $\tau_D = 0$ in the last

324

equation.

Classical Ziegler-Nichols procedure were used to do the controller parametrization and a refined tuning was achieved by simulation using the Integral Time-weighted Absolute Error (ITAE) as the performance criteria. The results are given as follows:

PI

$Kc = 0.895 \times 10^4$ l^2/mol.h

$\tau_I = 0.33h$

PID

$K_c = 0.5 \times 10^4 l^2/$ mol.h

$\tau_I = 1.4$ h

$\tau_D = 0.05$ h

A sampling time (T) of 0,2 h (12 min) was taken for on-line monitoring which is reasonable for industrial applications. This value was found, through extensive simulation, to be convenient taken into account the online system analysis limitations as well as closed-loop process stability. For example, Flow Injection Analysis (FIA) has been sucessfully used to monitor total protein concentration and enzyme activity during fermentation and downstrean processing (Papamichael *et al.*, 1990)

Figures 2, 3 and 4 show the PI and PID controller performance for perturbations (± 20% around reference values) in C_o, F_1 and F_r, respectively, when is assumed negligible time delay in the product measurement. It can be seen that variations in C_o and F_1 lead the system to have similar behaviour in terms of C_2. Both control laws present very good performance but a higher upset (maximum of 0.3%) occurs when the PID controller acts. For the recycle flowrate (F_r) changes, the free protein concentration in desorber (C_2) varies drastically reaching values around 2.5% higher than set-point and the control is obtained only after 4 hours.

As see in fig. 5, changes in the setpoint value (2% higher than the nominal value) are well supported by both, PI an PID controller, since the new setpoint is established in a short period of time (around 1 h).

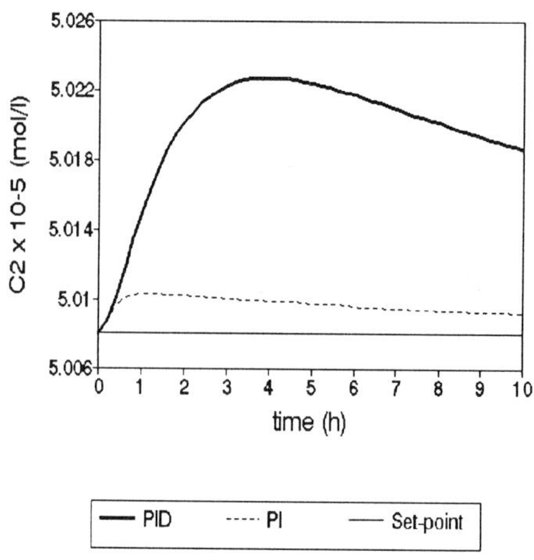

Figure 2 - PI and PID controller for perturbation in C_0 (20% higher).

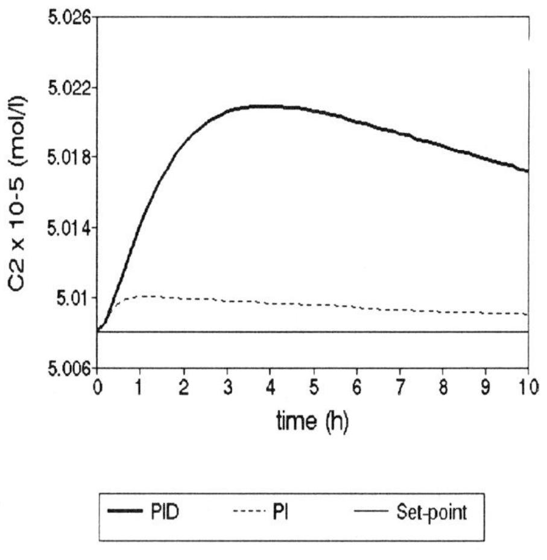

Figure 3 - PI and PID controller for perturbation in F_1 (20% higher).

On the basis of simulations, the feedback control algorithm with either PI or PID control law proved to be very effective in meeting the control objective even when drastic changes occur in all non-manipulated process

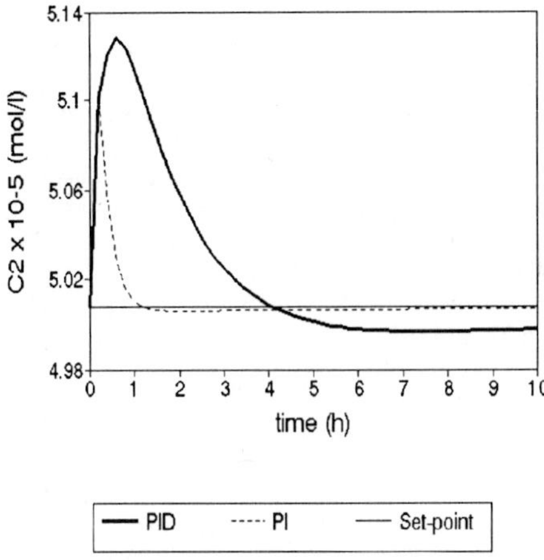

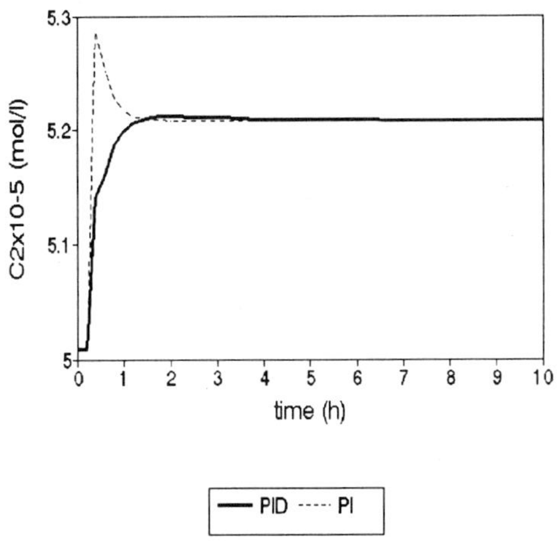

Figure 4 - PI and PID controller for perturbation in F_r (20% higher).

Figure 5 - PI and PID controller for changes in set-point (2% higher).

input as well as when the set-point is modified. Figure 6 shows a typical comparison between the closed loop and open-loop responses to a step change in the feed protein concentration. For the case without control action, the process is driven to a steady-state around 16% higher than the desired

conditions at the exit purified enzyme stream (C_2).

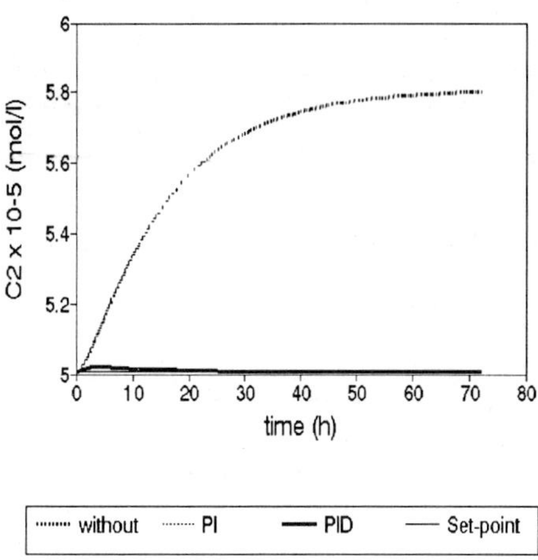

Figure 6 - Actions of PI and PID controllers and without control for perturbations in F_1.

Effect of Time Delay on Controller Performance

When industrial controller applications are required it is always usefulll to test the controller performance in the presence of delays in measurements. This has implications in the choice of the analytical technique to be used, especially when the state variable to be monitored is concentration.

In order to have information in a broad range of values, simulations of the controller performance were carried out for measurement delays of 0, 4.8, 7.2 and 9.6 minutes for the PI control and up to 12 minutes for case of PID control. The results are depicted in Figure 7 and 8 por PI and PID control laws, respectively. It can be concluded that the PI feedback controller has a good performance for concentration measurament delay of up to 7.2 minutes (aproximatelly 75% of the sampling interval). However, when a time delay of 9.6 minutes is considered the system presents a unstable behaviour.

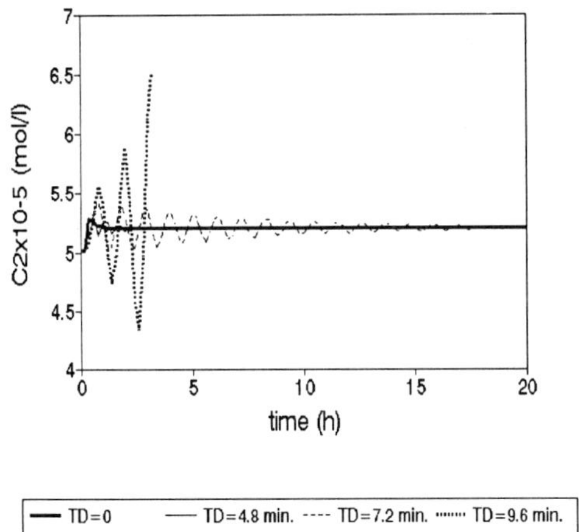

Figure 7 - Time delays in the PI controller for a change in the set-point.

On the other hand, the PID controller performance is very good (Figure 8) even for time delay of 12 minutes and this shows clearly the advantage of incorporating the derivative term in the control action.

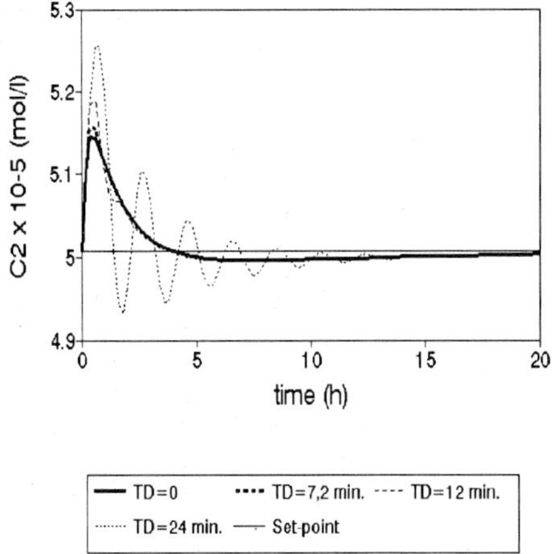

Figure 8 - Time delays in the PID controller for a perturbation in F_r

Feedback - Feedforward Controller

Basically, the limitation of pure feed-back controller is associated with the fact the control action only is active after measurement of the exit variable and its comparation with the desired

value (set-point). In some cases, inferential controller can be necessary but the same restriction in terms of feedback information is still present. One effective way to reduce the limitation is to implement a feedforward algorithm in addition of the feedback term. With this procedure it is possible to combine the advantages of both controller configurations.

This is the case for the process depicted in Figure 1, where the targeted protein concentration and flowrate in the feed stream are assumed to be measured on-line. Thus, a feedforward-feedback (FF) strategy can be implemented based on the state model obtained from eqs (1) to (4), with the time derivative set to zero. In this way, the steady state value of the manipulated variable (F_{2ss}) can be predicted from the measured disturbances, so that the following equation can be written:

$$F_{2_{SS(i)}} = f\left(F_{1_{(i)}}, C_{o_{(i)}}, F_{r_{(i)}} C_{2SP} \text{ and process parameters}\right)$$

$$(6)$$

and the PID feefback-feedforward controller in the discrete form, is given by:

$$F_{2_{(i)}} = \left[F_{2_{SS(1)}} - F_{2_{SS(i-1)}}\right] + F_{2_{(i-1)}} + \left[K_c 1 + \frac{T}{\tau_1}\frac{\tau_D}{T}.E_i\right] - \left[K_c 1 + 2\frac{\tau_D}{T}.E_{i-1}\right] + \left[K_c.\frac{\tau_D}{T}.E_{i-2}\right]$$

$$(7)$$

In order to verify the feed-back-feedforward controller algorithm performance, simulations were carried out assuming instantaneous action of the feedforward term as well as delays of 3, 6, 9 and 12 minutes in its action.

Figure 9 shows the results of the controller performance, including the conventional PID feedback, for changes in F_r (20% higher). It can be seen that when the feedforward acts immediatelly without delay, the set-point is obtained instantaneously. However, as should be expected, as the delay in the feedforward action is admited the set-point is not reached, but, even so the performance of the combined algorithm is better than when only feedback controler is used.

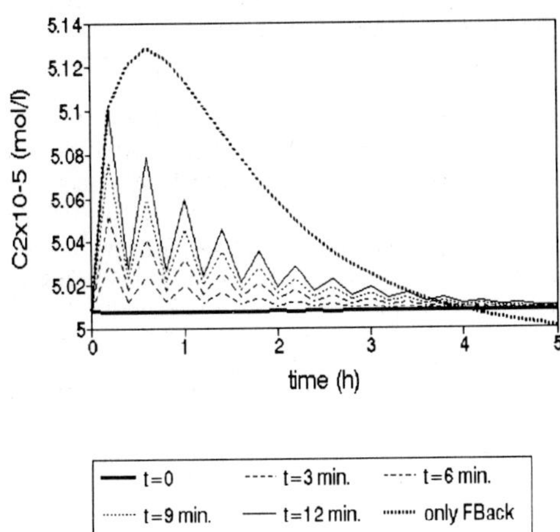

Figure 9 - FF contribution for the control system, with a F_r perturbation.

For the case of a 24 minutes delay in the concentration measurement in addition to a 6 minutes delay in the feedforward action, the control algorithm performs well, leading the set-point to be restablished after 6 hours of operation when the recycle flowrate is changed (20% higher) (Figure 10). This result is very encouraging since it shows that with the inplementation of the feedback-feedforward strategy a cheaper on-line (or even off-line) analytical technique for concentration measurement can be used while maintaining the system operation at high performance.

Conclusions

The mathematical model presented in this work provides important information on the system's dynamic and steady state behaviour, which otherwise would have required extensive experimental work.

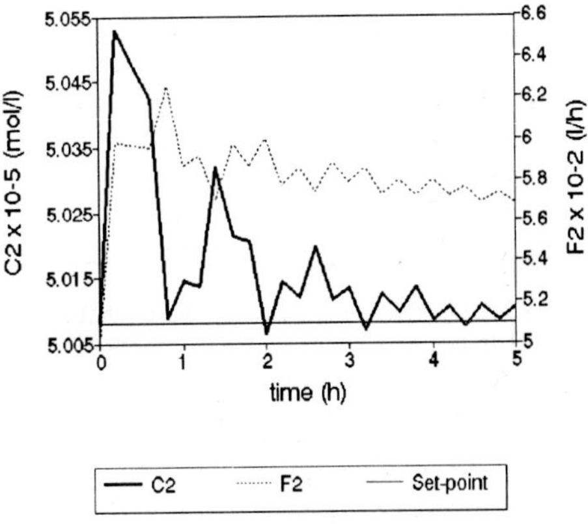

Figure 10 - Effects of a 6 minutes delay in FF action and 24 minutes delay in a FB.

On the basis of simulations, for the studied conditions, the free protein concentration in the desorber (C_2), the desired product, is effectively controlled by means of a feedback controller either PI or PID law, using the eluent flowrate as the manipulated variable.

The performance of the PI feedback is very good since the concentration measurement is provided no longer than 8 minutes. On the other hand, the inclusion of the derivative term in the control law allows concentration measurement delay to be 12 minutes while maintaining very good performance.

Model-based feedforward compensation showed to improve the performance of the controller, in case where the main disturbances can be measured on-line. In fact, the feedback-feedforward controller, used in discrete form, allowed very good process performance even where there exists 24 minutes of delay in concentration measurement and the feedforward acts with a delay of 6 minutes.

References

1. CHASE, H.A. ``Prediction of the performance of preparative affinity chomatography''. Journal of Chromatography, 297, 179-202, 1984.

328

2. COWAN, G.H.; GOISLING, I.S.; LAWS, J.F. and SWEETENHAM, W.P. "Physical and Mathematical Modelling to aid Scale-up of liquid chromatography", Journal of Chromatography, 363, 37-56, 1986.

3. PAPAMICHAEL, N., KRONER, K.H., SCHUTTE, H. and HUSTEDT, H. "Monitoring protein production whith on-line flow injection analysis, Paper presented at the 5th Euro-pean Conference on Biotechnologys, Copenhagen, 8-13 july, 1980.

4. PUNGOR, E.; AFEYAN., N.B.; GORDON, N.F. and COONEY, C.L. "Continuous Affinity-Recycle Extration: A novel protein separation techinique", Biotechnology; vol. 5, June, 604-608, 1987.

5. RODRIGUES, M.I.; ZAROR, C.A.; MAUGERI, F. and ASENJO, J.A. "Dynamic Modelling, Simulation and Control of Continuous Adsorption Recycle Extraction". Chemical Engineering Science, vol. 47, no. 1, 263-269, 1992.

6. RODRIGUES, M.I., PhD Thesis, FEA/UNICAMP, SP, BRAZIL, 1993.

Nomenclature

C_o	=	Feed protein concentration (mol/l)
C_1	=	Free protein concentration in adsorber (mol/l)
C_2	=	Free protein concentration in desorber (mol/l)
C_{2SP}	=	Set-point, product concentration (mol/l)
E_i	=	Error at i sampling interval, $(C_2 - C_{2SP})$ (mol/l)
F_1	=	Feed flowrate (ml/h)
F_2	=	Eluent flowrate (ml/h)
F_{2SS}	=	Steady state eluent flowrate (ml/h)
F_r	=	Recycle flowrate (ml/h)
k_1	=	Rate constant, forward adsorption reaction (ml mmol^{-1} h^{-1})
k_2	=	Rate constant, reverse adsorption reaction (h^{-1})
k_3	=	Rate constant, desorption reaction (h^{-1})
k_c	=	Proportional gain (l^2/mol.h)
q_m	=	Maximum adsorption capacity of gel (mol/l)
q_1	=	Bounded protein concentration in adsorber (mol/l)
q_2	=	Bounded protein concentration in desorber (mol/l)
t	=	Time (h)
T	=	Sampling interval (h)
V^1	=	Adsorber volume (ml)
V^2	=	Desorber volume (ml)
ε	=	Slurry voidage
τ_I	=	Reset time (h)
τ_D	=	Derivative time constant (h)
γ	=	Feed flow ratio (F_1/F_2)
ψ	=	Recycling rate (F_r/F_1)
τ_1	=	Residence time in the 1st stage
τ_2	=	Residence time in the 2nd stage
τ_p	=	Time constante (h)

Biodegradation of PCP in Contaminated Soil and in the Aqueous Phase

J.F. González[1] and W.-S. Hu[2]

[1]Fac. de Ingeniería, Universidad Nacional de Mar del Plata, Juan B. Justo 4302,
7600 Mar del Plata, ARGENTINA,
[2]Dept. of Chemical Engineering and Materials Science, Univ. of Minnesota,
Minneapolis MN 55455, U.S.A.

The decontamination of pentachlorophenol (PCP) laden soil was investigated. PCP was extracted from the soil followed by degradation by Flavobacterium sp. *Extractions were performed on whole soil with aqueous solutions, and the amount of PCP increased with NaOH concentration. The efficiency of extraction was found to be relatively low. The distribution of PCP in the particles was uneven. The smaller, slower settling particles contained most of the contaminant, while the larger particles, which constituted most of the soil mass were found to be contaminated to a lesser extent. We thus devised a scheme of separate treatment of the two fractions of the soil. The coarse, or sand, fraction was decontaminated by alkaline extraction. The PCP present in the extracts was then degraded by cells of* Flavobacterium sp. *The fraction of the smaller particles (i.e., fines) was treated by direct inoculation with cells of* Flavobacterium sp.

Pentachlorophenol (PCP) has been used as a wood preservative, insecticide and herbicide. A large amount of PCP was consumed by the wood-preserving industry (1, 2), and PCP contamination in soil poses a serious problem to the environment surrounding several wood treatment plants and saw mills (3,4). The contaminants presented in these soils are often transported into freshwater streams and lakes or in ground water by soil runoff. PCP has been found to be toxic to aquatic animals at concentrations as low as 0.5 μg/l (5).

Because of this toxicity, there is a need to decontaminate the PCP-laden soils. The possibility of eliminating PCP by biodegradation has been explored in a number of reports (6,7). One possible way of soil bioremediation has concentrated on the direct inoculation into the whole soil with bacteria capable of degrading the PCP (8,9). The results showed a large degree of variability. We explored a different approach of extracting PCP from soil followed by biodegradation of the extracts in a more homogeneous and controllable environment. This report describes our work on the

extraction of PCP from soil, soil fractionation, biodegradation of PCP and presents two simple models which take into account the loss of viability and the lag in PCP degradation.

<u>MATERIALS AND METHODS</u>

<u>Microorganism and medium</u> The *Flavobacterium* sp. was obtained from the Gray Freshwater Biological Institute of the University of Minnesota at Navarre, Minnesota. The medium was a modification of the minimal medium salts (MS) medium supplemented with 4.0 g/l of glutamate (10). The composition of the MS medium is as follows: KH_2PO_4: 0.17 g /l; K_2HPO_4: 0.65 g/l; $NaNO_3$: 0.5 g/l; $MgSO_4$: 0.1 g/l; $FeSO_4 \cdot 7H_2O$: 5.6 g/l.

<u>Cell preparation</u> *Flavobacterium* sp. were cultivated in a New Brunswick 2000 ml Multigen fermentor using a 1 l working volume. The agitation rate varied from 400 to 600 rpm. The pH was controlled at 7.4 $+/-$ 0.05 by the addition of 1.0 N NaOH or 1.0 N H_2SO_4 solutions using a ChemCadet pH

329

E. Galindo and O.T. Ramírez (eds.), Advances in Bioprocess Engineering. 329-335.
© 1994 Kluwer Academic Publishers. Printed in the Netherlands.

controller (Cole Parmer, Chicago, IL, USA). Growth was monitored by measuring the turbidity of the culture at 600 nm, using a Shimadzu UV-160 spectrophotometer (Shimadzu Scientific Instruments Inc., Columbia, MD, USA). One absorbance unit corresponds to $3.2*10^9$ cells/ml. In late exponential phase the cells were induced for PCP degradation with 50 mg/l of PCP. Induced cells were centrifuged, washed and re suspended in 5.0 mM potassium phosphate buffer (pH 7.4). The cells were then inoculated into the PCP solutions, the PCP containing extracts or slurries, as described later.

Soil and aqueous extraction of PCP
The soil, collected from a contaminated site in New Brighton, MN was obtained from BioTrol Inc. (Chaska, MN, USA). The soil contained a large amount of wood chips which were removed by sieving with a metallic mesh with an opening of 3 mm. PCP extraction from contaminated soil was carried out in 1 l glass beakers. The soil/water mixture was agitated at 380 rpm with a 9.0 cm wide Rushton-type impeller.

Degradation of PCP The degradation of PCP in solutions or soil extracts was carried out in 500 ml Erlenmeyer Flasks using 250 ml working volume.
     The degradation of PCP in slurries was carried out in a 1 l glass beaker using a magnetic bar at the bottom for stirring. The pH was controlled using a ChemCadet pH controller (Cole Parmer, Chicago, IL, USA) using a 1.0 N NaOH solution. In order to minimize evaporation, a rubber stopper was placed on top of the beaker. Water-saturated air was sparged from the bottom at 500 ml/min. No significant reduction in volume (other than that due to sampling) was observed. The temperature was controlled at 30°C ± 1°C using a heating tape.
     PCP concentration was measured in clear solutions by absorbance at 320 nm. In soil extracts and slurries, the PCP concentration was measured by high performance liquid chromatography (HPLC) using a Beckman 340 HPLC system under isocratic conditions. The mobile phase was 80% (v/v) of acetonitrile and 20% of 7.0 mM H_3PO_4.

The column used was a 25 cm C-18 reversed phase ODS column, with a flow rate of 2.0 ml/min. The PCP concentration in the samples was determined by comparison of their peak areas with those of external standards subject to the same treatment of the samples.

RESULTS AND DISCUSSION

Aqueous extraction of PCP from soil
     To estimate the extent to which PCP can be extracted from soil, extractions with tap water were initially performed. Two consecutive extractions on the same batch of soil only removed 20% of the contaminant (results not shown). The effect of NaOH en the extraction efficiency was then investigated. Three hundred grams of soil were mixed with 600 g of a 2.5 mM solution of NaOH. Extraction was carried out over a 2 h period. The contents were then allowed to settle and the supernatant was withdrawn slowly by siphoning with an 8 mm glass tube. Care was taken not to perturb the settled solids. Sodium hydroxide solution was then added to the same water/solids ratio. The same extraction procedure was repeated for the third extraction. The results are shown in Figure 1:

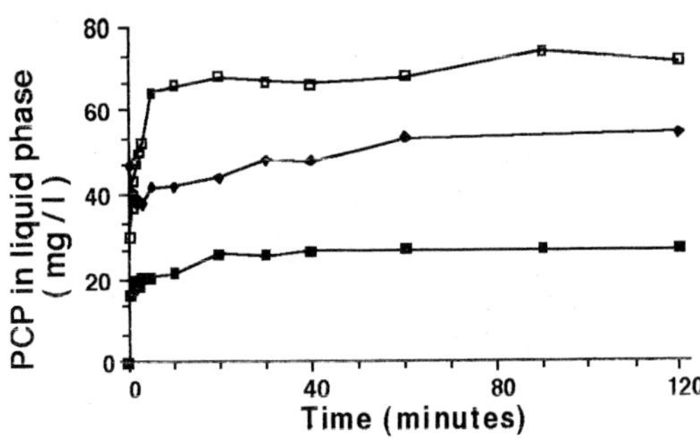

Figure 1: Kinetics of PCP extraction from soil in a cross-current scheme.
- □ 1st. stage, final pH 7.88
- ◆ 2nd. stage, final pH 8.38
- ■ 3rd. stage, final pH 9.33

     The amount of PCP extracted was lower in the second and third stages despite the higher pH reached. Overall, about 45% of the initial PCP

331

was transferred to the aqueous phase.
Replicate runs were made (data not
shown) with essentially the same
results. These and previous results
suggest that the decontamination of
the whole soil may not be carried out
effectively with the conditions used.
Although it might be possible to
extract more PCP from soil using more
concentrated solutions of NaOH, it was
deemed as unattractive due to the
excessive amount of alkali needed and
the possibility of forming stable
suspensions. This has been reported
to happen with humus-like components,
silt and clay particles, that readily
form a stable suspension with
extraction liquids at high pH (11).
This possibility might result in a
residual sludge which in turns needs
to be properly disposed of.

<u>Uneven distribution of PCP in soil</u>
In the extraction experiments
described above, the kinetics were
always fast in the first few minutes
after which the rate of PCP transfer
slowed down considerably. This
suggests that there exist more than
one mechanism by which the PCP
transfer takes place: one by which the
exposed part of the soil particles
rapidly reaches equilibrium, and
another in which pore diffusion takes
place. This may be due to the
heterogeneity in the soil particles,
since the particles of contaminated
soil are covered with the hydrocarbon
mixture used to dissolve the PCP in
the wood-impregnating process.

A sample of soil containing PCP
at 1450 mg/kg and oil/grease at 11.9
g/kg was mixed with an equal weight of
water and was stirred with the turbine
impeller for 2 h. The mixture was
transferred to a 100 ml plastic
graduated cylinder while being
stirred. The cylinder was then shaken
manually and then its contents were
allowed to settle overnight before
freezing them at -20°C. With its
contents frozen, the cylinder was cut
into slices to which the percent of
solids and PCP content were measured.
The results are shown in Figure 2.
The settled solids occupied the lower
quarter of the settled volume. The
supernatant contained a small amount
of unsettled solids (about 2% solids
content). At about 20 distance units

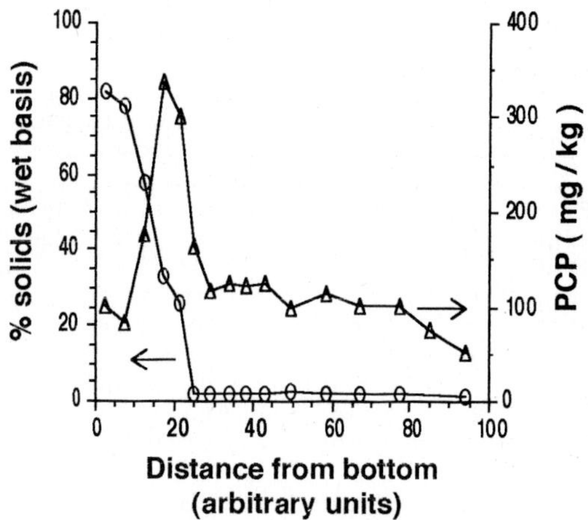

Figure 2: Distribution of solids (o)
and PCP (Δ) in settled slurry.
Experiment run in duplicate.

from the bottom, the percentage of
solids rises sharply. This indicates
the top of the settled solids. In the
bottom of the cylinder, the solids
constitute 80% of the mass. The PCP
concentration in the supernatant was
essentially constant, at about 100
mg/kg and was highest at about 20
distance units from the bottom. The
PCP concentration decreased again
towards the bottom to about the same
levels as in the supernatant. The
same results were obtained in a
duplicate run. This shows that the
PCP is distributed unevenly, with most
of the contaminant contained in the
smaller particles.

<u>Separate Treatment of the Soil
Fractions</u> It has been reported that
humus-like components , silt and clay
particles readily form a stable
suspension with the extracting liquid,
especially at high pH (11) Thus, if
PCP is to be extracted from these
particles, it is necessary to separate
these particles from the extract. If
a staged operation is to be employed,
the separation of these particles may
pose an even greater problem. The
larger particles in the soil are
easier to separate by settling and
therefore are more suitable to
decontamination by extraction. The
fine particles are difficult to
separate by settling. We thus
attempted to separate the soil into

two parts: one consisting of the relatively fast settling particles (sand fraction), the other (fines) consisting of the slowly settling particles. PCP in the sand fraction was extracted into the aqueous phase, followed by biodegradation by *Flavobacterium* sp. As for the fines, after they were separated from the sand, *Flavobacterium* sp cells will be inoculated directly to carry out the decontamination. In the laboratory-scale process that we used, a clear separation between the sand and fines could be observed and the separation of the sand posed no difficulties.

<u>Extraction of sand</u> To obtain the sand, soil samples were mixed with water and stirred with a turbine impeller. The slurry was allowed to settle and the sand was recovered from the bottom of the settled column. A continuous, counter-current three stage system, was simulated in the laboratory as shown in Figure 3.

A mixture (50% w/w) of sand and 0.1 M solution of NaOH was agitated for two hours in each stage and allowed to settle for 30 minutes. The supernatant was transferred to a second beaker containing fresh sand, while the original sand was extracted again with fresh NaOH solution. After extraction and settling in these two beakers, the contents were transferred in order to simulate the counter current scheme. The extraction proceeded as indicated by the arrows. Going downwards in the simulation scheme is equivalent to the passage of time in the start-up of an actual extraction system The composition of the streams gradually reaches a steady state value.

The speed with which the steady state of the simulation is approached is a function of the equilibrium reached in the stages. Further details of the simulation procedure can be found elsewhere (<u>12</u>)..The PCP concentration in the extracts (liquid phase) and raffinates (extracted solids) were measured The results of the simulation are shown in Figure 4.

The PCP content in the sand was very low, always below 30 mg/kg. The solid samples were analyzed after settling, but with extractant still present in the interstices.

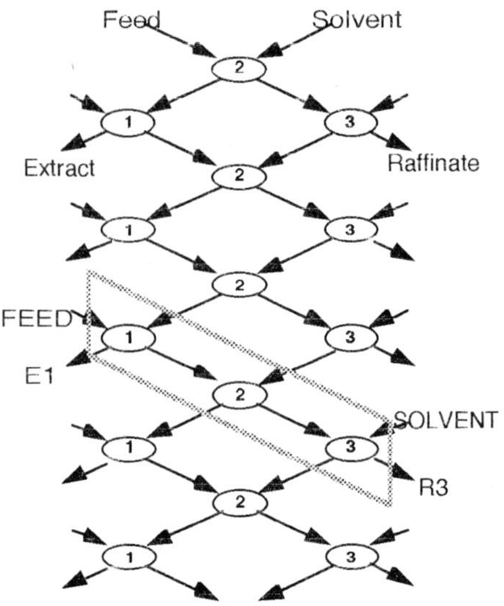

Figure 3: Scheme of a bench scale, three stages continuous counter-current simulation of an extraction. Each oval represents a mixing and settling operation.

It is therefore reasonable to assume that the sand itself was essentially decontaminated. The PCP present in the solvent at the output of the third stage reached a value of 630 mg/l in the 4th and 5th cycles of the simulation. These results indicate that the sand can be readily extracted and that the steady state in the simulation is reached in a relatively rapid manner.

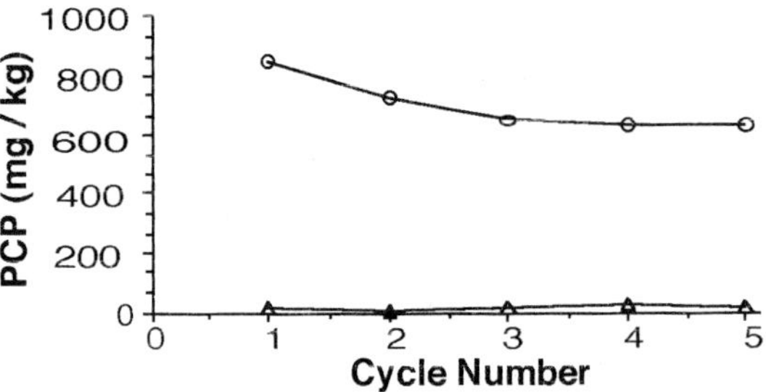

Figure 4: Results of simulation of countercurrent extracttion of sand
   o PCP content in extract (supernatant)
   Δ PCP content in raffinate (solids)
Cycle 1 corresponds to the first complete with three stages in Figure 3

<u>Treatment of the extracts</u> The extracts were then centrifuged at about 12000 g for 10 minutes, and the supernatant was used for subsequent biodegradation. The concentration of PCP in the extract was inhibitory to Flavobacterium (<u>13</u>). Thus, the extract was diluted with tap water to various PCP concentrations ranging from 30 to 130 mg/l and pH was adjusted to 7.4. Flavobacterium cells were inoculated at an initial cell concentration of about 0.3 absorbance units at 600 nm. The results are shown in Figure 5.

The PCP was degraded in all cultures. These results demonstrate that the sand, which constitutes most of the soil mass can be essentially decontaminated by alkaline extraction followed by biodegradation of PCP present in the extracts. The same results were obtained in duplicate runs (data not shown).

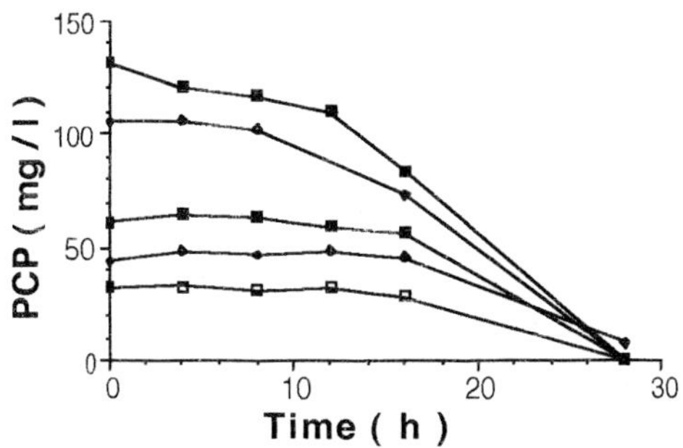

Figure 5: Degradation of PCP in extracts. Effect of the initial PCP concentration.

<u>Treatment of Fines by Direct Inoculation of Cells.</u> The fraction of fines from soil was subject to direct inoculation with Flavobacterium sp.

Six hundred grams of a 9% (w/w) slurry of soil fines and mineral salts medium were placed in a 1000 ml beaker. The initial cell inoculum was bout 0.1 absorbance units. Samples were taken periodically and analyzed for soluble and total PCP. The results are shown in Figure 6.

The soluble PCP decreased from 75 mg/kg at the start to undetectable (less than 1 mg/l) concentrations after 3 days. The total PCP decreased from 190 mg/kg to about 25 mg/kg in the same time period. One of the possible reasons for incomplete

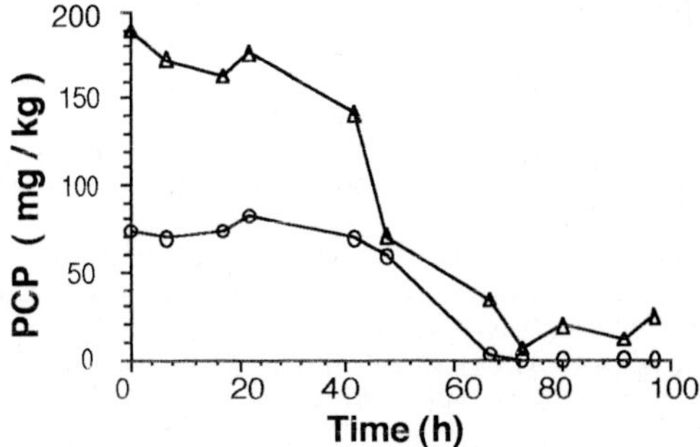

Figure 6: Total (▲) and soluble (○) PCP in a 10% fines slurry inoculated with Flavobacterium.

degradation of PCP in the soil particles is the strong adsorption of PCP . This has been reported previously (<u>14,15</u>) and has been proposed as a means of retaining the contaminants in the soil (<u>16</u>).

Another possible explanation for the incomplete degradation is the loss of viability of Flavobacterium cells. A viable count in Day 5 of the culture was less than 10^4 colony forming units (cfu/ml).

The reason for the decrease in Flavobacterium viability is not known. However, such a phenomenon of loosing viability was subsequently observed with other lots of soil and is probably rather common. We subsequently employed agar plates with MS and cellobiose as the sole carbon source for viable count. The use of cellobiose reduced the growth of indigenous flora and allowed the colonies of Flavobacterium to be enumerated. A slurry (15% w/w) of fines fraction was prepared from another lot of soil. Flavobacterium sp. cells were inoculated at about 10^8 cells/ml. The concentration of Flavobacterium decreased by more than four orders of magnitude in 5 h and no PCP degradation was observed (data not shown). In another experiment, the same slurry was heat-sterilized (121°C, 15 min.) to kill predators and other competing species that might be present in the soil. The same inoculation treatment was then performed. Cell concentration decreased from 1.3 $x10^7$ / ml to 2.2 $x10^6$/ml in 4 h, which was not as severe as in the case without heat treatment. The soluble PCP was completely degraded in 50 h from its original level of 90 mg/l (Figure 7). The loss of viability and its prevention by heat treatment could possibly be caused by disappearance of predators or removal of heat labile compound(s) toxic to *Flavobacterium*. In other studies of PCP degradation by *Flavobacterium* in MS medium, rapid loss of viability at high PCP concentrations was observed (<u>13</u>).

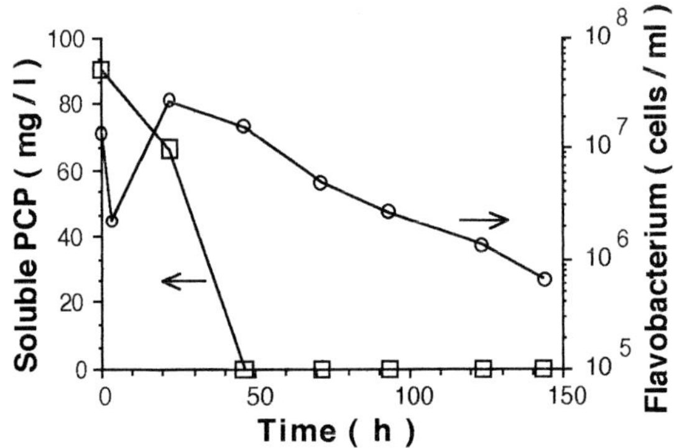

Figure 7: Treatment of sterilized soil 15% (w/w) slurries.
o Flavobacterium cells
□ Soluble PCP

It has also been shown that such loss of viability can be prevented by addition of glutamate to the medium (<u>13</u>). However, the nature of the effect of glutamate is not yet clear. furthermore, it is not known whether carbon sources other than glutamate will have a similar effect. If other carbon sources have a similar effect on retaining the viability , the results of the heat treatment reported above could also be explained by the possible release of compounds that can be utilized by *Flavobacterium* from soil by heating.

CONCLUSIONS

A scheme for soil treatment based on the separation of the soil in two major fractions: "sand" and "fines" was devised. The sand fraction could be essentially decontaminated by

extraction followed by biodegradation with *Flavobacterium* sp. inoculated in the extracts. The soil fines were also decontaminated with direct inoculation of cells; however, a residual amount of PCP was still present in the solids. The incomplete decontamination in the fines was thought to be caused, at least, partially by the loss of viability of *Flavobacterium* cells in the slurry.

LITERATURE CITED

1.- Hoos, R A W   Patterns of Pentachlorophenol Usage in Canada - An Overview. In: *Pentachlorophenol* (K. Ranga Rao, Ed.). Plenum Press, New York, USA. (1978).

2.- 1.- Cirelly, D P   In: *Pentachlorophenol* (K. Ranga Rao, Ed.). Plenum Press, New York, USA. (1978).

3.- Rao, Ranga K.   , In: *Pentachlorophenol* (K. Ranga Rao, Ed.). Plenum Press, New York, USA. (1978).

4.- Valo, R; V.Kitunen, M.S. Salkinoja-Salonen and S. Räisänen . . *Wat. Sci. Tech.* **17**:1381. (1985).

5.- Borthwick, P.W. and Schimmel, S.C. In: *Pentachlorophenol* (K. Ranga Rao, Ed.). Plenum Press, New York, USA. (1978).

6.- Stanlake G.J. and R.K.Finn. *Appl. Environ. Microbiol.* **44**:1421. (1982).

7.- Apajalahti, J.H. P. Karpanoja and M.S. Salkinoja-Salonen Rhodococcus *Intl. J. of Systematic Bacteriology* **36**:246. (1986)

8.- Crawford,R.L. and Mohn,W.W. *Enzyme Microb. Technol.* **7**:617. (1985).

9.- Edgehill R.U. and R.K.Finn. Eur. J. Appl. Microbiol. Biotechnol. **16**:179 1982.

10.- Watanabe,I. *Soil Sci. Plant Nutr.* **19**:109. (1973).

11.- Assink,J.W, Extractive Methods for Soil Decontamination: Operational Treatment Installation presented at the Netherlands. 2nd. International

Conference on New Frontiers for Hazardous Waste Management. Pittsburgh, Pennsylvania, September (1987).

12.- Alders, L. *Liquid-liquid extraction,* 2nd. Ed. Elsevier Publishing Company. Amsterdam. (1959)

13.- González, J.F. and W-S Hu . *Appl. Microbiol. Biotechnol* **35**:100. (1991)

14.- Boyd,S.A, J.Lee and M.M.Mortland, . *Nature* **333**:345.(1988).

15.- Schellenberg,K, C.Leuenberger and R.Schwarzenbach. Environ.Sci.Technol. **18**:652. (1984).

16.- Boyd,S.A, M.M.Mortland and C.T.Chiou, Soil Sci. Soc. Am. J. **52**:652.(1988).

17.- González, J.F. Pentachlorophenol Degradation by a *Flavobacterium* sp. and its Application to Soil Decontamination. Ph.D. Thesis, Univ. of Minnesota, Minneapolis, MN, U.S.A. (1990).

Biodenitrification Studies with a Bioreactor Operating in a Periodic Mode

J.-H. Wang, B.C. Baltzis, and G.A. Lewandowski

Department of Chemical Engineering, Chemistry, and Environmental Science,
New Jersey Institute of Technology, Newark, NJ 07102, U.S.A.

Biodenitrification of nitrite and nitrate/nitrite mixtures was studied with a pure culture of Pseudomonas denitrificans *(ATCC 13867), which is a strict anaerobe. The study involved theoretical analysis, and experimental verification of the biodenitrification process when a reactor operating in a periodic mode is employed. This reactor operates continuously in a cyclic mode, and each one of its cycles is made-up of three phases. During the first phase, the stream to be processed is fed to the reactor; reaction is initiated while the volume of the liquid hold-up of the reactor increases. During the second phase, reaction proceeds in the batch mode. During the third phase, part of the reactor contents are drawn-down till the volume of the reactor liquid hold-up returns to the value it had in the beginning of the cycle. At that point, the cycle starts repeating itself. A complete mathematical analysis of nitrite bioreduction is presented along with results from experiments performed with a 2 l reactor. Experimental results involve denitrification of water media containing either nitrite only, or mixtures of nitrate and nitrite. Theoretical predictions and experimental data are in excellent agreement.*

INTRODUCTION

Biological destruction of pollutants present in process waste streams is a very attractive alternative to other technologies, such as incineration and chemical catalytic oxidation. Biological treatment is both economical, and environmentally friendly as it occurs at ambient temperature. Biological treatment usually involves oxidation of organic compounds, but it can also involve reduction of pollutants such as nitrate and/or nitrite. These substances are present in waste streams of industrial processes such as munitions manufacturing. The volume of waste generated by these processes may be very large [1] and thus, proper process analysis and design may have a substantial impact on both the capital and operating cost of waste treatment facilities.

The major problem with biodenitrification is the fact that even when nitrite is not present in the original waste, it is produced during the process of nitrate reduction. Nitrite, when present at high concentration levels inhibits the reaction rate [2], and tends to accumulate in the reactor [3, 4], rather than being further reduced to nitrogen which is the desired final product. In cases where mixed cultures are used for denitrification of wastewater streams, a possible reason for nitrite accumulation can be population shifts as explained by Wilderer et al. [5]. These researchers have shown that under certain conditions the composition of the biocommunity changes in favor of facultative anaerobes which can reduce nitrate to nitrite only, and in the expense of the true denitrifying organisms. These results seem to suggest that either pure cultures of denitrifiers should be used, or the conditions of operation should be properly selected and monitored so that population shifts are prevented.

Nitrite concentrations in a bioreactor have been found to reach different maximum values even in cases where the biomass is assumed to have a constant composition, and thus the kinetics of the process remain unchanged. In such cases, the differences in the maximum nitrite concentration levels are simply the result of alternate approaches to bioreactor operation. Baltzis et al. [6], and Lewandowski and Baltzis [1], have

337

E. Galindo and O.T. Ramírez (eds.), Advances in Bioprocess Engineering. 337-344.
© *1994 Kluwer Academic Publishers. Printed in the Netherlands.*

shown that the rate at which a Sequencing Batch Reactor (SBR) is fed with the waste, as well as the ratio of the minimum to maximum reactor contents volume, have a profound effect on nitrite accumulation and complete denitrification. The aforementioned studies were based on the assumption that nitrate reduction follows Monod type kinetics [7], while nitrite reduction follows Andrews inhibitory type of kinetics [8].

The Sequencing Batch Reactor (SBR) mentioned above, is a reactor which has been widely used in aerobic treatment of liquid wastes, and in fewer anaerobic applications. In what could be called a classical SBR, one has a discontinuously operating reactor which processes each "batch" according to the following sequence. The waste is fed to the reactor slowly or rapidly during the fill-phase; reaction during this phase may or may not occur. Once the reactor is filled to a particular maximum volume, the react-phase begins, and during it the reactor operates in the classical batch mode. When the concentrations of the pollutants reach acceptable levels, mixing and/or aeration stops and the third phase in the sequence begins. During this settle-phase, the biomass is allowed to settle and then, in the fourth phase of the sequence, the effluent is drawn from the reactor. The effluent is the biomass free (or lean) part of the reactor contents at the end of the third phase. This mode of operation allows the biomass to be retained in the reactor at the end of the sequence. This sequence may be repeated immediately, or after a sometimes considerable idle period. These reactors are attractive when compared to the so called "activated sludge" continuous reactors because they do not require a separate clarifier, allow for cycling between aerobic and anaerobic conditions, present greater flexibility in meeting changes in the feed conditions, and allow for effective control of the quality of the discharge. Such SBRs have been extensively studied in the environmental literature [e.g., 9, 10, 11].

In our approach, we have eliminated the idle-phase from SBR operation, and -at a first step- we are not considering settling of biomass either. The latter has been done in order to separate reaction from physical phenomena such as settling, and investigate the dynamics of continuous cyclic operation, as well as its implications for reactor design. In earlier studies [12, 13] it has been shown that this type of operation leads to volumetric efficiencies greater than those of equivalent CSTRs, and that the complexity of the dynamics may allow for stable coexistence of competing microbial species, something impossible in CSTRs when the competition pattern is pure and simple [14].

The present work originated from earlier studies memtioned above [1, 6], which indicated that this continuous cyclically operated bioreactor has potential advantages regarding denitrification. The objective was to perform a complete analysis of the dynamics of the process, and subsequently validate experimentally the theoretically expected behavior. To avoid potential complications from microbial interactions, a pure culture was selected. The kinetics of denitrification were first investigated in shake flask experiments. Knowing the kinetics, we analyzed the process dynamics and then preformed experiments which validated the theory.

THEORY

The reactor considered in this study operates as follows. At the beginning of a cycle (t = 0) the reactor contains a liquid volume V_0. Feeding begins at a constant flowrate, resulting in a linear increase of the volume in the reactor, and stops at time t_1. The second phase of the cycle lasts from t_1 to t_2, and during it there is no inlet to, or outflow from the reactor thus, the volume of the reactor contents remains constant. At time t_2, the draw-down phase begins, again at a constant flow rate, and the volume drops linearly to V_0 (i.e., its value at t = 0) at a time t_3 which denotes the end of the cycle. The cycle is

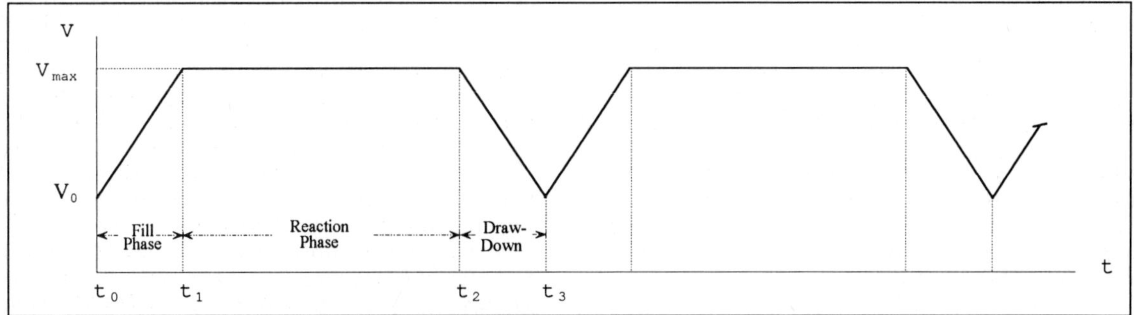

Figure 1: *Schematic representation of the variation of reactor contents volume during the cyclic operation of the reactor.*

then repeated. A schematic representation of this mode of operation is shown in Figure 1. Reaction occurs throughout the cycle, unless the concentration of nitrate and/or nitrite drops to zero before $t = t_3$. No settling of biomass is considered thus, a certain amount is lost during the third phase of the cycle. Unless the biomass lost is made-up for during the first two phases of the cycle, the culture is eventually washed out from the reactor.

The model equations for the general case where the inlet contains both nitrate and nitrite, constitute a set of mass balances for the biomass and the two pollutants of interest. The form of the equations depends on the phase of the cycle, and (in dimensionless form) they are as follows.

$$\frac{du}{d\theta} = \frac{(u_f - u)M}{\delta\sigma_1 + \theta} - \beta x g(u,z) \qquad (1)$$

$$\frac{dz}{d\theta} = \frac{(z_f - z)M}{\delta\sigma_1 + \theta} + \rho\beta x g(u,z) - \eta\beta x f(u,z) \qquad (2)$$

$$\frac{dx}{d\theta} = -\frac{xM}{\delta\sigma_1 + \theta} + \beta x\big[g(u,z) + f(u,z) - h(u,z)\big] \qquad (3)$$

Equations (1), (2), and (3) are for nitrate, nitrite and biomass, respectively. During the first phase of the cycle, i.e., for $0 \le \theta \le \sigma_1(1-\delta)$, the value of M is 1. For the remaining parts of the cycle, that is, for $\sigma_1(1-\delta) \le \theta \le 1-\delta$, the value of M is equal to zero. Functions g(u,z) and f(u,z) represent the specific growth rate of the biomass on nitrate and nitrite, respectively. Function

h(u,z) represents biomass loss due to maintenance. The particular form of the three functions mentioned above, depends on the particular culture which is used for carrying out the denitrification. For *P. denitrificans* (ATCC 13867) which was used in the experiments performed for the present study, the specific forms of the functions are,

$$g(u,z) = \frac{\varphi u}{1 + u + \gamma_1 u^2 + \varepsilon_1 uz} \qquad (4)$$

$$f(u,z) = \frac{z}{\omega + z + \gamma_2 z^2 + \varepsilon_2 uz} \qquad (5)$$

$$h(u,z) = \lambda_1 \Delta(u) + \lambda_2 \Delta(z) \qquad (6)$$

The form of functions (4) - (6), and the values of all kinetic parameters appearing in equations (1) - (6) were determined from an independent detailed kinetic study involving shake flask experiments. The results of that study have not yet been published, but the values of the kinetic parameters are shown in Table 1.

The model equations contain five parameters which can be varied, β, σ_1, δ, u_f, and z_f. The analysis of the model entails finding how the system behaves when these parameters assume different values. Since the input to the system changes periodically, the system responds in an oscillatory fashion and when transients decay, a stable periodic orbit is reached instead of a steady state. The system may reach different final orbits (known as limit cycles); for example, the culture may wash-out or establish itself in the reactor. The analysis then, is based on the bifurcation theory of forced systems, and is done by using special software packages. The general principles of this type of analysis have been discussed elsewhere [13], and the same methodology was used in the present study.

Table 1. Kinetic parameter values for the model equations

Parameter	Value
γ_1	0.46
γ_2	0.90
$\Delta(u)$	1, if $u > 0$
$\Delta(u)$	0, if $u = 0$
$\Delta(z)$	1, if $z > 0$
$\Delta(z)$	0, if $z = 0$
ε_1	0.096, if $z > 0.47$
ε_1	0, if $z < 0.47$
ε_2	4.80, if $z > 0.47$
ε_2	0, if $z < 0.47$
η	1.0
λ_1	0.084
λ_2	0.065
ρ	0.616
φ	0.71
ω	1.65

Figure 2 shows a bifurcation diagram for the case where only nitrite is fed to the reactor. This diagram shows the nitrite concentration at the end of a limit cycle, as a function of β, when all other parameters are fixed. For low values of β (region III), the concentration is equal to that in the incoming stream, something which implies that eventually the system cannot reach but the washout state. For values of β falling in region I, the culture establishes itself in the reactor, washout cannot occur, and the nitrite concentration at the end of the cycle drops as the hydraulic residence time increases. The interesting feature of this diagram, is region II. Theory predicts that for β values in this region, the system can reach two states under the same operating conditions: survival or washout. The actual state is determined by the conditions (concentrations) used during process startup. Theory also predicts that in region II, even when the culture is established and denitrification proceeds, relatively small perturbations can lead the system to washout. This diagram can help deciding what parameter values (e.g., β) should be used for attaining a stable pattern of operation. More information about the behavior of the system is contained in diagrams such as the one shown in Figure 3. In this case, two of the operating parameters are varied, and the behavior of the system predicted. Regions I, II, and III have the same meaning as in Figure 2.

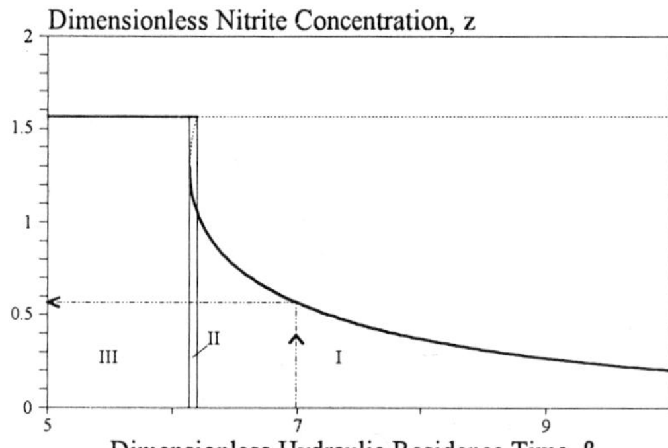

Figure 2: Bifurcation diagram for the case where $z_f = 1.56$, $u_f = 0$, $\delta = 0.5$, and $\sigma_1 = 0.1$. Regions I, II, and III are discussed in the text.

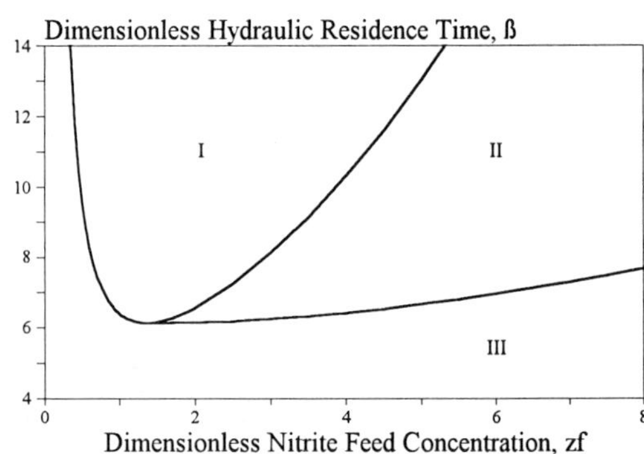

Figure 3: Predicted outcomes (see text) for the cyclically operated bioreactor when $u_f = 0$, $\delta = 0.5$, and $\sigma_1 = 0.1$.

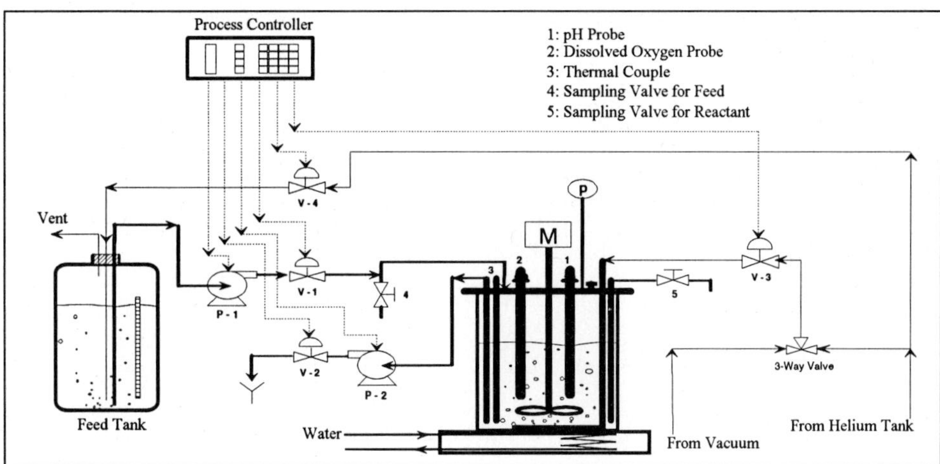

Figure 4: *Schematic diagram of the experimental unit*

EXPERIMENTAL RESULTS AND DISCUSSION

In order to verify the theoretical predictions, experiments were performed with a unit the schematic of which is shown in Figure 4. A time sequence controller was used, and allowed for automatic continuous operation. Experiments were run at 30°C, and the medium had enough buffering capacity so that the pH was maintained at 7.1 ± 0.1. The medium was prepared by slightly modifying the one proposed by Koike and Hattori [15]. Per 1L, it contained 1.5g K_2HPO_4, 5g KH_2PO_4, 5g $NaCl$, 0.2g $MgSO_4$, 0.2g $CaCl_2$, and trace amounts of a mixture containing $MnSO_4$, $CuSO_4$, $FeCl_3$, and $Na_2MoO_4 \cdot 2H_2O$. Methanol was used as the carbon source and was added to the medium so that the methanol to nitrate/nitrite ratio was 2 to 1 (molar). Nitrate and nitrite were added as KNO_3 and KNO_2, and their concentrations were monitored via Ion Chromatographic analysis. Biomass was monitored spectrophotometrically via optical density measurements.

In order to ensure that oxygen was not present in the system, the tank carrying the feed medium, and the medium in the reactor (before inoculation) were purged with helium gas at the start-up of each experiment. During the experiments, the reactor and the feed tank were perfectly sealed, and their contents were kept under a helium atmosphere of slightly positive pressure. As indicated in Figure 4, the unit was designed in a way that helium was automatically supplied to the reactor and the feed tank, whenever there was a slight loss of pressure. In addition, a dissolved oxygen probe was continuously immersed in the reactor. In the rare occasions when the probe indicated dissolved oxygen presence, the experiment was stopped.

Results from three experiments are presented here in Figures 5 - 7. These figures show experimental data, and model predicted concentration profiles shown as solid curves. The conditions under which these experiments were performed are shown in Table 2. In the same table the model parameter values needed for solving equations (1) - (6) are also shown; these values are needed in addition to the kinetic parameters given in Table 1.

Experiments 1 and 2 (Figures 5 and 6), involved nitrite only and correspond to Figure 2. Under the conditions of Experiment 1, theory predicts that the culture cannot eventually survive in the reactor. This is also the trend which is experimentally observed in Figure 5. The experiment was terminated after the fourth cycle because the biomass became inactive, possibly due to oxygen presence. For the first three cycles though, the agreement between experimentally measured, and model predicted concentration values is ex-

Table 2. Conditions for Experiments 1, 2, and 3

	Exp. 1	Exp. 2	Exp. 3
Maximum volume of reactor contents (L)	2.0	2.0	2.0
Minimum volume of reactor contents (L)	1.0	1.0	1.0
Nitrate concentration in feed stream (mg/L)	-	-	34.4
Nitrite concentration in feed stream (mg/L)	51.1	51.0	106
Nitrate in the reactor at start-up (mg/L)	-	-	23
Nitrite in the reactor at start-up (mg/L)	30.9	26.9	97.1
Biomass in the reactor at start-up (mg/L)	2.1	8.8	8.5
Total cycle time (h)	4.0	5.0	5.0
Time of fill phase (h)	0.4	0.5	0.5
Time of draw-down phase (h)	0.4	0.5	0.5
Feed flowrate (L/h)	2.5	2.0	2.0
β	5.59	6.99	6.99
u_f	-	-	1.08
z_f	1.6	1.59	3.3
u_o	-	-	0.72
z_o	0.97	0.84	3.0
x_o	0.21	0.89	0.86
δ	0.5	0.5	0.5
σ_1	0.5	0.5	0.5

cellent. Under the conditions of Experiment 2, theory predicts that the only possible outcome is survival of the culture with a nitrite concentration of about 18 mg/L (or $z = 0.56$) at the end of the cycle, as shown by the arrows in Figure 2. This is also what is experimentally obtained as shown in Figure 6. Again, experimental and model predicted concentration profiles are in very good agreement. In the case of Experiment 2, the feed contains nitrite at a level of 51 mg/L, so at the end of each cycle a 65% conversion is achieved.

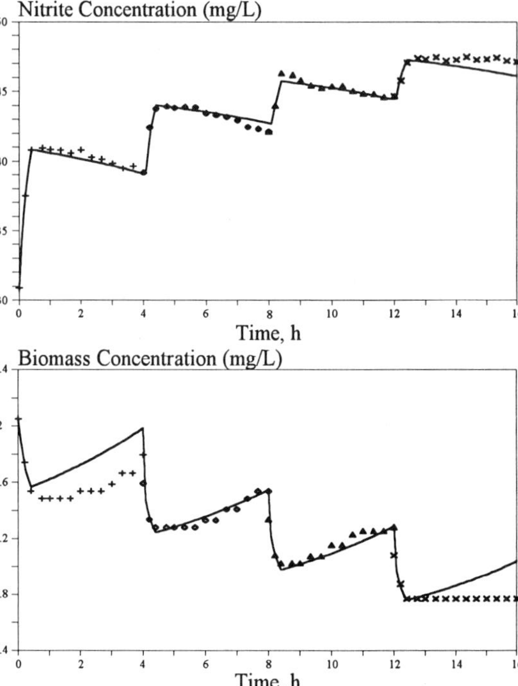

Figure 5. Experimental and model predicted concentration profiles for Experiment 1.

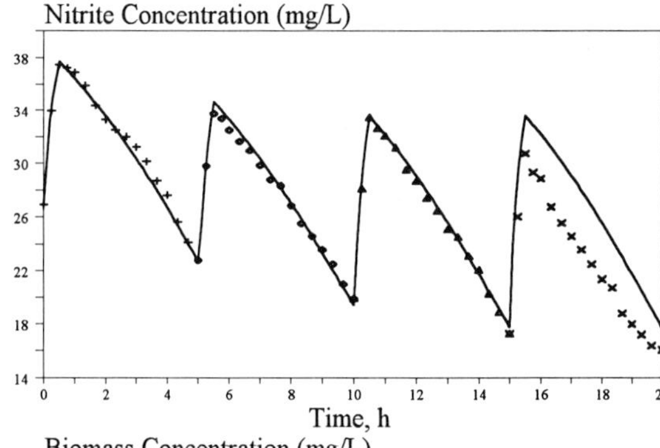

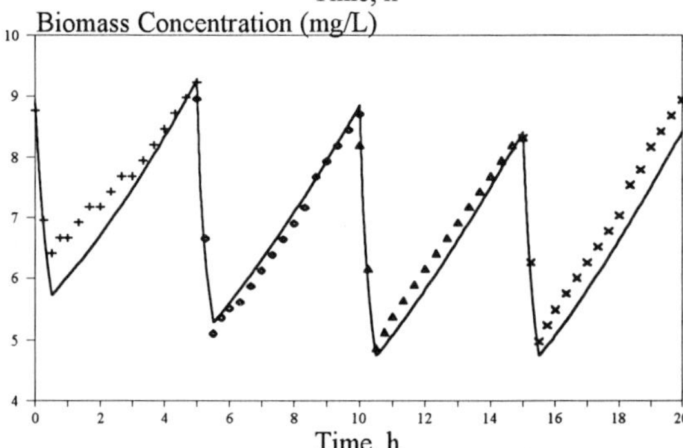

Figure 6. Experimental and model predicted concentration profiles for Experiment 2.

The theoretical analysis of the system when mixtures of nitrate and nitrite are to be biodegraded is much more complex, and is not shown here. It predicts though a dynamical behavior which is very interesting as multiple states of survival are predicted under the same operating conditions in some ranges of the parameter space. We show here results from an experiment involving a nitrate/nitrite mixture (Figure 7). The conditions were selected in order to confirm theoretical predictions of high nitrite accumulation in the system. Again the curves shown in Figure 7 represent model predicted concentration profiles. For nitrite and biomass the agreement is very good, while for nitrate the model predicts concentrations higher than those observed. In other experimental sets this is not the case. The high nitrite concentrations could had been avoided if the operating conditions, β or σ_1, had been selected differently.

Based on the results shown here, and many other data sets, we believe that we have a fully validated model of denitrification by a suspended culture in a cyclically operated reactor. This model, which could be solved for kinetic parameters corresponding to other cultures as well, can be used in process optimization studies.

NOMENCLATURE

x dimensionless biomass concentration; actual value divided by 9.92 mg/L

u dimensionless nitrate concentration in the reactor; actual value divided by 32 mg/L

u_f value of u in the stream fed to the reactor

z dimensionless nitrite concentration in the reactor; actual value divided by 32 mg/L

z_f value of z in the stream fed to the reactor

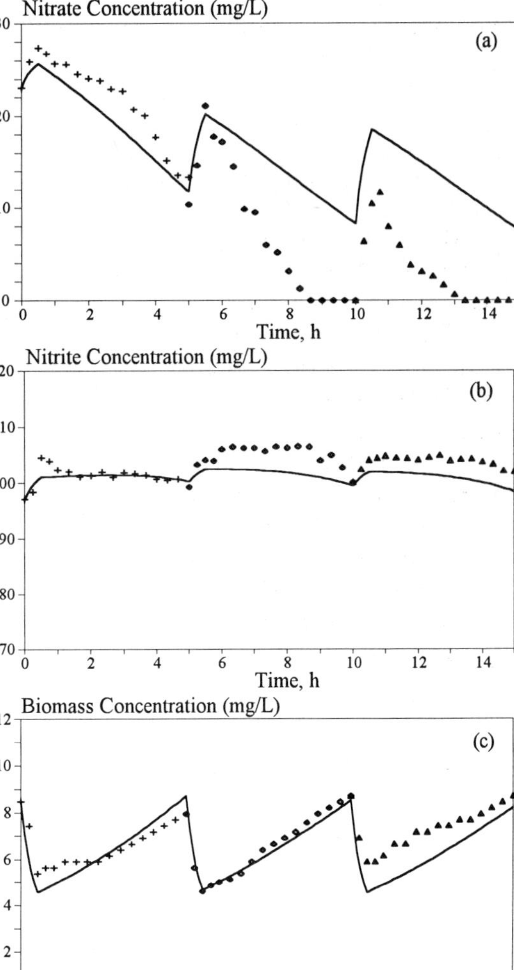

Figure 7. Experimental and model predicted concentration profiles for Experiment 3.

Greek symbols

β dimensionless measure of the hydraulic residence time; $1/\beta$ is a dimensionless measure of the dilution rate.

δ ratio of minimum to maximum volume of the reactor contents

θ dimensionless time

σ_1 fraction of the cycle time devoted to the first phase (fill)

LITERATURE CITED

1. Lewandowski, G. A. and B. C. Baltzis, *Chem. Eng. Sci.*, **47**, 2389 (1992).

2. Beccari, M., R. Passino, R. Ramadori, and V. Tandoi, *J. Water Pollut. Control Fed.*, **55**, 58 (1983).

3. Beltach, M. R. and J. M. Tiedje, *Appl. Envir. Microbiol.*, **42**, 1074 (1981).

4. Requa, D. A. and E. D. Schroeder, *J. Water Pollut. Control Fed.*, **45**, 1696 (1973).

5. Wilderer, P. A., W. L. Jones, and U. Dau, *Water Res.*, **21**, 2, 239 (1987).

6. Baltzis, B. C., G. A. Lewandowski, and S. Sanyal, "Sequencing Batch Reactor Design in a Denitrifying Application," in *Emerging Technologies in Hazardous Waste Management II*, D. W. Tedder and F. G. Pohland (Eds.), *ACS Symposium Series*, **468**, American Chemical Society, Washington, DC (1991).

7. Monod, J., *Recherches sur la Croissance des Cultures Bacteriennes*, Hermann et Cie., Paris (1942).

8. Andews, J. F., *Biotechnol. Bioeng.*, **10**, 707 (1968).

9. Chiesa, S. C. and R. L. Irvine, *Water Res.*, **19**, 471 (1985).

10. Irvine, R. L. and A. W. Busch, *J. Water Pollut. Control Fed.*, **51**, 235 (1979).

11. Silverstein, J. A. and E. D. Schroeder, *J. Water Pollut. Control Fed.*, **55**, 377 (1983).

12. Baltzis, B. C., G. A. Lewandowski, S.-H., Chang, and Y.-F. Ko, "Fill-and-Draw Reactor Dynamics in Biological Treatment of Hazardous Wastes," in *Biotechnology Applications in Hazardous Waste Treatment*, G. A. Lewandowski, P. M. Armenante, and B. C. Baltzis (Eds.), Engineering Foundation, New York (1989).

13. Dikshitulu, S., B. C. Baltzis, G. A. Lewandowski, and P. Pavlou, *Biotechnol. Bioeng.*, **42**, 643 (1993).

14. Fredrickson, A. G. and G. N. Stephanopoulos, *Science*, **213**, 972 (1981).

15. Koike, I. and A. Hattori, *J. gen. Microbiol.*, **88**, 1 (1975).

Dynamic Behavior of Activated-Sludge in Exponentially Fed-Batch Cultures Subjected to Step Perturbations

O.T. Ramírez, A. Aguilar-Aguila, and R. Quintero

Departamento de Bioingeniería, Instituto de Biotecnología,
Universidad Nacional Autónoma de México, A.P. 510-3, Cuernavaca, Morelos 62271, MEXICO

Laboratory scale wastewater treatment systems are often plagued by unattainment of true steady-states and the need for multiple systems operated simultaneously for extended periods of time. Such drawbacks become more critical when the experimental design includes changes in environmental or operating conditions which perturb the original steady-state. As an alternative, small samples from a laboratory scale model of an activated-sludge system (ASL) operating at steady-state were used as inoculum for an exponentially fed-batch culture operated in the variable volume mode (EFBC). Changes in the environmental and operating conditions were then made in the EFBC without perturbing the ASL. Such a system allowed the successful operation of the ASL for a year and the assessment of the effect of increasing dilution rates and step changes in NaCl concentration. The dynamic and steady state behavior the EFBC approximated that of a chemostat in a dilution range between 0.01 and 0.2 h $^{-1}$. Changes in NaCl concentration below 0.4 M resulted in a transient perturbation of the steady state followed by a rapid stabilization. Results on chemical oxygen demand, glucose, mixed liquor volatile suspended solids, particle count and particle mean volume in the EFBC compared to the ASL and batch culture are presented. The utility of the ASL-EFBC system for characterization of wastewater treatment systems is discussed.

The continuous bench-scale activated sludge reactor (ASL) has been traditionally used to assess the treatability of wastewater and determine the kinetic and stoichiometric coefficients of activated sludge processes (1). In such a method, a laboratory reactor is operated at a fixed dilution rate, with or without solids retention, and samples are taken until the reactor reaches the steady state. Steady state data at different dilution rates are then transformed and used to determine the various coefficients through linear regression methods such as the Lineweaver-Burke plots. For a constant dilution rate, a typical experiment requires at least 2 to 4 weeks of operation to achieve steady state conditions. Therefore, the simultaneous operation of four to five reactors is recommended to fully characterize a system (2). Although ASL are widely used for characterization purposes, they are labor and equipment intensive and can present other important drawbacks.

For instance, long operating periods can be needed to reach steady state in microbial populations of slow acclimatization to wastewater (3, 4). This is especially critical when perturbations to the system, either natural or induced, occur. Poor settling characteristics of the sludge (5, 6) and high turbulence in the settling chamber can result in the washout of the system when operated at high dilution rates, even if solids retention is employed. At low dilution rates, operational problems can also occur (7). For example, an accurate control of the food to microorganism ratio is difficult due to the very low flow rate needed to maintain the desired hydraulic and sludge residence time in small volume reactors.

Alternatively, exponentially fed batch culture (EFBC) can be used for characterization of waste waters. EFBC consist in feeding a medium to the reactor chamber in an exponentially increasing flowrate profile with respect to time.

345

E. Galindo and O.T. Ramírez (eds.), Advances in Bioprocess Engineering. 345-353.
© 1994 Kluwer Academic Publishers. Printed in the Netherlands.

Pioneering work by Yamane et al. (<u>8</u>) showed that EFBC operated in the variable volume mode, can closely approximate the behavior of continuous cultures. Since the ASL, with or without solids retention, is essentially a chemostat, it should be possible to use EFBC as an alternative to the ASL for characterization of wastewater treatment processes. Keller and Dunn (<u>9</u>) and Esener et al. (<u>10</u>) have shown that fed-batch cultures constitute a very useful experimental tool for modeling purposes and for characterization studies of microbial kinetics and energetics. Nevertheless, fed-batch operation has been exploited almost exclusively in its constant volume mode as a means to increase cell concentration and productivity in axenic cultures (<u>8</u>, <u>11</u>). To our knowledge, EFBC have not been used to investigate the behavior of mixed microbial populations in wastewater treatment systems. Compared to chemostats, variable volume EFBC are easier to operate at low dilution rates and no true washout occurs even at dilution rates above the maximum growth rate (<u>12</u>). Furthermore, the steady state condition is rapidly attained in EFBC and its transient behavior resembles that of a chemostat.

In this work, we report the behavior of activated sludge in fed-batch culture using a synthetic water as the feeding solution and supplied at an exponentially increasing rate. The effect of dilution rate and step changes in salt concentration was assessed. Comparison is made with a bench-scale ASL and with activated sludge in batch culture.

MATHEMATICAL FRAMEWORK

For a well-mixed fed-batch culture with sterile feed and no outflow, a cell balance over the bioreactor is given by:

$$\frac{d(xV)}{dt} = \mu(xV) \tag{1}$$

where x, V, t, and μ are cell concentration, culture volume, time, and specific growth rate, respectively. As described elsewhere (<u>12</u>), a constant and predetermined dilution rate, D, can be obtained in a fed-batch reactor if a variable, exponentially increasing feeding flow rate, $F(t)$, is set as:

$$\frac{dV}{dt} = F(t) = V_0 D \exp(Dt) \tag{2}$$

Where the subindex 0 refers to the time of feeding initiation. Combining equations (1) and (2) and rearranging yields:

$$\mu = D + \frac{d(Ln\, x)}{dt} \tag{3}$$

Similarly, a substrate balance yields:

$$\frac{dS}{dt} = D(S_f - S) - \frac{\mu x}{Y} \tag{4}$$

where S and S_f are the substrate concentration in the bioreactor and in the feed, respectively, and Y is the yield of biomass on substrate.

As shown by equations (3) and (4), the dynamic and steady state behaviors of an EFBC is the same of a chemostat, underscoring the utility of the former as an alternative to the later.

MATERIALS AND METHODS

<u>Analytical Techniques.</u> Chemical oxygen demand (COD), mixed liquor volatile suspended solids (MLVSS), settleable solids (SS), and oxygen consumption rate (OCR) were determined using standard methods (<u>13</u>). Glucose was determined with an enzymatic analyzer (YSI 2700, Yellow Springs, OH) and optical density (OD) was measured at 620 nm with a Beckman DU-650

spectrophotometer, (Fullerton, CA). Particle number and size distribution were determined with a Multisizer II electronic counter (Coulter Electronics, Hieleah, FL). For characterization of the ASL, COD was used as a measure of substrate concentration, and MLVSS, SS, and OCR were used as an estimate of biomass concentration and viability. Accuracy in the determination of COD and MLVSS in the EFBC and batch cultures can be lower than in the ASL due to the limited volume of sample available in the former cultures. Therefore, during characterization of the EFBC and batch cultures, glucose was used in addition to COD as a measurement of substrate concentration, and MLVSS was determined indirectly through conversion of OD measurements using a standard curve. In such cases, both glucose and COD are reported.

Synthetic Wastewater. The medium used for batch cultures and for feeding the ASL and EFBC contained, in mg/L: yeast extract, 150; glucose, 500; $(NH_4)_2HPO_4$, 200; $MgSO_4$, 30; $CaCl_2$, 5; and tap water, 100 mL/L. The average COD of the medium was 760 mg/L.

Activated Sludge System. The activated sludge system (ASL) consisted of a 25-liter acrylic cylindric vessel (30-cm diameter, 50-cm height), with four 3-cm baffles and a cover. It was agitated at 750 rpm with a 40 Watt mechanical stirrer (model 102, Talboys Eng., Montrose, PA), provided with two 5-cm Rushton turbines. Air was supplied to the vessel through an airstone at 1 vvm, measured on a glass-tube flow meter (Sho-rate 1350, Brooks, Hatfield, PA). The ASL was inoculated with an activated sludge population taken from a petroleum refinery wastewater facility operating with a sludge retention time of 14 days and a F/M ratio of 0.35 mg COD/mg MLVSS - day. Before inoculation, the sludge had been supplemented with acetic acid and then maintained in the synthetic water

described above. The synthetic wastewater was fed to the reactor with a peristaltic pump at a rate of 1.2 ml/min, resulting in an hydraulic retention time of 14 days. The overflow passed to a 15 liter pyramidal settler with an angle of 35°, where part of the settled sludge was recycled daily to the reactor in order to keep a food to biomass ratio (F/M) of 0.35 mg COD/mg MLSSV-day. The remaining sludge was disposed as surplus, and the overflow was collected in a vessel and analyzed weekly.

Exponentially Fed-Batch Culture. The exponentially fed-batch cultures were performed on a baffled 2-liter bioreactor Omni-Culture Bench-Top Fermenter (Vitris Co, Gardiner NY), equipped with two-6 blade Rushton turbines agitated at 600 rpm and operated as an open system at a controlled temperature of 29 °C. Dissolved oxygen (DO) and pH were measured with polarographic and glass electrodes, respectively (Ingold Electrodes Inc., Wilmington MA). All electrodes were steam sterilizable and their signals were amplified and sent to a data acquisition system for logging and control purposes. The data acquisition and control system consisted of a software written in Microsoft QuickBASIC (14) and a MacADIOS II data acquisition board (GW Instruments, Cambridge MA) connected to a Macintosh IIsi computer and interfaced to the bioreactor. The DO was controlled at 40% using a proportional - integral - derivative algorithm by varying the inlet gas composition (nitrogen and oxygen) with two mass flow controllers (Brooks 5850E) and maintaining the total gas flow rate constant.

The culture medium feeding rate was also controlled by the software via a microprocessor-based peristaltic pump (BioChem Technology, King of Prussia, PA) coupled to the computer's serial port (RS-232). The exponentially increasing feeding profile was established according to the mathematical analysis presented in the following section. Fed-batch operation was initiated after a period of batch growth. In order to reduce the transient phase, fed-batch operation in some experiments was initiated only after glucose was completely depleted (12). Glucose was measured off-line but its depletion could be inferred on-line as the time of sudden increase in the dissolved oxygen. For the various dilution rates tested, the initial culture volume ranged from 20 to 70 % of the volume at the end of the fed-batch phase. For comparison, batch cultures were also performed in the same bioreactor and conditions as the EFBC.

RESULTS AND DISCUSSION

Activated Sludge System. As shown in Figure 1, the ASL was successfully operated for almost a year. A slow transient start-up period was observed during the first 6 months, where COD removal efficiencies of around 80 % were obtained.

During such a period, an important reduction in the MLVSS and COD in the reactor occurred. Such a slow dynamic response of a ASL underscores the importance of designing alternative experimental tools for characterization of wastewater systems.

After 6 months of operation, a clear steady state was attained, with COD removal efficiencies higher than 95 %. Overall oxygen consumption rates fluctuated between 0.5 to 2 mg/L-h during the complete operation period. The improvement in sludge activity towards the next six months of operation could also be observed from an increasing specific oxygen uptake rate which reached its highest value at 10 mg_{O2}/g_{MLVSS}-hr. Such specific oxygen

uptake rate values are in agreement with those reported elsewhere (15). The well acclimated and active microbial population resulted in a stable system that tolerated step changes in the organic load such as the one at 290 days of operation.

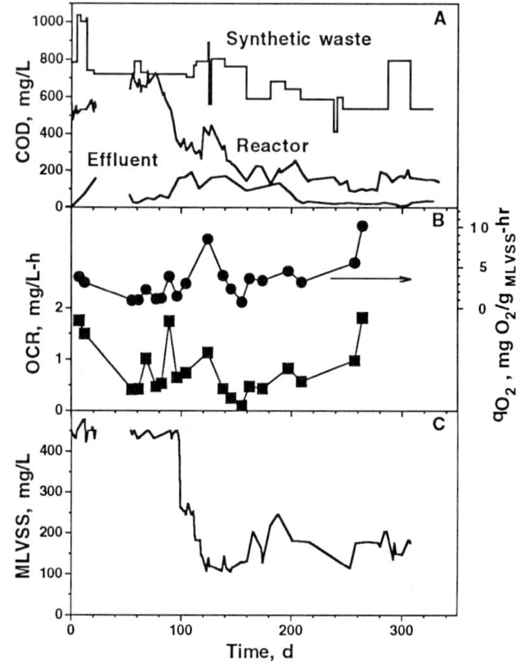

Figure 1. ASL, start-up and stabilization. A, COD in feed, reactor and effluent; B, Overall and specific oxygen uptake rate; C, MLVSS.

Table 1 presents a summary of the global performance of the ASL during the whole operation period. It can be seen that global parameters, such as sludge yield and COD remotion efficiency, are similar to full scale wastewater treatment plants (16, 17), demonstrating the similarity of the laboratory model with real systems.

Table 1.Global performance of the ASL.

Operation time:	320 days
Synthetic wastewater:	
Total influent volume	633 L
COD (range)	536-1038 mg/L
COD (mean)	678 mg/L
Aeration basin:	
COD (range)	88-737 mg/L
COD (mean)	336 mg/L
Settleable Solids (range)	15-45 ml/L
Settleable Solids (mean)	23 ml/L
Effluent:	
Total waste volume	618 L
COD (range)	11-191 mg/L
COD (mean)	73 mg/L
Surplus sludge:	
Total waste volume	15 L
Settleable Solids (mean)	98 ml/L
COD remotion efficiency	89.4 %
Sludge Yield	39 L SS/Kg COD

Exponentially Fed Batch Cultures. EFBC were inoculated with 400 mL of the settled fraction of a 1-L sample taken from the ASL. The sampled volume was less than 5% of the ASL volume, thus, the proposed experimental procedure allowed multiple experiments during short periods of time without perturbing the ASL. Figure 2 shows typical results of an EFBC operated at a dilution rate of 0.01 hr^{-1} and with DO controlled at 40 %. During the batch phase all the initial glucose and COD was consumed, the MLVSS and particle concentration increased exponentially, and the mean particle volume decreased after a small increase during the first two hours of culture. As seen in Figure 2E, an exponential increase in oxygen uptake rate also occurred. This was inferred from the increasing oxygen concentration in the inlet gas, set by the controller in order to keep a constant DO.

At 11 hours of operation exponential feeding, shown in Figure 2D, was initiated as prompted by the disruption of DO control. After a short transient period, MLVSS reached a steady state concentration. In comparison, glucose concentration and COD reached the steady state condition immediately after the fed-batch phase initiated. Particle concentration and mean particle volume showed slower dynamics towards steady state. While the former slowly increased after 10 hours into the fed-batch phase, the later slowly decreased. Such a behavior can reflect slow changes in the species distribution of the microbial community. For instance, changes in species distribution have been observed by Schmidt et al. (<u>18</u>) who reported smaller bacterial floc size in sequencing batch reactors (SBR) limited by glucose compared to SBR with glucose supplementation. As seen from the constant slope in total COD consumed and zero COD concentration (Figure 2B), a steady state in COD consumption rate at 5.24 mg$_{COD}$/L-h (see equation (4)) was also achieved.

To examine the dynamics of the EFBC after load perturbations, a step increase in NaCl in the range of 0.1 to 0.4 M was forced in the various cultures. Step perturbations in salt concentration were performed since Hisashi et al (19) have indicated the benefits of sea water addition for treatment of wastewater through activated sludge system. The step increase was performed by adding a known amount of salt directly to the reaction and feeding vessels after the steady state had been attained. Thus, the first fed-batch phase prior to the perturbation served as an internal control to asses the effect of the manipulated variable. As seen in Figure 2A, a step increase of 0.1 M NaCl at 41 h caused a sudden decrease in MLVSS. However, the steady state condition was rapidly recovered. A small increase in the specific substrate remotion rate could also be observed. Similar results were observed for the other concentrations of salt tested.

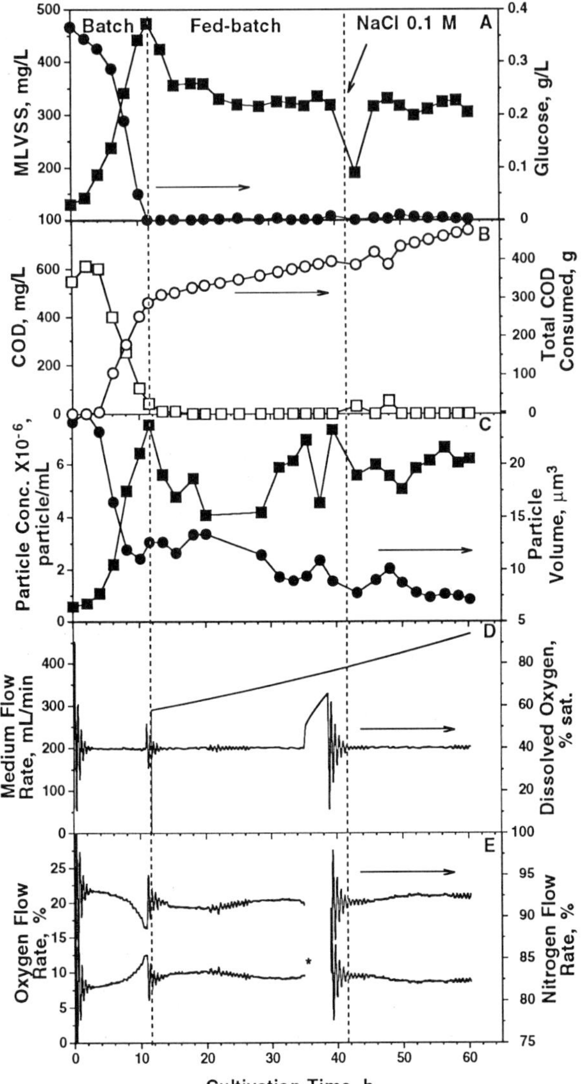

Figure 2. EFBC run at a dilution rate of 0.01 hr-1 and subjected to a step increase in salt conc.(0.1 M). A, MLVSS and glucose concentration; B, COD and total COD consumed; C, particle conc. and mean particle volume; D, medium flow rate and DO; E, inlet gas composition.
* indicates a disruption in the nitrogen supply.

Figure 3 presents a summary of the growth kinetics during the fed-batch phase for all the dilution rates tested. As seen from the constant slope, a constant specific growth rate

was attained in all cases. Furthermore, as seen in Table 2, the behavior corresponded to that which would have been expected for steady state in a chemostat, i. e., dilution rate being the same as the growth rate. However, in contrast to chemostats, stable operation was attained in the EFBC at dilution rates close or at the maximum specific growth rate. Chemostat operation in such a region would have resulted in washout of the culture. The maximum specific growth rate for the activated sludge, shown in Table 2, was 0.2 hr^{-1} and was obtained from batch culture data.

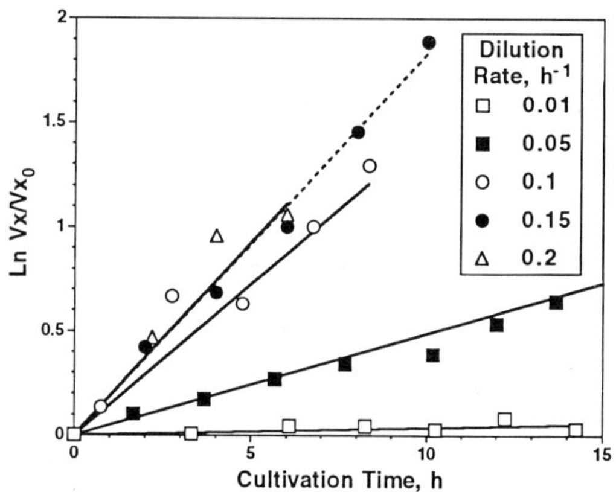

Figure 3. Summary of growth kinetics during fed-batch phase at increasing dilution rates. The subindex 0 refers to the conditions at feeding initiation. Growth rate was determined directly from the slope of each curve.

Table 2. Summary EFBC at increasing dilution rate.

D, hr^{-1}	μ, hr^{-1}
0.01	0.004
0.05	0.049
0.1	0.15
0.15	0.18
0.2	0.19
Batch (μ_m)	0.20

<u>Batch Cultures.</u> Figure 4 shows typical growth and substrate utilization kinetics in batch culture. From intensive parameters, such as MLVSS, COD, and glucose concentration, two clearly distinctive growth phases can be observed: an initial exponential phase and a second endogenous phase, after 8 hours of operation. During exponential growth MLVSS, particle volume and particle concentration rapidly increased, while glucose was completely depleted. The growth rate attained was 0.2 hr^{-1} which corresponds to the maximum growth rate (Table 2).

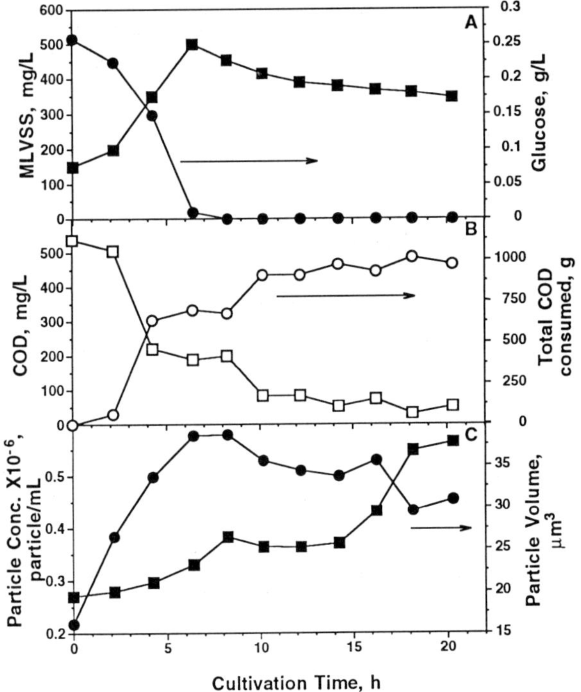

Figure 4. Behavior or activated sludge in batch culture. A, MLVSS and glucose concentration; B, COD conc. and total COD consumed; C, particle conc. and mean particle volume.

During the endogenous phase, MLVSS decreased and the residual COD was slowly degraded. It should be noted, however, that the endogenous phase can be further subdivided into two additional phases if single-cell parameters, such as mean particle volume, are analyzed (Figure 4C). Accordingly, an initial endogenous phase can be distinguished between 8 and 15 hr where particle concentration remains constant but mean cell volume decreases from 38 to 27 μm^3. Such behavior can be indicative of an invariant microbial population metabolizing internal reserves, i.e. a true endogenous phase. After 15 hr the particle concentration rapidly increased with a concomitant decrease in mean particle volume. Such a second endogenous phase could indicate severe substrate limitation resulting in cell disruption and accumulation of cell debris. Such results show that single-cell measurements can yield additional information not detectable through traditionally determined intensive parameters.

<u>CONCLUSIONS</u>

A laboratory scale model was design, constructed, and successfully operated for almost a year with global performance parameters and dynamic behavior that closely simulated real wastewater systems. The experimental set-up, in which step perturbations were tested in an EFBC inoculated with small sample volumes taken from an ASL, proved to be an important tool which allowed multiple experiments of slow processes to be performed in short periods of time. In particular, step changes in NaCl concentration below 0.4M resulted in a transient perturbation of the original steady state which later recovered to a new one.

It was shown that activated sludge could be cultivated in EFBC and that its dynamic and steady state behavior resembled that of a chemostat. Namely, a predetermined dilution rate could be fixed which resulted in a constant growth rate equal to the selected dilution rate. Furthermore, EFBC allowed simple operation at very low dilution rates and at, or close to, the maximum growth rate without washout. In addition, it was shown that determination of single-cell parameters can yield important information on the dynamic behaviour of mixed populations.

The results of this study indicate that the characteristics of the EFBC can be potentially useful in its application as a novel alternative in the filling stage of the Sequential Batch Reactors which are commonly used in wastewater treatment systems. It should be noted that the transient and steady state behavior of the EFBC is different from the one observed in other systems, such as the constantly fed-batch cultures (10) and the membrane bioreactors (19, 20).

NOMENCLATURE

COD	Chemical oxygen demand
D	Dilution rate
DO	Dissolved oxygen
MLVSS	Mixed liquor volatile suspended solids
OCR	Oxygen consumption rate
OD	Optical density
S	Substrate concentration
SS	Settleable solids
rpm	revolutions per minute
V	Volume
vvm	air volume/liquid volume - minute.
x	Biomass concentration
Y	Yield
μ	Specific growth rate
μ_m	Maximum specific growth rate

ACKNOWLEDGMENTS. Technical assistance by Ma. Elena Zamora is kindly appreciated.

LITERATURE CITED

1. Metcalf & Eddy Inc., Wastewater Engineering, McGraw-Hill Book Company, New York, **12** 481-571 (1972).

2. Ramalho, R.S., Introduction to wastewater treatment processes, Academic Press, New York, **5** 158-235 (1977).

3. Bermúdez, J.J., Jimeno, A., Cánovas-Diaz, M., Manjón, A. and Iborra, J.L., *Process Biochem.* 178-181 (1988).

4. Stephenson, D. and Stephenson, T., *Biotech. Adv.* **10** 549-559 (1992).

5. Karapanagiotis, N.K., Rudd, T., Sterritt, R.M. and Lester, J.N., *J. Chem. Tech. Biotechnol.* **94** 107-120 (1989).

6. Sheintuch, M., Lev, O., Einav, P. and Rubin, E., *Biotechnol. Bioeng.* **28** 1564-1576 (1986).

7. Adams, C.E. Eckenfelder, W.W. and Hovious, J.C., *Wat. Res.* **9** 37-42 (1975).

8. Yamane, T. and Shimuzu, S., *Adv. Biochem. Eng. Biotechnol.* **30** 145-194 (1984).

9. Keller R. and Dunn, I.J., *J. App. Chem. Biotechnol.* **28** 805-514 (1978).

10. Esener, A.A., Roels, J.A. and Kossen, W.F. *Biotechnol. Bioeng.* **23** 1851-1871 (1981).

11. Yee, L. and Blanch, H.W. *Bio/technol.* **10** 1550-1556 (1992).

12 Ramirez, O.T., Zamora, R., Quintero, R. and Lopez-Munguia, A. Exponentially fed-batch cultures as an alternative to chemostats: the case of penicillin acylase production by recombinant *E. coli*. *Enz. Microb. Technol.* (in press).

13. APHA-AWWA-WPCF, Standard methods for the examination of water and wastewater, 17th. edition., American Public Health Association, Washington D.C. **2** 71-87; **5** 10-17 (1989).

14 Aguilar-Aguila, A., Valentinotti, S., Galindo, E., and Ramírez, O.T., *Biotecnología* **5** (5-6) S130-S139 (1993).

15 Shamas, J.Y., Englande, A.J. *Wat. Sci. Tech.* **25** (1) 123-132 (1992).

16. Bowen, P.T., Magar, V.S., Otoski, R. and McMonagle, T. *Wat. Sci. Tech.* **25** (4-5) 281-287 (1992).

17. Wong, A.D. and Goldsmith, C.D., *J.W.S.T.* **20** (11-12-K) 131-136 (1986).

18. Schmidt, S.K., Smith, R., Sheler, D., Hess, T.F., Silverstein, J. and Radehaus, P.M. *Microb. Ecol.* **23** 127-142 (1992).

19. Hisashi, A., Minoru, I., Yasushi, K., Sadao Y. Method of treating waste water with activated sludge; adding sea water. International Patent Classification: C02C005/10 (1978).

20. Boulliot, P., Canales, A., Pareilleux, A., Huyard, A. and Goma, G. *J. Ferment. Bioeng.* **69** (3) 178-183 (1990).

21. Uribelarrea, J. L., Winter, J., Goma, G. and Pareilleux, A. *Biotechnol. Bioeng.* **35** 201-206 (1990).

Expansion Characteristics of Tapered Fluidized-Bed Bioreactors

C.-S. Wu and J.-S. Huang

Department of Environmental Engineering, National Cheng Kung University, Tainan 701, Taiwan, REPUBLIC OF CHINA

In addition to the derivation of theoretical models and the performance of sensitivity analysis, the bioparticles expansion tests of the conventional fluidized-bed bioreactor ($\phi = 0^o$) and the tapered fluidized-bed bioreactors ($\phi = 2.5^o, 5^o, 10^o$) were conducted to determine an optimal taper angle in this study. The most important findings indicated that the expanded volume of bioparticles and the expanded height of bioparticles in tapered fluidized-bed bioreactors are smaller than those in the conventional fluidized-bed bioreactor, and are decreased with the increasing taper angle. However, when the taper angle is greater than 5°, the reduction of both the expanded volume of bioparticles and the expanded height of bioparticles will not be very significant. In other words, taper angle of 5° could be very close to an optimal taper angle for designing a tapered fluidized-bed bioreactor.

At present, the conventional fluidized-bed bioreactor (abbreviated by CFB) is of great concern for treating medium-high strength industrial wastewaters. Nonetheless, the wash-out problem of bioparticles is rather easily occurred in this bioreactor, with an identical cross-section but a small operating range of flow rate, especially when the biogas generated in anaerobic reactors is attached onto bioparticles resulting in a small specific gravity of the bioparticle. Fortunately, the wash-out problem found in CFB can be overcomed by the tapered fluidized-bed bioreactor (abbreviated by TFB), Which has not only an unique geometrical configuration (i.e., narrower at the bottom and wider at the top) but small superficial velocity of fluid and drag force of bioparticles at the upper part of the bioreactor. However, most of recent studies on TFB mainly concern the relationship between the removal efficiency of a substrate and its volumetric loading (see Table 1). Studies regarding the selection of an optimal taper angle were still very scarce. Therefore, theoretical models describing the bioparticles'

expansion characteristics of the fluidized-bed bioreactors will be derived and sensitivity analysis will be performed. Besides, the bioparticles' expansion tests of CFB ($\theta = 0^o$) and TFB$_s$ ($\theta = 2.5^o, 5^o, 10^o$) will be also conducted to determine an optimal taper angle.

THEORY

Models for describing fluidization of bioparticles.

Assuming that the biofilm attached onto the bead shape activated carbon (abbreviated by BAC) (i.e., The bioparticle has a spherical shape.) has an uniform thickness. Thus, the wet density of the bioparticle (ρ_p) can be expressed as

$$\rho_P = \frac{\rho_m r_m^3 + \rho_f [(r_m + \delta)^3 - r_m^3]}{(r_m + \delta)^3} \tag{1}$$

According to Lewis & Bowerman (8), the relationship among the intersticial velocity of fluid (u_{in}), superficial velocity of fluid (u_s), bed porosity (ε) and particle terminal settling velocity in

355

E. Galindo and O.T. Ramírez (eds.), *Advances in Bioprocess Engineering.* 355-363.
© 1994 Kluwer Academic Publishers. Printed in the Netherlands.

Table 1. Literatures review on tapered fluidized-bed bioreactors.

Taper angle (°)	Volume (l)	Bottom width (cm)	Top width (cm)	Waste-water	Temp. (°C)	Loading (kg/m³-day)	Feed conc. (mg/l)	Hydraulic retention time (hr)	Reduction (%)	Remarks	Reference
2.7	2.4	2.50	7.60	Nitrate		180.0	3500	0.5	99	Anaerobic	Scott & Hancher (1)
1.4	3.0	2.54	7.62	Phenol		4.2-18.0	580-2200	2.2-3.9	99-56	Aerobic	Holladay et al. (2)
1.4	10.0	2.54	7.62	Phenol		8.7-37.4	45-260	0.2-0.8	99-87	Aerobic	Holladay et al. (2)
2.7	3.0	2.54	7.62	Thiocyanate	25	1.2-6.1	50-170	0.2	96-30	Aerobic	Lee et al. (3)
2.7	3.0	2.54	7.62	Phenol	25	2.0-80.6	9-260	0.1-0.3	99-29	Aerobic	Lee et al. (3)
3.4	0.8	3.00	6.00	Ammonium	30	12.0-19.5	100	0.1-0.2	98-75	Aerobic	Tanaka et al. (4)
12.4	1.0	0.32	9.50	Whey	35	7.7-19.5	2000-7000	2.4-9.4	90-70	Anaerobic	Boening & Larsen (5)
3.4	1.4	4.50	7.50	Nitrate	30	0.13	25	4.5	99	Anaerobic	Kurt et al. (6)
3.4	1.8	4.50	7.50	Whey	35	7.0-40.0	800-10000	2.4-4.6	97-63	Anaerobic	Denac & Dunn (7)
3.4	1.5	4.50	7.50	Molasses	35	10.0-40.0	1100-10000	2.4-4.3	97-70	Anaerobic	Denac & Dunn (7)

the fluidized-bed (v_i) is

$$u_{in} = \frac{u_s}{\varepsilon} = v_i \varepsilon^n \tag{2}$$

Richardson & Zaki (quoted by Khan & Richardson, (9)) pointed out that

$$v_i = u_t 10^{-\frac{d_p}{D}} \tag{3}$$

According to Khan & Richardson (10), the relationship between Re_t and Ga is

$$Re_t = (2.33 Ga^{0.018} - 1.53 Ga^{-0.016})^{13.3} \tag{4}$$

By definition

$$Re_t = \frac{\rho_l d_p u_t}{\mu} \tag{5}$$

$$Ga = \frac{\rho_l d_p^3 (\rho_p - \rho_l) g}{\mu^2} \tag{6}$$

Then, the value of u_t can be solved by using eq. (4) to eq. (6). In addition, Khan & Richardson (9) have derived an expression for n (in eq. (2)) and Ga from literatures that has the form

$$\frac{4.8 - n}{n - 2.4} = 0.043 Ga^{0.57}[1 - 1.24(\frac{d_p}{D})^{0.27}] \tag{7}$$

Models for expansion characteristics of bioparticles in TFB. When the superficial velocity of fluid at the bottom of the

bioreactor (u_{sO}) is greater than the critical superficial velocity of transport regime, the bioparticle at the bottom starts moving upward, as shown in figure 1. When the superficial velocity of fluid at a specific position (u_{sz}) in TFBs is equal to the bioparticle terminal settling velocity (v_i), the bioparticle ceases moving upward. Thus, Z_1 can be expressed as

$$Z_1 = \frac{1}{2 \tan\frac{\theta}{2}} (\frac{Q}{V_i})^{\frac{1}{2}} \tag{8}$$

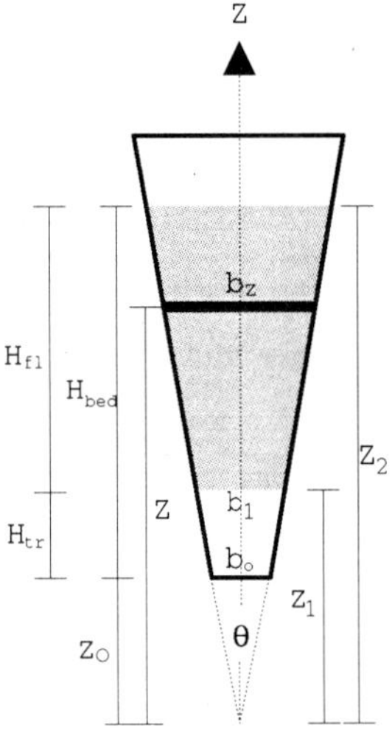

Figure 1. Geometric diagram of a tapered bioreactor.

And a continuity equation in liquid phase has the form

$$b_0^2 u_{so} = b_z^2 u_{in}\varepsilon_z \tag{9}$$

The equivalent diameter (D) at the position Z is

$$D = \frac{2b_z}{\sqrt{\pi}} \tag{10}$$

Thus, the porosity at the position Z (ε_z) can be solved by using eqs. (2), (3), (9), (14) and the geometric property of TFB.

$$\varepsilon_z = [\frac{z_0^2}{z^2}\frac{u_{so}}{u_t}10^{\frac{\sqrt{\pi}\,d_p}{4z\tan\frac{\theta}{2}}}]^{\frac{1}{n+1}} \tag{11}$$

In addition, total volume of bioparticles (V_p) can be expressed as

$$V_p = \frac{\pi}{6}d_p^3 WN \tag{12}$$

$$V_p = \int_{Z_1}^{Z_2}(1-\varepsilon_z)b_z^2 dz = 4\tan^2(\frac{\theta}{2})\int_{Z_1}^{Z_2}(1-\varepsilon_z)z^2 dz \tag{13}$$

When proper substitutions are made from eq. (11) to eq. (13), the following form can be obtained.

$$\frac{\pi}{6}d_p^3 WN = 4\tan^2(\frac{\theta}{2})\int_{Z_1}^{Z_2}[1-(\frac{z_0^2}{z^2}\frac{u_{so}}{u_t}10^{\frac{\sqrt{\pi}\,d_p}{4z\tan\frac{\theta}{2}}})^{\frac{1}{n+1}}]z^2 dz \tag{14}$$

The iteration using Newton-Raphson method is applied for solving Z_1 value in eq. (8). The Z_2 value can be solved by substituting for Z_1 value in eq. (14) and followed by the iteration using Simpson's rule. Thus, the volume fraction of the transport regime (ξ_{tr}) is

$$\xi_{tr} = \frac{V_{tr}}{V_{bed}} \tag{15}$$

The volume fraction of the fluidized regime (ξ_{fl}) is

$$\xi_{fl} = \frac{V_{fl}}{V_{bed}} = 1 - \xi_{tr} \tag{16}$$

The height fraction of the transport regime (ζ_{tr}) is

$$\zeta_{tr} = \frac{H_{tr}}{H_{bed}} \tag{17}$$

The height fraction of the fluidized regime (ζ_{fl}) is

$$\zeta_{fl} = \frac{H_{fl}}{H_{bed}} = 1 - \zeta_{tr} \tag{18}$$

The average bed porosity (ε_a) is

$$\varepsilon_a = \frac{(V_{fl}-V_p)}{V_{fl}} \tag{19}$$

In eqs. (15) and (16),

$$V_{tr} = \frac{4\tan^2(\frac{\theta}{2})}{3}(Z_1^3 - Z_0^3) \tag{20}$$

$$V_{bed} = \frac{4\tan^2(\frac{\theta}{2})}{3}(Z_2^3 - Z_0^3) \tag{21}$$

$$V_{fl} = V_{bed} - V_{tr} \tag{22}$$

Models for expansion characteristics of bioparticles in CFB. Assuming that bioparticles distributed in CFB are uniform (i. e., $\varepsilon = \varepsilon_z$). The equivalent diameter can be expressed as

$$D = \frac{2b_c}{\sqrt{\pi}} \tag{23}$$

From the continuity equation in liquid phase

$$b_c^2 u_{so} = b_c^2 u_{in}\varepsilon \tag{24}$$

Substituting for u_{in} in eq. (24) from eq. (2) gives

$$\varepsilon = (\frac{u_{so}}{v_i})^{\frac{1}{n+1}} \tag{25}$$

And $\quad V_p = b_c^2 H_{bed}(1-\varepsilon) \tag{26}$

Combining eqs. (12) and (26) for H_{bed} gives

$$H_{bed} = \frac{\pi d_p^3 WN}{6b_c^2(1-\varepsilon)} \tag{27}$$

In addition, $V_{bed} = b_c^2 H_{bed} \tag{28}$

From the afore-mentioned derivation, the specific surface area of the bioreactor(a) can be expressed as

$$a = \frac{\pi d_p^2 WN}{V_{bed}} \qquad (29)$$

<u>Sensitivity analysis.</u> On the basis of the following parameters values, θ = 5°, W = 500g, N = 14225 no. of BAC/g, ρ_m= 1.38 g/cm^3, b_o= 2.4 cm, b_c= 6 cm, δ = 90 μm, Q = 5000 l/day, the influence of the up and down variation of the afore-mentioned parameters values on the predicted values of the expanded height of bioparticles (H_{bed}) and expanded volume of bioparticles (V_{bed}) were determined by respectively letting the relative change of the parameter ($\triangle p/p$) and the relative change of both H_{bed} and V_{bed} calculated from the model as abscissa and ordinate. Thus, the sensitivity of H_{bed} and V_{bed} can be respectively expressed as

$$S_H = \frac{\frac{\Delta H_{bed}}{H_{bed}}}{\frac{\Delta P}{P}} \qquad (30)$$

$$S_V = \frac{\frac{\Delta V_{bed}}{V_{bed}}}{\frac{\Delta P}{P}} \qquad (31)$$

Figure 2 shows that, in TFBs, the influence of the variation of the taper angle (θ) and the particle diameter (d_p) on the predicted value of H_{bed} are more significant than the other parameters. Figure 3 shows that, in TFBs, the influence of the variations of the wet density of BAC (ρ_m) and d_p on the predicted value of V_{bed} is more significant than θ, and the rest parameters are less significant. Furthermore, figures 2 and 3 also reveal that the influence of the variation of θ on the predicted values of H_{bed} and V_{bed} are less significant when θ is greater than 5°. Figure 4 shows that, in CFB, ρ_m and column width (b_c) are the most sensitive parameters for the predicted values of H_{bed} and V_{bed}, d_p and flow rate (Q) are the next sensitive

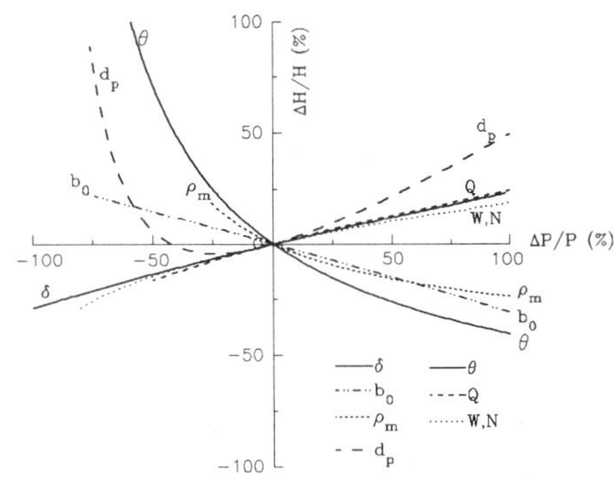

Figure 2. Sensitivity analysis of the model regarding bioparticles' expanded height in the tapered fluidized-bed bioreactor.

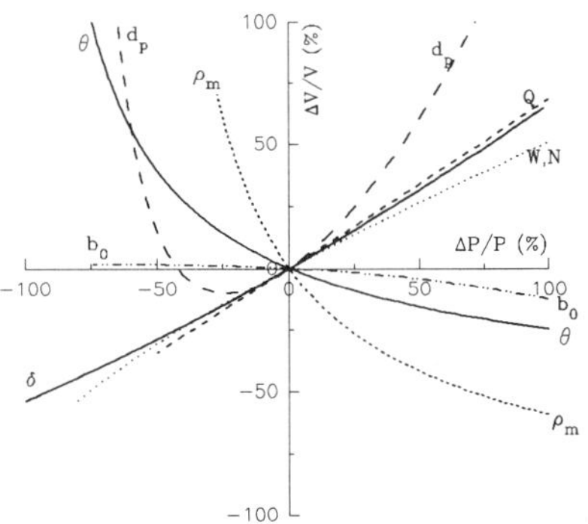

Figure 3. Sensitivity analysis of the model regarding bioparticles' expanded volume in the tapered fluidized-bed bioreactor.

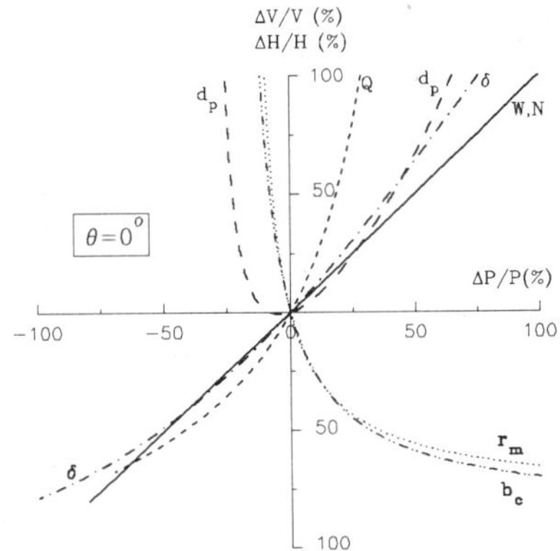

Figure 4. Sensitivity analysis of the models regarding expanded height and expanded volume in the conventional fluidized-bed bioreactor.

parameters, and the rest are the least sensitive parameters. Figure 4 also shows that, in CFB, the ratio of $\Delta H_{bed}/H_{bed}$ is equal to the ratio of $\Delta V_{bed}/V_{bed}$. Furthermore, the comparison among figures 2, 3 and 4 reveals that the influence of Q on $\Delta H_{bed}/H_{bed}$ and $\Delta V_{bed}/V_{bed}$ in CFB is more sensitive than that in TFB. This result simply implies that the operating range of Q in CFB is smaller than that in TFBs. In other words, the wash-out problem of bioparticles is very easily occurred in CFB.

MATERIALS AND METHODS

Experimental set-up. One set of column for CFB was constructed from plexiglass having dimensions of 6.0cm(L)$\times$6.0cm(w)$\times$163.8cm(H). Three sets of columns for TFBs with taper angles of 2.5°, 5°, 10° were constructed from plexiglass, as shown in figure 5, which had the same bed width at the bottom (i.e., 2.4cm), the different bed width at the top (i.e., 8.0cm, 10.3cm, 13.8cm) and the different bed height (i.e., 183.8cm, 141.9cm, 110.0cm). Each bioreactor has the same volume of 18.2 liters.

Experimental work for bioparticles' expansion test. The medium used in this study was the bead-shape activated carbon (BAC). After acclimation of the acetate synthetic wastewater with the seeded anaerobic sludge, the biofilm was attached onto BAC and ready for proceeding the study of bioparticles' expansion test. The experimental conditions of the bioparticle used in this study are shown in table 2. In order to avoid the interference from the biogas, the tap water instead of the acetate synthetic wastewater was introduced into the bioreactor. During the entire study, the flow rate was varied from high to low, and both the transport height and expanded height of bioparticles were measured.

Measurement of wet density of BAC (ρ_m) and wet density of biofilm (ρ_f). One grain of BAC, randomly selected from clean BACs, was measured for its radius(r_m) by a phase-constrast microscope. Then the clean BAC was put into the settling column with inner diameter of 8.3cm and height of 95cm for settling test. Thereafter, the required time for allowing the clean BAC to settle 70cm was measured. And ρ_m can be calculated by using the previously-mentioned settling equations (i.e., eqs. (1), (3), (4), (5) and (6)). The average value of ρ_m could be obtained by continuously repeating the afore-mentioned procedures for 20 different selected clean BACs. The average value of ρ_f could be obtained by the similar approach used for the measurement of ρ_m (Note: The biofilm thickness (δ) of bioparticles was ranged from 71μm to 231μm.).

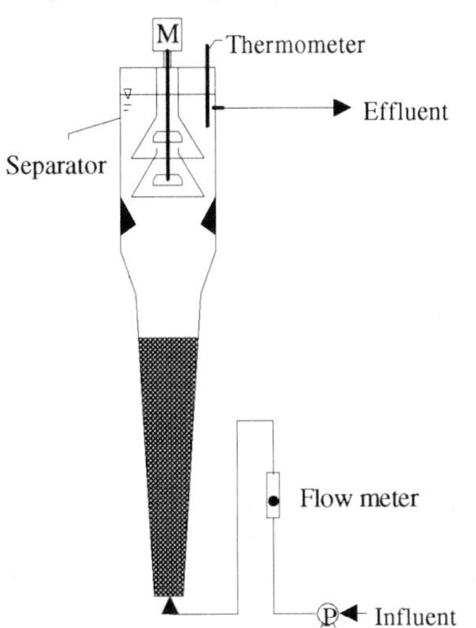

Figure 5. Schematic diagram of the tapered fluidized-bed experimental system.

Table 2. The experimental conditions for the bioparticles' expansion test.

θ (°)	r_m (μm)	d_p (μm)	δ (μm)	W (g)
0	220	463	11	375
2.5	262	556	16	126
5	224	573	63	176
10	238	479	2	752

RESULTS AND DISCUSSION

Experimental and simulated results from bioparticles' expansion test. In this study, the experimental conditions of the bioparticle are shown in table 2. The wet density of BAC (ρ_m) and wet density of biofilm (ρ_f) were respectively determined from the settling test, they were 1.38 g/cm^3 and 1.05 g/cm^3. Figure 6 shows that the expanded height of bioparticles (H_{bed}) was increased with the increasing flow rate in both CFB and TFBs. When the flow rate was raised to 4 l/min, H_{bed} in CFB would be greatly increased, resulting in a serious wash-out problem of bioparticles. However, at this same flow rate, H_{bed} in TFBs was decreased with the increasing taper angle. This simply implies that bioparticles are not easy to be washed out in TFBs. In addition, the changing rate of H_{bed} per unit flow rate in TFBs was smaller than that in CFB, and was decreased with the increasing taper angle. In other words, the operating flow rate in TFBs can be much larger than that in CFB. Under the operating range of the flow rate in CFB, bioparticles' transport phenomenon did not occur. In other words, the transport height of bioparticles(H_{tr}) is zero. Furthermore, the critical flow rate of bioparticles' transport regime for three sets of TFBs was nearly the same, ca. 1 l/min. This is because the bed width at the bottom of three sets of TFBs are the same. Meanwhile, at the same flow rate, the value of H_{tr} was decreased with the increasing taper angle. Moreover, at the same flow rate, the ratio of the expanded volume of bioparticles at this particular flow rate to that at the lowest flow rate (Vr) and the ratio of the expanded height of bioparticles at this particular flow rate to that at the lowest flow rate(Hr) in TFBs were smaller than those in CFB. Excepting that the value of Vr in taper angle of 5° was greater than that in taper angle of 2.5° (mainly because in taper angle the biofilm in taper angle of 5° was

thicker than that of 2.5°), the values of Vr and Hr in TFBs were decreased with the increasing taper angle. This result indicates that the expansion breadth in TFBs are smaller. In a word, there is a fairly good agreement of the bioparticles' expansion characteristics (i.e., the expanded volume and expanded height of bioparticles) between the simulated and experimental results.

Figure 7 shows that the volume fraction of the bioparticles' transport regime (ξ_{tr}) and the height fraction of the bioparticles' transport regime (ζ_{tr}) in TFBs were decreased with the increasing

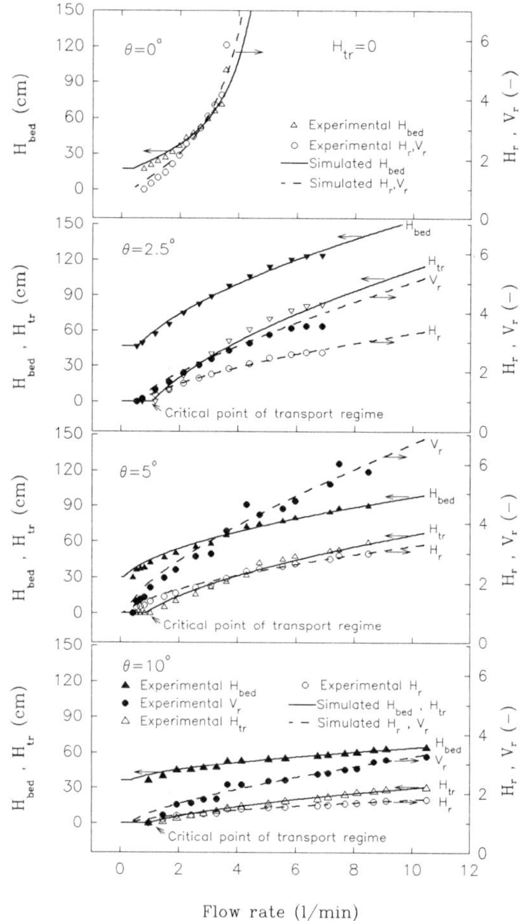

Figure 6. Experimental and simulated results of H_{bed}, H_{tr}, Hr, Vr vs. flow rate (Table 2 shows experimental conditions.).

taper angle. The reason for obtaining smaller values of ξ_{tr} and ζ_{tr} at larger taper angle is mainly because the value of u_s at lower position of the bioreactor is smaller than that of the critical superficial velocity of transport regime. Figure 8 shows that the average bed porosity (ε_a) in TFBs was decreased with the increasing taper angle, and ε_a influenced by the variation of the flow rate in TFBs was less than that in CFB. Results from figures 7 and 8 reveal that TFBs have much more space than CFB for retaining bioparticles.

In figure 9, at the same conditions of d_p (540 μm) and δ (90 μm), the simulated results further confirm that Vr, Hr and the changing rates of Vr, Hr per unit flow rate in TFBs were smaller than those in CFB, and were decreased with the increasing taper angle. However, the afore-mentioned influences became less significant when the taper angle was greater than 5°. Besides, figures 6 and 9 also reveals that the expanded volume of bioparticles (V_{bed}) and the expanded height of bioparticles (H_{bed}) in TFBs are smaller than those in CFB, and are decreased with the increasing taper angle. Nonetheless, the reductions of Vbed and Hbed will not be very significant when the taper angle of TFB is greater than 5°. In other words, taper angle of 5° could be very close to an optimal taper angle for designing a tapered fluidized-bed bioreactor.

SUMMARY AND CONCLUSIONS

The results obtained from theoretical models, sensitivity analysis and experimental tests clearly indicate that the expanded volume of bioparticles (V_{bed}) and the expanded height of bioparticles (H_{bed}) in TFBs (θ = 2.5°, 5°, 10°) are smaller than those in CFB, and are decreased with the increasing taper angle. Besides, there is a fairly good agreement of the expansion characteristics (i.e.,

the expanded volume and expanded height of bioparticles) between the simulated and experimental results. The simulated and experimental results also reveals that the operating range of the flow rate without incurring the wash-out problem of bioparticles in TFBs are much broader than that in CFB. The volume fraction of the bioparticles' transport regime (ξ_{tr}) and the height fraction of the bioparticles' transport regime (ζ_{tr}) in TFBs are decreased with the increasing taper angle. This implies that TFBs have much more space than CFB for retaining bioparticles. However, when the taper angle of TFB is

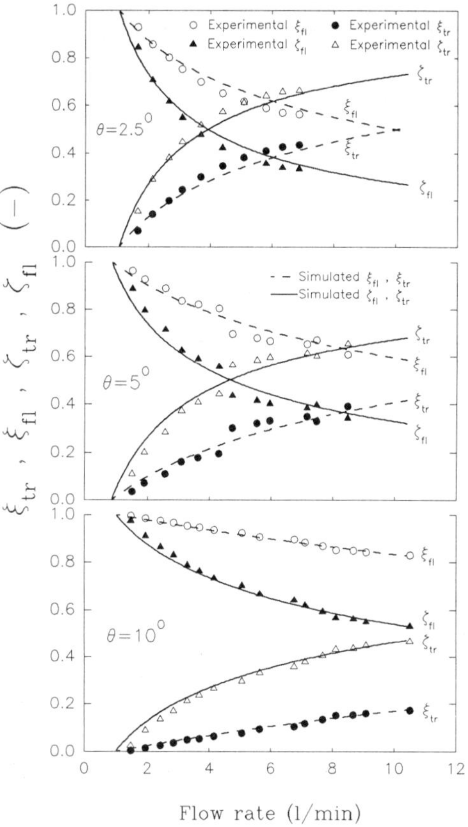

Figure 7. Experimental and simulated results of ξ_{tr}, ξ_{fl}, ζ_{tr}, ζ_{fl} at different flow rates in tapered fluidized-bed bioreactors (Table 2 shows experimental conditions.).

greater than $5°$, the reductions of V_{bed} and H_{bed} will not be very significant. In other words, taper angle of $5°$ could be an optimal taper angle for the designing purpose.

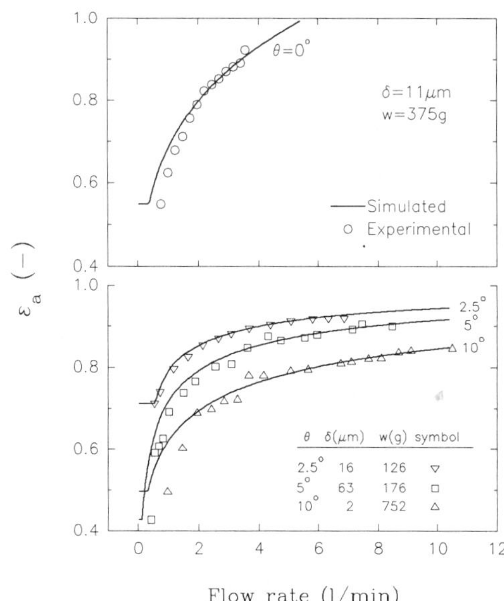

Figure 8. Experimental and simulated results of average bed porosity at different flow rates.

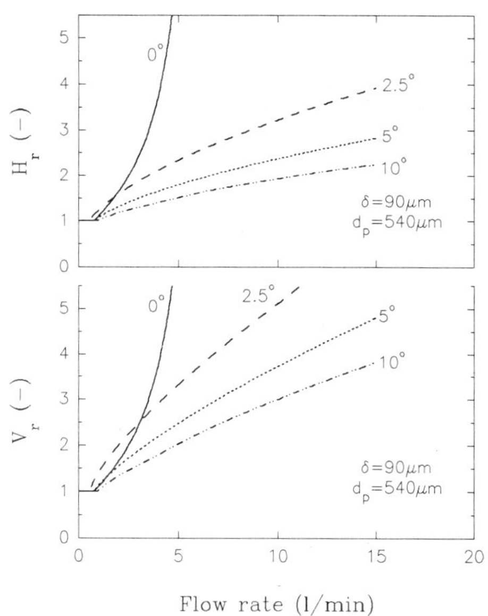

Figure 9. Simulated results of Hr and Vr at different flow rates.

ACKNOWLEDGEMENT

This research project (NSC-83-0410-E006-059) is financially supported by National Science Council, Republic of China in Taiwan.

NOMENCLATURE

D   equivalent diameter of bioreactor (L)

Ga   Galileo number $= \rho_1 d_p^3 (\rho_p - \rho_1) g / \mu^2$

H   height (L)

Hr   ratio of H_{bed} to H_{bed} which at the lowest flow rate (-)

N   number of particles per gram dry BAC=14225/g (M^{-1})

P   parameter

Q   flow rate $(L^3 T^{-1})$

Re_t Reynolds number based on $u_t = \rho_1 u_t d_p / \mu$

S_H $(\triangle H_{bed}/H_{bed})/(\triangle p/p)$ = sensitivity of bioparticles' expanded height (-)

S_V $(\triangle V_{bed}/V_{bed})/(\triangle p/p)$ = sensitivity of bioparticles' expanded volume (-)

V   volume (L^3)

Vr   ratio of V_{bed} to V_{bed} which at the lowest flow rate (-)

W   dry weight of media(BAC) which added into fluidized-bed bioreactor (M)

Z   axial distance from some position of tapered bioreactor to its hypothetical apex (L)

a   specific surface area (i.e., biofilm surface area per unit volume of bioreactor) (L^{-1})

b   bed width (L)

d   diameter (L)

n   index in equation (2)

r_m radius of media (L)

u   velocity of fluid (LT^{-1})

u_t terminal settling velocity of particle for large vessel ($d_p/D < 0.001$) (LT^{-1})

v_i terminal settling velocity of particle for single sphere (LT^{-1})

θ tapered angle of bioreactor (degree)

ρ wet density (ML^{-3})

δ thickness of biofilm (L)

ε bed porosity (-)

reactants and dissolved oxygen. Design guidelines have been developed for the oxygen requirements in terms of oxygen transfer coefficients, oxygen enrichment, and liquid recycle rate. Typical results for batch operation are given in Fig. 3.

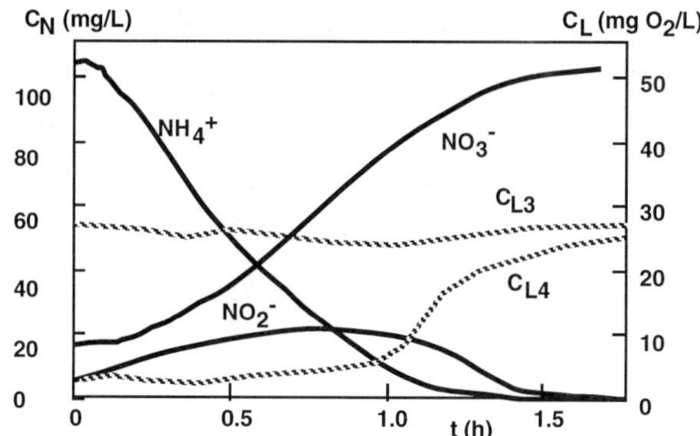

Figure 3. Batch nitrification experiment with pure oxygen, showing absence of oxygen limitation effects.

From theory, based on the dimensionless diffusion model equations, it was demonstrated that the stoichiometric ratio for the first step(3.5 mg O_2/mg NH_4^+-N) can be employed as a criterion to determine which substrate is limiting, oxygen or ammonia. For the present work, in the range of concentrations where limitation occurred, 4 mg/L NH_4^+-N and 14 mg/L O_2, the ratio of oxygen to ammonia in the bulk liquid determined which substrate was penetration-limiting, O_2 if < 3.5 and NH_4^+ if > 3.5. Simulations provided biofilm concentration profiles which demonstrated the role of the oxygen-ammonia ratio. Experiments indicated that, generally, high NO_2^- concentrations can be expected, which depend on the residence time, biofilm area, and oxygen concentration. This dependency was investigated with the model, as was the parametric sensitivity with respect to the saturation constants. Particularly important for the NO_2^- levels were the ratios of the saturation constants for oxygen. Simulations of the biofilm gradients are given in Fig. 4. Respirometer measurements established that the oxygen saturation constant for

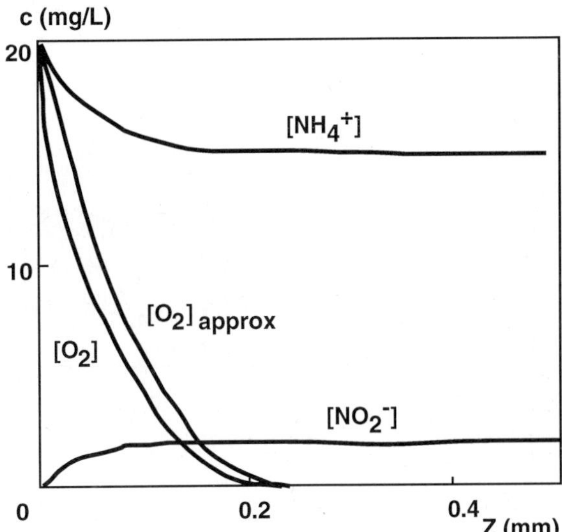

Figure 4. Steady state concentration gradients in the biofilm for constant concentrations in the reactor. For this case $[O2]R/[NH4^+]R = 1$, and the [O2] falls to zero midway in the film beyond which the other gradients become zero. The [O2] approx line is the zero-order kinetics approximation.

the NO_3^- forming step was highest. This led to NO_2^- accumulation at low oxygen levels. Dynamic pulse experiments were used to determine O_2 and NH_4^+ limitation (Fig. 5). As shown in Fig. 6, the limitation can be predicted from the O_2/NH_4^+ ratio. Repeated pulses of NH_4^+ caused a reproducible and stable response of the reactor. Oxygen uptake provided a reliable monitoring method.

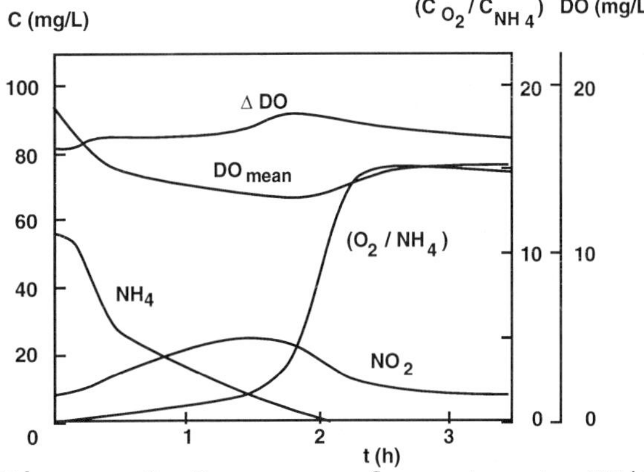

Figure. 5. Response of reactor to $NH4^+$ pulse under $NH4^+$ limiting conditions.

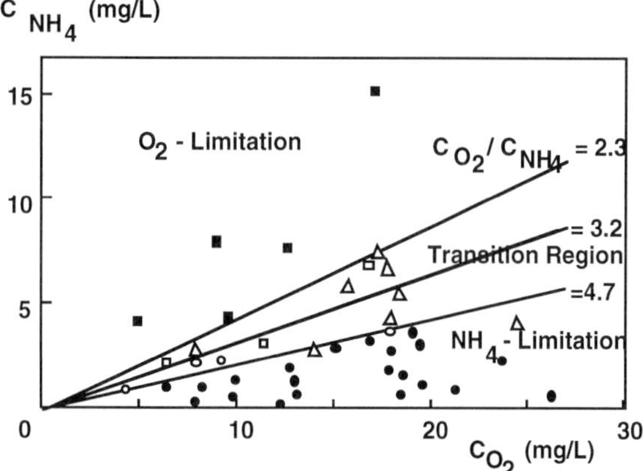

Figure 6. Regions of limitation as given by the concentrations of NH4+ and dissolved oxygen in the reactor.

The reaction rates were also measured in the fluidized-sand-bed biofilm reactor under a range of steady-state conditions with respect to bulk NH_4^+, NO_2^- and O_2 concentrations. In all experiments r_{NO2-} decreased more than r_{NH4+} under low oxygen conditions. This resulted in high NO_2^- effluent concentrations under low residence time conditions. In Fig. 7 the concentrations are seen as a function of flow rate (or residence time).

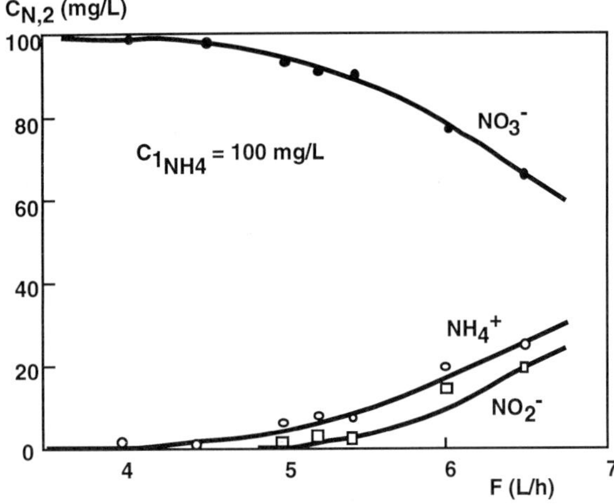

Figure 7. Outlet concentrations as a function of feed flow rate during continuous steady state operation.

Fig. 8 portrays the steady state concentrations as a function of dissolved oxygen concentration at a given flow rate and feed concentration.

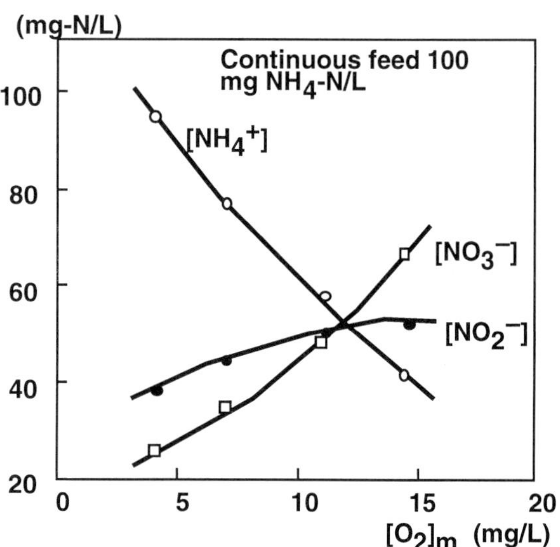

Figure 8. Steady state concentration data for NH_4^+, NO_2^- and NO_3^- as a function of dissolved oxygen concentration.

Figures 9 and 10 present the rate data in terms of a 1/2 order model for NH4 and O2, respectively. The 1/2 order model can be expected for the limiting substrate, as it is an approximation of the complex kinetics, whcih results from the zero-order approximation (see Fig. 4) and the first order diffusion process. In brackets the O2/NH4 ratios are given for the corresponding experiments. here it is clearly seen that NH4 limitation and O2 limitations correspond to the pulse experiments of Fig. 5

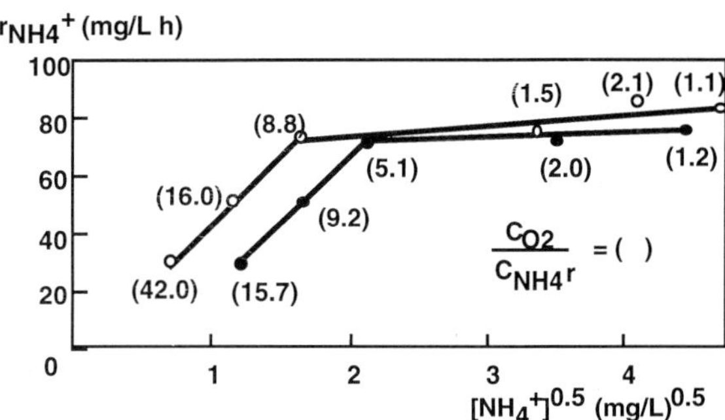

Figure 9. Experimental data showing half-order dependency on ammonia. The values for [O2]R/[NH4+]R are given in brackets.

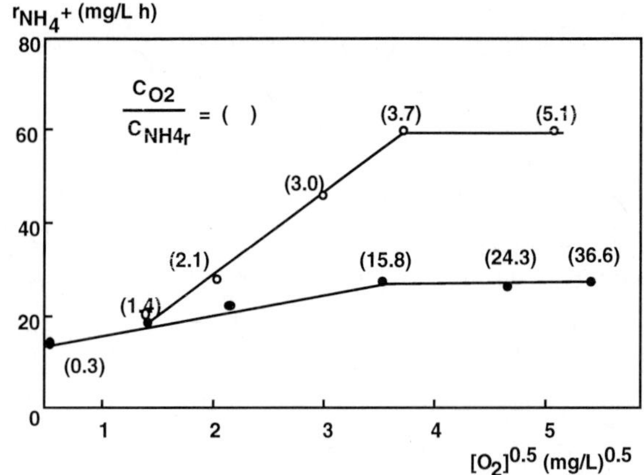

Figure 10. Experimental data showing half-order deopendency on oxygen. The values for $[O2]R/[NH4^+]R$ are given in brackets.

Table. 1 summarizes the rates of nitrification achieved.

Table 1. Maximum Rates of Nitrification

Oxidation steps	Based on total volume (mg N / L h)	Based on biomass (mg N / g h)
NH_4^+	84.5	134
NO_2^-	75.7	120

Biological Denitrification of Drinking Water Using Autotrophic Organisms with H_2 in a Fluidized-Bed Biofilm Reactor

Biological denitrification of drinking water was studied in a fluidized sand bed reactor using a mixed culture (6) Hydrogen gas was used as the reaction partner. The reaction kinetics were calculated with a double Monod saturation function. The K_S value for hydrogen was below 0.1% of saturation. No appreciable biofilm diffusion effects were detected. Reactor performance was a function of the culture's past history. Batch experiments always exhibited an accumulation of NO_2^-, but continuous experiments with a sufficiently long residence time always resulted in complete nitrogen removal. Rates of up to 23 mg N/L h, 25 mg N/g DW h, and 7.9 mg H_2/L h were achieved. Residence times of 4.5 h would be required for complete denitrification of water containing 25 mg NO_3^--N/L or approximately 1 h for every 5 mg/L.

Operation of a Three-Phase Biofilm Fluidized Sand Bed Reactor for Aerobic Wastewater Treatment

A biofilm fluidized sand bed column reactor (14 L) has been operated in the three-phase mode on a soluble glucose-yeast hydrolysate substrate in which the biofilm-sand phase (1-2.5 L) was suspended by direct aeration of the bed (1). Within two weeks a tight biofilm was formed whose activity resulted in a 90% reduction, with loads of 10.7 kg TOC/m^3 day. The residence time was 1 h. The biofilm remained intact during operation with high residence times (up to 23 h) over three weeks. Oxygen transfer coefficients varied with aeration rate and sand quantity between 0.02 and 0.04s^{-1} during nongrowth conditions; they decreased with increasing amounts of clean sand and were higher and relatively independent of the sand fraction with biofilm-covered sand. Aeration rates used in the 14 L reactor were 23-40 L/min (2.4-4.1 cm/s) and were sufficient to suspend 78-92% of the biofilm-covered sand. Clean sand was 50-75% suspended. Oxygen uptake rate varied between 15.4 and 23.1 mol/m^3 h.

Phenol Degradation in a Three-Phase Biofilm Fluidized Sand Bed Reactor

A previous three phase fluidized sand bed reactor design was redesigned by adding a draft tube to improve fluidization (2). The changes had little influence on the oxygen transfer coefficients (K_La) but greatly reduced` the aeration rate required for sand suspension. The resulting 12.5 L reactor was operated with 1 h liquid residence time, 10.2 L/min aeration rate, and 1.7-2.3 kg sand (0.25-0.35 mm diameter) for the degradation of phenol as sole carbon source. The K_La of 0.015 s^{-1} gave more than adequate oxygen transfer to support rates of 180 g phenol/h m^3 and 216 g oxygen/h m^3. The biomass-sand ratios of 20-35 mg volatiles/g gave estimated biomass concentrations of 3-6 g volatiles/L. Off-line kinetic measurements showed weak inhibition kinetics with constants of K_S = 0.2 mg phenol/L, K_{O2} =0.5 mg oxygen/L and K_I = 122.5 mg phenol/L.

Very small biofilm diffusion effects were observed. Dynamic experiments demonstrated rapid response of dissolved oxygen to phenol changes below the inhibition level. Experimentally simulated continuous stagewise operation required three stages, each with 1 h residence time, for complete degradation of 300 mg phenol/L h.

Packed- and Fluidized-Bed Biofilm Reactor Performance for Anaerobic Wastewater Treatment

Anaerobic degradation performance of a laboratory-scale packed-bed reactor (PBR) was compared with two fluidized-bed biofilm reactors (FBRs) on molasses and whey feeds (7). The reactors were operated under constant pH 7 and temperature 35 C conditions and were well-mixed with high recirculation rates. The measured variables were chemical oxygen demand (COD), individual organic acids, gas composition, and gas rates. As carrier, sand of 0.3-0.5 mm diameter was used in the FBR, and porous clay spheres of 6 mm diameter were used in the PBR. Startup of the PBR was achieved with 1-5 day residence times. Start-up of the FBR was only successful if liquid residence times were held low at 2-3 h. COD degradations of 86% with molasses (90% was biodegradable) were reached in both the FBR and PBR at 6 h residence time and loadings of 10 g COD/L day. At higher loadings the FBR gave the best performance; even at 40-45 g COD/L day, with 6 h residence times, 70% COD was degraded. The PBR could not be operated above 20g COD/L day without clogging. A comparison of the reaction rates show that the PBR and FBR performed similarly at low concentrations in the reactors up to 1 g COD/L, while above 3 g COD/L the rates were 17.4 g COD/L day for the PBR and 37.4 g COD/L day for the FBR. This difference is probably due to diffusion limitations and a less active biomass content of the PBR compared with the fluidized bed.

The results of dynamic step change experiments, in which residence times and feed concentrations were changed at constant loading, demonstrated the rapid response of the reactors. Thus, the response times for an increase in gas rate or an increase in organic acids due to an increase in feed concentration were less than 1 day and could be explained by substrate limitation. Other slower responses were observed in which the reactor culture adapted over periods of 5-10 days; these were apparently growth related. An increase in loading of over 100% always resulted in large increases in organic acids, especially acetic and propionic, as well as large increases in the CO_2 gas content. In general, the CO_2 content of the gas was very low, due to the large amount of dissolved CO_2 that exited with the liquid phase at low residence times. The performance of the FBR with whey was comparable to its performance with molasses, and switching of molasses to whey feed resulted in immediate good performance without adaptation.

Carrier Influence in Anaerobic Biofilm Fluidized Beds for Treating Vapour Condensate from the Sulphite Cellulose Process

The stability of the biofilm to adverse reactor disturbances is important for the design of anaerobic fluidized bed reactors, and it was therefore of interest to determine whether the adsorptive and porous properties of some potential carriers were important in biomass retention (8). The continuous anaerobic degradation of an acidic waste water from a cellulose process was investigated in six parallel reactors. The reactor volumes were from 750 to 800 mL and used varying amounts of carrier material as follows:

Reactor R1, Limestone
Reactor R2, Porous glass
Reactor R3, Quartz sand
Reactor R4, Act.carbon
Reactor R5, Act.carbon
Reactor R6, Coal

The reactors were operated for several months with vapour condensate from a sulphite cellulose process as feed. The performance and stability of the reactors with respect to degradation rate were tested for a range of loading conditions. Unbuffered, buffered and pH-controlled conditions were compared.

Four weeks after inoculation, nearly all the carriers appeared to be covered with a black biofilm. Electron microscope photos showed that activated carbon had a particularly thick microbial film. Using a loading of 10 kg COD/m^3 d with a residence time of 12 h, steady state conditions were reached after 70 days, as shown in Fig. 11.

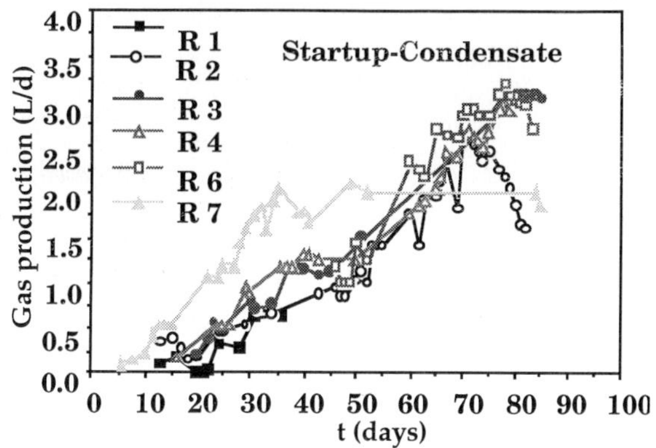

Figure 11. Gas production rates during startup

The reactor R1, containing limestone, was stopped because the carrier was too fragile, as was the reactor R5 with activated carbon, whose density was too low.

Experiments were conducted with non-buffered wastewater, with well-buffered wastewater, and with pH control. For non-buffered wastewater the effects on the pH of stepped loading (L), achieved by changing the hydraulic residence times (HRT), are shown in the Fig.12. The reactor R3 with quartz sand gave the best performance at the highest load, corresponding to the residence time of 4 hours.

In further experiments the wastewater was buffered, and the reactors were subjected to similar step increases in loading. From Fig. 13 it can be seen that the sand carrier systems behaved differently from the rest, exhibiting increased rates under loading conditions that caused the other reactors to fail or to begin to fail. This difference in the behavior of the sand system was reflected also in the pH effects; the pH drop at high loadings was much less for the sand carriers.

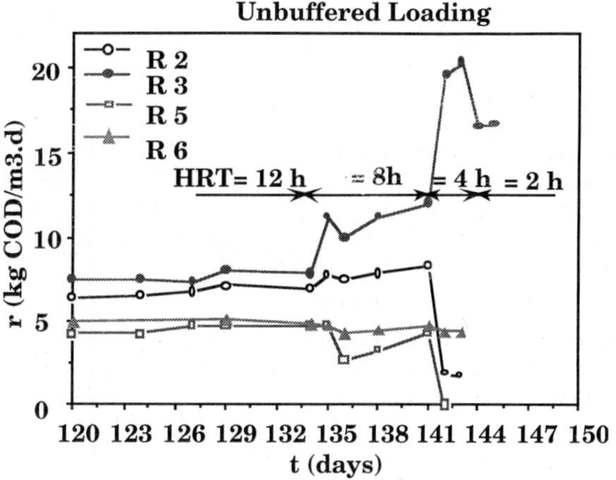

Figure 12. COD removal rates with unbuffered feed.

The inhibition of methanogens at low pH makes it desirable to operate high rate anaerobic systems with pH control. Reactor R3 was selected because of its superior performance and was subjected to step increases in loading, as shown in Fig. 14, where the gas production and

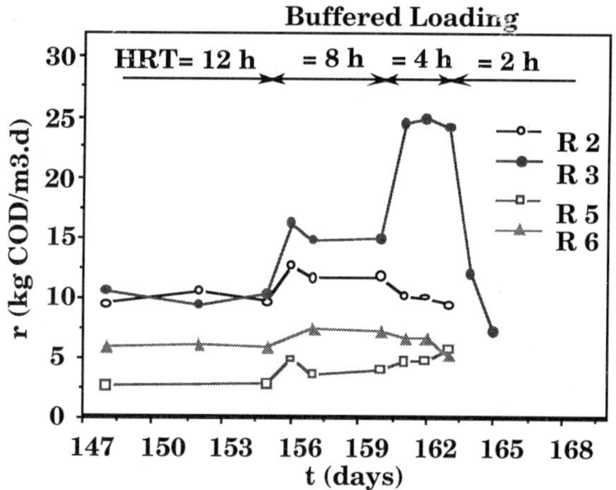

Figure 13. COD removal rates with buffered feed.

degradation rates are given. It is interesting that the degradation rates did not increase after Day 175, but that the gas productions did increase after each loading increase. Comparing values of the effluent concentrations with degradation rates indicated that the rates had reached a maximum value, governed by saturation kinetics.

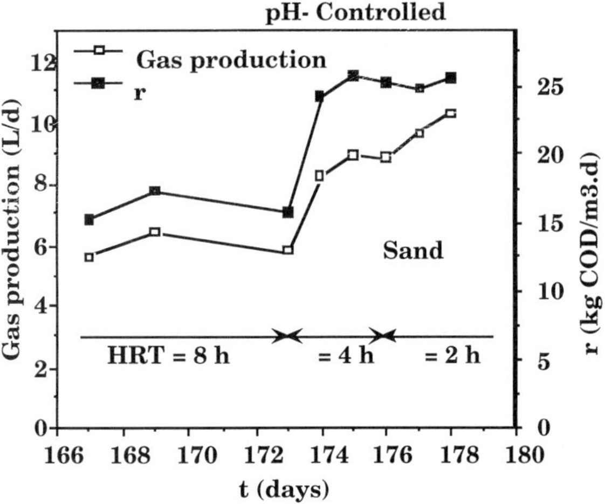

Figure 14. COD removal rates and gas production rates with pH control for the sand.

In photos of the biofilm-covered carriers, the surface roughness appears to have been important for the adhesion of biofilm on the sand, since the biofilm was located mainly in the crevices. The difference between the rather loose biofilm on activated carbon and the tighter biofilm on the sand was noteworthy.

These results show that the high-density, small-sized sand was the best carrier. Carrier porosity was not important, although surface roughness appeared to be important. No clear evidence of diffusional influence on the kinetics was observed. The highest rates and best stability were obtained with the sand carrier, which had the thinnest biofilm; these were 21 kg COD/m^3 d (unbuffered feed), or 25 kg COD/m^3 d (buffered feed and pH-controlled system). The sand system was relatively stable with respect to pH changes. With pH control, loading could be increased until a maximum reaction rate, as determined by the saturation kinetics, was achieved.

Absorption and Degradation of Dichloromethane in a Biofilm Fluidized Bed Reactor

This project involved a three-phase fluidized sand bed reactor for removing dichlormethane (DCM) from an air stream (9). . The organic was absorbed the air stream into the liquid phase and was degraded by an adapted wild culture. The overall reaction is as follows:

$$CH_2Cl_2 + O_2 \rightarrow CO_2 + 2HCl + Biomass$$

The production of HCl necessitates controlling pH by titration and regulating salinity control by a continuous liquid feed. Previously packed trickle-bed reactors have been studied, as seen in Fig. 15.

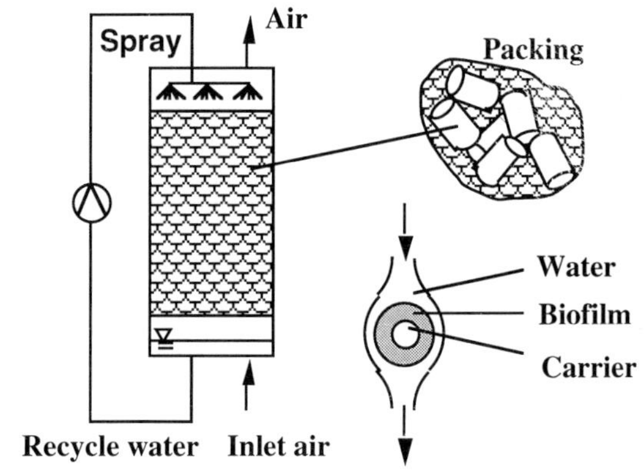

Figure 15. Trickle bed reactor for waste gas cleaning

Airlift Reactor and Instrumentation

In this study an airlift three-phase fluidized sand bed reactor was used, as shown below.

The conditions in the reactor are summarized in Table 2.

Table 2. Airlift Reactor Conditions

Volumes$(V_L + V_G)$	2.65 L
Carrier	Sand
Particle size	0.25 - 0.4 mm
Specific area	2000 m^2/m^3
Solids concentration	50 - 100 g/l
Expanded bed Volume	2.5 L
Expansion	ca. 2000 %
Biomass concentration	0.14 gTS/g sand
Aeration rate	1.47-5.89 L/min
Gas velocity	1.25-5.0 cm/s
Liquid velocity	to 10 cm/s
Feed rate	0.87 L/h
Residence time liquid	3 h
DCM conc.in feed air	0.5-30 mg/L
Temperature	30 °C
pH-controlled	7-8

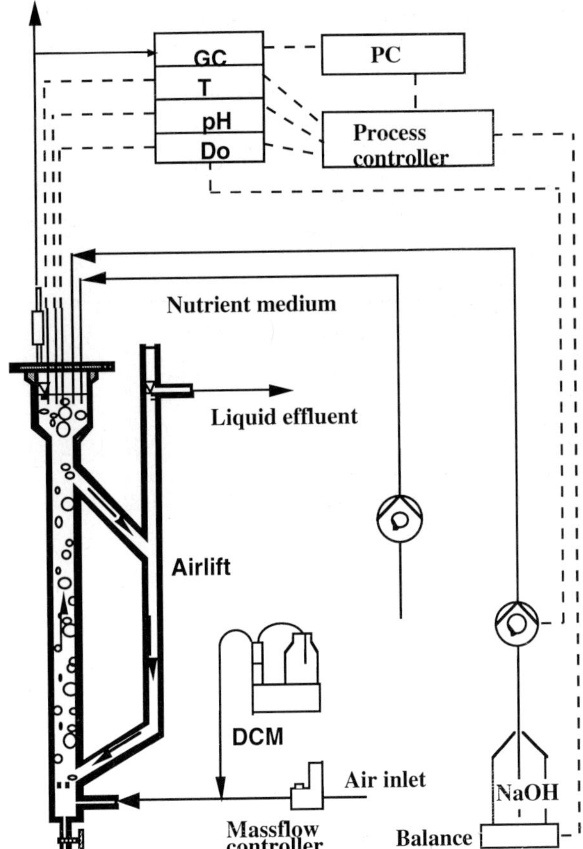

Figure 16. Airlift reactor and instrumentation.

The dependency of gas-related quantities, holdup and K_La, are given in Figs. 17 and 18. Gas holdup was directly influenced the gas-liquid transfer rate and was dependent on the amount of solid phase, as shown.

Steady-state experiments to determine the kinetics were made with the results for DCM and O_2 influence given in Figs. 19 and 20, respectively.

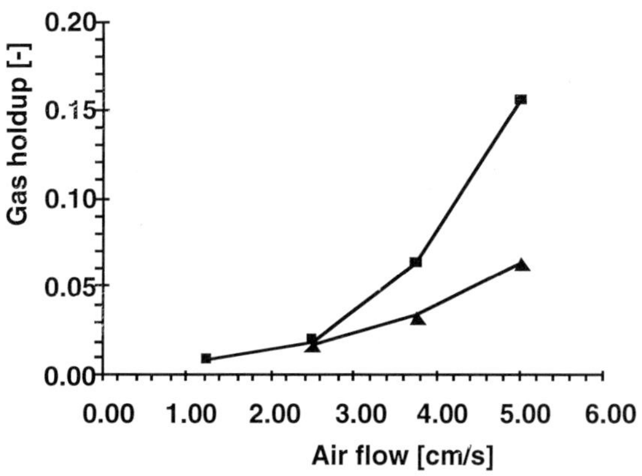

Figure 17. Gas holdup dependency on solids concentration and gas velocity $c_S = 0$ g / L [■] and $c_S = 100$ g / L [▲]

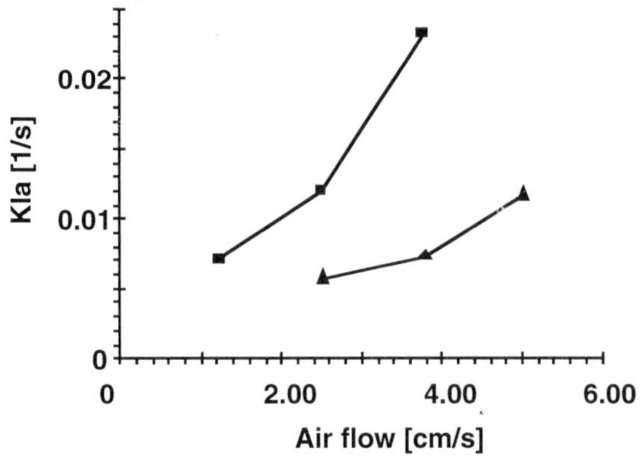

Figure 18. K_La for DCM with solids con-centration and superficial gas velocity. $c_S=0$ g/L [■] and $c_S=100$ g/L [▲]

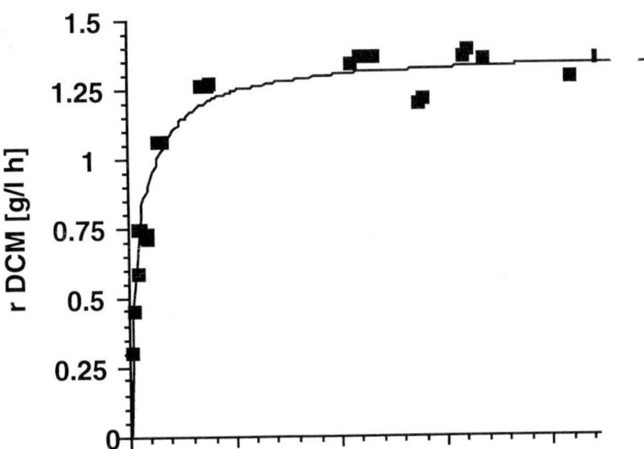

Figure 19. Degradation rate versus DCM concentration.

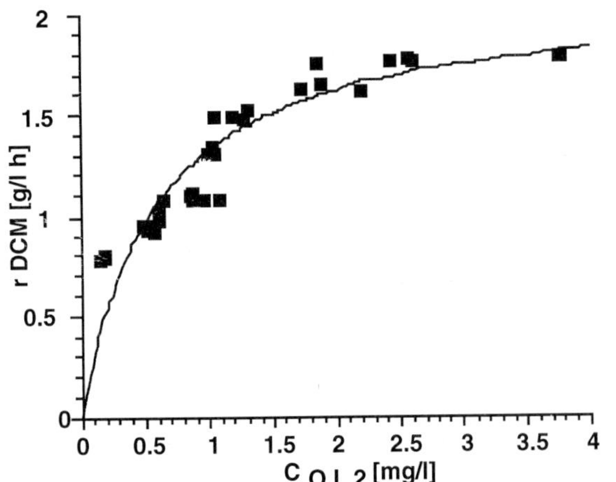

Figure 20. Degradation rate versus dissolved Oxygen concentration.

The rates were measured as a function of pH and temperature. Fig. 21 gives the pH influence.

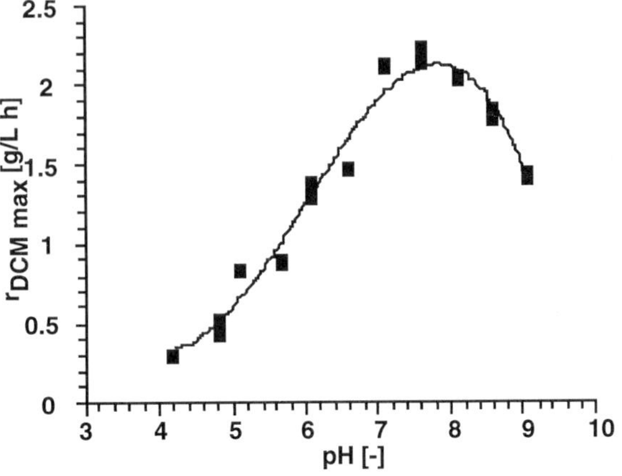

Figure 21. Influence of pH on maximum degradation rate.

The concentrations, and therefore the rates, were directly influenced by the operating parameters. Fig. 22 gives the influence of the liquid residence time, tau, and the corresponding Cl^- concentration on the rate.

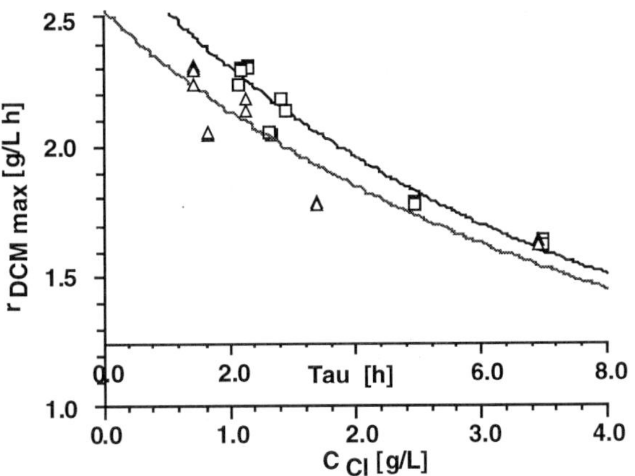

Figure 22. Influence of chloride C_{Cl} [□] and liquid residence time tau [Δ] on maximum degradation rate.

The load to the reactor is a measure of the DCM concentration and flow rate of the entering air. At low loading, the rate was proportional to the loading, as shown in Fig. 23. This is a DCM-limiting situation, for which DCM was very low in the liquid phase.

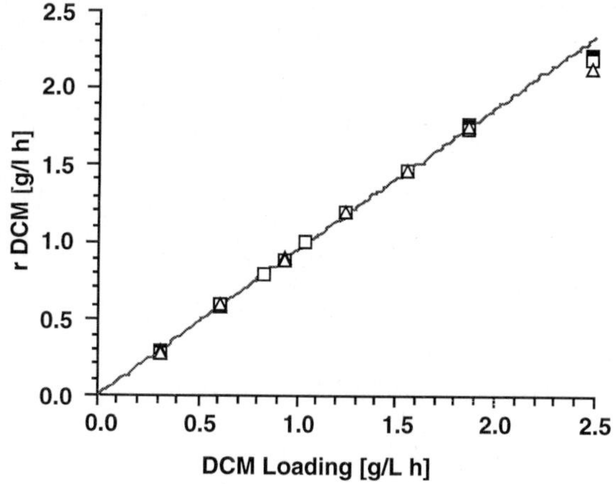

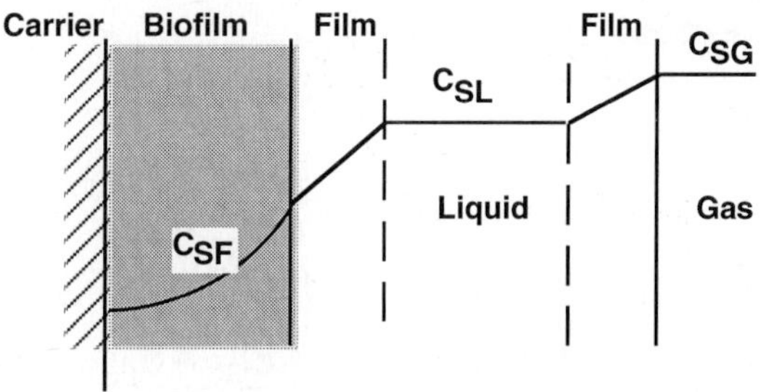

Figure 25. Transfer steps from gas to liquid to biofilm. Profiles at the solid-liquid interface.

Figure 23. Degradation rate versus loading rate.

The three phase reactor can be viewed for modelling purposes as shown in Fig. 26.

For a given air flow rate the DCM concentration in the inlet gas influenced the reactor conditions, as shown in Fig. 24.

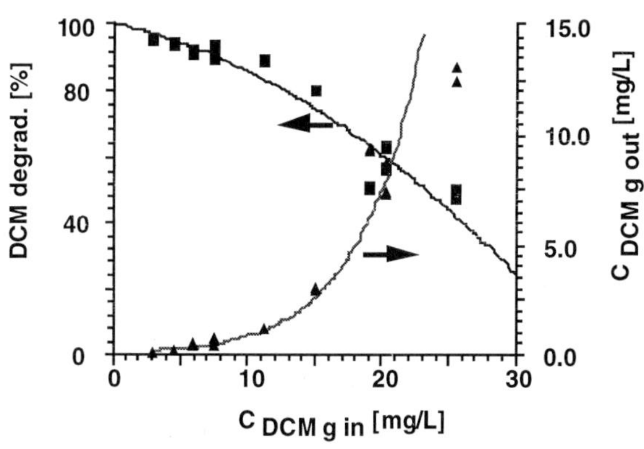

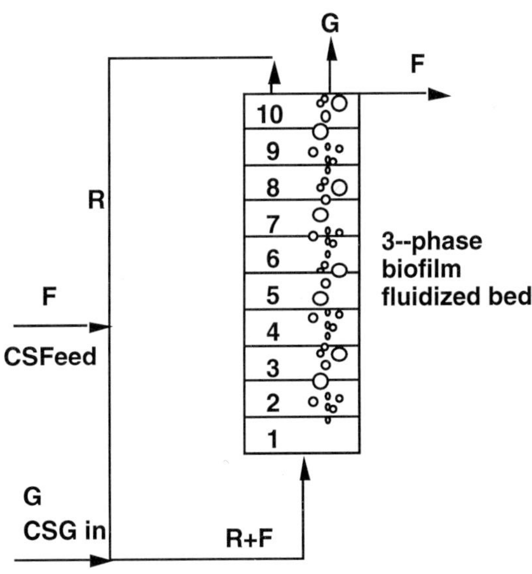

Figure 24. Percent degradation and outlet gas concentration versus inlet gas concentration

Figure 26. Three-phase fluidized bed for DCM-air treatment.

Model Development

The transfer of oxygen and dichloromethane from the gas phase to the liquid phase and from the liquid into the biofilm is depicted in Fig 25.

The flow can be modelled by a two-phase tanks-in-series approach in Fig. 27.

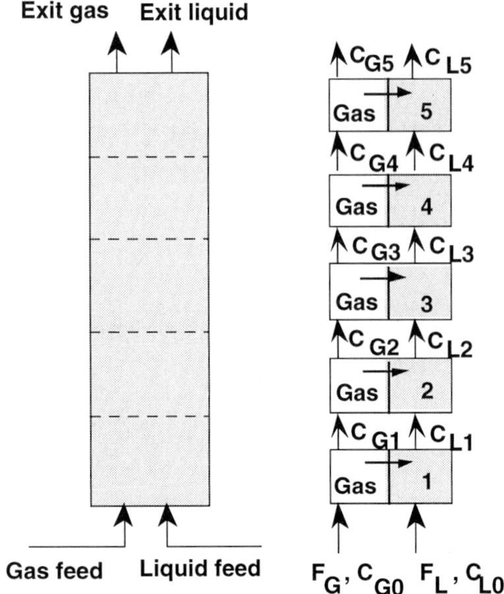

Figure 27. Stagewise model for the fluidized bed.

The follow is a summary of the mass balances for each component i in the gas and liquid phases:

DCM and oxygen in the gas phase:

$$V_G \frac{dC_{iGn}}{dt} = F_G(C_{iGn-1} - C_{iGn}) -$$

$$- K_La(C_{iLn}^* - C_{iLn})V_L$$

DCM, oxygen, chloride and hydrogen ions in the liquid phase with reaction and transfer:

$$V_L \frac{dC_{iLn}}{dt} = F_L(C_{iLn-1} - C_{iLn}) +$$

$$+ K_La(C_{iLn}^* - C_{iLn})V_L - r_{in}V_L$$

where the equilibrium is,

$$C_{iLn}^* = \frac{RT}{H_i} C_{iGn}$$

The reaction kinetics consider DCM, and oxygen limitation and inhibition by

Cl-, as well as the influence of pH and temperature as follows:

$$r_{DCMn} = r_{DCMmaxn} \frac{C_{DCMn}}{K_{DCM} + C_{DCMn}}$$

$$\frac{C_{On}}{K_O + C_{On}} \frac{K_{iCL}}{K_{iCL} + C_{CLn}}$$

The dependency of rate on temperature and pH was modelled as

$$r_{DCMmaxn} = r_{ODCMmaxn} \, F(pH) \, e^{-E/RT}$$

$$F(pH) = a \, pH^2 - b \, pH^3 - c \, pH + d$$

where the parameters a, b, c and d were fitted to the pH-rate data.

The oxygen uptake rate was related to the DCM uptake rate by a constant yield:

$$r_{On} = Y_{O/DCM} \, r_{DCMn}$$

The experimental process parameters needed to solve this model are the transfer coefficients for DCM and O_2, the Henry coefficients, and the temperature and pH dependencies for DCM and O_2.

The model was solved using a simulation language (ACSL-SimuSolv). In Figs.28 and 29 the contours are lines of constant rate.

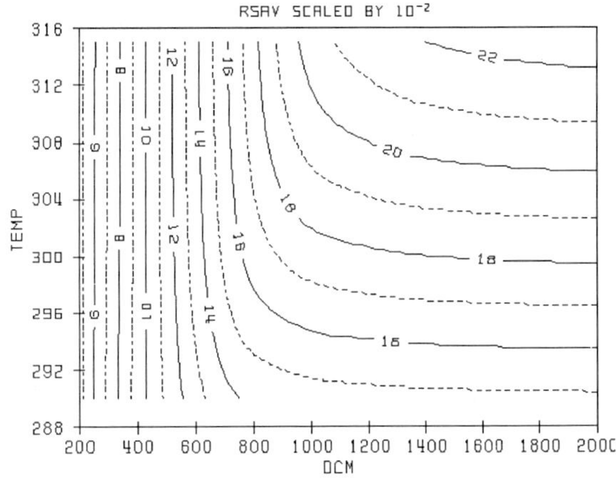

Figure 28. Degradation Rate as a function of temperature and the DCM loading.

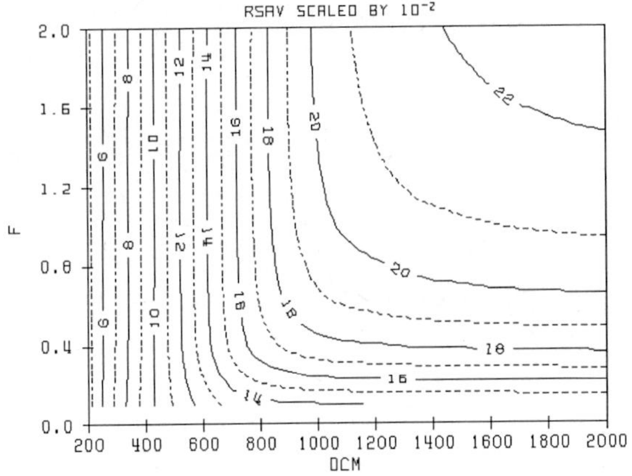

Figure 29. Degradation rate as a function of liquid feed rate and the DCM loading.

Figure 30 shows a comparison of response data with simulation.

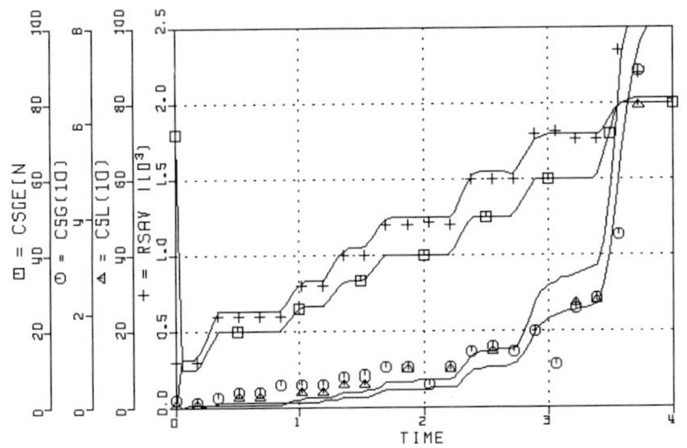

Figure 30. Response of column top to a step change in inlet gas concentration. Symbols: C_{GDCMin} [mg/L] [□], $C_{GDCM(10)}$ [O] [mg/L], $C_{LDCM(10)}$ [Δ] [mg/L], and $r_{DCMaverage}$ [+] [mg/L h] The solid lines are from simulation and the points are from experiment.

CONCLUSIONS

The examples discussed represent a wide variety of environmental bioprocesses, ie aerobic and anaerobic water and air treatment. Owing to the retention of highly active biomass on fluidized carriers, such as sand, the rates of reaction are orders of magnitude higher than would be expected with suspended culture. Experience has shown that with a mixed culture a biofilm can be obtained on rough or porous carriers for any biodegradable substance. For aerobic systems either a two-phase reactor with external aeration or a three-phase may be chosen. External aeration permits oxygen uptake measurements to be easily measured, which is valuable for monitoring and control.

ACKNOWLEDGEMENT

The author is grateful for the work of the many fine researchers who have contributed to this work. Financial support for this work was provided by the Sulzer Co., the ETH, and the Swiss government.

NOTATION

C	Concentration	mg/L
C_L	Dissolved oxygen	mg/L
DO	Dissolved oxygen	mg/L and % sat
F	Flow rate	L/h
K	Saturation const.	mg/L
R	Recycle flow	L/h
r	Reaction rate	kg/m3 h and mg/L h
S	Substrate conc.	mg/L
V	Volume	L and m^3
Y	Yield coeff.	(-)
Z	Distance	mm

Indices

1,2,3,4,and n refer to streams
F   refers to film
G   refers to gas
i   refers to component i
in   refers to inlet
L   refers to liquid
N   refers to nitrogen content
O   refers to oxygen
R   refers to reactor
S   refers to substrate
DCM, NH4, NO2, NO3 and Cl refer to the chemical species

LITERATURE CITED

1. Ryhiner, G, Petrozzi, S., and Dunn, I.J., *Biotechnol. Bioeng.*, **32**, 677 (1988).

2. Etzensperger, M., Thoma, S., Petrozzi, S., Dunn, I. J., *Bioprocess Engr.* **33**,175(1989)

3. Tanaka, H., Uzman, S., and Dunn, I.J.,*Biotechnol. Bioeng.* **23**, 1683 (1981)

4. Denac, M., Uzman, S., Tanaka, H., and Dunn, I.J. , *Biotechnol. Bioeng.* **25**, 1841 (1983)

5. Dunn, I.J., Uzman,S., and Studer, B., *Conserv. & Recycling*, **8**, 143 (1985).

6. Kurt, M., Dunn, I.J., and Bourne, J.R., *Biotechnol. Bioeng.* **29**, 493 (1987).

7 Denac, M. and Dunn,I.J., *Biotechnol. Bioeng.* **32,**159(1988).

8. Petrozzi, S., Dunn, I. J., Heinzle, E., and Kut, O. M., Can. J. Chem Engr., **69**, 527(1991).

9. Niemann, D., ETH Diss. No. 10025, Zurich, 1993

The Effect of Concentration and Hydraulic Shock Loads on the Performance of a Two-Stage High-Rate Anaerobic Wastewater Treatment System: Prediction and Validation

M. Romli[1], J. Keller[2], P.L. Lee[2], and P.F. Greenfield[2]

[1]Department of Agroindustrial Technology, Bogor Agricultural University, 16002, INDONESIA
[2]Chemical Engineering Department, The University of Queensland, 4072, AUSTRALIA

An improved version of a mechanistic mathematical model of anaerobic degradation processes has been developed. It was incorporated into three types of high-rate reactor configuration; namely, single-stage, two-stage, and two-stage with recycle. Implemented in a recently developed simulation package NIMBUS (1) the model was able to simulate the dynamic behaviour of anaerobic wastewater treatment systems. In the present study, the model has been used to predict the system response to a short-term concentration and hydraulic step changes at various volumetric organic loading rates. The comparison between the predicted and the measured data using a synthetic diluted molasses wastewater showed that reactor variables such as pH, alkali consumption rate, gas production and its composition, and effluent organic acids concentration were modelled reasonably well.

INTRODUCTION

Anaerobic treatment processes have been shown to offer many real benefits as a means for waste disposal in terms of environmental impact, energy, economic viability and material recovery. A major obstacle of the application of this technology is its reputation of being an unstable and difficult-to-control process. This notion, however, has been a result of a lack of understanding about the process. The development of a mechanistic model of anaerobic treatment processes may facilitate the documentation of the knowledge of the process and also serve as a reliable tool for design, optimisation and operational evaluation.

The mechanistic model of anaerobic wastewater treatment processes used in this study was initially developed by Costello (2). In the previous study, the model had been verified against the experimental data from single-stage high-rate anaerobic treatment configurations (3). Further modifications and improvements (4) have allowed the model to be used for the simulation of two-stage anaerobic wastewater treatment systems. This improved model has been shown to predict the two-stage reactor behaviour with and without recycle operation and to describe the effect of pH manipulation of the acidification reactor. In this present study, the model is used to study the effect of concentration and hydraulic shock loads on the performance of a two-stage high-rate anaerobic wastewater treatment system.

MATERIAL AND METHODS

Experiment. The experimental system consisted of a continuous stirred tank reactor (CSTR) as the acidification reactor and a fluidised sand bed reactor (FBR) as the methanogenic reactor (Figure 1). The liquid volume in the first stage was 1.64 litre and in the second stage 3.30 litre. Both reactors were operated at 35°C. Throughout the experiments the acidification reactor was controlled at a pH of 6.0 by automatic addition of sodium hydroxide, whereas the methanogenic reactor pH was only monitored. The recycle flow rate was set to zero (no recycle was applied). The feed used was a diluted molasses with the addition of nitrogen and phosphorous as well as trace elements to ensure good bacterial growth.

The experiments were conducted at a volumetric loading rate of 13 kg COD/m^3.day. After the system had reached steady state conditions with the overall COD removal of more than 90%, a step increase in feed concentration for 12 hours or feed flow rate for 6 hours was introduced. For a total time of at least 24 hours, reactor variables such as pH of the second reactor, alkali flow rate, gas generation rate, gas composition, and effluent organic acids concentration were measured at hourly intervals. A similar set of experiments were conducted at a higher volumetric loading rate which was 23 kg COD/m^3.day. The details of the reactor system, feed composition, and analytical methods were described elsewhere (4).

379

E. Galindo and O.T. Ramírez (eds.), Advances in Bioprocess Engineering. 379-384.
© 1994 *Kluwer Academic Publishers. Printed in the Netherlands.*

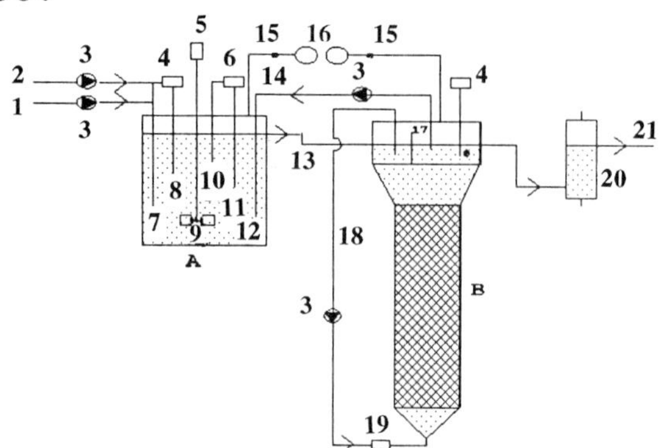

A. Acidification reactor
B. Methanogenic reactor

1. Feed stream
2. NaOH stream
3. Peristaltic pump
4. pH controller
5. Stirrer
6. Temperature controller
7. Feed inlet
8. pH electrode
9. Impeller
10. Heating element
11. RTD probe
12. Recycle inlet
13. Acid. effluent stream
14. Recycle stream
15. Gas sampling point
16. Gas meter
17. Weir
18. Recirculation stream
19. Water bath
20. Overflow system
21. Final effluent stream

Figure 1.   Schematic diagram of a two-stage
high-rate anaerobic wastewater treatment
system

Model and simulation. The model consisted of
a structured set of rate equations describing
the growth, production, and utilisation terms
for the major groups of anaerobic bacteria,
namely glucose (G), lactic (L), propionic (P),
butyric (B), acetic (A), and hydrogen (H)
bacteria.   These   equations   were   then
incorporated   into   the   overall   dynamic
equations for a two-stage high-rate reactor
system. A physico-chemical reaction system
was included in the model to calculate the pH
at any time given the concentrations of the
acidic and basic species in the reactor. The
molecular hydrogen inhibition and regulation
model similar to that developed by Mosey (5)
was also included to account for the effect of
hydrogen on anaerobic metabolic reactions.

The   model   was   implemented   in   a   recently
developed simulation package NIMBUS (1).   In
NIMBUS the model was set up in a modular form
according to the treatment plant flowsheet
(Figure 2).   To simplify the equations for the
bioreactor modules, the biomass calculations

were carried out in special biomass units, one
for each group of bacteria involved in the
degradation process.   To run the simulation
three types of model parameters, namely
experimental, physico-chemical and biological
parameters should be specified.   The physico-
chemical parameters were fixed according to
the   operating   temperature,   whereas   the
biological parameters were taken mainly from
the reported values in literature (6).

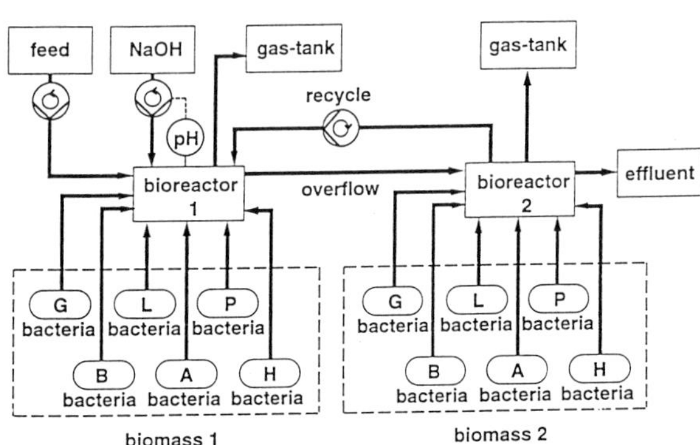

Figure 2.   Model configuration of a two-stage
anaerobic   wastewater   treatment   system   in
NIMBUS

RESULTS AND DISCUSSION

Effect on pH and alkali consumption

Figures 3 and 4 show the responses of alkali
consumption rate and pH of the methanogenic
reactor   to   a   step   increase   in   feed
concentration and a step increase in feed flow
rate respectively.   Both the experimental
results and the model predictions are
displayed. Note that the step increase is
started at time zero for a duration of shock
indicated by a vertical dotted line. As shown
in Figure 3, a 70% step increase in feed
concentration (initial organic loading of 13
kg $COD/m^3.day$)   immediately   doubles   the
consumption of alkali in the acidification
reactor from 24 to 45 ml/h.   The alkali
consumption rate remains at this level during
the period of disturbance and then drops
quickly to within 10% of the pre-shock level
in less than two hours. Since pH of the
acidification   reactor   is   controlled,   an
increase in organic acid production due to
shock load leads to an increase in the amount
of neutralising agent required to maintain the
ionic balance in the system.   A similar effect
is expected when a step increase in feed flow
rate is introduced as shown in Figure 4.

The other plot displayed in Figure 3 shows the
pH response of the methanogenic reactor to a
step increase in feed concentration.   As
expected,   effluent   pH   in   this   reactor
decreases due to more organic acids entering

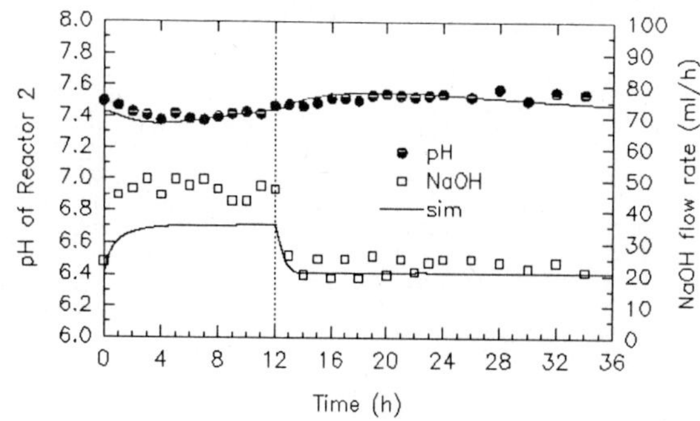

Figure 3. Effect of a step increase in feed concentration on alkali and pH

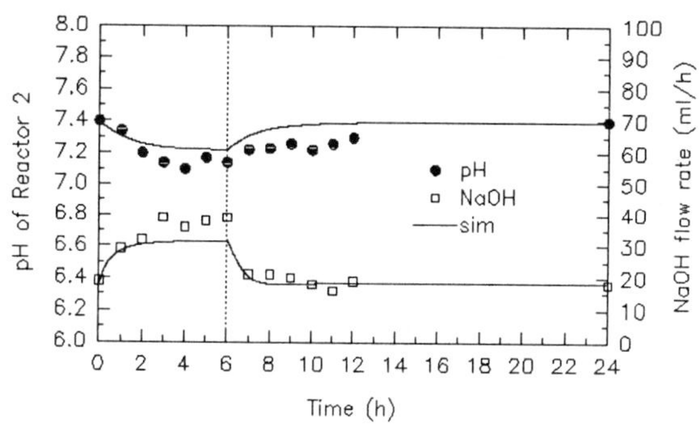

Figure 4. Effect of a step increase in feed flow rate on alkali and pH

Effect on gas production and composition

Effects of a step increase in feed concentration and feed flow rate on gas production rate are presented in Figures 5 and 6 respectively. The simulation plot in Figure 5 shows a slight increase in gas production rate of the acidification reactor during the step increase in feed concentration. A similar tendency is also indicated by the experimental data even though some fluctuations are observed. The large volume of the head space of the acidification reactor and its low gas generation rate make it difficult to measure variables of gas phase accurately. The opposite response of gas production is shown from the simulation plot in Figure 6. A step increase in feed flow rate slightly decreases the gas generation rate in the acidification reactor during the course of the disturbance. After the feed flow rate is returned to normal, the gas generation rate increases slightly and reaches its maximum, which is higher than the pre-shock value, in approximately three hours. The decreased gas production during the shock load is possibly due to a mechanism of hydrogen regulation where substrate degradation is directed toward lactic acid production which releases no gaseous products. As soon as the feed flow rate is returned to normal (liquid and thus solid residence times become longer), high concentration of substrate in the reactor is degraded at a normal rate, hence generating more gas in the first reactor.

In the methanogenic reactor, a 70% step increase in feed concentration increases gas generation rate by 55% as shown from the simulation plot in Figure 5. The experimental data show a similar qualitative trend but differ in the maximum value being 20% higher than the model prediction. This is most likely due to the violation in the assumption of using a constant gas transfer coefficient. It is argued that high bubble production during shock load period may enhance the gas transfer rate.

the reactor, and possibly also due to conversion of undegraded feed from the acidification reactor into organic acids. It is interesting to note that the decrease in pH is recovered before the feed shock load is terminated. This means that the excess production of organic acids has been utilised by methanogenic bacteria to produce neutral products (methane and carbon dioxide). A similar trend is shown in Figure 4, but quantitatively the step increase in feed flow rate causes a more severe effect than the step increase in feed concentration. The pH of the methanogenic reactor drops from 7.4 to 7.1 during the course of the disturbance and starts to recover only after the feed flow rate has been returned to normal. The comparison between experimental and simulation data shows that the model is able to predict the responses of both variable to a step increase in feed concentration and flow rate.

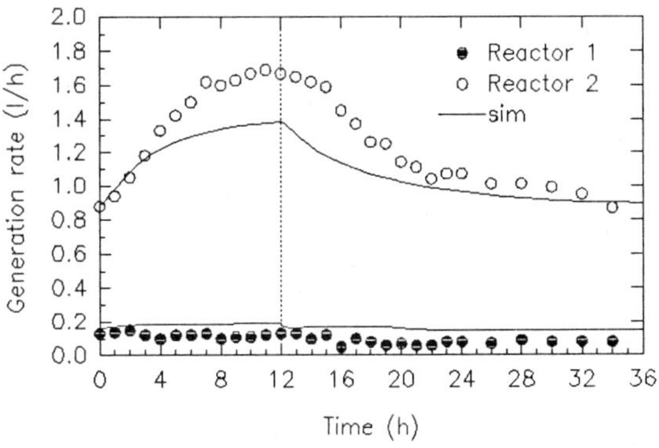

Figure 5. Effect of a step increase in feed concentration on gas production

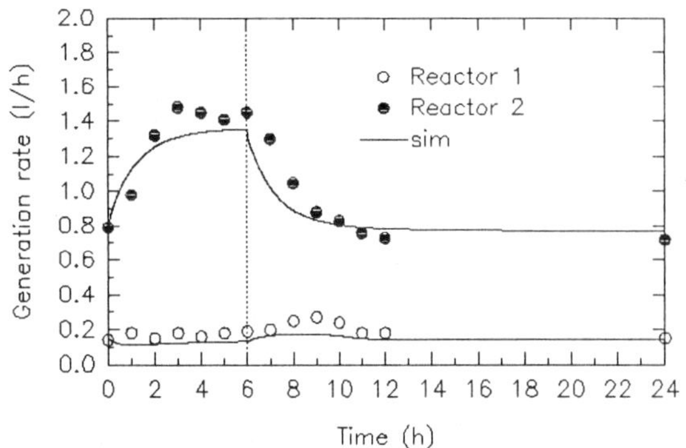

Figure 6. Effect of a step increase in feed flow rate on gas production

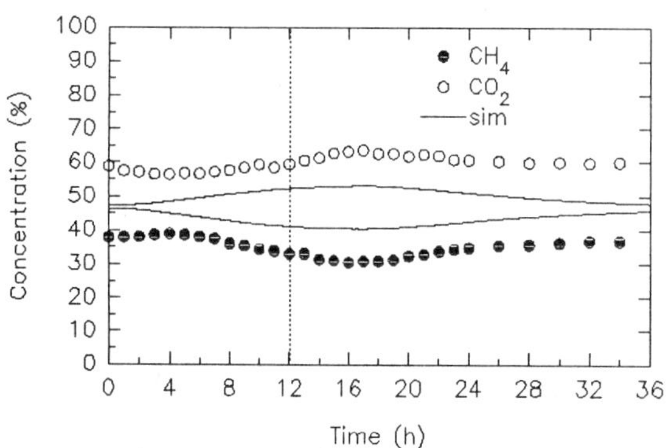

Figure 7. Effect of a step increase in feed concentration on gas composition of Reactor 1 at high loading rate

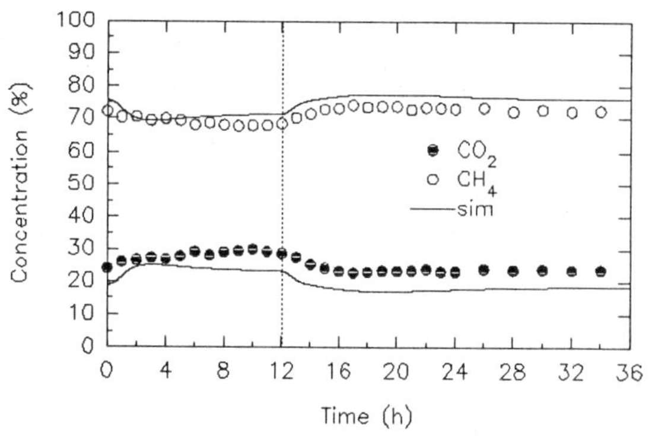

Figure 8. Effect of a step increase in feed concentration on gas composition of Reactor 2 at high loading rate

Effect of a concentration shock load on gas composition of the acidification reactor is presented in Figure 7. As expected, more carbon dioxide is produced with the increase in feed concentration, which therefore decreases the methane concentration. A similar trend is also demonstrated in the methanogenic reactor as shown in Figure 8. These figures are the responses of gas concentration to a step increase in feed concentration conducted at higher organic loading rate (23 kg COD/m^3.day). The responses to hydraulic shock load are found to be qualitatively similar to these results (figures not shown).

Effect on effluent organic acids

Effect of concentration shock load at a low volumetric loading rate on effluent organic acid concentration of the acidification reactor is shown in Figure 9. The experimental data of total organic acids represent the summation of all organic acids detected in the effluent samples including ethanol. Two species of unmodelled organic acids namely succinic and formic acids were occasionally detected in the system. These acids were typically detected at levels less than 5 C-mole/m^3 and were not found to accumulate significantly during the course of a disturbance. The presence of both acid species has been previously reported by a number of researchers.

It can be seen from Figure 9 that lactic acid is the only organic acid which accumulates significantly during the step increase in feed concentration. At normal operating conditions, as the figure indicates, the incoming glucose is converted mainly to acetic, propionic and butyric acids. The reason for the accumulation of lactic acid during shock loads could be explain as follow. The degradation of glucose into pyruvic acid leads to a reduction of one molecule of NAD (become $NADH_2$) for each molecule of pyruvate produced. The fermentation process has to be continued to produce final fermentation products with the main objective being to reoxidise the reduced NAD and desirably to generate additional ATP by substrate level phosphorylation to form, for example, acetic acid and butyric acid. These reactions release formic acid which accumulate in the liquid phase or split into hydrogen and carbon dioxide. When organic shock loads occur, the equilibrium state of oxidation and reduction of NAD during pyruvic acid formation may be affected due to the imbalance between the hydrogen removed and the electron acceptor available. This situation is overcome by directing the fermentation pathway towards production of lactic acid, since pyruvic acid itself is able to act as the hydrogen acceptor (7). Therefore, as shown in Figure 9, lactic acid keeps on accumulating during the course of disturbance, while acetic acid decreases. After the feed concentration has been returned to normal, lactic acid decreases rapidly and reaches the pre-shock level in less than 8 hours. On the contrary, acetic acid starts to increase and reaches the maximum level about 4

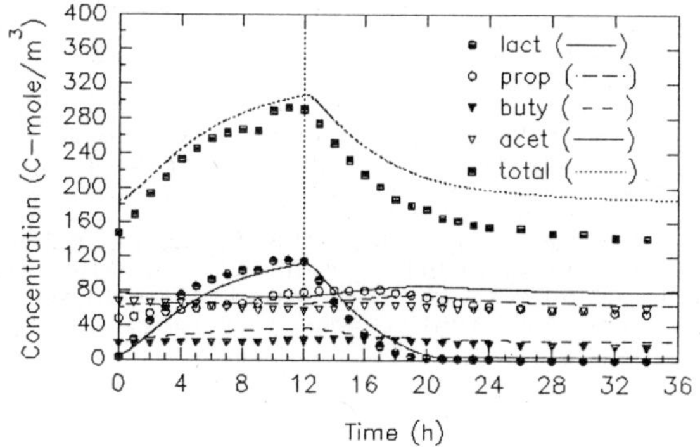

Figure 9. Effect of a step increase in feed concentration on effluent organic acid concentration in Reactor 1

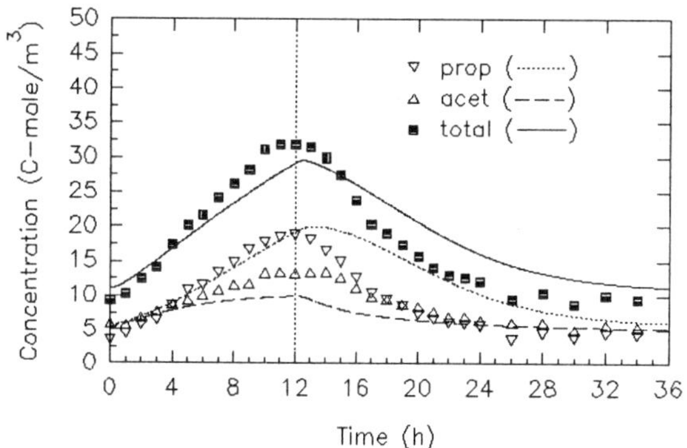

Figure 10. Effect of a step increase in feed concentration on effluent organic acid concentration in Reactor 2

hours after the disturbance has been terminated, followed by a subsequent decrease approaching the pre-shock value. It can be seen from the figure that the simulation predicts this behaviour very well.

Butyric acid response is also well predicted by the model. A step increase in feed concentration has only a marginal effect on the accumulation of butyric acid. The experimental results show a steady increase in propionic acid concentration during the step increase in feed concentration. This is expected since more lactic acid is available. Further accumulation of propionic acid after the period of disturbance is simply due to a result of lactic acid degradation which also

has increased the acetic acid production as mentioned before. Overall, a 12 hours step increase in feed concentration leads to an accumulation of total organic acids in the acidification reactor over the whole period of disturbance. The organic acid concentration returns to within 10% of the pre-shock level in less than 24 hours after the shock load is terminated.

Figure 10 shows the response of effluent organic acid in the methanogenic reactor to a concentration shock load. Only two species of organic acids are accumulated namely propionic and acetic acids. The figure shows the ability of the model to predict the response of effluent organic acids qualitatively and quantitatively.

Hydraulic shock loads led to a similar accumulation of total concentration of effluent organic acids. However, two species of organic acids namely propionic and acetic acids were usually found to behave in the opposite direction compared to the response to concentration shock loads. During the hydraulic shock loads, due to the decreased liquid as well as solid residence time, both organic acids dropped gradually. Their levels steadily returned to the pre-shock values after the shock load was terminated. The qualitative response of lactic and butyric acids was found to be similar to the one in the concentration shock loads (data not shown).

CONCLUSIONS

A number of experiments were conducted to study the dynamic behaviour of a two-stage high-rate anaerobic wastewater treatment system under shock load situations and to verify the improved version of mechanistic dynamic model of anaerobic degradation processes. The comparison between the simulated data and the experimental results shows the ability of the model to predict the dynamic behaviour of the system under shock load situations. Reactor variables such as pH, alkali consumption rate, gas generation rate and composition, and effluent organic acid concentration were modelled quite well. The ability of the model to predict the dynamic behaviour of the system in a various range of operating conditions (organic loading rates), therefore, provides a useful tool for design, optimisation, and operational evaluation purposes of two-stage anaerobic reactor configurations.

ACKNOWLEDGMENTS

The financial support provided by the IUC/WB XVII (to M.R.) and also by the Swiss National Science Foundation (to J.K.) is gratefully acknowledged.

LITERATURE CITED

1. Newell, R.B. and I.T. Cameron. *NIMBUS User's Manual*, CAPE Centre, The University of Queensland, Australia (1991).

2. Costello, D.J., P.F. Greenfield, and P.L. Lee, *Wat. Res.*, 25, 847 (1991).

3. Costello, D.J., P.F. Greenfield, and P.L. Lee, *Wat. Res.*, 25, 859 (1991).

4. Romli, M., "Modelling and Verification of a Two-Stage High-Rate Anaerobic Wastewater Treatment System", PhD thesis, The University of Queensland, Australia (1993).

5. Mosey, F., *Wat. Scie. Tech.*, 15, 209 (1983).

6. Pavlostathis, S.G. and E. Giraldo-Gomez, *Crit. Rev. Env. Ctrl*, 21 (5,6), 411 (1991).

7. Gaudy, Jr. A.F. and E.T. Gaudy, *Microbiology for Environmental Scientists and Engineers*, McGraw-Hill International Book Company, Tokyo (1981).

Investigation of Kinetic and Microbiological Features of UASB-Reactor Performance Under Various Organic Loading Rates

S. Kalyuzhnyi, V. Sklyar, and J. Rodríguez

Depto. de Biotecnología de Enzimas, Facultad de Ciencias Químicas, Universidad Autónoma de Coahuila, Saltillo, Coah., 25000, MEXICO

Kinetic and microbiological features of the performance of a laboratory UASB-reactor fed with synthetic wastewater were investigated under the increase of organic loading rates (OLR) in the range from 3.41 to 45.1 g COD l⁻¹ day⁻¹with detailed monitoring of the fed substrates and process intermediates, pH, TSS, VSS, specific biomass activities (acidogenic, acetoclastic, litotrophic) as well as distribution of number of the different groups of microorganisms through the reactor height. The optimal conditions for active granular biomass formation under the regimes investigated were found and the general regularities of the process were determined.For a stable UASB-reactor operation, overall average active biomass should not be less than 25 g VVS l⁻¹. In the sludge bed zone, this concentration should be higher than 60 g VVS l⁻¹. After changing the operating regime, the reactor should be maintained at new regime not less than 10-15 HRT. The increase of OLR should be made stepwise; passage to the next regime should be performed after achievement of quasi-steady-state on the previous regime with maximal possible organic matter conversion. Besides, in the operation of a technology using high-flow rate UASB-reactors, special attention should be paid to the efficiency of retention of very active biomass circulating in the sludge blanket zone.

One of the most perspective types of reactors used for anaerobic treatment of waste waters is UASB-reactor, suggested by Lettinga with co-workers at the beginning of 1980-th (1). High efficiency of this reactor is caused by maintaining inside the reactor the high concentration of active biomass mainly represented by good settled aggregates (granules and flocules), basically formed by methanogenic bacteria of *Methanotrix* genus. As a result of recent investigations dealing with .studying nature and structure of biomass granules formed, other principal groups of microorganisms involved in their composition were determined and the regularities of granules formation and their internal architecture were also elucidated to significant extent (1-6).

However, in spite of successes achieved, the data on kinetic regularities of nonstationary behavior of UASB-reactor under variation of the OLR and passage from one quasi-steady-state regime to another, as well as data on stratification of different groups of microorganisms within the reactor, are practically absent in the literature. But such information is very important for creation of fundamental ideas of anaerobic conversion as a whole, for elucidation of the process of formation and effective functio-

ning of granular anaerobic biomass, and also for quick start-up of the UASB-reactor and its stable operation.

The goal of this work is an investigation of kinetic and microbiological features of performance of laboratory UASB-reactor fed synthetic waste water under various OLRs with detailed detection of concentrations of fed substrates and process intermediates, determination of specific biomass activities, distribution of number of various microorganism groups through the reactor height for the following development of the mathematical model of the UASB-reactor.

MATERIALS AND METHODS

Reactor investigation. The investigations were carried out on laboratory UASB-reactor made from transparent plastic (diameter - 10 cm, height - 85 cm, total working volume - 2.7 l) equipped with sampling ports arranged in every 10 cm along the reactor height. The reactor was inoculated with 850 ml of suspended biomass from the pilot-scale anaerobic reactor for treatment of milk industry waste water. As a model waste water, the mineral media described in (7) with the adding

385

E. Galindo and O.T. Ramírez (eds.), Advances in Bioprocess Engineering. 385-390.
© 1994 Kluwer Academic Publishers. Printed in the Netherlands.

of glucose (2 g/l) and potassium acetate (from 2 to 5 g/l calculating on acetic acid) mixture as organic substrates was used. The influent pH value was 6.6-6.7, operating temperature was 35-37°C. OS concentration in liquid phase was translated into COD units proceeding from the substrate and the intermediate concentrations measured. Reactor operated under every OLR as long as biogas yield was stabilized and VFA concentration in effluent achieved a constant minimal value.

Analytical techniques. VFA, methane, carbon dioxide, and hydrogen concentration were measured using gas chromatography as previously describe (7). Determination of TSS, VSS, specific methanogenic biomass activities by acetate and H$_2$/CO$_2$ consumption and also acidogenic biomass activity by glucose consumption were carried out as described in manual (8).

Enumeration of bacteria. For microbiological research the sludge samples were taken from the different points along the reactor height during periods of time when the reactor achieved quasi-steady-state under corresponding OLR. For cultivation and determination of number of anaerobic bacteria the bicarbonate buffered Pfennig medium (9) with 0,002% of resasurin to control the redox potential was used (0,2 g/l of yeast extract, 2 ml/l of trace element Lippert solution (10), 10 ml/l of vitamin solution (11) were also added). Corresponding substrates were used for cultivation of different physiological groups of anaerobic bacteria: casein (5 g/l) - for proteolytic bacteria, glucose (1 g/l) - for saccharolytic bacteria, sodium acetate (3 g/l) - for acetate-utilizing methanogenes, mixture of hydrogen and carbon dioxide (80:20) - for hydrogen-utilizing methanogenes. The cultivation was carried out at 35°C in 50 ml serum bottles hermetically closed with rubber stoppers and aluminium covers. The method of 10-fold serial dilutions on corresponding substrates was used to determine number of anaerobic bacteria belonging to different physiological groups.

Scanning microscopy. Granules under investigation were placed into 0.15 M phosphate buffer (pH 7.0) with 5% glutaraldehyde during 16-18 h at 4°C for preliminary fixation. Dehydrated samples were dried at 50°C, stuck by electroconductive glue to the ob-

jective table, sprayed by gold in installation for ionic spray Spitter IFC-1100 and observed using ISM-1300 microscope (Japan).

Transparent thin section electron microscopy. For ultrathin section preparation the granules were fixed as described above. Strong fixation was carried out with osmic acid in accordance with Ritter-Kellenberger method (12). Thin sections were prepared using a microtome (LKB, Sweden), contrasted with 3% water solution of uranil-acetate and lead citrate. Preparations were observed using IEM-100C microscopy (Japan).

RESULTS AND DISCUSSION

Investigation of reactor performance under different OLR

After the start-up procedure the reactor was transferred on continuous regime of feeding. The initial OLR was 3.41 g COD/l·day. In two weeks the reactor achieved quasi-steady-state and OS concentrations in effluent decreased practically to zero and removal efficiency was close to 100% (Table 1). We use the term "quasi-steady- state" here and below because of a true steady-state for UASB-reactor is practically never achieved. Its long-term operation under the constant OLR doesn't lead to steady-state on biomass concentration - this concentration slowly but continuously increases.

Hereafter, for increasing of the OLR two methods were used: increasing of the influent flow rate and increasing of OS concentration in influent. Since growth rates of methanogenic bacteria are low (doubling time for bacteria of *Methanosarcina* genus - 20-30 h, for *Methanotrix* - 200-300 h), low flow rates were used at the beginning stage of the biomass adaptation to prevent its washing out. In order to avoid sharp decreasing of medium pH and other disturbances in the reactor performance, the OLR was increased slowly - step by step. The generalized results of our investigation of the UASB-reactor performance under different OLR are presented in Table 1. Analysis of the results of this study reveals some general regularities of achieving of quasi-steady-state of the UASB-reactor performance under any regime changing. In short, these regularities are following. Immediately after regime changing some disturbances in the reactor performance are observed (appearance of large gas bubble in sludge bed zone, partial colmatation, destruction of granules), probably, resulting from increased biomass lysis caused by temporary overloading of the reactor. Then after period of time to be equal to 3-6 hydraulic retention times (HRT), this undesirable phenomena are eliminated in general, but there is a rather high OS concentration in effluent and removal efficiency doesn't exceed 65-70%. Finally, the reactor performances becomes stable with removal efficiency close to 100% after period of time to be equal 12-15 HRT (on the average) from regime changing. In the cases of the OLRs lower than 21 g COD/l·day the quasi-steady-state of operation performance of the UASB-reactor is achieved faster - for 8-10 HRT. The process of stabilization of the reactor performance was most pronounced under the OLR of 45.1 g COD/l·day - it took above 30 HRT with relatively slow increase of removal efficiency from 69 to 97.3%.

Table 1. Operation performance of UASB-reactor under different OLR (quasi-steady-state)

OLR, g COD/l·day	HRT, h	S_{inf}, g COD/l	S_{eff}, g COD/l	removal effic.,%
3.4	22.5	3.21	0.005	99.8
4.4	17.6	3.21	0.005	99.8
5.5	18.6	4.28	0.01	99.7
6.4	20.0	5.35	0.01	99.8
6.7	19.2	5.35	0.01	99.8
10.2	12.6	5.35	0.02	99.6
14.2	12.7	7.49	0.31	95.8
17.6	10.2	7.49	0.29	96.1
20.4	8.8	7.49	0.02	99.7
23.3	7.7	7.49	0.23	97.0
28.6	6.3	7.49	0.02	99.7
31.4	5.7	7.49	0.30	96.0
34.0	5.3	7.49	0.12	98.4
45.1	4.0	7.49	0.21	97.2

The clear division of the reactor working zone into upper and lower parts corresponding to suspended biomass (sludge blanket) and granular biomass (sludge bed) was observed practically in all experiments. It is well seen from the data of Figure 1 (the boundary goes approximately between third and fourth sampling ports). And only under high flow rate (OLR of 45.1 g COD/l·day) the boundary between the zones is partially washed away.

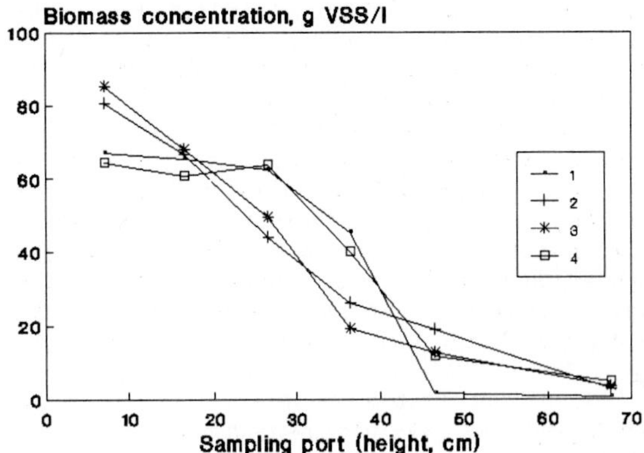

Figure 1. Distribution of biomass concentration through the reactor height under different OLR (g COD/l·day): 1 - 10.2; 2 - 20.4; 3 - 28.6; 4 - 45.1.

Thus, the study described above shows that, in general, granular biomass adapts quickly to regime changing and the OLR increase with high efficiency of the conversion process. Slow OLR increasing (during 4 months) leads to increasing of biomass concentration on the average by reactor volume up to 28-38 g VSS/l, in particular, in the sludge bed zone (on the average) - up to 60-70 g VSS/l (Figure 1). Also it should be noted that under the OLR of 45.1 g COD/l·day with the HRT of 4 h, the OS removal efficiency of 97% and the methane yield of 14.7 l/l reactor per day were achieved (Table 1) - this is a sufficiently high result for such type of the anaerobic reactors.

Distribution of biomass specific activities through reactor height

In the Tables 2-4 the results of investigations of distribution of specific biomass activities (acidogenic, acetoclastic, litotrophic) through the reactor height under different OLR are given. All data correspond to reactor performances under quasi-steady-state conditions.

In the sludge bed zone (first three sampling ports from bottom) the maximal acidogenic activities (by glucose consumption) were in the lowest sampling port that is quite regularly due to feeding of the main substrate (glucose) for this group of microorganisms is performed from below (Table 2). The maximal acetoclastic activity (methanogenic activity by CH_3COOH consumption) was at the first and second sampling ports (Table 3). At the next sampling port occurred in the granular biomass zone this activity decreased to some extent. It is due to the feeding influent has rather significant CH_3COOH concentrations (not less than 2 g/l).[3] Besides, acetic acid and other VFA are additionally formed under glucose decomposition. Litotrophic activity (methanogenic activity by H_2/CO_2 consumption) under high OLRs (more than 21 g COD/l·day) is maximal at lowest sampling port (Table 4), where due to acidogenic and obligate proton reducing bacterial activities the maximal amount of hydrogen is formed and necessity of its quick interspecies transport is arisen. At the next sampling ports this activity decreases to some extent. Under lesser OLR (10.2 and 20.4 g COD/l·day) this activity increases insignificantly through the reactor height, but its values are close to minimal for OLR of 28.6 and 45.2 g COD/l reactor per day. Probably, the litotrophic methanogenic activity values about 0.002-0.004 g CH_4/g VSS are characteristic for the granular biomass zone.

Table 2. Distribution of the specific acidogenic biomass activity (g glucose/g VSS·day) through the reactor height under different OLR (g COD/l·day).

OLR	Sampling ports (height, mm)					
	70	165	265	365	465	675
10.2	3.6	0.9	0.4	0.1	16.9	9.4
20.4	2.6	2.1	2.1	3.5	5.1	6.9
28.6	2.8	1.4	0.9	1.0	3.3	1.5
45.1	3.5	1.7	1.3	2.4	8.2	4.3

Table 3. Distribution of the specific acetoclastic biomass activity (g CH_4/g VSS·day) through the reactor height under different OLR (g COD/l·day).

OLR	Sampling ports (height, m)					
	70	165	265	365	465	675
10.2	0.06	0.05	0.03	0.02	1.21	0.81
20.4	0.02	0.04	0.03	0.05	0.05	0.26
28.6	0.03	0.07	0.03	0.13	0.14	0.32
45.1	0.05	0.04	0.04	0.03	0.12	0.20

In the sludge blanket zone (three sampling ports from the top of the reactor), where biomass concentration is lower (Figure 1), unexpected phenomenon is observed. Namely, all specific activities involved are increased up to two orders of the magnitude, being higher with reactor height increasing (Table 2-4). Thus, there is a simultaneous biomass concentration decrease and increase of its specific activities. It may be connected with the fact that the new microcolonies of young cells with markedly higher activity (in comparison with the cells occurred inside the granules) constantly forming on the granule surfaces are separated easily from this surface and rise into the upper reactor part by means of gaslifting effect. It should be also noted that this small biomass aggregates are precipitated to less extent than large granules and therefore they are scarcely settled into the sludge bed zone remaining mainly in the upper reactor part. Here they utilize the remain of VFA and are almost in fluidized state. Besides, the operating regimes with relatively long HRT (10 h and more) in comparison with regimes with relatively short HRT are accompanied by minimal hydraulic disturbances that makes difficult the gas microbubbles separation from the surface of small aggregates and also reduces speed of their sedimentation. It is supported (Tables 2-4) by large dispersion of specific microbiological activity values in the sludge bed and the sludge blanket zones for regimes with long HRT time, where the zone boundary is rather spread. It is especially well seen from data under OLR of 10.2 g COD/l·day (Table 2-4). Under this conditions the acidogenic activity in the upper reactor zone arises in 5 times, acetoclastic one - in 22 times, and litotrophic one - in 33 times.

Table 4. Distribution of the specific litotrophic biomass activity (g CH_4/g VSS·day) through the reactor height under different OLR (g COD/l·day)

OLR	Sampling ports (height, mm)					
	70	165	265	365	465	675
10.2	0.002	0.003	0.003	0.002	0.102	0.111
20.4	0.002	0.003	0.004	0.006	0.010	0.019
28.6	0.008	0.003	0.011	0.019	0.032	0.054
45.1	0.024	0.006	0.005	0.074	0.007	0.011

Thus, under elaboration of the technology using high-flow UASB-reactors, the attention should be paid to the efficiency of retention of the most active biomass circulating in the sludge blanket zone (this is already done empirically in modern modification of UASB-reactor - so called hybrid reactor (13)).

Microbiological investigation of biomass

Visual study of appearance of granules showed that in the lowest part of sludge bed zone there are mainly large and light ones whereas being taken higher, they are smaller and darker. On the average the granule diameter is of 1-2 mm. Often the granules are covered by slime. In the sludge blanket zone biomass is represented by amorphous light flocks and rarely very small black granules. Increase of the OLR leads to practically complete disappearance of more light granules in the sludge blanket zone of reactor.

Electron microscopy study of biomass samples taken under the OLR of 10.2 g COD/l·day demonstrated that almost through all the height of the reactor there are dense and well formed granules, being friable ones only in the upper reactor zone. They present both spherical and irregularly formed granules and consist, in general, of filaments of *Methanotrix* cells. Microorganisms with other morphology are rarely presented, *Methanosarcina* is practically absent. Besides, some granules, probably, contain iron sulfide giving them a black color. Small daughter granules are observed on the surface of many large granules.

The investigation of granules taken under the OLR of 20.4 g COD/l·day by electron microscopy showed that their composition was changed depending on sampling point. In granules from sludge bed zone *Methanotrix* cells predominate, *Methanosarcina* cells (large cocci) being in smaller amount. While in the samples taken from the sludge blanket zone, *Methanosarcina* cells are absent. Probably, it is caused by presence of traces of acetate in the sludge blanket zone. An affinity of *Methanosarcina* and *Methanotrix* for this substrate is markedly different. The predominance of *Methanotrix* is caused by its ability to utilize acetate in significantly lower concentrations in comparison with *Methanosarcina*.

Under the OLR of 28.6 g COD/l·day practically all granules have irregular form, being more friable than under lower OLRs. *Methanotrix* predominates in their composition with *Methanosarcina* inclusion, different rods and cocci are also presented. Probably, rise of acetate concentration stimulates an accumulation of acetate-utilizing methanogenes and following increase of gas formation leads to partial disruption of the structure of the granules.

The study of the structure of the granules under different OLRs performed by electron microscopy have demonstrated the presence of crystal inclusions, arising as a result of precipitation of mineral medium components. The surface of these crystals is overgrown by biomass and then they become the centers of granules formation.

The investigation of distribution of number of different microorganisms groups through the reactor height under regimes shows that maximal cells amount - up to 10^{11} cells/ml is in the sludge bed zone, and their number decreases up to the top of the reactor. A high concentration of proteolytic bacteria both in the lower (10^8-10^{12} cells/ml) and the upper part of reactor (10^7-10^{11} cells/ml) should be noted. It is evidently related with the presence of significant amount of bacteriolytical microorganisms responsible for lysis of microbial biomass. Short thin rods and small cocci were the predominant forms among proteolytic bacteria. A quantity of saccharolytic microorganisms in the reactor was at the range from 10^7 to 10^9 cells/ml in the both reactor zone. Under the increase of OLR the number of saccharolytic bacteria did not change markedly. The increase of OLR from 10.2 to 28.4 g COD/l·day led to elevation of number of methanogenes revealing the acetoclastic activity from 10^4 to 10^6 cells/ml. The number of methanogenic bacteria revealing litotrophic activity under growth on H_2/CO_2 mixture was rather high - 10^{10} cells/ml in the sludge bed zone and 10^7-10^9 cells/ml in the sludge blanket zone. In general, the overall quantity of methanogenic bacteria in the sludge bed zone is 2-3 orders of magnitude higher then in the sludge blanket zone.

CONCLUSIONS

Thus, the investigations described above permitted us to clarify the optimal conditions for active granular biomass formation for every regime studied and to find some general regularities of the process, namely: for stable UASB-reactor operation it is desirable to obtain an active biomass concentration on the average by the reactor volume - not less than 25 g VSS/l and particularly in the sludge bed zone -not less than 60 g VSS/l; after change of operating regime the reactor should be stood at new regime not less than 10-15 HRT; the increase of OLR should be made stepwise; a passage to the next regime should be performed after achievement of quasi-steady-state on the previous regime with maximal possible OS conversion. The process of anaerobic conversion is the most effective under maximal possible flow rates. Hence, working out a technology using high-flow rate UASB-reactors, a special attention should be paid to efficiency of retention of very active biomass circulating in the sludge blanket zone.

Microbiological study of reactor microflora showed a high number of different groups of microorganisms providing a complete substrates degradation to methane and carbon dioxide under various OLRs. The number of these microorganisms is on the average on 1-2 orders of magnitude lower at the upper part of the reactor in comparison with bottom of the reactor. Study of granules under all regimes investigated by electron microscopy demonstrated that a significant part their microbial biomass is represented by bacteria of *Methanotrix* genus. Besides, there were found a great amount of morphologically various bacterial forms distributed inside granules among *Methanotrix* filaments, often as microcolonies.

We plan to use above described data for the development of mathematical model of operation performance of the UASB-reactor.

NOMENCLATURE

COD	Chemical oxygen demand
HRT	Hydraulic retention time
OLR	Organic loading rate
OS	Organic substances
S_{eff}	OS concentration in effluent
S_{inf}	OS concentration in influent
TSS	Total suspended solids
UASB-reactor	Upflow anaerobic sludge blanket reactor
VFA	Volatile fatty acid
VSS	Volatile suspended solids

LITERATURE CITED

1. Lettinga G., van Velsen A.F.M., Hobma S.V., de Zeeuw W.J., and Klapwijk A. *Biotech. Bioeng.* **22**, 699 (1980).

2. Kalyuzhnyi S.V., Danilovich D.A., and Nozhevnikova A.N., *Anaerobic biological treatment of sewage*, VINITI Press, Itogi nauki i tekhniki, Ser. Biotekhnologiya, v.29, Moscow (1991) (in Russian).

3. Wiegant W.M., The "spaghetti theory" on anaerobic granular sludge formation, or inevitability of granulation, In: G.Lettinga, A.J.B. Zehnder, J.T.C. Grotenhius, L.W. and Hulshoff Pol (Eds.), *Granular anaerobic sludge: Microbiology and technology*, Pudoc, Wageningen, the Netherlands, p. 146 (1988).

4. Hulshoff Pol L.W., "The phenomenon of granulation of anaerobic sludge", Ph.D. Thesis, Agricultural University of Wageningen, the Netherlands (1989).

5. MacLeod F.A., Guiot S.R., and Costerton J.W. *Appl. Environ. Microbiol.* **56**, 1598 (1990).

6. Forster C.F. *J. Biotechnol.* **17**, 221 (1991).

7. Varfolomeyev S.D. and Kalyuzhnyi S.V. *Appl. Biochem. Biotechnol.* **22**, 331 (1989).

8. Lettinga G. and Hulshoff Pol L.W. (Eds) *International Course on Anaerobic Waste Water Treatment*, Agricultural University of Wageningen, the Netherlands (1990).

9. Pfennig N. *Zbl. Bakt. 1 Abt. Orig. Suppl.* **1**, 179 (1965).

10. Pfennig N. and Lippert K.D. *Arch. Microbiol.* **55**, 245 (1966).

11. Wolin E.A., Wolin M.Y., and Wolfe R.S. *J. Biol. Chem.* **238**, 2882 (1963).

12. Ryter A. and Kellenberger E.Z. *Naturforsch.* **13b**, 157 (1958).

13. Oleszkiewicz J.A., Hall E.R., Oziemblo J.Z. *Environ. Technol. Lett.* **7**, 445 (1986).

Reduction of Antiphysiological Compounds of Coffee Pulp Through Solid State Fungal Fermentation

C. Porres, J. F. Calzada, and D. Alvarez

Applied Research Division, Central American Research Institute for Industry (ICAITI), Av. La Reforma 4-47, zona 10, Ciudad de Guatemala, 01010, GUATEMALA

Caffeine, polyphenols and potassium present in coffee pulp have proved their negative effects on animal metabolic performance when they are included in the diet. The present research work was aimed to reduce caffeine and polyphenols in ensiled and aqueous extracted coffee pulp through solid state fermentation with the filamentous fungus strain Penicillum crustosum. *The bioconverted product presented reductions of antiphysiological compounds in the following ranges, related to the fermentation time: caffeine 52-88%, polyphenols 23-65%. The yield of fermented product was 89%. The reductions in dry and organic matter enzymatic digestibilities of coffee pulp were from 27% to 20% and 31% to 22%, respectively.*

The presence of caffeine, tannins and other polyphenols (mainly caffeic and chlorogenic acid) and potassium naturally ocurring in coffee pulp have detrimental effects upon animal growing when it is included in the diets.

Many works have been done to eliminate these antiphysiological factors through mechanical treatments such as screw pressing followed by aqueous extractions, alkaline extractions, grinding or heating. The resulting materials present only intermediate reductions of antiphysiological compounds (Gómez (1)).

Bergman, et al (2), describe a work in which a *Pseudomona aeruginosa* strain degraded caffeine via theobromine. During the metabolism of caffeine, formation of methylated catabolites of xanthine from methylxanthines were reported.

Schwimmer and Kurtzman (3), reported the isolation and the study of the metabolism of the *Penicillium crustosum* strain NRRL 5452, which efficiently utilized caffeine via theophyline as source of nitrogen in a clearly defined growth media with infusions prepared from commercial roast coffee. The rate of caffeine removal was estimated to vary between 0.24 to 0.28 mg/ml/day. It was found that coffee infusions inoculated with a coffee-derived culture (18 hours old) degraded 50% of the original caffeine contents after 6 hours of fermentation time.

Aquiahuatl, et al (4), reported the results obtained from their isolation and testing of strains naturally occurring in coffee plants. According to the author 280 strains of filamentous fungi were isolated from materials coming from coffee plantations of México. From the isolated and identified strains: *Aspergillus* and *Penicillium* were predominant, but other genera could be identified as *Fusarium, Trichoderma, Geotrichum* and some *Zygomycetes*. The paper presents a list of filamentous fungi with high capacity to degrade caffeine in liquid media containing such compounds as only source of nitrogen. The best caffeine degrading strain (in roasted coffee infusions) was a *Penicillium roqueforti* with 95% of reduction and a degradation velocity of 0.224 mg/ml/d followed by an *Aspergillus oryzae* (78% and a rate of 0.12 mg/ml/d). According to the author solid state fermentation of fresh coffee pulp with *Penicillium roqueforti* completely degraded the caffeine content after a period of 45 h.

A laboratory research work aimed to

E. Galindo and O.T. Ramírez (eds.), Advances in Bioprocess Engineering. 391-396.
© *1994 Kluwer Academic Publishers. Printed in the Netherlands.*

reduce antiphysiological compounds (caffeine and polyphenols) and increase protein content of coffee pulp through solid state fermentations with the strains: *Aspergillus niger A-10*, *Aspergillus oryzae* and *Sporotrichum pulverulentum* are described by Porres and co-workers (5). In relation to degradation of antiphysiological compounds in the fermented products it was concluded that the reductions of caffeine and polyphenols were: *A. niger* 13.2% and 3.1%, *A. oryzae* 23.3% and 49.7%, and *S. pulverulentum* 54% and 46% respectively. The expected high reduction levels were not reached with these strains.

Screening tests with a similar purpose, using 26 strains of basidiomycetes growing on coffee pulp during 30 and 60 days were described by De León (6). The strains which presented the highiest rates of reductions of caffeine and polyphenols during a growing period of 60 days were: *Phaenerochaete chrysosporium* 53.4% and 42.3% and *Phebia radiata* 52.8% and 52.1% respectively. Again, the expected high reduction levels were not reached.

The innovative aspect of the work presented here is the two-stage biological and mechanical reduction of the antiphysiological compounds. The first stage includes the pretreatment (silage, aqueous extraction and screw pressing). The second is basically the fermentation with the filamentous fungus strain *Penicillium crustosum*. The limiting factor for this technology is the presence of organic compounds (mainly organic acids) generated during silage. They can retard, and in some cases inhibit, fungal growth. However, it is not logical to scale-up a continuous industrial process based on seasonal raw materials. For this reason, silage is an imperative step of the process.

MATERIALS AND METHODS

Strains and inoculum. The fungi strains utilized in the experiments were:

* *Penicillium crustosum*, ICAITI 1159
* *Penicillium crysogenum*, ICAITI 1157
* *Penicillium melini*, ICAITI 1061

The innocula were prepared in sumerged culture using 400 ml shaker flasks with 125ml of growing medium prepared dissolving 30 g of Sabouraud dextrose broth per liter of water. The medium was sterilized at 121°C by 15 minutes. Mycelium was obtained from cultures maintained at low temperatures (1°C) in tubes

with agar. The culture was incubated at room temperature (19°C) for 3 days in a laboratory shaker. Final biomass concentration was estimated in 6 g of dry matter per liter.

Substrate. Fresh coffee pulp was screw pressed to reduce volume and moisture content from approximately 86% to 80%. Pressing of coffee pulp was carried out using a local manufactured screw press with a capacity of 1 ton of fresh material per hour. Afterwards 5% of sugar cane molasses (wet basis) was added to the material and ensiled during 99 days. After silage the material had a pH of 4.1.

Ensiled coffee pulp was extracted with cold water (temperature 20°C) in a rectangular tank of stainless steel (0.4 m height, 0.6 m width and 1.1 m long). Thirty liters of tap water were added to each 45.3kg of wet material and mixed intermittently for 30 minutes. After that, the pulp was screw pressed, sun dried and storage.

After pretreatments (silage, aqueous extraction and pressing) the content of antiphysiological compounds were reduced (as compared with those in fresh coffee pulp) as follows: caffeine from 0.71% to 0.22% and total polyphenols from 1.21% to 0.90%.

The substrate for fermentation was prepared by adding water to 9 kg of sun dried pretreated coffee pulp (moisture content approximately 16%) to achieve 65% of humidity and sterilized at 121°C by 15 minutes. Approximately, 21 kg of sterile substrate (moisture content 65%) were inoculated with 375 ml of sumerged culture of the fungus. The inoculated material was fermented in a solid state fermenter.

Fermenter. The fermenter consisted in a stainless steel ribbon blender (0.85 m long, 0.3 m width and 0.35 m height) with a volume of approximately 80 liter and driven by a 0.77 KW motor.

The fermenter had a central shaft with four paddles for mixing the material. At the bottom there was an orifice with a sliding door to discharge the product. With the bottom door open, a special part was adapted in order to feed air to the substrate at a flow-rate of 100 Ncm^3/s. All fermentation were carried out at room temperature (19-22°C) because the fermenter was not eqquiped with temperature control.

21 kg of substrate (65% of moisture content) occupied 70 liters of fermenter

volume (packing density 0.3 kg/l).

Before each fermentation the fermenter was disinfected with solutions of chlorine and quaternary ammonium and after that the apparatus was cleaned with live steam. The solid state fermentations were carried out at semi-aseptic conditions. The fermentations were maintained until sporulation was observed and this was the criteria for the fermentation time.

Organic mass losses after fermentation were calculated as the difference between the initial and the final dry weights of the pulp, expressed as percentage of the initial dry weight.

Reductions in chemical composition between the substrate and the bioconverted material were calculated based in the organic mass losses and the difference between initial and final values. The change was expressed as percentage of the original value (all expressed in dry basis).

Analytical techniques. Samples of pretreated coffee pulp and bioconverted coffee pulp were dried in a laboratory oven at 60°C during 24 h and after that, gounded and storage in plastic bags. The samples were analyzed for:

- Caffeine (Ishler (7)).
- Ash (AOAC 3.004 (8)).
- Nitrogen (AOAC 2.055 (8)).
- Total and condensed polyphenols (AOAC 9.110 (8)).

- Total sugars.
- In-vitro digestibilities (Clarke (9)).
- Lignin (Goering (10)).
- pH of silage.

RESULTS AND DISCUSSION

Strain selection. The contents of caffeine and polyphenols of the following materials: ensiled and aqueous extracted pulp and solid state fermented products (with different strains) are presented in Table 1.

These tests were aimed to select a microorganism with clear activity against caffeine and polyphenols and ability to grow on ensiled substrates. It is important to point out that fermentation times are retarded by the presence of organic acids in ensiled materials.

The selection criteria was reduction of antiphysiological compounds expressed in dry basis and as percentage of the total weight.

From the analysis of the data presented in the table it can be seen that only *Penicillium crustosum* was able to metabolize caffeine. The relative increment of caffeine percentage in others was due to weight losses through metabolism.

In screening test *Penicillium crustosum* reduce the initial contents of antiphysiological compunds of the substrate as follows: the caffeine content was reduced from 0.22% to 0.04% and total polyphenols from 0.90% to 49%.

Table 1. Selection of microorganisms with activity against antiphysiological compounds of coffee pulp.

STRAIN	CAFFEINE (%)	TOTAL POLYPHENOLS (%)
Substrate 1/	0.22	0.90
Bioconverted products:		
P. crysogenum 2/	0.32	1.12
P. melini 3/	0.43	0.77
P. crustosum 4/	0.04	0.49
Wild strain 5/	0.40	0.66

1/ Ensiled and aqueous extracted coffee pulp.
2/ Solid state fermentation with *P. crysogenum*, (fermentation time 216 h).
3/ Solid state fermentation with *P. melini*, (fermentation time 48 h).
4/ Solid state fermentation with *P. crustosum*, (fermentation time 155 h).
5/ Solid state fermentation with a wild strain (unidentified), (fermentation time 40 h).

This screening test allowed to find a strain with high capacity to degrade caffeine.

Chemical composition of converted pulp with _P. crustosum_. Contents of antiphysiological compounds before and after solid state fermentation of ensiled and extracted coffee pulp are shown in Table 2.

The mean organic mass loss due to respirations was 11.3%. The range of reductions in contents of antiphysiological compounds were: caffeine 51.5% - 88.0% and polyphenols 22.9% - 65.4% with a mean values of 74.4% and 42.5% respectively. The total amount of ash did not vary during fermentation.

In Table 3, the chemical composition of the materials above mentioned are shown.

The range of reductions of total sugars was 17.7% - 40.5% with a mean value of 32.7%. The digestibilities of total and organic matter decreased from 27.8% to 21.4% and 30.98% to 22.2% respectively.

Contents of nitrogen and lignin did not vary during fermentation. Slight changes were due to experimental errors (standard deviations of 0.16 and 1.06 respectively).

In spite of reductions in digestibilities on ensiled coffee pulp after fermentation, the bioconverted product presents attractive characteristics for animal feeding: protein 13%, caffeine less than 0.12%, acceptable reductions in polyphenols and moderated reductions in total sugar compared with the

Table 2. Contents of anitphysiological compounds in bioconverted coffee pulp with _Penicillium crustosum_ in 3 pilot assays.

FERMENTATION TIME (h)	CAFFEINE (%)	TOTAL POLYPHENOLS (%)	ASH (%)	YIELD (%)
Substrate (ensiled and extracted coffee pulp):				
0.00	0.22	0.90	6.22	----
Bioconverted products:				
137.0	0.12	0.78	7.40	89.0
142.0	0.04	0.35	7.13	89.0
168.5	0.03	0.62	6.97	88.1

Table 3. Chemical composition of bioconverted coffee pulp with _Penicillium crustosum_ in 3 pilot assays.

FERMENTATION TIME (h)	TOTAL SUGARS (%)	NITROGEN (%)	IVDMED (%)	IVOMED (%)	LIGNIN (%)
Substrate (ensiled and extracted coffee pulp):					
0.00	1.51	2.22	27.18	30.98	25.65
Bioconverted products:					
137.0	1.02	2.43	20.22	20.49	nd
142.0	1.01	2.13	22.91	23.89	29.63
168.5	1.41	2.53	20.91	22.28	27.58

original figures. A probiotic action of the biomass in the animal is also expected.

Kinetics of caffeine and polyphenols degradation. In Table 4, the data regarding degradation of caffeine and polyphenols and ash content during the fermentation time are presented.

From the analysis of these data it can be seen that only after 90 hours important reductions in antiphysiological compounds were observed.

After 170 h reduction in caffeine and polyphenols were 79.8% and 47.8% respectively. As it was mentioned previously, fungal growth is retarded by the presence of organic acids present in ensiled materials. It is estimated lower fermentations times could be obtained utilizing fresh pulp. One of the goals of this research was overcome some problems associated with fermentation of ensiled products.

In this Table 5, changes in chemical composition of pretreated coffee pulp during solid state fermentations are presented.

The in-vitro dry matter enzymatic digestibility (IVDMED) and the in-vitro organic matter digestibility (IVOMED) decreased in inverse relation to fermentation time. Thus reduction in IVDMED and IVOMED were approximately from 27% to 20% and 31% to 22% respectively (for fermentation times

Table 4. Kinetics of antiphysiological compounds degradation of coffee pulp during solid state fermentation with *Penicillium crustosum.*

TIME (h)	CAFFEINE (%)		TOTAL POLYPHENOLS (%)		ASH (%)
	CONTENT	RED.	CONTENT	RED.	
Substrate (ensiled and extracted coffee pulp):					
0.0	0.22	(0.0)	0.90	(0.0)	6.22
Bioconverted product:					
66.0	0.21	(8.7)	0.54	(42.6)	6.65
90.0	0.12	(48.7)	0.92	(3.9)	6.33
138.0	0.08	(67.0)	0.73	(26.3)	7.28
170.0	0.05	(79.8)	0.53	(47.8)	6.97

ABBREVIATION:
RED. % of reduction compared to initial value (dry basis).

Table 5. Changes in chemical composition of coffee pulp during solid state fermentations with *Penicillium crustosum.*

TYPE OF TREATMENT	TOTAL SUGARS (%)	NITROGEN (%)	IVDMED (%)	IVOMED (%)	LIGNIN (%)
Substrate (ensiled and extracted coffee pulp):					
0.0	0.83	2.05	27.18	30.98	26.22
Bioconverted product:					
66.0	2.83	1.91	24.81	31.09	26.78
90.0	1.68	2.04	22.34	22.41	23.51
138.0	1.16	2.27	20.52	21.31	28.55
170.0	1.08	2.32	19.81	22.11	31.54

among 0 to 170 h).

Slight variations due to experimental error were observed in the total amounts of nitrogen and lignin (standard deviations of 0.15 and 2.12), respectively.

At the beginning of the fermentation cycle the sugar content tend to increase up to a maximum value (increment of 79% with respect to the initial value) and then this figure was decreasing along the time (reduction of 36.6% with respect to the initial value).

CONCLUSIONS

1. The reductions of antiphysiological compounds through solid state fermentation of ensiled and aqueous extracted coffee pulp with *Penicillium crustosum* varied in the following ranges: caffeine 51.5% - 88.0%, and total polyphenols 22.9% - 65.4%. Total ash amount did not change with fermentation time, but the percentage increased as a result of weight losses.

2. The yield of bioconverted product from ensiled coffee pulp was 89%. Content of total sugar in bioconverted product tend to increase at the beginning of fermentation cycle but after approximately 90 h the content decreases to levels under the initial value. The total amounts of nitrogen and lignin did not vary during fermentation.

3. *Penicillium crustosum* on ensiled coffee pulp inoculated from submerged standard media cultures sporulated within a range of 137 - 169 hours.

4. Coffee pulp bioconverted with the fungus *Penicillium crustosum* produce an end product with promising characteristic for animal feeding (mainly ruminants).

ACKNOWLEDGMENT

This research was partially supported under Grant No. 936-5544.21. U.S.-Israel Cooperative Development Research Program, Office of the Science Advisor, U.S. Agency for International Development.

LITERATURE CITED

1. R. Gómez, "Processing of coffee pulp: chemical treatments", in Coffee pulp: Composition, Technology and utilization, Braham, J. E. and Bressani, R. (Eds.) IDRC publication, Ottawa, Canada (1979).

2. Bergmann, F., Ungar-Wdrow, H., Kwietny-Govrin, H., Goldberg, H., and Leon S. "Some specific reactions of the purine-oxidizing system of *Pseudomona aeruginosa*." Biochim. Acta 55: 512 (1962).

3. Schwimmer, S. and Kurtzman, R. H. "Fungal decaffeination of roast coffee infusions". Journal of Food Chemistry 37, 6, 921 (1972).

4. Aquiahuatl, M. A., Raimbault, M., Roussos, S. and Trejo, M. R. "Coffee pulp detoxification by solid state fermentation", in "Solid state fermentation in bioconversion of agro-industrial raw materials", Rimbault, M. (ed), Proceedings of the Seminar, ORSTOM-Montpellier (France), 25, 26 and 27 July 1988, 19-26 (1989).

5. Porres. C., De León, R., Gutierrez, E., Rolz, C. "Fermentación al estado sólido de pulpa de café", in Memorias del Simposio latinoamericano sobre biotecnología de biomasa y tratamiento de desperdicios. García, R. (ed), OEA/ICAITI Guatemala 1987, pp 416 - 422 (1987).

6. De León, R., Porres, C., Rolz, C. and Campos, M. "Crecimiento de hongos sobre pulpa de café", in Memorias del tercer simposio international sobre utilización integral de los subproductos del café. Porres, C., Rodas, C., and Calzada, J. (eds), PNUMA/OEA/ICAITI. Guatemala 1987, pp 76-84 (1987).

7. Ishler, N. H., Finucare, T.P., and Borker E., Anal. Chem. 20, 1162 - 1140 (1982).

8. Association of Official Analytical Chemists (AOAC). Official methods of analysis. Fourthteen edition, U.S.A., (1984).

9. Clarke, T., Flinn, P.C. and McGovan, A.A., Grass Forage Sci. 37, 1130-1140 (1982).

10. Goering, H.K. and Van Soest, P.J., Forage Fiber Analysis, USDA Agriculture hand-book, (1970).

Biological Removal of Hydrophobic Solvent Vapors from Airstreams

Z. Shareefdeen and B.C. Baltzis

Department of Chemical Engineering, Chemistry, and Environmental Science,
New Jersey Institute of Technology, Newark, NJ 07102, U.S.A.

Removal of benzene and toluene vapors from airstreams was studied in packed-bed biological reactors known as biofilters. The reactors, plexiglass columns 10 cm in diameter and 75 cm in height, were packed with a mixture of peat and perlite particles (volume ratio 2:3). A microbial consortium capable of completely mineralizing benzene and toluene was immobilized on the surface of the porous packing material, and formed a biolayer. Prehumidified air carrying either benzene or toluene was passed through the reactor. The solvent vapors and oxygen were transported to the biolayer where the contaminants underwent biodegradation. The experimentally observed maximum removal rates were 4.5 $g(m^{-3}$-packing$)^{-1}$ h^{-1} for benzene, and 24.8 $g(m^{-3}$-packing$)^{-1}h^{-1}$ for toluene. Under all operating conditions tried, toluene vapor was removed easier than benzene. Steady-state data are accurately predicted with a detailed mathematical model which accounts for possible limitation of the process by oxygen. However, calculations show that, when compared to hydrophilic compounds, biofiltration of hydrophobic substances is not much affected by oxygen.

INTRODUCTION

The use of biological processes for treatment of liquid industrial waste streams is a well established approach to solving problems of water pollution. Biological oxidation occurs at low temperature and requires inexpensive catalysts (microorganisms); thus, in most cases, it is an economical alternative to incineration and classical catalytic oxidation. Furthermore, when complete mineralization is achieved, bioprocesses treat pollutants without forming stable intermediates which may be more toxic than the original contaminants. Formation of toxic byproducts, such as dioxins, can occur when incineration is used. These advantages of waste-biotreatment have drawn the interest of industrial and academic researchers who hope to develop bioprocesses to deal with the problem of air pollution. Such a process is biofiltration.

Biofiltration employs microorganisms immobilized on porous solid particles which are placed, in packed-bed configuration, in a reactor known as a biofilter. Contaminated airstreams are passed through the biofilter, and as the pollutants are transported to the biolayer formed around the solid particles, they undergo biodegradation. The process is aerobic, and hence oxygen transport is also important. If the residence time is adequate, the airstream exiting the biofilter is pollutant free. Moisture is maintained in a biofilter through prehumidification of the airstream. If a water stream is recirculated through the bed, the term bioscrubber is used instead of biofilter. As the presence of a continuous liquid water phase introduces an extra mass transfer resistance, and also leads to possibly substantial pressure drop, the use of bioscrubbers seems warranted only for reasons of pH-control when chlorinated and/or nitro-compounds are to be treated.

The interest in biofiltration stems from the fact that in the recent years, environmental regulations regarding volatile organic compounds (VOCs) have become very stringent. VOCs, even if they are not classified as hazardous or toxic, are major contributors to the smog formation in the atmosphere. Biofiltration of VOCs was studied first in Europe by Ottengraf

397

E. Galindo and O.T. Ramírez (eds.), Advances in Bioprocess Engineering. 397-404.
© 1994 Kluwer Academic Publishers. Printed in the Netherlands.

and his co-workers, while in the very recent years a number of studies have been conducted or initiated in the USA. As this is a new field, there are a number of reports and papers in proceedings of meetings, but very few articles in journals [1,2,3,4,5,6].

The present work deals with hydrophobic solvent vapors, and more specifically with benzene and toluene. Both substances are classified as primary pollutants by the US EPA. They are present in gasolines [7], and they are also widely used as industrial solvents, as well as feedstocks for synthesis [8]. The objective of this study was not only to demonstrate feasibility of benzene and toluene vapor biofiltration, but to also mathematically describe the process with a model which would have predictive capabilities, and could be used as a basis for design calculations. It is believed that proper process modeling will facilitate scale-up and design of industrial scale biofilter units. It should be also emphasized that, as biofiltration involves mass transfer, kinetics and multiple phases (gas, biolayer, solid support), detailed and complete mathematical modeling of the process is a difficult undertaking which will require a lot of further work.

EXPERIMENTAL SET-UP AND PROCEDURES

A microbial consortium capable of completely mineralizing both benzene and toluene was obtained from the microbiology laboratory of Prof. R. Bartha (Rutgers University). This consortium was developed from contaminated soil samples. For the experiments with benzene, biomass was developed by growing an inoculum of the consortium in a simple mineral medium containing benzene as the sole carbon and energy source. After a small amount of biomass was grown in a shake-flask, it was transferred to a 10 liter sealed fermentor. The volume of the suspended culture was about 4L, allowing for excess oxygen concentration presence in the headspace of 6L. Benzene was added in small quantities over a period of two weeks. The culture was finally harvested through centrifugation, and was resuspended in fresh mineral medium not containing benzene before it was mixed with the solids.

The volume of the medium was calculated so that it would fill 50% of the pores of the packing material. The packing material consisted of a mixture of peat and perlite particles (2:3 volume ratio before mixing). The solids were autoclaved to ensure that no indigenous organisms present especially in the peat, would interfere with the biofiltration process. The solids were subsequently mixed with the prepared culture, and placed in a plexiglass column 10cm in diameter and 75cm in height. The biofilter was the part of the column which was filled with packing material. In the case of benzene biofiltration the height of the biofilter bed was 51cm, while for toluene it was 69cm. For the experiments with toluene, the same procedure was used except for the use of toluene instead of benzene in preparing the biomass from an inoculum of the original culture.

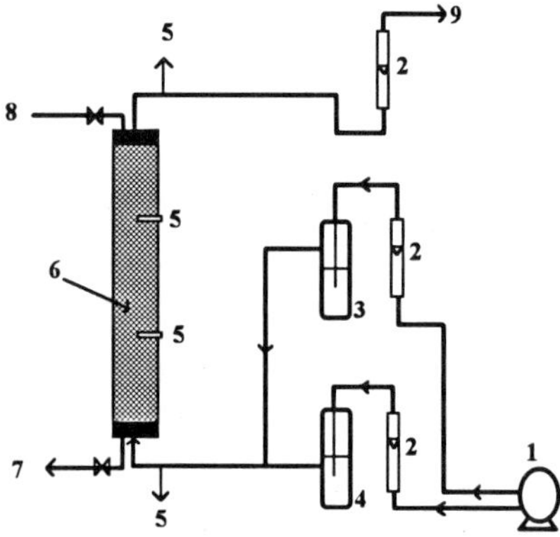

Figure 1. Schematic of the experimental packed-bed biofilter unit. The unit was 75cm high and 10cm in diameter. (1) air pump; (2) air flow meters; (3) water; (4) benzene or toluene; (5) sampling ports; (6) peat/perlite packing (7) valve which allows for pressure measurement and draining of water (8) valve which allows for pressure measurement and spraying of water (9) treated air (exhaust).

A schematic of the complete biofilter unit is shown in Figure 1. The main part of the airstream was passed through a water tank for humidification purposes. Another part of the external airstream was passed through a container carrying the solvent of interest. The two streams were subsequently combined to create the contaminated air. By varying the flowrates of the two streams before mixing, the space (or residence) time in the biofilter, as well as the concentration of the pollutant in the stream after mixing were controlled. Except at start-up, no nutrients were added to the biofilter bed. Furthermore, prehumidification of the airstream was enough to maintain adequate moisture levels in the bed. Only in rare occasions was it necessary to add some liquid water to the biofilter. Contaminant concentrations were measured at the inlet, exit, as well as at two locations along the bed. The side ports were at 35% and 80% of the biofilter bed height in the case of benzene, and 33% and 66% of the bed height in the case of toluene. Gas tight syringes were used for obtaining gas samples which were immediately subjected to GC analysis. The pressure drop along the column was monitored frequently, and was found to be negligible. In fact, even in columns which were in continuous operation for periods of over six months, the pressure drop never exceeded 0.25" water/m-bed.

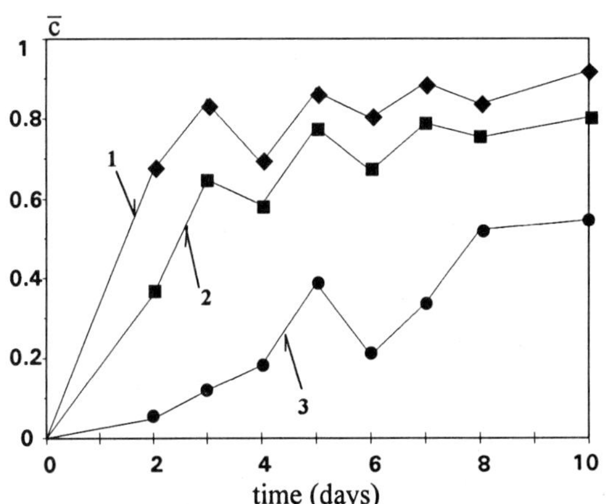

Figure 2: Characteristic transient response of biofilters during start-up; dimensionless concentrations at 0.35H, 0.8H, and at the exit of the biofilter (curves 1,2,3, respectively), as a function of time. Data from an experiment with benzene vapor fed at 0.43 g/m³, τ = 4.5 min, Vp = 4119 cm³, H = 51 cm.

Figure 2 shows a characteristic response of a biofilter during start-up. Initially, an amount of the contaminant is adsorbed on the solid packing material, or simply dissolved in the water retained in the pores of the solids. It is for this reason that initially no solvent vapor is detected at the exit of the unit. Eventually, after about eight days in the case shown in Figure 2, the unit reaches a steady state. The data shown in the results section are from steady state conditions.

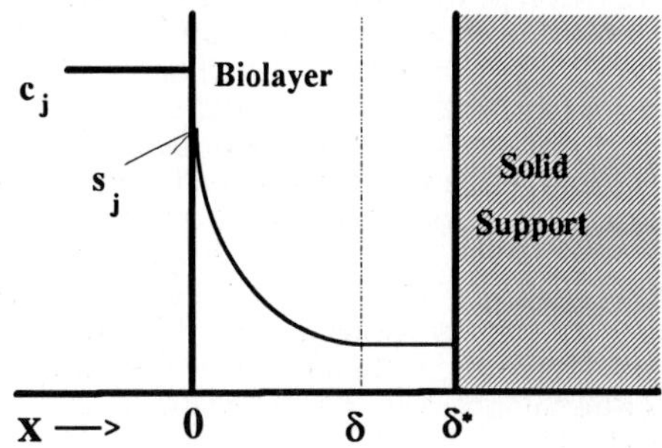

Figure 3. Schematic of the biofilm model concept at a cross-section along the biofilter column.

THEORY

A mathematical model which we had earlier introduced for describing steady state biofiltration of hydrophilic solvents, namely methanol [5], was used here and proved successful in describing steady state biofiltration of hydrophobic solvents as well. The basic concept of the model was originally introduced by Ottengraf and van den Oever [1], and is shown schematically in Figure 3. It is assumed that the biolayer thickness is small relative to the characteristic length of the solid particles, thus planar geometry can be used. The contaminant in the air and in the biolayer at the air/biolayer interface is in thermodynamic equilibrium as dictated by Henry's law. In our approach, the same is assumed to be true for oxygen. Due to diffusion and reaction in the biolayer, the concentration of the pollutant in the biolayer drops in the direction towards the biolayer/solid support interface. The concentration may drop to a zero value at a position δ of the actual biolayer, the thickness of which is δ^*. Thickness δ is the effective biolayer thickness, and in our approach it is defined as the position where the concentration of either the pollutant or oxygen (i.e., one of the two compounds needed in the bio-oxidation reaction), drops to zero. Mathematical calculations show that the value of δ is very small, and it never exceeds the value of 100μm. The biofilm density is assumed to be constant throughout the biofilter, while the diffusivity of a compound in the biofilm is taken as that in water corrected by a factor determined by a correlation proposed by Fan et al. [9]. The usual assumption that the kinetics of biodegradation in a biofilm are the same with those in suspended culture, is also made. Finally, plug flow of the air through the column is assumed.

The model is comprised of mass balances for the contaminant and oxygen, in the air and the biolayer. These equations are as follows.

$$\frac{dc_j}{dh} = \frac{A_{sj}f(X_v)D_jA}{F}\left[\frac{ds_j}{dx}\right]_{x=0} \tag{1}$$

$$\frac{dc_O}{dh} = \frac{A_{sj}f(X_v)D_OA}{F}\left[\frac{ds_O}{dx}\right]_{x=0} \tag{2}$$

$$\frac{d^2s_j}{dx^2} = \frac{X_v}{Y_jf(X_v)D_j}\mu_j(s_j)g(s_O) \tag{3}$$

$$\frac{d^2s_O}{dx^2} = \frac{X_v}{Y_{Oj}f(X_v)D_O}\mu_j(s_j)g(s_O) \tag{4}$$

with boundary conditions

$$c_j = c_{ji} \text{ and } c_O = c_{Oi} \text{ at } h = 0 \tag{5}$$

$$s_j = \frac{c_j}{m_j} \text{ and } s_O = \frac{c_O}{m_O} \text{ at } x = 0 \tag{6}$$

$$\frac{ds_j}{dx} = \frac{ds_O}{dx} = 0 \qquad \text{at} \quad x = \delta \tag{7}$$

The specific growth rates of the biomass on benzene and toluene were determined during an independent study involving shake-flask suspended culture experiments, and are as follows [10],

$$\mu_B(s_B) = \frac{0.68s_B}{12.22 + s_B} \quad (h^{-1}) \tag{8}$$

$$\mu_T(s_T) = \frac{1.50s_T}{11.03 + s_T + \dfrac{s_T^2}{78.94}} \quad (h^{-1}) \tag{9}$$

The functional dependence of the specific growth rate on oxygen, $g(s_O)$, is taken, [5], as

$$g(s_O) = \frac{s_O}{0.26 + s_O} \tag{10}$$

Except for A_{sj}, all model parameters are either measured or estimated from the literature (see Table 1). The value of A_{sj} is determined as follows. A number of data sets from different column experiments are fitted to the model by adjusting the value of A_{sj} so that the error between data and model predicted concentration values is minimized for all data sets, and all values within each set. This value of A_{sj} is then used for predicting concentration profiles of other column experiments (not used in the fitting approach), with the same substance. The model equations are solved by using either a shooting technique, or the method of orthogonal collocation for equations (3) and (4), and a 4th order Runge-Kutta method for equations (1) and (2). Details of the numerical approach are given elsewhere [5].

RESULTS AND DISCUSSION

Experiments with benzene vapor were performed under different inlet benzene concentrations, c_{Bi}, and space times, $\tau = V_p/F$. Each experiment was run for a period of at least two weeks. Table 2 shows the conditions of the experiments, and the measured exit concentration values of benzene. In the same table we show values of the observed removal rate, R_{exp}, which is defined as $(c_{Bi} - c_{Be})/\tau$. The load is defined as c_{Bi}/τ. The model predicted values for the exit concentration and the removal rate are also shown in Table 2. The value of parameter A_{SB} was determined as 23.3 m^{-1} by using two data sets, and was then used in predicting the values of concentrations in all other sets. One can easily see that the experimental and model predicted values of c_{Be} and R are very close in almost all cases. Judging the performance of a biofilter on the basis of the removal rate achieved, as is usually the case, may be misleading when the removal rate is not compared to the load. A better index would be the percent removal. Based on this index, one can say that the results of this study indicate that complete removal of benzene vapor requires substantial space times.

The agreement between experimentally measured and model predicted concentration values is good not only at the exit of the biofilter bed as shown in Table 2, but along the biofilter length (or height) as well. This is demonstrated in Figure 4. The model also predicts the concentration profiles of oxygen and benzene in the biolayer at any position along the biofilter length. Figure 5 shows such profiles at the middle point of a column. This graph shows that benzene is depleted much faster than oxygen in the biolayer. This was found to be always the case with both benzene and toluene, and it is opposite to what we had found earlier [5] with water soluble solvents. In the latter case, oxygen was depleted first and thus, it was determining the effective biolayer thickness. For the case shown in Figure 5, the effective biolayer thickness ($\theta = 1$) is predicted to be 53 μm, and is determined by the depletion of benzene.

Experiments were also performed with toluene vapor. In these experiments the inlet air toluene concentration was varied between 0.62 and 2.81 g/m^3, the space time from 2.7 to 8.6 min, and the load from 4.8 to 26.8 g/m^3-packing/h. The experimentally observed removal rates ranged from 4.8 to 24.8 g/m^3-packing/h. Removal rates of 20 g/m^3-packing/h have been also obtained by other investigators [1]. The minimum percent removal observed was 66.5%, while in many cases the percent removal was well above 90%. It appears that biofiltration of toluene is much easier when compared

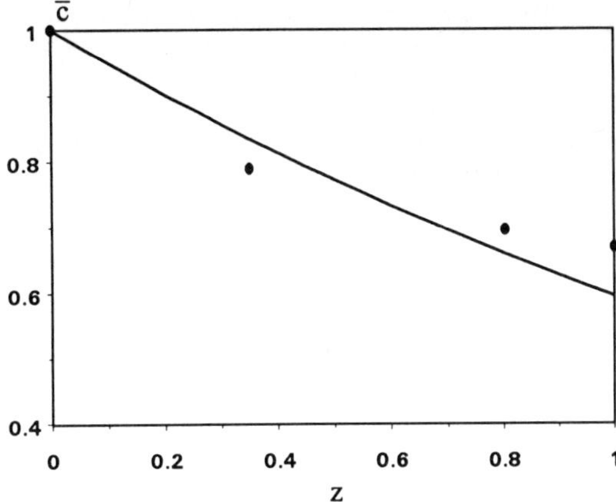

Figure 4 : *Benzene vapor concentration profile along the biofilter under steady state conditions : data and model prediction (curve). For this experiment, $c_{Bi} = 0.28\ g/m^3$, $\tau = 4.1$ min.*

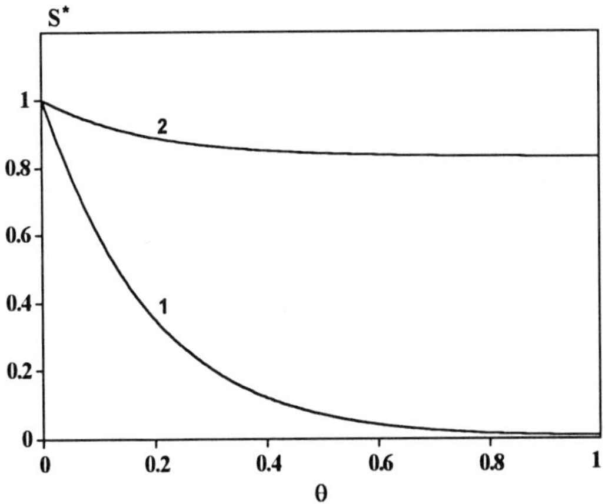

Figure 5 : *Model predicted concentration profiles in the biolayer at h = 0.5H. Curves 1 and 2 are for benzene and oxygen, respectively. These profiles are for the case where $c_{Bi} = 0.43\ g/m^3$, $\tau = 4.5$ min.*

Table 1. Model Parameter Values

A	19.6×10^{-4} m^2	PW*
A_{SB}	23.3 m^{-1}	PW
A_{ST}	40.0 m^{-1}	PW
c_{Oi}	275.0 g m^{-3}	[5]
D_B	1.04×10^{-9} m^2 s^{-1}	[11]
D_O	2.41×10^{-9} m^2 s^{-1}	[11]
D_T	1.03×10^{-9} m^2 s^{-1}	[11]
$f(X_v)$	0.195	[9]
m_B	0.23	[12]
m_O	34.4	[5]
m_T	0.27	[12]
X_v	100.0 kg m^{-3}	[5]
Y_B	0.708	[10]
Y_{OB}	0.336	PW
Y_{OT}	0.341	PW
Y_T	0.708	[10]

* PW: value measured or estimated in the course of the present work

Table 2. Steady state biofiltration of benzene vapors: Experimental data and model predictions.

τ	c_{Bi}	c_{Be} (exp)	c_{Be} (model)	Error (%)	Removal (%)	Load	R_{exp}	R_{model}	Error (%)
(min)		(g/m³)					(g/m³-packing/h)		
4.1	0.28	0.19	0.16	-15.8	32.1	4.1	1.3	1.6	23.1
4.5	0.43	0.23	0.25	8.7	46.5	5.7	2.7	2.5	-7.4
4.7	0.56	0.21	0.31	47.6	62.5	7.1	4.5	3.1	-31.1
2.7	0.13	0.09	0.09	0.0	30.8	2.9	0.9	0.8	-11.1
2.7	0.12	0.08	0.09	-11.1	33.3	2.7	0.9	0.8	-11.1
4.1	0.07	0.04	0.04	0.0	42.8	1.0	0.5	0.4	-20.0

to benzene. As in the case of benzene, toluene biofiltration data were subjected to model analysis. The agreement between experiments and theory was found to be remarkably good. Figure 6 shows a model predicted toluene concentration profile along with the actual measurements. In analyzing the toluene data, it was found that the value of A_{ST} is much higher than that of A_{SB} (see Table 1). It appears that when compared to benzene, since the consortium can grow easier on toluene, it can also better colonize the biofilter during start-up.

or a curve is passed through the data points by simple interpolation or some type of prediction. We believe that this approach is incorrect. Since variation in the load can be due to either a change in space time or in the inlet concentration, the data cannot fall on a single straight line or curve except if all were obtained under the same space time or inlet concentration. If both quantities were varied during the experiments, the data should fall in a region rather than on a curve (or line) in the removal rate-load space. We show this in Figure 7 for benzene, and Figure 8 for toluene. The boundaries of the regions (curves 1 and 2 in the graphs), are predictions under the minimum and maximum space time values used in our experiments. With one

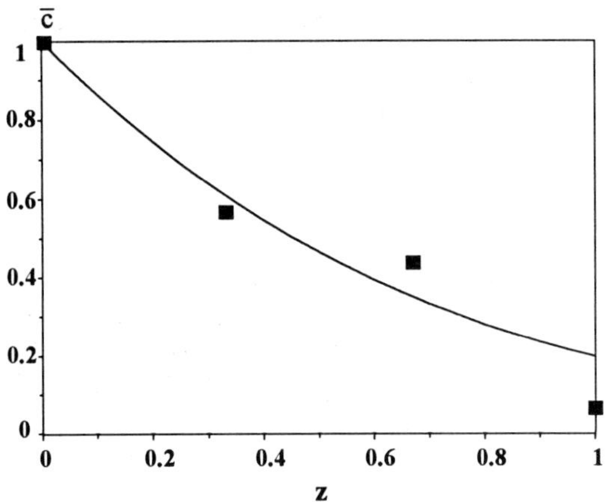

Figure 6 : Toluene vapor concentration profile along a biofilter column under steady state conditions when $c_{Ti} = 2.81$ g/m³, $\tau = 6.3$ min, $V_P = 5150$ cm³. Data are compared to the model prediction (curve).

Data from biofilters are presented in many cases in the form of a diagram showing the removal rate as a function of the load. Usually, a line

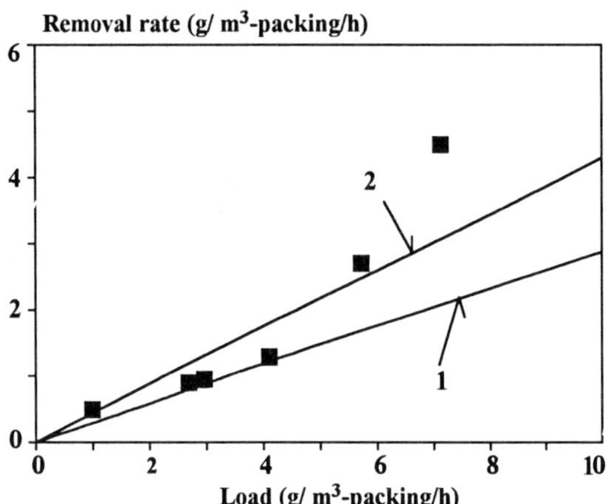

Figure 7 : Removal rate of benzene as a function of load. Data from experiments under various c_{Bi} and τ values. The two curves represent model predictions for the minimum and maximum τ values used in the experiments. For curve 1, $\tau = 2.7$ min; for curve 2, $\tau = 4.7$ min.

exception for benzene, all data points fall in the regions predicted by theory.

Regarding the model parameter values, the highest uncertainty seems to be associated with the biofilm density (X_V), and the biofilm surface area per unit volume of packing (A_{sj}). For this reason, we undertook a model sensitivity study on these parameters. As a base line, we used an experiment involving biofiltration of toluene. As

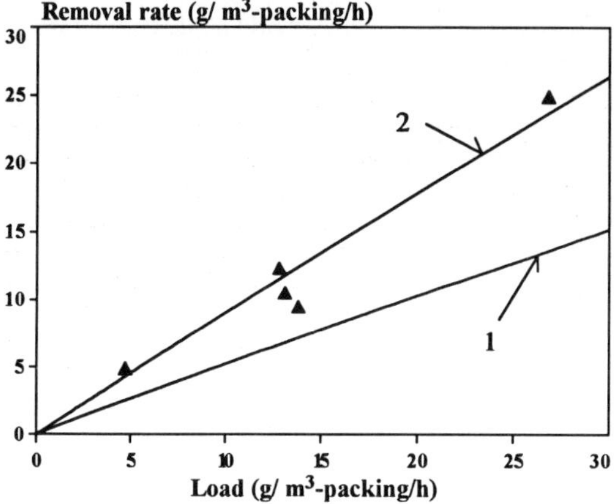

Figure 8 : Removal rate of toluene as a function of the load to the biofilter. Data from experiments under various conditions of c_{Ti} and τ. The curves represent model predictions for the minimum and maximum τ values used. For curve 1, $\tau = 2.7$ min; for curve 2, $\tau = 8.6$ min.

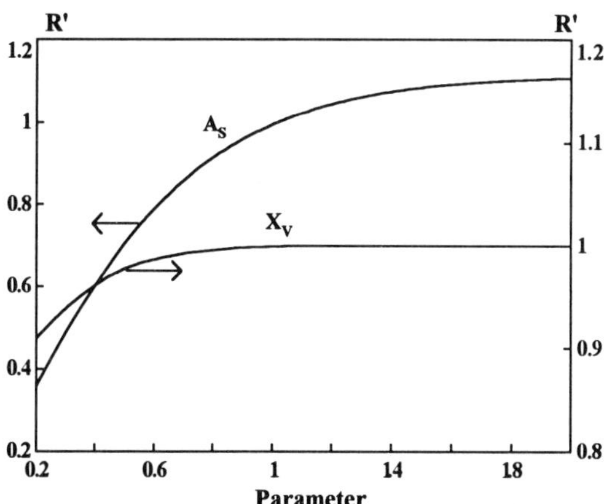

Figure 9 : Sensitivity of the model to the values of parameters A_S and X_V. Results of simulations compare the removal rate relative to a reference value when the parameters are changed relative to a reference value. The point of reference (1,1) corresponds to an actual biofiltration experiment with toluene for which $c_{Ti} = 1.65$ g/m³, $\tau = 7.7$ min, and the experimentally observed removal rate was 12.2 g/m³-packing/h.

discussed elsewhere [5], the values reported for X_V range from 20 to about 220 kg/m³. The value used in our calculations was 100 kg/m³. As Figure 9 indicates, for relative X_V values above 0.6, i.e., actual X_V values above 60 kg/m³, the predicted removal rate would be practically the same for any X_V value. If the real X_V value was between 20 and 60 kg/m³ the error in the removal rate would be less than 10%. The value of A_{sj} seems to be very important as shown also in Figure 9. It appears that if the real A_{sj} value is larger than the one estimated (relative value larger than 1), the impact on the prediction of the removal rate is less than 10%. On the other hand, if the real value is less than the estimated one, the error in predicting removal rates can be very substantial. It is for this reason that a careful approach needs to used in estimating A_{sj}.

CONCLUSION

Biofiltration can be used for removing benzene and toluene vapors from airstreams. Benzene appears to be harder to remove than toluene, at least with the culture used in this study. Steady state biofiltration can be nicely described and predicted with a model which accounts for reaction, as well as mass transfer of the pollutant and oxygen. Biofiltration of hydrophobic compounds appears to be less affected by oxygen limitation, when compared to treatment of hydrophilic compounds.

ACKNOWLEDGMENT

This work was supported through a grant from the Hazardous Substance Management Research Center (HSMRC), an NSF Industry/University Co-operative Research Center headquartered at the New Jersey Institute of Technology.

NOTATION

A : cross sectional area of the biofilter (m²)

A_{sj} : biolayer surface area per unit volume of packing (m⁻¹)

c_j : concentration of pollutant j in the air along the bed (g/m³)

c_{je} : value of c_j at h = H (g/m³)

c_{ji} : value of c_j at h = 0 (g/m³)

c_O : oxygen concentration in the air along the biofilter (g/m^3)

c_{Oi} : value of c_O at $h = 0$ (g/m^3)

$\bar{c}$: normalized concentration defined as c_j/c_{ji}.

D_j : diffusivity of compound j in water $(m^2\ h^{-1})$

D_O : diffusivity of oxygen in water $(m^2\ h^{-1})$

$f(X_V)$: diffusivity correction factor for the biofilm

F : flow rate of airstream $(m^3\ h^{-1})$

$g(s_O)$: functional dependence of specific growth rate on oxygen

h : position in the biofilter $h = 0$: entrance; $h = H$: exit

H : total height of the biofilter bed (m)

m_j : distribution coefficient of compound j between air and water

m_O : distribution coefficient of oxygen between air and water

R : removal rate $(g/m^3\text{-packing}/h^{-1})$

R' : relative value of removal rate

s_j : concentration of pollutant j in the biolayer (g/m^3)

s_O : concentration of oxygen in the biolayer (g/m^3)

s^* : normalized value of s_j or s_O ; actual value divided by that at $x = 0$

V_P : volume of the packing material (biofilter) (m^3)

x : position in the biolayer (m)

X_V : biofilm density $(kg\text{-dry cells } m^{-3})$

Y_j : biomass yield coefficient on compound j

Y_{Oj} : biomass yield coefficient on oxygen when compound j is the carbon source

z : dimensionless position in the biofilter $(z = h/H)$

Greek Symbols

δ : effective biolayer thickness (m)

δ^* : actual biolayer thickness (m)

θ : dimensionless position in the biolayer $(\theta = x/\delta)$

$\mu_j(s_j)$: specific growth rate of biomass on carbon source j

under excess oxygen conditions (h^{-1})

Subscripts

j=B : pollutant is benzene
j=T : pollutant is toluene

LITERATURE CITED

1.Ottengraf, S.P.P., and A.H.C., van den Oever, *Biotechnol. Bioeng.*, **25**, 3089(1983).

2.Leson, G., and A. M.Winer, *J Air Waste Manage. Assoc.* **41**, 1045(1991).

3.Diks, R. M. M. and S. P. P. Ottengraf, *Bioprocess Engineering,* **6**, 3,93 (1991).

4.Diks, R. M. M. and S. P. P. Ottengraf, *Bioprocess Engineering,* **6**,4,131 (1991).

5.Shareefdeen, Z., B. C. Baltzis, Y.-S. Oh, and R. Bartha, *Biotechnol. Bioeng.*, **41**, 512(1993).

6. Deshusses, M.A. and G. Hamer, *Bioprocess Engineering,* **9**, 4,141(1993).

7.Potter, T.L. "Fingerprinting Petroleum Products: Unleaded Gasolines," in: *Petroleum Contaminated Soils*, Vol. 3 , P.T. Kostecki and E.J. Calabrese (Eds.), Lewis Publishers, Chelsea, MI(1992).

8.Reisch, M.S.,*Chem. Eng. News, 70*, 15,16 (1992).

9.Fan, L.-S., R. Leyva-Ramos, K.D. Wisecarver, and B.J. Zehner, *Biotechnol. Bioeng.*, **35**, 279(1990).

10.Oh, Y.-S., Z. Shareefdeen, B. C. Baltzis, and R. Bartha, *Biotechnol. Bioeng.*(in press).

11.Reid, R.C., J.M. Prausnitz, and T.K. Sherwood, *The Properties of Gases and Liquids.*, McGraw-Hill, New York (1977).

12.Mackay, D. and W. Y. Shiu, *J. Phys. Chem. Ref. Data*, **10**, 4, 1175 (1981).

Toluene Removal from Air Stream by Biofiltration

M. Morales[1], F. Pérez[1], R. Auria[1,2], and S. Revah[1]

[1] Departamento. de Ingeniería de Procesos e Hidráulica, UAM-Iztapalapa, Apdo. Postal 55-534, 09340, México, D.F.; [2]ORSTOM (Institut Français de Recherche Scientifique pour le Développement en Coopération), Cicerón 609, Los Morales 11510, México D.F., MEXICO

Toluene removal was studied in a bench-scale biofilter. Continuous experiments were made during a period of four months. The biofilter was packed with peat previously inoculated with a mixed bacterial culture. A data acquisition system was installed to log temperature, air flow, relative humidity, carbon dioxide concentration, pressure drop and pH. Steady state was achieved in about 5 weeks. Maximal removal rate was 25 gtoluene$(m^3$ biofilter$)^{-1}h^{-1}$. Experiments were carried out at different toluene concentrations and two situations were identified: diffusion limitation (low concentration) and reaction limitation (high concentration). Biofilter behavior was adequately described by a simple model considering Monod kinetics and mass transfer between the biofilm and the gas phase. Nevertheless, parameter analysis showed a wide range of possible values for most of them.

Atmospheric contamination is a big global problem that humanity has to confront. An important part of the pollutants is produced by different industrial processes. Authorities in many countries have adopted stringent statutory regulations that limit emissions to protect public health and the environment.

Volatile Organic Compounds (VOC´s) are among the most common pollutants emitted by the chemical process industries. For example, in Mexico City, 70,000 ton./year of gaseous industrial wastes are emitted. Transformation industry (resins, glues, paints, inks, polymers, etc.) produce approximately 36,000 ton/year of solvents . These compounds cause an important health problem because they frequently are toxic, odorous and contribute to form ozone and photochemical smog.

Analysis of gaseous effluent characteristics and elimination process available must be made to select a VOC´s emission control[1]. Sometimes, biological waste gas treatments have

some specific advantages in comparison with other techniques (incineration, adsorption, absorption, etc.).

Biofiltration involves the use microorganisms to convert the pollutants to carbon dioxide, water and mineral salts. Therefore pollutants are not transferred to other phase but converted into less harmful oxidation products under mild conditions.

As environmental regulations for air emission become increasingly stringent, biofiltration is an economically promising alternative because it has low investment and operational costs [2,3]. It looks specially attractive to treat high flow rates with low concentrations of pollutants.

A biofilter is a packed reactor with an organic material (e.g., compost, peat, soil, etc.) on which a microbial population is immobilized forming a biofilm. The contaminated air flows through the filter bed and

405

E. Galindo and O.T. Ramírez (eds.), Advances in Bioprocess Engineering. 405-411.
© *1994 Kluwer Academic Publishers. Printed in the Netherlands.*

the pollutants are transported to the biofilm where they are degraded.

Historically, gas phase biotreatment has been employed for odor control at conventional treatment where H_2S or other odorous gases are produced. This technique is extensively used in the Netherlands where as many as 500 full scale biofilters are installed treating waste gas streams from diverse sources as chemical plants, print shops, fish frying, etc.[4]

Two fundamental aspects must be considered in biofilter operation: a)the microbial population and the filter material and b)the transport among phases (gas, solid, biotic, liquid). The selection of microorganisms for the biological degradation determines the ability of the biofilter to degrade a pollutant compound. Activated sludge from municipal sewage treatment plants has been successful used [5]. However, more recalcitrant compounds, such as phenol and benzene [6,7,8,9], require specific strains to be degraded. Studies on the filter material tend to improve filter characteristics, and composition of packing materials. In particular, the increase in material porosity has allowed reduced pressure drops and improved biological activity, and thus more efficient systems have been obtained [9,10,11,12,13]. Theoretical description of the phenomena involved in the operation of the biofilters has been reported [5,11]. Approaches based on the Monod kinetics have frequently use zero or first order kinetics to get analytical solutions. Proposed models involve different parameters with values that are often estimated or adjusted. The reported values show a great dispersion.

The present work aims to evaluate, the possibility of purifying waste gases polluted with toluene. This pollutant was selected because it is a common constituent of gaseous effluents from industries or shops that use or produce glues, paints, inks, resins, adhesives, etc. Toluene is listed as a hazardous air pollutant under the Clean Air Act, requiring the EPA to set emission standards. Furthermore, toluene is a toxic substance whose adverse effects on health are well documented.

The objective of this work is to study the gas phase toluene elimination by biofiltration at a bench scale and analyze the removal kinetics, the start-up and the stability of the system.

MATERIALS AND METHODS

Organisms and Medium. Packing material was peat. $Ca(OH)_2$ was added as a pH buffer at 0.04 $g/g_{dry\ peat}$. Peat was inoculated with a mixed culture obtained from activated sludge. Water, mineral salts, (Table 1), and the inoculum were added to the peat to bring the mixture to a final moisture content of 62%.

Table 1. Culture medium composition

Nutrient	g/l
K_2HOP_4	0.08
KH_2PO_4	0.20
$CaSO_4\ H_2O$	0.05
$MgSO_4\ 7H_2O$	0.50
$FeSO_4\ 7H_2O$	0.01
$(NH_4)_2SO_4$	1.00

Experimental Equipment. The equipment has a data acquisition system (RTI-820 Analog Devices and personal computer) to capture on line the different data.

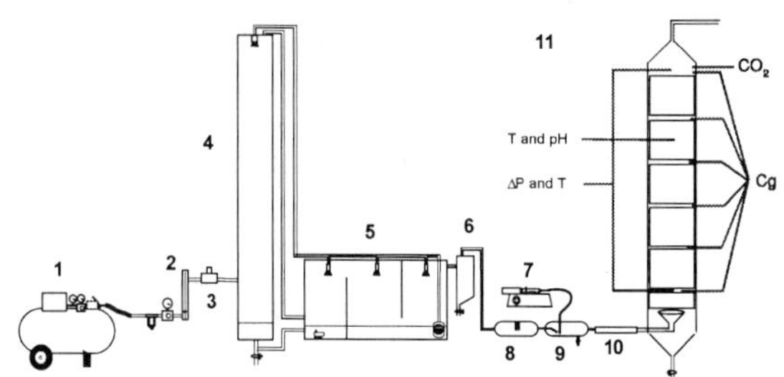

Figure 1. Biofiltration System: 1) Air-compressor, 2) Flowmeter, 3) Mass flow sensor, 4) Prehumidifier, 5) Humidifier, 6) Cyclone, 7) Syringe pump, 8) Relative humidity sensor, 9) Addition solvent 10)Static mixer, 11) Biofilter.

The bench scale biofilter that was used, is schematically shown in Figure 1. Air was supplied with a compressor and rate flow was measured with an electronic flowmeter. Air was humidified by water spraying and its relative humidity measured with a capacitive sensor (Humicor 6100, Coresi, France). Polluted gas was artificially produced by injecting liquid toluene to the airstream. It flows through a static mixer and was introduced by a diffuser to the bottom of the biofilter.

The biofilter consists of a 150 cm cylindrical acrylic column with an inner diameter of 14 cm. Packing material was distributed in 5 modules of 20 cm and each one is supported by a sieve plate to allow a homogeneous distribution of the gas flow through the filter bed.

Table 2. Operating Conditions

Packed bed volume	15.4 l
Packing density	0.3 g/cm^3
Air flow	26.03 l/min
Inlet concentration	241.82 ppmv
	0.7 g/m^3
Temperature	25 ^{0}C ±3
Residence time (empty)	35.48 sec
Average residence time	40 sec
pH	7
Humidity	62%
Dry peat mass	1.754 kg

The temperature of the airstream was measured by thermocouples in the inlet, outlet and in the packed bed. Sampling ports in each level allowed to determine toluene concentration by means of a gases IR analyzer (MIRAN-A FOXBORO USA) at 13.7 μm with a 6.25 m pathlength. CO_2 produced during toluene degradation is determined by an infrared analyzer (Milton Roy ZFP-9). Biofilter has two ports to measure pressure drop through the bed with a pressure transducer (Cole Palmer Model 7352-16). The pH was continuously monitored by a pH electrode (Omega model PHE 2385) inserted at the support.

The biofilter was started up and operated until steady state with the conditions described in table 2. Once the steady state was achieved, biofilter performance was studied under different air flow rate and toluene additions.

MATHEMATICAL DESCRIPTION OF PROCESS

In the work of Ottengraf and Van Den Oever [5] zero order biodegradation model was used to describe biofiltration of laquery emissions. A modification of that work is proposed introducing Monod kinetics.

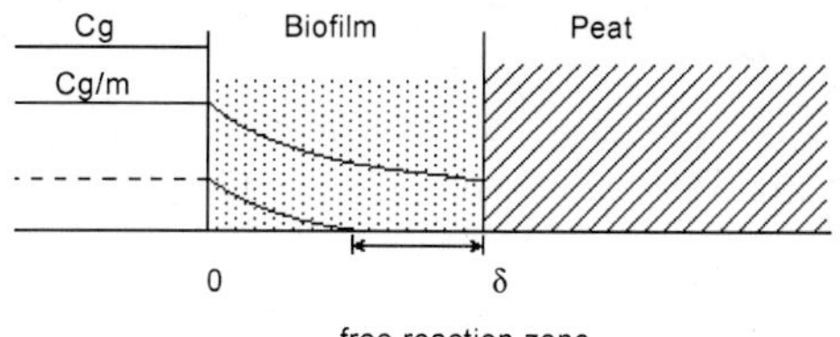

Figure 2. Biophysical Model

As airstream flows through the porous bed, the soluble components in the gases are transported into the moist bacterial film. Biodegradable compounds are then metabolized by bacteria. A schematic representation of the phases in the biofilter is shown in the figure 2. In steady state, the dimensionless differential equation describing the mass balance for a compound in the biological actived layer of the biofilter is:

$$\frac{d^2 C^*}{dx^{*2}} - \phi^2 \frac{C^*}{1 + C^*} = 0 \qquad (1)$$

where **C*** is a dimensionless concentration defined as **C|/Ks**, **x*** is a dimensionless coordinate **x/δ** and ϕ^2 is the Thiele modulus defined as **δ²Xμ/YD´Ks**.

With the boundary conditions:

$$x^* = 0 \quad \rightarrow \quad C^* = \frac{C_g}{mK_s}$$

$$x^* = 1 \quad \rightarrow \quad \frac{dC^*}{dx^*} = 0 \qquad (2)$$

A differential equation that describes the steady state mass balance for the gas phase compound at the position **h** in the filter bed is:

$$-U_g \frac{dC_g}{dh} = NA_s \qquad (3)$$

where **U_g** is the gas velocity, **A_s** is the area per unit volume of porous media, and **N** is the flux that is given by:

$$N = -D' \left(\frac{dC_1}{dx} \right)_{x=0} \qquad (4)$$

The procedure for solving the equations was the following: A finite difference approximation was used to convert equation 1 in an algebraic system. This was solved through the ZSPOW subroutine of the IMSL library which uses a modification of Newton´s method. A linear profile was used to initiate the calculus of the concentration profile in the biofilm to avoid negative values. Equation 1 was first solved at the inlet conditions from where the concentration profiles in the biolayer were determined. From these, the slope of concentration at x=0 was obtained and used in equations 3 and 4 to evaluate the gas phase concentrations at the position Δh. Here, equation 1 was solved and the procedure was repeated until the total height of the biofilter was reached.

Table 3. Parameter Values.

Parameter	Range of values	Ref.
μ (1/day)	0.6–9.9	5,14,15
Ks (g/m^3)	0.03–17.4	14,15
δ (μm)	40–100	13,16
X (g/m^3)	22,000–220,000	12
Y (g$_{cell}$/g$_{tol}$)	0.708–1.0	13,15
D´ (m^2/day)	0.73x10^{-4}–1.1x10^{-4}	5,13,15
m	0.27	5,13

To solve the model, the literature shows a wide variation in the parameters for toluene degradation, as it is shown in table 3. The work of Arvin and Arcangeli [15] is the only one that presents all parameters for toluene removal by bacteria in liquid phase.

RESULTS AND DISCUSSION

The adaptation period of the microbial population to toluene and the environmental conditions (mineral medium and peat) was approximately 5 weeks, at this time a steady state operation was reached. This period was longer than the 2 weeks reported elsewhere [5]. This may be due to the fact that not previous bacterial adaptation was made in this study. Figure 3 shows the global behavior of biofilter. Removal capacity slowly increased until a maximum of about 25 g/m^3/h, which is comparable to the 20 g/m^3/h reported by Ottengraf [5].

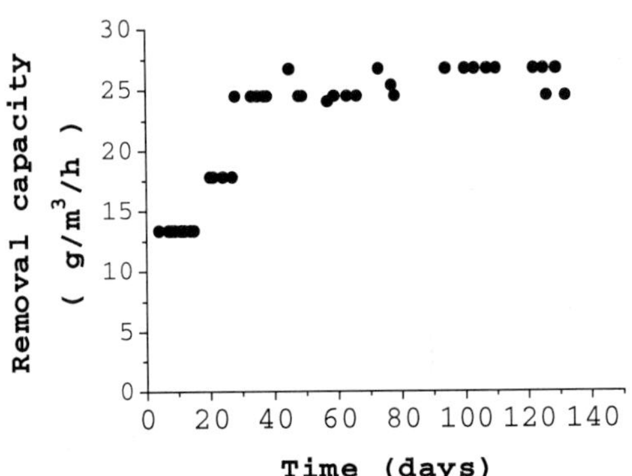

Figure 3. Evolution of Removal Capacity with time.

The biofilter proved to be very stable to small variations in environmental conditions such as temperature (which was not controlled) or inlet humidity. The system showed good recovery regarding operational accidents, (e.g., removal capacity was completely restored in 72 hours after a weekend electricity cut-off for 48 hours). These changes were logged by the data acquisition system.

Toluene elimination was determined by sampling the gas phase for each module. From figure 4 it can be concluded that the elimination occurs according to a zero order reaction. The profile shows a linear dependence to the height. The removal percentage was 35% with the conditions in table 2. Toluene removal efficiency about 99% has been obtained [11] but at very low concentrations (under 50 ppmv).

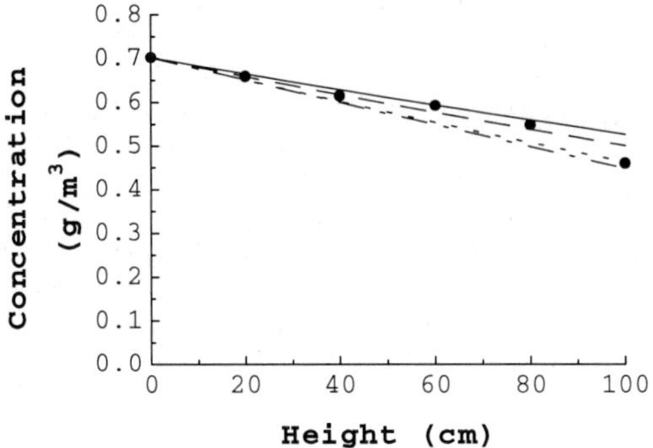

Figure 4. Concentration profile in the biofilter for different Thiele modulus values ——— 200, - - - 150, ···100, -·- 50, ● Data.

The biofilter was continuously operated during a period of about 4 months. The inlet toluene concentration was varied over a wide range to test the elimination capacity of the system under different conditions. These experiments were carried out in two different ways:

-at constant flow rate (26 l/min) the toluene supplied was varied from 0.34 to 5.10 ml/h.

-at constant toluene flow (1.17 ml/h) air flow was varied from 12 to 78 l/min.

As it has been pointed out in the theoretical model, the elimination capacity of the filter bed will no longer be constant but decrease when the toluene gas phase concentration is below a critical value (figure 5). Then diffusion limitation occurs and the biolayer is not fully active, therefore, penetration depth is smaller than the total layer thickness, hence, the maximal removal rate is not achieved (Figure 2). When reaction limitation is present, the biolayer is entirely active and the system reaches its maximal removal rate.

The analysis with Thiele moduli between 50 and 500 was made with fixed values of $D' = 1.1 \times 10^{-4}$ m^2/day, $Y=1$ g_{cell}/g_{tol} and $m=0.27$ [15]. The μ and the Ks were varied according to table 3. Selected biofilm thickness were 100 and 40 microns [13,16]. While many combinations could fulfill the selected Thiele moduli, only those that gave biofilm densities between 22,000 and 220,000 g/m^3 were retained. In the model the value of A_S was fitted.

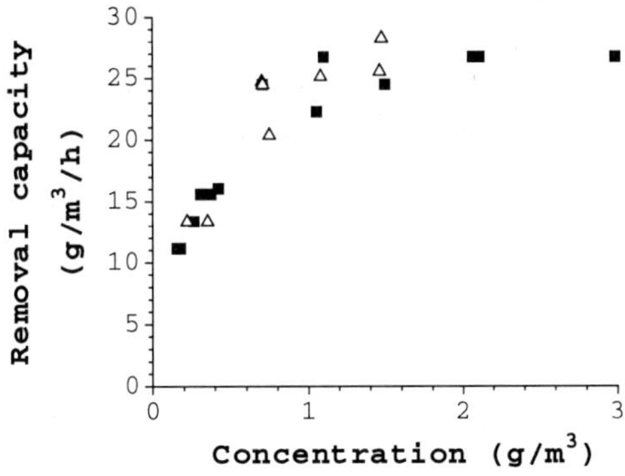

Figure 5. Removal capacity at different toluene inlet. ■ At variable toluene load, Δ At variable flow rate.

The concentration profile through the biofilter and the reactor removal capacity at different inlet toluene concentrations were simultaneously adjusted to the model. The results are depicted in figures 4 and 6. The resulting specific areas were from 18 to 175 m^2/m^3 for Thiele moduli between 50 to 200. The parameter values to represent experimental data are resumed in table 4.

Table 4. Adjustment parameter values. Ks= 0.05 g/m^3.

ϕ^2	$X\mu$ (g_{cell}/m^3day)	δ (μm)	A_s (m^2/m^3)
50	27,500	100	175
50	171,875	40	70
100	55,000	100	90
100	343,750	40	36
150	82,500	100	60
150	510,560	40	24
200	110,000	100	45
200	687,500	40	18

A linear relationship among parameters was observed: when the thickness biofilm is increased, the A_S value decreases proportionally. The quantity of microbial population is, under the studied conditions, the relevant parameter rather than the form in which it is distributed. To summarize, an overall zero order process could be observed in the biofilter.

In the operation of the biofilter an increase in the ΔP was observed (from 2.5 cm H_2O to 12 cm H_2O). This was specially observed in the first module and was related to an increase in the water content of support.

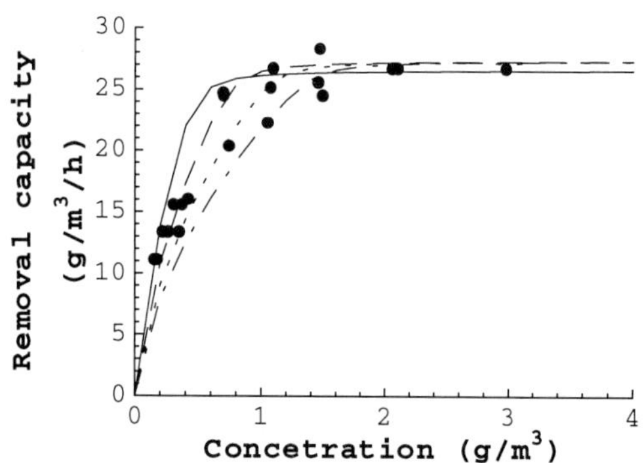

Figure 6. Behavior of Removal Capacity versus the influent gas concentration. Lines represent model at different values of Thiele modulus $-\cdot-$ 200, $\cdots$ 150, $---$100, $\longrightarrow$ 50, $\bullet$ Data.

CONCLUSIONS

Experiments with a bench scale biofilter have proved that toluene vapors can be efficiently removed from air. Maximal removal capacity in experimental system was 25 $g/m^3/h$. The system showed a great operational stability when breaks in toluene load and air flow occurred. The data acquisition system was useful since it permitted to obtain important information about system operation. An analysis of the involved parameter with the use of Monod kinetics was presented. According to the parameter values that adjust the experimental data, the global behavior of biofilter could be represented by an overall zero order model, as suggested by the work of Ottengraf [5]. Further experimental studies are been carried to increase the knowledge on the physical and microbial aspects on biofiltration.

NOMENCLATURE

A_S — Specific area (m^2/m^3)
C^* — Dimensionless concentration
C_g — Pollutant gas concentration (g_{tol}/m^3)
C_1 — Pollutant biofilm concentration (g_{tol}/m^3)
D' — Effective diffusion coefficient (m^2/day)
h — Height (m)
K_S — Saturation constant (g/m^3)
m — Distribution coefficient
N — Substrate flux (g_{tol}/m^2day)
U_g — Flow rate ($m^3/m^2/day$)
X — Biofilm density (g_{cell}/m^3)
x^* — Dimensionless coordinate
x — Coordinate (m)
Y — Yield coefficient (g_{cell}/g_{tol})
μ — Maximum growth rate (1/day)
δ — Thickness biofilm (μm)
ϕ^2 — Thiele modulus

LITERATURE CITED

1. Rinko J. and Taister M. Annual Meeting of APCA: 2-15 (1988)

2. Paul, P.G. and Castelijn, F.J., Edited by De Wall, K.J. Van Den Brink, W.J. Nihoff. Dordrecht. Neth.:231-237(1987).

3. Bohn, H.,. Chem. Eng. Prog. April: 34-40(1992).

4. Leson, G., Winner A.M., J. Air and Waste Manage. Assoc. 41: 8., 1045-1054 (1991).

5. Ottengraf S.P.P. and Van Den Oever, A.H.C., Biotechnol. Bioeng., 25: 3089-3102 (1983).

6. Claus, D. and Walker, N., J., Gen. Microbiol., 36: 107-122 (1964).

7. Rehm, H.J. and Ehrhardt, H.M., Appl. Microbiol. Biotechnol., 21: 32-36 (1985).

8. Dick B., Janssen, A.S., and Witholt, B. Innovations in Biotechnology, Edited by E.M. Mouwink and R.R. Van Der Meer: 169-178 (1984).

9. Zilli M., Converti A., Lodi A., Del Burghi M. and Ferraiolo G., Biotechnol. Bioeng., 41:693-699 (1993).

10. Anonymous. Document CLAIRTECH The Netherlands (1991).

11. Ergas S., Schroeder E. and Chang D. 86th Annual A&WMA Meeting and Exhibition (1993).

12. Shareefdeen Z, Baltzis B., Oh Y-S., Bartha R., Biotechnol. Bioeng., 41:512-524 (1993).

13. Baltzis B, Shareefdeen Z., 86th Annual Meeting and Exhibition (1993).

14 Alvarez P.J., Anid P.J., Vogel T.M. Biodegradation 2:43-51 (1991).

15. Arcangeli J.P. and Arvin E., Appl. Microbiol. Biotechnol., 37: 510-517 (1992).

16. Heijnen J.J., Van Loosddretch M.C., Mulder A. and Tijhuis L.,Wat. Sci. 26: 3-4, 647-654 (1992).

Invited paper

Metabolic Engineering of Cephalosporin Biosynthesis in *Streptomyces clavuligerus*

L.-H. Malmberg[1], A. Khetan[1], D. H. Sherman[2], and W.-S. Hu[1]

[1]Department of Chemical Engineering and Materials Science, [2]Department of Microbiology and Institute for Advanced Studies in Biological Process Technology, University of Minnesota, Minneapolis, Minnesota 55455-0132, U.S.A.

The biosynthesis of β-lactams is one of the most thoroughly studied antibiotic pathways. The availability of the characteristics and the time profiles of activities of enzymes involved in the biosynthesis allows one to critically evaluate the potential rate limiting steps in its production. Our approach to understanding the control of β-lactam biosynthesis has been pursued using a two stage strategy: 1) to predict the rate-limiting steps using a kinetic model and 2) to relax the rate limiting steps by engineering the biosynthetic pathway or by altering the kinetic parameters of the predicted key rate-limiting enzyme. Here we summarize the results of our kinetic analysis and the effect of amplifying the gene of a key limiting enzyme (lysine ε-aminotransferase (lat). The insertion of a single additional copy of lat into the chromosome resulted in major increase in antibiotic production.

A key to the success of metabolic engineering for enhance product formation is to predict the rate limiting steps involved in product biosynthesis. Such prediction is often hampered by the lack of knowledge on the biosynthetic pathway and its dynamic profile in the producing organism's life cycle. The wealth of information available on specific biosynthetic steps, including enzyme kinetic data, has allowed us to analyze rate-limiting reactions in the biosynthesis of β-lactam antibiotics from δ-(L-α-aminoadipyl)-L-cysteinyl-D-valine (ACV) tripeptide and its amino acid precursors (Fig. 1).

We have developed a structured kinetic model describing the production of cephalosporins in *Cephalosporium acremonium and Streptomyces clavuligerus* [1,2,3]. Using this model to simulate the production kinetics we demonstrated that the rate during batch culture fermentation paralleled those of experimental observation. Sensitivity analysis indicated that non-ribosomal condensation of ACV tripeptide from α-aminoadipic acid (α-AAA), valine, and cysteine, is the rate-limiting step in the pathway (Table1).

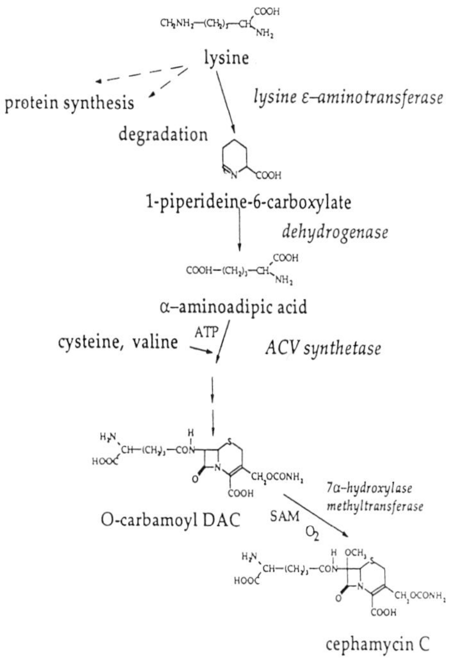

Figure 1 Biosynthetic pathway of cephamycin C form lysine, cysteine, and valine in s. clavuligerus.

413

E. Galindo and O.T. Ramírez (eds.), Advances in Bioprocess Engineering. 413-415.
© 1994 Kluwer Academic Publishers. Printed in the Netherlands.

It also predicted that precursor flux, which affects the formation of ACV tripeptide, may play an important role in controlling the rate of cephamycin C biosynthesis.

The precursor flux, specifically the biosynthesis of α-AAA, a branched pathway form primary to secondary metabolism appear to be limiting. The biosynthetic pathway of a α-AAA differs between fungi and actinomycetes. In fungi, this product is an intermediate of the lysine pathway whereas in actinomycetes α-AAA is a catabolic product of lysine (Fig.1). Studies have shown significant effects of primary metabolites on the production levels of β-lactam antibiotics. Addition of lysine in the fermentation culture of *Penicillium chrysogenum* repressed the level of antibiotics, attributed to the feedback inhibition of homocitrate synthetase by lysine. In contrast, addition of lysine and DL-meso-diaminopimelic acid (DAP) in *S. clavuligerus* stimulated cephamycin C production, possibly by providing a larger precursor pool for biosynthesis of α-AAA, or as a result of activation of aspartokinase, the first enzyme involved in lysine biosynthesis via the aspartate pathway. Furthermore, mutants resistant to the lysine analogue, S-(2-aminoethyl)-L-cysteine (AEC), produced increased levels of cephamycin C, possibly due to relaxed metabolic regulation of the lysine pathway and increased carbon flux to the biosynthesis of β-lactams. These findings along with our theoretical kinetic analysis suggest that biosynthesis of α-AAA plays a key role in controlling secondary metabolic pathway.

We thus hypothesize that relaxing the control of the biosynthesis of α-AAA will result in an elevated level of β-lactam biosynthesis in S. clavuligerus. Chromosomal integration of a key biosynthetic gene was used to augment carbon flux on production of cephamycin C[4]. Our strategy involved a unit increase in copy number of gene (*lat*) that encodes lysine ε-aminotransferase (LAT), the enzyme responsible for converting lysine to α-AAA, which represents the first dedicated secondary metabolic step in the β-lactam pathway. A targeted insertion strategy was used to generate a recombinant strain of *S. clavuligerus* (Fig. 2).

The recombinant strain produced cephamycin C and O-carbamoyl deacetylcephalosporin C at levels 5 fold higher than those observed in the wild-type (wt) strain (Fig. 3). Significantly, the production ratio of these two compounds is identical to wt which is consistent with an early step perturbation in the biosynthetic pathway. In addition, assay of lysine ε-aminotransferase revealed a 4 fold average increase in enzyme activity, whereas levels of ACV synthetase, whose structural gene is located just downstream of *lat*, remained essentially unchanged.

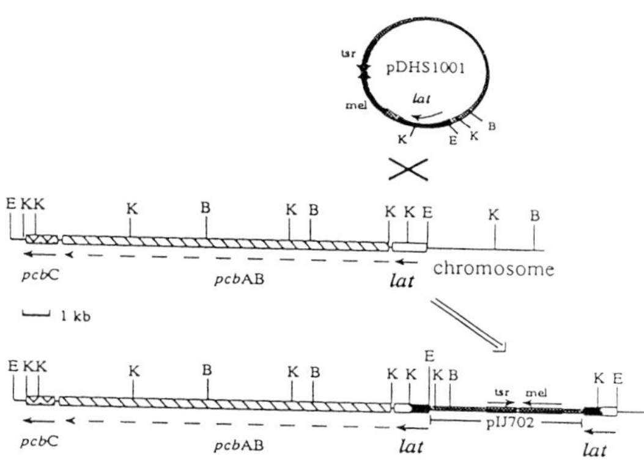

Figure 2 Integration of pDHS1001 into S. clavuligerus chromosome. pDHS1001 was inserted into the lat gene by a single crossover. The resulting LHM100 chromosome contains two copies of lat separated by plasmid pIJ702.

Our experimental finding strongly suggest that increasing precursor flux is an effective strategy for enhancing biosynthesis of cephamycin C. This unique combination of engineering analysis and molecular genetics may provide a general method to dissect mechanisms of metabolic control in the biosynthesis of important metabolites.

Table 1. Sensitivity analysis

	Sensitivity Coefficient
α-aminoadipic acid	0.70
cysteine	0.23
valine	0.30
ACS synthetase	1.00
cyclase	0.00
epimerase	0.00
expandase	0.00
hydroxylase	0.00

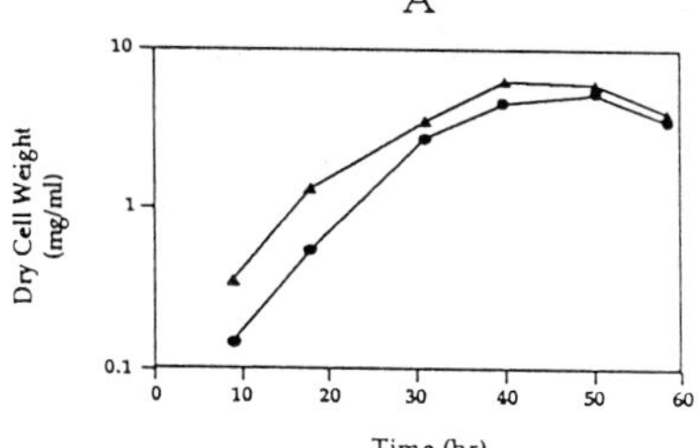

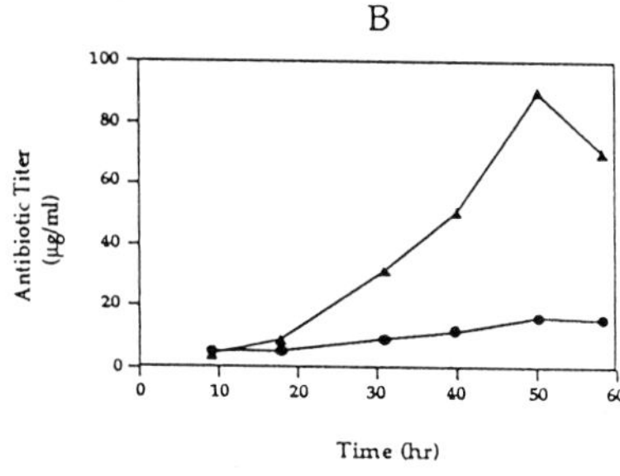

Figure 3. Growth (A) and antibiotic production (B) of LHM100 (△) and the wild-type (o) strains.

References

1. L.-H. Malmberg and W.-S. Hu, Biotechnol. Bioeng., 38 (1991) 941.

2. L.-H. Malmberg and W.-S. Hu, Appl. Microbiol. Biotechnol., 38 (1992) 122.

3. L.-H. Malmberg D.H. Serman, and W.-S. Hu, Ann. NY Acad Sci., 665 (1992) 16.

4. Malmberg, L.H., Hu, W-S. and Sherman, D.H. (1993) Precursor flux control through targeted chromosomal insertion of the lysine E-aminotransferase (*lat*) gene in cephamycin C biosynthesis. J. Bacteriol., 175, 6916-6924.

High Production of Lactic Acid from Metabolically Engineered *Saccharomyces cerevisiae*

L. Brambilla, D. Porro, E. Martegani, B.M. Ranzi, and L. Alberghina

Dipartimento di Fisiologia e Biochimica Generali, Sez. Biochimica Comparata, Università di Milano, Via Celoria 26, 20133, Milano, ITALY

The budding yeast Saccharomyces cerevisiae *is a safe and widely used host for the production of heterologous proteins. Nowadays, interesting perspectives from r-DNA applications also arise from the opportunity to modify metabolic pathways, yielding engineering yeast cells useful for chemicals production. In this paper we describe yeast strains that express the bovine muscle lactic dehydrogenase gene (LDH-A). These strains allow to carry out lactic acid fermentations with both high production (20 gl^{-1}) and high productivity (11 g l^{-1} h^{-1}) of lactic acid.*

In bacterial lactic acid fermentation, there is an inhibitory effect caused by the produced acid on the metabolic activities of the producing cells (i.e. Lactobacillus spp.)(7)(8). Beside the presence of lactic acid, lowering of the pH value inhibits cell growth (7). Therefore the addition of CaCO3, NaOH or NH4OH, to prevent lowering of pH, is a conventional operation in industrial processes (8)(9). These processes allow the production of lactate(s), maintaining pH at a constant value (about 5). However, as the solubility of lactate is low (9), at high concentration of this product the medium tends to solidify, complicating the fermentation behaviour and the following isolation procedures (9). Furthermore, additional operations are required to recover free lactic acid from its salt. All these problems could be overcome by producing lactic acid with microorganisms able to grow and survive S. cerevisiae represents one of the organisms of choice for industrial microbiology due to its easiness of genetic and physiological manipulations. It is used for the production of biomass, chemicals, ethanol as well as recombinant proteins, pharmaceutical agents and vaccines (1)(2)(3). Furthermore, the expression properties observed at the laboratory scale can be extrapolated to industrial scale.

The expression of heterologous enzymes in S. cerevisiae allows to modify its metabolic pathway, allowing the development of new strains utilizing new substrata or the production of new metabolites. In our laboratory we have developed metabolically engineered budding yeast cells for the production of biomass/ethanol (4)(5) and fructose 1,6 diphosphate (6) from waste materials such as lactose and whey. at low pH values, thus avoiding the production of lactate(s).

In this paper we describe the

E. Galindo and O.T. Ramírez (eds.), Advances in Bioprocess Engineering. 417-423.

© 1994 Kluwer Academic Publishers. Printed in the Netherlands.

expression of a bovine muscle lactic dehydrogenase gene (LDH-A) and the utilization of the recombinant yeast strains for the production of lactic acid. Since yeast cells are quite resistant to low pH values, no addition of bases is required and a satisfactory productivity of lactic acid was observed. Biotechnological perspectives are also discussed.

MATERIALS AND METHODS

Yeast Strains and Growth Conditions. S. cerevisiae GRF18 (MATa, leu2, his 3)(4), X4004-3A (MATa, lys5, trp1, met2, ura3) (4), and YSH 5.127-17C (Δ pdc1, Δ pdc5, Δ pdc6) (kindly provided by S. Hohmann)(10) strains were used in this work.

Batch cultures were run by shaking at 30 °C in rich (2% w v^{-1} peptone, 1% w v^{-1} yeast extract) or in minimal media containing 0.67% w v^{-1} Difco Yeast Nitrogen Base (YNB) without amino acids, supplemented with 50 μg ml^{-1} of the required supplements. Glucose (GLU) or Galactose (GAL) were used as carbon source.

Fed-batch fermentations were performed in a 2 l aerated stirred tank bioreactor equipped with temperature, dissolved oxygen, ethanol and pH controllers (11). During the fed-batch process the fermentations were aerated with a flow of 2 l min^{-1}. Foam was suppressed by occasional additions of drops of diluted and sterilized antifoam [poly(propylene glycol) 2000]. Temperature was fixed at 30 °C. Composition of the mineral medium and control of the bioprocesses were carried out as previously described (11).

Plasmids. Standard DNA manipulations were performed according to Maniatis et al. (12).

Plasmid pLAT1 was obtained by subcloning the bovine LDH-A cDNA (13)(about 1750bp)(kindly provided by N. Ishiguro) into pEMBLyex4 plasmid, downstream the UAS-GAL/CYC1 promoter sequences. pEMBLyex4 is a derivative of pEMBLyex2 (14),in which the XbaI site in the 2 μ region was eliminated. The obtained expression cassette UASGAL/CYC1-cDNA(LDH-A) does not result to be repressed by glucose and induced by galactose containing media, but high and comparable levels of expression have been observed using both the carbon sources. Such lack of transcriptional control has been ascribed to the long sequence (150bp) upstream to the first ATG codon.

Determination of cell number. Small samples of the cultures were sonicated and, after appropriate dilution with Isoton (Coulter Electronics, Harpenden, England), counted with a "Coulter Counter ZBI" equipped with a 100 μm orifice (15).

Growth rate was inferred by plotting cell number against time on a semi-logaritmic scale.

Determinations of Glucose, Ethanol, Galactose, Lactic acid and Lactic Dehydrogenase Activity. Glucose and ethanol determinations were acted as previously described (11).

Galactose and lactic acid were determined using Boehringer Kits (N.139084)(N.176303). LDH activity was determined using SIGMA Kit (DG1340-K), while total cell proteins determinations were carried out using a BIORAD Kit (N.500-0006).

RESULTS AND DISCUSSION.

Expression of LDH-A gene and lactic acid synthesis.

Cloning of a muscle bovine Lactate Dehydrogenase gene (LDH-A)(13) into S. cerevisiae cells (GRF18[pLAT1]) allows the introduction of a new and alternative pathway leading to the synthesis of lactic acid (Figure 1).

In batch cultures, production of both the heterologous enzyme and the new metabolite are not related to cellular growth. Absence of the carbon source (stationary phase of growth) seems to be related to the highest production of heterologous LDH-A enzyme (1.5 mU mg^{-1} of total cell proteins), while accumulation of lactic acid in the growth medium (up to 1.2 g l^{-1}) was obtained during the late exponential phase of growth.

Since yeast cells can efficiently grow and survive at low pH values (during the production of lactic acid the pH value dropped to 2.8; data not shown), production of lactate(s) can be avoided.

The growth condition reported in Figure 1 is not completely satisfactory because of the low lactic acid productivity (80 mg l^{-1} h^{-1}). Furthermore, high amounts of ethanol (up to 6 g l^{-1}; data not shown) are produced and this lowers the yield of lactic acid (10-15%; grams of lactic acid produced per grams of carbon source used).
Similar results have been obtained using host strain X4004 (data not shown).

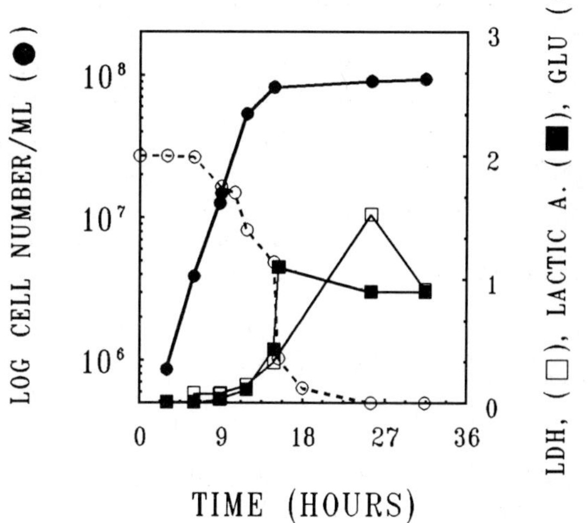

Figure 1. Lactic acid production during batch growth on YNB-GLU based medium of transformed GRF18[pLAT1] yeast cells.
Lactic dehydrogenase (LDH), mU mg^{-1} of total cell proteins; Glucose (GLU), g 100 ml^{-1}; Lactic Acid, g l^{-1}.

Production/Productivity Improvement.

Higher amount of lactic acid and an increased productivity were obtained by a modulation of the conditional operations of the bioprocess using a two-stage batch and fed-batch fermentations. Transformed GRF18[pLAT1] yeast cells were initially grown as shown in Figure 1; in stationary phase of growth (i.e., corresponding to the highest amount of heterologous LDH-A activity), additional glucose (60 g l^{-1}, final concentration) was added to the flask and lactic acid production was carried out without aeration and under a reduced agitation, Figure 2. Such condition allows a high production (12.5 g l^{-1}), with a higher yield

(30-40%), but still with an unsatisfactory productivity (285 mg l^{-1} h^{-1}). In fact, considering the duration and consequently the costs of an industrial fermentation, productivity probably represents a more relevant parameter than the absolute production.

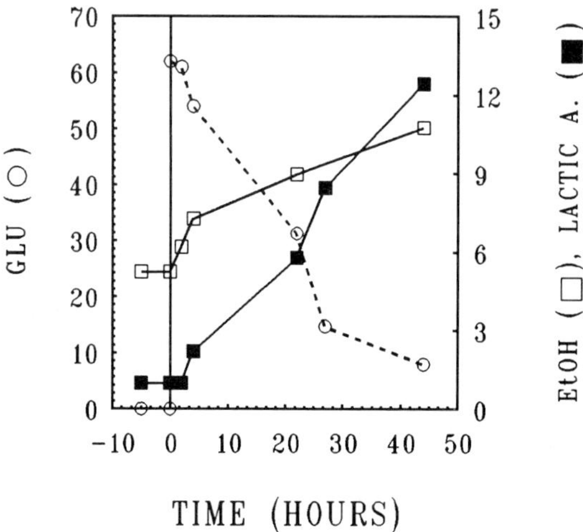

TIME (HOURS)

Figure 2. GRF18[pLAT1] yeast cells were pre-grown as shown in Figure 1. At time T=0, glucose was added (60 g l^{-1}, the initial concentration) and lactic acid production, ethanol and glucose levels measured. Glucose (GLU), g l^{-1}; Ethanol (EtOH), g l^{-1}; Lactic Acid, g l^{-1}.

A similar strategy was developed for transformed GRF18[pLAT1] yeast cells pre-grown to higher cell density by means of a computer controlled fed-batch fermentation, Figure 3. In this case, the production was as high as that shown in Figure 2, but with a more interesting productivity (till 11 g l^{-1} h^{-1} in the first hour of the process). Further addition of glucose allowed to increase the lactic acid

production up to 20 g l^{-1} (data not shown). Similar results have been obtained using the X4004 strain transformed with pLAT1 (data not shown).

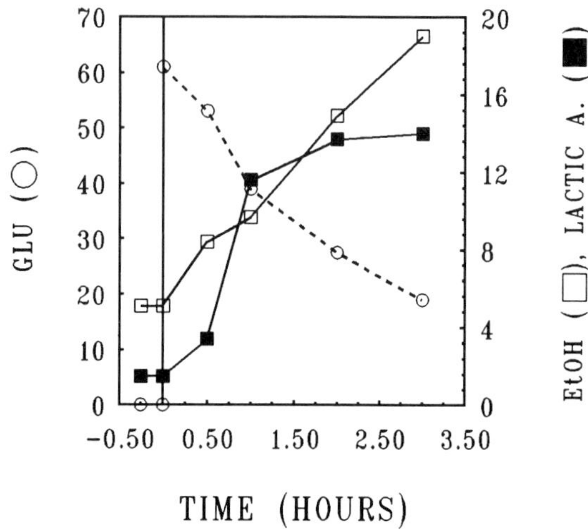

TIME (HOURS)

Figure 3. GRF18[pLAT1] yeast cells were pre-grown to high cell density (3.2x10^9 cells ml^{-1}) in computer controlled fed-batch culture. During this first phase, ethanol has been used as parameter controlling the addition of fresh mineral medium (11).
At time T=0, glucose was added (60 g l^{-1}, the initial concentration) and lactic acid production, ethanol and glucose levels measured. Glucose (GLU), g l^{-1}; Ethanol (EtOH), g l^{-1}; Lactic Acid, g l^{-1}.

Yield Improvement.

The production of ethanol could be limited by reducing the activity of pyruvate decarboxylase.
It has been previously reported that deletion of PDC1, PDC5 and PDC6 genes

results in very poor growth of yeast cells on glucose with a strong reduction of the overall PDC activity (10). The plasmid pLAT1 was used to transform YSH 5.127-17 strain (a pdc1, pdc5, pdc6 deleted strain); interestingly, transformed cells can be isolated simply through a selection for the ability to quickly grow on glucose. In fact, as a consequence of the cloned LDH activity a resumed fast growth rate (Td= 2.2 hours against 28 hours of the untransformed cells) was observed.

However, a very low activity of LDH and a related production of lactic acid were observed (data not shown).

A further and successfully attempt to improve the total yield of the process was based on the modulation/control of the metabolic flux rate. In fact, the Km of the LDH-A enzymes is 2×10^{-4} M (16) compared with 5×10^{-3} M for the yeast pyruvate decarboxylase endogenous enzyme (17). Thus, a low concentration of endogenous pyruvate should result in the production of lactic acid, limiting the production of ethanol. Theoretically this attempt could easily be performed by modulating the production of pyruvate (for instance, controlling the growth rate of the population by means of chemostat cultures as well as using nutrients supporting different glycolytic flow rates). For this purpose, transformed GRF18[pLAT1] yeast cells were grown in mixed glucose (0.3 % w v^{-1}) + galactose (2% w v^{-1}) media, Figure 4. This growth condition allowed, in the first phase, a fast production of biomass and ethanol (i.e., growth on glucose), and then a subsequent bioconversion of galactose in lactic acid with a yield as high as 70-80%.

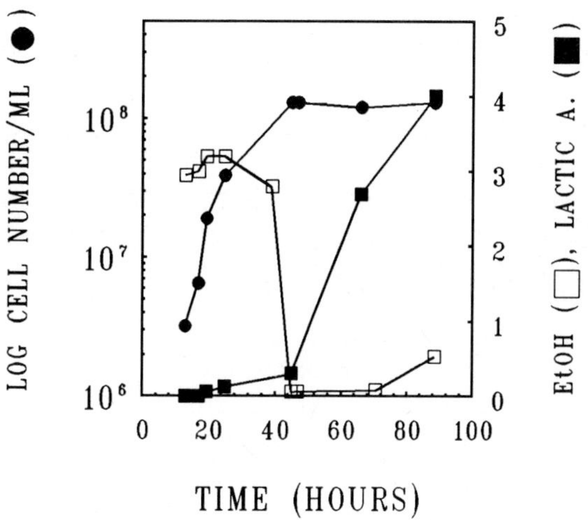

Figure 4. Lactic acid production during batch growth on YNB-based medium of transformed GRF18[pLAT1] yeast cells. Transformed cells were grown on a mixed GLU (0.3% w v^{-1}) + GAL (2% w v^{-1}). After depletion of glucose (20 hours of fermentation), cells began to metabolize galactose at a significant lower rate, resulting in production of lactic acid.

The behaviour described seems to be dependent upon the host strain employed. In fact, a higher growth rate on galactose and a lower yield have been obtained for X4004[pLAT1] transformed cells (data not shown).

Similar results have been achieved with GRF18[pLAT3D] transformed cells. Using this plasmid the expression of the heterologous gene was completely repressed by glucose and fully induced by galactose containing media (data not shown).

CONCLUSIONS.

In the last several years, our

laboratory has been engaged in developing systems for the production of heterologous proteins in yeast. These include both bacterial, plant and mammalian proteins (reviewed in Alberghina et al.(18)). Furthermore, rDNA technologies also allowed the transfer of new metabolic pathways interacting with the existing metabolic network. Such transfers have been utilized to obtain high production of biomass and ethanol from lactose/whey (the main wastes of the agro-alimentary industries) (4)(5) or for the production of lactic acid, as reported in this paper. The strategies described can not compete with the current productions of lactic acid from bacteria cells. For example, advanced fermentation techniques allow lactate(s) productions of 100-150 grams/liter (19) or productivity of 60-80 g l^{-1} hr^{-1} (20).
However, productions, productivities and yields obtained from metabolically engineered S. cerevisiae are sufficiently high and reproducible to find, in perspectives, interesting applications. In fact, the production of lactic acid from an heterologous host cells could solve the problems related to the production from bacteria cells. Such problems regard (i) inhibitory effects caused from the produced acid, (ii) strategies to prevent lowering of pH and (iii) purification procedures.

ACKNOWLEDGMENT.

The authors thank Prof.N. Ishiguro and Dr. S. Hohmann for providing LDH-A gene and YSH 5.127-17C yeast strain respectively.
Research supported by National Research Council of Italy, Special Project RAISA, sub-project 4, Paper N. 1451.

LITERATURE CITED.

1. Fiechter, A., G.F. Fuhrmann and O. Kappeli, *Adv. Microb. Physiol.*, **22**, 2031 (1981).

2. Buckholz, R.G., *Current Opionion in Biotechnology*, **5**, 538 (1993).

3. Ramonos, M.A., C.A. Scorer and J.J. Clare, *Yeast*, **8**, 423 (1992).

4. Porro, D., E. Martegani, B.M. Ranzi and L. Alberghina, *Biotechnol. Bioeng.*, **39**, 799 (1992).

5. Porro, D., E. Martegani, B.M. Ranzi and L. Alberghina, *Biotechnol. Letters*, **14**, 1085 (1992).

6. Compagno, C., A. Tura, B.M. Ranzi and E. Martegani, *Biotechnol. Bioeng.*, **42**, 398 (1993).

7. Hongo, M., Y. Nomura, and M. Iwahara, *Appl. Environ. Microbiol.*, **52**, 314 (1986).

8. Benninga, H. A history of lactic acid making, Kluvwer Academic Publishers (Dordrecht/Boston/ London) (1990).

9. Buchta, K., "Lactic acid", in Biotechnology, vol.3, p.409, Dellweg H. (Ed.), Verlag Chemie, Weinheim, Federal Republic of Germany (1983).

10.Hohmann, S., *J. Bacteriol.* **173**, 7963 (1991).

11. Porro, D., E. Martegani, A. Tura and B.M. Ranzi, *Res. Microbiol.*, **142**, 535 (1991).

12. Maniatis, T., E.F. Fritsch and J. Sambrook, (1982) <u>Molecular Cloning</u>. A Laboratory Manual. Cold Spring Harbor Laboratory, Cold Spring Harbor, New York (1982).

13. Ishiguro, N., S. Osame, R. Kagiya, S. Ichijo and M. Shinagawa, Gene, **91**, 281 (1990).

14. Baldari, C., J.A.H. Murray, P. Ghiara, G. Cesareni and C. L. Galeotti, *EMBO J.*, **6**, 229 (1987).

15. Alberghina, L., B.M. Ranzi, D. Porro and E. Martegani, *Biotechnol. Prog.*, **7**, 299 (1991).

16. Holbrook, J.J., A. Liljas, S. J. Seindel and M.G. Rossmann, "Lactate dehydrogenase," in <u>The Enzymes</u>, vol. 11, p. 191, Boyer, P.D. (Ed.), Academic Press, New York (1975).

17. Hohmann, S. and H. Cedeberg, *Eur. J. Biochem.*, **188**, 615 (1990).

18. Alberghina, L., M. Lotti, E. Martegani, B.M. Ranzi and D. Porro, Med. Fac. Landbouww. Rijksuniv. Gent 1993, in press, (1993).

19. Cheng, P., R.E. Mueller, S. Jaeger, R. Bajpai and E.L., Iannotti, *J. Ind. Microbiol.*, **7**, 27, (1991).

20. Mehaia, M.A. and M. Cheryan, *Enzyme Microb. Technol.*, **8**, 289 (1986).

Some Aspects of *Gibberella fujikuroi* Culture Concerning Gibberellic Acid Production

P. C. González[1], G. Delgado[1], M. Antigua[1], J. Rodríguez[1], P. Larralde[2], G. Viniegra[2], L. Pozo[3], and M. del C. Pérez[3]

[1]Cuban Institute for Research on Sugar Cane Byproducts (ICIDCA), Vía Blanca 804, P.O. Box 4026, S.M. Padron C. Habana, CUBA. [2]UAM - Iztapalapa, México; D.F, MEXICO
[3]Instituto de Cítricos y Frutales, Playa C. Habana, CUBA

Gibberellic acid (GA₃) is a plant growth regulator produced by Gibberella fujikuroi *in submerged fermentation. Studies concerning the effects of several carbon and nitrogen sources on GA₃ production were carried out. Sucrose-starch mixture was found to be the best carbon source. No significant differences between the nitrogen sources tested were observed.Two methods were compared for estimating μ of* Gibberella fujikuroi *grown in different proportions of glucose and starch. The specific growth rate of* Gibberella fujikuroi *estimated by an image processing technique (measuring tip extension and mean hyphal length in a Petri dish) was in close agreement with measurements performed in a stirred fermentor. In addition, submerged culture was found to be better than solid-state culture for the strain under study. 1 vvm and 700 rpm were found to be the conditions leading to the highest GA₃ production. Application of cell-free, HPLC characterized fermented broth in cuban "Tangor Ortanique" orchard resulted in a yield increase.*

Gibberellins, especially Gibberellic Acid (GA$_3$), constitute undoubtely the most important example of plant growth regulator produced from microbial origin, not only because of its world production volumes but also of its extensive practical uses in agriculture.

For gibberellin production, the fungi *Gibberella fujikuroi* (the perfect, sexual stage of *Fusarium moniliforme*), continues to be the most productive systems because non of the other microorganism were able to produce GA$_3$ and GA-like substances at commercially feasible levels.

In addition to strain selection and improvement of wild type strains, medium development, appropriate cultivation techniques, aeration-agitation scale up procedure and down stream processing operations are very important prerequisites for economical success in GA$_3$ production. As a result, former studies are required to make clear controversial points concerning fermentation working condition.

In this paper some studies for culture medium formulation in connection with carbon and nitrogen sources, oxygen transfer, specific growth rate (μ) determination using image processing thechnique and comparison between submerged and solid state cultures are presented. Finnally, yield increase resulted for the application of cell-free fermented broth in cuban orange plantation, are also offered.

MATERIALS AND METHODS

Microorganism: *Gibberella fujikuroi* IMI 58289 from the Commonwealth Mycological Institute, U.K, supplied as a gift by Dr. E. Cerda Olmedo, Univ.de Sevilla Spain. The strain were stored on agar Czapek slants at 4°C.

Culture techniques: Old cultures were reactivated on new agar-Czapek slants and grown for seven days at 29°C. Microconidia were washedout from these slants using sterile water and then used as inoculum for liquid Czapek medium diluted tenfold (v/v) and cultured in shake flasks with 250 rpm for 48 hr at 29°C. This mycelia suspension was used afterward as inoculum for fermenters, Petri dishes, shake flasks for carbon and nitrogen sources studies, and solid state fermentation with bagasse impregnated with culture medium.

Liquid media composition: The culture media were used with the following basal salt composition in g/L: KH$_2$PO$_4$, 5.0; MgSO$_4$.7H$_2$O, 1.0; solution with traces of Fe, Cu, Zn, Mn and MO, 2 ml/L. Carbon concentration 30.27 g/L and nitrogen concentration 0.44 g/L

425

E. Galindo and O.T. Ramírez (eds.), Advances in Bioprocess Engineering. 425-430.
© *1994 Kluwer Academic Publishers. Printed in the Netherlands.*

previously defined by Delgado et al (1). Initial pH value was adjusted to 3.0. For Image Processing studies, percentages of (glucose)/(starch + glucose) were established at: 25, 37.5, 50, 62.5, 75, 87.5 and 100%, with constant values of total carbohydrates (80 g/L).

Aeration-agitation studies: All experiments were carried out in 5 L fully instrumented BRAUN fermenters, provided with a polarographic electrode, for measuring of dissolved oxygen concentration (D.O %). Volumetric oxygen demand (rx= mg O_2 /L.s) and volumetric transfer coefficient (K_La = 1/s), were determined using the dinamic method reported by Taguchi and Humphrey (2), as a modification of the gassing-out technique.

Solid state fermentation with bagasse: Bagasse was impregnated with liquid medium to obtain a moisture content ·near to 70% after inoculation.
Cultures were carried out at pH 3.0 under non-aseptic conditions folowing the basic procedure previously described by Raimbault et al (3). Samples for biochemical assay were suspended in distilled water, homogenized and centrifuged. Reducing sugars and gibberellic acid were assayed in the supernatant.

Specific growth rate estimation on Petri dishes: Circular agar plates, having 10 cm diameter and 0.5 cm depth, were prepared with previous liquid media mixed with 1.5% agar. A small (0.5 cm diameter) round filter was placed in the dish center to assure the absorption of 10 μL of inocula from the shaker flasks. Solid cultures were grown by five-fold replication for 7 days at 29°C.
Each colony diameter was measured in five different directions every 24 hours. Daily average diameter was plotted vs. time and linear slopes were estimated by statistical regression (r=0.99). This parameter was called Vr and expressed in microns/h. After seven days, the colonies were observed with a Zeiss microscope (40x) provided with a video camera linked to an image processing system . The assembling consisted in a digitizing Matrox card and Biocom (France) software Imagenia 2000, running in an HP Vectra QS/20, AT, computer. Peripheral mycelial images were stored in the hard disk memory to be retrieved and processed later on. Retrieved images were cleaned up and improved by segmentation techniques. Distal hyphal lengths were measured on a VGH monitor using a "mouse" (electronic pantograph) previously calibrated with the grid of a red blood cell counting (Neubauer) chamber. Histograms for

n=30 measurements were recorded in order to evaluate the means and standard deviations. The mean hyphal length was called Le (microns); hyphal diameters were measured on same preparations. Specific growth rate was estimated as μ_1 in Eq.1: μ = [ln (2).Vr]/ [Lc.Ln (Lc/Lo)] where Lc= 2Le, Lo= 2D, D= hifal diameter, as described by C. P. Larralde et al (4).

Specific growth rate estimation in submerged cultures (μ_2): Triplicate measurements of suspended biomass in fermenters were made as follows, 30 mL broth samples were filtered through a previously weighed Whatman No 40 paper. Mycelial residues were oven dried at 60°C for 24 hr. Biomass concentration X or dry matter (D.M) were expressed as g/L. A semilog plot was made of average X values and the slope was estimated in the linear segment by a conventional regression program taking into account only the line segments with r>0.9. This slope was called μ_2.

General analytical methods: Filtrate after separation of biomass, was utilized for estimation of Reducing Sugars (R.S) by dinitrosalicylic acid reagent, Miller (5); nitrogen content by Kjeldahl method in a Automatic Kjeltec System (Tecator AB, Högänas, Sweden), Anon (6). Gibberellic acid determination was carried out using HPLC method, Barendse et al (7) in Phillips PU 4100 liquid chromatograph.

RESULTS AND DISCUSSION

Fermentation kinetic: Gibberellin production is a classic example of a secondary metabolite fermentation. This was corroborated by us in several trials carried out in 5 L fermentor by using *G. fujikuroi* IMI 58289 strain. In such trials exponential growth ceases when assimilable nitrogenous nutrient are exhausted from the medium. The metabolism of *G. fujikuroi* then switches over from trophophase to idiophase, which is characterized by the onset of formation of gibberellin and storage compounds.

Borrow et al (8), defined producing and non-producing phases of the gibberellin fermentation process. Exponential growth was observed during the *balanced phase* , and the uptake of carbon, nitrogen and other nutrients remain constant. In the following storage phase, when nitrogen is exhausted, the dry weight increases due to the accumulation of lipids, carbohydrates and polyols, Bruckner et al (9). Gibberellin production as well as other secundary metabolites start in this phase. The following maintenance phase is the main gibberellin or gibberellic acid-producing phase.

Nutrient media composition: For nutrient media optimization , the knowledge of biosynthesis regulation is main condition. A criterion for medium composition and other elements, is a fast-production high-concentration of gibberellins. The C:N ratios used in the production of GA_3 in the single-stage technique are calculated based on the total carbon and nitrogen sources and ranged from 6:1 to 188:1, Kumar (10). In a previous work, Delgado et al (1) an optimal C:N ratio for maximun level of GA_3 was found on 68.795 corresponding to 30.27 g/L of C and 0.44 g/L of N.

Different workers have used a wide variety of carbon sources for the production of GA_3 ; a distinction is made in slowly and readily utilizable carbon sources, Kumar et al (10) ; Borrow et al (11); Sanchez-Marroquín (12).

Sucrose, glucose, lactose, starch and starch / sucrose mixture as carbon sources, were evaluated by us in optimal carbon concentration. The influence of carbon source is offered in Table 1. Sucrose and starch lead to a high yield of GA_3, lactose significatively decreased GA_3 production by our *G.fujikuroi* strain. Nevertheless, a combination of two carbon sources with different uptake speed (starch/sucrose) yielded higher concentrations of GA_3 . In subsequent trials this mixture was used as carbon source.

Table 1. Effect of carbon and nitrogen sources on production of GA_3 by *G. fujikuroi*.

Carbon and nitrogen sources	Final concentration of GA_3 mg/L (7 days)
Glucose	40
Lactose	10
Sucrose	79
Starch	100
Sucrose/Starch	150
Ammonium sulfate	130
Urea	139
Ammonium nitrate	130
Yeast extract	135

A variety of organic and inorganic nitrogen were evaluated by different workers to study their effect on the production of GA_3, Kumar (10). It has been previously reported that ammonia is consumed rather than nitrate,

Harhash (13), despite no significant differences were found by us when ammonium sulfate, urea, ammonium nitrate and yeast extract were used in optimal nitrogen concentration. Those results are illustrated in Table 1. Subsequent experiments were conducted with ammonium nitrate.

Image processing technique for *G. fujikuroi* specific growth rate estimation: Evaluation of the specific growth rate (μ) for slow growing mycelia is a tediuos and time consuming operation. In addition, there is a risk of culture contamination and experimental errors during long periods of mycelial cultivation. Nevertheless, it seems to be important to measure μ value, to seek deeper into the physiology of some slow-growing organisms used for industrial fermentations, such as *G. fujikuroi*. An alternative way to evaluate μ is to use morphometric data from vegetative mycelia using image processing techniques of colonies of *G. fujikuroi* grown in Petri dishes, Viniegra et al (14).
Two methods were compared by us for estimating the μ of *G. fujikuroi*, grown on diferent proportion of glucose and starch. They were, μ_1 in Petri dish, using Image processing technique, and μ_2 in stirred fermenter by traditional method. Values of μ_1 and μ_2 were found in close agreement in the range of 0.04 and 0.09 h^{-1} (Fig. 1).

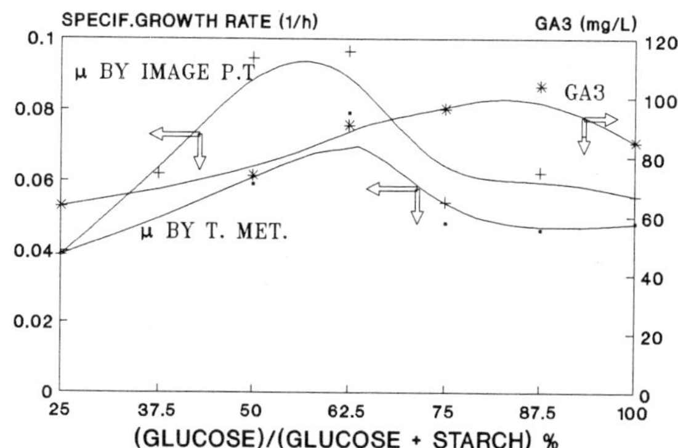

FIG.1: BEHAVIOR OF μ OF G.FUJIKUROI MEASURED BY TRADITIONAL METHOD AND BY IMAGE PROCESSING TECHIQUE

For μ_1 calculation Vr, Le and D values, were used. Only slight variation of Vr and D were observed for several glucose:starch ratio. In contrast significant diferences amonng Le values were found which provoke strong variation of μ_1 values. Apparently, *G. fujikuroi* has an adaptation mechanism to conserve Vr and D when growing in media with various proportions of glucose and starch. It should be pointed out the large statistical errors of Le distribution in each sample; but given the sample size (n=30) the confidence range of average values was less than 15%.

These trials made possible to define final carbon source composition of culture media. Although no coincidence was found between mayor GA_3 yield and the highest μ values as observed in Fig. 1.

Comparison between Solid State Fermentation (SSF) and Submerged Fermentation (Subm. F.) techniques: Some workers compared the production of GA_3 by SSF and Subm. F., Kumar et al (15). Based on equivalent carbohydrate contents, they found that in the case of solid medium (wheat bran) the accumulation of GA_3 was 1.6 times higher.

To compare SSF and Subm. F. for GA_3 production, several trials were carried out; Subm. F. is found to be better than bagasse-supported SSF, in the same culture media for our *G. fujikuroi* strain as shown in Fig. 2; in the other hand they yielded a completely different fermentation kinetics. Taking into account the economical advantages of SSF, it is necessary to carry out additional experiments with other supports.

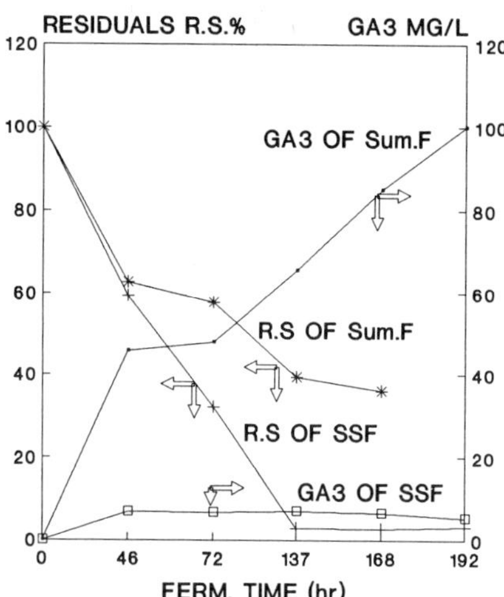

FIG.2: FERMENTATION KINETICS OF
G.fujikuroi BY SOLID-STATE AND
SUBMERGED TECHNIQUES.

Aeration-agitation: Though the effect of aeration-agitation is pronounced in the production of GA_3 in submerged fermentation, no kinetic studies directly related to the rate of production of GA_3 are available,

Jefferys (16). It is well known that a continous supply of oxygen is required for the production of GA_3, as the biosynthesis progresses through compounds with an increasingly level of oxidation. Differents aeration-agitation rates have been reported by other authors, Kumar et al (10).

To ellucidate the influence of oxygen transfer rate on fungae behavior, .different experiments were carried out in 5 L fermentors. Fig.3 illustrates dissolved oxygen profiles and GA_3 production for *G.fujikuroi* fermentation with different aeration strategies from 1 to 0.6 vvm, at a constant stirring speed 500 rpm. Dissolved oxygen levels, during the peak-demand period (40-50 h of fermentation time), dropped to the range 0-20% for both cases, nevertheless for 1 vvm D.O% dropped less than to 0.6 vvm, were as GA_3 production is about 1.3 times higher at the end of fermentation. Dropping of D.O % levels as low as close to cero values, which happens for 0.6 vvm, could have adverse effects on gibberellins-forming enzyme complex synthesis which onset just durind oxygen peak-demand period (40-50 h of fermentation time). Aeration of 1 vvm was used in a further experiment to compare different agitation strategies.

FIG.3: DISSOLVED OXYGEN PROFILES AND GA3
PRODUCTION FOR G.fujikuroi FERMENTATION
WITH DIFFERENT AERATION STRATEGIES

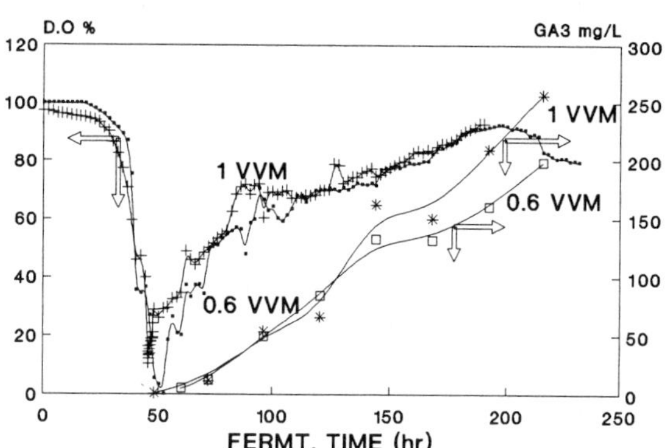

Is very well knowed that $K_L a$ is more influenced by the stirrer sped than the volumetric gas flow, this is particulary evident in the case of fermentation employing filamentous microorganisms, Alan Wiseman ed (17), what was corroborated in following results.

Fig.4 presents dissolved oxygen profiles and GA_3 production for *G.fujikuroi* fermentation with different agitation strategies (700 and 500 rpm) at 1vvm. The combination of 700 rpm/1vvm lead to higher levels of dissolved oxygen during the peak demand (70-80%) and concomitantly to an increase of GA_3 production. This combination was furtherly employed for process scale-up to 500 L fermentor, taking a constant impeller tip speed as scale-up criterion, a similar GA_3 concentration was achieved ranging from 250 to 300 mg/L.

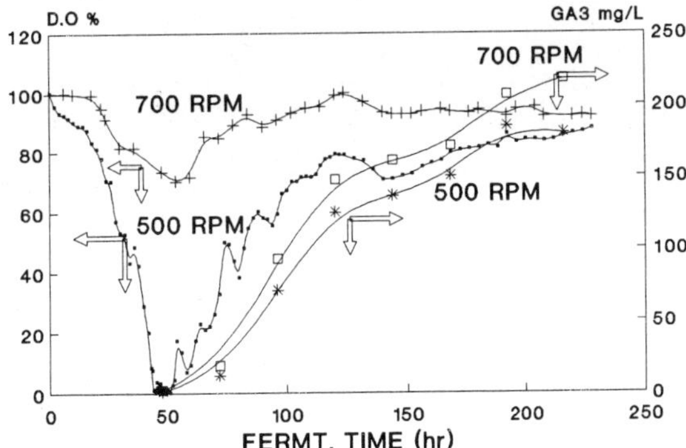

FIG.4: DISSOLVED OXYGEN PROFILES AND GA3 PRODUCTION FOR G.fujikuroi FERMENTATION WITH DIFFERENT AGITATION STRATEGIES

The behavior of rx and K_La (Fig. 5) showed that the maximum rx occur between 48 and 72 h when the fungae presented its higher biomass formation rate. When biomass reached its maximum concentration, the K_La values shows a tendency to decrease which was related to visual changes on reological properties of the system, although no broth viscosity measurements were made. Some authors reported that for filamentous fungi fermentation, as mycelial concentration increases, fluid viscosity likewise increases and aeration efficiency will therefore decrease, Alan Wiseman ed (17).

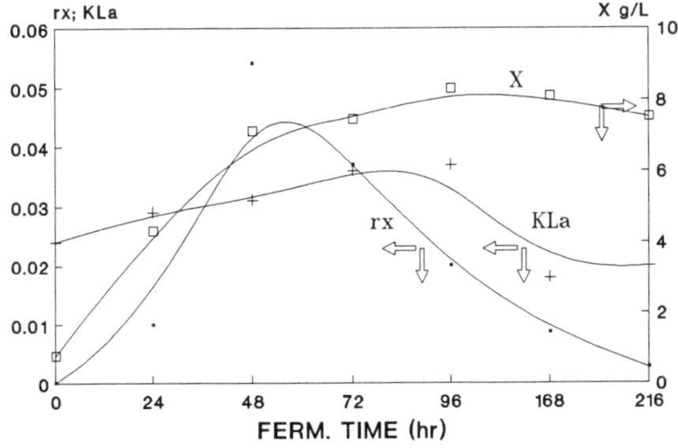

FIG.5: BEHAVIOR OF rx, KLa AND BIOMASS DURING G.fujikuroi FERMENTATION OT 700 RPM , 1 VVM

<u>Application of crude fermented broth in orange plantations</u>: In the US and Mediterranean countries, including Israel, GA_3 is commonly used in citrus cultivars in a large extent aiming for the increase of fruit size, number of fruits, etc.

The results of GA_3 application in cuban *"Tangor ortanique"* orchard by foliar spraying are summarized in Fig.6. It can be concluded that crude fermented broth produced in pilot plant 500 L fermentor, resulted to be in a final yield increase of 10 ton of fruit/ha, respect to the control. Moreover, yields were higher than those obtained with the foliar application of commercial russian GA_3 in the same concentration. This result may be explained by the presence, in crude fermented broth, of others plant regulator as kinetin previously detected by thin layer chromathography.

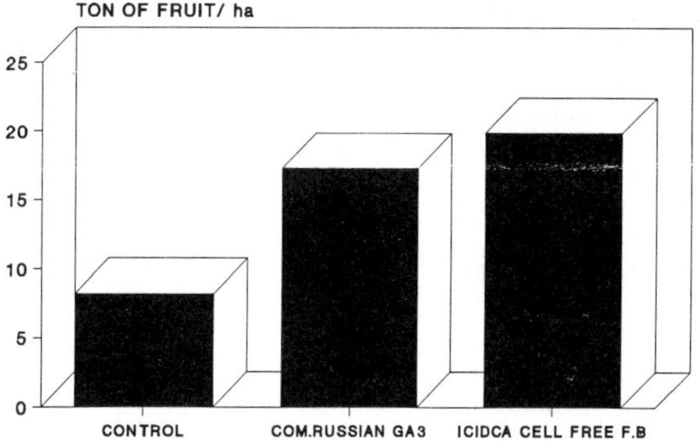

FIG.6: EFFECT OF FOLIAR GA3 APPLICATION ON FINNALLY YIELD OF CUBAN T. ORTANIQUE ORCHARD

<u>CONCLUSIONS</u>

The results from the present paper led to the optimization of carbon and nitrogen sources for gibberellic acid production in submerged fermentation. On the other hand, experimental assays permited the elaboration of primary criteria for process scaling up.

In addition the technical feasibility of image processing for specific growth rate determination was demonstrated. It seems to be convenient the development of a gibberellic acid determination technique in Petri dishes.

For an accurate conclusions it will be necessary to conduct further series of experiments in solid state fermentation.

430

NOMENCLATURE

Vr: Rate of tip extension (microns/h)
Le: Mean of hyphal length (microns)
μ_1: Specific growth rate by Image Processing Technique (1/h)
μ_2: Specific growth rate by Traditional Method (Submerged fermentation) (1/h)
Lc= 2Le : A critical hyphal length after which all leading hyphal branch out (microns)
L_0= 2D: Initial length of new branch (microns)
D: Hiphal diameter (microns)
rx: Volumetric oxygen demand (mg O_2/L.s)
K_La: Volumetric transfer coefficient (1/s).

REFERENCES

1. Delgado, G. González, P.and Rodriguez, J.A. Proc. III International Seminar on Sugar Cane By-Products, Havana, May 18-21, pp 128 (1993).

2. Taguchi, H. and Humphrey, A.E. J. Ferment. Technol. (Japan) 44, 881 (1966).

3.Raimbault, M. and Alazard, D. Eur. J. Appl. Microbiol. 9, 199-209 (1980).

4. C.P. Larralde; P.C. González-Blanco and G. Viniegra-González. Biotechnology Techniques (In press) (1994).

5. Miller, G.L. Anal. Chem. 31, 426-428 (1959).

6. Anon. Tecator A.B., Application Note 30/81 (1981).

7. Barendse, G.W.M. and Van De Werke; P.H. J. Chromatogr. 198, 449-455 (1980).

8. Borrow, A,; Brown, S.; Jefferys, E.G.; Kessel, R.H.J. and Swait, H.N. Can. J. Microbiol. 10, 407 (1964).

9. Bruckner, B. and Blechsmidt, D. Crit. Rev. Biotechnol. 11(2), 163-192 (1991).

10. Kumar, P.K.R. and Lonsane, B.K. Adv. Appl. microbiol. 34, 29-139 (1989).

11. Borrow, A.; Jefferys, E.G. and Nixon, I.S. U.S Patent 2, 906, 671 (1959b).

12. Sanchez-Marroquín, A. Appl. Microbiol. 11, 523-528 (1963).

13. Harhash, A.W. Acta Biol. Med. Ger. 17, 8-16 (1966).

14. Viniegra, G.; Saucedo, G.; López-Isunza, F. and Favela, E. IX International Biotechnology Symp. Abstract No. 339, Crystal City, USA (1992).

15. Kumar, P.K.R. and Lonsane, B.K. Biotechnol. Letters 9, 179 (1987).

16. Jefferys, E. G. Adv. Appl. Microbiol. 13, 283-323 (1970).

17. Alan Wiseman ed. Topic in Enz. and Ferm. Tech. Ellis Horwould Ltd, 223-266 (1979).

Continuous Culture to Produce Recombinant β–Galactosidase in *Bacillus subtilis*

C.A. Rincón, R. Quintero, and M. Salvador

Departamento de Bioingeniería, Instituto de Biotecnología, UNAM, Apdo. Post. 510-3,
Cuernavaca, Mor., 62271, MEXICO

A chromosomal-recombinant bacterial strain, Bacillus subtilis *BIBT10, was used to study the β-galactosidase production in continuous culture. The integrated genes consisted of the* Escherichia coli *lacZ gene controlled by the aprE promoter and the* Staphylococcus aureus *chloramphenicol resistance gene. The specific rate of β-galactosidase production was maximal at a specific growth rate of approximately 0.5 h⁻¹. A dilution rate of 0.4 h⁻¹ (in which less than 2% spores were found) was chosen to carry out extended production experiments. After 50 mean generation times, β-galactosidase production remained constant, indicating the high stability of this system. The β-galactosidase productivity in continuous culture, at a dilution rate of 0.4 h⁻¹, was 3.2 fold higher than that of batch cultures.*

Bacillus subtilis has been widely used for production of industrial enzymes. When compared with other bacteria, *B. subtilis* offers many potential advantages in the expression of foreign gene products. Mainly, it is non-pathogenic and non-toxigenic, its molecular genetics is well known, and there is experience in large-scale production. Although plasmids are likely to exhibit segregational and structural instability, recombinant gene expression has traditionally used self-replicating vectors. Alternatively, integration of recombinant genes in the resident chromosome has demonstrated to be a feasible technique in foreign protein production (1). In general, there is agreement that such sequences are stably maintained in the absence of selection (2), regardless of some cases where instability has been reported (3).

The existing reports of strains carrying foreign genes integrated in chromosome, where increased production of the enzyme of interest and/or the genetic construction stability are studied, include only plate and shaken flask experiments (1). The knowledge that has resulted from these studies is of the utmost importance and usefulness for scaling-up and industrial production. However, as experienced with recombinant plasmids, enzyme production and genetic stability can be affected by several fermentation conditions. Therefore, it is meaningful to study the behavior of this kind of recombinant strains in fermentors, in order to have more convincing and stronger data to support the scaling-up and large-scale production of such processes.

In our laboratory, we have been working with chromosomal-recombinant bacterial strains. In this paper we present the results obtained when one of such strains was grown in continuous culture. Some exploratory analyses of operating conditions for suitable production purposes were studied.

MATERIALS AND METHODS

Microorganism. The bacterial strain used was the previously reported *Bacillus subtilis* BIBT10 [Cm^R, *lacZ⁺*,

431

E. Galindo and O.T. Ramírez (eds.), Advances in Bioprocess Engineering. 431-435.
© *1994 Kluwer Academic Publishers. Printed in the Netherlands.*

degU32 (Hy)] (<u>1</u>). Upon transformation of *B. subtilis* by plasmid pAprlac2, not able to replicate in this host, the whole plasmid (carrying chloramphenicol resistance) was inserted in the resident chromosome presumably by a single recombination event between homologous, chromosomal and plasmid-borne subtilisin promoter sequences. The construction consisted in a gene fusion of the promoter sequence, the ribosome binding site, and twenty-four base pairs of the subtilisin structural *aprE* gene with *Escherichia coli lacZ* gene, followed by *Staphylococcus aureus* chloramphenicol acetyltransferase gene. The later has conserved its original promoter. After insertion, the incoming recombinant genes are flanked by *apr* promoter direct repeats (<u>1</u>).

Medium. The culture medium had the following composition: 0.8% (w/v) Nutrient Broth (Difco, Michigan), 13 mM KCl, 500 μM $MgSO_4 \cdot 7H_2O$, 1 μM $FeSO_4 \cdot 7H_2O$, 1 mM Na_2SO_4 and 1 μM $MnCl_2$. The medium was autoclaved at 15 lb/in^2 for 20 min. For solid media 1.4% (w/v) agar was added.

Growth conditions. Cultures were performed in a 2-liter BioFlo Model C30 fermentor (New Brunswick Scientific, New Jersey). Agitation was done by three 6-blade flat turbine impellers at 800 rpm with 3 baffle plates and aereation at 1 vvm. The pH was automatically maintained at 7.0 by 2 N H_3PO_4 solution. The temperature was controlled at 37 °C. A fresh loopful of a 14-hour culture was inoculated into a 500-mL flask with 100 mL of liquid medium, incubated for 10 h at 37 °C and 200 rpm of agitation, and used as inoculum for the fermentor. For continuous cultures the working volume was controlled at 1,470 mL. The medium feeding rate was controlled using a peristaltic pump and samples were withdrawn from cultures in steady state. Steady state was assumed when at least four residence times have passed and the optical density remained constant for two subsequent residence times. The cells were collected by centrifugation (11,750 *g*, 10 min, room temperature), washed once with 0.85% NaCl solution, and resuspended in the same solution immediately prior to analysis.

Analyses

<u>Protein determination</u>. Protein was determined according to the method of Lowry *et al.* (<u>4</u>).

<u>ß-Galactosidase</u>. Enzyme activity was assayed using a modification of the method of Miller (<u>5</u>). 10 μL of an appropriate dilution of the culture were added to 720 μL of Buffer Z. 10 μL of a 5% (w/v) lysozyme solution (Sigma, Missouri) were added and incubated 5 min at 37 °C to lysate cells. 100 μL of a 0.45% (w/v)o-nitrophenyl-ß-D-galacto-pyranoside (ONPG) solution (Sigma, Missouri) were added and the sample was incubated at 29 °C. The reaction was stopped with 150 μL of 1.2 M Na_2CO_3 and the absorbance was determined at 420 nm. One unit of ß-galactosidase activity was defined as the amount of enzyme required to hydrolyze 1 nmole of ONPG in 1 min at 29 °C.

<u>Spore staining</u>. Vegetative cells and spores were stained with the malachite green method (<u>6</u>) and counted in a Neubauer chamber.

RESULTS AND DISCUSSION

Effect of growth rate on ß-galactosidase production and spore formation in chemostats

As mentioned previously, the *lacZ* gene in *Bacillus subtilis* strain BIBT10 is controlled by the subtilisin promoter. It has been demonstrated that subtilisin production and spore formation in chemostat is higher at low specific growth rates (μ), and decreases when μ increases (<u>7</u>, <u>8</u>). Furthermore, it is generally accepted that production of extracellular enzymes by *Bacilli* is subject to catabolite repression (<u>9</u>). Thus, the ESP and ß-galactosidase (ß-gal) production (expressed both as enzyme activity per mg protein) in our strain can be expected to correlate with the specific growth rate in a negative way. The effect of μ on spore formation and vegetative cells, is presented in Figure 1. Such a behavior is consistent with that reported for a *B. subtilis* wild-type strain (8). As the specific growth rate increases, the spore population decreases. At a µ-value of 0.4 h^{-1} practically no spores were found (less than 2%).

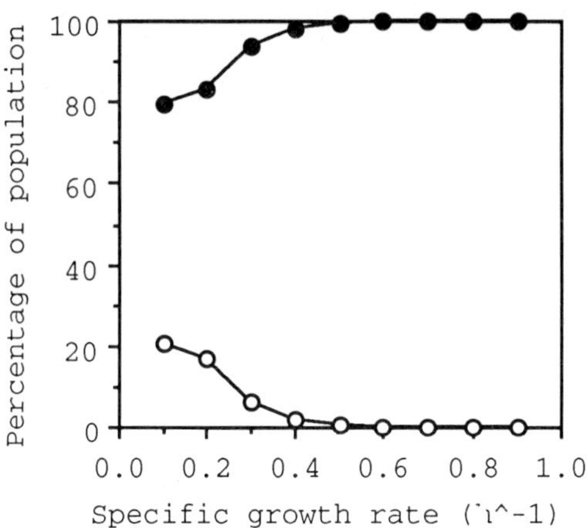

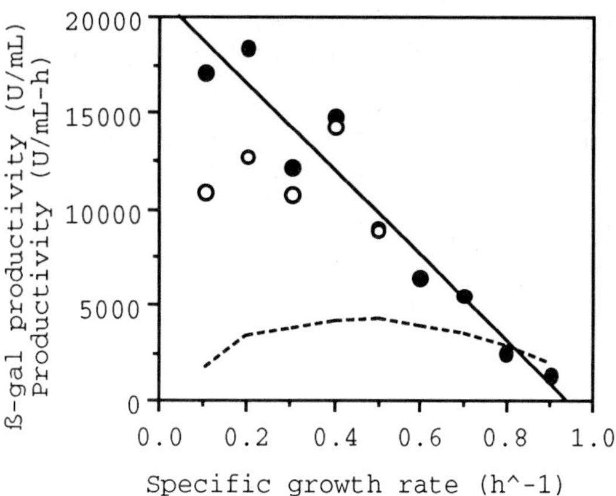

chemostat (7). When ß-gal productivity was obtained, it showed a maximum at a µ-value of approximately 0.5 h^{-1}. Above this growth rate, environmental and physiological conditions apparently repress ß-gal production.

FIGURE 1. Population distribution of vegetative cells (solid circles) and mature spores (empty circles) of *Bacillus subtilis* BIBT10 as a function of the specific growth rate in chemostat cultures.

ß-gal production was also affected by µ (Figure 2). The enzyme production showed a maximum at a specific growth rate of approximately 0.4 h^{-1}, and after that µ-value, it linearly decreased as µ increased. This indicates that there exists a growth rate where physiological conditions permit a maximal ß-gal production in continuous culture.

It has been demonstrated that mature spores do not produce subtilisin (10). In our system, there exists different ratios of spores and vegetative cells at dilution rates below 0.5 h^{-1}. Therefore, ß-gal production was corrected considering these ratios, as shown also in Figure 2.

This relation is similar to that reported for subtilisin production with a strain of *B. licheniformis* in

FIGURE 2. ß-galactosidase production considering total population (empty circles) and vegetative cells (solid circles), and productivity (dashed line) as a function of the specific growth rate during growth of *Bacillus subtilis* BIBT10 in chemostat cultures.

ß-galactosidase production in extended continuous culture at a single specific growth rate

In the chromosomal insertion of BIBT10 strain, a copy of subtilisin promoter is found at either extremity of the array. Therefore, it exists the possibility that the inserted genes can be excised by homologous recombination, and after the excision, the outgoing genes cannot autonomously replicate in *B. subtilis*. In sequential batch cultures without chloramphenicol, this strain has demonstrated to stably maintain the heterologous genes (expressed as ß-gal activity) for at least 50 generations (1). In order to examine the stability of the recombinant protein production in continuous culture, chemostat fermentations were carried out at a single dilution rate without chloramphenicol. A dilution rate of

0.4 h⁻¹ was chosen to permit relatively short fermentation times, high ß-gal production and nearly no spores present (see Figure 2). After about 50 generations, ß-gal production remained constant (Table 1). At the end of the culture, samples were withdrawn, plated in solid media without chloramphenicol and tested for recombinant genes loss in chloramphenicol (5 mg/L) and X-gal (Boehringer, Mannheim GmbH, Germany). All the colonies examined were able to grow in antibiotic and to produce ß-gal. We can conclude that under these conditions, the recombinant genes are stably maintained in chromosome, as reflected by a sustained ß-gal production throughout the fermentation process.

TABLE 1. Extended ß-galactosidase production of *Bacillus subtilis* BIBT10 in chemostat cultures at a specific growth rate of 0.4 h⁻¹.

No. of generations	Biomass concentration (OD at 525)	ß-gal production (U/mL)
2	4.75	9915
35	4.84	10004
47	4.73	9926

Specific rate of ß-galactosidase production in batch and continuous culture

A fermentation process is generally evaluated by the conversion yield and the overall productivity. The productivity of batch and continuous fermentors can be compared for the production of cell mass or for the product itself. For cell mass and some metabolites, the faster the organism grows the more favorable is the continuous over a batch process. To compare ß-gal productivity (expressed as enzyme units per mL per h) in batch and continuous culture, batch cultures were carried out without chloramphenicol.

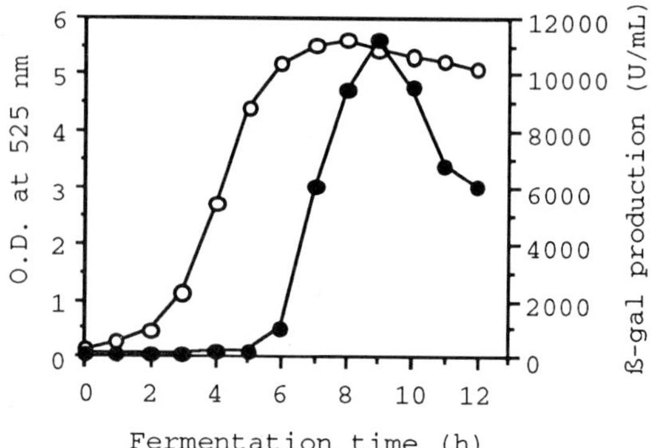

FIGURE 3. ß-galactosidase production (solid circles) and biomass concentration (empty circles) during batch culture of *Bacillus subtilis* BIBT10.

The ß-gal productivity was then compared with that of the chemostat cultures at 0.4 h⁻¹ of dilution rate. Figure 3 shows the typical profile of ß-gal production in batch culture. After exponential growth has concluded, a rapid increase in ß-gal production is observed. Such a behavior is a consequence of the onset of protease production. The production achieves a maximum of about 11,200 U per mL after 9 h fermentation time, corresponding to a ß-gal overall productivity of approximately 1,244 U/mL-h for the batch process. In continuous culture, the ß-gal productivity obtained at a µ-value of 0.4 h⁻¹ was 3,979 U/mL-h (9,948 U/mL from Table 1).

The calculated overall ratio of ß-gal productivity in continuous and batch cultures was approximately 3.2. The dilution rate for which the continuous culture productivity was obtained, is close to the µ-value at which ß-gal productivity is maximal. In terms of this, ß-gal production is clearly favored in continuous culture.

CONCLUSIONS

ß-gal production can be sustained for longer periods of time in continuous culture when compared with batch experiments. This may be a result of a stable genetic structure.

It is possible that using higher antibiotic concentrations, amplification and significantly higher ß-gal production levels can be attained as observed in batch experiments. The chromosomal insertion has proved to be stable for industrial production purposes, but some other conditions (antibiotic concentration, aereation, etc.) will have to be tested in order to demonstrate its full capacity. Additionally, continuous culture has proved to be an important tool in explaining fermentation conditions, as well as process development.

ACKNOWLEDGEMENTS

We thank Dr. Fernando Valle and Dr. Octavio T. Ramírez for help in several steps of the work and for reading the manuscript.

This work was partially supported by the grant No. 81257/016903 of the Consejo Nacional de Ciencia y Tecnología, México.

REFERENCES

1. **Salvador, M., R. Quintero and F. Valle.** Asian-Pacific J. Mol. Biol. Biotechnol. In press. (1994).

2. **Young, M. and S.D. Ehrlich.** J. Bacteriol., **171**, 2653-2656 (1989).

3. **Young, M.** J. Gen. Microbiol., **130**, 1613-1621 (1984).

4. **Lowry, O.H., M.J. Rosebrough, A.L. Farr, and R.J. Randall.** J. Biol. Chem., **193**, 265-275 (1951).

5. **Miller, J.H.** *Experiments in molecular genetics*, p. 352-355. Cold Spring Harbor Laboratory, Cold Spring Harbor, New York (1972).

6. **Gerhardt, P., R.G.E. Murray, R.M. Costilow, E.W. Nester, W.A. Word, M.R. Krieg, and G.B. Phillips.** *Manual of methods of general bacteriology*, p. 28. American Society for Microbiology, Washington, D.C. (1981).

7. **Frankena, J., H.W. van Verseveld, and A.H. Stouthamer.** Appl. Microbiol. Biotechnol., **22**, 169-176 (1985).

8. **Dawes, L.W. and J. Mandelstam.** J. Bacteriol., **103**, 529-535 (1970).

9. **Frankena, J., G.M. Koningstein, H.W. van Verseveld, and A.H. Stouthamer.** Appl. Microbiol. Biotechnol., **24**, 106-112 (1986).

10. **Nicholoson, W.L. and P. Setlow.** *Molecular biology methods for* Bacilli, p. 391-450. John Wiley & Sons, Ltd. Chichester, England (1990).

Rapid Ethanol Production from Sucrose Using *Zymomonas mobilis*

L.A. Kirk[1] and H.W. Doelle[2]

[1]Department of Microbiology, University of Queensland, St. Lucia 4072; [2]Microbiotech Pty Ltd, Maroochydore South, Queensland, AUSTRALIA

A new mutant strain of Zymomonas mobilis *UQM 2716 was developed which lost its ability to produce sorbitol from fructose. The mutant,* Zymomonas mobilis *ACM 3963 was co-immobilised with invertase on alginate. This combination allowed theoretical yields of ethanol from 100 and 150 g l^{-1} sucrose using either semi-defined media or sugarcane syrup within 3-5 h, respectively. The increased immobilised biomass concentration reduced batch fermentation times by 50-70%. The strain also improved significantly fed-batch fermentation with final yields of 14.3% v/v or 113 g l^{-1} and 13.8% v/v or 109 g l^{-1} ethanol for semi-defined media and sugarcane syrup, respectively.*

The batch fermentation of sucrose to ethanol by *Zymomonas mobilis* at 35°C has traditionally suffered losses in efficiency due to the production of the wasteful by-products sorbitol [1-5] and fructo-oligosaccharides [1,2,6]. These carbon diversions lead to poorer ethanol yields and increasing carbon balance discrepancies, which can be observed constantly at sucrose concentrations in excess of 150 g/l as well as high salt concentrations [7]. Sorbitol formation is the result of a pathway change due to the inhibition of fructokinase by high free glucose concentrations [8]. The enzyme glucose-fructose oxidoreductase simultaneously oxidizes glucose to gluconate and reduces fructose to sorbitol [9].

Sorbitol production can be reduced by the addition of selected salts, which often results in additional polymer (fructo-oligosaccharides) formation [10-13]. The formation of membrane-associated fructo-oligosaccharides appears to be a result of impairment of fructose uptake caused by either high substrate or high salt concentrations [13], which result in large losses in carbon balance calculations. The just recently identified polymer, 1-kestose, can be hydrolysed back to fructose upon the addition of invertase [10,11, 14,18].

Although invertase can effectively reduce fructo-oligosaccharides such as 1-kestose, the resulting improvement in sucrose hydrolysis increases either sorbitol formation or fructose accumulation.

We report here on a newly developed strain, which was found to be incapable of producing sorbitol from sucrose.

MATERIALS AND METHODS

Microorganisms, maintenance and preculture. The parent strain was *Zymomonas mobilis* UQM 2716, a laboratory strain developed from *Z.mobilis* UQM 2007 (NCIB 11199), and now deposited as ATCC 39676 (US Patent 4797360). The non-sorbitol-producing strain was a spontaneous mutant which arose from continuous subculturing on glucose medium of *Z.mobilis* UQM 2716. This strain was subsequently purified and has been deposited as *Z.mobilis* ACM 3963. Both strains were maintained as described previously [12]. The standard semi-defined medium for preculture contained 2 g/l each of $MgSO_4 \cdot 7H_2O$, $(NH_4)_2SO_4$, casein hydrolysate and yeast extract, plus 6 g/l KH_2PO_4, and 100 g/l glucose. The cultures were incubated at 32°C.

437

E. Galindo and O.T. Ramírez (eds.), Advances in Bioprocess Engineering. 437-440.

© *1994 Kluwer Academic Publishers. Printed in the Netherlands.*

<u>**Immobilisation of cells and invertase in alginate**</u>. The cell immobilisation technique has been described previously [19]. When 0.2 g/l invertase was co-immobilised with 2 g/l *Z.mobilis*, they were suspended in alginate prior to bead-making. The invertase used in these experiments (Sigma Grade V, Practical) was one where most of its activity was associated with the non-soluble portion and did not diffuse from the alginate bead.

<u>**Fermentation Conditions**</u>. Immobilised cells were inoculated into semi-defined media containing 6 g/l KH_2PO_4, 1 g/l $(CH_3COO)_2Ca$, 2 g/l each of $MgSO_4.7 H_2O$, $(NH_4)_2SO_4$, casein hydrolysate and yeast extract, and either 100, 150, or 200 g/l sucrose.

The sugarcane syrup was diluted to the appropriate sucrose concentration that the batch or fed-batch fermentations required. Also included was 6 g/l KH_2PO_4, and 1 g/l each of $MgSO_4.7 H_2O$ and $(CH_3COO)_2Ca$.

In all experiments, the pH was initially adjusted to 6.0-6.3, and maintained in the 5.0-5.5 range with 4M NaOH. All fermentations on sucrose were performed at 35°C.

<u>**Analytical Methoda**</u>. Sucrose, glucose, fructose and sorbitol were determined using High Performance Liquid Chromatography as described earlier [10,11,14,19]. Ethanol was estimated using the Ebulliometer [19]. Conversion efficiencies were calculated using the theoretical factors 0.511 for glucose and fructose and 0.538 for sucrose [19].

RESULTS AND DISCUSSION

The results outlined in Table 1 are a comparison between the parent strain UQM 2716 and the new mutant ACM 3963. Each strain was co-immobilised with invertase in alginate and fermentation was carried out in sucrose-based media. At the lower sucrose concentrations of 100 and 150 g/l, the mutant strain exhibited a 100% conversion efficiency from sucrose to ethanol. According to its constant glucose and fructose uptake rates [13], a concentration of 150 g/l sucrose had a proportionally longer fermentation time compared to the 100 g/l sucrose concentration. At 200 g/l sucrose, however, the conversion efficiency

dropped to 86% owing to the impairment of fructose utilization. This fructose accumulation also slowed down the invertase activity of hydrolysing and preventing further fructo-oligosaccharide formation. It should be noted that despite fructose accumulation, no sorbitol was formed.

In contrast, the parent strain produced sorbitol even at the lowest sucrose concentration resulting in only a 94% conversion efficiency. With increasing sucrose concentrations, the conversion efficiency decreased and reached identical values to the mutant strain at 200 g/l sucrose. It is of interest to note that the parent strain did not accumulate fructose as did the mutant, but sorbitol instead, the balance of carbon being a higher polymer formation. The total fermentation times of either strain are identical. The conversion efficiencies are very similar if not identical with previous experiments using free cells in batch fermentation [13], but fermentation time is severely reduced owing to the higher biomass used in immobilisation.

Table 2 exhibits the results of batch fermentations using sugarcane-syrup as substrate. A very similar trend could be observed. Whereas the mutant strain again exhibits a 100% conversion efficiency for 93 and 143 g/l sucrose containing syrup, the parent strain did not perform as well on the syrup as on the semi-defined medium. Free cell batch fermentations on sugarcane syrup were earlier reported [14,15] with conversion efficiencies in excess of 90%. However, the fermentation times were 50-70% above those of the immobilised cell fermentations.

Fed-batch fermentation techniques are often useful in obtaining a higher product formation when the organism is sensitive to the effects of high initial substrate concentrations [8].

The mutant strain was able to produce 113 g/l ethanol [14.3% v/v) from 216 g/l sucrose in less than 24 hours with an ethanol yield of 97%. This fermentation owes its efficiency to the lack of sorbitol formation. In contrast, the parent strain required 288 g/l sucrose to produce the same amount of ethanol, which gives only a 73% conversion efficiency. The sorbitol levels could be reduced by strictly maintaining glucose concentrations below 7 g/l [7], which is possible with free cell fed-batch fermentations [16,17], but difficult with such extremely short fermentation times.

In using equal amounts of sugarcane syrup (Table 4), the mutant strain confirmed its previously described superiority in conversion efficiency owing to no sorbitol formation and no polymer accumulation.

Table 1 - Fermentation with sucrose concentrations of 100, 150 and 200 g/l using *Z. mobilis* ACM 3963 (non sorbitol-producing) and *Z. mobilis* UQM 2716 (sorbitol-producing). Invertase, in the concentration of 0.2 g/l was co-immobilised with each strain.

Z.mobilis strain	Sucrose initial (g/l)	CE (%)	CCB (g/l)	Sorbitol final (g/l)	Ethanol final (g/l)	Sucrose final (g/l)	Glucose final (g/l)	Fructose final (g/l)	Fermentation time (hr)
ACM 3963	98	100	0	0	53	0	0	0	2-3
	153	100	0	0	82	0	0	0	4-5
	200	86	-7	0	93	0	0	22	6-7
UQM 2716	99	94	-1	7	50	0	0	0	2-3
	154	90	-3	12	75	0	0	0	4-5
	205	86	-13	17	95	0	0	0	6-7

Table 2 - Fermentation on sugar cane syrup with sucrose concentrations of 100, 150 and 200 g/l using *Z. mobilis* ACM 3963 (non sorbitol-producing) and *Z. mobilis* UQM 2716 (sorbitol-producing), both co-immobilised with 0.2 g/l invertase.

Z.mobilis strain	Sucrose initial (g/l)	CE (%)	CCB (g/l)	Sorbitol final (g/l)	Ethanol final (g/l)	Sucrose final (g/l)	Glucose final (g/l)	Fructose final (g/l)	Fermentation time (hr)
ACM 3963	93	100	0	0	50	0	1	0	2-3
	143	100	0	0	78	0	1	0	4-5
	193	83	-6	0	87	0	9	18	6-7
UQM 2716	99	89	-3	7	46	0	0	0	2-3
	144	89	-3	13	69	0	0	0	4-5
	194	86	-11	18	89	0	0	0	6-7

Table 3 - Comparision of fed-batch fermentations between *Z. mobilis* ACM 3963 (non sorbitol-producing) and *Z. mobilis* UQM 2716 (sorbitol-producing), with 0.2 g/l co-immobilised invertase.

Z.mobilis strain	Sucrose consumed (g/l)	EY (%)	CCB (g/l)	Sorbitol final (g/l)	Ethanol final (g/l)	Fermentation time (hr)
ACM 3963	216	97	-7	0	113	<24
UQM 2716	288	73	-31	51	113	<24

Table 4 - Comparision of fed-batch fermentations with sugar cane syrup between _Z. mobilis_ ACM 3963 (non sorbitol-producing) and _Z. mobilis_ UQM 2716 (sorbitol-producing), with 0.2 g/l co-immobilised invertase.

Z.mobilis strain	Sucrose consumed (g/l)	EY (%)	CCB (g/l)	Sorbitol final (g/l)	Ethanol final (g/l)	Fermentation time (hr)
ACM 3963	202	100	0	0	109	<24
UQM 2716	202	83	26	26	90	<24

CONCLUSIONS

The strain _Z.mobilis_ ACM 3963 has been shown to be of great commercial potential. Its unique characteristic of not being able to produce sorbitol, combined with the fructo-oligosaccharide hydrolysing properties of invertase, enables theoretical concentrations of ethanol to be produced from batch- and fed-batch fermentations of sucrose. Since ethanol is a low value product, the process for its production must be as efficient as possible.

LITERATURE CITED

1. Viikari,L. _Appl.Microbiol. Biotechnol._ **19**,252 (1984)

2. Viikari,L. _Appl.Microbiol. Biotechnol._ **20**,118 (1984)

3. Barrow,K.D., J.G.Collins, D.A.Leigh, P.L.Rogers and R.G.Warr, _Appl. Microbiol. Biotechnol._ **20**,225 (1984)

4. Leigh,D., R.K.Scopes and P.L.Rogers, _Appl. Microbiol. Biotechnol._ **20**,413 (1984)

5. Zacchariou,M. and R.K.Scopes, _J.Bacteriol._ **163**,863 (1986)

6. Viikari,L. and R.Gisler, _Appl. Microbiol. Biotechnol._ **23**,240 (1986)

7. Doelle,H.W. and P.F.Greenfield, _Appl. Microbiol. Biotechnol._ **22**,405 (1985)

8. Johns,M.R., P.F.Greenfield and H.W.Doelle, _Adv. Biochem. Eng./Biotechnol._ **44**,79 (1991)

9. Hardmann,M.J. and R.K. Scopes, _Eur.J.Biochem._ **173**,203 (1988)

10. Doelle,M.B., P.F.Greenfield and H.W.Doelle, _Appl. Microbiol. Biotechnol._ **34**,160 (1990)

11. Doelle,M.B., P.F.Greefield and H.W.Doelle, _Process Biochem._ **25**,151 (1990)

12. Kirk,L. and H.W.Doelle, _Appl. Microbiol. Biotechnol._ **37**,88 (1992)

13. Doelle,H.W., L.Kirk, R.Crittenden, H.Toh and M.B.Doelle, _CRC Crit. Revs. Biotechnol._ **13**,57 (1993)

14. Doelle,M.B. and H.W.Doelle, _J.Biotechnol._ **11**,25 (1989)

15. Doelle,H.W., L.D.Kennedy and M.B.Doelle, _Biotech. Lettrs._ **13**,131 (1991)

16. Kositanont,C., L.Edye and H.W.Doelle, _Microbios_ **61**,169 (1990)

17. Edye,L.A., M.R.Johns and K.N.Ewings, _Appl. Microbiol. Biotechnol._ **31**,129 (1989)

18. Doelle,M.B. and H.W.Doelle, _Appl.Microbiol. Biotechnol._ **33**,31 (1990)

19. Kirk,L.A., H.W.Doelle and R.I.Webb, _World J.Microbiol. Biotechnol._ **9**,366 (1993)

Improved Production of Proteases by *Brevibacterium linens* in Submerged Culture

S. Strauss[1], J. Zemanovic[2], and W. Hampel[1]

[1]Institute of Biochemical Technology and Microbiology, University of Technology Vienna; AUSTRIA, [2]Institute of Milk, Fats and Food Hygieny, Slovak Technical University, Bratislava; SLOVAKIA

Protease formation in submerged cultivations of Brevibacterium linens *was studied. The experiments were performed in a highly instrumented fermenter using a mineral salt medium containing 0.5% corn steep liquor (CSL) as sole carbon and nitrogen source. After 20 h of cultivation at 30 °C under an aeration rate of 0.33 vvm the organism produced a proteolytic activity of 33.1 U ml^{-1}. By continuously monitoring various cultivation parameters (dissolved oxygen, carbon dioxide and oxygen in effluent gas, and amount of acid added) and by regularly retrieving samples from the broth for off-line analysis of ammonia, α-amino nitrogen, amino acids and protease activity, a very clear picture of culture development could be obtained: respiration activity and substrate concentration indicated that the proteases were synthesized during linear growth and attained the maximum activity when entering the stationary phase. As amino acids were almost exclusively used for growth, mineral acid was added to maintain a constant pH, and consequently a considerable increase in the ammonium concentration was observed. With the exception of tyrosine and histidine, all amino acids were completely dissimilated. Further shake flask cultivation experiments proved that proteolytic activity can be increased up to 70 U ml^{-1} by supplying 15 g l^{-1} CSL.*

Brevibacterium linens constitutes the predominant part of the surface smear on a variety of aged and surface ripened cheeses such as Tilsiter, Limburger or Romadour, where it plays a major role in the ripening of these cheeses. During this process cheese proteins become cleaved at various sites of the protein network and the characteristic flavour of the cheese is developed by further reactions. Specific proteases are responsible for protein degradation and α-casein is the first fraction to be hydrolysed.

The first experiments to increase protease production in *B. linens* for preparative purposes were performed by Tokita and Hosono [1]. Very complex substances such as peptone, yeast extract, etc. were used as medium constituents in order to achieve maximal growth and to support enzyme formation. Furthermore, Zemanovic and Skarka [2] reported on the effect of several proteinic substances on enzyme synthesis and stated optimal conditions for submerged cultivations. Recently, Hayashi et al. [3] described an improved medium for protease production, isolated several proteolytic enzymes and showed their advantageous application to the ripening of cheddar cheese.

In the experiments described previously [4,5], we have studied protease production by *Brevibacterium linens* in labor pilot fermenters. Enhanced enzyme formation was achieved by the use of CSL as a carbon source. The results presented in this paper will elucidate the kinetics of microbial growth and enzyme formation. The aim of this work is to obtain information about optimal parameters for a technical production with respect to a possible application to the ripening process.

METHODS

Organism:

The strain *Brevibacterium linens* ATCC 9172 was used in the experiments.

Cultivation:

Cultivation was performed in a highly instrumented fermenter (WA 250; Chemap AG, Volketswil, Switzerland) at 800 rpm and 0.3 VVM aeration rate. Temperature and pH were controlled at the preset levels of 28°C and 7, respectively. Cultivation parameters monitored on-line and registered in intervals of 3 min by the computer-coupled fermentation system included temperature, pH,

441

E. Galindo and O.T. Ramírez (eds.), Advances in Bioprocess Engineering. 441-445.
© 1994 *Kluwer Academic Publishers. Printed in the Netherlands.*

dissolved oxygen, stirrer speed, oxygen and carbon dioxide in effluent gas and weight of neutralizing reagent (1M H_2SO_4) added to maintain constant pH. A PC-AT 386 SX was used with the FDS1 software for data acquisition and the IMM program package (ATS) for data analysis. The basic mineral salt medium for fermentation contained (g/10L): 2.7 KH_2PO_4, 8.9 $Na_2HPO_4*2H_2O$, 0.6 NaCl, 2 $MgSO_4$, 1.2 $(NH_4)_2SO_4$, 0.028 $FeSO_4$, 0.064 $CaCl_2*2H_2O$. Corn Steep Liquor (5g/L on basis of dry weight) was sterilised separately and added aseptically to the mineral solution.

The organism was grown for inoculation and shake flask culture experiments for 3 days on an orbital shaker (250 rpm) at 30°C in the above described media.

Analytical Methods:

Samples were withdrawn from the culture broth and centrifuged (10.000 g; 10min). The supernatant was submitted to the following determinations: Total proteolytic activity (TPA) was assayed according to Lin et al. [6], and activity of leucine aminopeptidase (LAP) according to Hennrich [7]. Both assays were performed at pH 7.5 and 38°C. Moreover, the concentrations of ammonia, alpha-amino-nitrogen and protein were estimated as described elsewhere (Reardon et al.[8]; Wylie and Johnson [9]; Bradford [10]). All analytical values stated are means of triplicate assays. Amino acids were determined by an automatic Amino Acid Analyzer (model AAA339; Mikroteckna; Praha, CFR).

RESULTS

Kinetics of growth and enzyme formation

In order to elucidate the kinetics of growth and enzyme formation the bacterium was cultivated in a highly instrumented fermenter on media containing 0.5% CSL in the basal mineral solution. Continuously acquired parameters as well as results from analytical determinations carried out in samples withdrawn from the culture fluid are outlined in Figure 1.
It could be observed that culture development proceeded in several distinct phases. After a short phase of adaptation (lag-phase) growth was almost exponentially without any increase in the proteolytic activity in the medium. But soon some metabolic limitations occurred as indicated by the respiration activity and the linear decrease in the α-amino nitrogen concentration. Nevertheless, protease formation started with only a small fraction of leucine aminopeptidase (LAP). Physiological alkaline compounds were the main source for catabolism as acid was added to the culture to maintain a constant pH. The decrease in the concentration of α-amino nitrogen and the increase of ammonium ions emphasized that the assimilation of free amino acids governed catabolism. Quantitative determinations of amino acids in the medium revealed that several neutral and basic amino acids such as alanine, leucine, valine, phenylalanine, lysine and arginine were the main substances (Table 1) for cell growth.

The effect of limitation in metabolic activity became more and more pronounced during the late growth phase starting at 13 hrs of cultivation. Stagnation and subsequent decrease in respiration activity was obviously caused by a reduced availability of metabolizeable materials. Amino acid analysis of samples from the medium proved the almost complete exhaustion of free amino acids, with the exception of only tyrosine, histidine, asparagine and leucine. Although these compounds were used for growth in the previous phases of the cultivation, a notable fraction remained unattacked by the bacterium. Nevertheless, proteolytic activity increased furthermore and attained a maximum level of 33 U/ml at 18 hrs of cultivation.

During the stationary phase adaptation to some minor constituents of the complex carbon source (CSL) occurred, which resulted in a further small growth phase. The total proteolytic activity in the medium was not affected by this effect but by increasing its fraction it influenced the activity of leucine aminopeptidase (LAP).

The cultivation was repeated twice showing reproducible results.

Influence of carbon source concentration on protease yield

Shake flask experiments showed that the

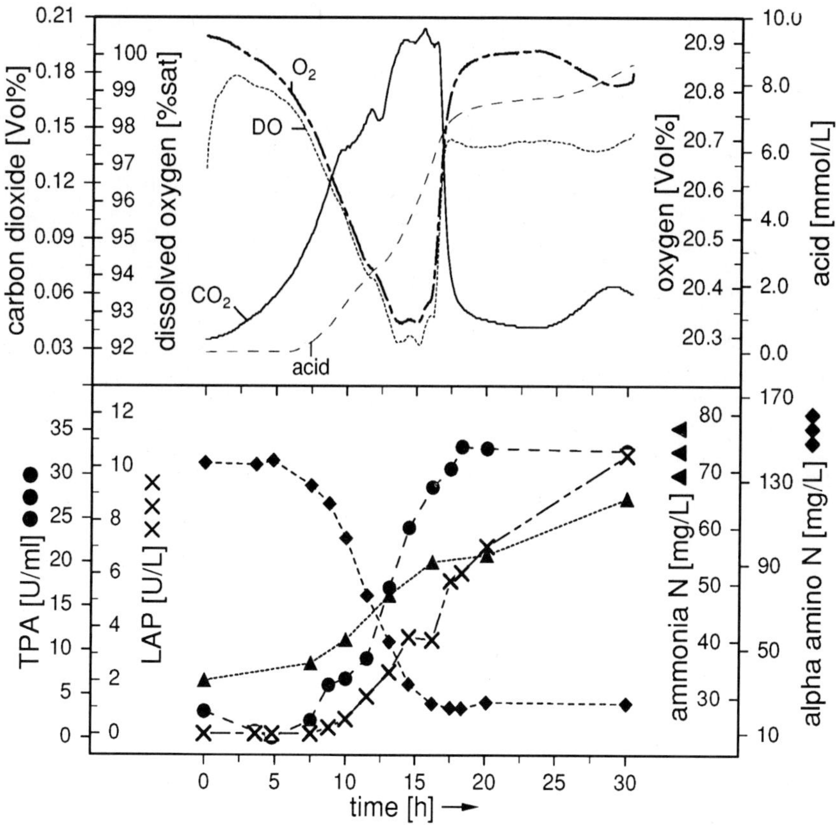

Figure 1. Cultivation of *Brevibacterium linens* on 5g/L CSL

Table 1. Concentrations of several Amino Acids [µmol/l] in the Cultivation Medium

Amino Acid	Cultivation time [hrs]						
	0.0	7.5	8.8	10.0	11.5	14.5	17.5
Asp	131	132	133	152	155	129	128
Thr	239	230	250	252	151	91	nd
Ser	475	392	341	274	210	98	nd
Glu	231	145	88	38	nd	nd	nd
Pro	650	650	550	520	390	140	nd
Gly	249	171	118	66	nd	nd	nd
Ala	1570	1350	1020	484	34	nd	nd
Val	685	639	627	584	519	143	nd
Met	211	184	162	138	126	nd	nd
Ile	328	274	256	249	211	59	nd
Leu	949	860	815	776	702	168	50
Tyr	361	340	331	328	232	67	117
Phe	412	392	387	296	236	nd	nd
His	334	228	187	139	55	37	30
Try	101	88	84	80	84	18	nd
Lys	533	475	419	379	342	nd	nd
Arg	453	359	325	277	202	nd	nd

total proteolytic activity could be increased considerably from 33 to 70 U/ml by increasing the concentration of CSL three times. A further increase in substrate concentration did not affect the yield of total proteolytic activity in the same way (Figure 2).

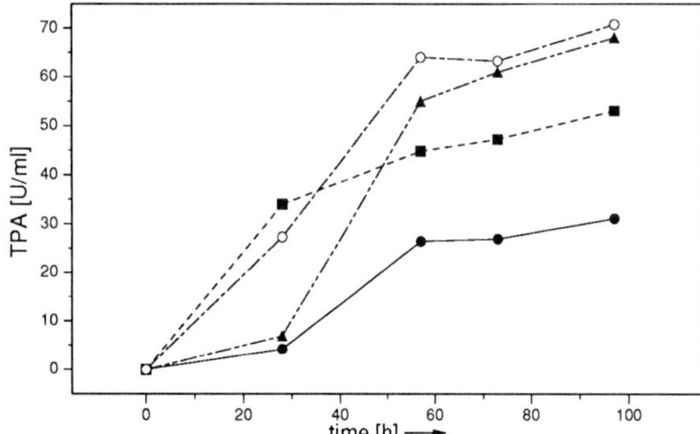

Figure 2. Influence of Substrate Concentration on the Amount of Proteolytic Enzyme formed. (● 5g/L CSL; ■ 10g/L CSL; O 20g/L CSL; ▲ 15g/L CSL)

DISCUSSION

Corn Steep Liquor in combination with basic mineral salts forms a culture media enabling an enhanced production of proteases by Brevibacterium linens. The most pronounced effect can be attributed to the supporting of rapid bacterial growth, which implies that proteolytic enzymes are synthesized at a high rate. Within one day enzyme activity levels can be attained which are comparable to those produced in a period of 3-4 days in previous experiments (Strauss et al. [4]) with other carbon sources such as malt extract, yeast extract, etc. Consequently, CSL seems to be an interesting carbon source for technical production of proteases. Moreover, CSL is a by-product of starch production from corn and easily available in adequate amounts.

The kinetic study revealed that most of the activity found in the medium at 18 hrs was formed during active growth. The growth of Brevibacterium linens is preferentially achieved by the utilization of free amino acids.

Amino acids are used extensively for growth by Brevibacterium linens. Active transport across the cell membrane is inducible and repressible. Some other components of the medium may have supporting or inhibiting effects on amino acid uptake; Hamouy et al. (1985) showed that the addition of lactate to the medium increased uptake of L-serine, whereas inhibition of transport occurred at 4% salt concentration. The inconsistencies in respiration activity during active growth may reflect such effects.

ACKNOWLEDGEMENT

This work was kindly supported by the Austrian Ministery of Science and Research (BMWuF)

NOTATION

CSL   Corn Steep Liquor
LAP   Activity of Leucine Aminopeptidase [U/L]
TPA   Total Protease Activity [U/ml]
DO   Dissolved Oxygen in culture broth [% sat]

LITERATURE CITED

1. Tokita, F. and Hosono, *Jap. J. Zootech. Sci.*, **43**, 39-48,(1972).

2. Zemanovic, J. and Skarka, B. "Culture media for extracellular proteinase production by Brevibacterium linens" in *Proc. 4th European Congress Biotechnol.1987*, Neijssel, O.M., van der Meer, R.R., Luyben, K.Ch.A.M., (Eds.), Elsevier Science Publ., Amsterdam, Vol.3, pp.541-544 (1987).

3. Hayashi, K., Cliffe, A. and Law, B.A., *Nippon Shokuhin Kogyo Gakkaishi* **37**, 737-739, (1990).

4. Strauss, S., Rauscher, U., Zemanovic, J. and Hampel, W. "Formation of Proteases by Brevibacterium linens," in *DECHEMA-Biotechnology Conference Series Vol.5A*, Kreysa,G., Driesel, A.J.,(Eds.), VCH Publ., New York; pp.181-184 (1992).

5. Strauss,S.,Kopecky,J.,Zemanovic,J. and Hampel,W.A., *Biocatalysis*, in press (1994)

6. Lin, Y., Means, G.E. and Feeney, R.E., *J.biol.Chem.*, **244** (4), 789-793,(1969).

7. Hennrich, N., Klochow, M., Lang, H. and Berndt,W. *Fed.Eur. Biochem.Soc.Lett.*,**2**(5), 278-280,(1969).

8. Reardon, J. and Foreman, J.A. and Seary R.L., *Clin.Chim.Acta*, **14** (3), 403,(1966).

9. Wylie, E.B. and Johnson, N.J., *Biochem.Biophys.Acta.*, **59**, 450,(1962).

10. Bradford, M.M., *Anal.Biochem.*,**73**, 248-254 (1976).

11. Hamouy, D., Boyaval, E., and Desmazeaud, M.J., *Milchwissenschaft*,**40**,133 (1985).

Production of K1 (Streptokinase) Using Batch and Fed-Batch Culture of Recombinant *E. coli*

R.E. Narciandi[1], F.J. Morbe[2], U. Knupfer[2], R. Wenderoth[2], and D. Riesenberg[2]

[1]Center for Genetic Engineering and Biotechnology, P.O. Box 6162, Cubanacan, Habana, CUBA;
[2]Hans Knoll-Institut fur Naturstoff-Forschung, Beutenbergstr. 11, D-07745 Jena, GERMANY

This paper discusses the effects of medium composition, host strains, cultivation strategy, and plasmid stability on the growth and expression of K1 (SK) recombinant fusion protein expressed in E.coli under the control of lpp-lac promoter. The production of recombinant K1 (SK) fusion protein, was compared in batch and fed-batch culture. In batch culture a high dependence of the protein expression level on the host strain and culture medium was observed. The highest expression level of K1 (SK) was obtained when the KS 474 (pINO 56) and TG 1 (pINO 56) strains were grown in synthetic medium, while in all cases a very low cell concentration was obtained. In order to increase the final cell concentration, a high cell density cultivation (HCDC) process was carried out. The expression of K1 (SK) using KS 474 (pINO 56) and TG 1 (pINO 56) strains was compared. A high difference related to the plasmid stability and cell growth was observed. By introducing fed-batch cultivation, the final cell dry weight and the specific expression level were increased significantly compared to batch cultivation. It was demonstrated that the expression form of K1 (KS) recombinant protein (soluble or insoluble) depended on the culture conditions.

Recombinant DNA technology for the expression of heterologous genes in microbial hosts offers an opportunity to produce large quantities of bioactive proteins, that would be difficult or impossible to derive otherwise.

Escherichia coli is well characterized regarding genetics, physiology and fermentation. It's manipulation by recombinant DNA technology is well developed. Therefore, E. coli has become a useful host for expression of many recombinant proteins. For recombinant E. coli, the desired outcome is not only a high expression of the target protein but also high cell density, thereby improving productivity and final product concentration.

As some fermentation processes are only economically attractive if high cell concentrations are possible, different culture systems have been developed in order to improve biomass accumulation. Simple batch fermentation of recombinant E. coli is unsuitable because of the necessity to supply high initial concentration of nutrients, which cause high growth inhibition, and yield low cell and product concentrations. Fed-batch culture is an efficient cultural technique of microorganisms to achieve high productivity of biomass and metabolites. This culture method have been often employed in many microbial industries, because of the possibility to control the concentration of all the nutrients (carbon source, nitrogen source, molecular oxygen, and mineral ions) necessary for growth, in the suitable ranges during the cultivation. As a result of this, high densities of biomass within short cultivation times can be achieved. Fed-batch technology is useful not only for the cultivation of microorganisms showing catabolic repression, but also to achieve high cell mass concentration, while maintaining an optimal culture condition. This paper discusses the effects of medium composition, host strains, cultivation strategy, and plasmid stability on the growth and expression of K1(SK) recombinant

447

E. Galindo and O.T. Ramírez (eds.), Advances in Bioprocess Engineering. 447-453.
© 1994 *Kluwer Academic Publishers. Printed in the Netherlands.*

fusion protein expressed in E. coli under the control of lpp-lac promoter.

MATERIALS AND METHODS

E. coli strains, plasmids and culture conditions. The bacterial strains used in this study were E. coli K12, TG 1 ; KS 474 ; KS 476 ; and RV 308. They were transformed with the pINO 56 plasmid, which codifies for the expression of the K1(SK) fusion protein under the control of lpp-lac promoters. The details of the genetic construction have been previously published (1).

Cultivation media. SOC medium (2) or Synthetic medium containing (in grams per liter): KH_2PO_4, 13.3 ; $C_6H_8O_7$ 1.7; $(NH_4)_2HPO_4$, 1; $MgSO_4$ $7H_2O$, 2.7; and 10 ml of trace metal solution (3), supplemented with thiamin-HCl, 20 ug/ml, ampicillin, 50 ug/ml, and initial glucose concentration of 35 g/l were used as culture medium. The feeding solution contained $MgSO_4$ $7H_2O$, 19.7 g/l, and glucose 700 g/l.

Cultivation conditions. The strains were stored in a 1 ml frozen glycerol stock vial at -70°C. It was aseptically transferred into a 500 ml flask with 50 ml of specific medium. This culture was grown overnight at 28°C and 200 rpm in a Infors AG CH-4103 Bottmingen incubator shaker and used as inoculum for all assays. Cultivations were performed in 500 ml flasks containing 50 ml of culture medium, at 28°C and 200 rpm, or in a 10 l fermenter (BIOSTAT ED10, equipped with DCU (Digital Measurement and Control System) and Micro-MFCS (Multi Fermentor Control System) from B. Braun Biotech International and exhaust gas analyzer from Hartmann Braun AG (Melsungen and Frankfurt, FRG), at a temperature of 28°C, gas flow rate of 10 l/min and initial stirrer speed 300 rpm. The pH of the medium was controlled at 6.8 with aqueous ammonia (25 %).

Analytical techniques. Glucose concentration of the culture medium was determined enzymatically with a glucose analyzer Medingen ESAT 6660 (Germany). The optical density (O.D) was measured at 550 nm. Cell dry weight (CDW), ammonia nitrogen and acetate were determined as a function of fermentation time as described by Riesenberg et al.(4). The percentage of plasmid harboring cells in the cell population was determined from the number of β-Lactamase positive colonies (5). Western-blot was performed on SDS gel transferred to nitrocellulose paper (6), using an antibody against commercial streptokinase. The soluble protein fraction obtained after ultrasonic treatment, was used to determine streptokinase biological activity by the chromogenic method (7). As standard, natural streptokinase (Sigma) was used.

RESULTS AND DISCUSSION

Selection of the host strain

In order to assess the effect of the host strains and medium composition on the K1(SK) protein expression and cell growth, different strains were transformed with pINO 56 plasmid and grown at shaken flasks level using SOC medium or Synthetic medium. In all cases the initial optical density was 0.1 O.D. (550 nm). At the early logarithmic phase (0.8 O.D (550 nm)), 1 mM of Isopropil-ß-D-thiogalacto-pyranoside (IPTG) (Promega, U.S.A.) was added as inductor of the expression system. In all shaken flasks assays, the samples were taken 4 hours after induction and the K1(SK) activity was determined.

The production level of K1(SK) varied widely depending on the E. coli host strains and the composition of the culture medium used (Table 1). Similar results have been previously reported by different groups upon studying the expression of recombinant proteins using different promoters, host strains, and culture medium (8,9). When the culture was performed in SOC medium, the cell growth was slightly higher than with synthetic medium, while in all cases the final cell concentration were very low.

Table 1. Effect of host strain and culture medium on the production of K1(SK).

Culture Medium	Strains	O.D (550 nm)	Specific Activity (U K1(SK)/1.O.D x 10^3)
Synthetic	TG1 (pINO56)	1.2	88.06
	RV 308 (pINO56)	1.5	19.94
	KS 474 (pINO56)	2.0	82.57
	KS 476 (pINO56)	3.6	9.44
SOC	TG1 (pINO56)	5.6	27.70
	RV 308 (pINO56)	3.9	23.58
	KS 474 (pINO56)	4.0	23.58
	KS 476 (pINO56)	7.0	14.31

At shaken flasks level, the K1(SK) fusion protein was expressed in soluble and insoluble form and in both cases a slight proteolytic degradation was observed by zymographic and western blot methods (data not shown) (1). The highest specific activity of K1(SK) from the soluble extract was obtained when the TG 1 (pINO 56) and KS 474 (pINO 56) strains were grown in synthetic medium yielding around 80-90 U K1(SK)/1.O.D x 10^3, which represent about 2% of expression in relation with the total cell protein. This value is 3.4 times higher than the specific activity obtained when the expression was performed using SOC medium (23-27 U K1(SK)/1.O.D x 10^3).

High Density Fed-batch fermentation

For large scale cultivation in particular, careful selection of a host strain in terms of plasmid stability and cell growth, in addition to protein production, is essential so as to attain stable and high-yield production. In order to increase the final cell concentration for the production of K1(SK), a high cell density cultivation (HCDC) process using two different host strains TG 1 (pINO 56) and KS 474 (pINO 56) was developed.

Principles of fed-batch fermentation. The process begans with the inoculation of a glycerol stock in a stirred tank reactor with glucose salt mineral medium. The HCDC was characterized by an exponential growth in batch phase at maximum specific growth rate, after a short adaptation step. The pO_2 decreased due to increased O_2 uptake by the growing culture, until it finally reached 20 % saturation. The pO_2 is further kept at this level by an exponential increase of the stirring speed. The pO_2 in the culture also increased sharply indicating the exhaustion of glucose. Glycerol from the inoculation stock was then metabolized. The fed-batch phase was initiated with the beggining of glucose feeding. The cells continued to grow further exponentially at a lower specific growth rate. The feeding rate of glucose was adjusted manually to keep the concentration of glucose below 20 g/l. When the stirrer speed reached the maximum level (1500 rpm), the air was enriched with pure oxygen. The expression of K1(SK) was induced at 10 g/l (CDW) by IPTG added to a final concentration of 1 mM.

Comparative analysis of cell growth and K1(SK) production in E.coli TG 1 (pINO 56) and KS 474 (pINO 56)

Fermentation of E.coli TG 1 (pINO 56). A typical growth kinetic and formation of recombinant K1(SK) during fed-batch cultivation of E. coli TG 1 (pINO 56) are shown in figure 1.

In the batch step, the culture grew exponentially with a specific growth

450

rate of 0.25 h^{-1}. When the glucose reached 10 g/l, the fed-batch phase was started by manual glucose feeding. After this moment _E. coli_ cells grew with a reduced rate of 0.16 h^{-1} under strict aerobic conditions at pO$_2$=20% (data not shown), while the biomass concentration increased from 10 to 61 g/l after 2.5 doubling time.

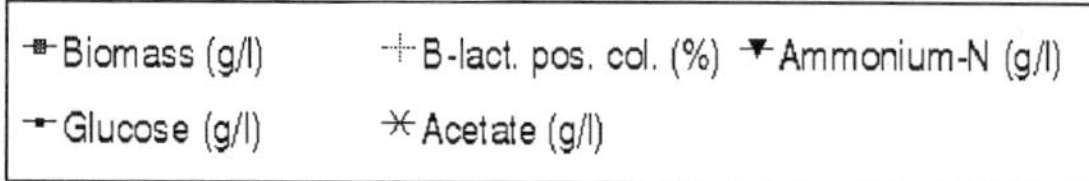

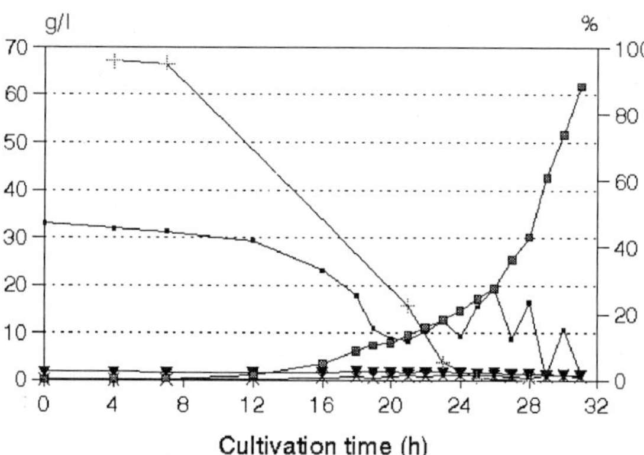

Figure 1. Kinetics of growth, consumption of glucose, plasmid stability and metabolites production during fed-batch cultivation of _E. coli_ (TG 1 pINO 56).

During all fermentation process using _E. coli_ TG 1 (pINO 56) ammonia was in excess in the culture medium (between 1.5-2.8 g/l) and was used as the nitrogen source, which correspond with that reported by Riesenberg _et al_ (_4_). Even significant changes in the concentration of glucose did not enhanced the formation of acetate, which is usually produced as the main by-product by _E. coli_ strains (_10_).

The acetate concentration remained at low levels (about 1 g/l). Both metabolites were used up after the consumption of initial glucose. The induction of the K1(SK) expression under the lpp-lac promoter by adding 1 mM IPTG, when the cell concentration was 10 g/l CDW, did not affected the cell growth. The fraction of plasmid harboring cells in the population was estimated as a function of fermentation time. During all fermentation process a high plasmid loss was observed. At the moment of induction, only 15 % of the cells contained plasmids. For this reason, the expression of K1(SK) was not detected by chromogenic method and western blot (data not shown).

Fermentation of _E. coli_ KS 474 (pINO 56). Figure 2 shows a typical high cell density fermentation with KS 474 (pINO 56). In batch phase, before IPTG induction, the cells grew exponentially with 0.27 h^{-1} specific growth rate, while the initial high glucose concentration was consumed. This behavior corresponded with that reported before for the TG 1 (pINO 56) strain. During this step the ammonium concentration increased up to 3 g/l and the acetate remained below 2 g/l.

The induction of K1(SK) expression was carried out by addition of IPTG (1 mM) at 10 g/l CDW. The cells continued growth during 4 more hours at low specific growth rate (0.16 h^{-1}) (Figure 2). After this moment only one doubling of biomass occurred and the final cell concentration was 20 g/l CDW. During fed-batch step, the ammonium concentration increased continually till a maximum of 5 g/l at 24 hour of fermentation, and later began to decrease until the end of cultivation, which demonstrated metabolic imbalances.

When the induction was performed, a rapid increase of acetate concentration, till 9 g/l was observed. This can be the responsible of growth inhibition. Although, according to Pan _et al_ (_11_), acetate concentration below 10 g/l have not been found to inhibit growth. The highest concentration of CO$_2$ (about 12%) (not shown) was still considerably lower than the data reported to be inhibitory for growth, i.e; 30-40% (<u>11</u>).

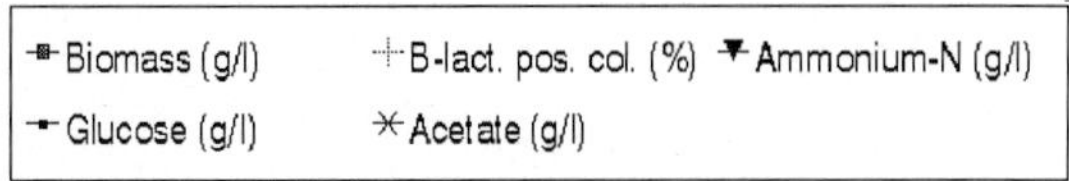

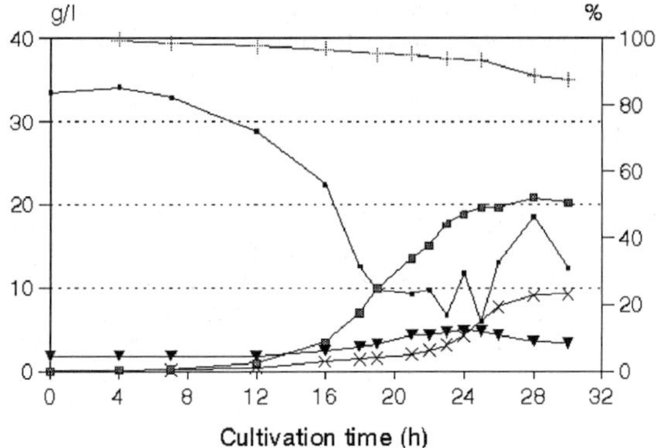

Figure 2. Kinetics of growth, consumption of glucose, plasmid stability and metabolites production during fed-batch cultivation of <u>E. coli</u> (KS 474 pINO 56).

As can be seen in figure 2, throughout all fermentation process almost no plasmid loss occurred, while the SK activity was not detected by chromogenic method on the soluble fraction.

Table 2 shows that the growth and the expression level of K1(SK) using KS 474 (pINO 56) varied with respect to cultivation strategy used.

Table 2. Effect of cultivation strategy on the growth and expression of K1(SK) using KS 474 (pINO 56).

	Batch	Fed-batch
Expression (%)	2	4
O.D	2	94
CDW (g/l)	0.43	20
Productivity (g CDW /1.h)	0.039	0.66

When the fed-batch method was used, the final CDW increased more than 46 times and the productivity of biomass production increased more than 16 times. Expression level was 2 times compared with batch culture.

In terms of K1 (SK) productivity, the fed-batch strategy was, then, more than 30 times higher.

K1(SK) expression

In order to know the localization of K1(SK) protein, samples corresponding to different fermentation times were taken. After ultrasonic treatment, the total cell protein and the insoluble material were analyzed by western blot using an antibody against commercial streptokinase.

Figure 3 (a) show that 4 hours before, and at the induction time, (lane 4 and 5 respectively) a slight basal expression of the heterologous protein was detected. The basal level production obtained in this conditions can be explained by the presence of the constitutive lpp promoter. The product formation increased directly after induction (figure 3 (a), lane 5 to 9). The highest K1(SK) expression (4%) estimated by western blot, was reached 6 hours after induction by IPTG, and maintained constant during all process. This shows that the production of the K1(SK) protein is fundamentally provided by the lac promoter. A slight proteolytic degradation was detected. Figure 3 (b) shows that all K1(SK) fusion protein was expressed on insoluble and inactive form, which do not correspond with the results in shaken flasks experiments, where all the K1(SK) produced was in a soluble and active form. The protein overexpression by IPTG induction and the toxicity of the heterologous protein on the host cell, may be the main cause of the results obtained in the fed-batch fermentations (<u>12</u>).

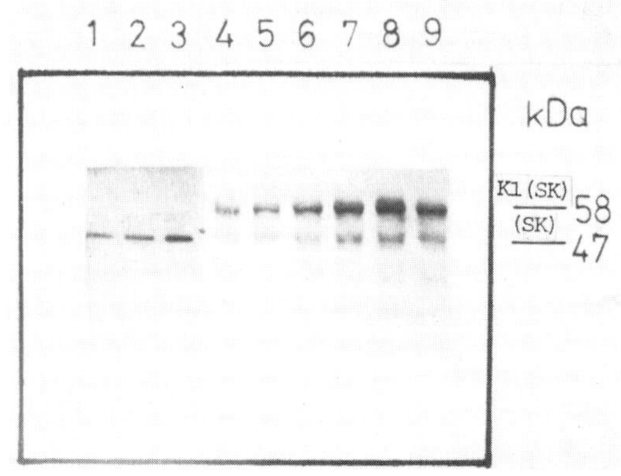

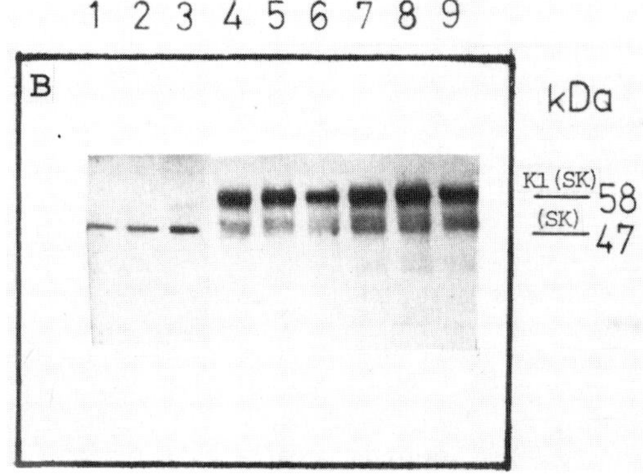

Figure 3. Kinetics of K1(SK) protein expression. Western blot analysis of K1(SK) protein samples. (a) Total cell proteins, (b) Insoluble fraction after U.S treatment. KS 474 (pINO 56) transformed _E. coli_ were grown in fed-batch culture at 10 l level in synthetic medium, supplemented with 35 g/l of glucose. The induction of K1(SK) expression was carried out at 10 g/l CDW by added 1 mM IPTG. The cells were cultured for 30 hours at 28°C. Samples corresponding at different cultivation time were analyzed by western blot using an antibody against commercial Streptokinase.
Lane 1,2,3. Commercial streptokinase (1 μg, 2 μg, 3 μg respectively). Lane 4. 4 hours before Induction with IPTG 1 mM. Lane 5. Induction moment. Lane 6 to 9. 2, 4, 6 and 8 hours after Induction.

CONCLUSIONS

In summary, the expression of K1(SK) is highly dependent on the host strain and the culture medium. As shown in the present paper, two _E. coli_ strains TG 1 (pINO 56) and KS 474 (pINO 56) yielded profound differences with respect to K1(SK) production, cell growth, metabolites production and plasmid stability, suggesting that host strain selection is crucial for the large and cost effective production of K1(SK). By introducing fed-batch cultivation, the final cell dry weight and the productivity were increased at less 46 and 16 times respectively, compared to batch cultivation, while the expression level increased 2 times. The expression form of K1(SK) recombinant protein depended on the culture conditions.

LITERATURE CITED

1. Malke, H. and Ferretti, J.J. Streptococcal Genetics (Eds, G. Dunny et. al.) Washington, D.C. Amer. Soc. Microbiol. 184 (1990).

2. Sambrook, J., Fritsch, E.F., and Maniatis, T. Molecular Cloning. A Laboratory Manual. Vol. 3. Cold Spring Harbor Laboratory, Cold Spring Harbor, N.Y.12. (1989).

3. Riesenberg, D., Schulz, V., Knorre, W.A., Pohl, H.D., Korz, D., Sanders, E.A., RoB, A., and Deckwer, W.D. J. Biotechnol. 20. 17 (1991).

4. Riesenberg, D., Menzel, K., Schulz, V., Schumann, K., Veith, G., Zuber, G., Knorre, W.A. Appl. Microbiol. Biotechnol. 34, 77 (1990).

5. Loser, C. PhD-Thesis. Technical University Dresden (1990).

6. Towbin,H.; Staehlin, T. and Gordon, J. Proc. Natl. Acad. Sci. USA. 76, 4350 (1979).

7. Kulisek, E.S., Holm, S.E., and Johnston, K.H. Anal. Biochem. 177, 78 (1989).

8. Luli, G.W. and Strohl, W.R. Appl. Environ. Microbiol. 56, 1004 (1990).

9. Narciandi R.E. and Delgado A. Biotechnol. Lett. 14, 1103 (1992).

10. Riesenberg, D. Current Opinion in Biotechnology. 2, 380 (1991).

11. Pan, J.G. Rhrr, J.S., Lebeault, J.M. Biotechnol. Lett. 9, 89 (1987).

12. Brosius, J. Gene. 2, 161 (1984).

Microbial Protein Production by Submerged Fermentation of Mixed Cellulolytic Cultures

I. Zabala[1], A. Ferrer[2], A. Ledesma[1], and C. Aiello[2]

[1]Laboratorio de Alimentos, Departamento de Química, Facultad Experimental de Ciencias, Universidad de Zulia, Grano de Oro, Módulo 2, Maracaibo; [2]Departamento de Química, Ciclo Básico, Facultad de Ingeniería, Universidad del Zulia, Av. Ziruma, Maracaibo, VENEZUELA

The hypercellulolytic mutant Trichoderma reesei *QM 9414 was cultured with* Monosporium sp., *a low cellulolytic microorganism, on Avicel to find out how it affected the biomass and enzyme production. Fermentations were carried out with both pure cultures and cocultures. The cocultures were prepared by inoculating* Monosporium sp. *at three different times to a well-developed* T. reesei *QM 9414 culture. The results showed an increase in biomass content on mixed fermentations (77%), with a higher production of extracellular proteins, but with a lower enzyme activity. The results of pH and reducing sugars suggest that this increase in biomass content was mainly due to the growth of* Monosporium sp.

The speed of degradation of lignocellulosic materials in controlled fermentations is usually slower than the one found in nature (6,8). This is mainly because fermentations are carried out with a pure culture of a lignocellulolytic microorganism instead of the great variety of microorganisms present in fertile soils (6,9). In addition, the biomass production is rather low, as it can be seen for two strong cellulolytic fungi such as *Trichoderma reesei* and *Chaetomium cellulolyticum* which produce a protein concentration of less than 0.07 g/l (4,15,16,17,21), and about 0.088 g/l (2,13,21) respectively in submerged fermentations.

The use of mixed cultures of lignocellulolytic microorganisms looks promising in increasing the protein content compare to pure cultures, and many of them have been reported to be more efficient in degrading lignocellulosic substrates and in producing high activity enzymes (3,4,5,14,18). Usually, only one of those is responsible of removing the byproducts of such

degradation, preventing catabolic repression.

In this work, *Trichoderma reesei* QM 9414 was used as a primary microorganism in a fermentation process of a cellulosic substrate. In order to increase its protein production, a second microorganism, *Monosporium sp.*, was added. *Monosporium sp.*, a filamentous fungus isolated from soil and vegetal material from Zulia State, Venezuela, has a weak cellulolytic activity (8). This strain was selected among 6 cellulolytic strains isolated from the same area, due to the considerable growth and a synergistic effect shown on streaked-plates and bottles co-cultured with *T. reesei* QM 9414, using filter paper and carboxymethylcellulose as substrates.

MATERIALS AND METHODS

Microorganisms. *Trichoderma reesei* QM 9414 and *Monosporium sp.* were kept on potato dextrose agar at 27°C for 14 days. In order to increase sporulation, they were transferred to a rice medium and

455

E. Galindo and O.T. Ramírez (eds.), Advances in Bioprocess Engineering. 455-460.
© *1994 Kluwer Academic Publishers. Printed in the Netherlands.*

incubated for 5 days at 27°C. The spore suspension was obtained by adding 100 ml of sterile distilled water. Then, it was standardized to get a final concentration of 1.0×10^6 spores per milliliter.

Culture Medium. Mandel and Weber's mineral solution (20) with 1% cellulose (as Avicel, Merck), was used. The urea needed in the medium was prepared separately and sterilized by filtration.

Fermentations. The fermentative processes were carried out in 125ml-erlenmeyer flasks with 50 ml of medium, in a shaker with controlled temperature and agitation, at 27°C and 100 rpm, by duplicate, during 5 days. The mixed fermentations started with *Trichoderma reesei* QM 9414, and then, spores of *Monosporium sp.* were added at different times.

Analytical Methods. The solids from the fermentation broth were separated by centrifugation at 9.200*g* for 30 minutes and dried. From these the biomass production was estimated as crude protein by a Microkjeldhal method (1). The cellulose content of the solids was determined substracting the biomass weight from the solids weight. The supernatant was analyzed to determine pH, extracellular protein by a Lowry method modified by Peterson et al.(10, 19), reducing sugars by the DNS method, and cellulase activity by the filter paper assay (1,12) with the 3,5-DNS acid reagent. The cellulase activity is expressed as μmol of reducing sugar released per minute and milliliter.

RESULTS

Figure 1 shows the results in biomass and cellulase production with both pure and mixed fermentations. The mixed fermentations started as a single process with a pure culture of *T. reesei*. *Monosporium sp.* was added at different times of fermentation, in order to determine the appropiate time to initiate the mixed fermentation. Figure 1b shows

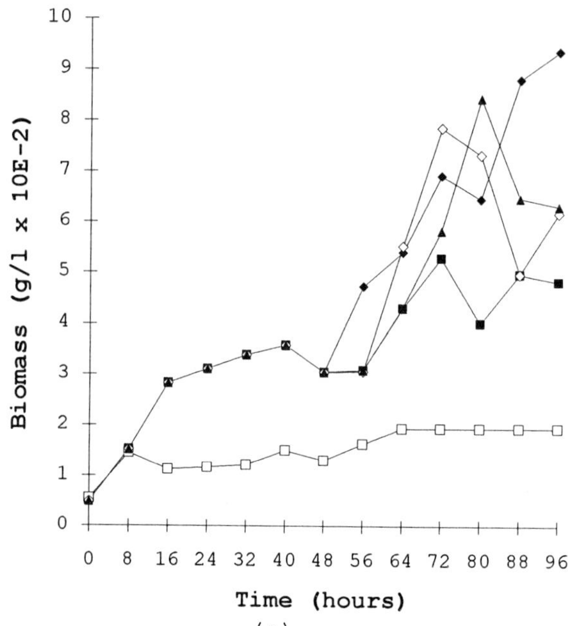

(a)

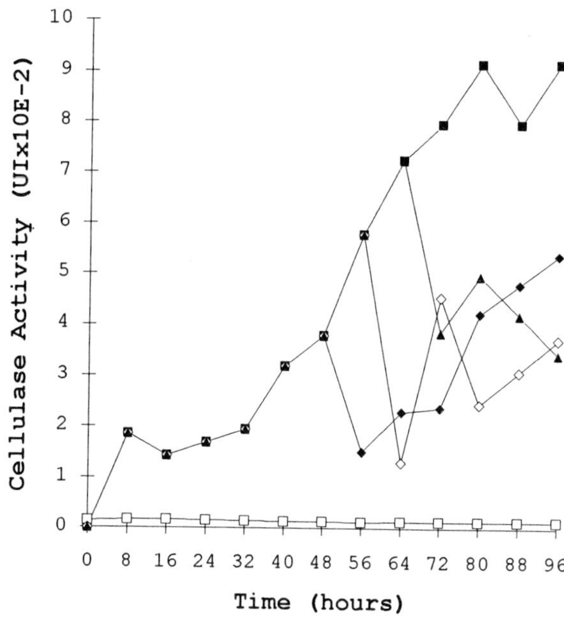

(b)

FIGURE 1. Production of Biomass and Cellulase in Submerged Fermentations of pure cultures and cocultures of *T. reesei* and *Monosporium sp.* a) Biomass Production; b) Cellulase Production; ■: *T. reesei*. □ : *Monosporium sp.* ◆: 48h Coculture. ◇: 56h Coculture. ▲: 64h Coculture.

the cellulase production. The biggest increment observed in the pure culture of *T.reesei* was between 48 and 56 fermentation hours (1.98 x 10^{-2}UI), followed by a second increment at 64 fermentation hours (1.47 x 10^{-2}UI). Those times were selected to start the mixed fermentations with *Monosporium sp.*, on the basis that *T. reesei*'s cellulase activity has already generated enough reducing sugars to support the growth of *Monosporium sp.* *T. reesei* as a pure culture, reaches a high cellulase activity that starts to increase after 32 hours, but the highest increments were observed at 48, 56 and 64 hours of fermentation. In contrast with the results obtained in biomass production, cellulase activity in mixed fermentations were lower than the one observed in single fermentations of *T. reesei*.

Figure 1a shows the biomass production in submerged fermentations with pure cultures and cocultures of *T. reesei* and *Monosporium sp*. It can be seen that all the mixed processes reached higher values of biomass than the single fermentations did. The best result in such processes was about a 77% increase in biomass content, obtained in the 48h-phasing fermentation.

Figure 2 shows the variation of pH. Those values indicate a very different behavior between single and mixed fermentations, so it could affect the expression of cellulolytic enzymes (15,16,17). Figure 3 and 4 show the results of extracellular protein and reducing sugars in both pure and mixed fermentation processes.

The increase of cellulase activities at 48, 56 and 64 hours of fermentation with pure cultures of *T. reesei* and cocultures with *Monosporium sp.* (Fig. 1b) coincide with the increase of extracellular protein observed at those times (Fig. 3) but cellulase activities of mixed fermentation were smaller, as compared to pure cultures of *T.*

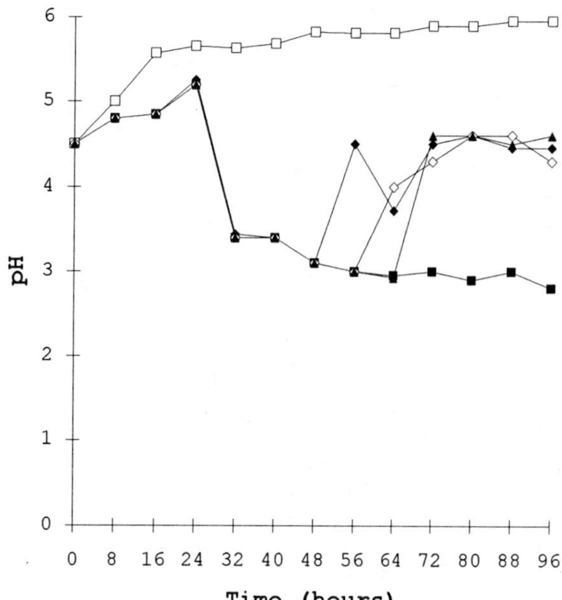

FIGURE 2. pH Variation in Submerged Fermentations of pure cultures and cocultures of *T. reesei* and *Monosporium sp.* ■: *T. reesei.* □ : *Monosporium sp.* ◆: 48h Coculture. ◇: 56h Coculture. ▲: 64h Coculture.

reesei (Fig. 1b) although extracellular protein was found in higher concentration with cocultures (Fig. 3). Pure cultures of *T. reesei* increased the medium pH during the first 20 fermentation hours (Fig. 2), which may be due to ammonia release by fungal urea breakdown coincident with the release of free sugars by cellulolysis (Fig. 4). Later on, biomass increases and uses ammonia (Fig. 1a), pH is reduced from 5 to 3 (Fig. 2) and sugar level is reduced to the initial value (1.14g/L). *Monosporium sp.* pure culture increases pH from 4.5 to nearly 5.5 during the first 20 hour-period of fermentation without further acidification of culture medium (Fig. 2) which is consistent with very little biomass increase after 20 hours (Fig. 1a) and very low levels of free sugars throughout, suggesting a very poor cellulolytic activity of this organism. All mixed fermentations showed an initial pH pattern similar to a pure *T. reesei* culture up to 40 or 60 hours (Fig. 2)

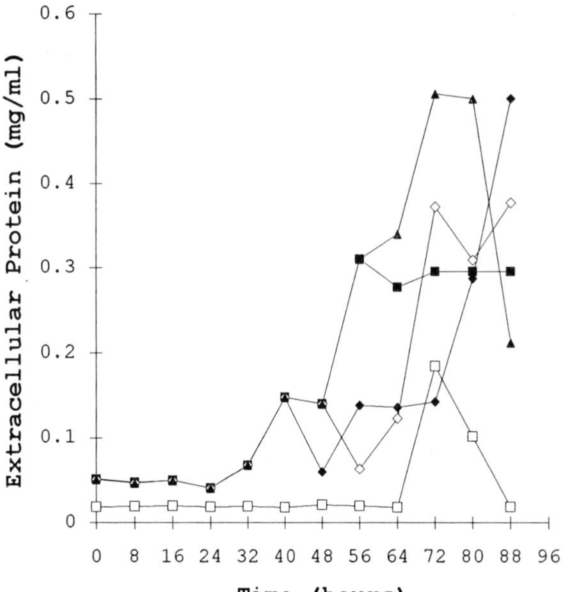

FIGURE 3. Extracellular Protein Production in Submerged Fermentations of pure cultures and cocultures of *T. reesei* and *Monosporium sp.* ■: *T. reesei*. ☐ : *Monosporium sp.* ◆: 48h Coculture. ✧: 56h Coculture. ⋏: 64h Coculture.

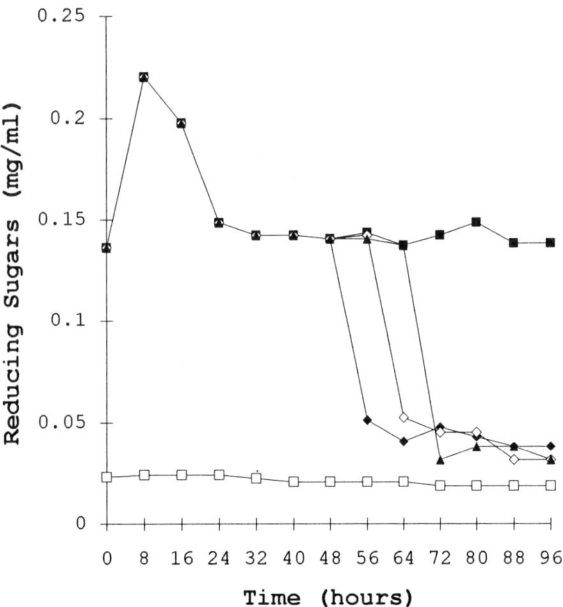

FIGURE 4. Reducing Sugars Production in Submerged Fermentations of pure cultures and cocultures of *T. reesei* and *Monosporium sp.* ■: *T. reesei*. ☐ : *Monosporium sp.* ◆: 48h Coculture. ✧: 56h Coculture. ⋏: 64h Coculture.

but with late pH increments which coincide with biomass production (Fig. 1a). Reducing sugar levels of mixed fermentations followed a similar pattern to pure *T. reesei* cultures but with a large decrease which coincided with the late biomass increase of all mixed fermentations.

DISCUSSION

The values of biomass obtained in fermentations of *T. reesei* are in agreement with values reported previously (2,4,18,21); it represents a protein productivity of 0.33 g per g of consumed cellulose (data not shown). However, the mixed fermentations showed an increase of this content, with best results in 48h-phasing fermentation processes.

The cellulase activity observed is in agreement with the one reported by Duff et al., (3) and slighty higher than the one described by Ghose et al.(4), and Panda et al.(15,16,17) for *Trichoderma reesei*, reaching a maximum value of 0.091 UI/ml. *Monosporium sp.* has very little cellulase activity, as some previous tests showed, with filter paper as the substrate (data not shown). The increments observed at 48, 56 and 64 hours of fermentation are consistent with the results of Figure 3, as the period of most liberation of extracellular protein. The fact that cellulase activities of mixed fermentations were smaller than those of pure cultures of *T. reesei* (Fig. 1b) but extracellular protein levels were larger in mixed fermentations (Fig. 3) may be related to pH differences between pure *T. reesei* cultures and mixed fermentations (Fig. 2) affecting the activities of cellulase excreted by *T. reesei* but inactive at the higher pH levels linked to biomass production in cocultures (Fig. 1a and 2). The early onset of free reducing sugars in cultures with *T. reesei* may be explained by a very active constitutive cellulase present in the spores and conidia of *T. reesei* and responsible of high concentration of oligosaccharides at a very early fermentation time as suggested by Kubicek (7). The concentration of those reducing sugars remains almost constant in the fermentation fluid during the process because of the slow assimilation rate of *T. reesei*. *Monosporium sp.*, on the other hand, keeps this concentration at very small values. The mixed processes, then, show a similar behavior once *Monosporium sp.* is introduced into the fermentation system: sugar concentration dramatically drops reaching very small values in a short time. This phenomenon has been described previously in mixed fermentations between a cellulolytic microorganisms and a non-cellulolytic one (11,18), explained by the rapid assimilation of sugars by the second microorganism: *Monosporium sp.* rapidly consumes the sugar generated by cellulases from *Trichoderma reesei*, causing the increment in biomass content observed.

In conclusion, biomass production of *Trichoderma reesei* in submerged fermentations of a cellulosic substrate has been approximately increased two fold by the utilization of a second microorganism, *Monosporium sp.*, in a mixed fermentation. In terms of biomass concentration, the best results were obtained when the second microorganism was inoculated at 48 hours of fermentation with *Trichoderma reesei*, even though cellulase activity in mixed fermentations was lower in all cases. The increase in protein content is due to fungal growth derived from sugar consumption by *Monosporium sp.* after its introduction into the fermentation system.

However, it was found that the yield of biomass per g of cellulose consumed at the end of the fermentation, was almost the same for the pure culture of *T. reesei* and the 48 hour coculture (0.33 g biomass/ g cellulose), what shows that the addition of *Monosporium sp.* increases the extent of cellulose degradation and therefore biomass production.

As the next step, submerged fermentations will be carried out to study the effect of the coculture on cellulase production when the pH is kept constant. In addition, the effect of the coculture on biomass production will be studied in solid-state fermentations.

LITERATURE CITED

1. AIELLO, C.; FERRER, A. "Fermentación de *T. reesei* QM9414. Efecto del tratamiento alcalino," Primer Simposio sobre Biotecnología. Astrodata, Caracas, p. 153 (1986).

2. CHAHAL, D.S.; MOO-YOUNG, M.; VLACH, D. *Mycologia*, **75**,597 (1983).

3. DUFF, S.; COOPER, D.; FULLER, O. *Enzyme Microbial Technol.*, **9**, 47 (1987)

4. GHOSE, T.K.; PANDA, T.; BISARIA, V. *Biotechnol. Bioeng.*, **27**, 1353 (1985).

5. HAN, Y.W.; DUNLAP, C.E.; CALLIHAN, C.D. *Food Technol.*, **25**, 32 (1971).

6. HECHT, V.; SCHUGERL, K.; SCHEIDING, W. *Eur.J. Appl. Microbiol. Biotechnol.*, **16**, 219 (1982).

7. KUBICEK, C. *J. Gen. Microbiol.*, **133**, 1481 (1987).

8. LEDESMA, A. Aislamiento de microorganismos celulolíticos aerobios de suelo y material vegetal en el estado Zulia. Thesis. Facultad Experimental de Ciencias, Universidad del Zulia. Venezuela (1990).

9. LEE, Y.H.; FAN, L.T. *Adv. Biochem. Eng.*, **17**, 101 (1980).

10. LOWRY, O.; ROSEBROUGH, N.; FARR, A.; RANDALL, R. *Anal. Chem.*, **16**, 190 (1965).

11. MARTY, D. *Biotechnol. Lett.*, **7**, 895 (1985).

12. MILLER, G. *Anal. Chem.*, **31**, 426 (1969).

13. MOO-YOUNG, M.; CHAHAL, D.S.; VLACH, D. *Biotechnol. Bioeng.*, **20**, 107 (1978).

14. MURRAY, W. *App. Environ. Microbiol.*, **51**, 710 (1986).

15. PANDA, T.; BISARIA, V.; GHOSE, T. *Biotechnol. Lett.*, **5**, 767 (1983).

16. PANDA, T.; BISARIA, V.; GHOSE, T. *Biotechnol. Bioeng.*, 30, 868 (1987).

17. PANDA, T.; BISARIA, V.; GHOSE, T. *Appl. Environ. Microbiol.*, **55**, 1044 (1989).

18. PEITERSEN, N. *Biotechnol. Bioeng.*, **17**, 1291 (1975).

19. PETERSON, G. *Anal. Biochem.*, **100**, 201 (1979).

20. REESE, E.T.; MANDEL, M.; WEISS, A. *Adv. Biochem. Eng.*, **14**, 181 (1972).

21. TSAO, G. *Process Biochem.*, **12**, 15. (1978).

Invited paper

Chitin as a Matrix for Enzyme Immobilization

A. Illanes

Escuela de Ingeniería Bioquímica, Universidad Católica de Valparaíso, Casilla 4059,
Valparaíso, CHILE

Chitin and its derivatives have been frequently used as a matrix for the immobilization of enzymes and cells. A thorough review on enzyme immobilization in chitin matrices is presented together with an assessment of their potential applications as industrial biocatalysts. Chitosan-immobilized fungal lactase is presented as a case study: the catalyst has been produced and used at pilot scale in the continuous hydrolysis of waste whey permeate to produce a sweet syrup for the dairy industry.

Much of the success for an immobilized enzyme as a process biocatalyst relies in the characteristics of the matrix. Desirable properties for a material to be used as an enzyme matrix are: large surface area per unit volume, high protein binding capacity, permeability, hydrophilicity, insolubility, chemical, mechanical, thermal and biological stability, rigidity, regenerability, flexibility for shaping, compatibility with product specifications and low cost. These characteristics are hardly met by a specific material and compromises among them have to be solved for each specific case. Chitin and their derivatives satisfy most of them, main drawbaks being its limited protein binding capacity and sensitivity to microbial chitinases. In the final balance, chitin and its derivatives are interesting materials to produce immobilized enzyme biocatalysts for industrial processes.

Chitin is a polymer of N-acetyl glucosamine units linked β1-4. The amino groups in chitin are acetylated. Deacetylated chitin is known as chitosan. In nature, a broad range of partially deacetylated polymers exists, the name of chitin or chitosan being set rather arbitrarily. A polymer in which most residues are acetylated is called chitin and the opposite holds for chitosan. The term chitosan will be preferred here, because any method used to extract chitin from its natural source produces some extent of deacetylation.

Chitin and chitosan, have been considered for long as very interesting and flexible raw materials and several applications have been proposed. Among those specifically related to biotechnology, recent applications in bioagriculture, waste treatment, downstream processing and biocatalysis are worthwhile to mention.

The amount of chitin is the highest in crustaceans, which constitute its main source as raw material for the industrial production of chitin and chitosan. Crustacean shells have about 25% of its dry matter in chitin and significant amounts of protein and pigment that can also be recovered if properly fractionated. The production of chitin from waste crab and king crab shells, with the recovery of pigmented protein has been evaluated in Chile. A total amount of 400 tons of chitosan and 300 tons of protein can be obtained, which represent 12 and 15% of total domestic potential market for these products.

CHITIN AS A MATRIX FOR ENZYME IMMOBILIZATION

Although this is a minor field of

461

E. Galindo and O.T. Ramírez (eds.), Advances in Bioprocess Engineering. 461-466.
© 1994 *Kluwer Academic Publishers. Printed in the Netherlands.*

application in quantitative terms, chitin and chitosan have been frequently reported as suitable carriers for the production of biocatalysts. Enzymes have been immobilized in chitosan by adsorption, covalently through bifunctional reagents such as glutaraldehyde, and also by entrapment. There is a considerable amount of information reported on enzyme immobilization on chitosan. An outline will be made aiming to evaluate chitosan as a carrier for enzyme immobilization.

Stanley et al. (1) immobilized fungal lactase by covalent linkage to chitosan through glutaraldehyde and also by simultaneous adsorption and crosslinkage with glutaraldehyde. Better results were obtained with the former system, in which 60 iu/g of carrier were obtained at a yield of 60%. Protein loaded in the carrier was 5 mg protein/g. The apparent Michaelis constant (K_m) of the immobilized enzyme was 86 mM, which is almost equal to the Michaelis constant of the soluble enzyme, 85mM, which means that the catalyst particle was free of diffusional restrictions.

Leuba and Widmer (2) immobilized various proteinases on Fluka chitosan (3.7 µmol-NH /mg). The matrix was prepared by reprecipitation and crosslinkage with epichlorhydrin and activation with glutaraldehyde. Immobilization yields were 15 and 17% for subtilisin and pronase and specific activities were 14 and 12.5 u/g of carrier respectively. Protein loaded was on the order of 30 mg/g, which is rather high for chitosan, but the catalytic efficiency of the immobilized enzymes were low, which is attributed to diffusional restrictions due to the high molecular weight of the substrate. Thermal stability, although modest, was significantly improved by immobilization.

Synowiecki et al (3) immobilized purified Fluka invertase (aproximately 200 iu/mg protein) on krill chitosan by adsorption and covalent binding with glutaraldehyde, obtaining specific activities of 1500 and 1200 iu/g of carrier respectively, corresponding to 7.5 and 6 mg protein /g carrier. Reported immobilization yield for the former was 45%. Although results were slightly higher for simple adsorption, immobilization in this case depended strongly on ionic strength, suggesting that binding is mainly of ionic nature. At ionic strengths higher than 0.01 M yield started to decrease down to 16% at ionic strength of 0.25 M.

Illanes et al. (4) immobilized crude invertase produced from autolysed bakers' yeast cells (31 iu/mg protein)) (5) on glutaraldehyde crosslinked chitin, obtaining a catalyst with 74 iu/g of carrier at a yield of 62%. Protein charge capacity was 3.3 mg/g of carrier and specific activity increased only to corresponding protein loads of 5 mg/g, the carrier being saturated with protein beyond that point. Enzyme half-life in the absence of substrate at 53 ºC was 25 days which is one order of magnitude higher than half-life for the soluble enzyme. Long-term packed-bed reactor operation with immobilized invertase at that temperature produced no elution of enzyme and operational half-life was estimated in 180 days. The carrier was regenerated three times with a loss in immobilization capacity of about 3% per cycle (6). All this demonstrates the feasibility of using chitosan as an industrial enzyme carrier. Chitosan was far superior than other carriers tested for invertase immobilization and diffusional restrictions were milder. Apparent K_m was 68.5 mM, as compared to 39.2 mM for the soluble enzyme. For invertase immobilized on activated charcoal, on the other hand, K_m was 185 mM. Optimum pH was displaced from 4.5 to 4.3 as a result of immobilization, which is in agreement with the net possitive charge of the carrier at those pHs.

Synowiecki et al (7) studied the effect of deproteinization and demineralization on the immobilization of different enzymes on krill chitosan. Results proved that the level of residual protein and minerals hinder enzyme immobilization. However, in certain cases immobilization was better when deproteinization was conducted at milder conditions. This can be attributed to the different degree of deacetylation produced by the alkaline conditions for protein extraction which affects covalent linkage through free amino groups in the chitosan molecule. In principle one can assume that higher specific activities should be obtained when the amount of free amino groups is higher. As shown for the case of invertase and amyloglucosidase, this was not the case, which can be attributed to steric hindrance and improper folding of the enzyme molecule produced by an excess of linkages.

Synowiecki et al (8) reported an extensive study on enzyme immobilization on chitosan by simple adsorption. Adsorption was produced mainly by ionic attraction between the amino groups of chitosan and the carboxyl groups of aspartic and glutamic acid residues, but also by hydrogen bonding, van der Waals forces and hydrophobic interactions. Although this system was simple and little denaturation ensued, catalyst stability was generally poor. These authors have tried to improve catalyst perfomance through activation with carbon sulfide which interacts with amino groups in chitosan, forming dithiocarbamino groups. The carrier had then both positive and negative charges interacting ionically with different amino acid residues in the enzyme molecule. Diastase immobilized by this scheme was almost as stable as that produced by covalent linkage through glutaraldehyde, but immobilization yield was higher, which makes carbon sulfide activation an interesting procedure. However, special caution has to be taken with ionic strength to avoid enzyme desorption.

Mitsutomi and Ohtakara (9) immobilized purified α galactosidase, on glutaraldehyde crosslinked colloidal chitosan at different enzyme to chitosan ratios. Specific activities ranged from 35 to 343 iu/g at yields from 46 to 70%. Residual enzyme increased from 13 to 39%, suggesting that the carrier was saturated with protein at the higher enzyme loads, which is surprising since highest protein load was only 0.52 mg/g carrier.

Pifferi et al (10) immobilized fungal - endopolygalacturonase (35 ui/mg protein) on monomaleyl and trimaleyl chitosan by simple adsorption and adsorption assisted by crosslinkage with glutaraldehyde. Specific activities from 1800 to 3100 iu/g of carrier, corresponding to 52 to 90 mg protein/g of carrier, were obtained at yields ranging from 19 to 32 %. Best results were obtained when immobilizing the enzyme to crosslinked trimaleyl chitosan. Catalyst half-life was 21 days at 25 ºC, twice the value for the soluble enzyme. Apparent K_m was 0.24 mM, while for the soluble enzyme K_m was 0.06 mM, meaning that in this case diffusional restrictions were observed, which is influenced by the high molecular size (30.000 daltons) of the substrate.

Kimura et al (11) tested different carriers for the immobilization of bacterial pullulanase, better results being obtained with chitosan beads. They reported specific activities of 85 and 135 iu/g of carrier at yields of 47 and 27% respectively, corresponding to a protein load of 3.6 mg/g for the former. Kusano et al (12) immobilized a partially purified bacterial pullulanse (88

iu/mg protein) on chitosan by adsorption and covalent linkage through bifunctional reagents like glutaraldehyde, hexamethylene diisocyanate (HD), toluene 2-4 diisocyanate (TD) and 4-4 diphenyl methane diisocyanate (DD). Yields were only moderate, around 20%, and enzyme losses were significant for the case of TD and DD. Protein load ranged from 10 to 20 mg/g. Apparent K_m for pullulanase immobilized on glutaraldehyde treated chitosan was 1.8 g/liter, which compares favorably with values for pullulanase immobilized on porous glass (K_m=6.4 g/liter) and on amberlite IRC-50 (K_m=6.8 g/liter). K_m for the soluble enzyme was 0.4 g/liter, meaning that the immobilized enzyme was subjected to diffusional restrictions, which were milder for the case of chitosan and quite insignificant if one considers the polymeric nature of the substrate. Chitosan immobilized pullulanase remained fully stable for 30 days at 60 ºC, while the same enzyme immobilized on amberlite IRC-50 lost half of its activity in that period of time.

Kim and Rhee (13) immobilized inulinase on glutaraldehyde activated chitosan and obtained at optimal conditions an immobilization yield of 23%, with specific activity of 77 iu/g of carrier. No data was presented on protein load. Specific activity was significantly higher for this carrier, as compared to others such as aminoethyl cellulose and diethylaminoethyl cellulose. Enzyme stability was also higher (estimated half-life was 24 days at 40º C) as was reactor productivity for the conversion of Jerusalem artichoke's inulin into fructose.

Braun et al (14) reported the immobilization of penicillin acylase from E.coli on reprecipitated chitosan shaped into powder, particles and beads and activated with glutaraldehyde. Enzyme was a crude preparation 0.25 iu/mg of protein. Yields and specific activities were markedly dependent on particle size, very good results being obtained for particles less than 0.5 mm in size. They studied the effect of protein load, results being rather surprising. Higher values exceed one gram of protein per gram of carrier, which is more than one order of magnitude higher than all others published. The authors declare their results are hard to compare with others, which is apparent from our analysis. Even though these results are spectacular in terms of protein load, enzyme yields and specific activities were high (from 20 to 200 iu/g) but not unusual, since the specific activity of the enzyme preparation was very small.

Enzyme losses were higher at smaller loads, which is hard to explain; however, enzyme yield decreased noticeably at loads higher than 800 mg protein/g of carrier, the sharp increase in residual activity meaning that the carrier was saturated beyond that point.

Ospina et al (15) compared several immobilized penicillin acylases in terms of their kinetic behaviour. Chitosan immobilized enzyme exhibited better kinetic properties than amberlite and epoxyacrylic resin immobilized enzymes and Novo's Semacylase. We have produced an immobilized penicilin acylase from Bacilus megaterium. Working with an enzyme with 15 iu/mg protein, obtained after broth microfiltration and ultrafiltration, an immobilized biocatalyst was obtained on activated chitosan with a yield of 25% and specific activity of 53 iu/g catalyst (16). Protein load was 7 mg/g catalyst, which is in the lower range of values reported for such matrix. Results are being improved by working with an enzyme further purified by selective adsorption into cellite and ion-exchange chromatography (17).

From the published information on immobilization of enzymes on chitosan-based matrices, some general statements can be made:

-Chitosan is a very flexible material which can immobilize enzymes by simple adsorption, by reticulation assisted adsorption, by covalent linkage through bifunctional and other activiting agents and even by gel entrapment.

-Ionic interactions between amino groups in the chitosan molecule and negative charges in the amino acid residues of the enzyme molecule are prevalent in adsorption, although van der Waal forces and hydrophobic interactions also play a role.

-Immobilization yields tend to be higher in adsorption than in covalent linkage due to the milder conditions in the former. Also specific activities are higher, but stabilities are rather poor and quite sensitive to ionic strength and pH, which is a serious drawback for reactor operation. Therefore, immobilization on activated chitosan through covalent linkage is the method of choice despite these disadvantages.

-When compared to other carriers, chitosan load capacity is rather low and can be estimated between 10 and 30 mg/g. Only one report presents protein loads of several hundreds mg/g, well over this range. Despite this modest capacity, enzyme yields are rather high, meaning that the catalytic expression of the immobilized protein is quite high. In fact, immobilization is mainly superficial and diffusional restrictions are low even with high molecular weight substrates, which is a definite advantage of chitosan as a matrix.

-Chitosan is quite stable, it can be used in prolongued operation and can be regenerated after enzyme decay. When shaped properly it is mechanically strong and can withstand harsh conditions as those found in industrial reactor operation.

-When compared with other carriers, chitosan is almost always selected because of higher yields, low diffusional restrictions, ease of handling and flexibility of configuration.

-The effect of impurities and the level of deacetylation affects significantly the immobilization capacity of chitosan. Residual protein and minerals decrease immobilization capacity but the level of deacetylation does not correlate with enzyme yield as expected, possibly because of improper enzyme folding and steric hindrances.

A thorough study on the efect of chitosan structure and purity on enzyme immobilization remains to be done.

IMMOBILIZATION OF FUNGAL LACTASE ON ACTIVATED CHITIN: A CASE STUDY

Lactase (β-D galactoside galactohydroase, EC 3.2.1.23) is a relevant enzyme for the dairy industry both in the production of delactosed milk and dairies and in the upgrade of waste whey and whey permeate. For the former, yeast neutral lactases (mainly from Kluyveromyces species) are used because of pH restraints. Acid fungal lactases (mainly from Aspergillus species) which are extracellular, simpler and more stable proteins are to be preferred for whey permeate hydrolysis. The potential for the economic production of hydrolyzed whey permeate is considered good, especially in those countries with limited production of conventional sweeteners, facing increasing environmental restrictions (18). Because of the high dilution and volumes of substrate, the use of immobilized enzymes is mandatory.

We have developed a process for the production of an industrial quality immobilized fungal lactase on activated chitosan (CIL). Soluble enzyme was a crude commercial lactase (ENZECO Fungal Lactase from A.oryzae, EDC, New York). Chitosan was produced from shrimp shells according to a modified Hackman procedure (19). The process considers an activation stage, in which the chitosan matrix is activated with glutaraldehyde, and an immobilization stage, in which the soluble enzyme is contacted with the activated chitosan matrix. Immobilization was studied at laboratory scale, where relevant variables were screened and then optimized in terms of specific activity and immobilization yield. Final results are presented in Figure 1. Specific activities over 400 iu/g at yields of 35% were obtained at optimum conditions (20). Better results were obtained when working with material of

lower degree of deacetylation (labeled "chitin" in Figure 1, as opposed to higly deacetylated material, labeled "chitosan"). This, although contrary to expected, has been also observed by Synowiecki et al (7) when working with invertase and glucoamylase. Kinetic behaviour of CIL was very good. The biocatalyst is esentially free of internal and external diffusional restrictions, and less sensitive to galactose inhibition than the soluble enzyme.

The process was scaled up to pilot plant level, immobilization being performed in situ (within the reactor. Results were highly reproducible with specific activities over 400 iu/g and yields close to 40% The biocatalyst was tested in the continuous operation of a pilot packed-bed reactor. Mechanical behaviour was excellent, no attrition or elution was observed and flow pattern was homogeneous, without channeling or significant void spaces within the catalyst bed. Enzyme operation half-life on whey permeate at 40ºC was 120 days as compared to 10 days for the soluble enzyme (20). As seen in Figure 2, chitosan immobilized lactase perfomed well when compared with leading commercial catalysts.

A preliminary economic evaluation of a chilean dairy facility for the production of lactase hydrolyzed whey permeate syrup and ultrafiltered protein was made based on the results obtained with CIL. Domestic market for such products was estimated in 1300 and 350 tons/year for syrup and protein respectively. Sale prices were estimated in US$ 0.3 and US$ 2.3 per kg. of syrup and protein respectively. Capital investment was US$

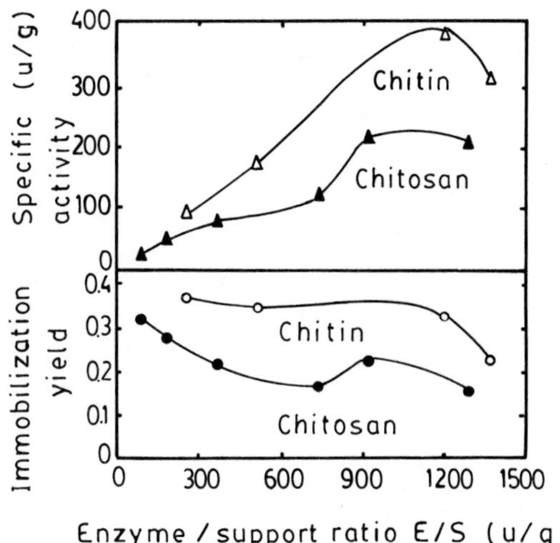

Figure 1. Immobilization yield and specific activity of immobilized lactase as a function of enzyme to support ratio (E/S).

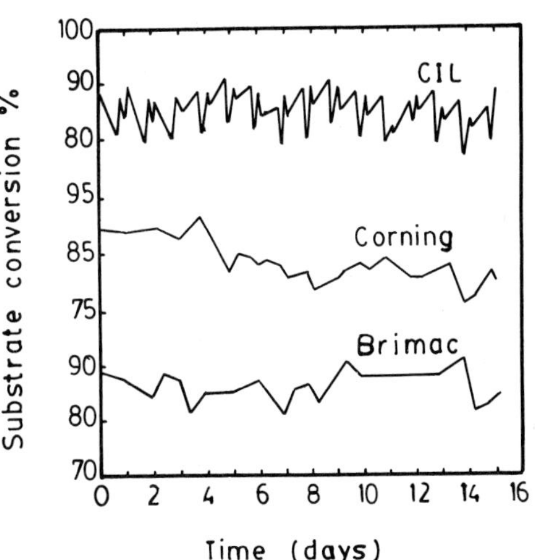

Figure 2. Long-term packed bed reactor performance with chitosan-immobilized lactase (CIL) and commercial immobilized lactase from Corning and Brimac. Operating conditions were 4ºC, pH 4.0, 72 ml/h of whey permeate, 40 g/l in lactose. Catalyst bed was washed daily with acetic acid solution to control bacterial contamination.

1.3 million and internal return rate of investment was 46% (20). However, the assumed value for the syrup may now be considered unrealistic when compared to commercial sweeteners in the market. Even so, if waste disposal is penalized, the developed technology might be an interesting alternative for permeate upgrade in Chile. The syrup has been tested successfully at the industrial level in products like milk pudding and toffees, and the product has been isomerized with immobilized commercial glucose isomerase to produce a sweetener syrup.

LITERATURE CITED

1. Stanley, W., Watters, G., Chan, B., and J. Mercer. Biotechnol.Bioeng. 17, 315 (1975).

2. Leuba, J. and F. Widmer. Biotechnol. Letters 1, 109 (1979).

3. Synowiecky, J., Sikorski, Z. and M. Naczk. Biotechnol.Bioeng. 23, 231 (1981)

4. Illanes, A., Chamy, R. and M.E. Zúñiga. Immobilization of invertase on crosslinked chitin, in Chitin in Nature and Technology, Muzzareli, R., Jeuniaux, C. and G. Gooday (Eds), Plenum Press, New York (1986).

5. Illanes, A. and Y. Gorgollón. Enzyme Microb.Technol. 7, 510 (1986).

6. Illanes, A., Zúñiga, M.E., Chamy, R. and M.P. Marchese. Immobilization of lactase and invertase on crosslinked chitin, in Bioreactor Immobilized Enzymes and Cells, Moo-Young, M. (Ed), Elsevier Applied Science London (1988).

7. Synowiecki, J. Sikorski, Z. and M. Naczk, Biotecnol.Bioeng., 23, 2211 (1981).

8. Synowiecki, J., Sikorska-Siondalska, A. and A. El-Bedawey, A. Biotecnol.Bioeng., 29, 352 (1987).

9. Mitsutomi, M. and A. Ohtakara. Agric. Biol.Chem. 12, 3153 (1984).

10. Pifferi, G., Tramontini, M. and A. Malacarne. Biotechnol.Bioeng. 33, 1258 (1989).

11. Kimura, T., Yoshida, M., Oishi, K., Ogata, M. and T. Nakakuki, Agric.Biol.Chem. 7, 1843 (1989).

12. Kusano, S., Shiraishi, T., Takahashi, S., Fujimoto, D. and Y. Sakano. J.Ferment. Technol. 68, 233 (1989).

13. Kim, C. and S. Rhee. Biotechnol.Letters 11, 201 (1989).

14. Braun, L., Le Chanu, P. and F. Le Goffic. Biotechnol.Bioeng. 33, 242 (1989).

15. Ospina, S., López-Munguía, A., González, R.L. and R. Quintero. J.Chem.Technol.Biotechnol. 53, 205 (1992).

16. Illanes, A., Ruiz, A., Vásquez, M. and R. Torres. Caracterización cinética de penicilina acilasa. Proc. 3rd Latinamerican Congress of Biotechnology, Santiago, p 223 (1993).

17. Torres, R. Illanes, A. and A. Ruiz. Comparación de dos estrategias de producción de penicilina acilasa. Proc. 3rd Latinamerican Congress of Biotechnology, Santiago, p 217 (1993).

18. Marwaha, S. and J. Kennedy. Internat. J.Food Sc.Technol. 23, 323 (1988).

19. Muzzarelli, R. Chitin, p 90, Pergamon Press, New York (1977).

20. Illanes, A., Ruiz, A., Zúñiga, M.E., Aguirre, C., O'Reilly, S. and E. Curotto. Bioproc.Eng. 5, 257 (1990).

Enzyme Reactor Performance Under Thermal Inactivation

A. Illanes, C. Altamirano, and O. Cartagena

Escuela de Ingeniería Bioquímica, Universidad Católica de Valparaíso, Casilla 4059, Valparaíso, CHILE

Enzymes are labile catalysts when used under harsh process conditions. Modelling enzyme inactivation under operation is therefore relevant for the design and evaluation of enzyme reactor performance. A model is presented to describe thermal enzyme inactivation under protection by reaction effecters and applied to develop equations to describe reactor performance. Two case studies are presented: batch reactor with immobilized penicillin acylase and packed-bed continuous reactor with immobilized lactase.

Thermal inactivation of enzymes has been a subject of considerable interest in the recent past (1). Temperature is a critical variable in enzyme reactor operation. It produces two opposed effects, increasing reaction rate, but also increasing the rate of enzyme inactivation. Therefore, an optimum temperature will always exist for an enzyme reactor and it will be time dependent. It is a fact that enzymes are more stable under operation, which has been attributed mainly to substrate protection. Except for substrate protection of enzyme inactivation, which has been frequently reported and studied (2), little consideration has been paid to other enzyme protecting agents during catalysis (3), despite its importance to assess reactor perfomance properly.

A generalized scheme for enzyme inactivation under effecter protection is presented and two cases are analyzed: 6 aminopenicillanic acid production with immobilized penicillin acylase in a recirculating batch reactor, and continuous hydrolysis of lactose with immobilized lactase in a packed-bed reactor.

MODELS FOR ENZYME INACTIVATION

First-order kinetics is the simplest and most used model to describe thermal enzyme inactivation. Several studies prop-

ose it as a suitable model to describe immobilized enzyme inactivation (4,5). However, more complex models have been proposed to describe it more rigorously, as is the case for parallel and series mechanisms. In the former, the enzyme is assumed to be heterogeneous, with two or more forms each inactivating by first-order kinetics, but at different specific rates (6); in the latter, the enzyme is considered to suffer a transition from a fully active stage to progressively less active stages down to a completely inactive form (7). These three models can be described by Equations (1) to (3). In the last two cases, two forms of enzymes (e_1 and e_2) were considered.

$$e = e_0 \exp(-k_D\, t) \qquad (1)$$

$$e = e_{10} \exp(-k_{D1} t) + e_{20} \exp(-k_{D2} t) \qquad (2)$$

$$e/e_0 = [1 - k_{D1}/(k_{D1} - k_{D2})\, e_1/e_0]\exp(-k_{D1} t) +$$
$$+ k_{D1}/(k_{D1} - k_{D2})\, e_1/e_2)\, \exp(-k_{D2} t) \qquad (3)$$

e represents the enzyme activity (volumetric or specific), e_0 its initial value and k_D the first-order enzyme decay constant, which strongly depends on temperature, usually in an Arrhenius mode.

E. Galindo and O.T. Ramírez (eds.), Advances in Bioprocess Engineering. 467-472.
© 1994 Kluwer Academic Publishers. Printed in the Netherlands.

ENZYME INACTIVATION UNDER PROTECTION

The protective effect of substrate was considered originally by O'Neill (8), who proposed a simplified model in which the enzyme first-order decay rate depended inversely on the substrate concentration. The subject was studied more recently by Chen and Wu (2), who considered different first-order decay rates for the enzyme and the enzyme-substrate complexes, in the reversible conversion of glucose into fructose by immobilized glucose isomerase.

We have extended Chen and Wu's analysis to consider potential protection of all effecters, that is, all substances that may interact with the enzyme molecule during catalysis. According to this, considering different first-order decay constants (α) for all possible enzyme complexes (E^*):

$$-de^*/dt = \alpha \, e^* \qquad (4)$$

For the case of the free enzyme: $\alpha = k_D$, and for the remaing enzyme complexes:

$$\alpha = k_D \, (1-n_i) \qquad (5)$$

where n_i represent the effecter protection factors. When $n_i = 1$, the enzyme is fully protected; when $n_i = 0$, the effecter does not protect and the effecter-enzyme complex decays at the same rate as the free enzyme.

Considering the reaction $S \rightarrow P_1 + P_2$, Figure 1 represents the scheme in which S, P_1 and P_2 are potential inhibitors, but also protecting effecters. Substrate inhibition is acompetitive, product inhibitio by P_1 is competitive and by P_2 is non-competitive.

The corresponding kinetic equation is:

$$v = \frac{k\,e\,s}{s(1+p_2/K_2+s/K') + K[1+p_1/K_1+p_2/K_2+p_1p_2/(K_1K_2)]} \qquad (6)$$

where v is the initial rection rate, s, p_1 and p_2 the molar concentrations of the respective species, k the catalytic constant, K the Michaelis constant for substrate and K_1, K_2 and K' the inhibition constants for P_1, P_2 and S respectively.

Defining the substrate conversion, X, as:

$$X = (s_0-s)/s_0 = p_1/s_0 = p_2/s_0 \qquad (7)$$

$$v(X) = \frac{k\,e\,(1-X)}{A + BX + CX^2} = k\,e_\sigma(X) \qquad (8)$$

$$A = 1 + K/s_0 + s_0/K' \qquad (9)$$

$$B = K/K_1 + K/K_2 + s_0/K_2 - 2s_0/K' - 1 \qquad (10)$$

$$C = s_0/K' + s_0K/(K_1K_2) - s_0/K_2 \qquad (11)$$

$$\sigma = \frac{v(X)}{k\,e} = \frac{(1-X)}{A + BX + CX^2} \qquad (12)$$

Material balance for the enzyme yields:

$$e = e_L + c + d + f + g + h + i \qquad (13)$$

where e_L is the molar concentration of the free enzyme and c, d, f, g, h and i are the molar concentration of enzyme complexes, according to Figure 1.

From Equations (4), (5) and (13):

$$de/dt = -k_D e_L - k_D(1-n_1)c - k_D(1-n_2)d -$$
$$- k_D(1-n_3)f - k_D(1-n_4)g -$$
$$- k_D(1-n_5)h - k_D(1-n_6)i \qquad (14)$$

$$de/dt = -k_D\,e\,[1 - \sigma(X)\,N(X)] \qquad (15)$$

$$N(X) = n_1 + n_2KX/[K_1(1-X)] + n_3KX/[K_2(1-X)] +$$
$$+ n_4s_0X/K_2 + n_5Ks_0X^2/[K_1K_2(1-X)] +$$
$$+ n_6s_0(1-X)/K' \qquad (16)$$

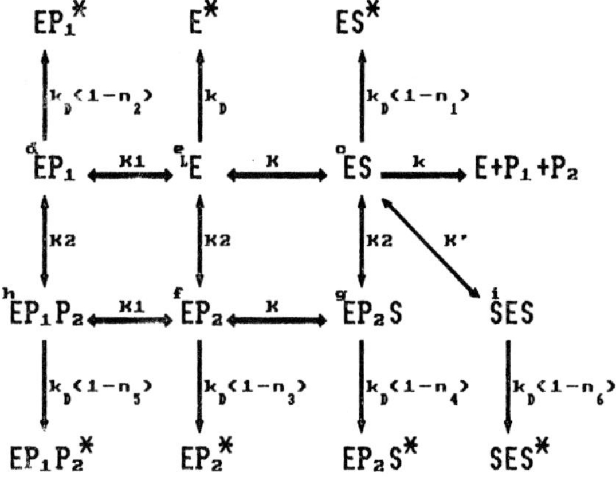

Figure 1. Scheme for reaction kinetics and enzyme inactivation in the conversion of S into P_1 and P_2. P_1 is a competitive inhibitor, P_2 a non-competitive inhibitor and S an acompetitive inhibitor. ($*$) represents the inactive forms of the enzyme species.

CASE 1. BATCH REACTOR PERFOMANCE WITH IM-
MOBILIZED PENICILLIN ACYLASE UNDER PROTECTED
THERMAL INACTIVATION.

Penicillin acylase (PA), (EC 3.5.1.11),
catalizes the hydrolysis of penicillin G
(PG) to 6 aminopenicillanic acid (6APA) and
phenylacetic acid (PhAA). PhAA is a competitive inhibitor, 6APA a non-competitive
inhibitor and PG has been frequently reported
to exert inhibition at high concentrations
(9, 10). Kinetics of PG hydrolysis by PA
can then be represented by Equation (6),
where S, P_1 and P_2 represent PG, PhAA and
6APA respectively, s, p_1 and p_2 their molar
concentrations and K, K_1, K_2 and K' the
Michaelis constant for PG and the inhibition
constants for PhAA, 6APA and PG.

The model for thermal inactivation is
now applied to the simulation of a batch
reactor operation with recirculation, which
is an appropriate configuration for an
enzyme inhibited by products and highly
sensitive to pH (11). For a batch reactor:

$$v = - ds/dt = s_0 \, dX/dt \qquad (17)$$

$$dX/dt = k \, e(t) \, \sigma(X) \, / \, s_0 \qquad (18)$$

$$de/dt = -k_D \, e \, [1 - \sigma(X) \, N(X)] \qquad (15)$$

Reactor performance is evaluated by the
simultaneous solution of differential Equations (15) and (18). Kinetic parameters
and operating conditions are presented in
Table 1 and correspond to Bacillus megaterium
PA immobilized in chitosan (12).

Table 1. Data for evaluation of batch reactor
performance with immobilized penicillin acylase.

Parameters

K	6.9	mM
K_1	16.5	mM
K_2	12.0	mM
K'	111.0	nM
k	$5.6 \cdot 10^{-4}$	h^{-1}

Operating Conditions

S_0	280	mM
e_0	33000	ui/l
X	0.96	
6APA	60	t/yr
Reaction	3.5	m
N batchs	300	

Results are presented in Figure 2,
assuming different combinations of protec-
factors n for the different enzyme effecters. Enzyme decay is plotted against total
reaction time, that is, the sum of times of
all batches. Time for each batch is determined to produce a substrate conversion of
0.96; therefore, the time for each batch
increases as the enzyme activity decays.
Final time in Figure 2 corresponds to that
required to perform the process task (60
tons of 6APA). As seen, when all n_i =1
(curve e), enzyme remains stable and total
time required is 1370 h. When all n_i =0
(curve a), enzyme decay is quite pronounced.
When n_1=0, all other n_i =1 (curve d), enzyme
decay is insignificant and total reaction
time is very much the same as in curve e.
The same holds for n_2 =0, n_3 =0, and n_6 =0, all
other n_i =1. This means that protection by
the effecters is not significant when forming
secondary complexes with the enzyme; this is
also the case for the tertiary complex SES.
When n_4=0, all other n_i =1 (curve c), enzyme
activity lost is close to 30% and total reaction time required is 1580 h. When n_5=0 all
other n_i=1 (curve b), enzyme activity lost is
close to 50% and total reaction time required
is 1780 h. This means that protection effect
is quite significant for the tertiary complexes $EP_2 S$ and $EP_1 P_2$. These conclusions
are valid for the kinetic data presented.
However, the same type of analysis, when
made for a commercial PA whose kinetic
parameters were different (the value of K
was almost the same, but K_1 and K_2 were much
higher in this case), led to other conclusions. In that case, protection by substrate was the most significant, both in ES
and $EP_2 S$ complexes (13). A sensibility
analysis was then made, considering different
K_1/K and $K_2/$ ratios. Results are presented
in Table 2, in terms of enzyme decay (e/e_0)
after 1200 h of reactor operation (roughly
one half-time for the unprotected enzyme).
Higher values for K_1 and K_2 correspond to
those of the commercial PA, lower values to
the chitosan-immobilized PA.

As seen, when the value of K is significantly lower than K_1 and K_2 , the protection
of substrate, both in secondary and tertiary
complexes is prevalent. When the value of K
is in the order of magnitude of K_1 and K_2 ,
substrate protection is less important and
in this case protection is significant only
in the tertiary complexes containing P_2 (EP_2
S; $EP_1 P_2$).

It can be concluded that the model
correctly predicts that the protection

effect is more significant for those effecters that interact with the enzyme more strongly (lower K values). The protecting effect reflects both in the amount of enzyme and the processing time required for a given process task. Since immobilized enzymes are expensive commodities, this aspect is certainly relevant in terms of production cost. Experimental determination of n_i values for PA is now underway.

Table 2. Model response assuming different values for K_1 and K_2. K=6.9 mM; K'=111 mM. Protection factor (n_i) identified is 0, all others are 1.

			e/e$_0$				
K (mM)	K_2 (mM)	n_1	n_2	n_3	n_4	n_5	n_6
330	240	0.79	0.97	0.98	0.87	0.97	0.74
82.5	60	0.88	0.97	0.96	0.76	0.91	0.85
16.5	12	0.98	0.95	0.95	0.74	0.69	0.97

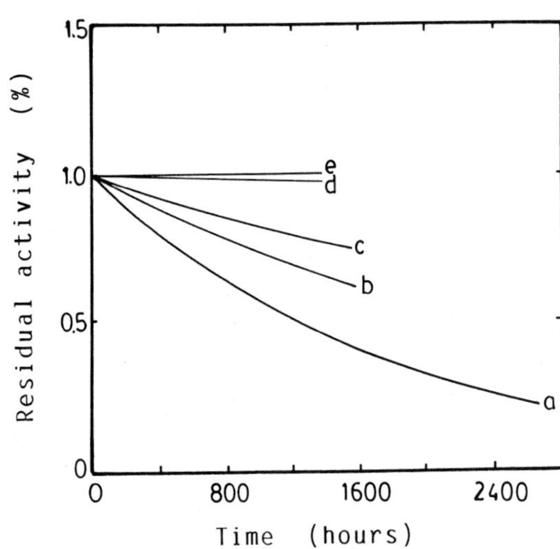

Figure 2. Enzyme decay during enzymatic hydrolysis of penicillin G with immobilized penicillin acylase in a batch reactor. (a): all n_i=0; (b) : n_5=0, all other n_i=1; (c) n_4=0, all other n_i=1; (d): n_i=0, all other n_i=1; (e): all n_i=1.

CASE 2. PACKED-BED REACTOR PERFORMANCE FOR THE CONTINUOUS HYDROLYSIS OF WHEY PERMEATE WITH IMMOBILIZED LACTASE.

Lactase (β-D-galactoside galactohydrolase, EC 3.2.1.23) catalyzes the hydrolysis of lactose to glucose and galactose. Immobilized fungal lactase can be used for the upgrade of whey permeate, which is currently a waste in the milk-processing industry (14). The enzyme is inhibited by galactose competitively and glucose has a mild mixed-type inhibition effect, which is not relevant for practical purposes (15). Therefore, Equation (6) reduces to:

$$ v = \frac{k\,e\,s}{s + K(1+p_1/K_1)} \qquad (19) $$

where S stands for lactose and P1 for galactose, s and p_1 being their molar concentrations, K the Michaelis constant for lactose and K_1 the inhibition constant for galactose.

In this case, Equation (8) applies with:

$$ A = 1 + K/s_0 \qquad (20) $$

$$ B = K/K_1 - 1 \qquad (21) $$

$$ C = 0 \qquad (22) $$

and Equation (16) simplifies to:

$$ N(X) = n_1 + n_2 KX/[K_1(1-X)] \qquad (23) $$

The model for thermal inactivation is now applied to the simulation of the continuous hydrolysis of whey permeate in a packed-bed reactor with immobilized lactase. For such reactor, under pseudo-steady state operation and piston-flow regime:

$$ F\,s_0\,dX = v(X)\,dV_R' \qquad (24) $$

$$ dV_R' = \varepsilon\,A_S\,dZ \qquad (25) $$

$$ dX/dZ = v(X)\,\varepsilon\,A_S\,/\,(F\,s_0) = $$

$$ = k\,e\,\sigma(X)\,\varepsilon\,A_S\,/\,(F\,s_0) \qquad (26) $$

where F is the feed flowrate, V_R', the reactor fluid volume, A_S its section, Z its height, and ε the liquid volume fraction in the catalyst bed.

Reactor performance is evaluated by the simultaneous solution of differential Equations (15) and (26). Kinetic parameters and operating conditions are presented in

Table 3, which correspond to fungal lactase immobilized in chitosan (15).

Results are presented in Figure 3, assuming different combinations of protecting factors n for the different enzyme effecters. Substrate conversion is plotted against reactor operating time. As seen, when n_1 (corresponding to ES) and n_2 (corresponding to EP_1) are 1 (curve d), enzyme remains stable and substrate conversion decreases very little. When both n_1 and n_2 are 0 (curve a), substrate conversion decreases sharply, approaching 0.4 after 300 days. By comparing curves b (n_1 =1, n_2 =0) and c (n_1 =0, n_2 =1), it can be concluded that in this case, protection by the product galactose is more important than substrate protection, which is in agreement with the general principle that the protection effect is more significant for those effecters that interact with the enzymes more strongly (lower K_i values). Profiles for enzyme decay through reactor length are presented in Figures 4 and 5 for all n_i =0 (unprotected system) and for n_1 =0 and n_2 =1 (full protection by galactose and no protection by lactose) respectively. In the first case, profiles are flat, but in the second case enzyme decay is differential, being slower near reactor outlet where galactose concentration is higher. Profiles for the fully protected system are similar to those in the second case, while for the case of n_1 =1 and n_2 =0 (full protection by lactose and no protection by galactose) profiles are similar to those in the first case.

Table 3. Data for evaluation of continuous packed-bed reactor performance with immobilized lactase.

Parameters

K	87	mM
K_1	20	mM
K_2	$2.9 \cdot 10^{-3}$	h^{-1}

Operating Conditions

s_0	110	nM
e_0	340	ui/l
X_0	0.90	
F	2.0	m^3/h
Reaction volume	3.3	m^3
ϵ	0.5	

It can be concluded that enzyme decay is differential through reactor height as the net result of substrate and product protection. If product protection is prevalent, as in this case, enzyme decay rate will decrease from inlet to outlet as product builds up and substrate is depleted. The opposite will occur when substrate protection is prominent. Again, consideration of effecter protection during catalysis affects reactor performance considerably and has to be taken into consideration for enzyme reactor design.

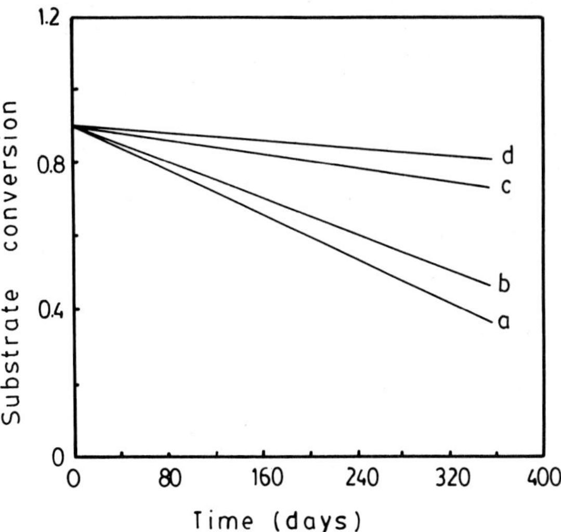

Figure 3. Substrate conversion decay during continuous hydrolysis of whey permeate with immobilized lactase in a packed-bed reactor. (a)): all n_i =0; (b): n_1 =1, n_2 =0 (c): n_1 =0, n_2 =1; (d): all n_i =1.

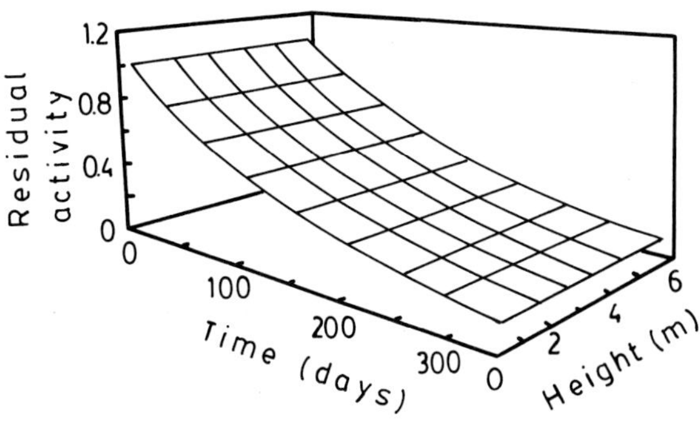

Figure 4. Profiles of enzyme decay through the catalyst bed in the continuous hydrolysis of whey permeate with immobilized lactase under non-protected conditions (n_1=n_2=0).

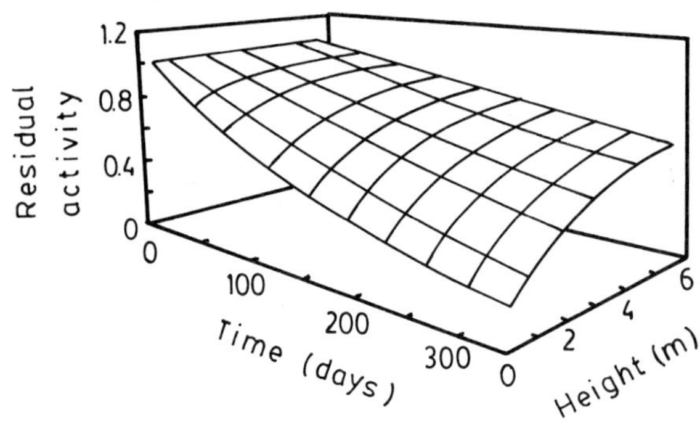

Figure 5. Profiles of enzyme decay through the catalyst bed in the continuous hydrolysis of whey permeate with immobilized lactase under full protection by galactose ($n_1=0$; $n_2=1$).

LITERATURE CITED

1. Henley, J. and A. Sadana. Biotechnol. Bioeng. 28, 1277 (1986).

2. Chen, K. and J. Wu. Biotechnol.Bioeng. 30, 817 (1987).

3. Alvaro, G., Fernández-Lafuente, R., Blanco, R. and J. Guisán. Enzyme Microb. Technol. 13, 210 (1991).

4. Erarslans, A. and H. Kocer. J.Chem. Technol.Biotechnol. 55 79 (1992).

5. Dubey, A., Bisaria, V., Mukhopadhyay, S. and T. Ghose.Biotechnol.Bioeng. 33,1311 (1989).

6. Dagys, R., Pauliokonis, A. and D. Kazlauskas. Biotechnol.Bioeng. 26, 620 (1984).

7. Henley, J. and A. Sadana. Enzyme Microb. Technol. 6, 35 (1984).

8. O'Neill, S. Biotechnol.Bioeng. 14, 473 (1972).

9. Park, M., Choi, C., Han, M. and B. Seong. Biotechnol.Bioeng. 24, 1623 (1982).

10. Warburton, D., Dunnill, P. and M. Lilly. Biotechnol.Bioeng. 15, 13 (1973).

11. Vandamme, E. Immobilized biocatalyst and antibiotic production, In: M. Moo-Young (Ed), Bioreactor ImmobilizedEnzymes and Cells, Elsevier AppliedScience, London, pp. 281-286 (1988).

12. Illanes, A., Ruiz, A., Vásquez, M. and R. Torres. Caracterización cinética de penicilina acilasa de Bacillus megaterium inmovilizada en quitina. Proc. 3rd. Latin-american Congress of Biotechnology, Santiago, Chile, p. 223 (1993).

13. Illanes, A., Cartagena, O. and R. Conejeros. Reactor design for penicillin acylase under thermal inactivation. Proc. Ninth International Biotechnology Symposium, Crystal City, USA, Nº 596(1992).

14. Illanes, A., Ruiz, A., Zúñiga, M., Aguirre, C., O'Reilly, S. and E. Curotto. Bioproc. Eng. 5, 257 (1990).

15. Illanes, A., Zúñiga, M. and A. Ruiz. Arch.Biol.Med.Exp. 23, 159 (1990).

Process Strategies in Enzymatic Racemate Resolution

J.L.L. Rakels, A.J.J. Straathof, and J.J. Heijnen

Department of Biochemical Engineering, Delft University of Technology, Julianalaan 67,
2628 RX Delft, THE NETHERLANDS

There is an increasing need to produce chiral building blocks, for the synthesis of pharmaceuticals, agro-chemicals and food-chemicals, rather than as racemates. Biocatalytic methods are suitable tools to accomplish the resolution of enantiomers. New strategies for enhancement of the enantiomeric purity of the product by enzymatic kinetic resolution were designed, modeled, and experimentally verified. Reaction kinetics, continuous process performance, and simultaneous multiple reactions and enzymes were considered.

THE STATE OF THE ART IN CHIRALITY

Chirality in biologically active molecules is of natural occurrence. Asymmetry is often involved in the interaction of pharmaceuticals, agrochemicals and food chemicals with a biochemical system of the living organism. There is a need for organic building blocks in order to synthesize such compounds. Chemical synthesis of a chiral building block usually results in a racemate, i.e. both enantiomers are produced in equal amounts. The chemical and physical properties of these isomers are mostly indistinguishable. The biological properties of the two enantiomers, however, may be entirely different [1,6,7,16]. In general only one of them is responsible for he desired activity. The second enantiomer may cause undesired inhibition, or, in the worst case, (toxic) side-effects. Nevertheless, a significant share of synthetic chiral pharmaceuticals and agrochemicals is still sold as racemate (Figure 1). The current situation is rapidly changing because in many countries product purity restrictions have been drastically adjusted towards avoidance of racemic drugs and preference for unpolluted single enantiomers in admission to the market [7,16,19].

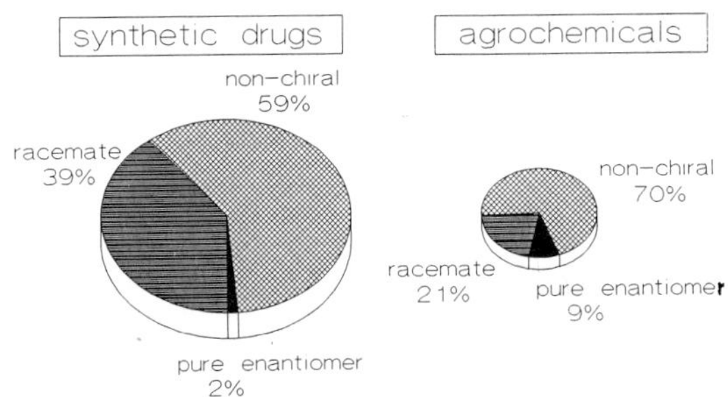

Figure 1. Estimates of commercialized synthetic drugs (total of >1500) and agrochemicals (total of 700) in 1990 by Athur D. Little Estimates [24]; see also [1,19].

Methods for the preparation of chiral building blocks

Various ways to accomplish the

E. Galindo and O.T. Ramírez (eds.), Advances in Bioprocess Engineering. 473-480.
© *1994 Kluwer Academic Publishers. Printed in the Netherlands.*

production of a pure enantiomer exist (6,11,18). They may be obtained from the "chiral pool", through transformation of prochiral substrates or by resolution of the enantiomers of a racemate, which is the main method for industrial synthesis at present. In contrast to the classical crystallization techniques, kinetic resolution is a relatively new technique in which one of the enantiomers of a racemate is more readily converted to product than the other. This may be realized by chemical or enzymatic methods. Its success depends on the difference in reaction rates of the two enantiomers. The utilization of a biocatalyst is a rational option when it is realized that we are dealing with biologically active compounds. Enzymes, being designed by nature, are characterized by their capability to catalyze reactions with high substrate specificity at mild and safe conditions. The number and variety of large-scale enzymatic processes is increasing, demonstrating that in many situations biotechnology offers a desirable solution. The future has been predicted to see an interesting race between biology, conventional chemistry and separation technology(6).

In this paper enzyme-catalyzed kinetic resolution will be addressed.

Enzymatic kinetic resolution of enantiomers

In this process, one enantiomer of the racemic substrate is preferentially converted, in the most favorable case yielding 50% product and 50% remaining substrate at the end of the reaction, both enantiomers in pure form. Unfortunately this ideal case is often no reality, because the enzyme exhibits a low or moderate enantioselectivity on non-natural substrates. In a regular biotechnological or chemical process one strives for maximum product yield, and tries to accomplish this in a minimal residence time in the reactor. However, for racemic compounds, first of all a sufficiently high enantiomeric purity should be aimed at. This is expressed as the enantiomeric excess, ee, which is a function of the concentrations c^R and c^S of the (R)- and (S)-enantiomer; if the (R)-enantiomer is required it is defined as:

$$ee = \frac{c^R - c^S}{c^R + c^S} \qquad (1)$$

For a racemate $ee = 0$, but if the (R)-enantiomer is 100% pure $ee = 1$. Commercial products usually have to meet very high purity demands (e.g. 99%). The manufacturer aspires to a maximum yield of sufficiently pure compound, and must therefore know the relationship between ee and yield. An important parameter within this respect is the enantiomeric ratio, E, which is a measure for the enzyme enantioselectivity. It was firstly defined by Chen et al.(4) for uni-uni reactions. For irreversible conversion of a racemic substrate, S, into a product, P, Equation (2) applies:

$$\frac{dc_S^R/dt}{dc_S^S/dt} = E \cdot \frac{c_S^R}{c_S^S} \qquad (2)$$

From integration of this equation, the relationship between the enantiomeric excess of the remaining substrate, ee_s, the degree of conversion, and the E-value may be generated. A graphical representation is depicted in Figure 2A. The enantiomeric purity of the initially racemic substrate increases when the kinetic resolution proceeds. When the desired purity is reached the reaction should be terminated, and the remaining pure substrate recovered. If the reaction proceeds too long, the yield of the desired pure compound decreases. The higher the enzyme enantioselectivity, the more favorable the resolution. From the product side the situation is rather different and more unfavorable (Figure 2B). Only a very high E-value of the enzyme may lead to a satisfactory ee_p. Even then, after 40-50% conversion the product purity collapses.

It can be concluded here that it is easier to obtain the remaining substrate than the reaction product in enantiomerically pure form. An enantiomeric purity of 100% for the remaining substrate can be attained, even when the enantioselectivity of the catalyst is modest, simply by adjusting the reaction conversion to a sufficiently high value (Fig. 2A).

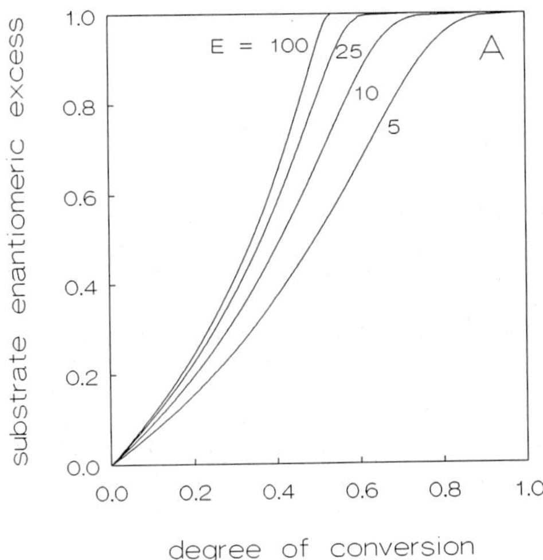

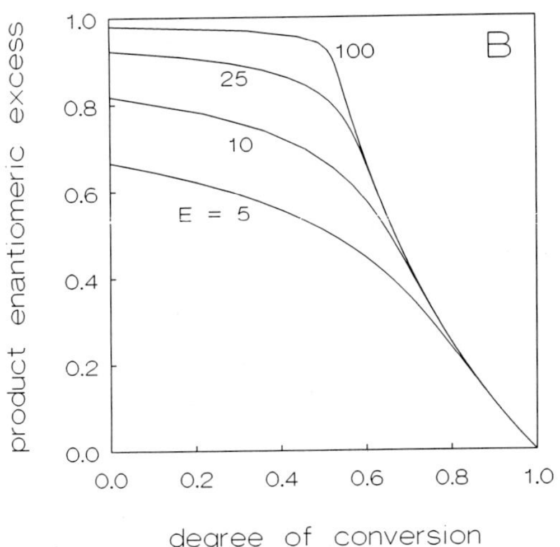

Figure 2. The effect of the enzyme enantioselectivity, E, on the *ee* of the remaining substrate (graph A) and the reaction product (graph B), as a function of the conversion ([4]).

Strategies to improve an enzyme-catalyzed kinetic resolution

It is obvious that the enantioselectivity of the enzyme should be high for practical purposes. Also, the catalyst should preferentially be applicable to a wide range of related compounds. It is very unattractive to develop a new catalyst for every new compound, since development costs are extremely high. Only a few cheap enzymes, which were in general not developed for applications in organic synthesis, are commercially available on a large scale. Examples are *Mucor mehei* lipase, used for cheesemaking, and the detergent enzyme, subtilisin. They are known to selectively catalyze the conversion of racemic alcohols, esters, carboxylic acids, amides, and also peroxides, phosphate and sulphate esters. Most of these compounds are synthetic substrates and enzyme enantioselectivities are often (too) low. The lack of a wide range of cheap, commercially available enzymes is a bottleneck in enantioselective resolution of enantiomers. In order to improve a kinetic resolution, several approaches may be adopted ([12]):

1. *Enzyme engineering*, by modification of the native structure of the enzyme. Activity and steric recognition properties of the active site may be altered by protein engineering ([26]) or chemical modification ([13]). Also immobilization may lead to a change in enzyme structure, influencing its selectivity ([10]). The screening for new enzymes should also be mentioned here ([3]).

2. *Substrate engineering*, i.e. modifying the substrate structure by varying the protecting group of a compound, thus accomplishing a relatively better fit of one enantiomer in the enzyme active site by changing steric requirements or polarity. Another possibility is changing the leaving group in ester hydrolysis, e.g. The applicability of this approach has been demonstrated in a number of cases ([12],[27]).

3. *Medium engineering*. Modification of the reaction conditions by use of different buffers, cosolvents, or by replacement of an aqueous medium for an organic medium or supercritical fluid, may result in increased enantioselectivity ([8],[9],[14]). Other, new problems, as reversibility of the reaction arose and have been addressed ([2],[5]). Also change of ambient pH, pressure and temperature ([17]) has been reported to affect the stereoselectivity of the enzyme.

The above mentioned strategies often focus on a particular specific reaction and are limited by lack of predictive models for selectivity enhancement. As

Table 1. Standard description of an enzymatic kinetic resolution, according to Equation (2), and possible alternatives.

Standard situation	Alternative
[substrate] $\gg$ [enzyme]	[substrate] $\approx$ [enzyme]
Homogeneous system	Heterogeneous system
Uni-uni kinetics	Bi-bi kinetics
Batch reactor	Continuous reactor
Irreversible reaction	Reversible reaction
Single reaction	Multiple reactions
Single enzyme	Multiple enzymes

long as enzyme-substrate-medium inter-actions are still poorly understood, these methods are largely empirical. This makes their application highly expensive and time-consuming. Molecular modeling may eventually lead to more organized research in enzyme, substrate and medium engineering in future. Thermodynamic models will be required to bring more structure in research on biocatalysis in non-conventional media.

In this paper another, more structured approach, for improve-ment of a kinetic resolution will be outlined.

<u>PROCESS ENGINEERING IN KINETIC RESOLUTIONS</u>

From a process engineering point of view, the search is not directed towards modification or development of substrates, enzymes or media, although these are definitely important process parameters, but employs the available tools, that is, easily available enzymes, substrates and solvents, and optimizes the process conditions by considering and (possibly) changing a standard situation. So far, this approach has hardly been addressed as a principle of improving a kinetic resolution process. The elegance of process engineering is its general applicability, which mainly resides in the generality of the concepts and the useful properties of the mathematical predictive models. To this end, the standard situation must be compared with possible alternatives. To this end, we will use Equation (2) as a starting point. It is a ratio of macroscopic balances for the (R)- and (S)-enantiomer of the substrate in a batch reactor. Transport phenomena are

supposed to be absent, and assumptions for the kinetic model are simple: A single substrate is irreversibly converted into a single product by a single enzyme, obeying uni-uni first order or Michaelis-Menten kinetics. Alternative concepts to this standard situation are listed in Table 2. We adopted this approach to improve enzyme catalyzed kinetic resolution processes.

In order to assess the various effects on the quality and yield of desired enantiomers, by changes in process design, one needs to quantify the reaction, which requires appropriate methodologies. For this, we will introduce two new methods.

<u>QUANTIFICATION METHODS</u>

We developed a more accurate extension of the widely accepted approach to determine the enzyme enantioselectivity from either substrate or product enantiomeric excess and the degree of conversion, as in figure 2 (<u>28</u>). It allows the calculation of E from ee measurements only. This is formally more correct than firstly calculating the degree of conversion from ee-values, and subsequently using it to calculate E. An example is shown in figure 3.

A second methodology asks for conversion measurements only. It deals with kinetic modeling of enzyme catalyzed reactions by integral progress curve analysis, and shows how to apply this technique to the kinetic resolution of enantiomers. We have shown that kinetic parameters for both enantiomers and the selectivity of the enzyme may be accurately and quickly obtained from the progress curve measurement of a racemate only (<u>29</u>). The bending of the curve around 50% conversion (when the fastest

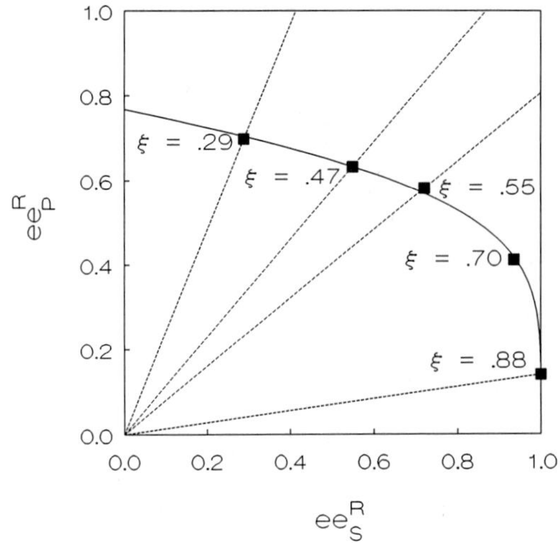

Figure 3 Experimental data (■) of the enantiomeric excess of the product (lactate, P) versus the substrate (2-chloropropionate, A) in a dehalogenation reaction. The curve represents the best fit, according to method of Rakels et al. (28), yielding $E = 7.0$.

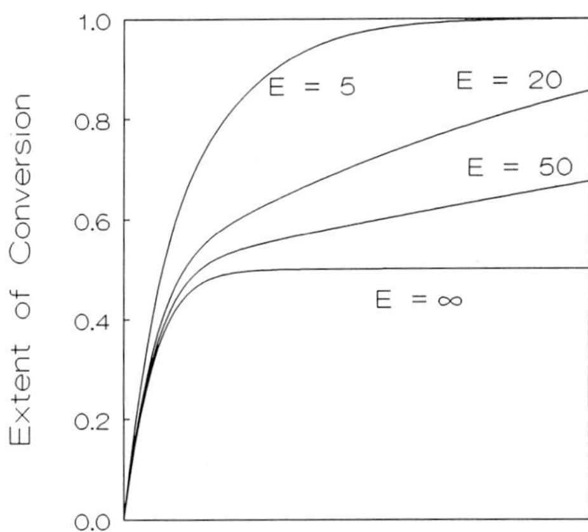

Figure 4. Effect of the enantiomeric ratio on the typical shape of a progress curve.

reacting enantiomer is mainly converted) is a measure for the enantioselectivity (see figure 4). We could apply this technique in various systems for model discrimination and verification, utilizing a covariance matrix analysis (29,31,33).

<u>PROCESS STRATEGIES</u>

<u>[substrate] ≈ [enzyme]</u>. When the initial substrate concentration is no longer much higher than the enzyme concentration, a situation which may for example occur inside a support particle for immobilized enzyme, the pseudo-steady state assumption for enzyme-substrate complexes is no longer valid. It has been shown that large deviations in ee-values as predicted from Chen's theory may be found (23). Generally, especially when non-immobilized enzyme is used, this exceptional situation will not occur. It will not be further discussed here.

<u>Heterogeneous system</u>. In a homogeneous system enzyme, substrate and product are completely soluble in the solvent. Heterogeneity occurs when multiple phases exist during reaction. If the enzyme is immobilized in a solid support, diffusional limitations between the reactant and the immobilized enzyme may be significant. Since conversion of the faster reacting enantiomer will be most restricted, the effective enantioselectivity will be reduced (15). Thus a kinetic resolution may be negatively effected by immobilization, assuming the intrinsic enzyme enantioselectivety to remain the same. A liquid-liquid two-phase system is also not appropriate for kinetic resolution (25). As a result of substrate partitioning between two phases, pseudo-equilibria may establish, causing a decrease in ee. Wu et al. (27) published a theoretical analysis of a three-phase enzyme membrane reactor system, with an organic and an aqueous liquid phase, and the enzyme immobilized on the membrane. Heterogeneity will not be further discussed here.

<u>Bi-bi kinetics</u>. Usually uni-uni first order or Michaelis-Menten kinetics are assumed to be valid, but we have pointed out that a proper kinetic analysis of many kinetic resolution processes leads to more complex models (bi-bi ping-pong or ternary complex), which predict that the enantiomeric product purity is dependent on chiral and non-chiral substrate and product concentrations (30). Thus, a good understanding of the

enzyme kinetics may lead to optimization possibilities of the kinetic resolution. This has been experimentally verified for carboxylesterase NP-catalyzed hydrolysis of racemic ethyl 2-chloropropionate (31), following ping-pong kinetics. The enantioselectivity increases when product inhibition plays a role. The non-chiral product (i.e. ethanol) was added as a cosolvent. The apparent value for E was be predicted by:

$$E^{app} = E \cdot \frac{[CH_3OH] / K_i^R + 1}{[CH_3OH] / K_i^S + 1} \qquad (3)$$

with K_i the methanol inhibition constant for the (R)- and (S)-enantiomer. At 60% conversion ee_S was increased from 0.65 to 0.70 in 4% ethanol, and to 0.75 in 20% ethanol. With *C. cylindracea* lipase an increase in *ee* from 0.35 to 0.48 in 20% ethanol was established, whereas the E-value of α-chymotrypsin dropped due to alcohol inhibition. These results support a new kinetic explanation of solvent effects.

<u>Continuous reactor</u>. There are two common alternatives to a standard batch performance of a kinetic resolution, namely reaction in a continuous stirred tank reactor (CSTR) or in a plug flow reactor. We have focussed on modeling of a kinetic resolution in an enzyme membrane reactor, which behaves as a CSTR. Although resolution reactions in continuous systems are gaining interest, the effects on product yield and purity when the standard batch reactor is replaced by a CSTR are poorly understood. We derived a mathematical description for conversion, ξ, and *ee* as a function of the enzyme enantioselectivity in such a system (32):

$$E = \frac{\xi - \dfrac{1}{(1 - ee_P)}}{\xi - \dfrac{1}{(1 + ee_P)}} = \frac{1 + ee_S}{1 - ee_S} \cdot \frac{1 + ee_P}{1 - ee_P} \qquad (4)$$

Comparison of a kinetic resolution in batch and CSTR is shown in figure 5. A continuous plug flow process should perform similar to a batchwise process, the length dimension of the tube replacing the time dimension of the batch (23).

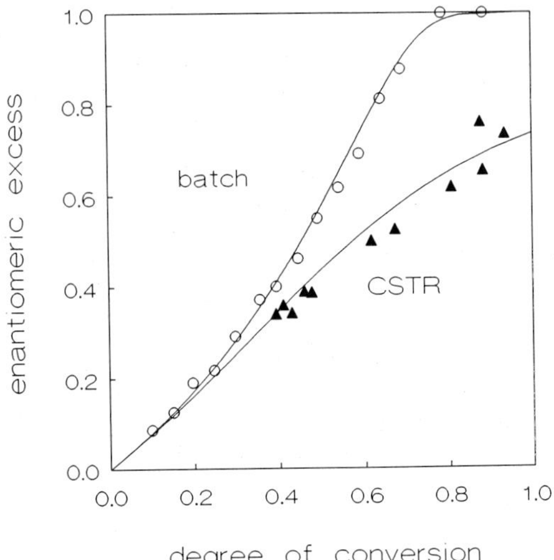

Figure 5. Carboxylesterase NP catalyzed kinetic resolution of racemic methyl 2-chloro-propionate in batch and CSTR. The curves show the model predictions for $E = 6.5$ and the symbols the experimental data.

<u>Reversible reaction</u>. Biocatalysis in organic media has opened up a new field of research, but reversibility of the reaction often plays a disturbing role in organic solvents, attainable yields and *ee*-values being lower than in irreversible reaction systems (5). Several attempts have been made to favor the thermodynamic reaction equilibrium in kinetic resolution of chiral alcohols in organic media (2), but the resolution of carboxylic acids seems to be more profound. We found a simple procedure to improve enantio-selective ester hydrolysis in organic solvents (34). The product carboxylic acid is pulled away by ion-pair formation using a non-reactive amine. This results in enhanced yield and *ee*, and eventually in facilitated work-up of the desired compound.

<u>Multiple reactions and enzymes</u>. Aiming at enhancement of product purity, multiple simultaneous resolutions may reinforce each other. The following possibilities may be distinguished: Two competitive parallel or two sequential reactions may be catalyzed either by a

single enzyme or by two compatible enzymes (22). We investigated the option that two enzymes simultaneously catalyze two reactions in sequence. Improved ee_p-values were attained in comparison to a single step reaction in carboxylesterase NP catalyzed hydrolysis of racemic 2-chloropropionate, followed by dehalogenation of the enantiomerically enriched 2-chloropropionate by DL-DEHALOGENASE into lactate (33). Compatibility of substrate and enzyme combinations is a prerequisite.

Competitive parallel kinetic resolution by a single enzyme is another approach. Now, the substrate enantiomer yielding the undesired product enantiomer is itself converted (with opposite enantioselectivity), into another product, by the same enzyme (Rakels et al., submitted for publication). These concepts compare favorable to conventional strategies, as product recirculation.

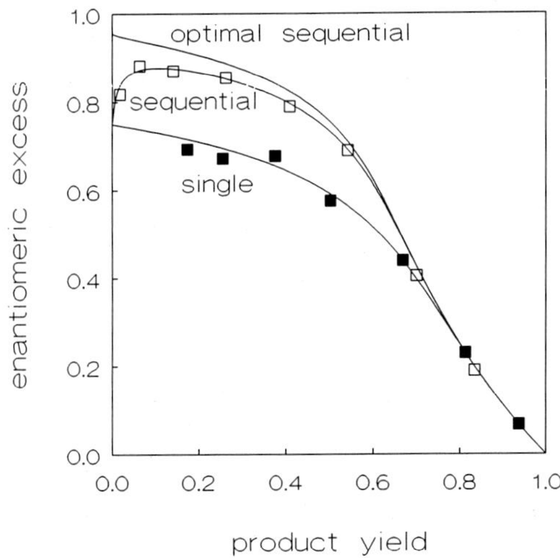

Figure 6 Results of the ee_p as a function of the yield for sequential kinetic resolution by two enzymes (□) and single resolution (■)

<u>CONCLUSION</u>

There are many degrees of freedom to improve an enzymatic kinetic resolution process, without necessarily changing substrate, enzyme or medium. We addressed a number of interesting possibilities, but many other variants and combinations of variants are thinkable. Knowledge of the enzyme kinetics appeared to be of tremendous value.

<u>LITERATURE CITED</u>

1. Ariëns, E.J., Wuis, E.W. and Veringa, E.J., *Biochem. Pharmacol.*, 1989, **37**, 9-18

2. Bevinakatti, H.S. and Newadkar, R.V. *Tetrahedron: Asym.*, 1993, 773-776.

3. Cheetham, P.S.J. *Enzyme Microb. Technol.*, 1987, **9**, 194-213.

4. Chen, C.-S., Fujimoto Y., Girdaukas, G., and Sih, C.J. *J. Am. Chem. Soc.*, 1982, **104**, 7294-7299.

5. Chen, C.-S., Wu, S.-H., Girdaukas, G., and Sih, C.J. *J. Am. Chem. Soc.*, 1987, **109**, 2812-2817.

6. Collins, A.N., Shel-drake, G.N., and Crosby, J., Chirality in industry, Wiley, New-York, 1992.

7. De Camp, W.H. *Chirality*, 1989, **1**, 2-6.

8. Dordick, J.S. In *Applied biocatalysis* vol 1; Blanch, H.W. and Clark, D.S., Eds; Marcel Dekker Inc.: New-York, 1991; pp 1-59.

9. Dordick, J.S. *Biotech-nol. Prog.*, 1992, **8**, 259-266.

10. Faber, K. *Tibtech*, 1993, **11**, 461-470.

11. Faber, K. Biotrans-formations in organic chemistry. Springer-Verlag, Berlin (1992).

12. Faber, K., Ottolina, G. and Riva, S. *Biocata-lysis*, 1993, **8**, 91-132.

13. Gu, Q.-M. and Sih. C.J. *Biocatalysis*, 1992, **6**, 115-126.

14. Klibanov, A.M. *Acc. Chem. Res.*, 1990, **23**, 114-120.

15. Matson, S.L., and Lopez, J.L. In: Frontiers in bioprocessing, S.K. Sikdar, M. Bier and P. Todd (eds.), CRC Press, Boca Raton, Florida, pp. 391-403 (1990).

16. Millership, J.S. and Fitzpatrick, A. *Chirality*, 1993, **5,** 573-576.

17. Phillips, R.S. *Enzyme Microb. Technol.*, 1992, **14,** 417-419.

18. Sheldon, R.A. Chiro-technology, Industrial synthesis of optically active compounds, Marcel Dekker, New-York (1993).

19. Shindo, H.S. and Caldwell, J. *Chirality*, 1991, **3,** 91-93.

20. Sih, C.J., and Wu, S.-H. *Topics Stereochem.*, 1989, **19,** 63-125.

21. Stinson, S.C. *Chem. Eng. News*, 1992, September 28, pp. 46-79.

22. Straathof, A.J.J, Rakels, J.L.L., and Heijnen, J.J. *Biocatalysis*, 1990, **4,** 89-104.

23. Straathof, A.J.J, Rakels, J.L.L., and Heijnen, J. J. *Ann. N. Y. Acad. Sci.*, 1992, **672,** 497-501.

24. Teeuwen, H. en Polastro, E.T. *Chemisch Weekblad*, 1990, **32,** 336-337.

25. Tol, J.B.A. van, Geerlof, A., Jongejan, J.A. and Duine, J.A. *Ann. N. Y. Acad. Sci.*, 1992, **672,** 462-470.

26. Wong, C-H., Chen, S-T., Hennen, W. J., Bibbs, J. A., Wang, Y.-F., Liu, J. L-C., Pantoliano, M. W., Whitlow, M. and Bryan, P.N. *J. Am. Chem. Soc.*, 1990, **112,** 945-953.

27. Wu, D., Belfort, G. and Cramer, S. *Ind. Eng. Chem. Res.*, 1990, **29,** 1612-1621.

28. Rakels, J.L.L., Straathof, A.J.J. and Heijnen, J.J. *Enzyme Microb. Technol.*, 1993, **15,** 1051-1056.

29. Rakels, J.L.L. Romein, B. Straathof, A.J.J. and Heijnen, J.J. *Biotechnol. Bioeng.*, 1994, **43,** 411-422.

30. Straathof, A.J.J., Rakels, J.L.L. and Heijnen, J.J. *Biocatalysis*, 1992, **7,** 13-27.

31. Rakels, J.L.L., Caillat, P., Straathof, A.J.J. and Heijnen, J.J. *Biotechnol. Progr.*, 1994, in press.

32. Rakels, J.L.L., Paffen, H., Straathof, A.J.J. and Heijnen, J.J. *Enzyme Microb. Technol.*, 1994, in press.

33. Rakels, J.L.L., Wolff, A., Straathof, A.J.J. and Heijnen, J.J. *Biocatalysis*, 1994, in press.

34. Rakels, J.L.L., Straathof, A.J.J. and Heijnen, J.J. *Tetrahedron: Asym.*, 1994, **5,** 93-100.

Design of Reaction Medium for Nonaqueous Biocatalysis: Analysis on Enzyme Hydration and Catalytic Activity in Organic Solvent

S.B. Lee, K.-J. Kim, and M. G. Kim

Department of Chemical Engineering, Pohang University of Science and Technology (POSTECH), P.O. Box 125, Pohang 790-330, KOREA

The effect of water on the enzyme hydration and on the catalytic efficiency of enzyme molecules have been analyzed in terms of the thermodynamic activity of water, which has been estimated by the NRTL or UNIFAC equations. When the amount of water bound to the enzyme was plotted as a function of water activity, the water sorption isotherms obtained from the water-solvent liquid mixtures were similar to the reported water-vapor adsorption isotherms of proteins. The results shown in the present study indicate that the adsorption of water from the organic media is not significantly dependent on the property of organic solvents. However, the catalytic activity of enzymes in nonaqueous solvents was found to be affected appreciably by the physicochemical properties of organic solvents even at the same water activity.

Recent finding that enzymes can be used in nonaqueous organic solvents greatly expands the potential scope and economic impact of biocatalysis(1-3). Enzyme reaction in nonaqueous solvents offers new possibilities for biotechnological production of many useful chemicals using reactions not feasible in aqueous media. In order to achieve maximum efficiency of the biocatalyst in organic solvent, however, the effect of microenvironmental factors that determine enzyme activity in nonaqueous environment should be further elucidated.

Since the majority of traditional studies of enzyme reaction have involved enzymes in aqueous environment, the knowledge on enzyme behavior in organic solvents is still far from complete understanding. In many cases the operating conditions such as the choice of organic solvent, optimum water content and the nature of the support material have been optimized via trial-and-error methods. Only recently has there been an effort to establish general rules for relating microenvironmental factors to the enzyme activity in nonaqueous media(4,5).

In this study, as one of our efforts for establishing such a rule, the effects of water on the enzyme hydration and on the catalytic efficiency of enzyme molecules have been analyzed in terms of the thermodynamic activity of water. For this purpose, the water activity has been estimated from the water content in organic solvents using the activity coefficient models described in the following section. After the present study has been completed independently, the author has found that similar ideas have been applied by Halling(6).

ESTIMATION OF THERMODYNAMIC WATER ACTIVITY

The activity of species i, denoted by a_i, is defined as:

$$a_i = f_i / f_i^{\circ} = x_i \, \gamma_i$$

where f_i and f_i° are the fugacities

481

E. Galindo and O.T. Ramírez (eds.), Advances in Bioprocess Engineering. 481-484.
© 1994 Kluwer Academic Publishers. Printed in the Netherlands.

in the mixture and in its standard state and x_i and γ_i are mole fraction and activity coefficient of component i, respectively. Deviation of the activity coefficient from unity reflects the discrepancy of the nature of the interaction between the solute and the solvent from an ideal solution. If vapor-liquid equilibrium (VLE) data are available from the literature(7), the activity coefficient can be calculated using the models such as the van Laar, Wilson, NRTL, and UNIFAC equations. Otherwise the activity coefficient can be estimated by the predictive methods, such as regular solution theory, ASOG, or UNIFAC equations.

In the case where VLE data exist the NRTL equation(8) is used in this study because the NRTL equation can predict the effect of temperature on activity coefficients for a moderate temperature range. If the difference in temperature between. VLE data and water-sorption experiments is significant, regular solution approximation has been applied. When VLE data are not available, the UNIFAC equation(9) has been used for the estimation of the thermodynamic activity of water. (see APPENDIX)

RELATIONSHIP BETWEEN WATER ACTIVITY AND ENZYME HYDRATION

The amount of water necessary for an enzyme to function in organic solvents is one of the major concerns of nonaqueous enzymology. Recently, several research groups have studied the role of water in nonaqueous enzyme catalysis.

Zaks and Klibanov(10) have investigated the dependence of the amount of water bound to the enzyme(WBE) on the water content of the solvent in which the enzyme is suspended. They found that the enzyme activities are approximately correlated with the WBE independent of the physicochemical properties of organic solvents.

Yamane et al.(11) have also studied the dynamic equilibrium between the

WBE and the amount of free water in various organic solvents. They proposed that general profiles of the adsorption isotherms of water significantly differ in water-miscible and in water-immiscible organic solvents. In water-miscible solvents the WBE approaches an upper-bound maximum while in water-immiscible solvents the WBE increases steadily until the amount of free water reaches its solubility limit in a particular solvent.

In this study, the aforementioned water adsorption data have been reanalyzed in terms of the thermodynamic activity of water instead of using the amount of water in the reaction medium. When the water contents, expressed in volume or weight percent, are converted to the water activity by the NRTL or UNIFAC equations as described earlier, the adsorption isotherms like the BET isotherm have been obtained (Fig. 1). It was found that the adsorption isotherms of water in the liquid mixtures were superimposable on the water-vapor adsorption isotherms. If all data points of Zaks and Klibanov(10) and Yamane et al.(11) were plotted as the amount of water bound to protein versus water activity, the adsorption isotherms were not

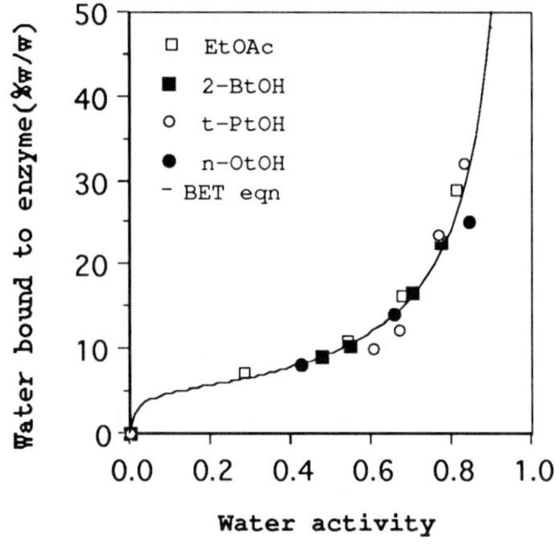

Figure 1. Water adsorption isotherm of alcohol oxidase at 298 K. Solid line represents the BET equation obtained by regression anlaysis.

significantly dependent on the kinds of organic solvents or the proteins (except ß-lactoglobulin) as shown in Fig. 2. The essential water which corresponds to the water at monolayer can also be estimated by the BET equation. In the case of chymotrypsinogen, for example, the BET monolayer water(W_m) was found to be 0.59 g/g-protein at 40°C, which compares well with the values obtained from the water-vapor adsorption experiment(12).

RELATIONSHIP BETWEEN WATER ACTIVITY AND ENZYME ACTIVITY

The analyses in the previous section indicate that the degree of enzyme hydration is primarily determined by the thermodynamic activity of water in the system. If the enzyme activities in organic media are, as suggested by Zaks and Klibanov(10), solely governed by the interactions of solvent with the water on the enzyme, the reaction rate at the given water activity might not be significantly affected by the nature or organic solvents. Such possibilities have been further examined by relating the catalytic activities of the enzyme to the thermodynamic activity of water.

When the data of Zaks and Klibanov(10) have been analyzed on the basis of thermodynamic water activity, it appears that the enzyme reaction rates at the given water activity differ appreciably depending on the nature of the solvents. As an example, the data for horse liver alcohol dehydrogenase are depicted in Fig. 3 where the reaction rate was plotted as a function of water activity instead of water content in organic solvents. Similar trends were also observed for alcohol oxidase and polyphenol oxidase.

A more profound effect of organic solvents on the catalytic activity of enzymes was exerted in the case where the polar solvents were used as the reaction medium. These were the cases for horseradish peroxidase(13), thermolysin(14), and penicillin acylase(15). The results for thermolysin shown in Fig. 4 clearly indicate the dependence of enzymatic activity on the property of organic solvents. The organic solvent might affect the interactions between the enzyme and the substrate by altering the intermolecular forces such as hydrogen bonding , electrostatic interaction, van der Waals, and hydrophobic forces. Hence, it is conceivable that both the amount of

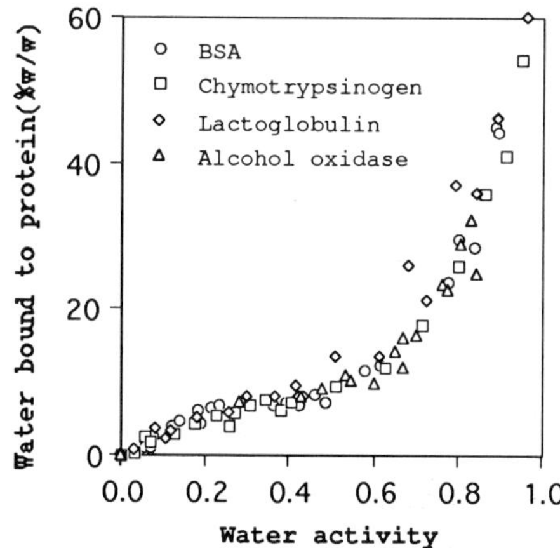

Figure 2. Water adsorption isotherms of proteins in water-organic solvent mixtures. The amounts of water bound to proteins are plotted against thermodynamic water activity instead of water content.

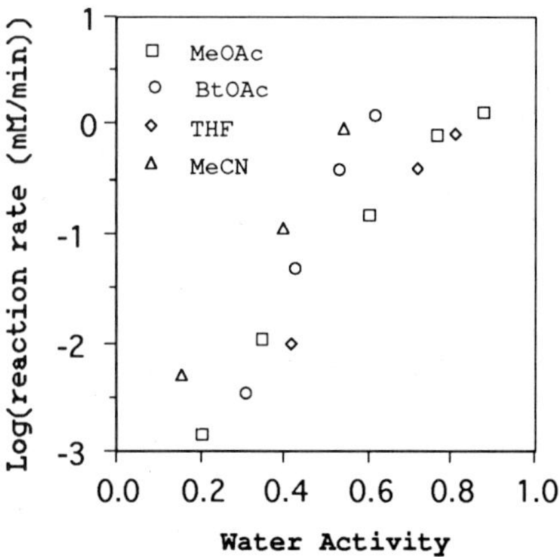

Figure 3. Relationship between reaction rate of alcohol dehydrogenase and water activity. Data were taken from Zaks and Klibanov(10).

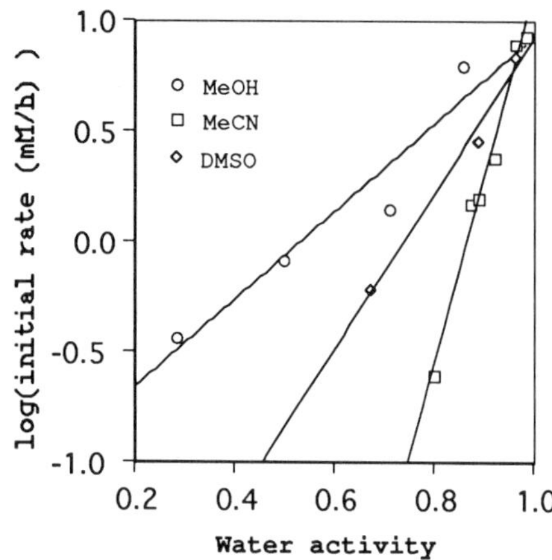

Figure 4. Relationship between reaction rate of thermolysin and water
activity. Data were taken from Hwang
(14).

the enzyme-bound water and the
property of the reaction medium
determine the enzyme activity in
organic solvents.

Further analyses on the interactions
of organic solvents with the
substrate and product and the enzyme
molecules are now in progress.

APPENDIX

NRTL activity coefficient model(8)
can be expressed as;

$$\ln\gamma_i = x_j^2 [\tau_{ji}(G_{ji}/(x_i+x_jG_{ji})^2 + \tau_{ij}G_{ij}/(x_j+x_iG_{ij})^2]$$

where $A_{ij}=g_{ij}-g_{jj}$, $\tau_{ij}=A_{ij}/R\,T$ (R=1.987
cal mol^{-1} K^{-1}), and G_{ij} = exp (-
$a_{12}\tau_{ij}$). Three NRTL equation
parameters (A_{12}, A_{21}, and a_{12}) were
obtained from the Vapor-Liquid
Equilibrium Data Collection(7).

UNIFAC activity coefficient model(9)
consists of two parts;

$$\ln\gamma_i = \ln\gamma_i(\text{combinatorial}) + \ln\gamma_i(\text{residual})$$

and each terms can be evaluated from
group contribution methods. The
volume and surface parameters, and
binary interaction parameters for
each pair of functional groups were
taken from the literature(16).

ACKNOWLEDGMENTS

This work was in part supported by
BioProcess ERC.

LITERATURE CITED

1. Klibanov, A. M., *CHEMTECH*, **16**, 354
(1986).

2. Zaks, A. and A. J. Russel, *J.
Biotechnol.*, **8**, 259 (1988).

3. Dordick, J. S., *Enzyme Microb.
Technol.*, **11**, 194 (1989).

4. Laane, C. and J. Tramper,
CHEMTECH, **20**, 502 (1990).

5. Mattiason, B. and P. Adlercreutz,
Trends in Biotechnol., **9**, 394
(1991).

6. Halling, P., *Biochim. Biophys.
Acta*, **1040**, 225 (1990).

7. *Vapor-Liquid Equilibrium Data
Collection*, DECHEMA Vol. 1 (1977);
Vol 1-a (1981); Vol 1-b (1988).

8. Renon, H. and J.M. Prausnitz,
AIChE J., **14**, 135 (1968).

9. Fredenslund, A., R. L. Jones and
J. M. Prausnitz, *AIChE J.*, **21**,
1086 (1975).

10. Zaks, A. and A.M. Klibanov, *J.
Biol. Chem.*, **263**, 8017 (1988).

11. Yamane, T., Y. Kojima, T. Ichiryu
and S. Shimizu, *Ann. N. Y. Acad.
Sci.*, **542**, 282 (1988).

12. Bone, S., *Biochim. Biophys. Acta*,
916, 128 (1987).

13. Epton, R., M. E. Hobson and G.
Marr, *Enzyme Microb. Technol.*, **1**,
37 (1979).

14. Hwang, K. A., MS Thesis, POSTECH,
Pohang (1992).

15. Kim, M. G., MS Thesis, POSTECH,
Pohang (1992).

16. Tiegs, D. ,J. Gmehling, P. Rasmus-
sen, and A. Fredenslund, *Ind.
Eng. Chem. Res.*, **26**, 159 (1987).

New Approaches to the Enzymatic Production of Oligopeptides: Synthesis of the "Delicious Peptide" and its Fragments

I. Gill, R. López-Fandiño, X. Jorba, and E. Vulfson

Department of Biotechnology and Enzymology, BBSRC Institute of Food Research, Earley Gate, Whiteknights Road, Reading RG6 2EF, U.K.

Short peptides are becoming increasingly important in view of their wide ranging biological activities which include antibiotic, hormonal, immunomodulating effects and sensory characteristics. The application of enzymes to the synthesis of short peptides offers numerous and well documented advantages but until recently little progress has been made in developing new routes which would avoid intermediate protection/deprotections steps. This communication describes the preparative syntheses of the "delicious" octapeptide and its functional fragments using a newly developed strategy for the sequential condensation of amino acid esters in organic solvents, and eutectic mixtures of substrates respectively.

The development of methods suitable for the large scale production of biologically active peptides is an actively investigated area. This is due to the rapidly increasing number of new structures which have been isolated and characterized in recent years (1-9). The advantages associated with enzymatic synthesis as compared to conventional chemical approaches are also widely recognized and documented (10-14). Indeed, the commercial feasibility of large scale protease catalyzed synthesis has been unambiguously proven in the manufacture of aspartame (15).

However, enzymatic methods generally suffer from the same drawbacks as solution and solid-phase chemical coupling in the sense that tedious protection-deprotection steps are required for the elongation of the polypeptide. This has significantly limited the scope of application of the enzymatic methods where the synthesis of more complex oligopeptide sequences is concerned. Recently, we demonstrated that the enzymatic assembly of an oligopeptide can be substantially simplified if *N*- unprotected amino acid esters are used as acyl donors (16). Thus, the synthesis of enkephalin precursors has been accomplished on a multi-gram scale in just three steps (16) and this methodology has been successfully applied to the design of a continuous bioreactor (17).

In this communication we wish to describe two new enzymatic strategies for the synthesis of oligopeptides, as exemplified by the preparation of the "delicious" octapeptide and its fragments. In particular, (i) the feasibility of the sequential assembly of the growing polypeptide chain from simple amino acid esters will be discussed and, (ii) the

synthesis of the functional fragments of the delicious peptide in a new bioreaction system, namely eutectic substrate mixtures (18-20), will be described.

Materials and Methods

Materials. Subtilisin Carlsberg (EC 3.4.21.14), thermolysin (EC 3.4.24.4), papain (EC 3.4.22.2), pronase E (EC undefined), chymopapain (EC 3.4.22.6), alaninamide hydrochloride, glycine ethyl ester hydrochloride, glycinamide hydrochloride, serine ethyl ester hydrochloride, leucine ethyl ester hydrochloride, aspartic diethyl ester hydrochloride, aspartic diallyl ester tosylate salt, glutamic diethyl ester hydrochloride, glutamic diallyl ester tosylate salt, N_α, N_ξ-di-CBZ-lysine, *N*- CBZ-glutamic acid, *N*- CBZ-serine, *N*- CBZ-aspartic acid, 2-(*N*-cyclohexylamino)ethane sulphonic acid (CHES), and *N*- tris-(hydroxymethyl)methyl-3-aminopropane sulphonic acid (TAPS) were obtained from Sigma Chemical Company. Diisopropylethylamine, 2-mercaptoethanol, dithiothreitol, thionyl chloride, allyl bromide, cesium carbonate, amberlite IRA400-OH and glycerol were obtained from Aldrich Chemical Company. Celite (GLC grade, 30-80 mesh) was obtained from BDH Ltd. All of the solvents used were of the highest purity available.

Preparation of substrates. *N*- acyl-amino acid ethyl esters and allyl esters were prepared from the corresponding free acids using the thionyl chloride and cesium salt methods respectively. The free bases of amino acid esters were prepared by neutralizing aqueous solutions of the corresponding hydrochloride or

485

E. Galindo and O.T. Ramírez (eds.), Advances in Bioprocess Engineering. 485-488.
© 1994 Kluwer Academic Publishers. Printed in the Netherlands.

tosylate salts with aqueous sodium hydroxide, followed by extraction with ethyl acetate and rotary evaporation. The free bases of amino acid amides were prepared by neutralizing ethanolic solutions or suspensions of the hydrochloride salts with amberlite IRA400-OH or ethanolic sodium hydroxide, followed by filtration and rotary evaporation.

<u>Preparation of immobilized enzymes</u>

Where immobilized enzymes were employed as catalysts, these were prepared using Celite as the support material as previously described (<u>17</u>).

<u>Reactions in eutectic mixtures</u>

In a typical reaction, the substrates and adjuvant were mixed thoroughly together and warmed to approx. 60^0C. After allowing to cool to room temperature, lyophilized or enzyme immobilized on Celite was added, and the mixture incubated in an open vial or beaker placed in a heating block maintained at 37^0C. Samples were taken at appropriate time intervals for HPLC analysis. Where a cysteine protease was used as the catalyst, 2-mercaptoethanol was included in the reaction mixture.

<u>Reactions in low-water media</u>

In a typical procedure, the substrates were dissolved in aqueous ethanol or a mixture of ethanol and acetonitrile. A solution of the enzyme dissolved in buffer, or enzyme immobilized on Celite was then added, and the mixture stirred on a heater-stirrer, or shaken in a shaker-incubator at 37^0C. Samples were taken at appropriate time intervals for HPLC analysis. Where a cysteine protease was used as the catalyst, dithiothreitol or 2-mercaptoethanol was included in the reaction mixture.

<u>HPLC analysis.</u>

HPLC analysis was carried out using an LDC Milton Roy CM4000 ternary gradient pump, connected with a Spectra Physics SP8450UV/VIS detector, an LDC Marathon autosampler, and an HP-Chem Station. Samples were quenched with pure methanol, centrifuged, then analysed on a 0.46x15 cm Nucleosil C18 / 3 μm column, at 45^0C, at
a flow rate of 1.0 ml/min. Methanol and 8:2 water : methanol, both containing 0.05 % v/v phosphoric acid, were used as the mobile phases.

<u>Characterisation of products.</u>

All products were fully characterised using FTIR spectroscopy (Perkin Elmer 1720-X FTIR spectrometer), FAB-MS (Kratos MS9/50TC spectrometer, glycerol matrix, xenon, 5-7 kV), and proton and carbon NMR (Jeol EX270-FT spectrometer, 270 MHz and 67 MHz, d_6-DMSO).

<u>Results and Discussion</u>

The delicious octapeptide, Lys-Gly-Asp-Glu-Glu-Ser-Leu-Ala, was first isolated from beef and

characterized by Yamasaki and Maekawa (<u>21</u>). This peptide possesses a "beef-soup" type flavour and its taste profile can be reconstituted from three fragments: Lys-Gly, Asp-Glu-Glu and Ser-Leu-Ala (<u>22</u>). The structure-taste relationship of the peptide is currently being investigated in order to define the minimal structural elements responsible for the sensory characteristics. However, the presence in the sequence of several multifunctional amino acids (which appear to be essential for the displayed activity), Asp, Glu, Lys and Ser, make this peptide and/or the above three fragments a challenging target for preparative synthesis and consequently, rather difficult to produce on a large scale.

One of the acknowledged advantages of enzymatic peptide synthesis is the dispensation with the requirement for the side chain protection of multifunctional amino acids which is clearly a problem associated with a conventional chemical synthesis of the delicious peptide. Following recent advances in the application of proteinases to the preparation of oligopeptides, we decided to investigate the feasibility of applying the enzymatic approach to the production of the delicious peptide and the three fragments which were shown to reconstitute the flavour.

Two independent approaches were chosen for the synthesis of the fragments and the complete peptide, namely the use of eutectic media and low-water organic media respectively. The choice of eutectic mixtures as a medium for performing the fragments syntheses was dictated by several advantages associated with this bioreaction system, particularly where the preparation of short peptides is concerned: (i) A high productivity of up to 0.8 g of product per gramme of reaction mixture can be achieved (<u>18</u>-<u>20</u>); (ii) A number of acidic and basic amino acid derivatives readily form eutectics in the presence of small quantities of organic solvents, termed adjuvants (<u>19</u>); (iii) Several proteases have been shown to retain their catalytic activity in eutectic mixtures (<u>19</u>); (iv) Non-hazardous solvents, such as ethanol, water and glycerol are effective in promoting the formation of eutectic mixtures from amino acid derivatives (<u>19</u>). The overall strategy for the preparative synthesis of all three fragments is shown in Figure 1, and the results obtained are summarized in Table 1 (lines 1-3).

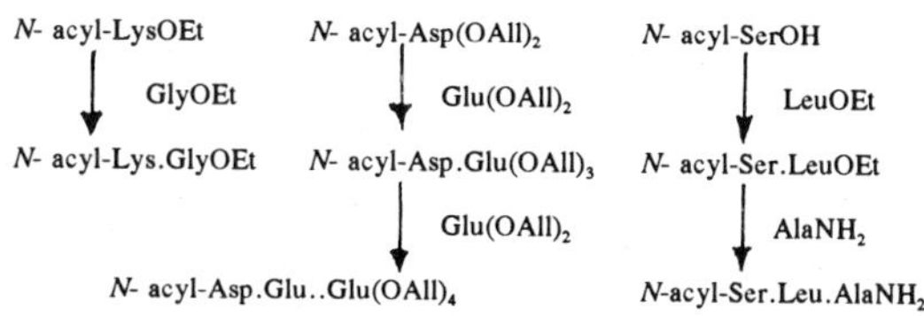

Figure 1. Enzymatic synthesis of precursors of the delicious peptide fragments. All the reactions were performed in eutectic mixture of substrates in the presence of up to 30% (w/w) of adjuvants.

From Figure 1, it can be seen that precursors of the three fragments were readily synthesized by the sequential addition of amino acid esters to the suitably protected N-terminal amino acid esters. These syntheses were performed in liquid or semi-liquid eutectic mixtures of the substrates, containing small quantities of adjuvants such as water, ethanol and glycerol, and proteases such as papain, chymopapain and thermolysin. The overall protocols for the preparation of each fragment were very facile, without any requirement for intermediate protection / deprotection steps, and the syntheses were readily scaled up to produce multi-gramme quantities of the desired peptides. Furthermore, the use of eutectic media as a bioreaction system allowed the attainment of high productivities of up to 0.5 g per gramme of reaction mixture.

Where the synthesis of the complete octapeptide was concerned, because of the unprecedented length of the target molecule, a novel sequential approach to oligopeptide synthesis had to be developed, which would minimize the requirement for intermediate protection / deprotection / activation steps. To investigate the feasibility of such a sequential strategy, we examined whether it was possible to synthesize the N- and C-terminal tetrapeptide fragments, Lys-Gly-Asp-Glu and Glu-Ser-Leu-Ala, via the stepwise addition of amino acid esters to suitably protected N-terminal amino acid derivatives. This strategy is presented in Figure 2, and the results are summarised in a Table 1 (lines 4-5).

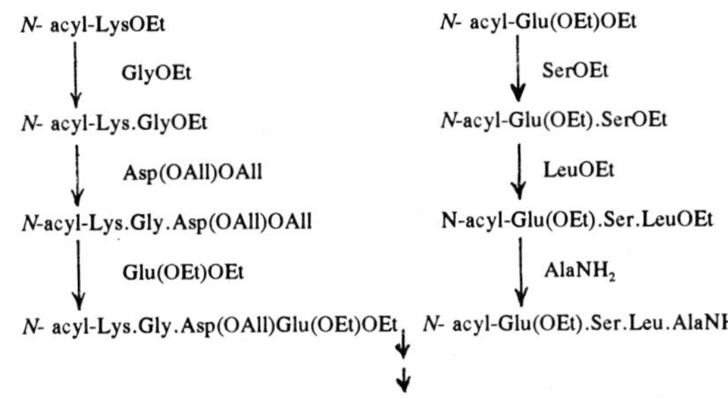

Figure 2. The sequential assembly of the delicious peptide. All reactions were performed with thiol and serine proteases in ethanol-water mixtures.

It is evident from Table 1, that the designed approach was indeed successful, and that the considerable simplification that was achieved by the use of the sequential strategy allowed the synthesis of each tetrapeptide to be completed in only three steps. The reactions were conducted in low-water organic media, using a number of free and immobilized proteases, including chymopapain and pronase E as catalysts. The final fragment condensation has been successfully carried out, and is currently being optimized and scaled up(not shown).

Product	Scale (mmol)	Overall Yield(%)
Lys.GlyOH	10	42
Asp.Glu(OH).Glu(OH)OH)	10	30
Ser.Leu.AlaNH₂	8	55
N- acyl-Lys.Gly.Asp(OAll).Glu(OEt)OEt	7	34
N- acyl-Glu(OEt).Ser.Leu.AlaNH₂	5	34

Table 1. The substrates used for the syntheses are described in Figures 1 and 2. All the reactions were conducted at 37^{0}C using equimolar quantitites of substrates for the preparation of Lys. GlyOH and Ser.Leu.AlaNH₂ Two times excess of the nuclophile was employed in all other syntheses. The deprotection of the final oligopeptides products was performed using conventional methods (*23*) . All the products were obtained in 95% + purity and fully characterised (see Materials and Methods)

To the best of our knowledge, the delicious peptide is one of the longest oligopeptide prepared to date solely by enzymatic coupling. This advance has become possible only because of the development of an entirely new approach for the sequential assembly of the growing polypeptide chain.

This methodology provides a new and potentially powerful strategy for the synthesis of peptides, and in particular, should prove valuable in the case of larger multifunctional oligopeptides, for which facile and economic synthetic routes are at present unavailable. In conclusion the delicious peptide and its functional fragments have been prepared by sequential addition of amino acid esters on a multi-gram scale using conventional low water systems and eutectic mixtures of substrates respectively.

Literature Cited

1. Grenby, T.H. *Trends Food Sci. Technol.*, 2, 2 (1991).

2. Kirimura, J., Shimizu, A., Kimizura, A., Ninomiya, T. and Katsuya, J. *J. Agric. Food.Chem.*, 3, 31 (689).

3. Kleinkauf, H. and Von-Doehren, H. *Crit. Rev. Biotechnol.*, 8, 1 (1988).

4. Kurihara, Y. and Nirasawa, S. *Trends Food Sci. Technol.*, 5, 37 (1994).

5. Meisel, H. and Schlimme, E. *Trends Food Sci. Technol.*, 2, 41 (1990).

6. Mills, E. N. C., Alcocer, M. J. C. and Morgan, M. R. A. *Trends Food Sci. Technol.*, 3, 64 (1992).

7. Rizo, J. and Gierasch, L. M. *Ann. Rev. Biochem.*, 61, 387 (1992).

8. Snyder, S. H. and Innis, R. B. *Ann. Rev. Biochem.*, 48, 755 (1979).

9.   Tager, H. S. and Steiner, D. F. *Ann. Rev. Biochem.*, 45, 510 (1974).

10. Fruton, J. S. *Adv. Enzymol. Rel. Areae Mol. Biol.*, 53, 239 (1982).

11. Kitaguchi,H.and Klibanov,A.M.*J.Amer. Chem. Soc.*, 111, 9272 (1989).

12. Kullmann,W.*Enzymatic peptide synthesis*, CRC Press, Boca Raton, Florida (1987).

13. Morihara, K. *Trends Biotechnol.*, 5, 164 (1987).

14. Wong, C. -H., Matos, J. R., West, J. B. and Barbas, C. F. *Developments in Industrial Microbiology.*,*J. Ind. Microb.*, 29, 171 (1987).

15. Gross, A. *Food Technol.*, 45, 96 (1991).

16. Gill, I. and Vulfson, E. N. *J. Chem. Soc. Perkin Trans. 1*, 6, 667 (1992).

17. Richards, A. O'L., Gill, I. and Vulfson, E. N. *Enz. Microb. Technol.*, 15, 928 (1993).

18. Gill, I. and Vulfson, E. N. *J. Amer. Chem. Soc.*, 115, 3348 (1993).

19. López-Fandiño, R., Gill, I. and Vulfson, E. N. *Biotechnol. Bioeng.*, 43, 11 (1994).

20. López-Fandiño, R., Gill, I. and Vulfson, E. N. *Biotechnol. Bioeng.*, 43, 11 (1994).

21. Yamasaki, Y. and Maekawa, K. *Agric, Biol. Chem.*, 42, 1761 (1978).

22. Tamura, M., Nakatsuka, T., Tada, M. and Kawasaki, Y. *Agric. Biol. Chem.*, 53, 319 (1989)

23. T.W.Green and P.G.M. Wuts. Protective groups in organic synthesis, Wiley, Pub., New York. (1991),

Aspergillus sp. 2M1 Xylanase: Production, Characterization and Application in the Pulp and Paper Industry

N. Durán[1], E. Curotto[3], E. Esposito[2], C. Aguirre[3], and R. Angelo[1,3]

[1]Inst. Quimica, Biological Chemistry Laboratory, C.P. 6154, Campinas, S.P. CEP 13081-970,
[2]Chem Eng. Faculty, Universidade Estadual de Campinas, S.P., BRAZIL; [3]Inst. Química,
Universidad Católica de Valparaiso, Casilla 4059, Valparaiso, CHILE

Aspergillus *sp. 2M1 produces high levels of β -xylanases and β-xylosidases. Different carbon sources as inductors of xylanases were used. Production of xylanases were evaluated in a combined packed-bed/air-lift bioreactor (PBAL) in free and immobilized form and with a stationary tray bioreactor (STB). In the free form, xylanase activity was similar to that in Erlenmeyer flasks, but in the immobilized form higher values were observed. Similar xylanases activities with the STB were found. In order to understand xylanases stabilities, chemical modifications were done. Among others, 1-ethyl-3(3-dimethyl aminopropyl) carbodiimide (EDC) affected the xylanase activity. On the basis of this study, it was concluded that one or more carboxyl groups are in the active site of the xylanase from* Aspergillus *sp. 2M1. Xylanases induced by oat, birch and pinus xylans exhibited their optimum pH at 5.5 and at 55°C. At this temperature the half-lives of the xylanase activities were 11.8, 7.2 and 4.4 min for pinus, birch and oat, respectively. Xylanases induced by birch xylan were applied for a pre-bleaching experiment. Pre-bleaching of pinus Kraf pulp followed by a short oxygen-peroxide sequence, resulted in a very high selectivity as compared with commercial xylanases.*

INTRODUCTION

Production and characterization of the hydrolytic enzymes of fungi and others microorganisms is of practical importance in the search for new enzymes for industrial use. The pulp and paper industry is interested in the application of biotechnology in order to substitute chlorocompounds of the traditional pulp bleaching processes at least partially by a xylanase treatment (1). Treatment of radiata pine Kraft pulp with xylanase enhanced its bleachability, resulting in saving in chlorine chemicals. A fully bleached pulp, which had been treated with xylanase showed only minor changes in handsheet properties (2). Xylanase used in a single treatment of unbleached softwood Kraft brownstock pulp, had the same delignification and brightness capability as 5 to 7 Kg/t of chlorine dioxide (1). Similar results with hardwood Kraft pulp (3) and softwood (4) were obtained. Recently using PULPZYME HA (Novo Ind.) and pure xylanase on black spruce pulp, it was found that the pre-bleaching effect was associated with a marked drop in xylan degree of polymerization in the pulp. CARTAZYME HS (Sandoz) pretreatment of Kraft pulp enable a significant reduction in elemental chlorine needed to achieved a high brightness pulp. Mill scale enzyme treatment was published (5).

All of these enzymatic treatments can be used to develop new chlorine-free chemical bleaching sequences, such as with oxygen and/or hydrogen peroxide (6). Recently, the bleaching by a sequence free of chlorine by using oxygen reinforced with hydrogen peroxide (PO) (7) on organosolv pulps was used (8).

Xylanases are produced by different microorganisms (9-12). One potential source of cellulase-free xylanase could be certain strains of *Aspergillus* fungi. In previous works production of xylanases were described for strains of several *Aspergillus* species (12,13), but, in general all are producers of cellulases (12). However, recently we have isolated an *Aspergillus sp.* (2M1 strain) with no cellulase or protease and very low beta-glucosidase (0.003 U/mL) activities (11,14).

Due to the industrial importance of xylanases, besides the productivity, there are many concerns with stability and the active site behaviour of these enzymes with different chemicals (15). In other cases the carboxylic groups were found as the important moiety in this kind of inactivation (16).

The aim of this work is to study the xylanases production, comparing two types

489

E. Galindo and O.T. Ramírez (eds.), Advances in Bioprocess Engineering. 489-494.
© 1994 Kluwer Academic Publishers. Printed in the Netherlands.

of reactors (a PBAL and STB bioreactors), chemical and thermal stabilities, some of the characteristic of the *Aspergillus sp.* 2M1 xylanases (ASPERZYME 2M1) and their application in a pulp pre-bleaching process followed with an oxygen-peroxide short sequence.

MATERIALS AND METHODS.

Chemicals. Birch and oat spelt xylans, and 3,5-dinitro-salicylic acid (DNS) were from SIGMA. Pinus xylan was prepared as described previously (17).

Collection: From nine samples of wood and vegetal soil in decomposition from the X and VII Region in Chile the *Aspergillus sp.* (2M1 strain) was selected.

Growth conditions: A) Solid Medium. The isolated fungus was cultivated in agar plates containing Vogel medium (11) with 1% of birchwood xylan as the carbon source, following previous work (11). B) Liquid Medium. 50 mL of Vogel solution in the presence of 1% xylan as the carbon source in a 250 mL Erlenmeyer in an orbital shaker at 150 rpm at 28°C during 4 days was used (11).

DNS reagent for the modified Bailey's method: 1.6 g NaOH and 30 g of sodium potasium tartrate in 50 mL of distilled water was stirred under heating and added slowly 1 g of DNS and after cooling, it was completed 100 mL with distilled H_2O_2.

DNS reagent for the standard Bailey's method (18): The DNS reagent was prepared following a published method (11).

Enzymatic assays: The ß-xylanase was determined by measuring the reducing sugars by the standard Bailey's method (18) or the modified Bailey's method. The xylanase activity was measured as the xylose liberated per minute (umol/min, IU). The beta-xylosidase activity was determined by standard methods (11).

Temperature stability: The remaining ß-xylanase activity was determined after preincubation of the enzyme at 30-55°C for different times, in 100 mM phosphate buffer, pH 6.0 The enzyme inactivation kinetics first order reaction constants (k_i) were determined.

pH effect on xylanase activity: Xylanase activity was measured in a range of pH of 4.5-8.0 using 50 mM acetate-phosphate buffer at 50°C.

Chemical modifications of xylanases: Chemical modifications were done using several reagents under reaction conditions wich are described in table 4 (16).

Enzymatic treatment: Enzyme prebleaching was performed at 2.5 consistency at pH 4.5 (acidified by sulfuric acid) with 5 U/g (ASPERZYME 2M1) of dry pulp for 3 h at 50°C.

Pulps: Unbleached *Pinus radiata* Kraft was provided by ARAUCO Celulosa, Arauco, Chile. The bleaching process was carried out as previously published (7,19). Viscosities were measured by the capillarity viscometer method (TAPPI Standard T-230 OS-76) and Kappa number by the published method (20).

Bioreactors: A) Packed-bed/air-lift (PBAL): The experiments with the selected strain **Aspergillus sp.** 2M1 were carried out in a modified packed bed reactor (Figure 1) (21,22). Cubic particle of Nylon foam (0.5 cm side) were used as packing material (3 g per 500 mL).

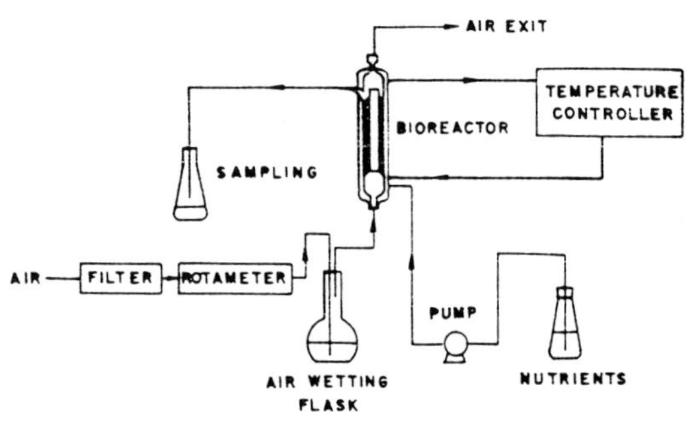

Figure 1. Experimental set-up (Packed-bed/air-lift bioreactor).

Operation Procedure: 500 mL Vogel medium as described in liquid culture (Erlenmeyer) was used, and inoculum (a half-Petri plate of mycelium) was fed into the reactor (23). The bioreactor was operated for 4 days. After this period the xylanolytic enzymes were measured in the culture medium. Air flow at a rate of 0.43 L/min was maintained. B) Stationary Tray Bioreactor (STB): A schematic diagram of stationary tray bioreactor used in this work is shown in Figure 2.

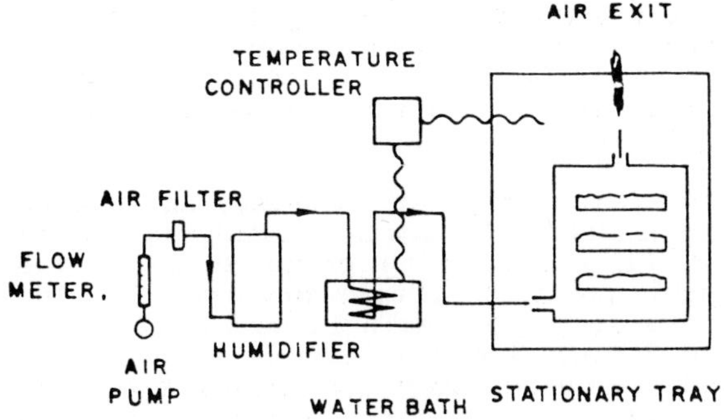

Figure 2. Experimental set-up (Stationary tray bioreactor).

(24).The STB was conducted in stationary shallow tryas, where the substrate was distributed in 5 cm deep layers in a cabinet wherein optimum growth parameters were maintained for obtaining high yield of enzyme production. The STB was humidified as described in Fig.2. Metallic trays with a perforated bottom were employed. The perforation was provided to afford adequate aeration of the undersurface of the substrate in the trays. <u>Operation Procedure</u>: The isolated fungus was cultivated in agar trays containing Vogel medium (11) with 1% of birchwood xylan as the carbon source, following the same procedures as previously described (11). The bioreactor was operated 4 days. The humity was maintained at 75%.

RESULTS AND DISCUSSION

The xylanases levels induced by different carbon sources, in liquid cultures, gave 127 U/mL, 64 U/mL and 67 U/mL with birch, oat spelt and pinus xylans, respectively.

Recently we have studied the induction of xylanases on different pH and different buffer nature (14,17). Now, the temperature effect on the xylanolytic activities induced by oat spelt xylan and birch xylan in a liquid culture was studied. The results are shown in Figure 3. In assay period of 5 min the optimum enzymatic activity in both cases was observed at 50-55°C. Under the same enzymatic assay, the pH of maximal xylanolytic activity was 5.5, either on oat spelt or birch xylans (Figure 4).

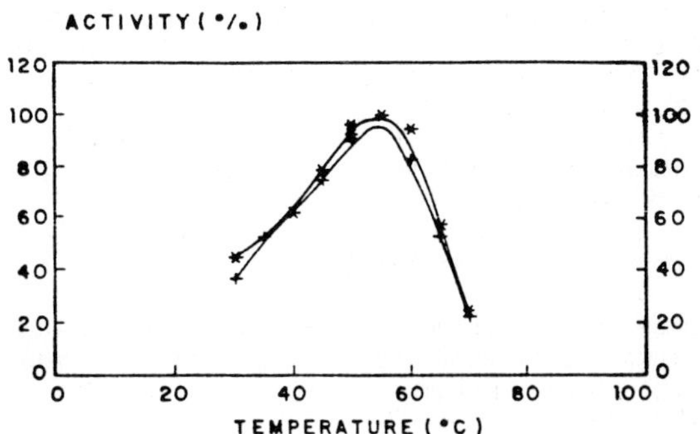

Figure 3. Temperature effect on the xylanase activity induced by birch and oat xylans: (-+-) birch; (-*-) oat.

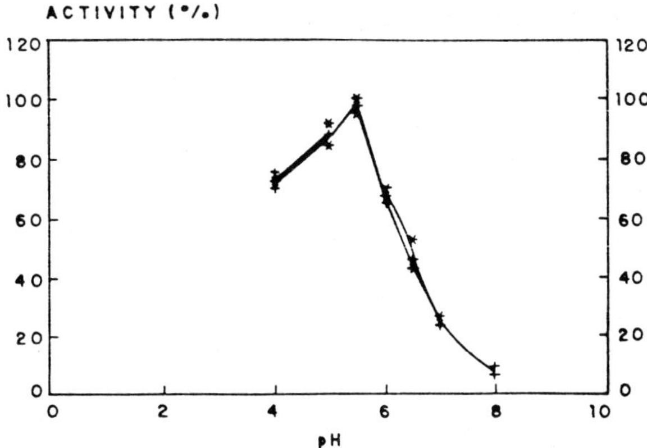

Figure 4. pH effect on the xylanase activity induced by birch and oat xylans: (-+-) birch; (-*-) oat.

The isolated fungus was cultivated in Petri plates from which extracts of the crude enzyme were taken for ß-xylanase, ß-xylosidase, cellulase, and protease analyses (11). The results of the activities are presented in table 1 . The assays were carried out directly on the extract obtained from the plates, being xylanase activity higher in the standard Bailey's method than in the modified one. Protease activity was not detected. However, the measurement were done at the 4th day. In the submerse culture, some of the extracts showed proteolytic activities around the 7th days of culturing (not showed). Cellulase activity was not found in any of the described culture in which the carbon source used was birch wood xylan. The enzymatic activity detected by the described method also is expressed in the submerse fungal culture in liquid medium at the same temperature, pH, culture medium, carbon sources and growth periods (table 1).

Table 1. Comparison of the enzymatic activities found in the solid (Plates) and liquid medium.

Sol.cult. Agar plat.	β-Xyl.(a) U/mL	Ut/g (b)	β-Xylo. U/mL	Ut/g (b) x 10³
2M1(d)	33(127)(c)	1891	11(13)(f)	0.5
2M1(e)	75(74)	2590	26(g)	-

a) Birch xylan as substrate; b) Total Units/g agar-fungi dried weight; c) Liquid culture activities in parenthesis; d) Modified Bailey's method. e) Standard Bailey's method; f)Ref. 14; g)Ref.11

The 2M1 strain exhibited higher xylanase activity in the liquid culture than in the solid one, when the modified Bailey's method was applied. However, no difference in the xylanase activity measured by the Bailey's methods was found. Optimum conditions for the xylanase activity in the presence of either oat spelt and birch xylans were 55°C and pH 5.5.

Production of xylanases using a PBAL bioreactor gave similar values (100 U/mL) than in a batch conditions (Erlenmeyer) using birch xylan as inductor. However in the immobilized form, 60 U/mL were obtained (result not showed). In a preliminary experiment with a STB bioreactor showed to be slightly more efficient than the PBAL bioreactor in its free form.

One time we selected conditions for the xylanases production and in order to understand the stability of these induced xylanases, the inactivation rate constants were studied following the method described previously (11,25). At 40°C the $t_{1/2}$ in birch, oat and pinus xylans were 20.4, 9.37 and 38.5 h, respectively. At 55°C the $t_{1/2}$ in birch oat and pinus xylans were 7.2, 4.4 and 11.8 minutes, respectively. The enzymes were reasonable stable at 40°C, but at 55°C a rapid denaturation was observed.

At the temperatures studied the xylanase produced by oat spelt xylan was the most unstable. This is an indication that different carbon sources induced different xylanases (26).

The instability at 55°C is inadequate for industrial purpose, but if glycerol in 50% is added to crude extract, the stability is improved notoriously, increasing their half-lives in 7.2 min to 2.6 h and 4.4 min to 1.5 h when the inductor were birch and oat spelt xylans, respectively. Figure 5 exhibited this comparison in the presence of glycerol.

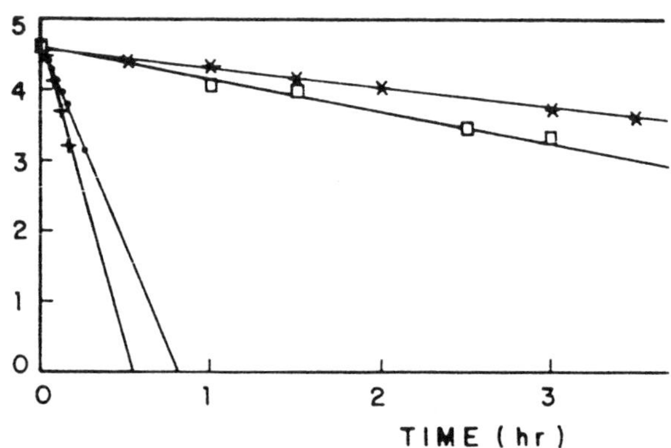

Figure 5. Glycerol effect (50%) on the xylanase stability at 55°C induced by birch and oat xylans: (- . -) birch without glycerol; (-*-) birch with glycerol; (-+-) oat without glycerol; (-□ -) oat with glycerol.

In order to understand better the induced xylanases by birch xylan, chemical modifications were done using several reagents under reactions conditions which were generally similar to those use for cellulase, lysozyme and for xylanases from *Streptomyces sp.* (16). The different functional groups of the enzymes were modified with the following reagents: Carboxyl groups, with 1-ethyl-3-(3-dimethylaminopropyl)carbodiimide (EDC); indole groups, with N-bromosuccinimide (NBS); imidazole groups, with methylene blue; guanidine groups, with 2,3-butanedione; phenol groups, with N-acetylimidazole; and thioether groups, with chloramine T.

The chemicals that were found to affect xylanase activity were EDC for the carboxylic groups and NBS at high concentration for oxidation of indoles (table 2). Carbodiimide reacts with sulfhydryl groups of cysteines, aromatic hydroxyl groups of tyrosine, and carboxyl groups alike. In the case of EDC affecting an enzyme activity, it would be possible that cysteine and tyrosine residues have a relationship to the enzyme activity like a carboxyl group does. Due to the absence of modification of cysteine and tyrosine residue with methylene blue and N-acetylimidazole treatment, respectively, it is concluded that these two amino acids were not related to the enzyme activity. At 0.1 mM effect slightly the xylanase activity (38% reduction) but tryptophan was obviously affected under these conditions. At 0.5 mM of NBS, the activity of the crude xylanases were reduced to less than 14 % on account of the destruction of protein which means the cleavage of peptide bonds by the high concentration of NBS, but not the modification of the active sites of the enzymes. These results are similar as published recently by Marui et al. ([16]) with purified xylanases from *Streptomyces sp.*

Since, the stability and chemical behaviour of xylanase induced by birch xylan from *Aspergillus sp. 2M1.* was known, the industrial potentiality was tested. Then, a prebleaching treatment on Kraft pulp was afforded with this enzyme, which was named ASPERZYME 2M1.

Table 2. Effect of chemical reagents on xylanase activities

Reagents	Residual Act. (%)	Reagent (mM)	Reaction Time(Min)
EDC(a)	29	100.0	60
NBS(b)	62	0.1	30
NBS	14	0.5	30
Methylene Blue (c)	99.7	25.0	60
2,3-Butane-dione (d)	98	10.0	60
N-Acetyl imidazole(e)	100	5.0	60
Chloramin T (f)	97.4	1.7	20
Iodoacetate (g)	100	11.3	75

a) EDC, containing 0.1 M acetate buffer pH 4.75 and 0.7 M NaCl. b) NBS, in 0.1 M acetate buffer, pH 4.0. c) Meth.blue, in 25 mM phosphate buffer pH 7.0. d) 2,3-butanedione, in 50 mM borate buffer pH 8.5. e) N-acetylimidazole, in 50 mM barbital buffer pH 7.5. f) Chloramin T, in 0.1 M Tris buffer pH 8.5. g) Iodoacetate, in 0.1 M citrate buffer, pH 6.0.

Table 3 shows the results with our xylanase, ASPERZYME 2M1 (5 U/g of dry pulp) acting on Kraft pulp. A 20% of Kappa value reduction and a high fibres protection was found (less viscosity reduction). A very high selectivity ratio (2.20) as compared with those of commercial CARTAZYME (Sandoz) (1.51) and PULPZYME (Novo Ind.) (1.65)([19]) was determined.

Table 3. Xylanase-oxygen/peroxide sequence (a).

KRAFT PULP:Enzyme treat.:pHi 4.5; pHf 5.4

STAGE	Kappa Value (%)	Delig. Effic. (%)	Visc. (cps)	Red. Visc. (%)	Select.Ratio XPO/PO
Contr.	23.1	–	30.9	–	–
PO	18.3	20.8	18.5	40.0	0.52
X-PO	14.6	36.8	21.1	31.7	1.16 2.2

a) ASPERZYME 2M1 5 U/g.

In summary, a Packed-bed/Air-lift Bioreactor (PBAL) with immobilized fungi and the Stationary Tray Bioreactor (STB) appeared as efficient for xylanase production. A more detailed study with STB is actually under progress. ASPERZYME 2M1 which has a good stability and is produced with no cellulases presence, improved the reduction of Kappa number and a significative fibre protection in the pulp. This results open the possibilities of new studies on alternative pulping and bleaching process with no pollutant reagents.

REFERENCES.

1.   Tolan, J.S. and R.V. Canovas. *Pulp Paper Can.* 93, 39 (1992).

2.   Allison, R.W., T.A. Clark and S.H. Wrathall. *APPITA* 46, 269 (1993).

3.   Bajpai, P., N.K. Bhardwaj, S. Maheshwari and P.K. Bajpai. *APPITA* 46, 274 (1993).

4.   Senior, D.J. and J. Hamilton. *TAPPI J.* 76, 200 (1993).

5.   Scott, R.P., F. Young and M.G. Paice. *Pulp Paper Can.* 94, 57 (1993).

6.   Kantelinen, A., B. Hortling, M. Ranua and L. Viikari. *Holzforschung* 47, 29 (1993).

7.   Parthasaraty, V.R. *TAPPI J.* 73, 243 (1990).

494

8. Ruiz, J., N. Rojas, J. Mena, S. Urizar, J. Freer, E. Schmidt, S. Quadri and J. Baeza. *Proc. 2nd Braz. Symp. Chem. Lignins & Other Wood Comp.* (N. Durán and E. Esposito, Eds) FAPESP Ed., Campinas, S.P., Brazil. 3, 120 (1992).

9. Milagres, A.M.F. and N. Durán. *Progress Biotechnol.* 7, 539 (1992).

10. Wong, K.K.Y. and J.N. Saddler. (1992). *Crit. Rev. Biotechnol.* 12, 413 (1992).

11. Curotto, E., C. Aguirre, M. Concha, A. Nazal, V. Campos, E. Esposito, R. Angelo, A.M.F. Milagres and N. Durán. *Biotechnol. Tech.* 7, 821 (1993).

12. Bailey, M.J. and L. Viikari. *World J. Microbiol. Biotechnol.* 9, 80 (1993).

13. Attili, D., A.P. Macedo and E. Esposito. *Proc. 3th.Braz. Symp. Chem. Lignins & Other Wood Comp.* (D. P. Veloso, E.A. Do Nascimento and R. Ruggiero, Eds.), Belo Horizonte, M.G. Brazil. 4, 000 (1994).

14. Curotto, E., M. Concha, C. Aguirre, S. O'Reilly, S., A. Ferreira and N. Durán. *Proc. VIII Workshop in Bioorganic Chemistry (Brazil-Chile)*(H. Mansilla, P. Pacheco and J. Villaseñor, Eds.)(Univer. Talca Publ), Chile. 88 (1991).

15. Khasin, A., I. Alchamanati and A. Shoham. *Appl. Environ. Microbiol.* 56, 1725 (1993).

16. Marui, M., K. Nakanishi and T. Yasui. *Biosci. Biotech. Biochem.* 57, 662 (1993).

17. Curotto, E., C. Aguirre, R. Angelo, E. Esposito and N. Durán. *Proc. 3th. Braz. Symp. Chem. Lignins & Other Wood Comp.* (D. P. Veloso, E.A. Do Nascimento and R. Ruggiero, Eds.), Belo Horizonte, M.G. Brazil. 4, 000 (1994).

18. Bailey, M.J., P. Biely and K. Poutanen. *J. Biotechnol.* 23, 257 (1992).

19. Ruiz, J., S. Urizar, J. Freer, J. Baeza, J. Rodriguez and N. Durán. *Proc. 3th. Braz. Symp. Chem. Lignins & Other Wood Comp.* (D. P. Veloso, E.A. Do Nascimento and R. Ruggiero, Eds.), Belo Horizonte, M.G. Brazil. 4, 000 (1994).

20. Berzins, V. *Pulp Paper Mag. Can.* 67, 206 (1966).

21. Cammarato, M.C. and G.L. Sant'Anna. *Environm. Technol.* 13, 65 (1992).

22. Durán, N., E. Esposito and V.P. Canhos. *In* Cellulosics: Pulp, Fibre and Environmental Aspects. (J.F. Kennedy, G.O. Phillips and P.A. Williams, Eds.) Ellis Horwood Ser. Polym. Sci. and Technol. Publ. Chapter 73, 493 (1993).

23. Esposito, E., N. Durán, J. Freer, J. Baeza and L. Innocentini-Mei. *Proc. CHEMPOR'93, Intern. Chem. Eng. Conference*, Porto Portugal, 201 (1993).

24. Weiland, P. *In* Treatment of Lignocellulosics with White Rot Fungi. (F. Zadrazil and P. Reiniger, Eds.) Elsevier Appl. Sci. Comission of the European Communities. 64 (1988).

25. Ferrer, I., E. Esposito and N. Durán. *Enzyme Microb. Technol.* 4, 402 (1992).

26. Curotto, E., M. Concha, V. Campos, A.M.F. Milagres and N. Durán. *Appl. Biochem. Biotechnol.* in press (1994).

Optimization of Cell Harvesting and Assay Procedures for Reductive Biotransformations in Obligate Anaerobes

E.T. Davies and G.M. Stephens

Department of Chemical Engineering, UMIST, Manchester, M60 1QD, U.K.

Caffeate reductase is an enoate reductase which is produced by the acetogenic, obligate anaerobe, Acetobacterium woodii. *The function of the enzyme appears to be the reduction of aromatic enoates, which are utilised as alternative electron acceptors to carbon dioxide, the usual electron acceptor for this organism. The substrate specificity of this enzyme seems to differ from that of other enoate reductases. Therefore, caffeate reductase may be useful for biotransformations which cannot be achieved at present. The objective of this work was to develop an assay for caffeate reductase so that the substrate range and characteristics of the enzyme could be investigated. Using cells harvested from a chemostat culture, an anaerobic harvesting procedure and an assay for caffeate reductase in whole cells were developed. An electron donor was found to be essential for the reduction of caffeate, and for this fructose was used. The optimum buffer pH was found to be 7.0, and the optimum caffeate concentration was 2.5 mM. At higher caffeate concentrations a decrease in reduction rate was observed. The substrate range of the enzyme appeared to be fairly wide, as cinnamaldehyde and cinnamyl alcohol were reduced, as well as cinnamate, p-hydroxycinnamate and caffeate.*

Reductases are commonly employed in the production of chiral compounds, important building blocks in synthetic chemistry (1). The use of enzymes is advantageous for this purpose as high degrees of stereo- and regio-specificity can be obtained (2). However, only a few reductases are known to catalyze the reduction of carbon-carbon double bonds (3). Work on the enoate reductases from *Clostridium* species has demonstrated that such enzymes are extremely useful in synthetic chemistry (4, 5, 6). It would be useful to identify and characterise more enzymes which can catalyse the reduction of enoates and other classes of unsaturated compounds, since these may be exploitable for new biotransformations which cannot be achieved using existing enzymes.

The homoacetogenic obligate anaerobe, *Acetobacterium woodii*, can reduce various phenylacrylates in a reaction catalysed by a novel enzyme, caffeate reductase (7). Normally, A. woodii uses the reduction of carbon dioxide to acetate as the electron-accepting process for the fermentation of various organic electron donors (8,9), but phenylacrylate reduction serves as a supplementary electron-

accepting process (10). In addition, the reduction of phenylacrylates enables the organism to generate ATP via electron transport-dependent phosphorylation (11, 12). The substrate range of caffeate reductase seems to differ from that of the clostridial enoate reductases (3,5,7), and, for this reason, we are investigating the use of this enzyme for biotransformations. The first step in this research was to develop and optimize an assay for the reductase in whole cells. Since A.woodii is an obligate anaerobe, it was also necessary to develop strictly anaerobic procedures for preparing cell suspensions for use in the assay.

MATERIALS AND METHODS

The organism used was *Acetobacterium woodii* DSM1030. Cells were maintained in an anaerobic, fructose-limited chemostat culture with 14.3mM fructose in the presence of caffeate. In some experiments, the cultures were grown with 28mM fructose (excess fructose) and this is mentioned in the corresponding figure legends. However, this resulted in lower rates of caffeate reduction and this growth condition was not used routinely. For

495

E. Galindo and O.T. Ramírez (eds.), Advances in Bioprocess Engineering. 495-499.
© *1994 Kluwer Academic Publishers. Printed in the Netherlands.*

496

biotransformation assays, cells were harvested from the chemostat using an Atmos bag which had been flushed with nitrogen, to avoid exposing the cells to air, and transferred to an anaerobic cabinet. Cells were collected and washed by filtration, followed by centrifugation in a minifuge and resuspension in the assay buffer. Caffeate reductase activity was determined by adding fructose as an electron donor and caffeate (or another substrate) to initiate the reaction. The residual concentration of the unsaturated substrate was measured by determining the UV absorbance at 312nm. Reduction rates given are average values for duplicate assays.

RESULTS

Development of an Anaerobic Harvesting Procedure for Preparing Washed Cell Suspensions.

The main challenge in developing a procedure for preparing cell suspensions was to maintain strictly anaerobic conditions. After removal from the chemostat, the cells had to be harvested and washed in an anaerobic cabinet. However, there is very little space in an anaerobic cabinet and this placed restrictions on the type of equipment which could be used. Therefore, it was only possible to harvest the cells by filtration or by centrifugation in a microcentrifuge. The two techniques were compared. Cells harvested by centrifugation reduced caffeate at a rate of $9.93\mu mol.min^{-1}.g^{-1}$, whereas cells harvested by filtration exhibited a rate of $8.21\mu mol.min^{-1}.g^{-1}$ indicating that there was little difference between the effectiveness of the techniques. However, it took longer to prepare the cell suspensions by centrifugation (95 min) than by filtration (60 min) due to the small volume of each centrifuge tube (1.5ml). Since the difference in caffeate reductase activity was relatively small, filtration was used for collecting cells in subsequent work.

Development of an Anaerobic Assay for Caffeate Reductase

A variety of physical and chemical factors were considered likely to influence caffeate reductase activity in cell suspensions, especially the composition of the assay mixture.

Aromatic compounds often inhibit the growth or metabolism of microorganisms, and therefore it was important to test for caffeate toxicity (figure 1). The enzyme activity increased with caffeate concentrations up to 2.5mM but then reduced to about 50% of the maximum level with 7mM caffeate. This suggested that the optimal caffeate concentration for use in the assays was 2.5mM. The reduction in enzyme activity at higher caffeate concentrations suggests that either the substrate, or the product, or both, was toxic. Therefore, it was also important to test for hydrocaffeate toxicity. However, there was no evidence for hydrocaffeate inhibition even when hydrocaffeate was added at concentrations up to 5mM, so that hydrocaffeate toxicity would be unlikely at the optimal caffeate concentration.

An alternative explanation for the reduced reduction rates at high caffeate concentrations was that the supply of the electron donor had become limiting. The presence of an electron donor in the assay mixture was found to be essential for the reduction of caffeate, since there was no enzyme activity in the absence of fructose. However, there was little variation in caffeate reductase

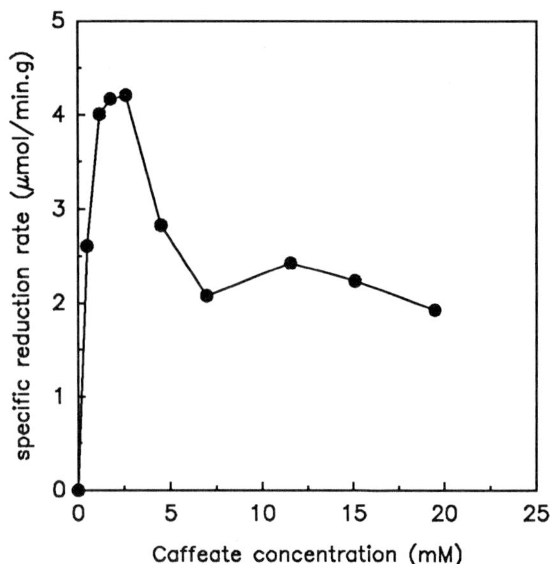

Figure 1 The effect of initial caffeate concentration on the activity of caffeate reductase in harvested cells. Cells grown with 28mM fructose.

activity when the fructose concentration was varied between 5mM and 100mM (Figure 2), suggesting that the electron donor was in excess.

A physical factor which may affect reduction rates is mass transfer in the assay suspension, since contact must be made between the enzyme and the substrate. Continuous stirring of the assay mixture using a magnetic stirrer increased the enzyme activity slightly. Stirred suspensions reduced caffeate at a rate of $8.21\mu mol.min^{-1}.g^{-1}$ while suspensions agitated only by inversion before withdrawing samples had a reduction rate of $7.52\mu mol.min^{-1}.g^{-1}$. Therefore, the cell suspensions were stirred continuously during subsequent assays.

The composition of the buffer was also optimized. Acetogenesis via the acetyl CoA pathway in *A. woodii* is sodium dependent, and a sodium pump is involved in the generation of ATP during the reduction of CO_2 to acetate (13). A trans-membrane electrical potential is also necessary in the phosphorylation of ADP linked to caffeate reduction, but a sodium pump is not thought to be involved (12). Efficient reduction of caffeate should depend upon the diversion of reducing power from CO_2 reduction to the caffeate reductase. The availability of sodium ions might influence the activity of the pathway for CO_2 reduction and this was tested by comparing the rate of caffeate reduction in cells washed and resuspended in sodium and potassium phosphate buffers. Suspensions prepared in sodium phosphate buffer had a slightly higher initial rate of caffeate reduction ($9.17\mu mol.min^{-1}.g^{-1}$) than cells resuspended in potassium phosphate buffer ($8.55\mu mol.min^{-1}.g^{-1}$), but caffeate reduction continued for longer in the presence of potassium ions (figure 3). This demonstrated that there was no advantage in using sodium-containing buffers. The higher initial reduction rate in the sodium-containing buffer may be due to an increase metabolic activity. However, in the presence of sodium ions, cells may favour acetogenesis over caffeate reduction, which could explain the subsequent decrease in reduction rate.

The pH of the buffer was also influential, with the pH optimum for reduction being 7.0 (figure 4). The enzyme activity was reduced substantially at lower pH values. Cultures of *A. woodii* normally produce

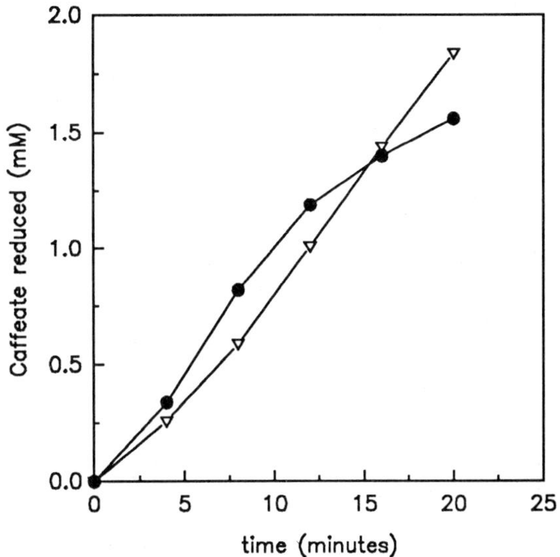

Figure 2 The effect of initial fructose concentration on the activity of caffeate reductase in harvested cells. Cells grown on 28mM fructose.

Figure 3 The effect of washing and resuspending cells in sodium buffer (●) and in potassium buffer (v) on the activity of caffeate reductase in harvested cells.

acetic acid during growth, which causes the pH of the culture to fall to values at which caffeate reductase activity would decline. However, the provision of caffeate in growing cultures reduces the amount of acetate produced (11), and this may explain why the enzyme functions better in less acidic conditions.

The validity of the assay which had been developed was assessed by comparing the rates of caffeate reduction in growing cultures and in the cell suspensions. When cells were grown in the chemostat with 28mM fructose and 1mM caffeate, the rate of reduction ($q_{caffeate}$) was 4.82μmol.min^{-1}.g^{-1}. This may be compared with the maximum reduction rate in harvested cells (V_{max}, calculated from the data shown in Fig 1) which was 5.31μmol.min^{-1}.g^{-1}. This demonstrates not only that the assay provided an accurate measure of caffeate reductase activity, but also that the harvesting procedure did not affect the integrity of the cells.

Substrate Range Of Caffeate Reductase

It was found that caffeate reductase was not strictly an enoate reductase as the range of substrates reduced was not confined to acids (table 1). Thus, cinnamaldehyde and cinnamyl alcohol were both reduced, with the percentage reduction of the aldehyde and alcohol being significantly higher than that for cinnamate after 15 and 30 minutes.

The presence and number of hydroxyl groups was also found to have an effect on the reduction of substrates. Cinnamate was reduced at a significantly slower rate than 4-hydroxycinnamate, which in turn was reduced at a much lower rate than caffeate (3,4-dihydroxycinnamate). This suggests that the presence of hydroxyl groups are not essential for the activity of the enzyme, but that the rate is enhanced with hydroxylated substrates.

CONCLUSIONS

The study of reductases in the past has been concentrated on enzymes from aerobic or facultatively anaerobic organisms, such as yeasts, and in well characterised anaerobes, such as the clostridia (14, 3, 4). The investigation of novel enzymes from other obligate anaerobes has been hampered by the difficulty of culturing the organisms and in subsequently assaying enzyme activity. In order to overcome the difficulties of working with obligate anaerobes, a new method of cell harvesting has been developed. This was found to result in reduction rates in harvested cells that were comparable with those in growing cultures. The assay is also a relatively simple and quick process.

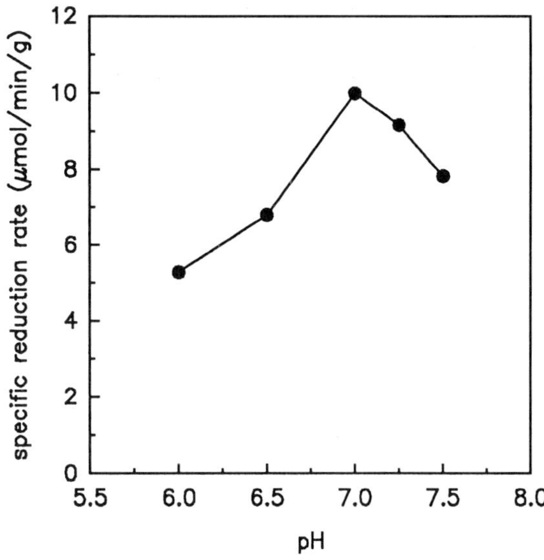

Figure 4 The effect of buffer pH on the activity of caffeate reductase activity in harvested cells.

Table 1 The reduction of a range of substrates by harvested cells grown in the presence of 1mM caffeate. All substrate concentrations were 2.5mM.

Substrate	% reduction after:-	
	15min	30min
caffeate	48.7	91.0
4-hydroxycinnamate	16.0	65.2
cinnamate	12.8	33.2
cinnamaldehyde	87.2	88.8
cinnamyl alcohol	39.6	60.4

It has also been shown that caffeate reductase is able to catalyse the reduction of aromatic enoates, aldehydes and alcohols. Other enoate reductases from anaerobic bacteria only catalyse the reduction of enoates and unsaturated aldehydes. The relatively high rate of reduction of cinnamaldehyde suggests that phenylacrylates may not be the physiological substrates of this reductase. Since the substrate range differs so markedly from other enoate reductases, and cofactors or mediators are not required, the caffeate reductase of *A. woodii* may be industrially useful. Further work will be done to determine how different growth conditions and electron donor availabilities affect caffeate reductase activity. The substrate range and the stereospecificity of the enzyme will also be investigated in greater depth.

ACKNOWLEDGEMENT

This work was funded by the Science and Engineering Research Council.

LITERATURE CITED

1. Whitesides, G.M. and Wong, C.-H., *Angew. Chem. Int. Ed. Engl.*, **24**, 617 (1985).

2. Leuenberger, H.G. and Wirz, B., *Chimia*, **47**, 82 (1993).

3. Simon, H., Bader, J., Günther, H., Neumann, S., Thanos, J., *Angew. Chem. Int. ed. Engl.*, **24**, 539 (1985).

4. Bader, J. and Simon, H., *Arch. Microbiol.*, **127** (1980).

5. Bühler, M., Giesel, H., Tischer, W., Simon, H., *FEBS Lett.*, **109**, 244 (1980).

6. Bostmemrun-Desrut, M., Kergomard, A., Renard, M.F., Veschambre, H., *Agric. Biol. Chem.*, **47**, 1997 (1983).

7. Bache, R. and Pfennig, N., *Arch. Microbiol.*, **130**, 255 (1981)

8. Fuchs, G., *FEMS Microbiol. Rev.*, **39**, 181 (1986).

9. Diekert, G., *FEMS Microbiol. Rev.*, **87**, 391 (1990).

10. Drake, H., "CO_2, reductant, and the autotrophic acetyl-CoA pathway: Alternative origins and destinations," in *Microbial growth on C1 compounds*, Murell, J.C. and Kelly, D.P. (eds.) published by Intercept Ltd. (Andover) (1993).

11. Tschech, A. and Pfennig, N., *Arch. Microbiol.*, **137**, 163 (1984).

12. Hansen, B., Bokranz, M., Schönheit, P., Kröger, A., *Arch. Microbiol.*, **150**, 447 (1988).

13. Heise, R., Müller, V., Gottschalk, G., *J. Bacteriol.*, **171**, 5473 (1989)

14. Ward, O.P. and Young, C.S., *Enzyme Microb. Technol.*, **12**, 482 (1990)

A Scalable Method for the Purification of Recombinant Human Protein C from the Milk of Transgenic Swine

W.N. Drohan[1], T.D. Wilkins[2], E. Latimer[2], D. Zhou[2], W. Velander[3], T.K. Lee[1], and H. Lubon[1]

[1]Holland Laboratory, American Red Cross, Rockville, MD 20855; [2]TechLab, Inc., Blacksburg, VA 24061; [3]Department of Chemical Engineering, Virginia Polytechnic Institute, Blacksburg, VA 24061, U.S.A.

Transgenic pigs which produced from 0.1 to 1.0 mg ml⁻¹ of recombinant human Protein C (rHPC) in milk were generated. A process for the purification of rHPC from milk has been developed which used selective precipitation by polyethylene glycol (PEG), enrichment of rHPC by barium/citrate precipitation, viral inactivation by solvent/detergent treatment and fractionation of rHPC on an ion exchange column. The overall recovery of rHPC by this process was 24-35%. rHPC was 95% pure and had enzymatic and anticoagulant activities similar to those of plasma derived HPC. These results demonstrate that transgenic pigs can be efficient "bioreactors" for the production of rHPC, and that the purification of the protein from milk by conventional methods can be developed for economical manufacture.

Bioreactors have recently progressed from conventional cell culture systems and fermentors to transgenic animals. The milk, blood or urine of transgenic livestock are considered to be vehicles for the production of large quantities of therapeutic proteins. The limitations of transgenic protein production include small number and low viability of embryos, limited understanding of embryo biochemistry and physiology, low transgene integration rates, a lack of embryonic stem cells for any of the livestock species and high animal costs (1). Despite this, transgenic livestock animals have been developed expressing human α-1-antitrypsin, tissue plasminogen activator and Factor IX in milk, and hemoglobin in blood (2,3).

We have used transgenic pigs (4) as model farm animals to produce a naturally occurring human anticoagulant, Protein C (HPC; 5), a member of the plasma vitamin K-dependent protein family which are responsible for hemostasis (7). Protein C undergoes extensive co- and posttranslational modification during its biosynthesis in the human liver. These modifications include glycosylation, γ-carboxylation and β-hydroxylation of specific amino acids, and endoproteolytic processing to remove the pre- and propeptides, and a basic dipeptide that connects the light and heavy chains (6). These modifications have complicated the production of large amounts of functional HPC and other vitamin K-dependent proteins in cell culture systems. Protein C, Prothrombin, Factor X, Factor VII, Factor IX and Protein S share the common feature of containing domains with 9-12 γ-carboxyglutamic acid (Gla) residues, as well as other sequence homologies. Although there are selective methods to isolate Gla-containing proteins, their low concentration in plasma and high degree of homology make the production of these proteins as therapeutics challenging. As a result, an expensive immunoaffinity column has been used for retrieving HPC from human plasma (8). Even so, the supply of Protein C from plasma cannot meet the projected demand which is around 100 kg per year. Thus, high-level production of vitamin K-dependent proteins in the milk of transgenic livestock animals may solve these problems. In addition, this reduces the risk of potential contamination with human blood-borne viruses.

Recombinant human Protein C (rHPC) has been expressed in porcine milk at levels ranging from 0.1 mg/ml to 1 mg/ml which is significantly higher than the level present in human plasma, 0.004 mg/ml (4, 5). We present herein a scalable method for the purification of rHPC from the milk of transgenic pigs.

<u>MATERIALS AND METHODS</u>

<u>Transgenic swine.</u> The construction of the hybrid gene containing the HPC cDNA inserted into the KpnI site of the first exon of mouse whey acidic protein gene, the method of preparation of DNA used for micro-injection, the generation and identification of transgenic swine have been described earlier (4). In this study, we used the milk of transgenic pigs 83-1 and 29-2. Milk let down was induced by intramuscular administration

501

E. Galindo and O.T. Ramírez (eds.), Advances in Bioprocess Engineering. 501-507.
© *1994 Kluwer Academic Publishers. Printed in the Netherlands.*

of 20-30 IU of oxytocin and milk was collected by hand milking, from day 5 through day 60 of lactation. Whole milk was then diluted with an equal volume of deionized water and processed immediately. Generally, 1.5 to 2 L of milk were used per run.

Protein analysis. The denatured protein samples were separated on 10% or 8-16% gradient polyacrylamide gels at 30 mA for about two hours and silver stained (9). Protein concentrations were determined either by the method of Bradford (10) or by absorbance at 280 nm, using an extinction coefficient of 1.45 for a 1 mg/ml solution of Protein C. A 6 kDa fragment from the light chain region was generated by chymotryptic digest (11) and amino-terminal sequence analysis was performed on an automated Hewlett Packard G1005A protein sequencing system.

Determination of Protein C activity. Protein C was activated for ten minutes by Protac, a glycoprotein isolated from the venom of southern copperhead snake, Agkistrodon contortrix (American Diagnostics). The amidolytic activity of APC was determined by the hydrolysis of a tripeptide substrate, Glu-Pro-Arg-p-nitroanilide (S-2366, Kabi) at room temperature (12). One unit was defined as the amount of protein required to produce 1 μmole of product (p-nitroanilide, pNA) in one minute using an extinction coefficient of 9620 M^{-1} cm^{-1} for pNA at 405 nm. The kinetic parameters were derived from double reciprocal plots of the initial rate as a function of substrate concentrations ranging from 0.167 to 1.0 mM.

The anticoagulant activity of rHPC was determined as a function of the activated partial thromboplastin time (APTT), as described (13). The samples were incubated with Protein C deficient plasma (George King Biomedical) at 37^0C followed by the addition of the APTT reagent (Organon Technika) and Protac. The reaction was initiated by the addition of $CaCl_2$. The clotting time was recorded by an MLA Electra 900 Coagulometer. Duplicates were run for each test sample. One unit of anticoagulant activity was defined as equivalent to 4 μg of human Protein C present in 1 ml of plasma.

RESULTS AND DISCUSSION

Transgenic pigs may become the livestock animals of choice for the production of therapeutic proteins in milk. Pigs are usually not raised for milk, but it is possible to collect between 200 to 400 L from a lactating sow in a year. In the production of rHPC by transgenic pigs, the expression level usually increased during lactation, as shown in Figure 1 for pig 83-1, which consistently made about 0.1 to 0.5 g/L. Sow 29.1 produced about 0.2 to 1.0 g/L (4). As such, a herd of 100 pigs expressing rHPC at an average of 0.5 g/L can produce approximately 15 kg of the recombinant protein in one year. For a cell culture system producing 0.01 g/L/day to provide this

amount, the manufacturer would need to generate 4,000 L of culture media daily.

Producing human therapeutic proteins in the milk of transgenic animals has some significant advantages, however, isolating the desired recombinant protein from milk can be challenging (14). Milk is a complex colloidal mixture consisting of fat globules, casein micelles and whey components. The recombinant protein could be distributed amongst these different phases. In addition, the colloidal nature of milk makes solid-liquid separation difficult. Since skimming the milk greatly decreases the viscosity making it more manageable, we decided to use that as the first step.

At 6.8 g/100 g, pig milk contains more fat than cow milk, therefore we have explored different techniques of fat removal. Filtration could be a method of choice. We experimented with different combinations of filter materials such as paper, glass fiber and nylon. We also evaluated filter aids such as Celite and Hyflo Super Cel and found that although some combinations could remove the fat, the filters clogged quickly and would be difficult to scale up. We decided to use centrifugation to skim the milk, a process that can be adapted to industrial scale centrifuges, although it is not as convenient as filtration. We found that most of the rHPC remained in the aqueous phase, after skimming twice by centrifugation at 3000 X g for 25 minutes and straining through cheesecloth. The concentration of rHPC in skim milk was usually between 0.1 to 1.0 mg/ml.

Next was a volume reduction step in which we attempted to precipitate virtually all the rHPC and a minimum of other proteins, leaving most of the soluble components such as small peptides and sugars behind. After examining conditions involving buffers, pH, temperature, and concentration of different precipitating agents, we found that addition of polyethylene glycol (PEG) in phosphate buffered saline (PBS) to a final concentration of 17% in skim milk allowed us to reproducibly precipitate about 80% of rHPC and total protein. After centrifugation at 7000 X g for 25 minutes and washing the pellet with 17% PEG, we were able to remove most of the β-lactoglobulin, α-lactalbumin, serum albumin and other whey components, such as lactose and small peptides, from the precipitate, Figure 2. The pellet with rHPC was stored frozen at -20^0C. At this stage, the process may be halted for several weeks to accumulate larger quantities of milk protein for large-scale purification.

About 40% of the rHPC was consistently associated with the casein micelles, and solubilization of the micelles greatly improved the recovery of rHPC. The 17% PEG pellet was solubilized in 120 mM sodium citrate, 100 mM EDTA, pH 7.0, which chelates Ca^{++} ions and dissociates the casein micelles, releasing rHPC. The buffer was then changed to 20 mM Tris, 1 M NaCl, pH 7.0, which enhanced the solubility of rHPC.

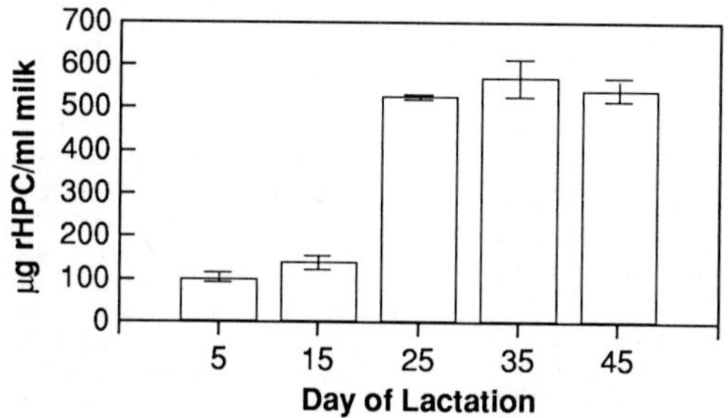

Figure 1. Expression of rHPC during lactation of pig 83-1.
Levels of rHPC in milk were determined using a "sandwich" ELISA, with a rabbit anti-HPC antiserum for the immunocapture of rHPC, followed by detection with a goat anti-HPC antiserum and a rabbit anti-goat antiserum conjugated to horseradish peroxidase (4). The values shown represent mean ± S.E.M., for three determinations.

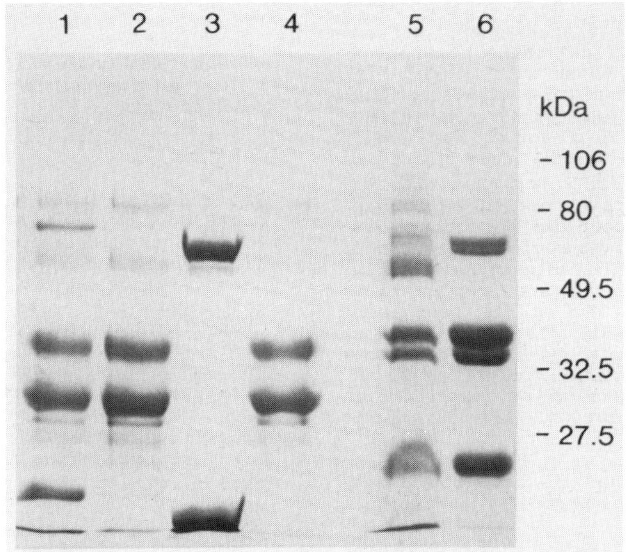

Figure 2. SDS-PAGE analysis of proteins at different stages of purification. Approximately 20 μg of protein were reduced with 50 mM dithiothreitol, electrophoresed on a 10% SDS-PAGE gel and stained with Coomassie Blue R-250. Lane 1, skim milk; Lane 2, 17% PEG pellet; Lane 3, 17% PEG supernatant; Lane 4, 12% PEG pellet; Lane 5, barium/citrate eluate after dialysis; Lane 6, 7.5 mM $CaCl_2$ eluate from the DEAE column.

Subsequent addition of PEG in 1 M NaCl to a final concentration of 12% precipitated a significant portion of the caseins and other milk proteins while the rHPC remained in the supernatant, Figure 2, lane 4 and Table 1. The removal of caseins at this stage facilitates the subsequent barium/citrate extraction of rHPC, as caseins are strong divalent metal binding proteins and will also be precipitated in the process.

Since the final product was to be a parenterally administered therapeutic, the incorporation of a viral inactivation step was necessary. Tri-(n-butyl) phosphate (TNBP) and Triton X-100, at 1% each, have been successfully used for the large scale viral inactivation of HPC from human plasma. The solvent/detergent mixture disrupts the membranes of viruses that have lipid envelopes. The result is either complete structural disruption or destruction of the cell receptor recognition site on the viral coat. In both cases, the viruses become noninfectious (15). Here we were interested in determining if the incorporation of a viral inactivation step would interfere with the subsequent purification process. We found that the addition of solvent/detergent after the 12% PEG precipitation did not affect the overall recovery or purity of the final product (Table 1). We also found that this unit operation could be conveniently carried out overnight at 4^0C or be placed after the barium/citrate precipitation step.

After the viral inactivation step, sodium citrate was added to a final concentration of 12 mM, followed by dropwise addition of 1 M $BaCl_2$ to a final concentration of 80 mM. The solution was then stirred for 30 minutes and centrifuged at 7000 X g for 15 minutes. Barium binds to the γ-carboxylated glutamic acid (Gla) residues of rHPC and precipitates it as rHPC-barium-citrate complex. This was a very efficient method for selecting Gla containing proteins and provided a 47-fold purification of rHPC, Table 1. Proteins eluted from the barium citrate complex with 20 mM Tris, 200 mM EDTA, pH 8.0 contained approximately 30% rHPC. This eluate was suitable for ion-exchange chromatography after it was dialyzed into 50 mM Tris, 20 mM EDTA, pH 7.4. The process may be stopped here and the protein solution may be kept overnight at 4^0C.

An anion exchange column was used to remove residual impurities. The protein solution was filtered through a 0.45 μm membrane and loaded onto a DEAE-Sepharose column equilibrated with the above buffer. The capacity of the column was 25 mg protein per ml of resin. We tested different conditions for loading and found that maximum rHPC binding occurred at a NaCl concentration less than 80 mM or an EDTA concentration less than 50 mM. The presence of EDTA appears to prevent aggregation of rHPC, making the DEAE chromatography more consistent. After the column was washed sequentially with 50 mM Tris and 50 mM Tris with 50 mM NaCl, pH 7.4, rHPC could be eluted with either a NaCl

gradient from 0.2 to 0.3 M, or with $CaCl_2$ at 5, 7.5, 10 and 25 mM in 50 mM Tris, pH 7.4, Figure 3. This "pseudo-affinity" method allows the separation of fully γ-carboxylated (9 Gla) rHPC, from partially γ-carboxylated (6-7 Gla) rHPC (16). We performed a similar chromatographic step and obtained a 40% recovery of functionally active rHPC, eluted at 5 to 10 mM $CaCl_2$ from the column. The overall recovery of the process was usually 24-35%. An example of the recovery from a run is shown in Table 1 and Figure 3.

SDS-PAGE analysis of proteins from three different lots is presented in Figure 4. Under reducing conditions, rHPC migrated as single chain, heavy chain and light chain forms. The molecular weights of the rHPC polypeptides were 1-2 kDa lower than those of HPC. These differences are probably due to the varying carbohydrate composition of the human and recombinant proteins (9). The presence of multiple heavy chain forms suggests different levels of glycosylation at the three putative N-glycosylation sites. The amount of single chain, ~30%, is greater in rHPC than in HPC, ~10%, which may be the result of incomplete proteolytic processing due to over-expression of the protein in the mammary gland.

Kinetic analysis of lots R18, R19 and R20 showed that the activated rHPC had better affinity for the tripeptide substrate Glu-Pro-Arg-p-nitroanilide, S-2366, Table 2, but the turnover of the substrate to product was slower. The catalytic efficiency (k_{cat}/K_m) of rHPC was comparable to that of its plasma counterpart.

The amidolytic and anticoagulant activities of rHPC were about 75-85% that of HPC, Table 2. Amino-terminal sequence analysis of the light chain showed that there were negligible amounts of glutamic acid residues present at the putative Gla sites. Since the phenylthiohydantoin (PTH) derivatives of Gla residues are not extracted during Edman degradation, this suggests that all the nine residues were γ-carboxylated in purified rHPC, figure 5 and that the slightly lower specific activity was not due to the absence of γ-carboxylation. At present, the cause of the slightly lower activity of rHPC is unknown.

In conclusion, our studies demonstrate that a human anticoagulant factor can be produced by transgenic livestock animals and purified by conventional methods in quantities suitable for large-scale production. The purification process developed can be scaled up to 1000 to 5000 L batches, as in the processing of human plasma. Our observations provide a strong impetus for the further development of trangenic pigs as bioreactors for the production of pharmacologically active proteins in milk.

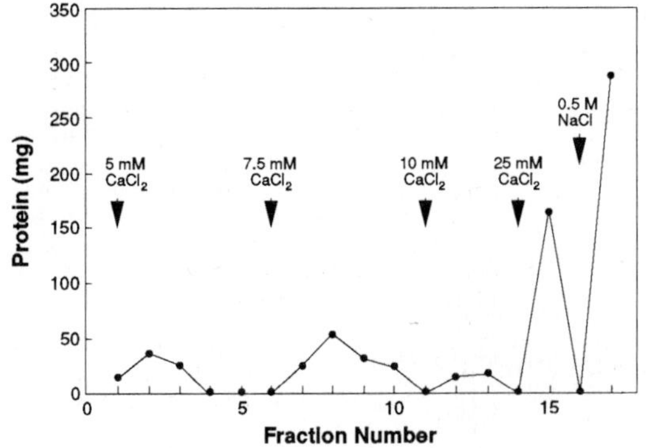

Figure 3. Separation of rHPC on a DEAE-Sepharose column.

The eluate from the barium citrate complex was applied to a DEAE-Sepharose column (2 x 20 cm). Approximately 1 g protein in 50 mM Tris-HCl, 20 mM EDTA, pH 7.4 was loaded. The column was washed with the above buffer, followed by 50 mM Tris-HCL, pH 7.4 and 50 mM Tris-HCl, 50 mM NaCl, pH 7.4. The proteins were eluted with 5, 7.5, 10 and 25 mM CaCl₂ in 50 mM Tris-HCl, 50 mM NaCl, pH 7.4 and with 0.5 M NaCl. Fractions were monitored by reading the absorbance at 280 nm. The presence of rHPC in the elution peaks was confirmed by SDS-PAGE and amidolytic assays. Fractions containing functional rHPC at purity greater than 90% eluted with 5 to 10 mM CaCl₂ and were pooled for further analysis.

Table 1. Purification of rHPC from the milk of transgenic swine

Description	Total rHPC (g)[a]	Total Protein (g)[b]	rHPC/ Protein (%)[c]	Fold purity
1. Skim milk	1	158	0.63	1
2. 17% PEG pellet	0.756	130	0.58	0.92
3. Citrate/EDTA supernatant	0.452	98	0.46	0.73
4. 12% PEG supernatant	0.439	16.6	2.64	4.18
5. Viral inactivation	0.570	ND[d]	ND	ND
6. BaCl₂/EDTA eluate & dialyzed	0.506	1.7	29.8	47.0
7. DEAE eluates				
5 mM CaCl₂	0.077	0.078	99	156
7.5 mM CaCl₂	0.126	0.133	95	150
10 mM CaCl₂	0.035	0.039	90	142

[a] rHPC concentration was determined by ELISA in steps 1 to 6 and by A_{280} values in step 7.

[b] Determined by the method of Bradford in steps 1 to 6 and by A_{280} in step 7.

[c] In step 7, the ratio of rHPC to total protein was determined by densitometry of a SDS-PAGE gel.

[d] Not determined because the solvent/detergent mixture interferes with the assays.

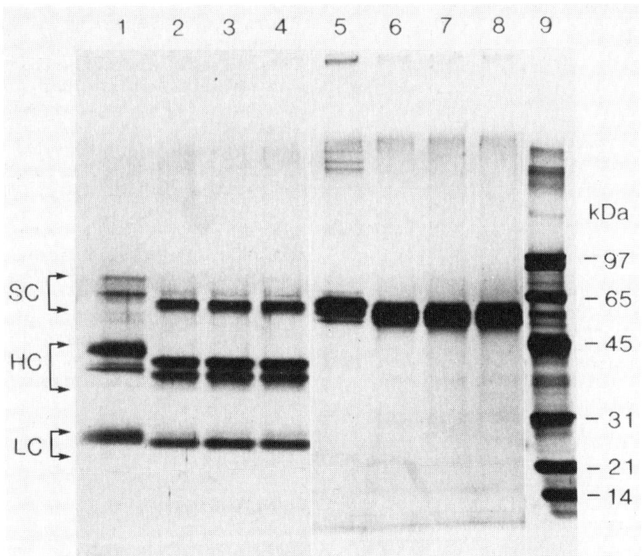

Figure 4. SDS-PAGE analysis of plasma derived HPC and three lots of purified rHPC.
Protein samples, 0.3 μg each, were subjected to SDS-PAGE on a 8-16% gradient gel and visulaized by silver staining.
Lanes 1-4 and 9 were run under reducing conditions and lanes 5-8, non-reducing conditions. Lanes 1 and 5, HPC; lanes 2 and 6, rHPC from run 18; lanes 3 and 7, rHPC from run 19; lanes 4 and 8, rHPC from run 20; lane 9, molecular weight standards. SC: single chain; HC: heavy chain; LC: light chain.

Table 2. Functional activity of HPC and rHPC
These data represent mean values from duplicate determinations.

Enzyme	Amidolytic (units/mg)	K_m for S-2366 substrate	Anticoagulant (units/mg)
HPC	36.4	0.32	250
R18	25.5	0.2	186
R19	26.8	0.2	199
R20	23.2	0.2	205

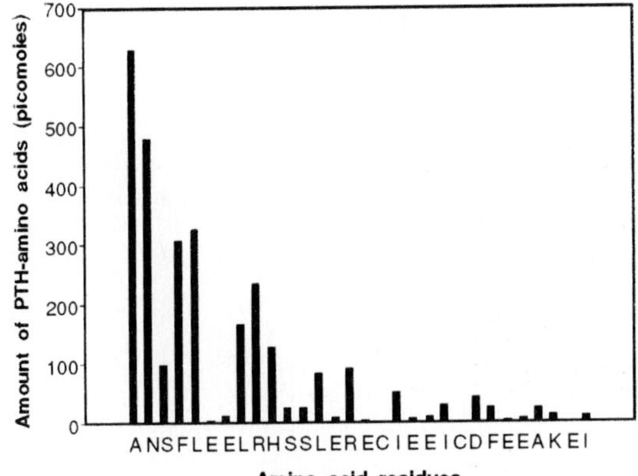

Figure 5. Amino terminal analysis of the light chain of rHPC.

Acknowledgements: The authors thank Drs. B.L. Williams and F. Gwazdauskas for the generation of transgenic pigs, Ms. M.H. Wroble for technical support, Dr. L. Medved and Ms. M. Migliorini for the N-terminal sequence of the light chain, Ms. K. Hyre for typing and Drs. N. Bangalore and R. Paleyanda for critical review of the manuscript.

LITERATURE CITED

1. Wall, R.J., H.W. Hawk and N. Nel, *J. Cell. Biochem.*, 49, 113 (1992).

2. Clark, A.J., H. Bessos, J.O. Bishop, P. Brown, S. Harris, R. Lathe, M. McClenaghan, C. Prowse, J.P. Simons, C.B.A. Whitelaw and I. Wilmut, *Bio/Technology* 7, 487 (1989).

3. Hodgson, J., *Bio/Technology*, 10, 863 (1991).

4. Velander, W.H., R.L. Page, T. Morcol, C.G. Russell, R. Canseco, J.M. Young, W.N. Drohan, F.C. Gwazdauskas, T.D. Wilkins, and J.L. Johnson, *Ann. N.Y. Acad. Sci.* 665, 391 (1992).

5. Esmon, C.T., *J. Biol. Chem.* 264, 4743 (1989).

6. Yan B.S., B.W. Grinnell and F. Wold, *Trends in Biol. Sci.* 14, 264 (1989).

7. Furie, B. and B.C. Furie, *Blood*, 75, 1753 (1990)

8. Kang, K., D. Ryu, W.N. Drohan and C.L. Orthner, *Biotech. Bioeng.*, 39, 1086 (1992).

9. Drohan, W., D-W. Zhang, R. Paleyanda, R. Chang, M. Wroble, W. Velander and H. Lubon, *Transgenic Res.*, in press (1994).

10. Bradford, M., *Anal. Biochem.* 72, 248 (1976).

11. Esmon, N.L., L.E. DeBault and C.T. Esmon, *J. Biol. Chem.*, 258, 5548 (1983).

12. Odegaard, O.R., K. Try and T. R. Andersson, *Hemostasis* 17, 109 (1987).

13. Martinoli, J.L. and K. Stocker, *Thromb. Res.* 43, 253 (1986).

14. Wilkins, T.D. and W. Velander, *J. Cell. Biochem.*, 49, 333 (1992).

15. Horowitz, B., M.E. Wiebe, A. Lippin and M.H. Stryker, *Transfusion*, 25, 516 (1985).

16. Yan, B.S, P. Razzano, Y.B. Chao, J.D. Walls, D.T. Berg, D.B. McClure and B.W. Grinnell, *Bio/Technology*, 8, 655 (1990).

A Laboratory Study on the Behavior of *Thiobacillus ferrooxidans* during Pyrite Bioleaching in Percolation Columns

M.G. Monroy[1], M.A. Dziurla[2], B.-T. Lam[2], J. Berthelin[2], and P. Marion[3]

[1]Inst. Metalurgía, Univ. Aut. San Luis Potosí, A.P. 581, 78240 San Luis Potosí, S.L.P., MEXICO;
[2]Centre de Pédologie Biologique, C.N.R.S.-U. Nancy I.B.P. 5, and [3]Lab. Environ. et Minéralurgie,
C.N.R.S.-E.N.S. de Géologie Appliquée de Nancy, B.P. 40, 54501 Vandoeuvre les Nancy, FRANCE

There is currently a great and increasing interest to recover precious metals from low-grade refractory sulfide gold ores by bioleaching in heaps and dumps. In order to optimize the efficiency of this oxidation process, it is necessary to evaluate the influence of biological parameters controlling sulfide oxidation in percolation devices. A detailed laboratory study in small columns containing simulated sulfide ores (pure natural pyrite FeS_2 mixed with silica sand) was performed to estimate the bacterial distribution and oxidation rates during sulfide bioleaching process. The aim of this study was to establish a more accurate relationship between sulfide oxidation and the growth and activity of the bacteria (Thiobacillus ferrooxidans). Results showed that there were three bacterial classes, depending on their attachment to mineral surfaces: class 1, non-adhering bacteria or free bacteria in the leaching solution; class 2, bacteria adhering poorly to mineral surfaces and bacteria located in the interstitial medium and class 3, bacteria adhering strongly to mineral. After inoculation, the number of free bacteria decreased appreciably as a consequence of the attachment of bacterial cells to mineral surfaces. Although at least 80% of the total bacterial population was fixed throughout the process, the bacterial oxidation rates of pyrite in percolation devices showed a closer correlation with the growth of free bacteria and bacteria located in the interstitial medium than with the growth of adhering bacteria. The number of adhering bacteria did not significantly change through the column axis during the process (between 10^5 and 10^6 cells (mg mineral)$^{-1}$.

Precious metals from refractory sulfide gold ores are mainly associated to sulfides commonly gold-bearing pyrite (FeS_2) or arsenopyrite (FeAsS) [1-3]. To recover the precious metals from these refractory ores, the oxidation of gold-bearing sulfides is necessary to release the chemically combined gold or the dispersed native gold micro-inclusions. The biooxidation of refractory sulfide gold ores and concentrates has been extensively studied during the recent past years as a commercial option to oxidize the gold-bearing sulfides, previously to chemical leaching of gold by conventional cyanidation [4-6]. Although the bioleaching of coarse ores in columns has been shown to be effective in laboratory and pilot

tests as a pretreatment of low-grade refractory gold ores [7-9], certain biological, operational and mineralogical factors can determine finally the feasability of a bioleaching process in heaps [10]. Moreover, comparison of results from different works is difficult, due to the fact that different refractory sulfide gold ores and microbial strains have been used. Therefore, to better define the role of the biological and mineralogical factors, laboratory optimized conditions (pure minerals and culture collection bacterial strains) are required.

In this work, the monitoring of bacterial populations and sulfide oxidation kinetics during the bioleaching of simulated sulfide ores

509

E. Galindo and O.T. Ramírez (eds.), Advances in Bioprocess Engineering. 509-517.
© 1994 *Kluwer Academic Publishers. Printed in the Netherlands.*

in columns was performed in order to establish a relation between growth and activity of *Thiobacillus ferrooxidans* and the pyrite and arsenopyrite oxidation in percolation devices. The sulfide oxidizing bacteria *Thiobacillus ferrooxidans* has been chosen for this work because it is the main bacterial species involved in the sulfide bioleaching process.

MATERIALS AND METHODS

Bacteria and Growth Medium. The chemiolithotrophic and acidophilic *Thiobacillus ferrooxidans* was provided by Deutsche Sammlung Mikroorganismen (DSM 583). The growth medium for the bacterial culture and the leaching experiments consisted of 0.4 g KH_2PO_4, 0.4 g $MgSO_4.7H_2O$, 1.0 g $(NH_4)_2SO_4$ per liter, adjusted to pH = 1.80 with 1N H_2SO_4 solution. This medium was sterilized at 120°C during 20 min.

Simulated Sulfide Ores. Simulated ores were prepared with silica sand (Fontainebleau sand, 100-150 μm) intimately mixed with pure pyrite, in order to have 2 wt.% of sulfide in the ore. The sulfide minerals used for the experiments were prepared from large pure natural crystals. The pyrite crystals were originated from a sedimentary deposit in Logroño, Spain. The crystals were separately and carefully ground and sieved in order to obtain the fraction smaller than 2mm to be used. Pyrite crystals have minor impurities, i.e., inclusions of gypsum, and no presence of non-stoechiometric zones within crystals. The chemical composition of pyrite samples obtained by quantitative electron microprobe (Q.E.M.) and wet analysis are given in Table 1.

Table 1. Chemical composition of pyrite

	wet chemical analysis	Q.E.M. chemical analysis
Fe	43.66 %	46.51 %
S	49.15 %	52.68 %
As	0.33 %	0.40 %
Total	93.14 %	99.59 %

Experimental Devices. The columns used in this study (Small Columns) "SC Percolation Device" were made of a glass column (300 mm high x 50 mm diameter) and had a perforated rubber stopper in the bottom. Each column was packed with 80 g of silica sand (Fontainebleau sand, 100-150 μm) at the bottom, 306 g of simulated sulfide ore in the middle and 120 g of silica sand at the upper part. The three beds were separated by a polyamide woven sieve 100 μm porosity (UGB, France). Leaching solution was fed by gravity to a separate reservoir containing initially 400 ml of culture medium, which was circulated to the top of the column by a variable flow peristaltic pump (Ismatec) at the flow-rate of 100 $l.h^{-1}.m^{-2}$ (figure 1).

In all cases, dry simulated ore was gently poured into the columns to minimize stratification of sulfide particles in the column profile. Silica sand and sulfide ore were not sterilized before inoculation. Once a constant flow-rate was established (12-48 h), the percolation set-up was inoculated at the top of the column with an actively growing culture to have an initial population of 5.10^7 bacteria.g^{-1} of mineral. The columns were mounted in a darked and heated room at a controlled temperature of 30°C. Each type of column was set up and run in duplicate.

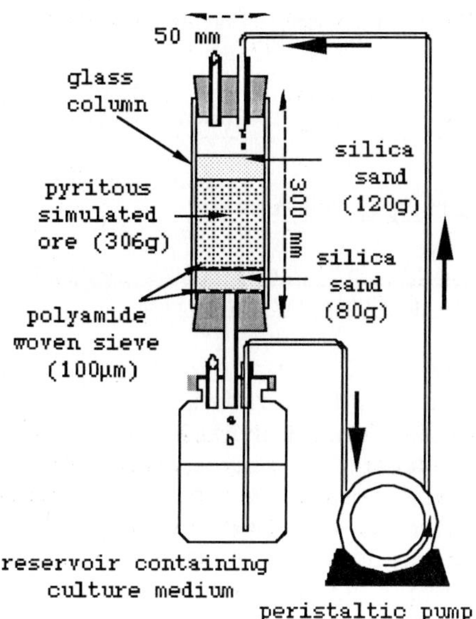

Figure 1. SC Percolation device used for leaching tests.

Enumeration of adhering and free bacteria.

The number of free bacteria was measured by direct microscopical counting using a Thoma cell on the leaching solution collected from the reservoir. The Tween 80 treatment method described by Dziurla *et al.* [11] allowed to estimate the number of bacteria adhered at mineral surfaces when their number was greater than 10^3 bacteria.mg^{-1} of mineral. During and at the end of the biooxidation experiments, mineral samples of about 300 mg were collected from the columns with an ethanol cleaned spatula. Three sites located at the top, in the middle and at the bottom of columns were selected for this bacterial estimation. Mineral samples were initially washed several times with 1 ml of culture medium and shaked for 1 min. in a vortex. Washing solutions were collected together in order to enumerate immediately the released bacteria by direct microscopical counting. After completion of the washing, samples were resuspended twice into 1 ml of 10% [vol/vol] Tween 80 solution, shaked twice over 1min., settled during 10min. prior to performing direct microscopic counts into Tween treatment solutions.

The washing and Tween treatments used to release the bacteria adhering to mineral surfaces demonstrated the presence of three bacterial classes, depending on their attachment to mineral surfaces : <u>Class 1</u>, non-adhering bacteria or free bacteria (*i.e.* not associated with solids), counted directly in leaching solution; <u>Class 2</u>, weakly-adhered bacteria and bacteria located in the interstitial medium, counted together in washing solutions; and <u>Class 3</u>, strongly-adhered bacteria, only released and counted by the Tween 80 treatment method.

Tween 80 treatment was necessary to release the bacteria adhering strongly to mineral. Ten to one hundred times more of bacteria were released by this chemical and physical treatment in comparison to simple washing. In all the leaching tests in column, the simple washing with culture medium removed only a small proportion of the bacterial population attached to mineral surfaces. This result is not in agreement with previous reports which found that washing with medium is a suitable method to recover bacteria from mineral material [12].

ATP analysis.

The ATP quantification of cells coming from bioleaching tests was performed by the very sensitive luciferin-luciferase assay [13]. ATP was extracted from cells with NRB^R reagent (Lumac Ltd., Lumac Systems AG), after lysis with a solution Tris (50 mM)-EDTA (4 mM) at pH 7.75. The ATP content of the extracts was measured in triplicate by the luciferin - luciferase luminescence reaction using Lumit PM^R purified enzyme. The light output was measured for 10 s on a Lumacounter (Dynatech, France), immediately after addition of the enzyme.

Solution and solid analysis. The total iron, arsenic and sulphur concentrations in the leaching solutions were determined by inductive coupled plasma analysis (ICP Jobin Yvon JY32) using an internal Sr, HCl standard (0.6% of matrix). Corrections for evaporation during the experiment were carried out before sampling by the addition of distilled water in the reservoir. pH measurements were effected in the solution from the reservoir by a Tacussel Eh-pH meter (PHN 78) and a glass combined electrode (Ag/AgCl, Ingold) whereas Eh measurements were conducted by mean of a platinum wire electrode. Metals contents in the sulfides and in the bioleached residues were determined by acidic dissolution of solid samples using a micro-wave furnace, followed by ICP analysis (Jobin Yvon JY70 PLUS). Quantitative electron microprobe analysis of pyrite crystals were performed on a SX Cameca apparatus, using the following analytical conditions : accelerating voltage 20 kV, 20 nA, counting time 10 s.

Microscopic observations. Mineral samples for microscopic observations of corrosion patterns in sulfide grains were collected in the columns and dried under O_2 free atmosphere to preserve the surface and the corrosion pits. In presence of precipitates formed on the surface grains during the bioleaching process, samples were washed with a 6N HCl solution prior to microscopic examination. Other mineral samples collected from the columns were directly treated with osmium tetra-oxide [vap] during 14 days in order to examine the presence of bacterial cells adhering to mineral grains. Microscopic observations were then carried out on a scanning electron microscope F.B. Cambridge Stereoscan equipped with a Si-Li Princeton energy dispersive spectrometer.

RESULTS AND DISCUSSION

Dynamics of bacterial population during pyrite bioleaching

The chemical and biological phenomena established during the pyrite biooxidation in stirred reactor [14] were also observed during the bacterial leaching of a simulated ore containing pyrite in percolation columns. These phenomena can be summarized as follow.

Initially, the number of free bacteria decreased appreciably after inoculation (figure 2). At this time, the strongly - adhered bacteria represented 98 per cent of the total bacterial population in the percolation system (figure 3). This rapid decrease of the number of free bacteria was attributed to the attachment of bacterial cells to mineral surfaces. It is well known that in the recirculating systems, the cells quickly colonize the mineral surfaces, due to the large surface / volume ratio of the inorganic phase rather than due to an interaction between bacteria and mineral particles [15].

Eight days after inoculation, the number of the bacteria class 1 and class 2 began to increase (figure 2). The pH began to decrease slightly and the redox potential increased and reached a stable value between 800 and 850 mV/SHE (figure 4). In addition, the iron solubilization began also to increase slowly (figure 5). This step corresponds to the exponential growth phase of the free bacteria in the percolation suspension and to the beginning of the biological dissolution of pyrite. At the end of the exponential growth phase, the free bacteria reached a number of 5.10^7 bacteria.ml^{-1}, which remained stable in the course of the bioleaching experiment.

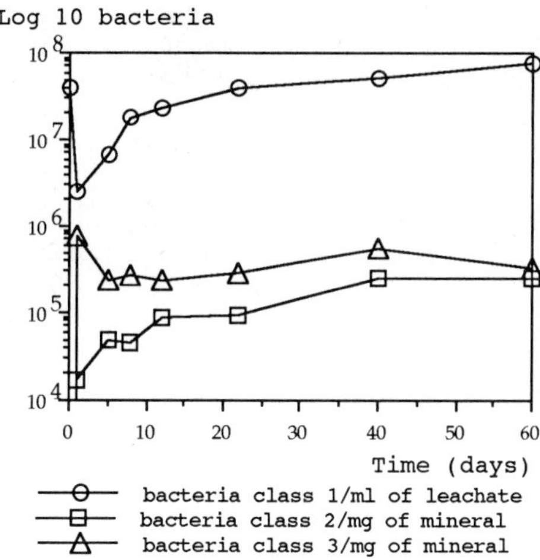

Figure 2. Growth of three bacterial classes during the bioleaching process of a pyritous simulated ore in SC percolation device.

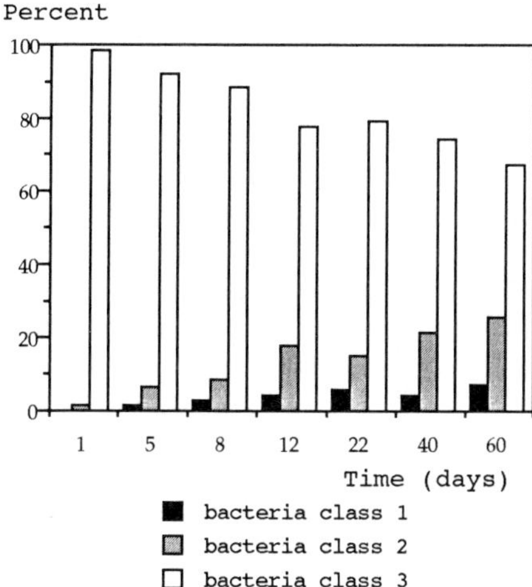

Figure 3. Percentage of three bacterial classes during the bioleaching process of a pyritous simulated ore in SC percolation device.

Even if the number of free bacteria in solution reached a maximal concentration at 20 days, the number of bacteria located in the interstitial liquid (bacterial population here considered within class 2) increased slightly and continuously, whereas the number of strongly-adhered bacteria remained stable at about 3.10^5 bacteria.mg^{-1} of mineral. This step was dominated by a strong increasing of the iron amount in solution and a constant decreasing of the pH (figures 4 and 5).

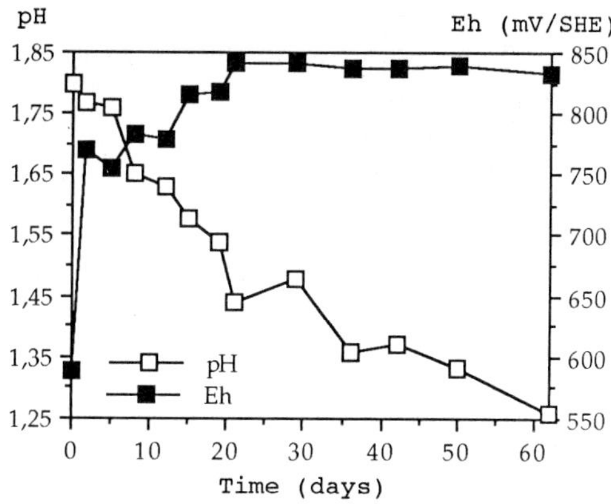

Figure 4. Evolution of pH and redox potential during the bioleaching process of a pyritous simulated ore in SC percolation device.

Comparison of bacterial growth and iron oxidation rates

The bacterial oxidation of pyrite in columns involved a very close correlation between the number of bacteria class 1 and class 2 and the iron amount in solution (figure 5). This correlation is not observed with the growth of strongly-adhered bacteria. At the end of the bioleaching, the percentage of strongly-adhered bacteria decreased from 95 to 75%.

The fact that the iron solubilization start simultaneously with the increase of free bacteria in the leachate and in the interstitial medium suggests a main role of the bacteria class 1 and class 2 in the continuity of the bioleaching process in percolation devices. The ATP estimation of cells coming from a bioleaching test in column get well with this last hypothesis : the free bacteria (in the leaching solution and in the interstitial medium) had a higher concentration of ATP than the strongly-adhered cells in the exponential growth phase (table 2).

Table 2. ATP concentrations of bacterial populations (*Thiobacillus ferrooxidans*) during the bioleaching process of a pyritous simulated ore in SC percolation device.

bacterial classes	ng ATP.bacteria^{-1}	
	exponential growth phase	stationnary growth phase
free bacteria	$1.04 \ 10^{-9}$	$2.75 \ 10^{-10}$
adhered bacteria	$5.49 \ 10^{-10}$	$6.51 \ 10^{-10}$

The relative abundance of the strongly-adhered bacteria (figure 6.A) could allow the initial sulfide oxidation by the direct bacterial mechanism. This bacterial class seems to be responsible for the initial apparition of corrosion patterns in sulfide particles (figure 6.B) and consequently of the initial increase of both total iron and ferrous ion amounts in solution.

Ferrous ions in the leachate, produced in this way, are subsequently oxidized by the free bacteria and the exponential growth phase begins. The large quantity of ferric ions produced by the activity of free bacteria could oxidize the

sulfide particles. Thus, the important and homogeneous porosity developed in sulfide grains along the

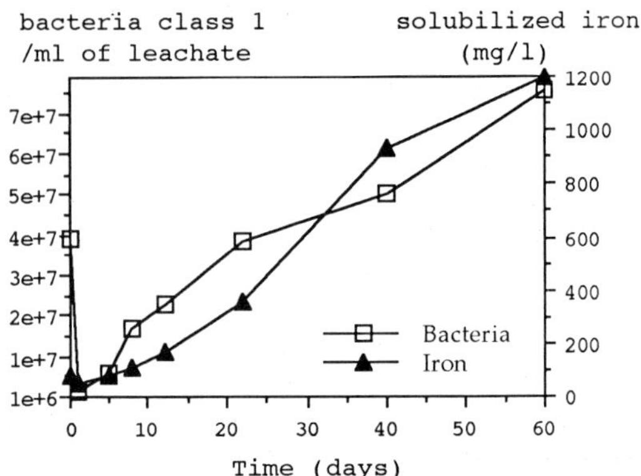

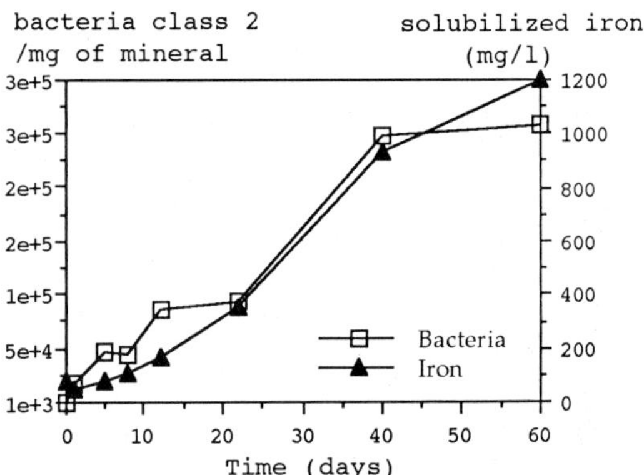

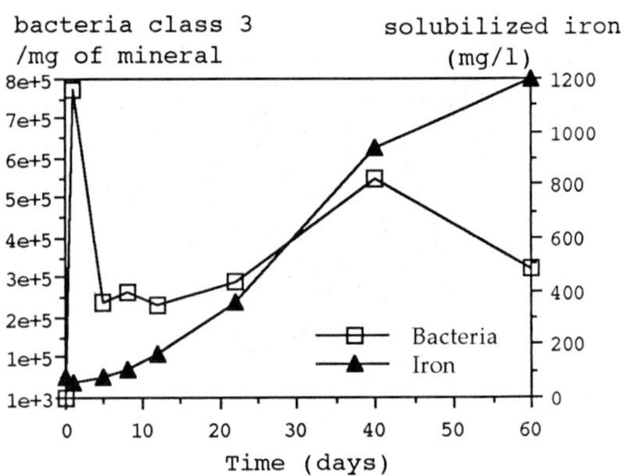

Figure 5. Bacterial growth and dissolution of total iron during the bioleaching process of a pyritous simulated ore in SC percolation device.

mineral profile (figures 6.C and 6.D) involved certainly both bacterial and chemical oxidation processes with the beginning of the exponential growth phase. The results described here seem to be supported by previous works [14] showing that the pyrite oxidation by *Thiobacillus ferrooxidans* concerns a succession of different steps which involve particular behaviors of mineral and biological phases.

It can be underlined that the spatial distribution of the strongly-adhered bacteria was homogeneous along the mineral profile (figure 7), showing only weak variations in the total amount of attached cells throughout the bioleaching process (between 3 and 8.10^5 bacteria.mg^{-1} of mineral). It is interesting to note that there was a similar number of bacteria adhering strongly to the silica sand particles in comparison to bacteria adhering to the pyritous simulated ore. This result is in agreement with those of DiSpirito *et al.*[16] and Solari *et al.*[17] which showed that adhesion of *Thiobacillus ferrooxidans* to quartz is only two to four times weaker than to pyrite. The amount of bacteria adhering strongly to mineral surfaces in percolation devices was found to be ten to one hundred times higher than the attached cells number in stirred leaching systems. With respect to the free bacteria in the interstitial medium, the observed decrease of cell number downward

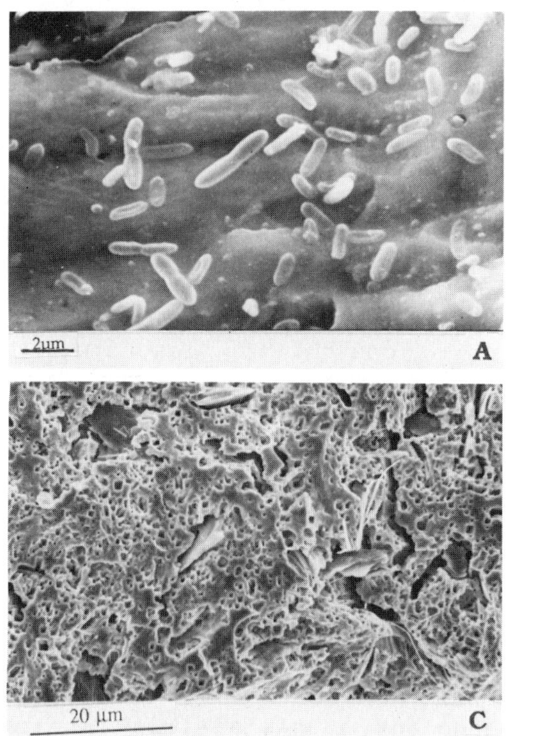
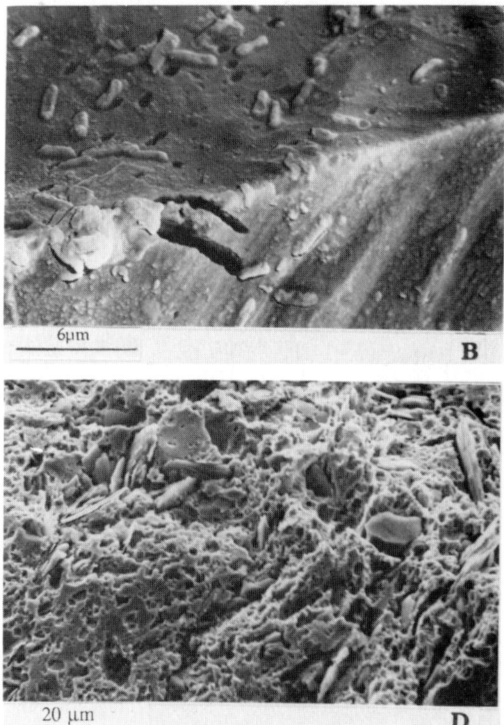

Figure 6. SEM photographs (secondary electrons) of the pyrite surfaces during the bioleaching of a pyritous simulated ore in SC percolation device : (A) bacteria adhering strongly to pyrite surfaces after 4 days; (B) apparition of corrosion patterns in a pyrite grain by the activity of bacteria adhering strongly after 18 days; (C) deep corrosion pores in a pyrite grain coming from the up of the mineral profile after 60 days; (D) deep corrosion pores in a pyrite grain coming from the bottom of the mineral profile after 60 days.

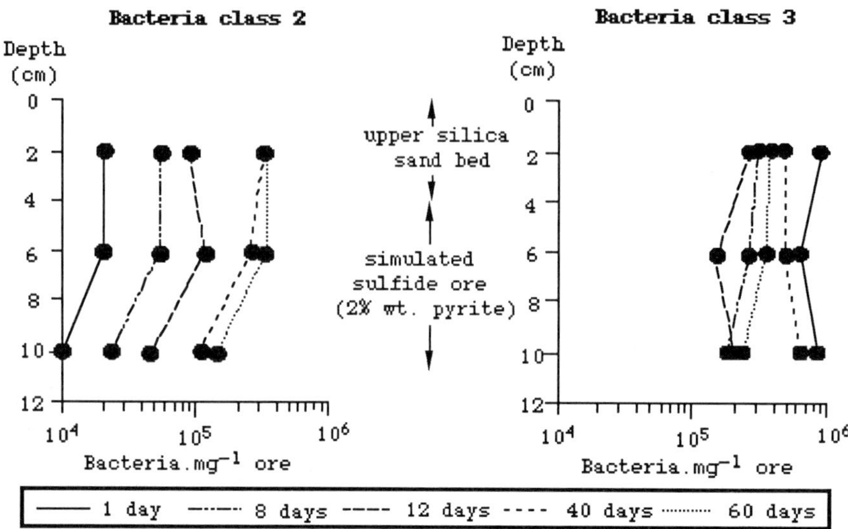

Figure 7. Growth and distribution of bacterial populations within SC percolation device during the bioleaching process of a pyritous simulated ore.

(figure 7) could be interpreted as a result of a diminution in the oxygen concentration, although we do not have evidence to support this hypothesis.

Only a few works have been reported on the evolution and the activity of the bacterial populations during the sulfide biooxidation in percolation systems. Data presented here have showed a particular influence of each class of bacterial populations on sulfide oxidation kinetics. The presence and amount of each bacterial class might thus condition the kinetics of bioleaching and gold recovery during the treatment of refractory sulfide gold ores in heaps.

CONCLUSIONS

Even if the desorption method used here to estimate the distribution of bacterial populations in leaching devices is not sensitive and specific, the information obtained gives an answer to some questions concerned with the mechanism of the sulfide bioleaching process in percolation systems. In conclusion, the great number of bacteria adhering strongly to sulfide minerals allows to begin the oxidation process by direct attack whereas the free bacteria in solution and in the interstitial medium assure the continuity and advance of the process by promoting high concentrations of ferric ions in the leachate. The behavior (the presence, the amount and the activity) of the three bacterial classes might thus have a high influence on sulfide oxidation kinetics and gold release during the bioleaching treatment of refractory gold ores in percolation systems.

ACKNOWLEDGMENTS

French Ministry of Research and Technology supported this research (Grant 89-R-06704). Mexican Council of Science and Technology (CONACyT) supported the PhD studenship to M. Monroy. G. Belgy (C.P.B., Nancy) conducted leached solution analyses and A. Kholer (Nancy University) provided S.E.M. observations.

LITERATURE CITED

1. Cathelineau, M., M.C. Boiron, P. Holliger, P. Marion and M. Denis., "Gold-rich arsenopyrites : crystal-chemistry, gold location and state, physical and chemical conditions of crystallization", In *The Geology of Gold Deposits : The perspective in 1988*, Economic Geology Monograph 6, 328-341 (1989).

2. Marion, P., M. Monroy, P. Holliger, M.C. Boiron, M. Cathelineau, F.E. Wagner and J. Friedl, "Gold-bearing pyrites : A combined ion microprobe and Mössbauer spectrometry approach", In M. Pagel and J. Leroy (Ed.), *Source, Transport and Deposition of Metals*, 677-680, Balkena, Rotterdam (1991).

3. Cook, N.J. and S.L. Chryssoulis, *Can. Mineral.*, **28**, 1-16 (1990).

4. Torma, A.E., "A Review of Gold Biohydrometallurgy", In G. Durand, L. Bobichon and J. Florent (Ed.), *Proc. 8th Int. Biotechnology Symp.*, Paris, 1158-1168 (1989).

5. Lawrence, R.W., "Biotreatment of Gold Ores", In H.L. Ehrlich and C.L. Brierley (Ed.), *Microbial Mineral Recovery*, 127-148, McGraw-Hill, Inc., New York (1990).

6. Lindström, B.E., E. Gunneriusson and O.H. Tuovinen, *Crit. Rev. Biotech.*, **12**, 1/2, 133-155 (1992).

7. Brierley, J.A. and L. Luinstra, "Biooxidation-Heap Concept for Pretreatment of Refractory Gold Ore", In A.E. Torma, J.E. Wey and V.I. Lakshmanan (Ed.), *Biohydrometallurgical Technologies Volume 1 : Bioleaching Processes*, 437-448, TMS, Pennsylvannia (1993).

8. Mihaylov, B.V. and J.L. Hendrix, "Gold Recovery from a Low-Grade Ore Employing Biological Pretreatment in Columns", In A.E. Torma, J.E. Wey and V.I. Lakshmanan (Ed.), *Biohydrometallurgical Technologies Volume 1 : Bioleaching Processes*, 499-511, TMS, Pennsylvannia (1993).

9. Harrington, J.G., R.W. Bartlett and K.A. Prisbrey, "Kinetics of Biooxidation of Coarse Refractory Gold Ores" In J.B. Hiskey and G.W. Warren (Ed.), *Hydrometallurgy : Fundamentals, Technology and Innovation*, 691-708, SME, Michigan (1993).

10. Monroy, M., P. Marion, J. Berthelin and G. Videau, "Heap-bioleaching of simulated refractory sulfide gold ores by *Thiobacillus ferrooxidans* : a laboratory approach on the influence of mineralogy", In A.E. Torma, J.E. Wey and V.I. Lakshmanan (Ed.), *Biohydrometallurgical Technologies Volume 1 : Bioleaching Processes*, 489-498, TMS, Pennsylvannia (1993).

11. Dziurla, M.A., M. Monroy, B.T. Lam, S. Picquot and J. Berthelin, "A Method to Estimate the Acidophilic (*Thiobacillus ferrooxidans*) Attached to Pyrite and their Comparison to non-adhering Bacteria on Solid Media", presented at the *6th Int. Symp. on Microb. Ecology*, Barcelona (1992).

12. Southam, G. and T.J. Beveridge, *Appl. Environm. Microbiol.*, **58**, 1904-1912 (1992).

13. Jenkinson, D.S. and J.N. Ladd, "Microbial biomass in soil : measurement and turnover", A.E. Paul and J.N. Ladd (Ed.), *Soil Biochemistry vol. 5*, 415-471, Marcel Dekker, New York (1981).

14. Mustin, C., J. Berthelin, P. Marion and P. de Donato, *Appl. Environm. Microbiol.*, **58**, 1175-1182 (1992).

15. van Loosdrecht, M.C.M., J. Lyklema, W. Norde and A.J.B. Zehnder, *Microbial. Ecology*, **17**, 1-15 (1989).

16. DiSpirito, A.A., P.R. Dugan and O.H. Tuovinen, *Biotechnol. Bioeng.*, **25**, 1169-1173 (1983).

17. Solari, J.A., G. Huerta, B. Escobar, T. Vargas, R. Badilla-Ohlbaum and J. Rubio, *Colloids and Surfaces*, **69**, 159-166 (1992).

Purification of G6PDH from Unclarified Yeast Cell Homogenate Using Expanded Bed Adsorption (EBA) with STREAMLINE™ Red H-E7B

Y.K. Chang, G.E. McCreath, and H.A. Chase

Department of Chemical Engineering, University of Cambridge, Pembroke Street,
Cambridge CB2 3RA, U.K.

The use of expanded beds of STREAMLINE™ Red H-E7B in a STREAMLINE 50 EBA apparatus for the direct extraction of glucose-6-phosphate dehydrogenase (G6PDH) from unclarified preparations of disrupted bakers' yeast has been investigated. It has been demonstrated that such crude feedstocks can be applied to the bed without prior clarification steps (centrifugation and microfiltration). The ability of combining clarification, capture, and purification in a single step will greatly simplify downstream processing flowsheets and reduce the costs of protein purification. The purification of G6PDH from unclarified yeast homogenate was chosen as a model system outlining the typical features of direct broth extraction. Procion Red H-E7B STREAMLINE™ adsorbent has been assessed for its performance in expanded bed procedures. The absorbent was shown to be successful in achieving isolation of G6PDH from unclarified homogenate with a purification factor of 12 and yield of 86% in a single step process. STREAMLINE™ Red H-E7B could be cleaned using simple clean-in-place procedures without affecting either adsorption or the bed expansion properties of the adsorbent. The results clearly demonstrate the potential of expanded bed technology in achieving a significant reduction in the number of processing steps required to purify proteins from crude feedstocks. In addition, the use of biomimetic dyes as affinity ligands seems to be an ideal compromise between selectivity and robustness.

Downstream processing flowsheets for the recovery of proteins often involve streams which contain particulate material. Such streams include:

• Fermentation broths containing whole cells.

• Preparations of broken cells containing cell debris.

The flowsheets are also likely to contain adsorption and/or chromatography steps for the purification of proteins. However, problems often occur when particulate-containing liquids are applied to packed beds of adsorbents as flow through the bed is likely to become blocked as a result of the bed acting as a filter for the particulates. Although there is a possibility of using simple stirred tank adsorption procedures in the presence of particulates, these are likely to be inefficient as the method constitutes at the most a single equilibrium stage and the handling of the adsorbent can present problems. Hence, there is a need to develop a new unit operation in which adsorption separation procedures can be carried out with feedstocks that contain particles. Novel technologies have been described for the direct extraction of proteins from unclarified cell homogenates (<u>1</u>, <u>2</u>)

The application of Expanded Bed Adsorption (EBA) to direct extraction results from the expansion of the adsorbent bed when solution is applied to the bottom of the bed in an upward direction. Once a minimum liquid flow velocity has been exceeded, the bed starts to expand and gaps occur between the beads (i.e. the external voidage of the bed increases). Particulates can now pass freely through the bed whilst soluble proteins can be adsorbed to the adsorbent bed. The terminal velocity of adsorbents developed for use in conventional packed bed procedures are low as a result of their small particle size and similar density to the liquid phase. When used in EBA procedures with high viscosity of fermentation broths or cell homogenates, beds of these adsorbents expanded to large extents even at low superficial velocities and bed stabilities are noticeably affected. Hence, larger beads of a greater density are required for direct recovery of proteins in industrial processes. Pharmacia Biotech has manufactured a specialised matrix (STREAMLINE™) for use in expanded bed procedures. It is possible to operate STREAMLINE™ adsorbents at 8 times the

519

E. Galindo and O.T. Ramírez (eds.), Advances in Bioprocess Engineering. 519-525.
© *1994 Kluwer Academic Publishers. Printed in the Netherlands.*

flow rate possible with Sepharose FF (3). This feature enables feedstocks to be processed at faster flow rates (100-300 cm/h) more suited to industrial implementation of this technique.

Since the discovery of the interaction of Cibacron blue with various enzymes, there has been a considerable expansion in the application of reactive dyes to protein purification (4). Dye ligand chromatography has become popular because the ligands are inexpensive, easy to couple to matrices and extremely stable. Numerous examples of protein purification using dye ligands have been published (4). Demonstration of the direct extraction of G6PDH from cell homogenates in which no clarification or pre-purification using Red H-E7B as a selective ligand has been published previously (5). In this paper, we have developed expanded bed affinity chromatography (EBAC) for direct extraction of G6PDH from unclarified yeast cell homogenate using the biomimetic dye Procion Red H-E7B coupled to the STREAMLINE™ matrix.

Use of expanded beds for the adsorption of proteins from crude feedstocks

The principle and operation of expanded bed adsorption for the direct extraction from a crude homogenate has been described previously (2). The protocol for the operating EBA is very similar to that undertaken in a packed bed procedure:

• The adsorbent is expanded to create a stable bed by passing buffer through the column.
• The particulate-containing stream is applied to the expanded bed until the capacity for the target protein(s) is substantially utilised.
• Particulates are removed from the bed by a wash procedure performed in the expanded mode.
• Elution is carried out with the bed returned to a packed configuration.

• The bed is subjected to clean-in-place procedures before the cycle of operation is repeated.

Creation of a stable expanded bed using purpose designed apparatus

Fluidized beds are traditionally thought of as a well-mixed system with back-mixing of both the liquid and adsorbent phases. Such a situation will give comparatively poor adsorption performance as the bed will constitute, at the most, a single equilibrium stage. Previous approaches to obtaining a stable bed have involved the use of magnetic particles with stabilisation by an external magnetic field (6). However, a very stable bed can be obtaining in an easier manner by using an adsorbent with a defined distribution of particle sizes. Under these circumstances, the larger, heavier particles will be confined to the bottom of the bed. This property confers stability on the bed and avoids the undesired circulation of adsorbent particles with associated back-mixing of liquid. Flow characteristics through the expanded bed are similar to those through a conventional packed bed. Adsorption performance is very similar to that which would be achieved with the same amount of adsorbent in a packed bed mode at the same flow rate (3,7).

MATERIALS AND METHODS

Materials.

The underivatised STREAMLINE™ matrix was a gift from Pharmacia Biotech (Uppsala, Sweden). Bakers' yeast was purchased in the form of pressed blocks from a local supplier. All other chemicals used were obtained from Sigma (Poole, UK).

Preparation of dye ligand STREAMLINE™ adsorbent.

STREAMLINE™ matrix (500 g) was added to 500 ml of dye solution containing 2.5 g Procion Red H-E7B. The gel suspension was mixed by rotary mixer at room temperature for 40 min in order to allow complete diffusion of the dye into the particles. Then, 100 ml of 22% (w/v) NaCl solution was added and mixing continued at room temperature for a further 1.5 hr. At the end of this period, Na_2CO_3 (5 g) was added and the slurry heated to 60°C and mixing continued for 6 hr. After immobilisation was complete, unreacted dye was washed away from the dyed adsorbent with a sequence of solutions containing distilled water, 1 M NaCl in 25% ethanol, 6 M urea in 0.5 M NaOH, and 50 mM phosphate buffer, pH6.0 (1500 ml each).

Measurements of bed expansion characteristics.

The characteristics of bed expansion were measured in a STREAMLINE 50 column at 4°C at increasing superficial velocity as described previously (2).

Preparation of disrupted bakers' yeast homogenate.

Large scale disruption of *Saccharomyces cerevisiae* cells was carried out as described previously (8). The total protein concentration and G6PDH activity of disrupted cell preparations were determined as previously described (5).

Direct extraction of G6PDH from unclarified yeast cell homogenate.

The isolation of G6PDH from bakers' yeast cell homogenates was carried out at 4-8°C in a STREAMLINE 50 column. STREAMLINE™ Red H-E7B was poured into the column to give a settled bed height of 20.0 cm. Then the top adapter of the column was positioned at about 2 times the settled bed height (45 cm), to minimise the volume of liquid above the expanded bed. Two Pharmacia syringe P-6000 pumps (pump A and pump B) were used in the EBA system and the liquid flow rate was controlled , either manually or automatically, using the Pharmacia FPLC Manger LCC 500 CI process controller. Pump A was used to pump buffer during equilibration, and washing stages. Pump B was used to pump the eluent solutions. In addition, a peristaltic pump C was used to pump crude homogenate onto the expanded bed. Initially, the adsorbent bed was washed with 0.05 M phosphate buffer (pH 6.0), using pump A at a flow rate of 147 cm/h, until 5 bed volumes of buffer had passed through the bed and the bed had expanded to the desired height of 40.0 cm. Pump A was then switched off and simultaneously pump C was started in order to apply crude yeast cell homogenate. To maintain a constant extent of bed expansion and to prevent

the bed from compacting against the top adapter in the column, the velocity of flow was reduced to 52 cm/h during application of the feedstock. The optical density of the outlet stream was recorded using a UV-1 monitor at 280 nm. As soon as the absorbance in the column outlet stream began to rise as a result of the cells breaking through the bed (580 ml), samples of the liquid in the outlet stream were collected, and stored at 4-8°C. Samples were taken at regular intervals (20 ml) during the runs. When 900 ml of the homogenate had passed through the bed, pump C was switched off and pump A was switched on to deliver the wash solution to the column. The wash solution of 25% glycerol (v/v) in 0.05 M phosphate buffer (pH 6.0), was used to remove cells and cell debris from the bed. The velocity of flow was increased during washing from 52 cm/h to 92 cm/h to maintain the same extent of bed expansion until the homogenate was observed, on a qualitative basis, to have been washed out from the bed (only a single bed volume of 25% glycerol/buffer solution was required). The bed was then returned to a packed bed mode and 0.05 M phosphate buffer (pH 6.0) was used to wash glycerol from the bed in a downward direction at 152 cm/h until the absorbance had dropped to zero. First, elution of the bed was carried out at 15 cm/h using 10 mM NADP/0.1 M triethanolamine-HCl, pH8.0 (50 ml). 0.1 M tiethanolamine-HCl, pH8.0 was then applied until the optical density in the outstream had returned to zero. Second, 2 M NaCl/0.1 M triethanolamine-HCl, pH 8.0 at 25 cm/h was used to elute non-G6PDH proteins until the absorbance returned to a steady minimum value. Each sample collected (25 ml) during the experiment was analysed for its G6PDH activity and its total protein content. Cleaning-in-place was carried out in a packed bed mode by using a cocktail containing 60% ethanol, 0.5 M NaOH, and 4 M urea at 50 cm/h (2-3 bed volumes).

RESULTS AND DISCUSSION

Characteristics of Bed expansion

The variation of the expansion of the bed with flow rate is a function of the viscosity and density of liquid flowing through it. Bed expansion characteristics were found to follow the Richardson & Zaki correlation (<u>9</u>). The experimentally determined terminal velocities (U_t) of particles agree well with calculated theoretical values from the Stokes' equation (<u>10</u>) (Table 1).

The normal operational temperature is 4-8°C with cell homogenate systems and the viscosity of homogenate is high due to high biomass content and because of certain substances released during cell lysis (predominantly nucleic acids). Figure 1 illustrates the effect of viscosity on the degree of bed

Table 1. Richardson & Zaki parameters for STREAMLINE™ Red H-E7B in a STREAMLINE 50 column at 4°C. STREAMLINE™ Red H-E7B has a mean diameter of 186 μm and a density of 1180 kg/m³. The table gives the fluidizing solutions used in experiments to study variation in the height of the expanded bed as a function of liquid flow the bed. The experimental values of the parameters U_t and n are obtained by fits of the data to Richardson & Zaki equation. Theoretical values of the terminal velocities of the adsorbent beads are also calculated from Stokes' equation.

Fluidizing Solution	Solution Density (kg/m³)	Solution Viscosity (mPas)	U_t (Experiment) (mm/s)	U_t (Stokes' Eq.) (mm/s)	Richardson-Zaki exponent (n)
0.05 M phosphate pH 6.0	1026	1.213	2.46	2.45	4.79
25% glycerol	1100	4.349	0.49	0.36	4.94

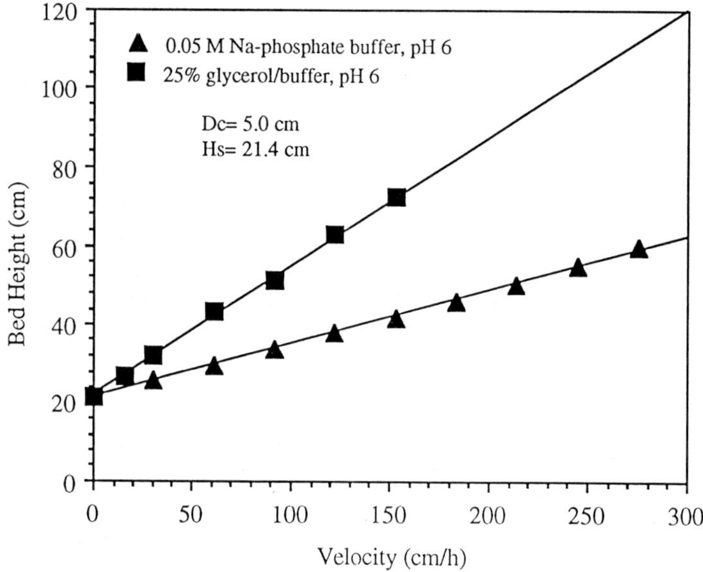

Figure 1. Bed expansion characteristics for STREAMLINE™ Red H-E7B. The characteristics of expansion of a bed of STREAMLINE™ Bed H-E7B were determined in a STREAMLINE 50 column at 4°C. The height of the settled bed of adsorbent was 21.4 cm. Bed expansion was performed in the presence of the following solutions: (▲) 0.05 M sodium phosphate buffer, pH 6.0; (■) 25% (v/v) glycerol/buffer, pH 6.0.

expansion in a model experiment with 25% glycerol/buffer system at 4°C. The bed expansion of STREAMLINE™ Red H-E7B in the presence of 25% glycerol is greater than that in buffer only. The superficial velocity of flow required to give 100% bed expansion at 4°C in buffer alone was 3 times greater than that in 25% glycerol.

Direct extraction of G6PDH from a yeast cell homogenate

The potential for the use of EBA in direct extraction has been shown in the purification of G6PDH from a crude bakers' yeast homogenate using STREAMLINE™ Red H-E7B. Large scale disruption of yeast cells was achieved by milling with 500 μm glass ballotini in a high speed mixer at 4800 rpm. The purification of G6PDH from unclarified homogenate was carried out using an 5 cm diameter expanded bed of STREAMLINE™ Red H-E7B (400 ml, 2x height of settled bed). The results of a typical experiment are shown in Figure 2.

A steep rise in optical density at 280 nm was observed after 580 ml of homogenate had been applied to the column which indicated the front of the particulate material in the liquid leaving the bed. Total protein broke through the bed nearly at the same time. However, G6PDH activity did not break through the bed until 820 ml of the homogenate had been applied to the bed. This is due to the high selectivity of dyed adsorbent for G6PDH under these conditions. At the end of the adsorption stage, residual particulate material was completely washed out from the expanded bed using 25% (v/v) glycerol in phosphate buffer, pH 6.0. Figure 2 demonstrates that only

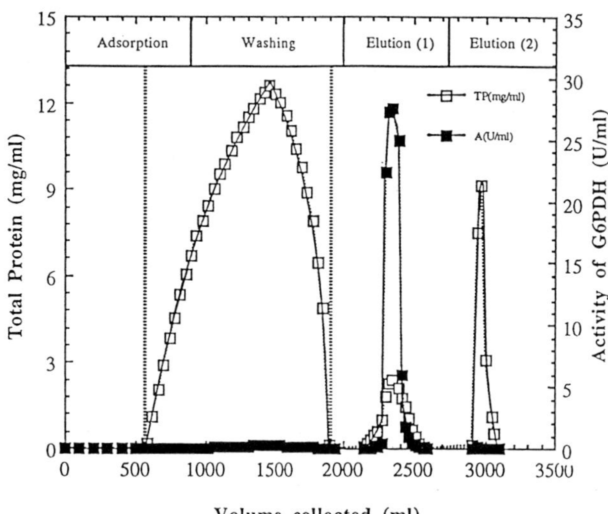

Figure 2. Expanded bed affinity chromatography of G6PDH on STREAMLINE™ Red H-E7B. The chromatographic protocol used to purify from a feedstock of disrupted yeast cells broken by a high speed mixer was described in the text. Samples collected throughout the experiment were assayed for total protein content (□), and G6PDH activity (■). The optical density at 280 nm of the stream leaving the bed was monitored continuously (⋮) and was measured of the presence of particulates in this stream. Washing was performed with 25% glycerol in 50 mM phosphate buffer, pH 6.0, followed by elution with 10 mM NADP in 0.1 M triethanolamine-HCl, pH 8.0, followed by elution with 2 M NaCl in 0.1 M triethanolamine-HCl, pH 8.0. The various stages of the procedure were performed at different flow rates as described in the text.

a single expanded bed volume of the wash solution was required to remove particulate material completely from the bed as evidenced by the rapid decline in optical density of the outlet stream by this point. Total protein was also observed to fall to zero by the end of the wash phase. Following washing, the bed was returned to a packed bed mode and eluted firstly with 10 mM NANP/0.1 M triethanolamine-HCl, pH 8.0, which was shown to elute G6PDH completely, and secondly with 2 M NaCl to elute non-G6PDH proteins. There appeared to be no particulate material in the eluent as judged by the lack of turbidity and the absence of particles on examination of the liquid under a microscope. G6PDH was found to have been recovered with a purification factor of 12 and yield of 86% (Table 2). A similar chromatographic protocol was also carried out with clarified yeast cell homogenate prepared by removed of particulate by centrifugation. The clarified solution (41.2 mg/ml, 9.88 U/ml, 3 ml) was loaded to Pharamacia HR 10 column packed with STREAMLINE™ Red H-E7B (6.6 ml) at a flow rate of 152.8 cm/h. An overall purification of 6.4 for G6PDH was achieved in 87.6% yield under similar process conditions (data not shown). Comparing with the results showed that the adsorption performance of G6PDH was not severely affected by the presence of whole yeast cells and cell debris. Reproducibility studies confirm that the bed expansion and

adsorption characteristics do not change as a result of the subsequent cleaning-in-place (CIP) procedures (data not shown). Overall, the results indicated that the use of stable expanded bed adsorbents is an effective method for the adsorption of proteins from crude feedstocks without prior removal of particulate matter.

CONCLUSIONS

The ability to achieve clarification, capture, and purification in a single step will greater simplify downstream processing flowsheets and reduce the cost of protein purification. Hence, the use of expanded bed adsorption techniques should generate more favourable process economics when introduced commercially. This is particularly important at a time when the biotechnology industry is undergoing a transition from laboratory to process scale development. In this paper, STREAMLINE™ Red H-E7B was shown to be successful in achieving isolation of G6PDH from unclarified homogenate with a purification factor of 12 and yield of 86% in a single step process. The omission of centrifugation or filtration steps would certainly result in more cost effective process with the opportunity of increased yields of product.

Table 2. Purification of glucose-6-phosphate dehydrogenase (G6PDH) from bakers' yeast cell homogenate. G6PDH was purified from bakers' yeast homogenate using an expanded bed of STREAMLINE™ Red H-E7B. The experimental protocol was as described in the text and the chromatogram from the run was shown in Figure 2. Samples collected during chromatography were assayed for total protein content and G6PDH activity and allowed the purification table to be calculated.

Purification Step	Volume (ml)	Total Activity (U)	Total Protein (mg)	Specific Activity (U/mg)	Purification Factor	Yield G6PDH (%)
Crude homogenate	900	3060	12807	0.239	(1.0)	(100)
Flow through	900	131	6318	-	-	-
Washing	400	65	4092	-	-	
Elution (1)	445	2623	915	2.864	12.0	86
Elution (2)	192	9	1621	-	-	-
Total recovery (%)		92.42	101.18			

ACKNOWLEDGEMENTS

We would like to thank Pharmacia Biotech for the generous provision of chromatographic materials. Also, YKC greatly acknowledges the Mingchi Institute of Technology (Taiwan, ROC) for financial support.

LITERATURE CITED

1. Chang,Y.K., G.E.McCreath, N.M.Draeger and H.A. Chase, *Trans IChemE*, **71**(Part A), 299 (1993).
2. Chase, H.A. and N.M. Draeger, *J. Chromatogr.*, **597**, 129 (1992).
3. Chang,Y.K. and H.A.Chase, *Procedings from I.Chem.E.Res.Event.*, University College London, London, January, p. 168, (1994).
4. Kopperschläger,G., H.J.Böhme and E. Hofmann , in *Advances in Biochemical Engineering,* (Ed.Fiechter A.),Springer Verlag, Berline, **25**, 101(1982).
5. McCreath,G.E., H.A.Chase and C.R. Lowe, *J.Chromatogr.*, **659**, 275 (1994).
6. Chetty, A.S. and M.A. Burns, *Biotechnol. Bioeng.*, **38**, 963 (1991).
7. Chase, H.A. and Y.K. Chang, *Presented at 22nd FEBS Congress*, Stockholm, Sweden, July, 1993.
8. McCreath,G.E., *PhD Thesis*, University of Cambridge, 1993.
9. Richardson,J.F.and W.N.Zaki, *Trans. Inst. Chem. Eng.*, **32**, 35 (1954).
10. Kunii,D. and O.Levenspiel, *Fluidization Engineering*, R.E. Krieger, Pub., Malabar (1984).

Purification of Recombinant Hepatitis B Surface Antigen (rec-HBsAg) from *P. pastoris*: A Process Development Study

L. Pérez[1], S. López[2], A. Beldarraín[1], D. Arenal[1], and E. Pentón[1]

[1]Center for Genetic Engineering and Biotechnology, [2]Center for Biological Productions, Havana, CUBA

The aim of this study was to purify rec-HBsAg cloned and expressed in P. pastoris *starting from a semipurified material. A sequence of six steps was developed where the separation was achieved by the combined effect of immunoaffinity, ion exchange and molecular exclusion chromatographies. The downstream process was scaled up 250 to 500 fold and resulted in a rec-HBsAg purification ≥ 97% with an overall yield of 40% depending on the quality and characteristics of the starting material.*

The yeast *P. pastoris* has already been used for the expression of various heterologous proteins (<u>1</u>), including HBsAg, in a genetic construction in which the gene coding for this protein was cloned integrated into the chromosome of a mutant strain under the regulation signal of the enzyme Alcohol Oxidase I promoter (<u>1</u>,<u>2</u>). This makes possible the activation of the system by methanol to produce HBsAg from 2 to 3% of the soluble protein. However, the complete purification of the antigen from this source had not yet been disclosed (<u>1</u>,<u>2</u>).

HBsAg is a well known protein component of the surface of the hepatitis B virus (HBV), which forms particles with some 100 subunits of 24 KD. These particles are stabilized by disulfide bonds (<u>3</u>), and contain carbohydrates and lipids (<u>1</u>,<u>4</u>).

The aim of this work was to provide a chromatographic purification process for obtaining highly purified and particulated rec-HBsAg with an adequate yield, starting from a clarified and filtered preparation derived from a crude homogenate of the *P pastoris* recombinant strain. The antigen so prepared was incorporated as the main active ingredient of a vaccine which is now a commercially available product, registered and distributed in several countries.

MATERIALS AND METHODS

The starting material was clarified and purified in previous steps by acid precipitation and adsorption / desorption on a matrix of diatomaceous earth (<u>5</u>), followed by its concentration and filtration.

The chromatographic media employed were: **a.** Immunoaffinity matrix made with antihepatitis B monoclonal antibodies (MAb) coupled to CnBr-activated Sepharose, supplied by the Hybridomas Division (CIGB, Cuba). The

527

E. Galindo and O.T. Ramírez (eds.), Advances in Bioprocess Engineering. 527-534.
© 1994 Kluwer Academic Publishers. Printed in the Netherlands.

MAb employed is class 1 IgG obtained as described (6); **b.** DE-52 Cellulose from Whatman, UK.; **c.** Analytical and preparative PW 5000 columns from Tosohass, Japan and **d.** Sephadex G-25 M, Pharmacia and Biotechnology, Sweden.

All chemical reagents used were analytical grade. Tris (hydroxymethyl aminomethane), ethylendiaminetetra-acetic acid (EDTA), Sodium chloride, Sodium deoxicolate, Potassium thiocyanate and Sodium phosphate were purchased from Merck (Darmsstadt, Germany).

The concentration of rec-HBsAg was determined by an ELISA method (7), using sheep polyclonal antibodies against HBsAg and a conjugate of these antibodies with horse-radish peroxidase. The reference HBsAg preparation was calibrated using a primary standard preparation obtained from the National Institute for Biological Standards and Control (UK).

The protein concentration was determined by the Bradford Coomassie Brilliant Blue method (8), the lipids were determined by the phospho-vainilline method (9), the DNA determinations were done by a hybridization procedure (10), carbohydrates were determined by the anthrone method (11) and the endotoxins using the LAL test (12). All chromatographic equipment and columns were purchased from Pharmacia and Biotechnology (Sweden).

The rec-HBsAg purity is defined as the ratio between the quantity (mg) of antigen measured by ELISA and the quantity (mg) of proteins as measured by the Bradford method, expressed in %. The % of lipids, carbohydrates and DNA removed is calculated for each step by subtracting the % of recovery from 100.

RESULTS AND DISCUSSION

Downstream Process

The downstream process presented in Figure 1 is carried out at room temperature starting from a semipurified material. Figure 1 shows the final downstream process adopted, it has six chromatographic and two non chromatographic steps.

Table 1 shows the characterization of the starting material including the average values of the main contaminants (host proteins, carbohydrates, lipids and DNA) found in three different batches. The process developed took into consideration these four type of contaminants to reduce them to the minimum acceptable quantities regulated by WHO (13).

Table 1. Characterization of starting material.

Properties	Values
Conductivity	19 mS/cm
pH	7.2
HBsAg: content	0.214 mg/mL
Ip	5.6
Mw	1200 Kd
Protein conc.	1.100 mg/mL
Lipids conc.	1.438 mg/mL
Carbohydrates conc.	1.746 mg/mL
DNA conc.	0.200 μg/mL

The semipurified starting material has a relatively high rec-HBsAg content, it is highly soluble, stable at room temperature and heated up to 60°C for 2 hours at the range of pH used during the process, which is carried out with the same buffer system: 20mM Tris-HCl pH 7.2 with 3mM EDTA, unless otherwise specified.

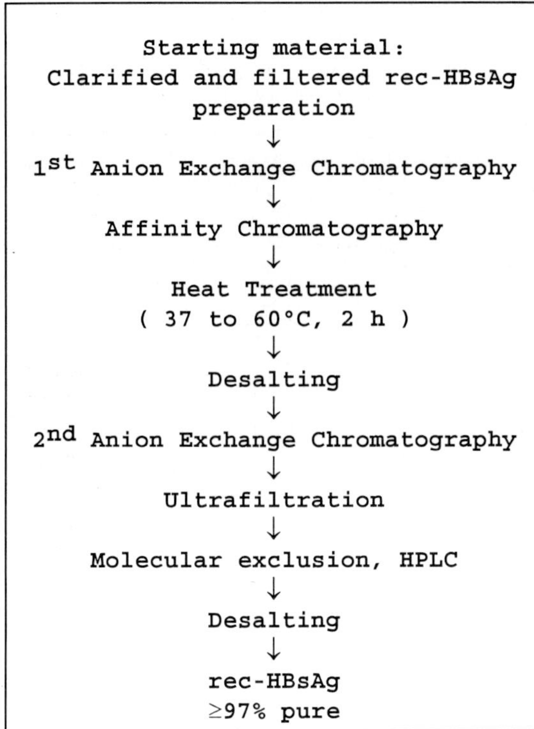

Starting material:
Clarified and filtered rec-HBsAg
preparation
↓
1st Anion Exchange Chromatography
↓
Affinity Chromatography
↓
Heat Treatment
(37 to 60°C, 2 h)
↓
Desalting
↓
2nd Anion Exchange Chromatography
↓
Ultrafiltration
↓
Molecular exclusion, HPLC
↓
Desalting
↓
rec-HBsAg
≥97% pure

Figure 1. Downstream process for the purification of rec-HBsAg from *P.pastoris*

The conditions, aims and results of each step may be briefly summarized as follows:

Step I. First anion exchange chromatography.
Negative respect to rec-HBsAg, which passes through the column while high molecular weigh DNA, pigments, and some host proteins remain fixed on the bed. The running buffer used for this step contains NaCl (C=28 mS.cm^{-1}) and the regeneration buffer 2M NaCl.

Step II. Affinity chromatography.
Affinity chromatography in CnBr activated Sepharose coupled to anti hepatitisB MAb which recognizes antigenic hepatitis B surface antigen particles. This is the most selec-

tive single step which purify the antigen more times (Table 2). Adsorption conditions C= 70mS.cm^{-1}.

Table 2. Results of the Downstream Process in terms of purity and recovery of each individual step. S.M.: starting material, I: Anion exchange No.1, II: Affinity chromatography, IV: Desalting, V: Anion exchange No.2, VI: Ultrafiltration, VII: Molecular exclusion, VIII: Final Desalting.

Step	HBsAg mg/mL	Purity %	HBsAg partial %	Yield total %
S.M.	0.214	20	–	100
I	0.158	22	98	98
II	0.693	94	64	63
IV	0.465	94	102	64
V	0.692	97	74	47
VI	9.315	97	90	43
VII	3.570	98	95	41
VIII	1.116	98	97	39

Desorption is carried out by the chaotropic agent 3M KSCN in elution buffer which rendered the best rec-HBsAg yield with a purity higher than 90%. The reduction of carbohydrates and DNA was more than 99% and lipids elimination was around 86% as shown in Table 3.

Step III. Heat treatment of rec-HBsAg in elution buffer with 3M KSCN.
After immunopurification this step promotes subunit aggregation and improves the recovery of 22 nm parti-

particles of HBsAg in the next ion exchange step.

Table 3. Clearance of non protein contaminants in the three principal separation steps during Downstream process. S.M.:starting material, II: Affinity chromatography eluate, V: Ion exchange No.2 eluate, VII: Molecular exclusion eluate.

Step	Lipids g	Lipids Rmvd (%)	Carbohyd. g	Carbohyd. Rmvd (%)	DNA mg	DNA Rmvd (%)
S.M.	658.9	-	109.4	-	23.8	-
II	91.2	86.2	1.0	99.0	0.04	99.8
V	56.2	38.4	1.1	0.0	<0.00	87.5
VII	10.6	81.1	1.0	6.4	<0.00	0.0
Overall	-	98.4	-	99.1	-	99.9

Step IV. Gel filtration chromatography in a Sephadex G-25M column. Desalting step for conditioning the material previous to the second ion exchange. Running buffer with $C = 2$ $mS.cm^{-1}$.

Step V. Ion Exchange in DE-52 Cellulose chromatography.
This step increases the purification of the antigen up to 95% (Table 2), lower more than 87% of the DNA to values below 10 pg / 20 μg antigen and reduce 38 % of lipids (Table 3).
Adsorption conditions, running buffer with $C= 2$ $mS.cm^{-1}$
Washing conditions, running buffer plus NaCl to $C= 10$ $mS.cm^{-1}$
Desorption conditions, running buffer plus NaCl to $C= 40$ $mS.cm^{-1}$

The traces of free IgG which might be incorporated during immunopurification are removed in the washing step before the elution.

Step VI. Concentration and Diafiltration by ultrafiltration.
Carried out using hollow fiber cartridges (cut off 0.1μm) to reduce around 10 times the volume and reduce KSCN ion traces present in the preparation.

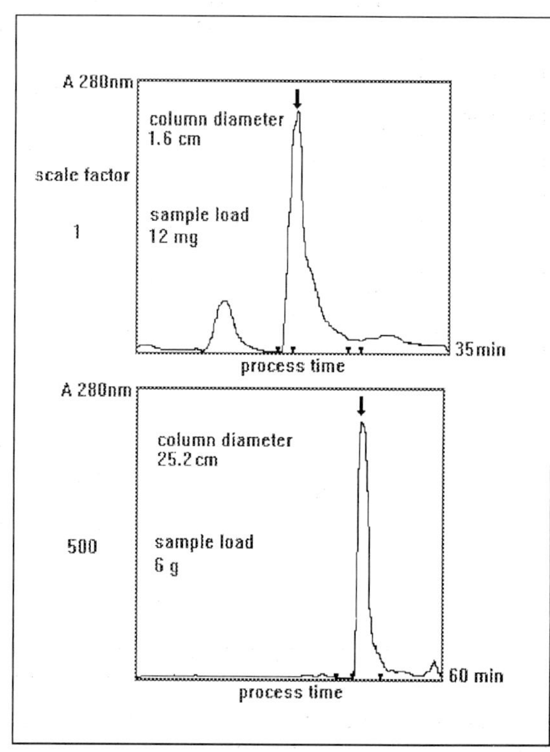

Figure 2. Ion exchange on DE-52 Cellulose records for laboratory and production scale. The arrow ↓ indicates rec-HBsAg peak.

Step VII. Molecular exclusion in PW 5000, HPLC system (Tosohass column). The column is equilibrated in 20 mM Tris-HCl pH 8.0 buffer, containing 0.05 % Sodium deoxicolate and 0.2 M NaCl. The presence of detergent during the operation allows the separation of non properly particulated forms of HBsAg and removes 75% the endotoxin content, leading to a preparation with less than 3 pg of endotoxins per μg of pure rec-HBsAg.

This chromatographic separation is also useful to remove 80% of the lipid (Table 3) content of the initial material of this step, leaving only around 1 μg lipids per μg of HBsAg, which corresponds to non-removable lipids.

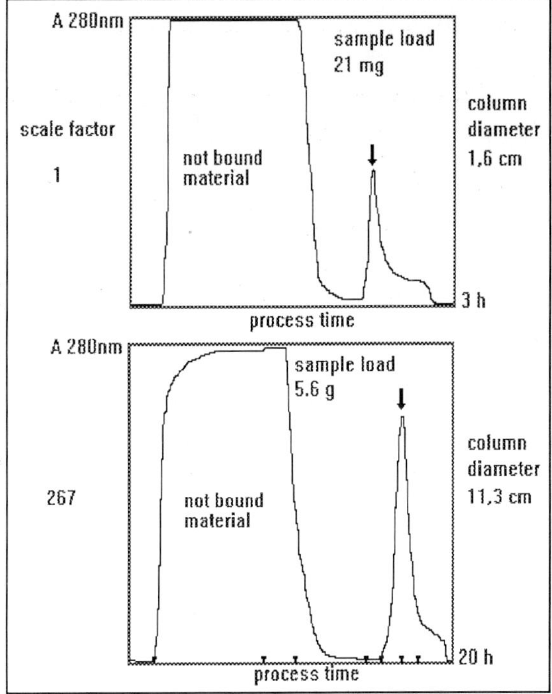

Figure 3. Immunoaffinity records for laboratory and production scale. The arrow ↓ indicates rec–HBsAg peak.

Step VIII. Final formulation of the HBsAg in the storage buffer using gel filtration in Sephadex G-25 M.

The results of the Downstream process in terms of purity, recovery and antigen concentration are shown in Table 2 detailing each step.

Scale up procedure

Scaling up denotes an increase in the volume handled in a production cycle. An essential feature of the scale up procedure is the introduc-tion of new and larger pieces of apparatus, whose effects on chromatography separations must be analyzed and balanced out.

The way chosen for scaling up was to increase proportionally a group of parameters such as the sample load, column diameter, matrix volume and volumetric flow rate while maintaining the bed height, linear flow rate, sample concentration and the ratio sample load: bed volume (14). This principle was successful in all cases except for the affinity chromatography.

Table 4. Scale up results in Anion Exchange on DE-52 Cellulose chromatography.

Parameter	Lab scale	Pilot scale	Produc-tion
Bed volume, mL	12	236	6000
Bed height, cm	6	12	12
Bed diameter, cm	1.6	5.0	25.2
Flow rate, cm/h	50	50	50
Sample load, mg	12	240	6000
Scale up factor	1	20	500
Yield %	80	75	75
Process time, min	35	60	60

After the purification scheme was optimized on a laboratory scale, the whole process was scaled up to the final processing volume. The first step was scaled up to a pilot scale, and then to the production scale. The typical laboratory, pilot and production scale parameters for the three more effective chromatographic steps,. are described in Tables 4, 5 and 6, including the scale up factors, % of recovery and process timing.

Table 4 presents the results obtained for the anion exchange step, where all general rules for scaling up were applied. The results were excellent in terms of performance of the medium. Figure 2 shows the record chart obtained for the minimum and maximum scales.

For affinity chromatography, (see Table 5), it was not possible to increase the bed diameter maintaining the bed height constant as usual, because the rec-HBsAg recovery decreased proportionally. This failure may be explained by the slow access of the target molecules to the inside of the bed, as far as the diffusion of macromolecules, such as HBsAg, may take several seconds (15), and in practice, only the surface layers of the immunosorbent are used. The records for analytical and production scale are shown in Figure 3. From Table 5 it can be derived that to duplicate the bed diameter it was necessary to increase 8.5 times the bed height although the sample: matrix ratio had to be lowered 1.6 times to obtain the same recovery.

Table 5 . Scale up results in Affinity chromatography on MAb-Sepharose matrix.

Parameter	Lab scale	Pilot scale	Production
Bed volume, mL	14	140	6000
Bed height, cm	7	7	60.0
Bed diameter, cm	1.6	5.0	11.3
Flow rate, cm/h	35	35	35
Sample load, mg	21	215	5600
Scale up factor	1	10	267
Yield %	80	75	75
Process time, h	3	6	20

Scaling up molecular exclusion was relatively easy because the separation was achieved in PW 5000 (Tosohass column) only by adjusting the main parameters in each scale. In this case, the jump was done directly from laboratory to preparative scale.

Figure 4 shows a typical record where it can be observed that for a scale up factor of 125 the bed diameter increased only 6 times while the recovery and process time remained constant (Table 6).

Table 6. Scale up results of molecular exclusion chromatography on PW 5000 columns.

Parameter	Lab scale	Production
Bed volume, mL	26	1 178
Bed height, cm	60	60
Bed diameter, m	7.5	50
Flow rate, cm/h	40	40
Sample load, mg	10	1 250
Scale up factor	1	125
Yield %	95	95
Process time, h	2.5	2.5

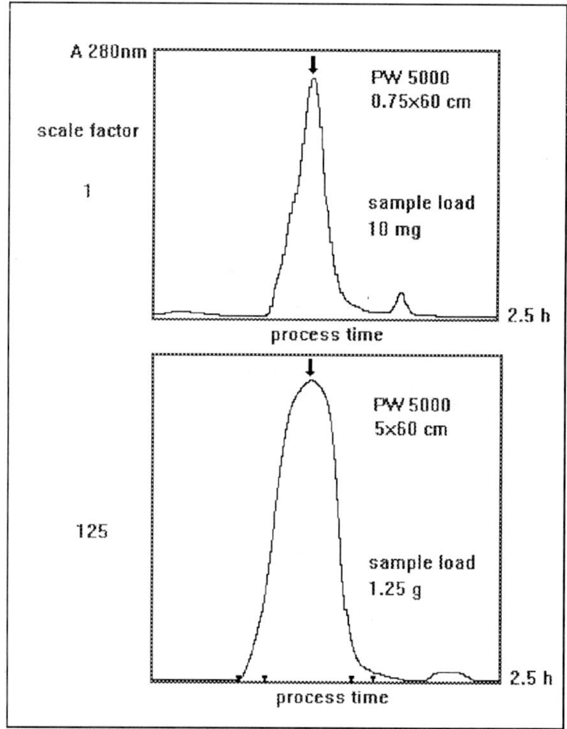

Figure 4. Records for molecular exclusion HPLC. The arrow ↓ indicates HBsAg peak. The second peak was found to be non properly particulated rec-HBsAg.

CONCLUSIONS

The downstream process developed has a logical sequence of steps and renders a rec-HBsAg preparation compatible with the high purity required for its intended use. The processes are reproducible, easily handled and hygienyzed and were completely scaled up to the present production scale.

Figure 5 is a typical HPLC pattern of the purified preparation, note that a single antigen peak is obtained with a retention time of 75.96 ± 2.34 min. and a purity ≥ 97% as estimated from the area below the curve and 97 ± 3% by densitometric scanning of the Coomassie blue stained electrophoretic gel (data not shown).

Regarding DNA, lipids and carbohydrates the process remove more than 98% of these contaminants from the starting material.

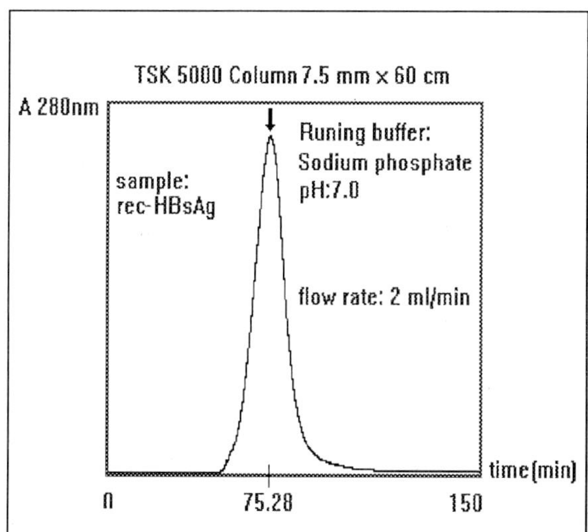

Figure 5. Records of molecular exclusion HPLC analysis to the final preparation obtained by the procedure described, developed in Quality Controls Laboratories.

NOMENCLATURE

C	Conductivity
Mw	Molecular weight
Ip	Isoelectric point

LITERATURE CITED

1. Muzio, V.L, et al., patent, publication number 0480525A2.,Euro. Pat. Aplic., (1993).

2. Craig, W.S., R.S. Siegel, patent 89106753, Euro. Pat. Aplic.,(1990).

3. Wampler, D.E., E.D. Lehman, J. Borger, W.J. McAleer, and E.M. Scolnick, Proc. Nat. Acad. Soc., **82**, 6830, USA (1985)

4. Gavillanes, F., J. Gómez Gutierrez, M. Aracil, and D.L. Peterson, Biochem. Journal., **265**, 857 (1990).

5. Agraz, A., Y. Quiñones, E. Pentón and L. Herrera. "Adsorption- Desorption of Recombinant Hepatitis B Surface Antigen (r-HBsAg) from *P. pastoris* on a Diatomaceous Earth Matrix: Optimization of parameters for purification", Biotechnology and Bioengineering **42**, 1238 (1993).

6. Fontirrochi, G., M.E. Pérez, and C. Duarte, Biotecnología Aplicada. **10**, 1, 24, Cuba (1993).

7. González, A., A. Alerm, A. Salgado, V. Ramírez, and T. González, "Cuantificación del HBsAg en muestras Biológicas con fines asistenciales y preparativos", Biotecnología Aplicada. **10**, 2, 104, Cuba (1993).

8. Bradford, M.M., Anal. Biochem., **72**, 248 (1976).

9. Woodman, D.D. and C.P. Price, Clin. Chem. Act., **38**, 39 (1972).

10. Sambrook, J., E.F. Fristsch and T. Maniatis, Molecular Cloning Laboratory Manual, 2th ed., USA, (1989).

11. Trevelyan, W.E. and J.S. Harrison, Biochem. Journal. **23**, 1824 (1952).

12. Levin, J., and Bang, F.B. Bulletin John Hopkins Hospital, **115**,

265, USA (1964).

13. Word Health Organization, Techn. Report Series, 786 (1989).

14. Sofer, G.K., Nyström, L.E. Process Chromatography., 55, London (1989).

15. Duncan L. and S. Pepper, Lab. Meth. in Immunology, **II**, 218 (1990).

Evaluation of Affinity Chromatographic Methods in the Purification of Recombinant Streptokinase

N. Pérez, P. Rodríguez, L. Hernández, E. Muñoz, D.R. Orta, and S. Pérez

División de Trombolíticos, Centro de Ingeniería Genética y Biotecnología, Apartado 6162, La Habana 6, 10600, CUBA

Streptokinase is an extracellular protein produced by several strains of streptococci. It is able to convert blood plasminogen to plasmin. In this paper we describe the purification of recombinant streptokinase using monoclonal antibodies and a combination of two affinity matrixes: human plasminogen-Sepharose 4B and monoclonal antibody anti-streptokinase-Sepharose 4B. An homogenous preparation of recombinant streptokinase with specific activity around 50,000 IU mg[-1] and purity higher than 93% was obtained.

Streptokinase (SK) is used as an agent for the therapy of thrombotic disorders, it activates the human proteolytic system. Several methods have been described for the preparation of this protein obtained from natural sources (7), (2), (14), (15) and by recombinant DNA techniques (8). In some cases ion exchangers have been used together with other purification procedures. The use of an affinity column of canine plasmin (plm)-Sepharose was described by Castellino et al., 1976 (5), furthermore, Rodríguez et al., 1992 (13) described the results obtained for the purification of natural and recombinant Streptokinase (rSk) on NPGB[1] -human plasminogen (hPlg)-Sepharose.

In this paper we describe the purification of rSk with Monoclonal antibodies (MAbs) immobilized to Sepharose. Our

procedure gives a homogeneous preparation with a molecular weight of 47 000 and between 91 and 94% of purity. The specific activity was higher than 47 000 IU/mg of protein in all cases.

Two affinity chromatographic procedures were successfully combined in order to obtain an homogeneous preparation of rSk. MAbs-Sepharose and NPGB-hPlg-Sepharose were used as affinity matrixes.

MATERIALS AND METHODS

Anti rSk monoclonal antibodies 122, 130 and 139 type IgG1 were supplied from Hybridomas Division (CIGB, Cuba), the Mabs were purified from mouse ascitic fluid and subjected to affinity chromatography using Protein A-Sepharose 4B (Pharmacia, Sweden), according to the manufacturer suggestions (1).

---

[1] p'nitrophenyl-p'guanidine-benzoate

E. Galindo and O.T. Ramírez (eds.), Advances in Bioprocess Engineering. 535–540.
© 1994 Kluwer Academic Publishers. Printed in the Netherlands.

Human plasminogen was obtained according to the method of Deutsch and Mertz, 1970 (6), using an affinity column of Lysine-Sepharose (Pharmacia LKB Biotechnology, Upssala, Sweden).

Monoclonal antibodies solution were dialyzed against NaHCO3 0.1 M pH 8.3 and coupling to Cianogen Bromide (CNBr) activated Sepharose 4B (Pharmacia Fine Chemicals, Sweden) (1), 5 mg MAbs/ml gel of each immune absorbent.

Protein analysis

Protein concentration was determined by the method of Bradford (3) . Streptokinase enzymatic activity was measured by the chromogenic substrate S-2251 method (9). SDS-Polyacrylamide gels electrophoresis 12.5 % were conducted by the method of Laemmli , 1970 (10).

Western blot analysis was performed by procedures of Towbin et al., 1979 (12) and Burnett ,1981 (4). Inmunodot Blot was performed by modifications of the method of Moeremans et al., (1985) (11).

Purification of rSK using MAbs columns.

Mabs affinity chromatography experiments were done with gels packed in three analytical columns of 2 ml each. The flow rate (20 mL/h) was controlled with a peristaltic pump.

The absorbance of the eluate was recorded with a Uvicord SII LKB (Pharmacia, Sweden). Starting material used in these assays were obtained from fermentation of E. coli cells W 3110 transformed with the expression vector for SK prepared in our laboratory (8). The cellular disruption was carried out in a French press and the streptokinase were solubilized with 6 M Urea in 4, 1 M Tris HCl

pH 7 buffer.. The rSk represented in this starting material was 20 % of the proteins.

The starting material was gel-filtered to remove the urea and renaturalize the protein in Sephadex G-25, previously equilibrated with 0.02 M Sodium Phosphate pH 7.4 buffer The elution (5 ml) was applied to a flow rate of 20 ml/h to the MAbs columns. They were washed with 15 ml of the same buffer, containing 0.5 M NaCl. The elution procedure was performed with 0.2 M Glycine pH 2.5. Fractions of 3 ml were collected. The rSk obtained from each column was dialyzed against distilled water. Western blotting and protein SDS PAGE were done with those preparations.

Purification of rSk combining two affinity chromatography.

A column of NPGB-hPlg-Sepharose with 20 ml of gel was used to purify 50 ml of the starting material containing 142 mg of total protein, this step was done using Rodríguez et al., method (13). A 15 ml fraction with 30 mg of protein was collected. It was gel filtered in Sephadex G-25 with, 0.01 M Tris-HCl pH 7.5 buffer containing 0.15 M NaCl, to remove the Glycine and it was passed through the column of MAb 130-Sepharose. Conditions for MAbs columns were similar to the above experiment. A fraction of 4 ml with 12 mg of protein was collected.

RESULTS AND DISCUSSION

The results obtained in the purification assays using MAb 122-Sepharose, MAb130 Sepharose and MAb 139-Sepharose are shown in table 1. The total protein of the starting material was about 3.5 mg in the assays with each MAbs; the amount of purified protein obtained in the elution steps, was around 0.7 mg in each assay. The specific activity of starting materials was about 14 000 IU/mg and the purified material obtained had a specific activity from 47 000 to 49 000 IU/mg of protein in each case. Recovered protein was 30% of the total starting protein (results not shown). The absorption capacity calculated for these columns was from 65 to 70 µg of protein/mg of MAb immobilized. The final purified rSK had 91 to 94% of purity, (table 1). It was judged comparing the intensity between the dots of the *E. coli* protein applied in serial dilutions and the samples of each purification assay blotted over the nitrocellulose membrane. The Immunodot blot was developed with an anti host protein antiserum (figure 1).

Samples of protein eluted of each purification assay were applied on Polyacrylamide-gel-electrophoresis 12.5 % (figure 2a). A major component of MW 47 000 was detected and other proteolytic fragments of the major protein with MW from 45 000 up to 30 000 Da. These components of degradation can appear during the purification process (5). This results are confirmed by the Western blot using antiserum anti rSK and antiserum anti *E. coli* (figure 2b and 2c).In these Western blot we can observe that rSK keep the same degradation products before and after purification but the *E. coli* proteins contaminants desapear after this purification step.

From these results we conclude that anti-streptokinase MAb affinity columns purify .the initial sample from *E. coli* contaminant. Results obtained with the three columns were very similar, but we considered to use the MAb 130 in the rest of the experiments by the yield and absorption capacity.

Chromatography on Human Plasminogen Sepharose and Monoclonal Antibody 130.

We began with 50 ml of starting material which had 7 042 IU/mg of specific activity, this material was purified using a matrix of hPlg Sepharose, 15 ml with 30 mg of purified protein was collected. This preparation shows 31 345 IU/mg of specific activity, and the purity evaluated by anticoli immunodot was about 80 % (figure 3). The calculated purification factor was 4.45 fold. The overall purification factor for all process was 6.75 fold (table 2). We consider that the purification factor could be even higher if we change some conditions, as the ionic strength of washes in the step of purification with MAb.

Results of the *E. coli* contaminant in the different purification steps are shown in the immunodot assay of figure 3. On column 1, serial half dilution of an elution sample from hPlg-Sepharose were applied (starting with 11 µg). On columns 2 and 3 two samples from the MAbs Sepharose purification step were applied, those samples start with 4.3 µg and consecutive half dilutions were applied. *E. coli* contaminant standard curve was applied from line **F** to line **H** the initial value of the curve is 4.7 µg of protein and them half dilution from the previous one. The final purity of the rSK was estimated between 93% and 96.5%, by comparison of the intensity of the signals between the dots of the standard curve and dots in the anticoli immunodot assay (figure 3).

The specific activity obtained for the final purified rSK was 47 535 IU/mg.

The figure 4 shows 12.5% SDS PAGE with samples of this purification process. On this experiments shown that the molecular integrity of the rSk obtained with 2 affinity

Table 1.
Comparative evaluation of three MAbs on the purification of rSK

Gel	Specific Activity*	Yield* (%)	Purification ratio	Abs. Capacity* (SK/MAbs)	Final Purity (%)
MAb-122	50,000	60	3	68	91
MAb-130	47,500	75	3.2	73	94
MAb -139	48,000	67	4.2	65.7	94

* Values are the mean of two experiments

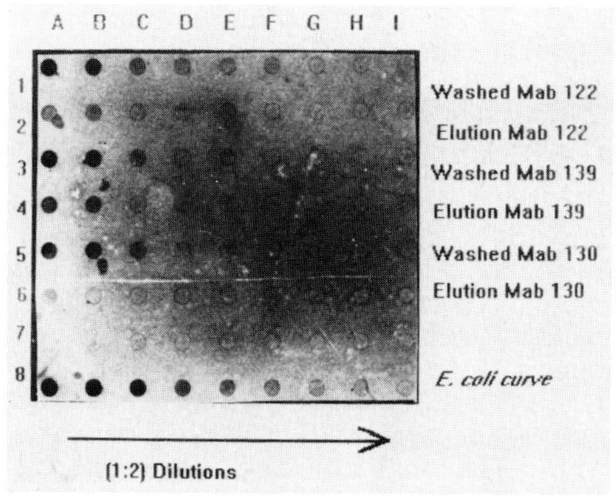

Figure 1. Immunodot blot anti *E. coli* proteins performed with the samples of each purification assay (A to I each column correspond to a half dilution of the precedent), **Line 1**, Washed of the MAb 122: column, well (A) 2.2 µg; **Line 2**, Elution of the MAb 122: column, well (A) 8.2 µg; **Line 3**, Washed of the MAb 139: column, well (A) 1.68 µg; **Line 4**, Elution of the MAb 139: column, well (A) 4.4 µg; **Line 5**, Washed of the MAb 130 column, well (A) 2.48 µg; **Line 6**, Elution of the MAb 130 column, well (A) 4 µg. **Line 8**, Controls of the *E. coli* contaminant proteins, . Well (A) 4.7 µg.

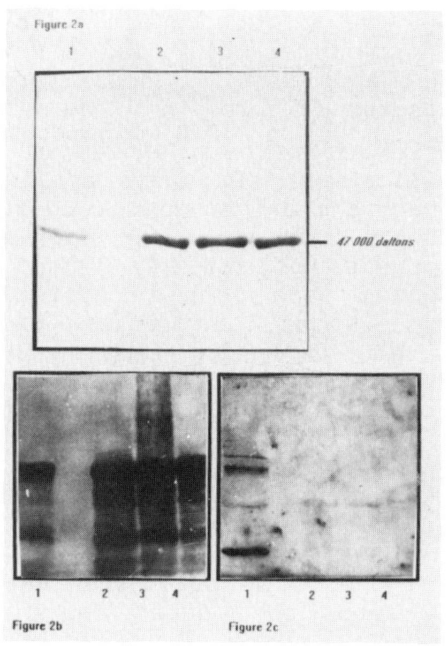

Figure 2. Western Blotting analysis of eluates obtained by purification processes on each MAb column 139, 130 and 122, lines 2, 3 and 4 respectively. Starting material of rSk was applied on line 1 (10 µg of protein was applied in each line). (2a) SDS PAGE gel electrophoresis stained with Coomassie; (2b) Western Blot with antiserum anti rSk (2c) Western Blot with antiserum anti-*E. coli* contaminant proteins.

Table 2

Results of the two consecutive affinity steps for the rSK purification

Step	Volume (ml)	Protein* (mg)	Activity* (IU)	Specific Activity* (IU/mg)	Purification ratio	Purity (%)
Starting Material	50	142	1,000,000	7,042		~20
hPlg-Sepharose	15	30	940,850	31,362	4.45	~80
MAb-130 Sepharose	4	12	570,480	47,540	6.75	93 - 96.5

* Values are the mean of two experiments

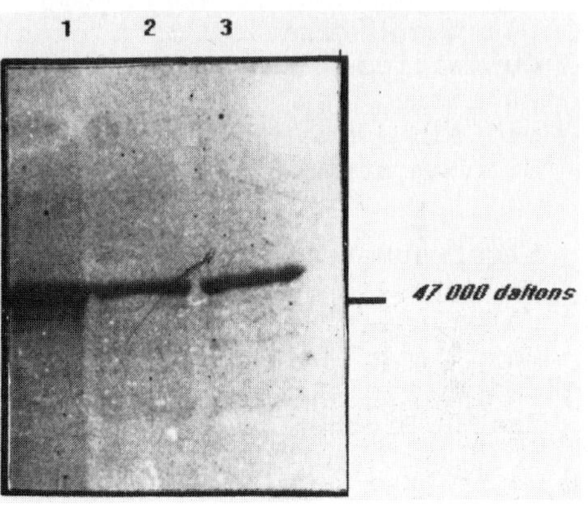

Figure 4. SDS PAGE of samples from different steps in the 2 affinity chromatography purification assay of rSk. Line 1 Sample from hPlg-Sepharose (9 µg), Line 2 and 3 samples of the elution of the MAb 130-Sepharose column (7 µg in each one).

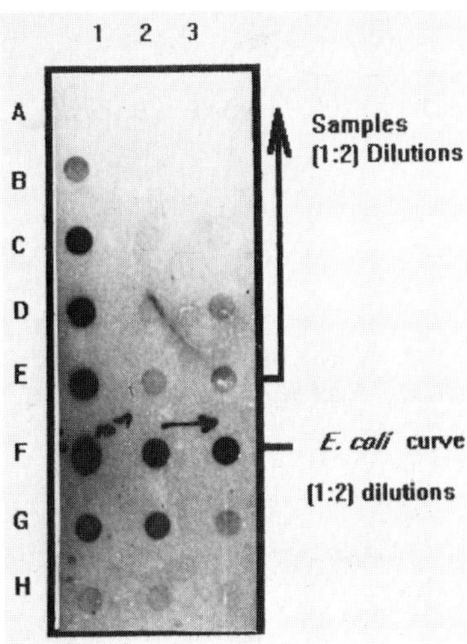

Figure 3. Immunodot blot performed with the samples obtained from the combined affinity chromatographic process.. The samples were applied with 1/2 dilution from well E to well A, as it is indicated by the larger arrow. Column 1, samples eluted from hPlg-Sepharose column (11 µg applied in line E); columns 2 and 3 samples eluted after the MAb-Sepharose column (4.3 µg applied in line E). The *E. coli* contaminant proteins standard curve was applied with 1/2 dilution on lines F to H, the first well (F1) has 4.7 µg.

purification step are similar than in the previous experiments (figure 2)

This is the first report about the use of one step immunoaffinity chromatography for the purification of recombinant Streptokinase with high purity. We conclude that it is possible to obtain, using one step MAb affinity chromatography, rSK with very similar purity than those obtained using two affinity chromatography steps. One step method is recommended in terms of yield and simplicity (compare 60 to 75% yield against around 57% in the case of two steps method) and because the final product in both cases have similar specific activities. (Tables 1 and 2).

REFERENCES

1. **-Affinity Chromatography Principles and Methods.** Pharmacia. Sweden. (1983).

2. -Blatt W. F., H. Segal and J. L. Gray (1964). Purification of Streptokinase and Human Plasmin. Their Interaction. **Throm. Diath. Haemorrh** 11:393-402

3. -Bradford, M. M. (1976). A rapid and sensitive method for the quantification of microgram quantities of protein utilizing the principle of the protein-dye binding **Analytical. Biochemistry.** 72: 248-254.

4. -Burnett N. N. (1981). Electrophoretic transfer of proteins from SDS-polyacrylamide gels to unmodified nitrocellulose and radio iodinated protein A. **Anal. Biochemistry,** 112: 195-203.

5. -Castellino F. J., J. M. Sodetz, W. J. Brockway and G. E. Siefring, Jr. (1976). Streptokinase (21). **Methods in Enzimology.** 45: 244-257.

6. -Deutsch, D. G. and E. T. Mertz (1970). Plasminogen: purification from Human plasma by affinity chromatography. **Science.**170:163-167.

7. -Dillon H. C. and L. W. Wannamaker (1964). Physical and inmunological differences among streptokinases. **J. Exptl. Med:** 121-151

8. -Estrada M. P.,L. Hernández, A. Pérez, P. Rodríguez, R. Serrano, R. Rubiera, A. Pedraza, G. Padrón, W. Antuch, J. de la Fuente and L. Herrera. 1992 High level expression of Streptokinase in *Escherichia coli*. **Biotechnology 10,** 10: 1138-1142.

9. -Hernández L., P. Rodríguez, A. Castro, R. Serrano, M. P. Rodríguez, R. Rubiera, M. P. Estrada, A. Pérez, J. de la Fuente and L. Herrera. (1990). Determination of streptokinase activity by quantitative assay. **Biotecnología Aplicada 7,** 2: 153- 160.

10. -Laemmli U. K. (1970). Cleavage of structural proteins during the assembly of the head of the bacteriophage T4 . **Nature 227:** 227-680.

11. -Moeremans M., Daneels G. and De Mey J. (1985). Sensitive colloidal metal (gold or silver) staining of protein blots on nitrocellulose membranes. **Anal Biocnem. 145:** 315-321.

12. -Towbin H., T. Stahelin and J. Gordon (1979). Electrophoretic transfer of proteins from polyacrylamide gels to nitrocellulose shuts. **Proc Natl. Acad. Sci USA, 76:** 4350-4354.

13. -Rodríguez P.,L Hernández,E Muñoz, A Castro, J. De la Fuente and L. Herrera. 1992. Purification of Streptokinase by affinity chromatography on immobilized acylated human plasminogen. **BioTechniques, 12,** 3: 424-429.

14. -Taylor F. B. ; Jr., and J. Botts (1968). Purification and Characterization of Streptokinase with Studies of Streptokinase Activation of Plasminogen. **Biochemistry 7** :233-242.

15. -Taylor F. B., and R. H. Tomar (1969). Streptokinase 63. **Methods Enzimol. :** 807-821.

IN APPRECIATION

We want to express our gratitude to the referees of the manuscripts submitted to this publication

Arturo Aguilar-Aguila
Alejandro Alagón
Grant Allen
Richard Auria
Alejandro Azaola
Paulina Balbás
Basil Baltzis
Eduardo Bárzana
Hugo Barrera
Ursula Bilitewski
Edmundo Brito
Francisco Calzada
Robert Carrier
Lidia Casas
Luis Covarrubias
Gopal Chotani
Bruce Dale
Wolf-D. Deckwer
Mayra de la Torre
Roberto de León
Mario Díaz
Irving Dunn
Guadalupe Espín
Amelia Farrés
Ernesto Favela
Luis Flores
Enrique Galindo
Mariano García-Garibay
Luis Garrido
Alfonso Gómez

Simón González
Guillermo Gosset
Martin Griot
Leopoldo Güereca
Eduardo Gutiérrez
Jean Pierre Guyot
Roberto Guzmán
Raman H. Siva
Davis Hubbard
Arthur Humphrey
Andrés Illanes
Blanca Jiménez
Nazmul Karim
Chris Kent
Dhinakar Kompala
Nicholas Kossen
Ilangovan Kuupusamy
Guillermo Larios
Arye Lazar
Hermilo Leal
Robert Lencki
Felipe López-Isunza
Agustín López-Munguía
Ma. Teresa Lucas
Andrew Lyddiatt
Alfredo Martínez
Petia Mijaylova
Oscar Monroy
Murray Moo-Young
Adalberto Noyola

Alberto Ochoa
Efren Parada
Martín Patiño
Lourival Possani
Rodolfo Quintero
Octavio T. Ramírez
Govind Rao
Mattias Reuss
Sergio Revah
Carlos Rolz
Sevastianos Roussos
Miguel Salvador
Sergio Sánchez-Esquivel
Alfredo Sánchez-Marroquín
Sergio Sánchez-Ruiz
Susana Saval
Leobardo Serrano
J.D. Shewale
Wei-Shou Hu
Geoffrey Slaff
Gloria Soberón
Xavier Soberón
Alberto Tecante
Colin Thomas
Eugenio Tisselli
Fernando Valle
Rafael Vázquez
Gustavo Viniegra
Zhibing Zhang

UNIVERSITY
COLLEGE LONDON
LIBRARY